Systems Analysis and Design in a Changing World

John W. Satzinger
Southwest Missouri State University

Robert B. Jackson
Brigham Young University

Stephen D. Burd
University of New Mexico

Systems Analysis and Design in a Changing World, **John W. Satzinger, Robert B. Jackson, and Stephen D. Burd.**

Senior Editor	Jennifer Normandin
Associate Publisher	Kristen Duerr
Production Editor	Debbie Masi
Developmental Editor	Karen Hill
Associate Product Manager	Amanda Young
Marketing Manager	Susan Ogar
Text Designer	Ann Small
Cover Designer	Efrat Reis

For more information, contact Course Technology, One Main Street, Cambridge, MA 02142; or find us on the World Wide Web at www.course.com.

For permission to use material from this text or product, contact us by

- Web: www.thomsonrights.com
- Phone: 1-800-730-2214
- Fax: 1-800-730-2215

ISBN 0-7600-5879-2

Printed in the United States
2 3 4 5 CRKEN 03 02 01 00

DEDICATION

To JoAnn, Brian, and Kevin—JWS

To Anabel and my children for their continued support—RBJ

To Dee, Amelia, and Alex—SDB

PREFACE

It is not easy to develop information systems in today's rapidly changing environment, but the satisfaction and rewards for a job well done are substantial. This text is designed for students and practitioners of analysis and design who are up to the challenge of trying to do it right.

Systems analysis and design is a practical field that relies on a core set of concepts and principles, as well as what sometimes seems an eclectic collection of rapidly evolving tools and techniques. Learning analysis and design today therefore requires an appreciation of the tried-and-true techniques widely embraced by experienced analysts plus the mastery of new and emerging tools and techniques that new graduates are increasingly expected to apply on the job. This text was developed for use in undergraduate and graduate courses in systems analysis and design where instructors and students do not want to ignore the past or hide from the future.

Objectives and Vision

This text was developed by a team who shared a commitment to produce an analysis and design text that was different—a text that is flexible and streamlined, yet comprehensive and deep. We were guided by the belief that the text must be flexible enough to appeal to instructors emphasizing more traditional approaches to systems analysis and design as well as those emphasizing the latest object-oriented techniques. We did not want to oversimplify the problem of system development, yet we felt that it was time to reevaluate whether many of the topics and features included in competing analysis and design texts were still essential. At the same time, many new developments are affecting systems analysis and design, and we wanted to include key trends—package solutions, enterprise resource planning (ERP), components, the Internet, and so on.

We also wanted the text to teach the key concepts and techniques, not just describe them. Therefore, we focus on fundamentals of lasting value and then show how these fundamentals apply to all approaches to development. Then we explore traditional structured analysis and design and object-oriented analysis and design in depth. Flexible and streamlined? Comprehensive and deep? We think you will agree these objectives have been achieved with this text.

Innovations

This text is unique in its integration of key systems modeling concepts that apply to both the traditional structured approach and the newer object-oriented approach—events that trigger system activities and objects/entities that are part of the system's problem domain. We devote one chapter to event partitioning and modeling key objects/entities. After completing that chapter, instructors can emphasize structured analysis and design or object-oriented analysis and design, or both. The object-oriented approach is not added as an afterthought—it is assumed from the beginning that everyone should understand the key object-oriented concepts. The traditional approach is not discarded—it is assumed from the beginning that everyone should understand the key structured concepts.

The traditional approach presented in this text is based on modern structured analysis and design as refined by McMenamin and Palmer, Ed Yourdon, and Meilir Page-Jones. Modern structured analysis is an integrated, model-driven approach that includes event partitioning, data modeling with entity-relationship diagrams (ERD), and process modeling with data flow diagrams (DFDs). Modern structured design is also based on event partitioning and uses the structure chart for software design. Database design using a relational database management implementation is featured. Instructors who emphasize the structured approach to development will be satisfied by the presentation and depth of coverage in this text.

The object-oriented approach presented in this text is based on the Unified Modeling Language (UML) from the Object Management Group as originated by Booch, Rumbaugh, and Jacobson. A model-driven approach to analysis starts with use cases and scenarios and then defines classes of objects and object interactions. We include requirements modeling with use case diagrams, class diagrams, sequence diagrams, collaboration diagrams, and statechart diagrams. Design models are also discussed in detail, with particular attention to design class diagrams and statechart diagrams that are used to design methods. Our database design chapter covers two approaches to object persistence—a hybrid approach using relational database management and a pure approach using object-oriented database management systems (OODBMS). Instructors who emphasize the object-oriented approach will not be disappointed by the presentation and depth of coverage in this text.

Additional concepts and techniques are included in response to the realities of system development today. First, rapid application development and component-based development are covered in depth. Second, package solutions and enterprise resource planning (ERP) are described as alternatives to custom development.

Features and Pedagogy

The text uses an ***integrated case study*** of moderate complexity—Rocky Mountain Outfitters—to illustrate key concepts and techniques on a specific system project. An overview of the strategic systems plan for Rocky Mountain Outfitters (RMO) is presented in Chapter 1 to place the system project in context. It is a planned project of strategic importance to RMO, and the need to integrate the new system with legacy systems and other planned systems is emphasized from the beginning. The planned system architecture provides for rich examples—a client-server Windows-based component as well as an e-commerce component with direct customer interaction via the Internet. Details about the case are integrated directly into each chapter to make a point or to illustrate a concept—just-in-time examples—rather than isolating the case study in separate sections of the chapters. The same system project is used to illustrate traditional and object-oriented models and solutions, so both approaches can be understood and directly compared.

Short stand-alone ***opening case studies*** describe a real-world situation relevant to the material in each chapter. A variety of companies and situations are included to provide the reader with a broad view of the problems and opportunities found in the real world.

Each chapter includes extensive ***figures and illustrations*** designed to clarify and summarize key points and to provide examples of models and other deliverables produced by an analyst. ***Margin definitions*** of key terms are placed in the text when the term is first used. End-of-chapter material includes a ***detailed summary***, an ***indexed list of key terms***, and a list of ***additional resources*** and references. Each chapter also includes ample ***review questions***, problems and exercises to get the student ***thinking critically***, a collection of ***experiential exercises*** involving additional research or problem solving, and end-of-chapter ***case studies*** that invite students to practice completing analysis and design tasks appropriate to the chapter. Some cases extend from chapter to chapter to build on earlier concepts and techniques.

Organization and Use

The text includes fifteen chapters organized into four parts—far fewer chapters than competing texts. Additionally, many chapters are modular—sections can be skipped without loss of continuity based on the objectives for the course. Some chapters are entirely optional. The *Instructor's Manual* includes a discussion of different approaches to using the text in analysis and design courses, including suggested course outlines for instructors emphasizing the traditional structured approach or the object-oriented approach and for instructors teaching graduate courses on analysis and design.

Part 1: The Modern Systems Analyst

Chapter 1 discusses the work of a modern systems analyst, including a streamlined discussion of systems and the role of the systems analyst as a problem solver in a modern business organization. Chapter 2 moves right to the heart of the course—a systems development project—introduced while describing the system development life cycle (SDLC), project planning, feasibility assessment, and project management. Students are drawn quickly into the RMO project so that the material has a meaningful context. Chapter 3 then asks, Now that we have a project, what do we have to do to get this system built? That is, what are the methodologies, models, tools, and techniques that can be used to develop systems? We make it clear that there are a variety of approaches to system development and that today's analysts need to be familiar with all of them. Even if students specialize in one approach in their course or later in their job, they should be able to distinguish among the structured approaches, information engineering, and the object-oriented approach in a meaningful way.

Part 2: Systems Analysis Tasks

Part 2 moves ahead with systems analysis techniques. Chapter 4 covers investigating system requirements, including gathering information and interviewing system owners and users. Chapter 5 covers modeling system requirements—using our approach, which includes event partitioning and modeling objects/entities, as described earlier. Chapter 6 continues requirements modeling using the traditional approach, including data flow diagrams (DFDs), data flow definitions, process descriptions, and so on. A few additional models that an analyst might encounter are also presented. Chapter 7 continues the discussion from Chapter 5 using the object-oriented approach to requirements. Instructors can simply choose to emphasize Chapter 6 or Chapter 7 to focus the course on either the traditional or the object-oriented approach, or both. Chapter 8 presents an overview of technical environments that affect the generation of alternative system solutions. Then, a comprehensive guide to generating and evaluating alternatives is presented, including the reality that a package solution is always a viable option.

Part 3: System Design Tasks

Chapter 9 covers system design, including both traditional structured design models and object-oriented models. The instructor can include either or both sections. Chapter 10 covers database design—relational, hybrid, and object-oriented databases. Again, the instructor can emphasize any or all approaches. Chapter 11 discusses system inputs and outputs, with particular attention to system controls. Chapter 12 covers human-computer interaction, and we include general principles and concepts of dialog design in addition to using UML diagrams to model the dialog.

Part 4: Implementation and Support

System implementation is increasingly technology specific, and because of the diverse development environments in the real world, we decided to streamline the discussion of implementation. However, we did include two chapters on important alternative approaches to implementation. Although the text emphasizes iteration and prototyping throughout, we include a comprehensive discussion of rapid application development and component-based development in Chapter 13. Similarly, although package solutions are discussed as viable alternatives throughout, we include a detailed discussion of packages and enterprise resource planning (ERP) in Chapter 14, including specific examples from SAP. Chapter 15 provides an overview of implementation and support that addresses traditional technology and object technology.

Available Support

The text includes a package of proven supplements for instructors and students. The *Instructor's Resource Kit* includes *Course Presenter*, *Course Test Manager*, *Electronic Instructor's Manual*, and *Figure Files*.

Course Presenter includes a PowerPoint lecture presentation in outline form with figures for each chapter. *Course Test Manager* can be used to create paper or LAN-based tests with multiple-choice, fill-in-the-blank, and/or essay questions. The *Electronic Instructor's Manual* includes suggestions and strategies for using the text, test bank questions, answers to review questions, and suggested solutions to chapter exercises and cases. *Figure Files* allow instructors to create their own presentations using figures from the text.

Many instructors like to include software for students to use for exercises and course projects, and this text offers many bundling possibilities. A popular option to use for drawing diagrams is Visio Professional 5.0. Some instructors like to emphasize CASE tools, and Course Technology can bundle several popular CASE tools with the text, including Oracle Designer, Visible Analyst Student Edition, and others. Contact Course Technology for the latest information.

Credits and Acknowledgments

This project was launched following some initial brainstorming between publisher Kristen Duerr of Course Technology and author John Satzinger. We agreed that an analysis and design text required a major commitment from the publisher and from a team of authors to be competitive. It was agreed that no one person or even pair of authors could complete a text that met the objectives—flexible and streamlined, yet comprehensive and deep. Therefore, Course Technology took an active role in assembling a team of authors who shared the vision. The senior editor brought in to manage the project was Jennifer Normandin, who had a major role in bringing the authors together and shaping the direction and final form of the text. This text truly could not have been completed without the commitment of Kristen Duerr and the leadership and energy of Jennifer Normandin. We thank them for everything.

Another essential member of the team is developmental editor Karen Hill of Elm Street Publishing Services. She collected and digested the comments and reactions of reviewers, provided guidance and design for the features and chapter pedagogy, suggested improvements and refinements to the organization and content, read each draft of each chapter from the perspective of the student to help us be consistent and clear, and (to the extent it is humanly possible) edited the chapters to provide a consistent style.

We also want to thank some other key people for their specific contributions—Richard A. Johnson of Southwest Missouri State University for writing Chapter 14 on packages and ERP, William Baker for contributing material on presentation techniques, and Mario Busjra for designing the RMO Logo and Web site. Many other colleagues and friends at SMSU, BYU, the University of New Mexico, and elsewhere contributed to and supported our work in one way or another. Special thanks also to Lavette Teague, Lorne Olfman, and Paul Gray for guidance and inspiration.

Many other people were involved in the production of this text. Amanda Young of Course Technology was always available and helpful. Debbie Masi, the production editor, came through for us to help sort out the subtleties of diagrams that probably appeared rather primitive to the artists. The production team went the extra mile to be true to the diagramming conventions and standards to the extent that we could define them.

Last, but certainly not least, we want to thank the reviewers who worked so hard for us, beginning with an initial proposal and continuing all the way through to the completion of the text. It is not easy to review a work in progress, and we were lucky enough to have reviewers with broad perspectives, in-depth knowledge, and diverse preferences. We listened very carefully, and the text is much better as a result of their input. The reviewers included:

Robert Beatty, *University of Wisconsin*, Milwaukee
Paul H. Cheney, *University of Central Florida*
Jon D. Clark, *Colorado State University*
David Little, *High Point University*
Ellen D. Hoadley, *Loyola College in Maryland*
Robert Keim, *Arizona State University*
Rebecca Koop, *Wright State University*
James E. LaBarre, *University of Wisconsin—Eau Claire*
George M. Marakas, *Indiana University*
Roger McHaney, *Kansas State University*
Bruce Neubauer, *Pittsburgh State University*
Mary Prescott, *University of South Florida*
Robert Saldarini, *Bergen Community College*

All of us involved in the development of this text wish you all the best as you take on the challenge of analysis and design in a changing world.

- John Satzinger
- Bob Jackson
- Steve Burd

CONTENTS

PART 1 | *The Modern Systems Analyst*

PART 2 | *Systems Analysis Tasks*

PART 3 | System Design Tasks

APPENDICES

PART 1

The Modern Systems Analyst

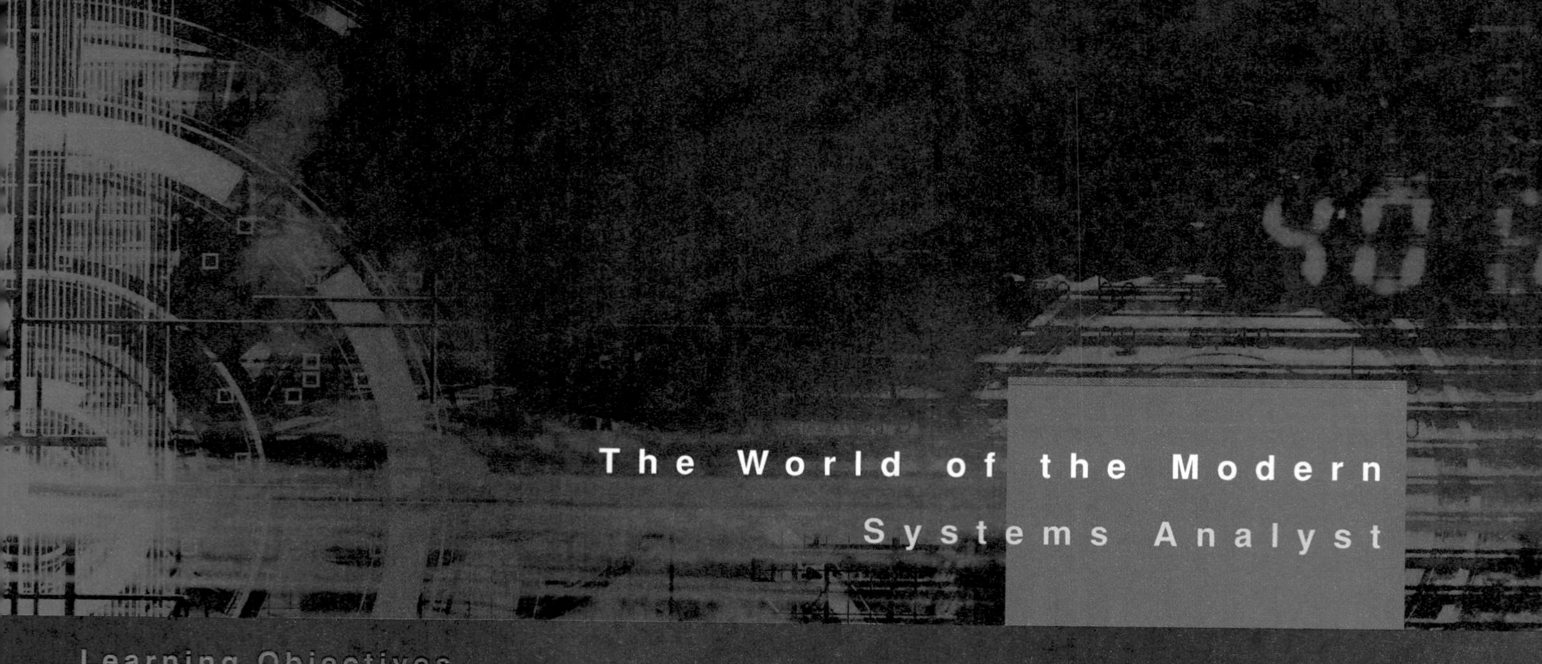

The World of the Modern Systems Analyst

Learning Objectives

After reading this chapter, you should be able to:

- Explain the key role of a systems analyst in business
- Describe the various types of systems an analyst might work on
- Explain the importance of technical, people, and business skills for an analyst
- Explain why ethical behavior is crucial for a systems analyst's career
- Describe the many types of technology an analyst needs to understand
- Describe various job titles and places of employment where analysis and design work is done
- Discuss the analyst's role in strategic planning for an organization
- Describe the analyst's role in a system development project

Chapter Outline

The Analyst as a Business Problem Solver

Systems that Solve Business Problems

Required Skills of the Systems Analyst

The Environment Surrounding the Analyst

The Analyst's Role in Strategic Planning

A Strategic Systems Plan for the Rocky Mountain Outfitters Case

The Analyst as a System Developer (The Heart of the Course)

A SYSTEMS ANALYST AT RUTHERFORD MANUFACTURING

Mary Wright thought back about her two-year career as a programmer analyst. She had been asked to talk to visiting computer information system (CIS) students about life on the job. "It seems like yesterday that I finally graduated from college and loaded up a U-Haul to start my new job at Rutherford," she began.

Rutherford Manufacturing is a sporting goods manufacturer located in St. Louis. Rutherford sells mainly to retail store chains but recently began selling directly to consumers over the Internet. This, and other competitive changes in manufacturing and distribution, made information systems particularly important to Rutherford.

"At first I did programming, mainly fixing things that end users wanted done. I completed some training on C++ and object-oriented analysis to round out my experience," explained Mary. "The job was pretty much what I had expected at first," she continued, "until everything went crazy over the IMDS project."

The Integrated Manufacturing and Distribution System (IMDS) project had been part of the company's information systems plan drawn up the year before. Edward Duke, the CEO of Rutherford Manufacturing, pushed for more strategic planning overall at the company from the beginning, including having a five-year strategic plan for information systems. IMDS was scheduled for the third or fourth year of the plan, but suddenly priorities changed. One of the larger retail customers of Rutherford went out of business. Then another retailer decided to eliminate sporting goods altogether. Additionally, sporting goods, like virtually all consumer goods, were beginning to be sold over the Internet.

Rutherford's sales dropped 15 percent in one quarter. A decision was made to make a major commitment to the Internet. The IMDS project was the key, launched immediately with the objective of radically improving manufacturing and distribution processes, mainly using Internet technology and direct sales.

"It seemed like the IMDS project was the only thing the company cared about," continued Mary. "I was assigned to the project as the junior analyst assisting the project manager, so I got in on everything. Suddenly I was in meeting after meeting, and I had to digest all kinds of information about manufacturing and retailing, like I was in business school. I met with manufacturing supervisors and marketing managers. I traveled all over to visit supplier locations (including a four-day trip to Seoul, Korea, on about two days notice!). I interviewed technology vendor representatives and consultants. I spend a lot of time at my computer, too, writing reports, letters, and memos; not programming!"

"We have been working on the project for seven months now, and every time I turn around, Mr. Duke, our CEO, is saying something about how important the IMDS project is to the future of the company—to the employees and to the stockholders. Mr. Duke is in on many of our meetings, and he even sat next to me the day I presented a list of key requirements for the system to the top management team. This is not at all the way I thought it would be."

OVERVIEW

Information systems are crucial to the success of modern business organizations, and new systems are constantly being developed to make businesses more competitive. The key to successful system development is thorough systems analysis and design. **System analysis** means understanding and specifying in detail what the information system should do. **System design** means specifying in detail how the many components of the information system should be physically implemented. This text is about systems analysis and design techniques used by a **systems analyst**, a business professional who develops information systems.

systems analysis
the process of understanding and specifying in detail what the information system should do

systems design
the process of specifying in detail how the many component parts of the information system should be physically implemented

systems analyst
a business professional who uses analysis and design techniques to solve business problems using information technology

This chapter describes the world of the systems analyst—the nature of the work, the knowledge and skills that are important, and the types of systems and special projects an analyst works on. First, the analyst's work is described as problem solving for an organization, so the problem-solving process followed by the analyst is described. Next, since most problems an analyst works on are solved in part by an information system, it is important to review the types of information systems involved. A systems analyst is a business professional that requires extensive technical, business, and people knowledge and skills, so these are reviewed. Next, the office technologies worked with and the variety of work places and positions where analysis work is done are surveyed. Sometimes an analyst works on special projects like strategic planning and enterprise resource planning.

Finally, this chapter introduces Rocky Mountain Outfitters (RMO), a large, regional sports clothing distributor headquartered in Park City, Utah. They are following a strategic information systems plan that calls for a new customer support system that will integrate phone orders, mail orders, and direct customer orders via the Internet. The Rocky Mountain Outfitters case is used throughout the text to illustrate analysis and design techniques.

The Analyst as a Business Problem Solver

Systems analysis and design is, first and foremost, a practical field. Analysts must certainly know about computers and computer programs. They possess special skills and develop expertise in programming. But they must also bring a fundamental curiosity to explore how things are done and the determination to make them work better.

Developing information systems is not just about writing programs. Information systems are developed to solve problems for organizations, as the opening case demonstrated, and a systems analyst is often thought of as a problem solver rather than a programmer. So, what kind of problems does an analyst typically solve?

- Customers want to order products any time of the day or night. So the problem is being able to process those orders round the clock without adding to the selling cost.
- Employees want paychecks automatically deposited in their accounts with options for withholding money for 401(k) plans. So the problem is being able to process payroll transactions that handle these options.
- Management wants to know the current financial picture of the company first thing every morning. So the problem is how to collect and present the information to them.
- Shipping needs to know what shipments to arrange for each day so the department can get them ready. So the problem is how to get the details about the day's shipments to the department.
- Production needs to plan very carefully the amount of each type of product to produce each week. So the problem is how to estimate the dozens of parameters that affect production and then allow planners to explore different scenarios before committing to a specific plan.

Information system developers work on problems such as these—and many more. Their focus is the business and how to make it run efficiently and effectively. All programming eventually done for the system contributes to solving the business problem, but these are not just programming problems.

How does an analyst solve problems? Systems analysis and design focuses on understanding the business problem and outlining the approach to be taken to solve it (see Figure 1-1). Usually the solution takes advantage of information technology—an information system. But that is just part of the story.

To thoroughly understand the problem, the analyst must learn everything possible about it—who is involved, what business processes come into play, what data need to be stored and used, what other systems would be affected when solving this problem. Then the analyst needs to confirm for management that the benefits of solving the problem outweigh the costs. Sometimes it would cost a fortune to solve the problem. Is it worth it to try to solve it at all?

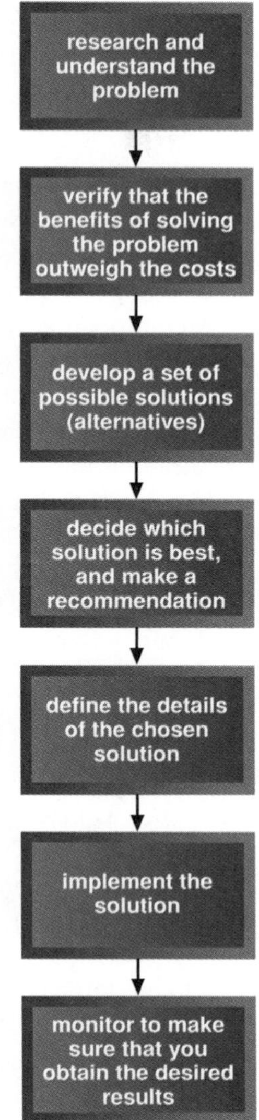

FIGURE 1-1
The analyst's approach to problem solving.

If solving the problem is feasible, the analyst develops a set of possible solutions. Each possible solution needs to be thought through carefully. Then the analyst needs to decide, in consultation with management, which possible solution is the best alternative overall. Usually, an information system is part of each alternative, but there can be many different configurations. The analyst must answer the following questions about each possibility:

- What should it do?
- What are the component parts?
- How should they be configured?
- What technology should be used to build the different components?
- Who should build them?

There are always many different alternatives to consider, and the challenge involves selecting the best—that is, the solution with the fewest risks and most benefits. The analyst needs to consider whether alternatives for solving the problem are cost effective, but they must also be consistent with the corporate strategic plan. Does the solution contribute to the basic goals and objectives of the organization? An alternative may solve the problem in the short run, but it can cause other problems down the line. Should this alternative be considered at all? Analysts must weigh the potential solutions and make tough decisions.

Once the systems analyst has decided which alternative to recommend and management has approved the recommendation, the details must be worked out. Here the analyst is concerned with creating a blueprint (design specifications) for how the new system will work. Systems design specifications cover databases, user interfaces, networks, operating procedures, conversion plans, and, of course, program modules. Once the design specifications are complete, the actual construction of the system can begin.

An information system can cost a lot of money to build and install, perhaps millions of dollars. So detailed plans must be drawn up. It is not unusual for 10 to 15 programmers to work on programs to get a system up and running, and those programmers need to know exactly what the system is to accomplish—thus, detailed specifications are required. We present in this text the tools and techniques that an analyst uses during system development to create the detailed specifications. Some of these specifications are the result of systems analysis and some are the result of systems design.

Although this text is oriented toward potential systems analysts, it also provides a good foundation for others who will become involved with a business problem that could be solved with the help of an information system. Managers throughout business must become more and more knowledgeable about how to use information technology to solve business problems. Many general business students take a systems analysis and design course to round out their background. Many graduate programs such as master of business administration (MBA) and master of accountancy (M'Acc) programs have technology tracks that include courses that use this book. Remember that systems analysis and design is not just about developing systems, but it is about solving business problems using information technology. So even though they never build an information system, managers need to gain expertise in these concepts to be effective in their jobs.

Systems That Solve Business Problems

We described the systems analyst as a business problem solver. We said that the solution to the problem is usually an information system. Before we talk about how you learn to be a systems analyst, let's quickly review what an information system is.

Information Systems

system
a collection of interrelated components that function together to achieve some outcome

information system
a collection of interrelated components that collect, process, store, and provide as output the information needed to complete business tasks

subsystem
a system that is part of a larger system

A **system** is a collection of interrelated components that function together to achieve some outcome. An **information system** is a collection of interrelated components that collect, process, store, and provide as output the information needed to complete a business task. Completing a business task is usually the "problem" we have talked about.

A payroll system, for example, collects information on employees and their work, processes and stores that information, and then produces paychecks and payroll reports (among other things) for the organization. A sales management system collects information about customers, sales, products, and inventory levels, and then processes and stores the information. Then information is provided to manufacturing so the department can schedule production.

What are the interrelated components of an information system? There are several ways to think about components. Any system can have **subsystems**. A subsystem is a system that is part of another system. So the subsystems might be one way to think about the components. For example, a customer support system might have an order-entry subsystem that creates new orders for customers. Another subsystem might handle fulfilling the order, including shipping and back orders. A third subsystem might maintain the product catalog database. The view of a system as a collection of subsystems is very useful to the analyst. These subsystems are interrelated components that function together.

supersystem
a larger system that contains other systems

Every system, in turn, is part of a larger system, called a **supersystem**. So the customer support system is really just a subsystem of the production system. The production system includes other subsystems, such as inventory management and manufacturing. Figure 1-2 shows how one system can be divided, or decomposed, into subsystems, which in turn can be further decomposed into subsystems. This approach to dividing a system into components is referred to as **functional decomposition**.

functional decomposition
dividing a system into components based on subsystems that in turn are further divided into subsystems

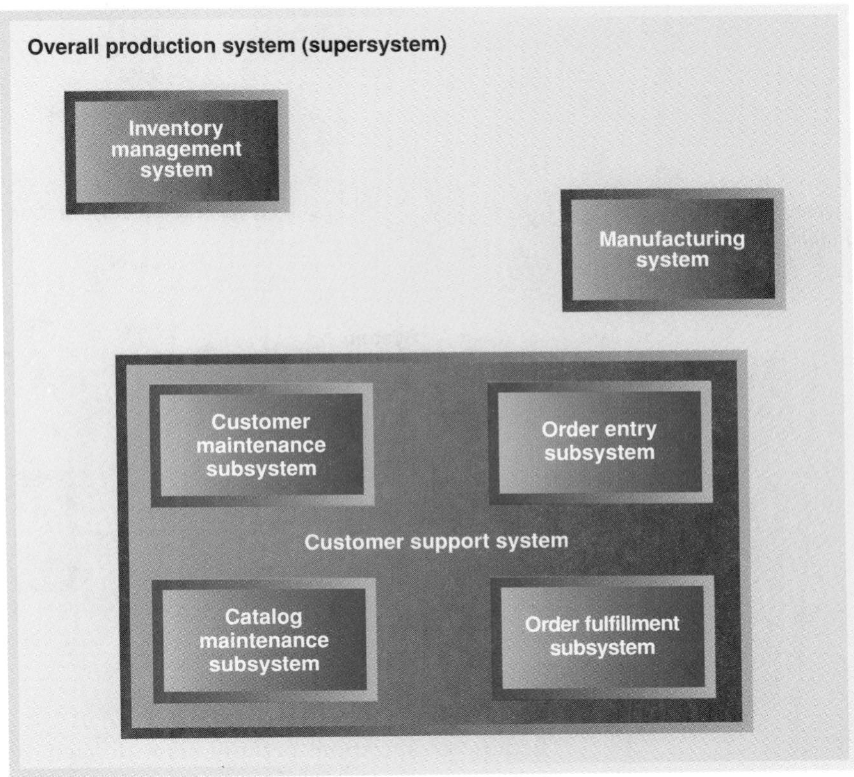

FIGURE 1-2
Information systems and subsystems.

Another way to think about the components of a system is to list the types of things that interact. For example, an information system includes hardware, software, inputs, outputs, data, people, and procedures. This view is also very useful to the analyst. The hardware, software, inputs, outputs, data, people, and procedures are interrelated components that function together in a system, as shown in Figure 1-3.

system boundary
the separation between a system and its environment that inputs and outputs must cross

automation boundary
the separation between the automated part of a system and the manual part of a system

Every system has a boundary between it and its environment. Any inputs or outputs must cross the **system boundary**. Defining what these inputs and outputs are is an important part of systems analysis and design. In an information system, people are also key components, and these people do some of the work accomplished by the system. So there is another boundary that is important to a systems analyst—the **automation boundary**. On one side of the automation boundary is the automated part of the system, where work is done by computers. On the other side is the manual part of the system, where work is done by people (see Figure 1-4).

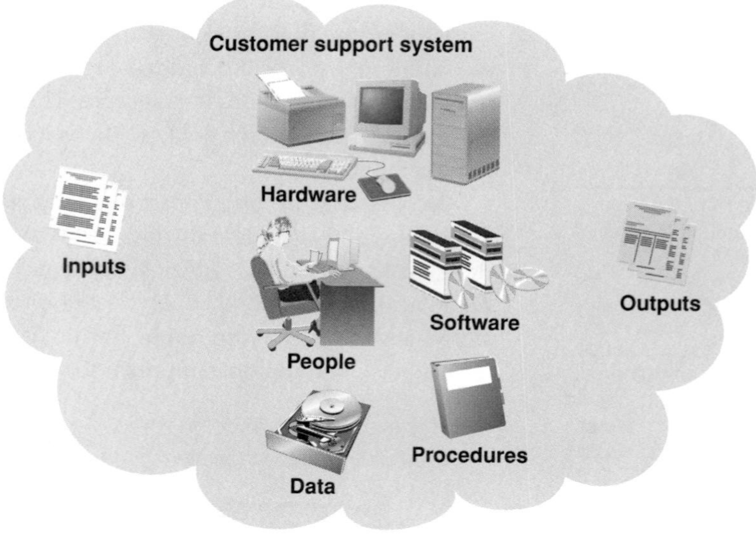

FIGURE 1-3
Information systems and component parts.

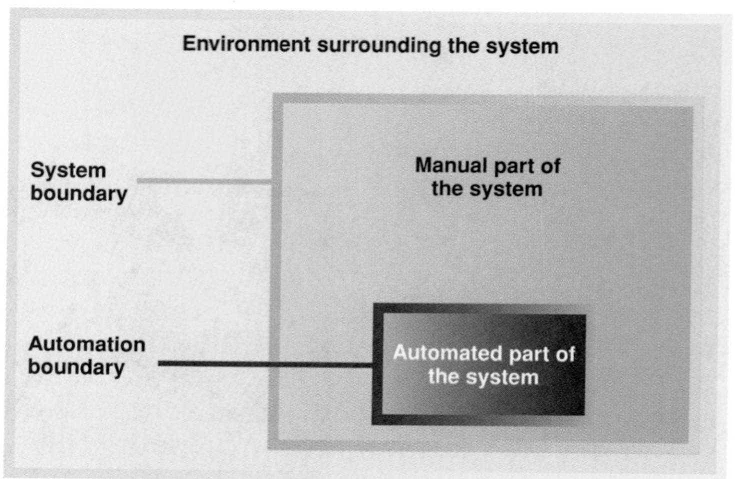

FIGURE 1-4
The system boundary versus the automation boundary.

Types of Information Systems

Because organizations perform many different types of activities, there are many different types of information systems. The systems found in most businesses include transaction processing systems, management information systems, executive information systems, decision support systems, communication support systems, and office support systems (see Figure 1-5). You learned about these types of systems in your introductory information systems course, so we will only briefly review them here.

FIGURE 1-5
Types of information systems.

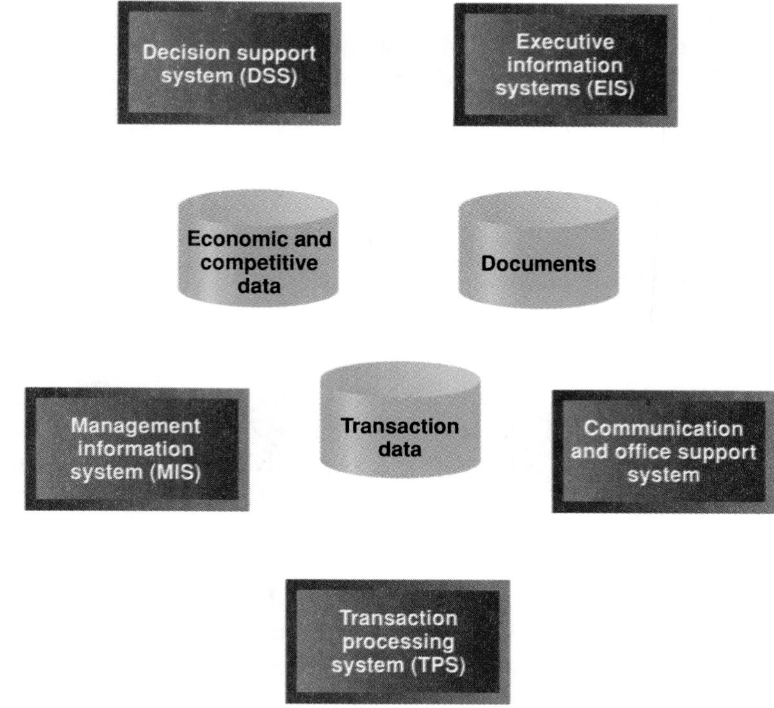

transaction processing systems (TPS)
information systems that capture and record information about the transactions that affect the organization

management information systems (MIS)
information systems that take information captured by transaction-processing systems and produce reports that management needs for planning and control

executive information systems (EIS)
information systems that provide information for executives to use in strategic planning

decision support systems (DSS)
support systems that allow a user to explore the impact of available options or decisions

communication support systems
support systems that allow employees to communicate with each other and with customers and suppliers

Transaction processing systems (TPS) capture and record information about the transactions that affect the organization. A transaction occurs each time a sale is made, each time supplies are ordered, each time an interest payment is made. Usually these transactions create credit or debit entries in accounting ledgers. They eventually end up on accounting statements used for financial accounting purposes, such as the income statement. Transaction processing systems were among the first to be automated by computers. Newer transaction processing systems use state-of-the-art technology and present some of the greatest challenges to information systems developers. They also present some of the greatest competitive advantages and returns on investments for companies. The newer transaction processing systems are called on-line transaction processing systems—OLTP.

Management information systems (MIS) are systems that take information captured by transaction processing systems and produce reports that management needs for planning and controlling the business. Management information systems are possible because the information has been captured by the transaction processing systems and placed in organizational databases.

Executive information systems (EIS) provide information for executives to use in strategic planning. Some of the information comes from the organizational databases, but much of the information comes from external sources—news about competitors, stock market reports, economic forecasts, and so on.

Decision support systems (DSS) allow a user to explore the impact of available options or decisions. Sometimes this is referred to as "what if" analysis, because the user asks the system to answer questions such as, "What if sales dip below $100 million during the third quarter and interest rates rise to 7.5 percent?" Financial projections made by the DSS can then explore the results. Some decision support systems are used to make routine operational decisions, such as how many rental cars to move from one city to another for a holiday weekend based on estimated business travel patterns.

office support systems
support systems that help employees create and share documents, including reports, proposals, and memos

Communication support systems allow employees to communicate with each other and with customers and suppliers. Communication support now includes e-mail, fax, Internet access, and video conferencing.

Office support systems help employees create and share documents, including reports, proposals, and memos. Office support systems also help maintain information about work schedules and meetings.

Required Skills of the Systems Analyst

Systems analysts (or anyone doing systems analysis and design work) need a great variety of special skills. First, they need to be able to understand how to build information systems, and this requires quite a bit of technical knowledge. Then, as discussed previously, they have to understand the business they are working for and how the business uses each of the types of systems. Finally, the analyst needs to understand quite a bit about people and the way they work, since people will use the information systems.

Technical Knowledge and Skills

It should not be surprising that a systems analyst needs technical expertise. Even if an analyst is not involved in programming duties, it is still crucial to have an understanding of different types of technology—what they are used for, how they work, and how they are evolving. No one person can be an expert at all types of technology; there are technical specialists to consult for the details. But a systems analyst should understand the fundamentals about:

- Computers and how they work
- Devices that interact with computers, including input devices, storage devices, and output devices
- Communications networks that connect computers
- Databases and database management systems
- Programming languages
- Operating systems and utilities

tools
software products used to help develop analysis and design specifications and completed system components

A systems analyst also needs to know a lot about tools and techniques for developing systems. **Tools** are software products that help develop analysis or design specifications and completed system components. Some tools used in system development include:

- Software packages such as Microsoft Access and PowerBuilder that can be used to develop systems
- Integrated development environments (IDEs) for specific programming languages, such as Sun Java Workshop or Microsoft C++
- Computer-aided system engineering (CASE) tools that store information about system specifications created by analysts and sometimes generate program code
- Program code generators, testing tools, configuration management tools, software library management tools, documentation support tools, project management tools, and so on

techniques
strategies for completing specific system development activities

Techniques are used to complete specific system development activities. How do you plan and manage a system development project? How do you complete systems analysis? How do you complete a systems design? How do you complete implementation and testing? How do you install and support a new information system? Much of this text explains how to use specific techniques for planning, analysis, and design. But it also covers some aspects of implementation and support. These techniques include:

- Project planning techniques
- Systems analysis techniques
- Systems design techniques
- System construction and implementation techniques
- System support techniques

Business Knowledge and Skills

Other knowledge and skills that are crucial for an analyst include those that apply to understanding business organizations in general. After all, the problem to be solved is a business problem. What does the analyst need to know? The following are examples:

- What activities and processes do organizations perform?
- How are organizations structured?
- How are organizations managed?
- What type of work goes on in organizations—for example, finance, manufacturing, marketing, customer service, and so on?

Systems analysts benefit from a fairly broad understanding of businesses in general, so they typically study business administration in college. In fact, computer information systems (CIS) or management information systems (MIS) majors are often included in the college of business for that reason. The accounting, marketing, management, and operations courses taken in a CIS or MIS degree program serve a very important purpose of preparing the graduate for the workplace.

Systems analysts also need to understand the type of organization for which they work. Some analysts specialize in a specific industry for their entire career—one might specialize in manufacturing, retailing, financial services, or aerospace. The reason for this is simple: It takes a long time to understand the problems of a specific industry. An analyst with deep understanding of a specific industry can solve complex problems for companies in the industry.

It is also important to understand a specific company. Often, just knowing the people who work for a company and understanding subtleties of the company culture can make a big difference in the effectiveness of an analyst. It takes years of experience working for the company to really understand what is going on. The more an analyst knows about how an organization works, the more effective he or she can be. Some specifics the analyst needs to know about the company include:

- What the specific organization does
- What makes it successful
- What its strategies and plans are
- What its traditions and values are

People Knowledge and Skills

Because analysts often work on development teams with other employees, systems analysts need to understand a lot about people and possess many people skills. An analyst spends a lot of time working with people, trying to understand their perspectives on the problems they are trying to solve. It is critical that the analyst understand:

- How people think
- How people learn
- How people react to change
- How people communicate
- How people work (in a variety of jobs and levels)

An analyst must use a variety of people skills to get the required information and to influence and motivate people to cooperate. These types of skills include interpersonal skills and communication skills. Because analysts come into contact with so many people throughout an organization, they have a unique opportunity to influence the organization as a whole.

A Few Words about Integrity and Ethics

One aspect of a career in information systems that is often underestimated by students is the importance of personal integrity and ethics. A systems analyst is asked to look into problems that involve information in many different parts of an organization. The information might be very private, especially if it involves individuals, such as salary, health, and job performance information. The analyst must have the integrity to keep this information private.

But the problems the analyst works on can also involve confidential corporate information, including proprietary information about products or planned products, strategic plans or tactics, and even top-secret information involving government military contracts. Sometimes security processes or specific security systems employed by the company can be involved in the analyst's work.

Some system developers work for consulting firms that are called in to work on specific problems for clients. They are expected to uphold the highest ethical standards when it comes to private proprietary information they might encounter on the job. Any appearance of impropriety can destroy an analyst's career.

The Environment Surrounding the Analyst

Types of Technology Encountered

Most students have used and are familiar with personal computers. In many university MIS or CIS degree programs, students complete modest course projects that run on PCs. But not all businesses function with desktop systems, and sometimes students don't recognize how different the large-scale systems are that they will work on when they are out in the real world.

Many basic systems now are run with personal computers on the desktop, but behind the scenes, these computers may be connected to data centers through very complex networks. When a company talks about an on-line order-processing application, it might be talking about a system with thousands of users spread over hundreds of locations. The database might contain hundreds of tables with millions of records in each table. The system might have taken years to complete, costing millions of dollars. If the system fails for even an hour, the company could lose millions of dollars in sales. Such a system is critical for the business. The programmers and analysts who work to support and maintain the system work in round-the-clock shifts, often wearing beepers in case of a problem. The importance of these systems to business cannot be overstated.

Different configurations of information systems that future analysts may encounter, which are discussed in later chapters, include:

- Desktop systems
- Networked desktop systems that share data
- Client-server systems
- Large-scale centralized mainframe systems
- Systems using Internet, intranet, and extranet technology

Just as an organization's business environment continually changes, so does the technology used for its information systems. The rapid change in technology often drives needed changes. Thus, it is important for everyone involved in information system development to upgrade their knowledge and skills continually. Those who don't will be left behind.

Typical Job Titles and Places of Employment

We have discussed the work of a systems analyst in terms of problem solving and system development. But it is important to recognize that many different people do systems analysis and design work. Not all have the job title of systems analyst. Sometimes analysis and design work is done by end users assigned to the project to provide expertise. Often analysis and design work is combined with other tasks (such as programming or end-user support).

Some job titles you may encounter are:

- Programmer analyst
- Business systems analyst
- System liaison
- End-user analyst
- Business consultant
- Systems consultant
- System support analyst
- System designer
- Software engineer
- System architect

Sometimes systems analysts might also be called project leaders or project managers. Be prepared to hear all kinds of titles for people who are involved in analysis and design work.

People who do analysis and design work are also found at organizations of all sizes—small businesses, medium-sized regional businesses, national Fortune 500 businesses, and multinational corporations. The type of technology used and the nature of the projects differ quite a bit based on the size of the organization. In addition, some businesses have very centralized information systems departments or divisions, while others have smaller information systems units that serve specialized parts of the organization. And the size of the organization may not always correspond to the size of the system. So some analysts work at smaller companies with a large company feel, while others work at a large company with a small company feel.

But not all analysts work directly for the company with the problem to solve. Analysis and design work is accomplished in many different work arrangements, including:

- Programmer analysts working for the company (part of the time is spent in analysis and design work)
- Systems analysts working for the company (who specialize in analysis and design)
- Independent contractors (who work as analysts or programmer analysts)
- Outsource provider employees (who work on- or off-site on a specific project)
- Consultants (who work for a consulting firm on a specific project)
- Software development firm employees (who work on developing and supporting packages)

Analysis and design work is also done in other situations. Naturally, companies like Microsoft, Sun Microsystems, and Netscape hire system developers to create or adapt software packages such as Office 2000 and operating systems such as Windows 2000.

These developers do analysis and design work, too, but the problem to be solved is generally different. Computer science graduates are more likely to work on Windows 2000 or Office 2000, where their technical skills can be used to best advantage. In this text, though, we assume the analyst is working on information systems that solve business problems, not on operating systems or software packages.

The Analyst's Role in Strategic Planning

We have described a systems analyst as someone who solves specific business problems by developing or maintaining information systems. The analyst might also be involved with senior managers on strategic management problems—that is, problems involving the future of the organization and plans and processes to ensure its survival and growth. Sometimes an analyst only a few years out of college can be summoned to meet with top-level executives and even asked to present recommendations to achieve corporate goals. How might this happen?

Special Projects

First, the analyst might be working to solve a problem that affects executives, such as designing an executive information system. The analyst might interview the executives to find out what information they need to do their work. An analyst may be asked to spend a day with an executive or even travel with an executive to get a feel for the nature of the executive's work. Then the analyst might develop and demonstrate prototypes of the system to get more insight into the needs of the executives.

Another situation that could involve an analyst in strategic management problems is a business process reengineering study. **Business process reengineering** seeks to alter radically the nature of the work done in a business function. The objective is radical improvement in performance, not just incremental improvement. Therefore, the analyst might be asked to participate in a study that looks carefully at existing business processes and procedures and then to propose information system solutions that can have a radical impact. Many tools and techniques of analysis and design are used to analyze business processes, redesign them, and then provide computer support to make it work.

business process reengineering
a technique that seeks to alter the nature of the work done in a business function with the objective of radically improving performance

Strategic Planning Processes

Most business organizations invest considerable time and energy completing strategic plans that typically cover five or more years in the future. During the **strategic planning** process, executives ask themselves such fundamental questions about the company as where they are now, where they want to be, and what they have to do to get there. A typical strategic planning process can take months or even years, and often plans are continually updated. Many people from throughout the organization are involved, all completing forecasts and analyses, which are combined into the overall strategic plan. Once a strategic plan is set, it drives all of the organization's processes, so all areas of the organization must participate and coordinate their activities. Therefore, a marketing strategic plan and a production strategic plan must fit within the overall strategic plan.

strategic planning
a process which executives try to answer questions about the company such as where they are now, where they want to be, and what they have to do to get there

Information Systems Strategic Planning

One major component of the strategic plan is the **information systems strategic plan**. Information systems are so tightly integrated with an organization that nearly any planned change calls for new or improved information systems. Beyond that, the information systems themselves often drive the strategic plan. For example, with the rapid developments in the Internet, many new companies have come into existence

information systems strategic plan
the plan defining the technology and applications the information systems function needs to provide to support the organization's strategic plan

(Amazon.com, eToys.com, etc.) and many others have changed the way they compete because of the new technology. In other cases, the possibilities presented by information systems technology have led to new products and markets with more subtle impacts. Information systems and the possibilities that they present play a large role in the strategic plans of most organizations.

Although information systems plans are always part of the company strategy, information systems strategic planning sometimes involves the whole organization. Usually at the recommendation of the chief information systems executive, top management will authorize a major project to plan the information systems for the entire organization. Unlike ongoing planning, a specific information systems strategic planning project might be authorized every five years or so, depending on the changes in the industry or company.

In developing the information systems strategic plan, members of the staff look at the organization overall to anticipate the problems that need solving rather than reacting to systems problems as they come up. Several techniques help the organization complete an information systems strategic planning project. A consulting firm is often hired to help with the project. Consultants can offer experience with strategic planning techniques and can train managers and analysts to complete the planning project.

Usually managers and staff from all areas of the organization are involved, but the project team is generally led by information systems managers with the assistance of consultants. Systems analysts often become involved in collecting information and interviewing people.

Many documents and existing systems are reviewed. Then the team tries to create a model of the entire organization in terms of the business functions it performs. Another model, one that shows the types of data the entire organization creates and uses, is also developed. The team examines all of the locations where functions are performed and data are created and used. From these models, the team puts together a list of integrated information systems for the organization to carry out its business functions, called the **application architecture plan**. Then, given the existing systems and other factors, the team outlines the sequence needed to implement the required systems.

applications architecture plan
a description of the integrated information systems needed by the organization to carry out its business functions

Given the list of information systems needed, the team defines the **technology architecture plan**—that is, the types of hardware, software, and communications networks required to implement all of the planned systems. The team must look at trends in technology and make commitments to specific technologies and possibly even technology vendors.

technology architecture plan
a description of the hardware, software, and communications networks required to implement planned information systems

In the ideal world, a comprehensive information systems planning project solves all of the problems faced by information systems managers. Unfortunately, the world continues to change at such a rate that plans must be continually updated. Unplanned information system projects come up all the time, and priorities must be continually evaluated.

Enterprise Resource Planning (ERP)

An increasing number of organizations are addressing their information systems requirements using an approach called **enterprise resource planning** (ERP), which commits to using an integrated set of software packages for key information systems. Software vendors such as SAP and PeopleSoft offer comprehensive package solutions for companies in specific industries. To adopt an ERP solution, the company must carefully study its existing processes and information needs and then determine which ERP vendor provides the best match. These ERP solutions are so complex that it takes a major commitment to research the options, involving nearly everyone in the information systems department and throughout the organization. The ERP solutions are also very expensive, in initial cost and in support cost. Much change is involved, for management and for staff. Once the decision is made to adopt an ERP approach, it is very difficult to return to

enterprise resource planning (ERP)
a process in which an organization commits to using an integrated set of software packages for key information systems

the old ways of doing business and the old systems that were replaced by the ERP solution. An analyst involved in an ERP project will have to draw heavily on technical, business, and people skills and can play a key role in the project.

A Strategic Systems Plan
for Rocky Mountain Outfitters

To demonstrate the important systems analysis and design techniques in this text, we use a system development project for a company named Rocky Mountain Outfitters (RMO) as a continuing example. RMO recently completed an information systems strategic planning project with the help of a consulting firm. The outcome of the plan included a list of information system development projects that the company needed to complete to reach its growth and profit objectives over the next five years. This section provides some background on the company and summarizes the overall information systems plan. Beginning with the next chapter, we will focus on one of the needed information systems—the customer support system. The reason for developing the customer support system, though, was identified in the strategic plan of the company.

Overview of Rocky Mountain Outfitters

RMO began in 1978 as the dream of John and Liz Blankens of Park City, Utah. Liz had always been interested in clothing and had worked her way through college by designing, sewing, and selling winter sports clothes to the local ski shops in Park City. She continued with this side business even after she graduated, and soon it was taking all of her time. It was also about this time that she and John Blankens were married. He had worked for several years for a retail department store chain after completing an MBA. Together they decided to see whether they could expand the business and reach a larger customer base.

The first step in their expansion involved direct mail-order sales to customers via a small catalog. As interest in the catalog increased, the Blankenses sought out additional lines of clothing and accessories to sell along with their own product lines. They also opened a retail store in Park City.

By the late 1990s, Rocky Mountain Outfitters had grown to a large, regional sports clothing distributor in the Rocky Mountain and Western states. The states of Arizona, New Mexico, Colorado, Utah, Wyoming, Idaho, Oregon, Washington, Nevada, and the eastern edge of California had seen tremendous growth in recreation activities. The market for both winter and summer sports clothes had also exploded. The sports of skiing, snow boarding, mountain biking, water skiing, jet skiing, river running, jogging, hiking, ATV biking, camping, mountain climbing, and rappelling had all seen a tremendous increase in interest in these states. Of course, all of these activities required appropriate sports clothes. RMO expanded its line of sportswear to respond to this market. It also added a line of casual and active wear to round out its offerings to the expanding market of active people.

Rocky Mountain Outfitters now employs over 600 people and produces almost $100 million annually in sales. The mail-order operation is the major source of revenue, at $70 million. Retail sales remained a modest part of the business, with sales of $2.5 million at the Park City retail store and $5 million at the recently opened Denver store. Several years ago, the Blankenses added a phone-order operation that now accounts for $20 million in sales. To the customer, this was a natural extension, but to RMO, considerable changes in systems were required to handle the phone orders on-line. Although the mail-order and phone-order operations share the same catalog, their systems and management remain separate. Integrating and expanding the systems that support mail-order and phone sales are a high priority in the RMO strategic systems plan.

Organization and Locations

Rocky Mountain Outfitters is still managed on a daily basis by John and Liz Blankens. John is president, and Liz is vice president of merchandising and distribution (see Figure 1-6). Other top managers include William McDougal, vice president of marketing and sales, and JoAnn White, vice president of finance and systems. The systems department reports to JoAnn White.

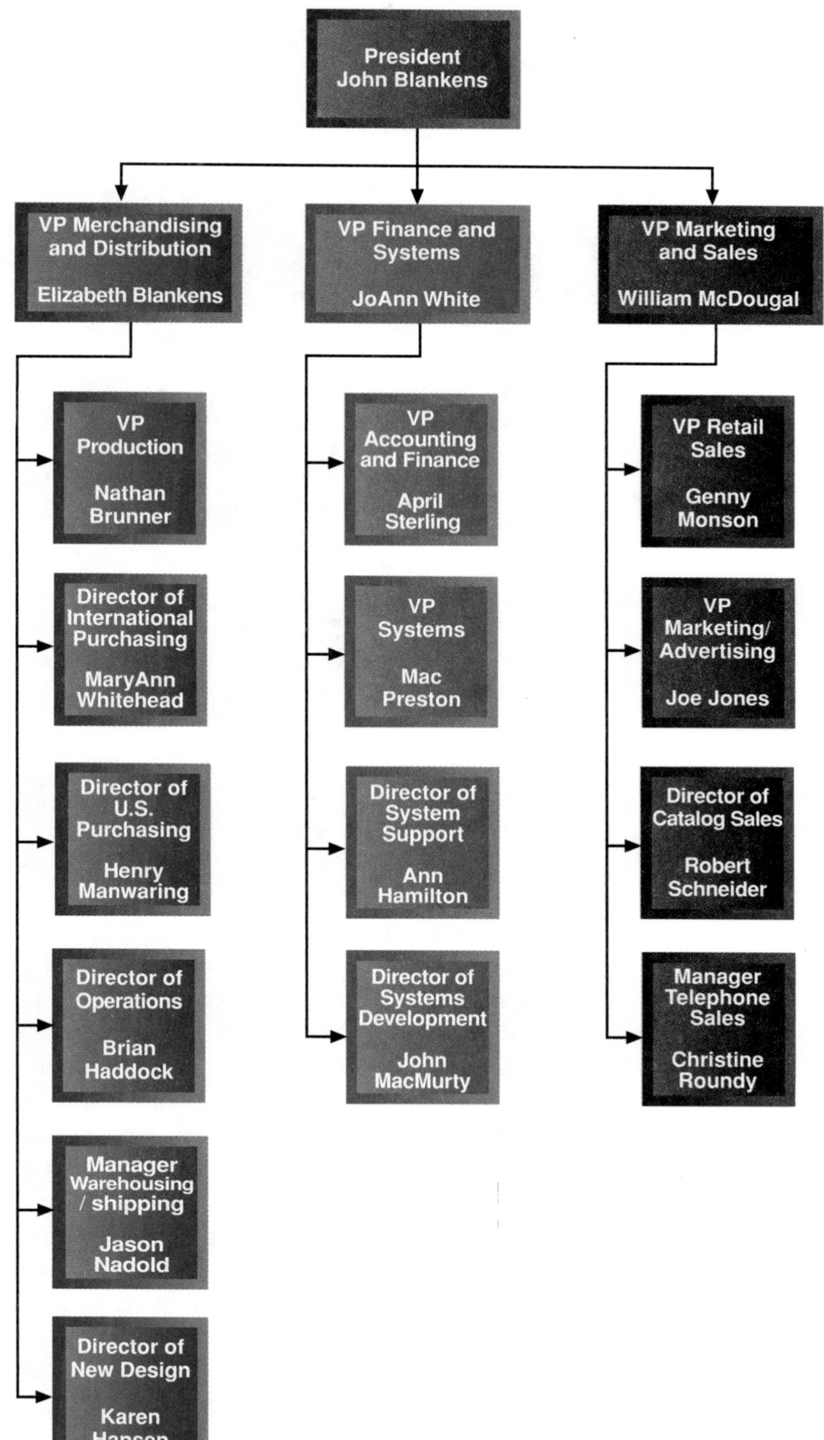

FIGURE 1-6
Rocky Mountain outfitters' organizational structure.

One hundred and thirteen employees work in human resources, merchandising, accounting and finance, marketing, and information systems in the corporate offices in Park City, Utah. There are two retail stores: the original Park City store and the newer Denver store. Manufacturing facilities are located in Salt Lake City and more recently Portland, Oregon. There are three distribution/warehouse facilities: Salt Lake City, Albuquerque, and Portland. All mail-order processing is done in a facility in Provo, Utah, employing 58 people. The phone-sales center, employing 20, is located in Salt Lake City. Figure 1-7 shows the locations of these facilities.

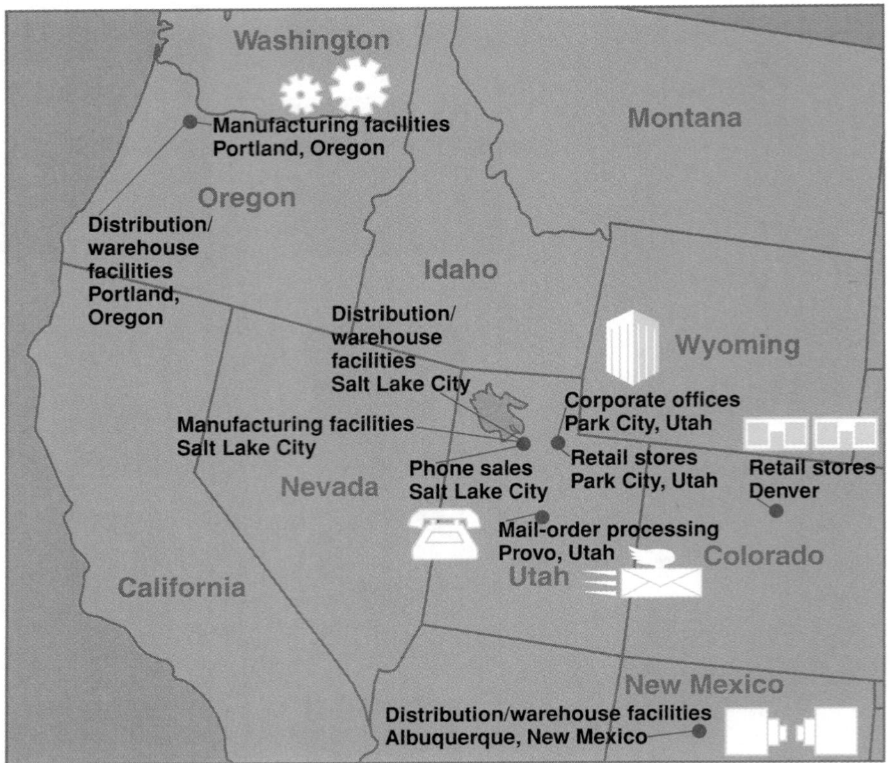

FIGURE 1-7
Rocky Mountain Outfitters' locations.

Information Systems Department Organization

The Information Systems department is headed by Mac Preston, an assistant vice president, and there are nearly 50 employees in the department (see Figure 1-8). Mac's title of assistant vice president reflects a promotion following the successful completion of the information systems planning project. He is not quite equal to a full vice president, but the position is considered increasingly important to the future of the company. Mac reports to the finance and systems vice president, whose background is in finance and accounting. The information systems department will eventually report to John Blankens directly if Mac has success implementing the new strategic information systems plan.

Mac organized information systems into two areas—system support and system development. Ann Hamilton is director of system support. System support involves such functions as telecommunications, database administration, operations, and user support. John MacMurty is director of system development. System development includes four project managers, six systems analysts, 10 programmer analysts, and a couple of clerical support employees.

FIGURE 1-8
Information systems department staffing.

IS staffing

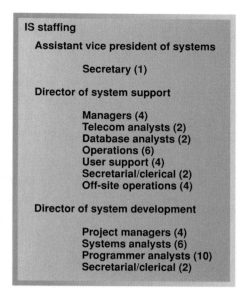

Assistant vice president of systems

Secretary (1)

Director of system support

Managers (4)
Telecom analysts (2)
Database analysts (2)
Operations (6)
User support (4)
Secretarial/clerical (2)
Off-site operations (4)

Director of system development

Project managers (4)
Systems analysts (6)
Programmer analysts (10)
Secretarial/clerical (2)

Existing Systems

Most of the computer technology and information systems staff at Rocky Mountain Outfitters is located at the data center in Park City. A small mainframe computer runs the inventory, mail-order, accounting, and human resource functions by connecting to the distribution and mail-order sites by dedicated telecommunication links. The manufacturing sites have dial-up capability to the mainframe.

Office functions at the home office, distribution sites, and manufacturing sites are supported by local area networks with file servers. Retail stores run a point-of-sale package with a local server and dial-up batch updates to the inventory system on the mainframe. The phone sales center has a small local area network with a client-server order-processing application running under Windows. Batch inventory updates are made to the mainframe.

The existing information systems technology includes:

- Retail Store Systems: A retail store package with point-of-sale and batch inventory update.
- Office Systems: A local area network with office software/e-mail in Park City and other sites.
- Merchandising/Distribution: A mainframe application developed in-house using COBOL/CICS with some DB2 relational database and VSAM files.
- Mail Order: A mainframe application developed in-house using COBOL. Mail-order clerks in Provo use dedicated terminals.
- Phone Order: A modest Windows application developed using Microsoft Access as a quick solution to customer demand for phone orders. It is a multiuser file-server design that is not integrated well with merchandising/distribution and has reached capacity.
- Human Resources: An application developed in-house for payroll and benefits running on the mainframe.
- Accounting/Finance: A mainframe package from a leading accounting package vendor, implemented 10 years ago.

The Information Systems Strategic Plan

The strategic thrust of RMO is to build more direct customer contact to improve service and to expand the marketing presence beyond the Western states. One strategy is to expand the phone-order capability, and the other is to add direct customer access through the Internet. The Information Systems department recognizes it has to plan for rapid growth and be prepared to enhance systems that support customer interaction.

The strategic systems planning project resulted in the following technology architecture plan, application architecture plan, and time frames for implementation (see Figure 1-9):

FIGURE 1-9
Rocky Mountain Outfitters' application architecture plan.

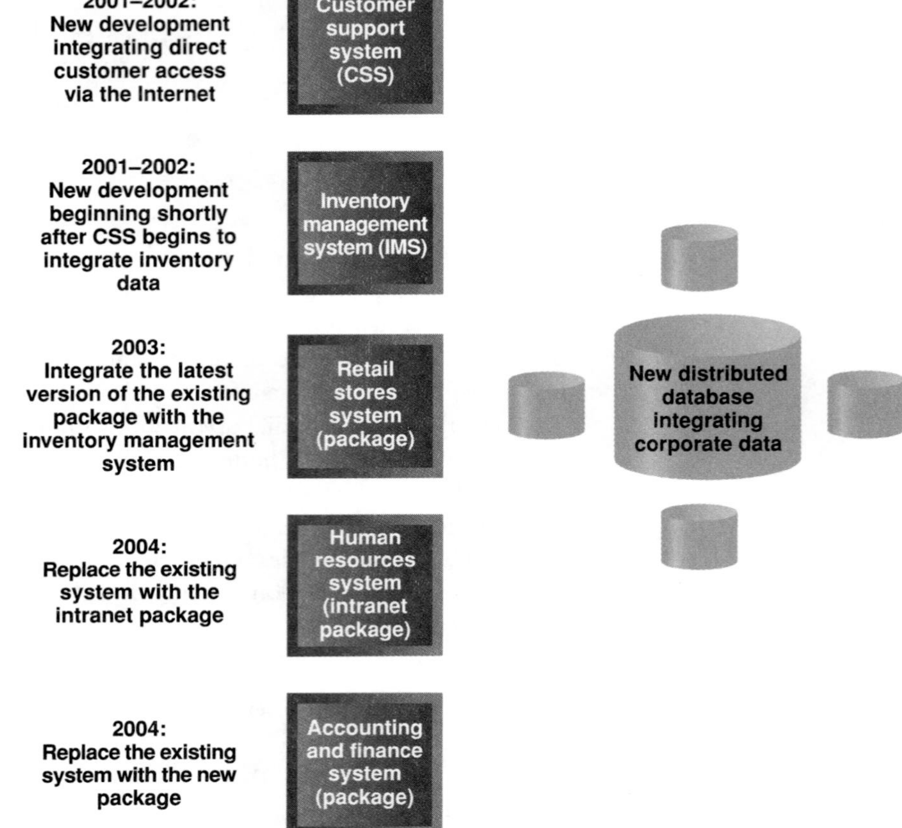
2001–2002:
New development integrating direct customer access via the Internet
Customer support system (CSS)
2001–2002:
New development beginning shortly after CSS begins to integrate inventory data
Inventory management system (IMS)
2003:
Integrate the latest version of the existing package with the inventory management system
Retail stores system (package)
2004:
Replace the existing system with the intranet package
Human resources system (intranet package)
2004:
Replace the existing system with the new package
Accounting and finance system (package)
New distributed database integrating corporate data

Technology architecture plan

- Move business applications to the client-server architecture, reserving the mainframe for database and telecommunications functions to allow incremental as well as rapid growth in capacity.
- Move toward conducting the business via the Internet, first with a web site, then with some customer and supplier transaction processing support linked to the internal systems and databases.
- Anticipate the eventual move toward intranet solutions.

Application architecture plan

- Accounting/finance: purchase a package solution with the client-server architecture.
- Human resources: purchase a package solution, possibly an intranet application, to maximize employee access to human resources (HR) forms, procedures, and benefits information.

- Customer support system: design an order-processing and fulfillment system that integrates the three order-processing requirements: mail order, phone order, and direct customer access.
- Inventory management system: design a merchandising and inventory system that can seamlessly integrate with customer support in all strategic areas: mail order, phone order, direct customer access, and retail store systems.
- Retail store systems: integrate store management systems with the inventory management system.

Time frames for implementing application architecture plan

- First, implement the customer support system (CSS).
- Second, implement the inventory management system, starting shortly after the customer support system is started. This can be a phased implementation, starting with improving the fulfillment/distribution component used by CSS, followed by improving the vendor management and other merchandising components.
- Third, integrate retail store systems with the inventory management system.
- Finally, upgrade the human resources system and the accounting/finance system.

Two Major New Systems

The first two projects planned include the customer support system and the inventory management system. These two systems actually resulted from some reengineering of the processes followed at Rocky Mountain Outfitters. Rather than just replace or marginally improve the existing systems, staff redefined how to group together the business processes logically. A few more details on each of these systems follow.

The Customer Support System The first system to be addressed is the customer support system. Rocky Mountain Outfitters has always prided itself on its customer orientation. One of the core competencies of RMO has been the ability to develop and maintain customer loyalty. John kept himself current on important business concepts, and he was fully aware of developing effective business processes that had a customer focus. So even though this system included all of the required sales and marketing components, he wanted everyone to understand that its primary objective was to support RMO's customers.

The customer support system includes all functions associated with providing product to the customer, from order entry to arrival of the shipment. Some of the functions that are included in the system include:

- Order entry
- Customer service/catalog
- Order tracking
- Shipping
- Back ordering
- Returns
- Sales analysis

Some wish lists associated with the system included the following: With the growth of the Internet, Rocky Mountain Outfitters considered it critical that it also provide Internet shopping capabilities. Catalog customers should be able to order either by telephone, through the mail, or over the Internet. All catalog items would also be available through a sophisticated Rocky Mountain Outfitters web catalog. Order entry needed to support the graphical, self-help style of the Internet, as well as a streamlined, rapid-response approach required for telephone and in-house clerks. All other types of sales and customer services also require both types of interfaces.

Such things as returns, back orders, and order status need to be accessible both on the web and to in-house operators. RMO wanted this system to be a sophisticated, high-support system. The project team would have to assess these possibilities carefully when proposing a solution.

The Inventory Management System The customer support system is described in detail later in the text, but it is important to understand the inventory management system because it has to support all those "behind the scenes" activities that ensure that merchandise is available and ready for sale. Even though Rocky Mountain Outfitters manufactures many of its lines of clothing, it does purchase most items from other manufacturers. The inventory management system includes the following functions:

- Vendor management
- Manufacturing requirements planning
- Normal volume ordering and reordering
- Emergency out-of-stock ordering
- Cost tracking
- Aging of inventory stock
- Turnover analysis
- Warehouse goods tracking

Even though RMO is not quite large enough to justify a completely automated warehouse, it was decided that the system should be built with row and bin information included in the inventory file in anticipation of the day when such information would be necessary.

Most of the functions supported by the inventory management system do not require immediate real-time responses. The exception may be the emergency out-of-stock ordering feature. Most of these functions can be performed at night, when the computer systems are not so busy.

Other ideas for the inventory management system include maximizing any benefits from electronic commerce either through the Internet or through electronic data interchange (EDI) capability.

The Analyst as a System Developer
(The Heart of the Course)

We have discussed many roles a systems analyst can play in an organization, including strategic planning and helping identify the major information systems projects that the business will pursue. However, the main work of an analyst is working on a specific information system development project. This text is about planning and executing an information systems project—in other words, working as a system developer. The text is organized around this theme. In this section, we provide an overview of the text.

Part 1: The Modern Systems Analyst

The first part of the text describes the work of the systems analyst. This chapter (Chapter 1) describes the nature of the analyst's work in terms of types of problems solved, the required skills, and the job titles and places where an analyst might work. We hope it is clear so far that the analyst does much more than think about and write programs. The rest of the text is organized around the problem-solving approach we described at the beginning of the chapter.

Not only does the analyst get involved with business problems, but the analyst can work on very high-level strategic issues and with people at all levels of the organization relatively early in his or her career. In this chapter we also described Rocky Mountain Outfitters and its strategic information systems plan. The rest of the text will focus on one of the planned new systems—the customer support system—and its development.

Chapter 2 gets to the heart of the system development project by describing how a project is planned and managed. The system development life cycle (SDLC) is introduced as the technique for managing and controlling the project. Other project management tools and techniques are also introduced, including feasibility studies, project scheduling, and project staffing. An information systems project is just like any other project in these respects. It is important for an analyst to understand the role of project management. Specific issues also arise when planning an information systems project, and an analyst needs to be familiar with them and how they relate to the larger context of the activities of project planning for an information systems project.

Chapter 3 focuses on the variety of approaches available for developing an information system. A variety of tools, techniques, and methodologies are discussed, including the traditional structured approach and the newer object-oriented approach. System developers should be familiar with the fundamental concepts of both approaches. This text covers both approaches throughout, pointing out where they are similar and where they are different. Systems analysis and design activities are also described in Chapter 3, and the same basic activities are followed no matter which approach to system development is used.

Part 2: Systems Analysis Tasks

Chapters 4 through 8 cover systems analysis in detail. Chapter 4 discusses techniques for gathering information about the problem to be solved by the new system so the system requirements can be defined. The variety of people who are affected by the system are also discussed. All of these people need to be interviewed and kept up to date on the status of the project. Techniques such as prototyping and walkthroughs are introduced to help the analyst communicate with everyone involved.

Chapter 5 introduces the concept of models and modeling to record the detailed requirements for the system in a useful form. When discussing an information system, two key concepts are particularly useful: "events" that cause the system to respond, and "things" the system needs to store information about. These two concepts, events and things, are important no matter which approach to system development you are using—either the traditional structured approach or the object-oriented approach. The entity-relationship diagram (ERD) is introduced as a model for showing the things affected in the traditional approach, and the class diagram is introduced as a model of things in the object-oriented approach.

Chapters 6 and 7 continue the discussion of modeling system requirements, at which point the traditional structured approach begins to look different from the object-oriented approach. Chapter 6 covers the traditional approach to requirements, which focuses on processes. Data flow diagrams (DFDs), structured English, and data flow definitions are emphasized. Chapter 7 covers the object-oriented approach to requirements, which focuses on objects and their interactions. Case diagrams, statechart diagrams, and sequence diagrams are emphasized. System developers should be familiar with both approaches, but it is important to recognize that in a given system development project, one approach or the other will be used. These two approaches to requirements are not used simultaneously.

Chapter 8 begins with an overview of the technical environments that might be used to implement the system. Then techniques for generating alternatives for actually implementing the system are presented. Each alternative is described and evaluated carefully for feasibility. Then the best alternative is recommended to management. The final approval of the recommended alternative is a key decision point for the project.

Part 3: System Design Tasks

Once one of the alternatives is chosen, work on the actual design details begins. Chapters 9 through 12 cover detailed design issues. Chapter 9 provides an overview of systems design, including the types of models used with the traditional approach (structure charts and pseudocode) and the object-oriented approach (class diagrams and sequence diagrams).

Chapter 10 describes the issues involved when designing the database for the system, using either a relational database, an object-oriented database, or a hybrid approach that combines relational databases with object technology.

Chapter 11 covers designing controls, inputs, and outputs, including types of reports that are typically produced on-line and on paper. Information systems controls are discussed, including the importance of assuring that inputs are accurate and complete and that processing is done correctly. Techniques for protecting the system from unauthorized access are also discussed.

Chapter 12 discusses human-computer interaction, including guidelines for developing user-friendly systems. The chapter covers Windows graphical user interfaces and newer web-based interfaces, and these design concepts apply to both the traditional approach and the object-oriented approach.

Part 4: System Implementation and Support

This text emphasizes systems analysis and design using a fairly traditional view of the SDLC. There are other approaches to developing systems with which you should be familiar that focus more on implementation. Rapid application development (RAD) and component-based development are discussed in Chapter 13. Implementing a software package instead of developing a system is almost always a viable alternative, and Chapter 14 describes software packages and the comprehensive approach to package solutions referred to as enterprise resource planning (ERP).

Finally, no matter how the system is obtained, a major part of the project is making the system operational and keeping it that way. Chapter 15 describes the fourth and fifth phases of the SDLC: system implementation and system support. The analyst's role in implementing the system includes quality control, testing, training users, and making the system operational (conversion). Maintenance and support of the system can continue for many years, involving fixing problems and enhancing the system over time.

Often, a new programmer analyst is involved in maintenance and support of an existing system. Maintenance and support of the system are also the most expensive parts of the project, and decisions made during analysis and design can have a big impact on the ease of maintenance and the overall cost of the system over its lifetime.

Summary

A systems analyst is someone who solves business problems using information systems technology. Problem solving means looking into the problem in great detail, understanding everything about the problem, generating several alternatives for solving the problem, and then picking the best solution. Information systems are usually part of the solution, and information systems development is much more than writing programs.

A system is a collection of interrelated components that function together to achieve some outcome. Information systems, like any other systems, contain components, and an information systems outcome is the solution to some business problem. Information systems components can be thought of as subsystems or as types of things that interact, such as hardware, software, inputs, outputs, data, people, and procedures. There are many different types of systems that solve organizational problems. These include transaction-processing systems, management information systems, executive information systems, decision support systems, communication support systems, and office support systems.

A systems analyst needs a variety of knowledge and skills. These include technical, business, and people knowledge and skills. Integrity and ethical behavior are crucial to the success of the analyst. Analysts encounter a variety of technologies that often change rapidly. Systems analysis and design work is done by people with a variety of job titles, including systems analyst but also programmer analyst, systems consultant, and system engineer, among others. Analysts also work for consulting firms, as independent contractors, and for companies that produce software packages.

A systems analyst can become involved in strategic planning by working with executives on specific systems, by helping with business process reengineering projects, and by working on information systems strategic plans. Sometimes an information systems strategic planning project is conducted for the entire organization, and analysts often are involved. The Rocky Mountain Outfitters strategic planning project described in this chapter is an example.

Usually the systems analyst works on one systems development project, one that solves a business problem identified by strategic planning. That is the emphasis in the rest of this text: how the analyst works on a system development project, completing project planning, systems analysis, systems design, systems implementation, and system support activities.

Key Terms

applications architecture plan, p. 15
automation boundary, p. 8
business process reengineering , p. 14
communication support systems, p. 10
decision support systems, p. 9
enterprise resource planning (ERP), p.15
executive information systems (EIS), p.9
functional decomposition, p.7
information system, p.6
information systems strategic plan, p. 15
management information systems (MIS), p. 9
office support systems, p. 10
strategic planning, p. 14
subsystem, p. 7
supersystem, p. 7
system, p. 6
system analysis, p. 3
system boundary, p. 8
system design, p. 3
systems analyst, p. 3
techniques, p. 10
technology architecture plan, p. 15
tools, p. 10
transaction processing systems (TPS), p. 9

Review Questions

1. Give an example of a business problem.
2. What are the main steps followed when solving a problem?
3. Define *system*.
4. Define *information system*.
5. What are the types of information systems found in most organizations?
6. List the six fundamental technologies an analyst needs to understand.

7. List four types of tools the analyst needs to use to develop systems.
8. List five types of techniques used during system development.
9. What are some of the things an analyst needs to understand about businesses and organizations in general?
10. What are some of the things an analyst needs to understand about people?
11. What are some of the types of technology an analyst might encounter?
12. List 10 job titles that involve analysis and design work.
13. How might an analyst become involved with executives and strategic planning relatively early in his or her career?

Thinking Critically

1. Describe a business problem your university has that you would like to see solved. How can information technology help solve it?
2. Describe how you would go about solving a problem you face. Is the approach taken by a systems analyst, as described in the text, any different?
3. Many different types of information systems were described in this chapter. Give an example of each type of system that might be used by a university.
4. What is the difference between technical skills and business skills? Explain how a computer science graduate might be strong in one area and weak in another. Discuss how the preparation for a CIS or MIS graduate is different from computer science.
5. Explain why an analyst needs to understand how people think, how they learn, how they react to change, how they communicate, and how they work.
6. Who needs greater integrity to be successful, a salesperson or a systems analyst? Or does every working professional need integrity and ethical behavior to be successful? Discuss.
7. Explain why developing an information system requires different skills if the system is client-server versus large-scale centralized mainframe architecture.
8. How might working for a consulting firm for a variety of companies make it difficult for the consultant to understand the business problem a particular company faces? What might be easier for the consultant to understand about a business problem?
9. Explain why a strategic information systems planning project must involve people outside the information systems department. Why would a consulting firm be called in to help organize the project?
10. Explain why a commitment to enterprise resource planning (ERP) would be so difficult to undo once it has been made.

Experiential Exercises

1. It is important to understand the nature of the business you work for as an analyst. Contact some information systems developers and ask them about their employers. Do they seem to know a lot about the nature of the business? What types of classes did they take to prepare themselves (for example, banking classes, insurance classes, retail management classes, hospital administration classes, manufacturing technology classes)? Do they plan to take additional courses? If so, which?
2. Think about the type of position you want (working for a specific company, working for a consulting firm, or working for a software package vendor). Do some research on each type by looking at companies' recruiting brochures or web sites. What do they indicate are the key skills they look for in a new hire? Are there any noticeable differences between consulting firms and the other organizations?

Case Study

Association for Information Technology Professionals Meeting

"I'll tell you exactly what I look for when I interview a new college grad," Alice Adams volunteered. Alice, a system development manager at a local bank, was talking with several professional acquaintances at a monthly dinner meeting of the Association for Information Technology Professionals (AITP). AITP provides opportunities for information systems professionals to get together occasionally and share experiences. Usually a few dozen professionals from information systems departments at a variety of companies attend the monthly meetings.

"When I interview students, I look for problem-solving skill" continued Alice. "Every student I interview claims to know all about COBOL and Java and SQL and XYZ or whatever the latest package is," claimed Alice. "But I always ask them one thing: 'How do you generally approach solving problems?' And then I say, "I know you know how to program, but what would you say are the greatest problems facing the banking industry these days?'"

Jim Parsons, a database administrator for the local hospital, laughed. "Yes, I know what you mean. It really impresses me if they seem to appreciate how a hospital functions, what the problems are for us—how information technology can help solve some of our problems. It is the ability to see the big picture that really gets my attention."

"Yeah, I'm with you," added Sam Young, the manager of marketing systems for a retail store chain. "I am not that impressed with the specific technical skills an applicant has. I assume they have the aptitude and some skills. I do want to know how well they can communicate. I do want to know how much they know about the nature of our business. I do want to know how interested they are in retail stores and the problems we face."

"Exactly," confirmed Alice.

1. Do you agree with Alice and the others about the importance of problem-solving skills? Discuss.
2. Should you research how a hospital is managed before interviewing for a position with an information systems manager at a hospital? Discuss.
3. Do you think it really makes a difference whether you go to work for a bank, a hospital, or a retail chain? Discuss.

Further Resources

For a comprehensive review of information systems concepts, see:

Carol Frenzel, *Management of Information Technology*. Course Technology, 1998.
Effy Oz, *Management Information Systems (2nd ed.)*, Course Technology, 2000.
Kathy Schwalbe, *Information Technology Project Management*. Course Technology, 1999.
Ralph M. Stair and George W. Reynolds, *Principles of Information Systems* (4th ed.), Course Technology, 1999.

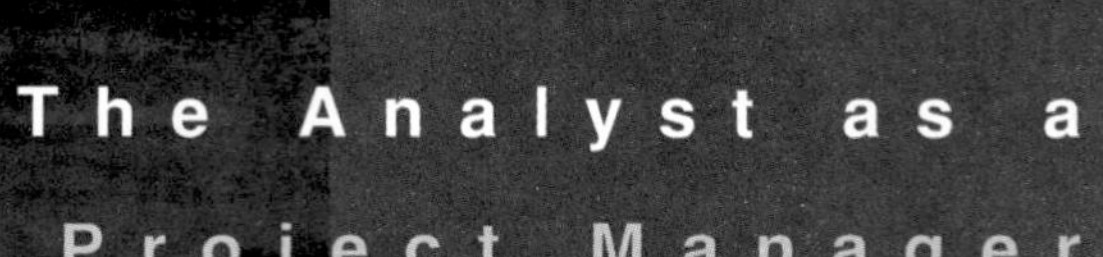

The Analyst as a Project Manager

Learning Objectives

After reading this chapter, you should be able to:

- Explain the purpose and various phases of the systems development life cycle (SDLC)
- Explain the responsibilities of a project manager
- Explain the activities in the project planning phase of the SDLC
- Describe how the scope of the new system is determined
- Develop a cost/benefit analysis of a proposed project
- Develop a project schedule using PERT and Gantt charts

Chapter Outline

Systems Development Projects and the SDLC

Project Management

Project Initiation

The Project Planning Phase

BLUE SKY FAMILY OF MUTUAL FUNDS: MANAGING THE IRA PROJECT

Gary Johnson was just putting the final touches on his presentation for tomorrow's executive review meeting. He was the leader for a small team of three systems analysts who had been working frantically over the last month to plan the development of a new system and to prepare the report. The three had worked closely to gather the information, do the evaluation, and prepare the report and presentation prior to launching the project. He thought back to six weeks ago, when he first got word of this new assignment.

Gary had been the project manager on a successful development project that was in the final stages of testing before going into production when he received the assignment to investigate the feasibility of a new system. He spent the first week after being assigned the new project turning over his current management responsibilities to the new project leader. The second week he located two other experienced systems analysts to work with him. It took almost the whole week to get their schedules reworked so that they could work with him for the month it would take to develop the new project plan. Finally, after two weeks, he and his team were able to begin work in earnest.

The objective of the new system is to support the sales and administration of retirement account investments for the Blue Sky Family of Mutual Funds. Blue Sky offers many different types of mutual funds to the public, including investments in investment retirement accounts (IRAs). IRAs are a special breed of mutual fund that have very strict information and reporting requirements. Part of the strategic plan for Blue Sky was a project to replace the current information system supporting IRAs. The project was scheduled to begin in six months. However, new legislation had just opened a possibility for Blue Sky to offer a new type of retirement mutual fund, the Roth IRA. To maintain its competitive position, Blue Sky needed to begin offering this new IRA beginning next January. Suddenly, the need for a new system was accelerated. The strategic plan was modified to indicate that development needed to begin immediately.

Even though the system was almost certain to be approved and funded by senior management, Gary still thoroughly evaluated the project's feasibility. His first activity had been to define the problem precisely and determine exactly what the scope of the project should be. Obviously, he couldn't estimate the size of the project unless he knew what the system needed to include, so he developed a list of business benefits of the new system and a detailed list of functions that the system had to provide. This list of functions, called the system capabilities, helped him begin to estimate the cost of developing the system as well as the expected financial benefits. He then began to develop a plan and schedule for the project based on the required system capabilities and some preliminary brainstorming the three team members had done to identify possible system alternatives.

As Gary worked on the schedule, he assigned one of the other analysts to confirm the feasibility of the project and to investigate potential problems that could cause difficulties in the development and deployment of the new system. The third analyst began calculating the costs and benefits to ensure that the new system was still economically feasible. All three team members had to coordinate their efforts because the results were interdependent.

Over the last several days, they had put the results into a final report and prepared for the executive presentation. It always made Gary nervous to present before the executives of the firm, especially for a project that had such high visibility and importance to the company. Although he was nervous, he was also confident. He and the team had done a thorough and professional job. He had already begun planning for his next task, which was to begin staffing the project and investigating the system requirements in much more detail.

OVERVIEW

Chapter 1 described the business environment, with its insatiable need for information systems in today's competitive and rapidly paced global economy. That chapter also identified the role of systems analysts and their role in information technology (IT) and IT strategic planning. You also learned about the various types of information systems. In this chapter, we begin to narrow the focus to explain the specifics of how an information system is developed within a company. The objective of this textbook, and the course that you are taking, is to teach you state-of-the-art concepts, techniques, and skills that you will use in the development of new information systems for today's organizations. This chapter continues with this objective by introducing and explaining the concept of the systems development life cycle. The systems development life cycle is the general term that is used to describe the method and process of developing a new information system.

This chapter also introduces the principles of project management. Project management encompasses the skills and techniques that are necessary to successfully plan and manage the development of a new system. As a knowledge worker and problem solver, you will need both technical skills and management skills to be a contributing member of a systems development team. This chapter provides you with the fundamentals of project management, and later chapters further elaborate key management principles associated with the various phases of the project.

The last major section in the chapter describes the activities of the project planning phase. The planning process for a new project entails several important steps, such as defining the scope of the project, comparing the estimated costs and anticipated benefits of the new system, and developing a project schedule. The section explains these specific steps and teaches the skills associated with the steps.

Systems Development
Projects and the SDLC

project
A planned undertaking that has a beginning and an end, and which produces a predetermined result or product

Chapter 1 explained that systems analysts solve business problems. Obviously, to make this problem-solving work productive, it needs to be organized and goal-oriented. An analyst accomplishes that result by organizing this work into projects. A **project** is a planned undertaking that has a beginning and an end, and produces a predetermined result or product. The term *systems development project* describes a planned undertaking, which is normally a large job that produces a new system. In our case, the system being developed is a computer-based information system. Some development projects are very large, requiring hundreds of hours of work by many people and spanning several calendar years. For a systems development project to be successful, an analyst must have detailed plans. Success depends heavily on having an organized, methodical sequence of tasks and activities that culminate with an information system that is reliable, robust, and efficient.

The Systems Development Life Cycle

phase
A division of the SDLC in which similar activities are performed

The development of any new information system normally requires three major sets of activities: analysis activities, design activities, and implementation activities. Analysis activities are those that provide a thorough understanding of the business's information needs and requirements. The focus of analysis is on the business needs, not on any particular computer technology. Design activities are those that define the architecture and structure of a new system to satisfy those requirements. It is during design that analysts begin the process of conceptualizing a computer-system solution. Implementation is the actual construction, testing, and installation of a functioning information system. In information system terms, each set of activities is a **phase**. Thus, a development project has an analysis phase, a design phase, and an implementation phase. Figure 2-1, which shows the phases of information systems development, adds two more phases. The first phase, project planning, consists of activities that are required to initiate, plan, and obtain approval for the project. It is normally a short phase, but it is critical to the overall success of the project.

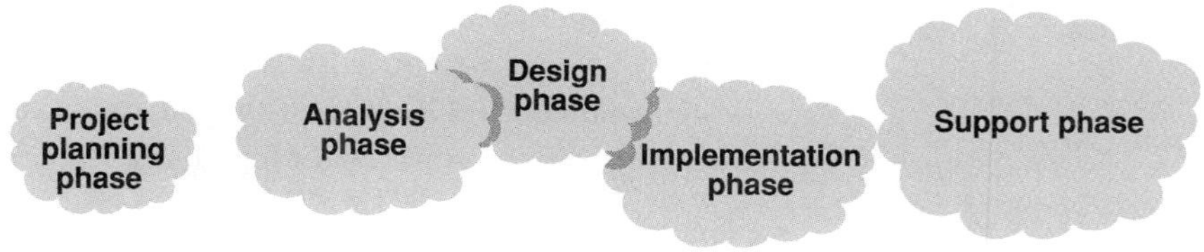

FIGURE 2-1
Information systems development phases.

The last phase, support, is really not part of project, but it is a phase in the total life of the information system. The support phase includes all the activities that maintain and enhance the information system during its productive lifetime. Obviously, the support phase encompasses a much longer period of time than all the other phases combined. It also usually is the most expensive.

systems development life cycle (SDLC)
A method of systems development that consists of planning, analysis, design, implementation, and support phases

This method of systems development is called the **systems development life cycle (SDLC)**. The activities of every project can be classified into these five phases. This classification of the activities of a systems development project is the most fundamental. However, in today's complex world of systems development, there are many variations on the basic five-phase structure illustrated in Figure 2-1. For example, different development approaches include prototyping, rapid application development (RAD), and joint application development (JAD). The differences in the various approaches are either in the set of activities composing each phase or in the method of carrying out the activity. But all information systems development requires planning, analyzing the need, designing a solution, and implementing the final system. As you gain an understanding of the objective of each phase, and the various alternative activities, you will be able to adjust to any type of systems development. This chapter introduces the activities of a standard, conventional SDLC. Later chapters elaborate the conventional approach and explain the various alternative approaches to systems development.

An SDLC is more than simply organizing the project into phases. The idea of an SDLC includes principles of management, planning, organization, problem solving, coordination, control, division of labor, scheduling, and so forth. Since the primary objective of this textbook is to help you develop the skills to be a successful systems analyst on a development project, the principles of systems development are explained in the context of the activities associated with systems development.

It is important to note that all successful large development projects use some type of organized systems development method. Small projects also require analysis, design, and implementation, but may not be as rigorous or include as much planning. The concept of relating similar activities into phases and defining the steps is what makes the SDLC a method for development. Without the structure and organization provided by an SDLC method, development projects are at severe risk for missed deadlines, escalating budgets, and ultimately a low-quality system. As a methodology, an SDLC provides the structure, method, controls, and checklist needed to ensure successful development.

The First Four Phases of the Systems Development Life Cycle

To begin to understand information systems development, you first must identify the main objectives and primary activities within each of the first four phases of the systems development life cycle.

planning phase
The initial phase of the SDLC whose objective is to scope and plan the project

The Planning Phase The primary objectives of the **planning phase** are to identify the scope of the new system, ensure that the project is feasible, develop a schedule, allocate resources, and budget for the remainder of the project. The following are the five activities in the project planning phase:

- Define the problem
- Confirm project feasibility
- Produce the project schedule
- Staff the project
- Launch the project

Probably the most important activity of the planning phase is to define precisely the business problem and the scope of the required solution. At this stage in the project, you will not know all of the functions or processes that will be included within the system. However, it is important to identify the major uses of the new system and the business problems that the new system must address.

The next major element is to confirm that the project is feasible. Many projects are initiated as part of an enterprisewide strategic plan. Within the plan, each project must be feasible on its own. Feasibility analysis includes financial, operational, technical, and schedule feasibility. Each of these types of feasibility analysis is explained in more detail later in the chapter.

The next two activities, producing the project schedule and staffing the project, are almost always done together. A detailed project schedule with tasks, activities, and required staff is developed. Fortunately, excellent methods and tools are available to provide support for this activity. Large projects require elaborate schedules with specific identifiable milestones and control procedures. A critical task is identifying the necessary human resources and a plan to acquire those resources at the required times during the project.

Finally, the total plan for the project is reviewed with upper management and the project is initiated. Initiation of the project entails allocating funds, assigning project members, and obtaining other necessary resources such as offices and development tools. An official announcement often communicates that the project has begun.

analysis phase
One phase of the SDLC whose objective is to understand the user's needs and develop requirements

The Analysis Phase The primary objective of the **analysis phase** is to understand and document the business needs and the processing requirements of the new system. Analysis is essentially a discovery process. The key words to drive the activities during analysis are discovery and understanding. There are six primary activities that are considered part of this phase:

- Gather information
- Define system requirements
- Build prototypes for discovery of requirements
- Prioritize requirements
- Generate and evaluate alternatives
- Review recommendations with management

problem domain
The area of the user's business for which a system is being developed

Gathering information is a fundamental part of analysis. During this activity, the systems analysts meet with users to learn as much as possible about the **problem domain**. The problem domain is the area of the user's business that needs an information system solution and is being researched. Analysts obtain information about the problem domain by observing the users performing the business procedures; by interviewing and asking questions of the users; by reading existing documents about procedures, business rules, and job responsibilities; and by reviewing existing automated systems. Gathering information is the core activity for discovery and understanding.

It is not sufficient simply to gather information. The information obtained must be reviewed, analyzed, and structured so that an overall understanding of requirements for the new system can be developed. This is called defining the system requirements, and the primary technique that is used is to draw diagrams that express the processing requirements for the new system. The objective of the diagrams is to model the processing requirements for the new system.

One important activity to help to gather and understand the requirements is to build a prototype of pieces of the new system for the users to review. In developing requirements for new systems, it is often easier for users to express their needs by reviewing working prototypes of processing alternatives. In much the same way as "a picture is worth a thousand words," a prototype can elicit valuable insights from end users.

As the processing requirements are uncovered, each must be prioritized. There are always more requests for automation support than the budget or resources can provide. Thus, the most important needs must be identified and given priority in the new systems development. As requirements are prioritized, various alternative solutions are researched. Analysts investigate such issues as how many of the high-priority requirements can be included, how state-of-the-art the new system should be, and whether it should be built from scratch or bought and adapted.

Finally, one of the alternatives is selected and recommended to upper management. A recap of the results of the analysis phase activities is provided and decisions about an alternative are finalized.

design phase
The phase of the SDLC in which the system and programs are designed

The Design Phase The objective of the **design phase** is to design the solution system. High-level design consists of developing an architectural structure for software programs, databases, the user interface, and the operating environment. Low-level design entails developing the detailed algorithms and data structures that are required for program development. The design phase utilizes as its input the information obtained during the analysis phase. Seven major activities must be done during design. Design activities are closely interrelated and generally have substantial overlap.

- Design and integrate the network
- Design the application architecture
- Design the user interfaces
- Design the system interfaces
- Design and integrate the database
- Prototype for design details
- Design and integrate the system controls

The network consists of computer equipment, network, and operating system platforms that will house the new information system. Many of today's new systems are being installed in network and client-server environments. Design includes configuring these network environments. Sometimes the design uses the existing operating environment and strategic IT plans. At other times, substantial work must be done to develop an operating environment to provide the level of service required by the new system.

application
The portion of the new information system that satisfies the user's needs in the problem domain

The **application** is the portion of the new information system that satisfies the user's needs in the problem domain. In other words, the application provides the processing functions for the business requirements. Designing the application consists of taking the model diagrams of the requirements that were developed during analysis and designing the appropriate computer programs.

The user interface is a critical component of any new system. During the analysis activities, some of the prototyping activity may have defined elements of the user interface. These elements are all combined during the design phase to produce an integrated user interface consisting of forms, reports, screens, and sequences of interactions.

Most new information systems must also communicate with other, existing systems. The design of the method and details of these communication links must be precisely defined.

Databases and information files are an integral part of information systems for business. The model diagrams of the information requirements defined during analysis are used to design the database that will support the applications of the new system. At times the database for the specific system must be integrated with information databases of other systems already in use.

During the design phase it is often necessary to verify the workability of the proposed design. One method to accomplish this is to build working prototypes of parts of the system to ensure that it will function correctly in the operating environment. Analysts can test and verify alternative design strategies by building prototypes of the new system. Sometimes, prototypes can be saved and used as part of the final system.

Finally, every system must have sufficient controls to protect the integrity of the database and the application program. Because of the highly competitive nature of the global economy and the risks associated with technology and security, every new system must include mechanisms to protect the information and assets of the organization. These controls should be integrated into the new system while it is being designed, not after it has been constructed.

implementation phase
The phase of the SDLC in which the new system is programmed and installed

The Implementation Phase During the **implementation phase**, the final system is built, tested, and installed. The objective of this phase is not only to have a reliable, working information system but also to ensure that the users are all trained and that the business is benefiting. All the prior activities come together during this phase to culminate in an operational system. Six major activities make up the implementation phase:

- Construct software components
- Verify and test
- Develop prototypes for tuning
- Convert data
- Train and document
- Install the system

Construction of the software can be achieved through various techniques. The conventional approach is to write computer programs using a language such as Visual Basic or Java. Other techniques, based on development tools and preexisting components, are becoming popular today. The software must also be tested. The first kind of testing verifies that the system actually works. Additional testing is also required to make sure that the new system meets the needs of the system's users.

During implementation, the analysts may build additional prototypes. These prototypes are used to verify different implementation strategies and to ensure that the system is tuned so that it can handle the volume of transactions after it is placed in production.

Almost every new system replaces an existing system, either a completely manual system or an earlier automated system. Normally, the existing information is important and needs to be converted to the format required in the new system. Converting the data often becomes a small project of its own, with analysis, design, and implementation to clean and convert the data to the new system.

No system is successful unless the users understand it and can use it appropriately. A critical activity during implementation is to train the users on the new system so that they will be productive as soon as possible.

Finally, the actual changeover is the culminating activity in this phase. The new equipment must be in place and functioning, the new computer programs must be installed and working, and the database must be populated and available. The individual pieces of the new system must be up and running before the system can be used for its intended purpose. In today's widely dispersed organizations, the system must frequently be installed in many locations and integrated throughout the organization.

support phase
A phase that occurs after the system is installed

help desk
The availability of support staff to help the users with any technical or processing problem associated with an information system

The Support Phase The objective of the **support phase** is to keep the system running productively during the years following its initial installation. The activities during this phase fall into two categories: (1) providing support to end users and (2) maintaining and enhancing the computer system. End-user support is an ongoing effort that is carried out through **help desks** and training programs. A help desk consists of technical personnel who can provide help when users have questions or problems. It may or may not be a physical location such as a desk. Enhancements and upgrades to the system occur throughout the life of the system to maintain and extend its productive life.

System maintenance can also be classified into different types of activities, from simple program error corrections (also known as bug fixes) to comprehensive enhancements and upgrades. Again, the range of activities is quite broad. Within the support phase, additional development projects may occur. During your career, you may have the opportunity to participate in several upgrades.

Scheduling of Project Phases

waterfall method
A method of executing an SDLC where one phase leads (falls) to the next phase

Figure 2-1 may have given the impression that a project progresses through analysis, then design, and then implementation. In fact, during the 1970s and 1980s, analysts tried to make projects proceed from one phase to the next. That was called the **waterfall method**, based on the metaphor that when one phase was finished, the project team dropped down to the next phase.

Analysts no longer attempt to use the waterfall approach. The activities are still classified as planning, analysis, design, and implementation, but many activities of the phases overlap and are done concurrently. Figure 2-2 illustrates how the phases of a typical project might overlap. But, why does overlap occur?

The answer is efficiency. At the time the team is analyzing needs, it may be thinking about and designing various forms or reports. To help understand the needs of the users, the team may want to design part of the final product. The team frequently throws part

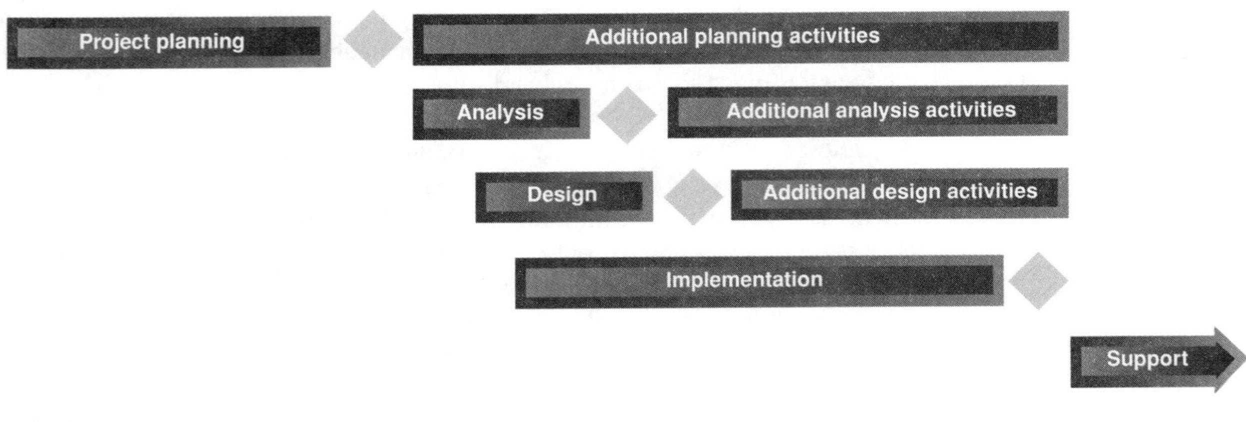

FIGURE 2-2
Overlap of systems development activities.

of the early design away, but also retains many good components. In addition, many components of a computer system are interdependent, which necessitates that the team perform portions of analysis and design at the same time.

Then why not overlap all activities completely? The answer is dependency. Some activities naturally depend on the results of prior work. The team cannot complete the design until the analysts have some idea of the basic nature of the problem, so some analysis must happen before the team can design. It would also be inefficient to write program code before the team has an overall design structure, because it would have to throw too much of the code away. Experience has proven that the most successful projects have schedules similar to the one shown in Figure 2-2. That kind of schedule provides flexibility, yet requires timely completion of identified milestones.

Another principle of systems development is iteration. Iteration is the process of looping through the same development activities multiple times, sometimes at increasing levels of detail, sometimes at increasing levels of accuracy. In many cases, systems development is much like peeling an onion. For example, the analysis phase is a discovery phase to understand the business problem. The team first investigates the processing needs at a broad level (the outside layer). Then the team iterates through the various processing needs at more and more detail (the inside layers) to be sure that the analysts truly understand all facets of the business information needs.

Another version of iteration occurs as you obtain more detail in one area of a business and discover that it has ramifications in previously studied areas. Sometimes this is called the "ripple effect," where a solution to a problem affects a prior decision or problem. This is a normal part of systems development, and you should use it as a tool to improve the quality of your work.

The Project Team

The project team is probably the most important component of a successful systems development project. In his classic book *The Mythical Man-Month*, Frederick Brooks indicates that a successful project team should function much like a surgical team, with each member of the team performing a specialized task critical to the whole. Although the analogy between a project team and surgical team does not hold for all aspects, a project team does need to be balanced, and its members bring various skills and abilities to the project. The size and make-up of the team also changes during the life of the project. Figure 2-3 illustrates a typical staffing scenario for the several phases of a project.

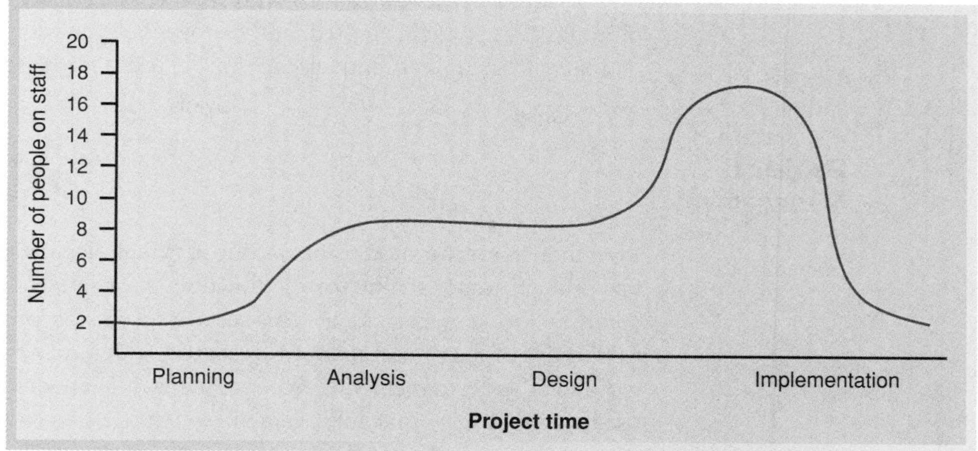

FIGURE 2-3
Staffing levels of a typical project.

During the project planning phase, the team consists of only a few members, typically a project manager and one or two experienced systems analysts. The activities of this first phase are primarily concerned with determining the scope of the required solution, assessing feasibility, and performing a cost/benefit analysis. As explained in the following sections, these activities are called management activities because they are concerned with planning, scheduling, controlling, and evaluating the project. Experienced analysts and developers usually have the required management expertise to participate in this phase.

The analysis phase requires team members with good analytical skills and strong problem domain knowledge. Remember that the problem domain is the focus for the new system, such as a point-of-sale system or an inventory management system. The project team during this phase adds systems analysts, business analysts, and at times even selected key users. If the project is for a large system, the project team may be divided into smaller teams to focus on different areas of the organization.

Design is a more specialized activity requiring the addition of members with technical expertise, such as networking and database experts. If the systems analysts are also experienced in technical issues, they may carry out the design activities alone. It is not unusual, however, to add a few technical experts to the team to develop the design. As in the analysis phase, during the design phase several teams may work together to design different components of the final system. This is especially true where various subsystems, platforms, or databases are involved.

During implementation, more programmers and quality-control team members are usually added. The project team is usually largest during the implementation phase. Programmers work with the existing team members to convert the design into actual code. Quality-control persons are involved with the development of test plans and test data. They also are active in testing the system to ensure that it will function under extreme or adverse conditions. Also during implementation, additional users may be added to the team as project team members with the responsibility to conduct training. These people are frequently knowledgeable users who learn the system first and then act as trainers and coaches for the rest of the user community.

Project Management

Even though every project team has one person designated as the project manager who has primary responsibility for the functioning of the team, all experienced members contribute to its management. This section focuses on project management from the perspective of the project manager, recognizing that many management tasks are delegated to other team members. There are many different definitions of *management*. One useful definition is that management is getting things done through other people. Project management is a special type of management. Remember the definition of a project: a planned undertaking that has a beginning and an end, and which produces a predetermined result or product. We extend this definition of management with the following definition of project management: **Project management** is the organizing and directing of other people to achieve a planned result within a predetermined schedule and budget.

project management
Organizing and directing of other people to achieve a planned result within a predetermined schedule and budget

The question at this point is, "what makes a successful project manager?" As with all key persons who are able to get things done, a good project manager is worth his or her weight in gold. As you progress in your career, you should note project management skills you observe in others as well as those you learn by your own experience. One place to start is with the set of skills described in Chapter 1 for a systems analyst. However, good project management requires more than just those skills. A good project manager knows how to plan, execute the plan, anticipate problems, and adjust for variances. Project management skills *can* be learned. Those who aspire to managing projects are proactive in a self-improvement plan to learn the necessary skills.

A project manager reports to and works with several groups of people. The **client** is a person or group of people who will be paying for the development of the new system—the customer. The client releases funds and ultimately approves the project. For in-house development, the client can be an executive committee or a particular vice president who is funding the project. For large, mission-critical projects, an **oversight committee** may be formed. This committee consists of clients and other key executives who have a vision of the strategic direction of the organization and have a strong interest in the success of the project. Experience in directing large development projects is helpful but certainly not a prerequisite for a member of the committee. **Users**, on the other hand, are the people who will actually be using the new system. In some cases, the client and user are the same people. More frequently, however, they are not. The user typically provides information about the detailed functions and operations of the new system. The client may also provide some input on the required functions, but primarily approves and oversees the project and provides the funding. Figure 2-4 depicts the various groups of people involved in a development project.

client
The person or group who funds the project

oversight committee
Clients and management who review and direct the project, like a board of trustees for a company

user
The person or group who will use the new system

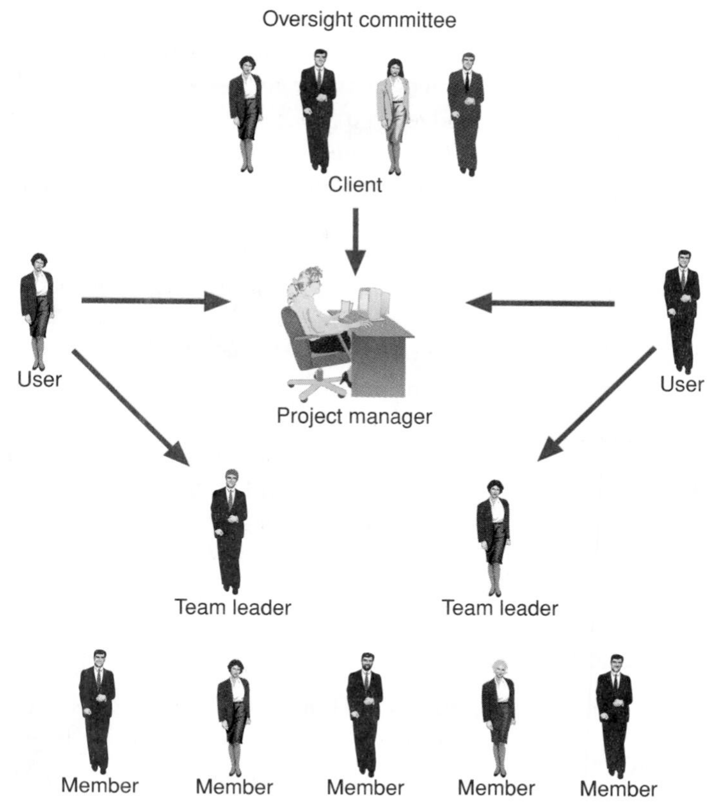

FIGURE 2-4
Participants in a systems development project.

Appendix A provides a comprehensive discussion of the responsibilities of a project manager. This section provides a brief overview of project management, summarizing two key points: first, planning and organizing, and second, directing the project.

Planning and Organizing

The first set of tasks of project management consists of those that plan and organize the project. The activities of the project planning phase identify the major components of planning and organizing the project. As you will see in the section on project planning, a project manager must be able to define the scope of the problem precisely and then develop a plan to carry out the project. This plan must include a detailed task list and schedule for the completion of the tasks. Superimposed on the task list are the resources, particularly personnel, required. The project manager also defines the milestones and interim deliverables that are used to monitor progress.

The project schedule identifies and schedules all the activities in every phase of the SDLC. By looking at the project schedule, the reader can tell what is to be done, who is to do it, and what the result of the activity is going to be.

The project manager also organizes the work and the workers into groups to accomplish the identified tasks effectively. A comprehensive and detailed plan is a critical element to a successful project.

Directing—Executing, Monitoring, and Controlling

Once the project begins, the project manager is responsible to see that the plan is executed and that the project progresses on schedule. One aspect of executing the plan is to monitor the completion of the various tasks. As they are completed, new assignments are made to team members for upcoming tasks. At frequent intervals, the project manager meets with the oversight committee to review the schedule, the budget, and progress.

The project manager also uses various techniques to keep the project on schedule in the face of obstacles and unanticipated problems. Since each problem is unique to the specific project, system, resource, management, and business environment, the project manager must draw from experience and creativity. The mettle of a successful project manager is measured in large part by his or her ability to anticipate and correct these problems.

Project controls include all activities a project manager uses to monitor and control the project. The project schedule includes quality-control and testing activities to ensure that the components are well developed. Milestones are also included in the schedule. **Milestones** are points in the project where critical pieces of the project are completed and can be identified by a specific piece of documentation or a specific status review meeting.

milestone
An event in the project schedule to denote the completion of a predetermined portion

Project Initiation

Projects are initiated in various ways. The optimal method is through the long-term information systems strategic plan. In the last chapter, you learned about strategic planning and how it identifies and directs the overall efforts of the organization to maintain a competitive position. These major strategic plans result in specific projects. Projects initiated in this way are sometimes described as top-down projects.

Projects may also be initiated from the bottom up. Individual department managers or process managers are close to the daily work within a company. They often identify information system problems within their respective areas. Sometimes these needs are brought to the attention of the strategic planning committee and integrated into the overall business strategy. At other times there is an immediate need that cannot wait for the strategic plan to be updated, such as a new sales commission schedule or a new productivity report. In these cases, the process managers will request initiation of individual development projects.

Finally, projects can be initiated to respond to outside forces. One common outside pressure is legislative changes that require new information and external reporting requirements such as changes in tax or labor law. Legislative changes can also change the range of services and products that an organization can offer in the market, expanding or contracting the type of business an organization can conduct. Changes such as these can affect the strategic plan, resulting in an urgent need for new systems. Examples include recent changes in the telecommunications industry, with cable TV and telephone companies vying for opportunities to provide cellular services, Internet access, and personalized entertainment.

However the project is initiated, it usually requires an initial review to ensure that the benefits outweigh the costs and risks of development. Thus, the first activities of almost every project are those that precisely define the business problem, determine the scope of the project, and analyze feasibility. Project managers group all of these initiating activities together in the project planning phase.

Project Initiation for Rocky Mountain Outfitters

As was noted in the last chapter, Rocky Mountain Outfitters' (RMO's) strategic thrust is to build more direct customer contact, to improve service, and to expand the marketing presence beyond the Western states. As part of this strategic plan, RMO has decided to begin the project to develop the customer support system.

The project begins with the assignment of a project leader and one experienced systems analyst. Barbara Halifax is the project manager for this activity. She has been with RMO for several years. Prior to joining RMO, she worked for the information systems consulting division of a large accounting firm. Her experience in consulting gave her broad exposure to many different companies and systems. Senior management at RMO has complete confidence in her abilities to manage this project. Steven Deerfield, a senior systems analyst, has also been assigned to this project. Barbara and Steven have worked together before and have very compatible work styles. This project is a critical component of RMO's long-range strategic plan, and these systems analysts are two of the best in the company.

An overview of the customer support system was given in Chapter 1. Recall that the primary objective of the system is to support RMO's objective of building customer loyalty. The system is to help in this objective by supporting all types of customer services, including ordering, returns, and on-line catalogs for both telephone sales and new Internet capability. Customers not only must have access to the on-line catalog of RMO products either via a telephone sales representative or the Internet but must also be able to see their past purchasing history. Finally, there are several "bells and whistles" that managers at RMO would like to include in the system to support their vision of RMO customer service.

The following section describes the activities of the project planning phase, using examples from the RMO project. Again, these important activities are management activities to plan, organize, schedule, and obtain approval for the project. Even though only two members of the team have been assigned at this point, Barbara and Steven have extensive experience and excellent management skills.

The Project Planning Phase

The project planning phase, as depicted in Figure 2-5, consists of the activities that are required to get the project organized and started. As discussed earlier in the chapter, these activities are:

- Define the problem
- Confirm project feasibility
- Produce the project schedule
- Staff the project
- Launch the project

FIGURE 2-5
Activities of the project planning phase.

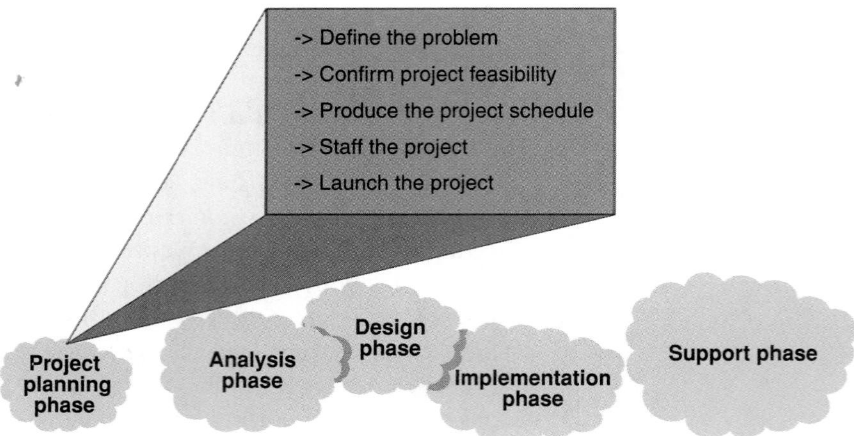

You will notice that several of these activities correlate to components of project management; this is because project management encompasses project planning.

The project planning phase is usually staffed with two or three highly experienced systems analysts, one of whom serves as the project manager. The other systems analysts are experienced developers with strong analytical skills as well as experience in managing and controlling projects. The team members first assigned frequently become the core team leaders around whom the rest of the team is built. At the successful conclusion of this phase, the project will have assigned resources, schedules, and a budget.

To help you learn about the activities of the project planning phase, we use our running case of Rocky Mountain Outfitters as the example, first explaining the purpose of the activity and then showing how the activity is done for RMO.

Defining the Problem

Defining a problem is one of the most important activities of the project. The objective is to define precisely the business problem to be solved and thereby determine the scope of the new system. The activity defines the target that you want to hit; if the target is ill defined, then all subsequent activities will lack focus, creating problems throughout the project.

The first task within this activity is to review the business needs that originally initiated the project. As with RMO, if the project was initiated as part of the strategic plan, then the planning documents are reviewed. If the project originated from departmental needs, then key users are consulted to help them understand the business problem. Another source of business needs is a list of benefits that the organization expects to achieve from a new information system, often called **business benefits**. Figure 2-6 is an example of the statement of need and the list of business benefits for RMO. Notice that the statement of need is a brief paragraph describing the business problem that initiated the project. The list of business benefits contains the results that the organization hopes to realize from a new system.

business benefits
The benefits that accrue to the organization; often measured in money

FIGURE 2-6
Statement of need and business benefits for the customer support system.

Description

Catalog sales began at Rocky Mountain Outfitters as a small experiment that soon developed into a rapidly growing division of the company. Support was initially provided by manual procedures with some simple off-the-shelf programs to assist in order taking and fulfillment. As a result of opportunities in the market, RMO initiated a comprehensive strategic planning activity. Part of the new strategic plan is the development of new information systems to support the long-term objectives of RMO for sales, customer support, and inventory. This project, to develop a new customer support system, is the first of several critical systems that are to be built.

Anticipated Business Benefits

- Reduce errors caused by manual processing of orders
- Expedite order fulfillment through quicker order processing
- Maintain or reduce staffing levels in mail-order and phone-order processing
- Open a new sales channel through the Internet
- Increase turnover by tracking sales of popular items and slow movers
- Increase customer loyalty through extensive customer support and information

The second task in this activity is to identify, at an abstract or general level, the expected capabilities of the new system. For this task, the focus shifts from looking at the problem from the business perspective to looking at the problem from the viewpoint of the required system capabilities. The objective is to define the scope of the problem in terms of the requirements of the information system that can solve it. Although at first this may not appear to be defining the problem, it is a necessary part of understanding the total scope of the project. The expected system capabilities are documented and combined with the previous list of business benefits. Figure 2-7 is an example of the list of system capabilities for the customer support system for RMO. Note the difference between this list and the previous list. The system capabilities are the mechanisms that will enable the organization to achieve the business benefits.

FIGURE 2-7
List of system capabilities for the customer support system.

The customer support system shall meet the following objectives:

- Be a high-support system with on-line customer, order, back order, and returns information
- Support traditional telephone and mail catalog sales with rapid-entry screens
- Include Internet customer and catalog sale capability, including purchase and order tracking
- Maintain adequate database and history information to support market analysis
- Provide a history of customer transactions for customer queries
- Be able to handle substantial increases in volume (300 percent or more) without degradation
- Support 24-hour shipment of new orders
- Coordinate order shipment from multiple warehouses
- Maintain history to support analysis of sales and forecasting of market demand

context diagram
A graphical diagram showing the scope of a system

Frequently, as part of identifying system capabilities, the project team also develops a diagram that shows the primary users of the system and the information the users and the system exchange. This **context diagram** provides an explanation of the scope of the problem from the viewpoint of the information needs of the system users. Figure 2-8 is an example of a context diagram for the RMO customer support system.

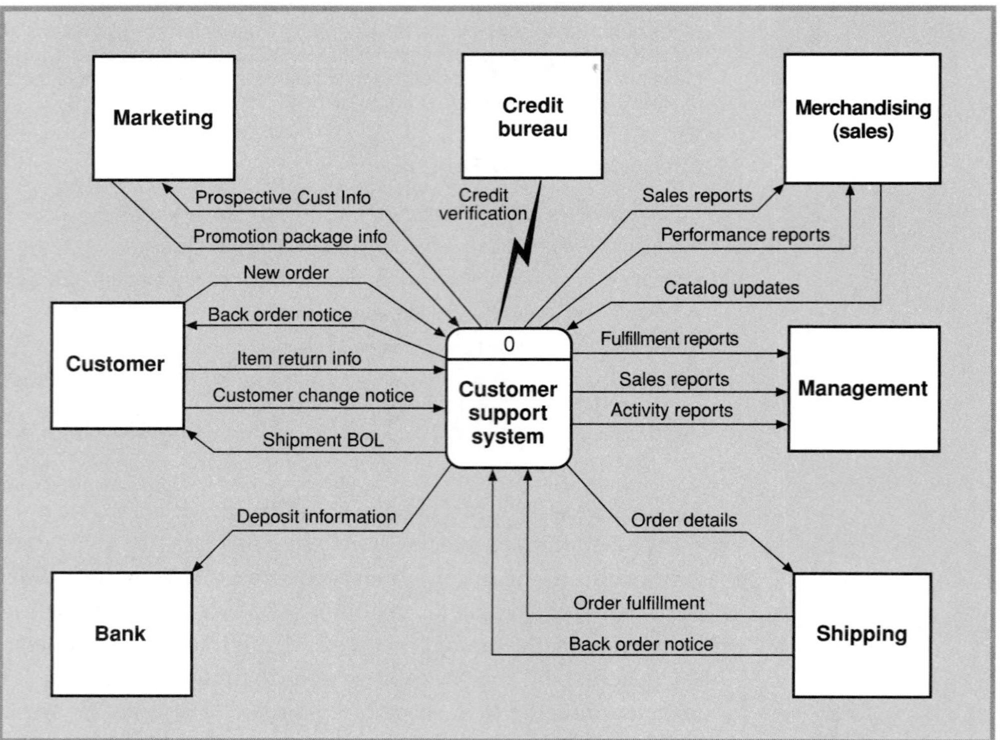

FIGURE 2-8
Context diagram for the RMO customer support system.

The rectangle with rounded corners in the middle of the diagram represents the customer support system itself. The squares are the external agents that provide information to the system or that receive information from the system. The lines with arrowheads are the major inputs and outputs to and from the system. This diagram only identifies the major information flows into and out of the system. The objective is to get an overview of a proposed solution without getting involved in the details.

Note that the context diagram is also used during the analysis phase. During the project planning phase, the diagram helps define the scope of the problem. This diagram becomes a starting point for the more detailed investigation done during analysis.

The key question to be answered when defining the problem activity is, "Do we understand what we are supposed to be working on?"

Defining the Problem at RMO Barbara and Steven, the CSS project team, developed the lists of business benefits and system capabilities and the context diagram after talking to William McDougal, vice president of marketing and sales, and his assistants. Chapter 4 explains more about interviewing users and eliciting important information. The important point to note here is to obtain information from and involve people who will be using the system and who will get the most benefit from it. They can provide insight to ensure a system does what it needs to do.

One additional task is required to complete problem definition. A preliminary investigation of alternative solutions is also conducted to reassess the assumptions the project team made when the project was initiated. Since the schedule and budget for the remainder of the project assume a particular development approach, it is critical to make those assumptions explicit so that all participants understand the constraints on the project schedule and so that the team can perform an accurate feasibility analysis.

For example, if an off-the-shelf program is identified as a possible solution, then part of the schedule during the analysis phase must include tasks to evaluate the program against the needs. If the most viable solution appears to be a new system developed completely in house, then detailed analysis tasks are planned and scheduled.

While Barbara was finishing up the problem definition statement, Steven did some preliminary investigation on possible solutions. He researched the trade magazines, the Internet, and other resources to determine whether there were sales and customer support systems that could be bought and installed rapidly. Although he found several, none seemed to have the exact match of capabilities that RMO needed. He and Barbara, along with William, had several discussions about how best to proceed. They decided that the best approach was to proceed with the analysis phase of the project before making any final decision about solutions. They would revisit this decision, in much more detail, after the analysis phase activities were completed. For now, Barbara and Steven began developing a schedule, budget, and feasibility statement for the new system.

Confirming Project Feasibility

When a project is initiated, it is assumed that a new system is feasible to develop and install. As more insight is gained during problem definition, team members should confirm that the project actually is feasible. During project feasibility, the project manager answers such questions as, "Are the expected benefits reasonable?" and "Are the assumed costs realistic?"

The objective in assessing feasibility is to determine whether a development project has a reasonable chance of success. The team assesses the original assumptions and identifies other risks that could jeopardize the project's success. After identifying those risks, the team establishes plans and procedures to minimize their effects. However, if there are serious risks, the team must discover and evaluate them as soon as possible.

Developing a list of potential risks is fairly difficult, which is another reason experienced systems analysts are involved in planning. They have encountered and dealt with problems and know where risks are likely. Generally there are five areas of risk for a new system that are considered when confirming project feasibility:

- economic feasibility
- organizational and cultural feasibility
- technological feasibility
- schedule feasibility
- resource feasibility

Economic Feasibility Economic feasibility consists of two tests: (1) Is the anticipated value of the benefits greater than projected costs of development, and (2) does the organization have adequate cash flow to fund the project during the development period? Even though the project may have received initial approval, final approval usually requires a thorough analysis of the development costs and the anticipated financial benefits. Obviously, a new system must increase income, either through cost savings or by increased revenues. A determination of the economic feasibility of the project always requires a thorough cost/benefit analysis.

cost/benefit analysis
The analysis to compare costs and benefits to see whether the investment in the development of a new system will be more beneficial than costly

Developing a **cost/benefit analysis** is a three-step process. The first step is to estimate the anticipated development and operational costs. Development costs are those that are incurred during the development of the new system. Operational costs are those that will be incurred after the system is put into production. The second step is to estimate the anticipated financial benefits. Financial benefits are the expected annual savings or increases in revenue derived from the installation of the new system. Third, the cost/benefit analysis is calculated based on detailed estimates of costs and benefits. The most frequent error in cost/benefit analysis is lack of thorough definition of costs and benefits.

Development Costs

Although the project manager has final responsibility for estimating the costs of development, senior analysts always assist with the calculations. Generally project costs come in the following categories:

- Salaries and wages
- Equipment and installation
- Software and licenses
- Consulting fees and payments to third parties
- Training
- Facilities
- Utilities and tools
- Support staff
- Travel and miscellaneous

Salaries and wages are calculated based on the staffing requirements for the project. As the project schedule and staffing plans are developed, the project manager can estimate the cost for salaries and wages. Figure 2-9 is an example of the estimated cost for salaries for the RMO customer service system. The numbers in this table were calculated by identifying personnel who are needed for the project and the length of time they are to be assigned to the project.

FIGURE 2-9
Supporting detail for salaries and wages.

SUPPORTING DETAIL FOR SALARIES AND WAGES

Team member	Salary/wage for project
Project leader	$101,340.00
Senior systems analyst	$90,080.00
Systems analyst	$84,980.00
Programmer analysts	$112,240.00
Programmers	$58,075.00
Systems programmers	$49,285.00
Total salaries and wages	**$496,000.00**

Each of the other categories of costs requires detailed calculations to determine the estimated costs. Detailed estimates of equipment, software licenses, training costs, and so forth can be calculated. These details are then combined to provide an estimate of the total costs of development. Figure 2-10 is a summary table of all of the costs. Again, each line in the summary table must be supported with details such as those given in Figure 2-9.

FIGURE 2-10
Summary of development costs for RMO.

SUMMARY OF DEVELOPMENT COSTS FOR RMO	
Expense category	**Amount**
Salaries/wages	$496,000.00
Equipment/installation	$385,000.00
Training	$78,000.00
Facilities	$57,000.00
Utilities	$152,000.00
Support staff	$38,000.00
Travel/miscellaneous	$112,000.00
Licenses	$18,000.00
Total	**$1,336,000.00**

Sources of Ongoing Costs of Operations

Once the new system is up and running, normal operating costs are incurred every year. These annual operating costs must also be accounted for in calculating the costs and benefits of the new system. Generally, project managers do not include the normal costs of running the business in this cost. Only the costs that are directly related to the new system and its upkeep and maintenance are included. The following list identifies the major category costs that may be allocated to the operation of the new system:

- Connectivity
- Equipment maintenance
- Computer operations
- Programming support
- Amortization of equipment
- Training and ongoing assistance (help desk)
- Supplies

Figure 2-11 summarizes the annual operating costs for RMO. As with the development costs, each entry in the summary table should be supported with detailed calculations. This figure represents the costs that are anticipated for the RMO system; other organizations may have a different set of operating costs.

FIGURE 2-11
Summary of annual operating costs for RMO.

SUMMARY OF ANNUAL OPERATING COSTS FOR RMO	
Recurring expense	**Amount**
Connectivity	$60,000.00
Equipment maintenance	$40,000.00
Programming	$65,000.00
Help desk	$28,000.00
Amortization	$48,000.00
Total recurring costs	**$241,000.00**

Sources of Benefits

The project manager and members of the project team can determine most of the development and operational costs. However, the benefits relate to the user and the client, so they must determine the value of the anticipated benefits. Members of the project team can and do assist, but project team members should never attempt to determine the value of benefits by themselves.

Benefits usually come from two major sources: decreased costs or increased revenues. Cost savings come from greater efficiency in the operations of the company. Areas to look for reduced costs include the following:

- Reducing staff due to automating manual functions or increasing efficiency
- Maintaining constant staff with increasing volumes of work
- Decreasing operating expenses, such as shipping charges for emergency shipments
- Reducing error rates due to automated editing or validation
- Ensuring quicker processing and turnaround of documents or transactions
- Capturing lost discounts on money management
- Reducing bad accounts or bad credit losses
- Reducing inventory or merchandise losses due to tighter controls
- Collecting accounts receivables more quickly
- Capturing income lost due to "stock outs" with better inventory management
- Reducing the cost of goods with volume discounts and purchases
- Reducing paperwork costs with electronic data interchange and other automation

This is just a sampling of the myriad benefits that can accrue. Unlike development costs, there are no "standard" benefits. Each project is different and so are the anticipated benefits. Figure 2-12 is an example of the benefits that RMO expects from the implementation of the new customer support system.

FIGURE 2-12
Sample benefits for RMO.

SAMPLE BENEFITS FOR RMO		
Benefit/cost saving	**Amount**	**Comments**
Increased efficiency in mail-order department	$125,000.00	5 people @ $25,000
Increased efficiency in phone-order department	$25,000.00	1 person @ $25,000
Increased efficiency in warehousing/shipping	$87,000.00	
Increased earnings due to web presence	$500,000.00	Increasing at 50% /year
Other savings (inventory, supplies, etc.)	$152,000.00	
Total annual benefits	**$889,000.00**	

Financial Calculations

net present value (NPV)
The present value of dollar benefits and costs for an investment such as a new system

Companies use a combination of methods to measure the overall benefit of the new system. One popular approach is to determine the **net present value (NPV)** of the new system. The two concepts of net present value are (1) all benefits and costs are calculated in terms of today's dollars (present value) and (2) benefits and costs are combined to give a net value. In other words, the future stream of benefits and costs is netted together and then discounted by a certain factor for each year in the future. The discount factor is like an interest rate, except it is used to bring future values back to current values. Appendix B provides detailed instructions, using RMO's customer support system as the example, on calculating the various economic feasibility measures. You should read through this section first to get an overview and then walk through Appendix B to ensure you understand the details.

FIGURE 2-13
Net present value, payback, and return-on-investment calculations for RMO.

Figure 2-13 is copy of the RMO net present value (NPV) calculation done in Appendix B (Figure B-1). In this case, the new system gives an NPV of $3,873,334 over a five-year period using a discount rate of 10 percent.

	RMO cost benefit analysis	Year 0	Year 1	Year 2	Year 3	Year 4	Year 5	Total
1	Value of benefits	$ -	$ 889,000	$ 1,139,000	$ 1,514,000	$ 2,077,000	$ 2,927,000	
2	Discount factor (10%)	1	0.9091	0.8264	0.7513	0.6830	0.6209	
3	Present value of benefits	$ -	$ 808,190	$ 941,270	$ 1,137,468	$ 1,418,591	$ 1,817,374	$6,122,893
4	Development costs	$(1,336,000)						$(1,336,000)
5	Ongoing costs		$ (241,000)	$ (241,000)	$ (241,000)	$ (241,000)	$ (241,000)	
6	Discount factor (10%)	1	0.9091	0.8264	0.7513	0.6830	0.6209	
7	Present value of costs	$ -	$ (219,093)	$ (199,162)	$ (181,063)	$ (164,603)	$ (149,637)	$(913,559)
8	PV of net of benefits and costs	$(1,336,000)	$ 589,097	$ 742,107	$ 956,405	$ 1,253,988	$ 1,667,737	
9	**Cumulative NPV**	**$(1,336,000)**	**$(746,903)**	**$ (4,796)**	**$ 951,609**	**$2,205,597**	**$ 3,873,334**	
10	**Payback period**		**2 years + 4796 / (4796 + 951,609) = 2 + .005 or 2 years and 2 days**					
11	**5-year return on investment**		**6,122,893 - (1,336,000 + 913,559) / (1,336,000 + 913,559) = 172.18%**					

payback period, or breakeven point
The time period at which the dollar benefits offset the dollar costs

return on investment (ROI)
A measure of the percentage gain received from an investment such as a new system

Another method organizations use to determine whether an investment will be beneficial is to determine the **payback period**. The payback period, sometimes called the **breakeven point**, is the point in time at which the increased cash flow (benefits) exactly pays off the costs of development and operation. Appendix B also provides the details necessary for this calculation. Figure 2-13 illustrates the calculations for the payback period. A running accumulated net value is calculated year by year (see line 9). The year when this value becomes positive is the year in which payback occurs. In the RMO example, this happens within the third year.

The **return on investment (ROI)** is another economic measure used by organizations. The objective of the ROI is to calculate a percentage return needed (like an interest rate) so that the costs and the benefits are exactly equal over the specified time period. Figure 2-13 includes an ROI calculation for RMO; Appendix B provides details. The time period can be the expected life of the investment (that is, the productive life of the system) or it can be an arbitrary time period.

For RMO, assuming a five-year benefit period, the ROI is 172.18 percent. In other words, the investment in the development costs returned over 172 percent on the investment for a period of five years. Obviously, since the system is generating benefits at that point in time, if you assume a longer lifetime, such as 10 years, you would get a much higher ROI.

Intangibles

tangible benefits
Benefits that can be measured or estimated in terms of dollars and that accrue to the organization

intangible benefits
Benefits that accrue to the organization but that cannot be measured quantitatively or estimated accurately

The previous cost/benefit calculations depend on quantifiable costs and benefits. However, in many instances, both costs and benefits cannot be measured and values cannot be determined. If an estimated dollar value can be placed on a benefit or a cost, it is treated as a **tangible benefit** or cost. Where there is no reliable method of estimating or measuring the value, then it is considered an **intangible benefit**. In some instances, the intangible benefits far exceed the tangible costs, at least in the opinion of the client, and the system is developed even though the dollar numbers do not indicate a good investment.

Examples of intangible benefits include the following:

- Increased levels of service (in ways that cannot be measured)
- Increased customer satisfaction (not measurable)
- Survival (a standard capability common in the industry, or common to many competitors)
- The need to develop in-house expertise (such as with a pilot program with new technology)

Examples of intangible costs include the following:

- Reduced employee moral
- Lost productivity (perhaps inestimable)
- Lost customers or sales (during some unknown period of time)

Even though the intangibles do not enter into the calculations for NPV, ROI, or payback, they should be considered and, in fact, may be the deciding factor on whether the project proceeds or not.

Sources of Funds

The cost/benefit analysis is done in conjunction with the development of the project budget, which we explained earlier. Economic feasibility is concerned with a positive result from the cost/benefit analysis and the source of funds for the systems development. Development projects can be financed in various ways. Frequently, new information systems are financed using a combination of current cash flows and long-term capital. The project team may not be involved in obtaining the financing for the project. However, the results of the cost/benefit analysis will greatly influence the financing decisions.

Organizational and Cultural Feasibility As discussed in Chapter 1, each company has its own culture, and any new system must fit within that culture. There is always the risk that the new system may depart so dramatically from existing norms that it cannot be successfully deployed. The analysts involved with feasibility analysis should evaluate organizational and cultural issues to identify potential risks for the new system. Such issues might include the following:

- A currently low level of computer competency
- Substantial computer phobia
- A perceived loss of control by staff or management
- Potential shifting of political and organizational power due to the new system
- Fear of change of job responsibilities
- Fear of loss of employment due to increased automation
- Reversal of longstanding work procedures

It is not possible to enumerate all the potential organizational and cultural risks that exist in organizations. The project management team needs to be very sensitive to feelings and areas of reluctance within the organization in order to identify and resolve these risks.

Again, the question analysts ask for operational feasibility is, "What items might prevent the effective use of the new system and the resulting loss of business benefits?"

Once the risks are identified, then positive steps can be taken to counter them. For example, additional training sessions can be held to teach new procedures and provide increased computer skills. Higher user involvement in the development of the new system can increase user enthusiasm and commitment.

Technological Feasibility Generally, a new system brings new technology into a company. At times the new system stretches the state-of-the-art of the technology. Other projects may use existing technology but combine it into new, untested configurations. Finally, even existing technology can be new if there is a lack of expertise within the company. If an outside vendor is providing capability, it is usually assumed to be an expert in the area in which it provides service. However, even an outside vendor can suffer from technological risks.

The project management team needs to assess the proposed technological requirements and available expertise. When these risks are identified, the solutions are usually fairly straightforward. The solution to technological risks include conducting additional training, hiring consultants, or hiring more experienced employees. In some cases, the scope and approach of the project may need to be changed to ameliorate technological risk. The important point is that a realistic assessment will identify technological risks early and permit corrective measures to be taken.

Schedule Feasibility The development of a project schedule is always an activity with high risk, and schedule feasibility means the project can be completed on time. Every schedule requires many assumptions and estimates about the project. For example, the needs, and hence the scope, of the new system are not well known; the estimated time to research and finalize requirements must be estimated; and the availability and capability of team members are questionable.

Another frequent risk in the development of the schedule occurs when upper management indicates that a new system must be deployed within a certain time. The tendency is then to build the schedule to show that it can be done. Unfortunately, this usually spells disaster. The schedule should be built without preconceived completion dates. Once the schedule is completed, then comparisons can be made to the requested dates to see whether time tables coincide. If not, then corrective measures, such as reducing the scope of the project, can be taken to increase the probability of the project's timely completion.

One of the objectives of defining milestones during the project schedule is to permit the project manager to assess the ongoing risk of the schedule slipping. If milestones begin to be missed, then corrective measures can possibly be taken early. Contingency plans can be developed and carried out to reduce the risk of further slippage.

Sometimes there is an absolute deadline for a project that simply cannot be missed, most notably the recent Y2K projects that had to be completed before January 1, 2000. Another example of an absolute deadline applies to the Olympics in Park City, Utah (up the street from RMO) in 2002. If the systems supporting the Olympic games are not completed on time, they are useless. Sometimes fixed deadlines are crucial to organizations in less obvious ways. If a university's course registration system is not implemented on time during the spring, then it cannot be used for fall semester. If a direct customer ordering system on the Internet, like the customer support system at RMO, is not completed by early October, then the whole season of holiday sales are at risk.

Resource Feasibility Finally, the project management team must assess the availability of resources for the project. The primary resource consists of the members of the team. Development projects require the involvement of systems analysts, system technicians, and users, and one risk is that the required people may not be available to the team when needed. An additional risk is that the people who are assigned may not have the necessary skills. Once the team is functioning, there is the continual risk that members may leave the team. This can happen internally if other special projects arise within the organization, or externally if qualified team members are hired by other organizations. Although the project manager usually does not like to think about this risk, people with skills are in short supply and sometimes do leave projects.

The other resources required for a successful project include adequate computer resources, physical facilities, and support staff. Generally, these resources can be made available, but the schedule can be affected if there are delays in the availability of these resources.

An assessment of each of these five areas of feasibility is an important part of the project planning phase. The key question to be answered when completing feasibility analysis is, "Is it still feasible to begin working on this project?"

Developing the Project Schedule

Before we discuss the details of a project schedule, let's clarify three terms: task, activity, and phase. Fundamentally, a phase is made up of a group of related activities, and an activity is made up of a group of related tasks. A task, then, is the smallest piece of work that is identified and scheduled. Activities are also identified, named, and scheduled. For example, suppose a project manager is scheduling the design phase. Within the design phase, the manager identifies activities such as designing the user interface, designing and integrating the database, and completing the application design. Within the "designing the user interface" activity, the analyst might identify individual tasks such as designing the customer entry form and designing the order-entry form. Thus, the task, activity, and phase breakdown provides a three-level hierarchy.

During the project planning phase, it is not possible to schedule every task in the entire project because it is too early in the project to know all of the tasks that will be necessary. Frequently, a project management team builds a project schedule that identifies the major activities of the project, as we did when we explained SDLC at the beginning of the chapter. Experienced systems analysts can usually develop a schedule of the major activities. Then a detailed list of tasks is developed for each phase of the project shortly before its commencement.

Appendix C provides a detailed example of the development of a project schedule for the RMO project. This section discusses the major steps required to develop a schedule. You should read this section to get an overall understanding of the process and then work through Appendix C to understand the details and learn how to develop a schedule.

Developing a project schedule is a four-step process. First, using the list of activities identified in the standard SDLC as a guide, project members identify all of the individual tasks for each activity. Other activities that are unique to the specific project, and their corresponding tasks, are also identified.

The second step is to estimate the size of the task—that is, the number of human resources, the person-days required, the calendar time required, and any other special resources that may be needed. Figure 2-14 is a portion of the table developed in Appendix C (Figure C-2) showing the tasks, along with the estimated time and resources required. Even though this figure is just a sample, notice how detailed and comprehensive it is. Developing the task list and estimating the amount of work for each task usually requires considerable thought.

	TASK LIST WITH RESOURCE ESTIMATES Phases, activities, and tasks	Optimistic duration (days)	Pessimistic duration (days)	Most probable (days)	Number of resources
1.0	Project planning phase				
1.1	Define the problem				
1.1.1	Meet with users	1	4	2	2
1.1.2	Determine scope	1	3	2	2
1.1.3	Write statement of business benefits	1	2	1	1
1.1.4	Write statement of need	1	2	1	1
1.1.5	Define statement of system capabilities	1	2	1	1
1.1.6	Develop context diagram	1	2	1	1
1.2	Confirm project feasibility				
1.2.1	Identify intangible costs and benefits	1	1	1	2
1.2.2	Estimate tangible development and operational costs	1	3	2	2
1.2.3	Estimate tangible benefits and do cost/benefit	1	3	2	2
1.2.4	Calculate NPV, ROI, and payback	1	1	1	1
1.2.5	Evaluation organizational and cultural feasibility	1	1	1	1
1.2.6	Evaluate technical feasibility	1	2	2	2
1.2.7	Evaluate required schedule	1	2	1	1
1.2.8	Evaluate resource availability	1	1	1	1
1.3	Produce the project schedule				
1.3.1	Develop work breakdown schedule	2	4	3	2
1.3.2	Estimate resources, durations, and precedences	1	3	2	2
1.3.3	Develop PERT chart and Gantt chart	1	3	2	2
1.4	Staff the project				
1.4.1	Develop a resource plan for the project	1	1	1	2
1.4.2	Identify and request technical staff	1	2	1	1
1.4.3	Meet with users, identify and request user staff	1	3	2	1
1.4.4	Organize project team	1	1	1	1
1.4.5	Conduct team-building exercises	2	5	3	2
1.4.6	Conduct initial skills training	2	6	4	2
1.5	Launch the project				
1.5.1	Prepare presentation materials	1	1	1	1
1.5.2	Make executive presentation	1	1	1	1
1.5.3	Set up project facilities and support resources	1	5	3	2
1.5.4	Conduct official kick-off meeting	1	1	1	1
2.0	Analysis phase				
2.1	Gather information	30	120	60	5
2.2	Define system requirements	30	100	60	5
2.3	Build prototypes for discovery of requirements	25	80	50	2
2.4	Prioritize requirements	5	15	10	2
2.5	Generate and evaluate alternatives	5	15	8	2
2.6	Review recommendations with management	2	5	3	2

FIGURE 2-14
A task list with resource estimates.

The third step is to determine the sequence for the identified tasks. The team accomplishes this step by determining which tasks are predecessors for a given task. The sequence is sometimes obvious. At other times, the team sequences the tasks simply to schedule the work in an orderly fashion.

The final step is to schedule the tasks themselves, taking into account that resources cannot be double-scheduled. Although the scheduling of the tasks is not always simple, techniques and tools are available to help create a project schedule.

PERT/CPM
A method of scheduling a project based on individual tasks or activities

A **PERT/CPM** chart is a charting method that shows the relationships among tasks. PERT/CPM stands for *Project Evaluation and Review Technique/Critical Path Method*. By defining which tasks can be done concurrently and which ones must be done serially, project managers can determine the total expected duration of the project. The longest path, from the first task to the last task, of dependent tasks is called the **critical path**, because if any of the tasks on that path slip, then the entire project schedule will slip. Other tasks not on the critical path usually have some slack time. The slack time for a task is the amount of time that the task can slip without affecting the schedule. A PERT/CPM chart graphically shows these dependencies, the critical path, and slack times. It is an effective tool to develop a project schedule.

critical path
The path on a PERT chart that indicates the shortest completion period for the project

By showing which tasks can be done concurrently, a PERT chart also assists in assigning staff. It is always a juggling act to balance the availability and work load of the team members with the dependent and independent tasks. Figure 2-15 is a PERT chart for the RMO customer support project. Each box represents a scheduled task. Within the box is the number of the task, its name, its duration, its projected start and end dates, and its slack time. The tasks on the heavy line are critical path tasks.

Gantt chart
A bar chart which represents the tasks and activities of the project schedule

Another charting technique that is used for project management and control is a **Gantt chart**. A Gantt chart contains basically the same information as a PERT chart but presents it in a different format. A Gantt chart presents the tasks as horizontal bars where the vertical tick marks are calendar days and weeks. Thus, a Gantt chart does not show the dependencies quite as graphically, but it does show calendar information well. Figure 2-16 illustrates the same RMO project schedule in a Gantt chart format.

Planning Phase
1 | 30 days
Mon 1/10/0 | Fri 2/18/00

Define the problem
2 | 8 days
Mon 1/10/0 | Wed 1/19/0

Produce the project schedule
18 | 7 days
Wed 1/26/0 | Thu 2/3/00

Meet with users
3 | 2 days
Mon 1/10/0 | Tue 1/11/0

Determine Scope
4 | 2 days
Wed 1/12/0 | Thu 1/13/0

Write Statement of Business Benefits
5 | 1 day
Fri 1/14/00 | Fri 1/14/00

Write Statement of Need
6 | 1 day
Mon 1/17/0 | Mon 1/17/0

Define Statement of System capabilities
7 | 1 day
Tue 1/18/0 | Tue 1/18/0

Develop Context Diagram
8 | 1 day
Wed 1/19/0 | Wed 1/19/0

Identify Intangible Costs and Benefits
10 | 1 day
Thu 1/20/0 | Thu 1/20/0

Estimate tangible development and
11 | 3 days
Fri 1/21/00 | Tue 1/25/0

Calculate NPV, ROI and Payback
13 | 1 day
Wed 1/26/0 | Wed 1/26/0

Evaluation Organizational &
14 | 1 day
Thu 1/27/0 | Thu 1/27/0

Estimate tangible benefits and do
12 | 2 days
Fri 1/21/00 | Mon 1/24/0

Evaluate Technical Feasibility
15 | 2 days
Thu 1/27/0 | Fri 1/28/00

Evaluate Required Schedule
16 | 1 day
Wed 2/2/00 | Wed 2/2/00

Evaluate Resource Availability
17 | 1 day
Mon 2/7/00 | Mon 2/7/00

Develop Work Breakdown Structure
19 | 3 days
Wed 1/26/0 | Fri 1/28/00

Estimate resources, durations, precedence
20 | 2 days
Mon 1/31/0 | Tue 2/1/00

Develop PERT chart and Gantt chart
21 | 2 days
Wed 2/2/00 | Thu 2/3/00

Staff the project
22 | 11 days
Fri 2/4/00 | Fri 2/18/00

Develop a resource plan for the project
23 | 1 day
Fri 2/4/00 | Fri 2/4/00

FIGURE 2-15
Partial PERT chart for the customer support project.

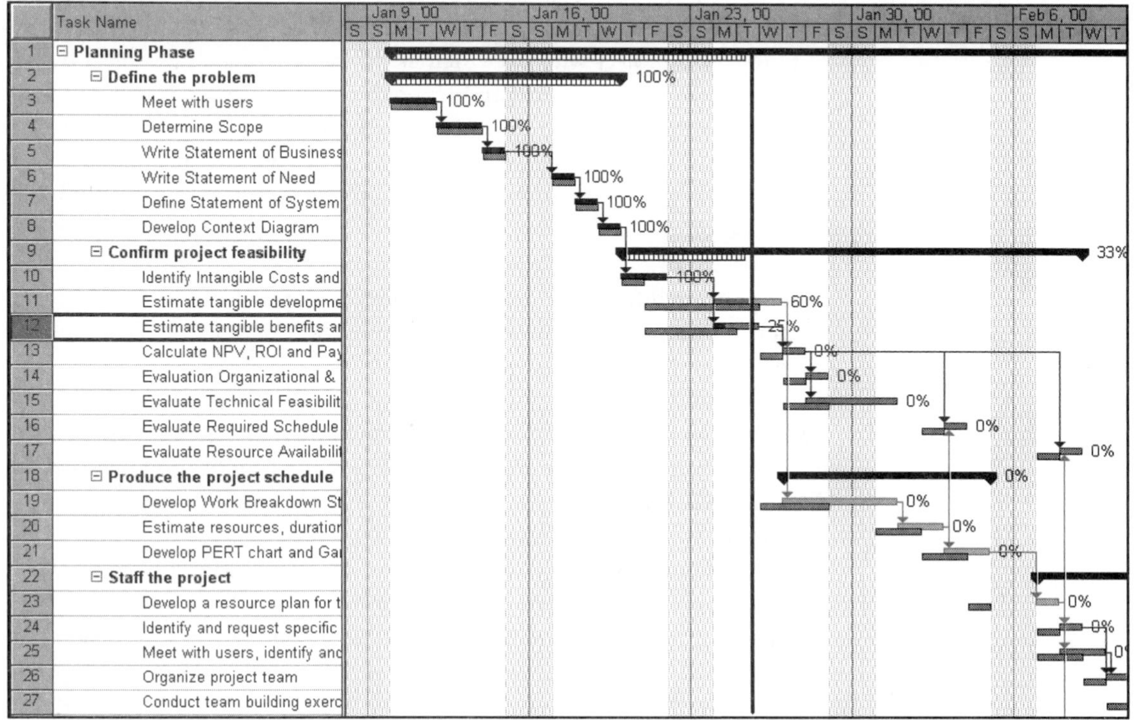

FIGURE 2-16
Gantt chart for the customer support project.

PERT charts are quite useful in building the initial project schedule, but Gantt charts are more useful in tracking the project after it has begun. As tasks are completed, they can be marked with a different color so progress is easy to track. It is a simple matter to observe which tasks are to the left of today's date and have not been colored. Those tasks are behind schedule. Tasks to the right of today's date and that have been colored are ahead of schedule.

More detailed information on these scheduling techniques, especially how to build schedules using these tools, is given in Appendix C. The key question to be answered when completing this activity is, "Can the project be completed on time given the available resources?"

Staffing the Project

The responsibility for staffing falls primarily on the project manager. This activity demonstrates the project manager's organizational and teamwork skills. There are five tasks within this activity:

- Develop a resource plan for the project
- Identify and request specific technical staff
- Identify and request specific user staff
- Organize the project team into work groups
- Conduct preliminary training and team-building exercises

Based on the tasks identified in the project schedule, a detailed resource plan can be developed. In fact, the schedule and the resource requirements are usually developed concurrently. If a tool such as Microsoft Project is being used to build the schedule, then the resources required for each task are part of the total schedule. In developing the plan, the project manager must recognize (1) that resources are usually not available as soon as requested and (2) that a period of time is needed for a team member to become acquainted with the project.

Once the plan has been developed, then specific people can be identified and requested to become part of the team. The team needs both technical staff and user staff. The technical staff consists of the systems analysts, the programmer analysts, the network specialists, and other technicians. Technical staff members frequently move from project to project whenever their skills are required. The project manager often meets with the director or vice president of information systems to identify and schedule the necessary resources. In some instances, additional technical staff may be needed, so the human resources department may become involved in finding and recruiting new people. Even though finding technical members for the team is standard procedure, it does require some time to find and assign all of the required team members.

The user staff are people within the organization who are assigned to the team. Sometimes it is difficult to get users assigned to a team full time. Assignment to a project team is not part of the normal job progression for a user. However, projects do progress more smoothly if a few full-time team members can represent the user community and act as liaisons.

On small projects, the project team may all work together. However, a project team larger than four or five members usually is divided into smaller working groups. Each group will have a group leader to coordinate the tasks assigned to the group. The responsibility for this task falls on the project leader.

Finally, training and team-building exercises are conducted. Training may be done for the project team as a whole, such as when new technology, a new database, or a new programming language is used. In other cases, individual training may be needed for new team members who are unfamiliar with the tools and techniques being used. Appropriate training for both technical people and users should be conducted. Team-building exercises are especially important when team members have not worked together before. The integration of user members of the team with technical people is an important consideration in developing effective teams and workgroups.

The key question to be answered when completing this activity is, "Are the resources available, trained, and ready to start the project?"

The Rocky Mountain Example Three more people were assigned to RMO's customer support system project when it reached the staffing stage: Ming Lee, who is an experienced systems analyst, and two junior programmer/analysts, Jennifer and Jason. The schedule called for the commencement of the analysis phase and the related fact-finding activities, which Chapter 4 discusses.

Launching the Project

After the previous activities are complete, then it is time to launch the project. The scope of the new system is defined, the risks have been identified, the project is feasible, a detailed schedule has been developed, and team members have been identified and are ready to start. Two final tasks usually occur at this point. First, the oversight committee is finalized and meets to give final go-ahead for the project, including the release of the necessary funds. Second, a formal announcement is made, through the standard communications channels in the organization, that gives credence to the project and solicits cooperation from all involved parties in the organization. In other words, the project gets the blessing and visible support of the senior executives of the organization. No project should begin without these two events.

The key question to be answered when completing this activity is, "Are we ready to start?"

Summary

The focus of this chapter is on project management activities. Three major themes are covered: (1) the systems development life cycle (SDLC), (2) the various components of project management, and (3) the project planning phase.

The development of a new system requires an organized, step-by-step approach called the systems development life cycle. The SDLC defines the overall approach to systems development. The SDLC is divided into four development phases and one support phase. For development, the phases are project planning, analysis, design, and implementation. Each of these phases is further divided into activities, and the activities are divided into individual tasks. In this way, project managers can identify and schedule the individual work tasks that make up the project.

Project management is the organizing and directing of other people to achieve a planned result within a predetermined schedule and budget. The six primary areas of responsibility of a project manager are (1) provide leadership and vision, (2) plan and organize the project, (3) organize and manage the project team, (4) estimate costs and develop the project budget, (5) monitor and control the project schedule, and (6) ensure the quality of the final result.

The project planning phase is carried out primarily by the project manager and one or two other senior analysts. Many of the responsibilities of the project manager are carried out within the activities of the project planning phase. This phase consists of five activities: (1) defining the problem, (2) confirming project feasibility, (3) producing the project schedule, (4) staffing the project, and (5) launching the project.

Key Terms

analysis phase, p. 32
application, p. 34
business benefits, p. 42
client, p. 38
context diagram, p. 44
cost/benefit analysis, p. 45
critical path, p. 54
design phase, p. 33
Gantt chart, p. 54
help desk, p. 35
implementation phase, p. 34
intangible benefits, p. 50
milestone, p. 40
net present value (NPV), p. 48

oversight committee, p. 38
payback period, or breakeven point, p. 49
PERT/CPM, p. 54
phase, p. 30
planning phase, p. 32
problem domain, p. 32
project, p. 30
project management, p. 38
return on investment (ROI), p. 49
support phase, p. 35
systems development life cycle (SDLC), p. 31
tangible benefits, p. 50
user, p. 38
waterfall method, p. 35

Review Questions

1. List and explain the activities of the project planning phase.
2. What are the six activities of the analysis phase?
3. Describe the purpose of each phase of the SDLC.
4. Define *project management.*
5. Describe the types of feasibility analysis.
6. What is the purpose of a cost/benefit analysis?
7. Explain the difference between a PERT chart and a Gantt chart.
8. List five or six possible sources of tangible benefits from the installation of a new system.
9. List four or five sources of development costs.
10. What is meant by the *critical path*?
11. What is the purpose of a system context diagram?
12. Describe the six major responsibilities of the project manager.
13. Describe the seven activities of the design phase.
14. Describe the four activities of the implementation phase.

Thinking Critically

1. Write a one-page paper that distinguishes among the fundamental purposes of the analysis phase, the design phase, and the implementation phase.
2. Given the following narrative, make a list of expected business benefits.
Especially for You Jewelers is a small jewelry retailer in a college town. Over the last couple of years, Especially for You has experienced a tremendous increase in its business. However, its financial performance has not kept pace with its growth. The current transaction-processing system, which is partially manual and partially automated, does not track customer billing and receipts sufficiently, and Especially for You has difficulty determining why its costs are so high. In addition, Especially for You runs frequent specials to attract customers. It has no idea whether these specials are profitable or whether they generate additional sales. Especially for You also wants to increase repeat sales to existing customers, so it needs a customer database. Especially for You wants to install a new direct sales and accounting system to help solve these problems.
3. Given the following narrative, make a list of system capabilities.
The new direct sales and accounting system for Especially for You Jewelers is an important element in the future growth and success of the jewelry company. The direct sales portion of the system needs to keep track of every sale and be able to access the inventory system for cost data to provide a daily profit and loss report. The customer database needs to be able to produce purchase histories to assist management in preparing special mailings and special sales to existing customers. Detailed credit balances and aged accounts for each customer would help solve the problem with the high balance of accounts receivables. Special notice letters and credit history reports would help management reduce accounts receivables.
4. Build a PERT/CPM chart based on the following list of tasks and precedences to build and test a screen form for a new system. Identify the critical path.

Task	Description	Duration	Precedence
0	Start	0	—
1	Meet with user	2	0
2	Review existing forms	1	0
3	Identify and specify fields	3	1, 2
4	Build initial prototype	2	3
5	Develop test data (valid data)	4	3
6	Develop error test data	2	5
7	Test prototype	3	4, 6
8	Make final refinements	3	7

5. Many students have jobs. Suppose that you worked in a dentist's office and were asked to develop a system to keep track of patient appointments. How would you start? What would you do first? What kinds of things would you try to find out first? How does your approach compare to what this chapter has described?

Experiential Exercises

1. Download a trial version of Microsoft Project and build a project schedule based on the following activities. Print out both a PERT chart and a Gantt chart. (Note: The set of files is about 25 MB.) Download a trial version of Microsoft Project and build a project schedule based on the following activities. Print out both the PERT chart and the Gantt chart. (Note: The set of Microsoft Project files is about 25 Meg.)
   Below is a list of tasks for a student to have an international experience by attending a university abroad. You can build schedules for several versions of this set of tasks. For the first iteration, assume that all precedence tasks must finish before the succeeding task can begin (simplest version). For a second iteration, identify several tasks that can begin a few days before the end of the predecessor task. For a third iteration, modify the second version so that some tasks can begin a few days after the beginning of a predecessor task. Also insert a few overview tasks such as Application tasks, Preparation tasks, and Travel and Arrival tasks. Be sure to state your assumptions for each iteration.
2. Build a project plan to show your progress through college. Include the course prerequisite information. If you have access to Microsoft Project or a similar project management tool, enter the information in the tool.
3. Using information from your organizational behavior classes or other sources, write a one-page paper on what kinds of team-building and training activities might be appropriate as the project team is expanded for the analysis phase.
4. Research and write a two-page paper on why systems development projects fail.
5. Ask a systems analyst about the SDLC that he or she uses at work. If possible, ask the analyst to show you a copy of a project schedule.

Task ID	Description of Task	Duration	Precedence
1.	Obtain Forms from international exchange office	1 day	none
2.	Fill out and send in foreign university application	3 days	1
3.	Receive approval from foreign university	21 days	2
4.	Apply for scholarship	3 days	2
5	Receive notice of approval for scholarship	30 days	4
6.	Arrange Financing	5 days	3, 5
7.	Arrange for housing in dormitory	25 days	6
8.	Obtain passport and required visa	35 days	6
9.	Send in pre-registration forms to university	2 days	8
10.	Make travel arrangements	1 day	7, 9
11.	Determine clothing requirements and go shopping	10 days	10
12.	Pack and make final arrangements to leave	3 days	11
13.	Travel	1 day	12
14.	Move into dormitory	1 day	13
15.	Finalize registration for classes and other university paperwork	2 days	14
16.	Begin classes	1 day	15

Case Study: Custom Load Trucking

It is time for Stewart Stockton's annual performance review. As Monica Gibbons, an assistant vice president of information systems, prepares for the interview, she reviews Stewart's assignments over the last year and his performance. Stewart is one of the up-and-coming systems analysts in the company, and she wants to be sure to give him solid advice on how to advance his career. She knows, for example, that he has a strong desire to become a project manager and accept increasing levels of responsibility. His desire is certainly in agreement with the needs of the company.

Custom Load Trucking (CLT) is a nationwide trucking firm that specializes in rapid movement of high-technology equipment. With the rapid growth of the communications and computer industries, CLT is feeling more and more pressure from its clients to be able to move its loads more rapidly and precisely. Several new information systems are planned that will enable CLT to schedule and track shipments and trucks almost to the minute. However, trucking is not necessarily a high-interest industry for information systems experts. With the shortage in the job market, CLT has decided not to try to hire project managers for these new projects but to build strong project managers from within the organization.

As Monica reviews Stewart's record, she finds that he has done an excellent job as a team leader on his last project. His last assignment was as a combination team leader/systems analyst on a four-person team. He had been involved in systems analysis, design, and programming, as well as managing the work of the other three team members. He had assisted in the development of the project schedule and had been able to keep his team on schedule. It also appears that the quality of his team's work was as good as, if not better than, other teams on the project. She wonders what advice she should give him to help him advance his career. She also wonders whether now is the time to give him his own project.

1. Do you think the decision by CLT to develop its own project managers from the existing employee base is a good one? What advice would you give to CLT to make sure that it has strong project management skills in the company?
2. What kind of criteria would you develop for Monica to use to measure whether Stewart (and other potential project managers) is ready for project management responsibility?
3. How would you structure the job for new project managers to ensure, or at least increase the possibility, of a high-level of success?
4. If you were Monica, what kind of advice would you give to Stewart about managing his career and obtaining his immediate goal to become a project manager?

Further Resources

Howard Eisner, *Essentials of Project and Systems Engineering Management.* John Wiley & Sons, Inc., 1997.

James P. Lewis, *Team-Based Project Management.* American Management Association, 1998.

Sanjiv Purba, D. Sawh, and B. Shah, *How to Manage a Successful Software Project.* John Wiley & Sons, Inc., 1995.

Felix Redmill, *Software Projects: Evolutionary vs. Big-Bang Delivery.* John Wiley & Sons, Inc., 1997.

Walker Royce, *Software Project Management: A Unified Framework.* Addison-Wesley, Inc., 1998.

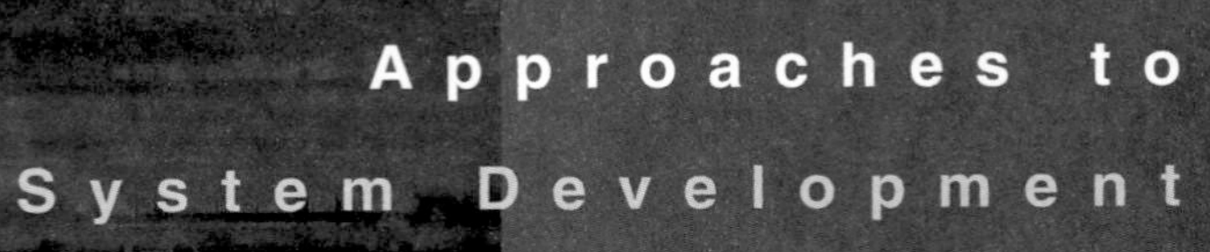

Approaches to System Development

Learning Objectives

After reading this chapter, you should be able to:

- Explain the differences between a model, a tool, a technique, and a methodology
- Describe the three overall approaches used to develop information systems
- Describe some of the variations to the system development life cycle (SDLC)
- Explain how computer-aided system engineering (CASE) tools are used in system development
- Describe the activities of the system analysis life cycle phase
- Describe the activities of the system design life cycle phase

Chapter Outline

Methodologies, Models, Tools, and Techniques

Three Approaches to System Development

System Development Life Cycle Variations

Computer-Aided System Engineering (CASE) Tools

Analysis and Design Phases in More Detail

DEVELOPMENT APPROACHES AT AJAX CORPORATION, CONSOLIDATED CONCEPTS, AND PINNACLE MANUFACTURING

Bob, Mary, and Kim, graduating seniors, were discussing their recent interview visits to different companies that recruited CIS majors on their campus. All agreed they learned a lot by visiting the companies, but they also all felt somewhat overwhelmed at first.

"At first I wasn't sure I knew what they were talking about," cautiously volunteered Bob. During his on-campus interview, Bob had impressed Ajax Corporation with his knowledge of data modeling. When he visited the Ajax home office data center for the second interview, the interviewers spent quite a lot of time describing the company's system development methodology, and presented a fairly detailed demonstration of a CASE tool the company uses.

"They started out listing a bunch of acronyms I wasn't sure I had heard of: BSP, BAA, and things like that," Bob continued. "They said they had used ADW and then IEF, but now they were using Sterling Software CoolTools to develop all new systems. I was thinking, 'Great. I only know COBOL and some Visual Basic.' I felt much better when someone explained to me they were using the information engineering approach and had been for about 10 years. They had changed CASE tools a few times, though. When they showed me the data model in the CASE tool and then generated some COBOL programs, I was thinking, 'Okay, I know this.'"

"I know what you mean," said Mary, a very talented programmer who knew just about every new programming language available. "Consolidated Concepts went on and on about things like OMT and UML and some people named Booch, Rumbaugh, and Jacobson. But then it turned out that they were using the object-oriented approach to develop systems, and they liked the fact that I knew C++ and Java. No problem once I got past all of the terminology they used. They said they'd send me out for training on Rational Rose, a CASE tool for the object-oriented approach."

Kim had a different story. "A few people said to forget everything I learned in school." Kim had visited Pinnacle Manufacturing, which had purchased a complete development methodology called IM One from a small consulting firm. Most employees agreed it works fairly well. The people who had worked at the manufacturer for quite a while thought IM One was totally unique, and they were quite proud of it.

"Well, that got my attention," continued Kim, "but then they started telling me about their SDLC, about data flow diagrams, about events, and about entity-relationship diagrams, and things like that." Kim had recognized many of the key concepts in the methodology were fairly standard models and techniques from the structured approach to system development.

Bob, Mary, and Kim all agreed that there was much to learn in these work environments, but also that there are many terms used to describe the key concepts they learned in school. They were all glad they focused on the fundamentals in their CIS classes and that they had been exposed to a variety of approaches to system development.

OVERVIEW

As the experiences of Bob, Mary, and Kim demonstrate, there are lots of ways to develop an information system, and developing an information system is very complex. People who develop systems rely on a variety of aids to help them with every step of the process. The system development life cycle (SDLC) described in Chapter 2 provides an overall framework for the process of system development. But the developer relies on many more concepts for help, including methodologies, models, tools, and techniques. It is very important to understand what these concepts are before discussing system development in any detail.

This chapter reviews three approaches to system development—the structured approach, the information engineering approach, and the object-oriented approach. Some variations of the SDLC are also described. Additionally, an analyst needs computer support to complete work tasks. This support is provided through computer-aided system engineering (CASE) tools, so some of these tools are described. Most of the models, tools, and techniques discussed in this chapter are used during the analysis and design phases of the SDLC, and the chapter concludes with a description of the activities carried out during systems analysis and systems design no matter which approach to system development is used.

At Rocky Mountain Outfitters, one of Barbara Halifax's initial jobs as the project manager for the customer support system project is to make decisions about the approach used to develop the system. All of the options described in this chapter are open to her. We will not describe her final decisions, though, because we use the customer support system example throughout this text when we present more details about all approaches.

Methodologies, Models, Tools, and Techniques

A systems analyst has a variety of aids to assist in analysis and design. Among them are methodologies, models, tools, and techniques. We discuss each of these in the following sections.

Methodologies

system development methodology
Comprehensive guidelines to follow for completing every activity in the system development life cycle, including specific models, tools, and techniques

A **system development methodology** provides guidelines to follow for completing every activity in the system development life cycle, including specific models, tools, techniques, and a system development life cycle. Some methodologies are homegrown, developed by systems professionals in the company based on their experience. Some methodologies are purchased from consulting firms or other vendors.

Some methodologies (whether homegrown or purchased) contain written documentation that can fill a bookcase. Everything the developers might do at any point in the project is defined, including how documentation should look and what reports to management should contain. Other methodologies are much more informal—general descriptions of what should be done are contained in one document. Sometimes the methodology a company adopts is "just follow some sort of methodology," but such freedom of choice is becoming rare. Most people want the methodology to be flexible, though, so that it can be adapted to many different types of projects and systems.

If a methodology contains instructions about how to use models, tools, and techniques, then you must understand what models, tools, and techniques are.

Models

model
A representation of some important aspect of the real world

Anytime people need to record or communicate information about something in the real world, it is very useful to create a model, and a model in information system development has the same purpose as any other model. A **model** is a representation of some important aspect of the real world. Sometimes the term *abstraction* is used because we abstract an aspect of particular importance to us. Consider a model of an airplane. To talk about the aerodynamics of the airplane, it is useful to have a small model that shows its overall shape in three dimensions. Sometimes a drawing showing the cross-sectional details of the wing of the plane is what is needed. In another case, a list of mathematical characteristics of the plane might be necessary to understand how the plane will behave. All of these are models of the same plane.

Some models are physically similar to the real product. Some models are graphical representations of important details. Some models are abstract mathematical notations. Each emphasizes a different type of information. In airplane design, aerospace engineers use lots of different types of models. Learning to be an aerospace engineer involves learning how to create and use all of the models. It is the same for an information system developer, although models for information systems are not yet as standardized or precise as aerospace models. But system developers are making progress. First, it is important to recognize that the field is very young and many senior analysts were self-taught. More importantly, though, an information system is much less tangible than an airplane—you can't really see, hold, or touch it. Therefore, the models of the information system can seem much less tangible, too.

What sort of models do developers make of aspects of an information system? Models used in system development include representations of inputs, outputs, processes, data, objects, object interactions, locations, networks, and devices, among other things. Most of the models are graphical models, which involve drawing a representation using agreed-upon symbols and conventions. These are often called *diagrams* and *charts*. You have probably drawn models showing program logic using flow charts. Much of this text involves showing you how to read and create a variety of models that represent an information system.

Another kind of model important to develop and use is a project planning model, such as PERT or Gantt charts, as shown in Chapter 2. These models show a representation of the system development project itself, highlighting the tasks and the task completion dates for the project. Another model related to project management is a chart showing all of the people assigned to the project. Figure 3-1 lists some models used in system development.

FIGURE 3-1
Some models used in system development.

Some models of system components

- Flow chart
- Data flow diagram (DFD)
- Entity-relationship diagram (ERD)
- Structure chart
- Use case diagram
- Class diagram
- Sequence diagram

Some models used to manage the development process

- PERT chart
- Gantt chart
- Organizational hierarchy chart
- Financial analysis models – NPV, ROI

Tools

tool
Software support that helps create models or other components required in the project

A **tool** in the context of system development is software support that helps create models or other components required in the project. Tools may be simple drawing programs for creating diagrams. They might include a database application that stores information about the project, such as data flow definitions or written descriptions of processes. A project management software tool, like the one described in Chapter 2, is another example of a tool used to create models. The project management tool creates a model of the project tasks and task dependencies.

Tools have been specifically designed to help system developers. Programmers should be familiar with integrated development environments (IDEs) that include many tools that help with programming tasks—smart editors, context sensitive help, and debugging tools. Some tools can generate program code for the developer. Some tools reverse-engineer old programs—taking the code and generating a model from the code so the developer can figure out what the program does in case the documentation is missing (or was never done).

CASE tool
Computer-aided sytem engineering tool designed to help a systems analyst complete development tasks

The most comprehensive tool available for system developers is called a **CASE tool**. CASE stands for computer-aided system engineering tool. These tools are described in more detail later in this chapter. Basically, CASE tools help the analyst create the important system models, and then they automatically check the models for completeness and compatibility with other system models. Finally, the CASE tools can generate program code based on the models. Figure 3-2 lists types of tools used in system development.

FIGURE 3-2
Some tools used in system development.

Project management application
Drawing/graphics application
Word processor/text editor
Computer-aided system engineering (CASE) tools
Integrated development environment (IDE)
Database management application
Reverse-engineering tool
Code generator tool

Techniques

technique
A collection of guidelines that help the analyst complete a system development activity or task

A **technique** in system development is a collection of guidelines that help the analyst complete a system development activity or task. This often means step-by-step instructions for creating a model, or it might include more general advice for collecting information from system users. Some examples include data-modeling techniques, software-testing techniques, user-interviewing techniques, and relational database design techniques.

Sometimes a technique applies to an entire life cycle phase and includes creating several models and other documents. The modern structured analysis technique (discussed later) is an example. Even strategic system planning techniques discussed in Chapter 1 and project planning techniques discussed in Chapter 2 fit this definition. Figure 3-3 lists some techniques commonly used in system development.

FIGURE 3-3
Some techniques used in system development.

Strategic planning techniques
Project management techniques
User interviewing techniques
Data-modeling techniques
Relational database design techniques
Structured analysis technique
Structured design technique
Structured programming technique
Software-testing techniques
Object-oriented analysis and design techniques

How do all these components fit together? A methodology includes a collection of techniques that are used to complete activities within each phase of the system development life cycle. The activities include completion of a variety of models as well as other documents and deliverables. As with any other profession, system developers use software tools to help them complete the activities. Figure 3-4 shows the relationships among the components of a methodology.

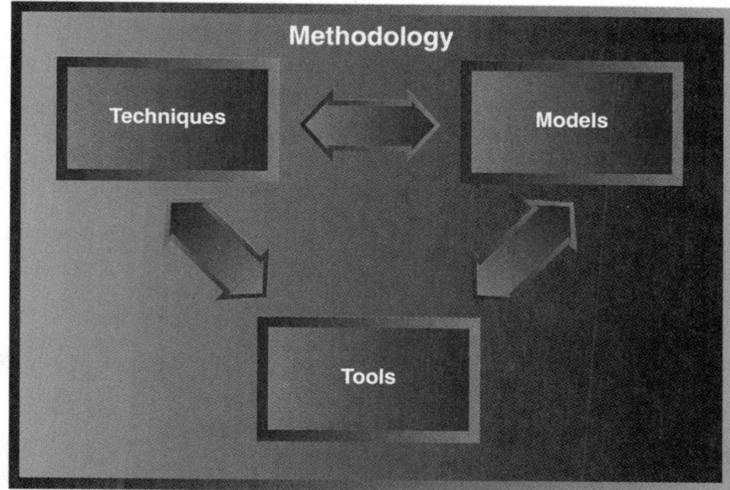

FIGURE 3-4
Relationships among components of a methodology.

Three Approaches to System Development

System development is done in many different ways. This diversity can confuse new employees when they go to work as system developer. Sometimes it seems every company that develops information systems has its own approach. Sometimes different development groups within the same company use different approaches, and each person in the company may have his or her own way of developing systems.

Yet, as you have already seen in the opening case, there are many common concepts. In virtually all development groups, some variation of the system development life cycle is used, with phases for project planning, analysis, design, implementation, and support. Additionally, virtually every approach uses models, tools, and techniques that make up an overall system development methodology.

All system developers should be familiar with three very general approaches to system development because they form the basis of virtually all methodologies. These three approaches are called the structured approach, the information engineering approach, and the object-oriented approach. This section reviews the major characteristics of all three approaches and provides a bit of history.

The Structured Approach

structured approach
System development using structured programming, structured analysis, and structured design techniques

Structured programming, structured analysis, and structured design are the three techniques that make up the **structured approach** to system development. Sometimes these techniques are collectively referred to as the structured analysis and design technique (SADT). The structured programming technique, developed in the 1960s, was the first attempt to provide guidelines to improve the quality of computer programs. You most certainly learned the basic principles of structured programming in your first programming course. The structured design technique was developed in the 1970s to make it possible to combine separate programs into more complex information systems. The structured analysis technique evolved in the early 1980s to help make the requirements for the computer system clearer to the developers before they designed the programs.

structured program
A program or program module that has one beginning and one ending, and each step in the program execution consists of sequence, selection, or repetition constructs

Structured Programming Quality programs not only produce the correct outputs each time the program runs but also allow other programmers to easily read and modify the program later. And programs need to be modified all the time. A **structured program** is one that has one beginning and one ending, and each step in the program execution consists of one of three programming constructs:

- A *sequence* of program statements
- A *decision* where one set of statements or another set of statements executes
- A *repetition* of a set of statements

Figure 3-5 shows these three structured programming constructs.

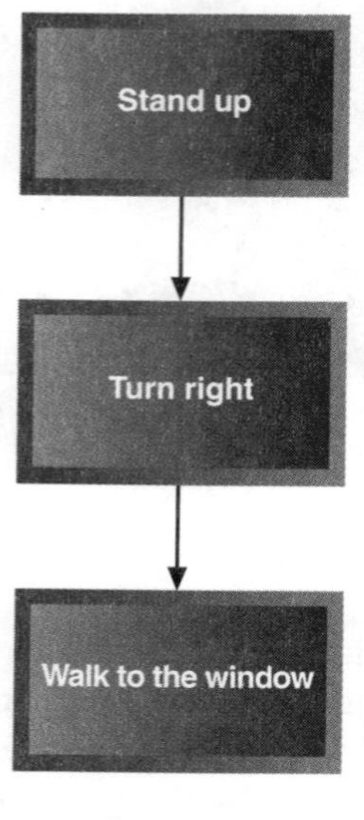

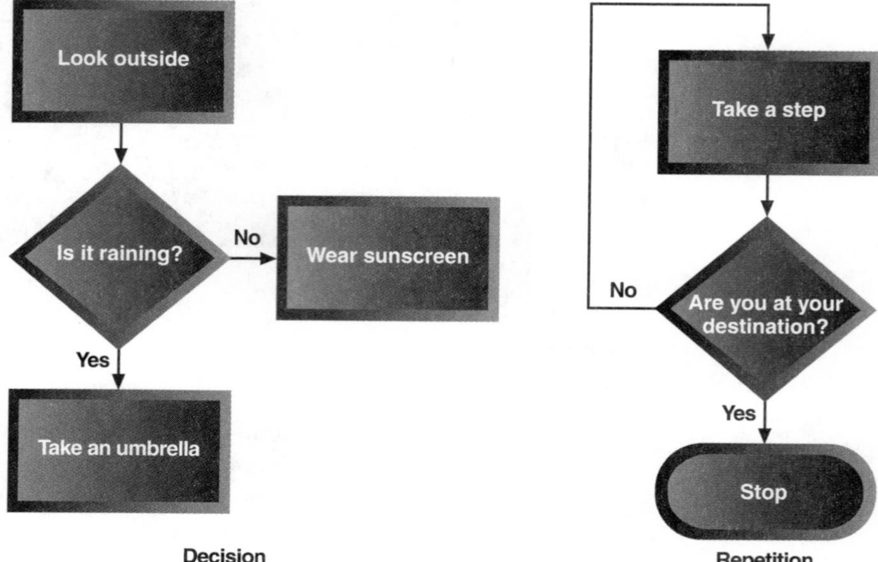

FIGURE 3-5
Three structured programming constructs.

Before these rules were developed, programmers made up programming techniques as they went along, which resulted in some very convoluted programs. Most programmers were happy if the programs ran at all, and they were even happier if the programs produced the right outputs. But following these simple rules made it much easier to read and interpret what a program does.

top-down programming
Dividing more complex programs into a hierarchy of program modules

Another concept related to structured programming is top-down programming. **Top-down programming** divides more complex programs into a hierarchy of program modules (see Figure 3-6). One module at the top of the hierarchy controls program execution by "calling" lower-level modules as required. Sometimes the modules are part of the same program. For example, in COBOL, one main paragraph calls another paragraph using the Perform keyword. In Visual Basic, a statement in an event procedure can call a general procedure (like a subroutine). Each program module (paragraph or procedure) is written using the rules of structured programming (one beginning, one end, and sequence, decision, and repetition constructs).

Sometimes separate programs are produced that work together as one "system." Each of these programs uses top-down programming and the structured programming rules. But the programs themselves are organized into a hierarchy, as with top down programming. One program calls other programs. When the hierarchy involves multiple programs, such an arrangement is sometimes called *modular programming*.

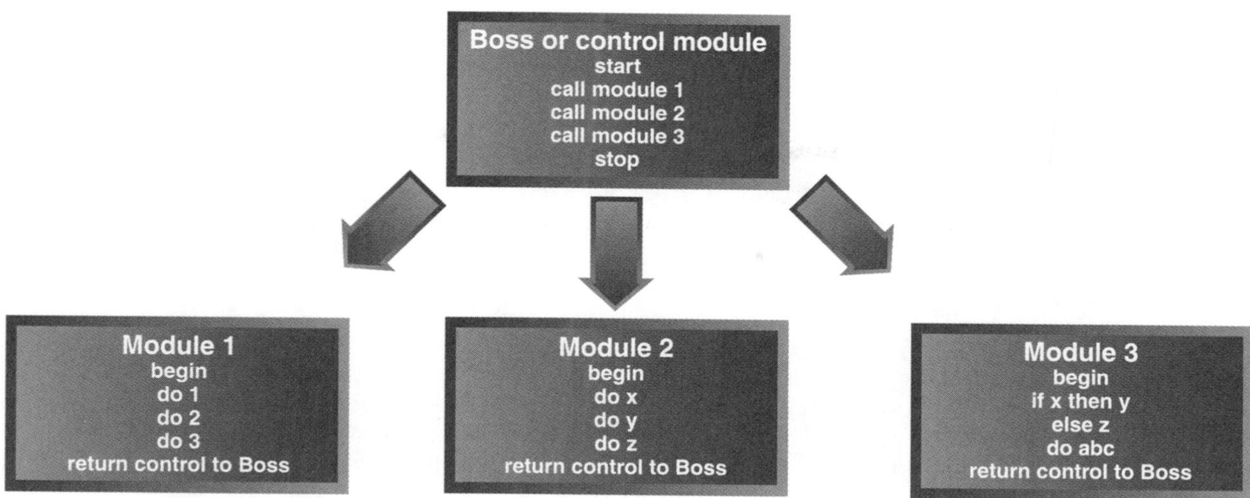

FIGURE 3-6
Top-down or modular programming.

structured design
A technique providing guidelines for deciding what the set of programs should be, what each program should accomplish, and how the programs should be organized into a hierarchy

structure chart
A graphical model showing the hierarchy of program modules produced in a structured design

Structured Design As information systems continued to become increasingly complex through the 1970s, each system involved many different functions. Each function performed by the system might be made up of dozens of separate programs. The **structured design** technique was developed to provide some guidelines for deciding what the set of programs should be, what each program should accomplish, and how the programs should be organized into a hierarchy. The modules and the arrangement of modules are shown graphically using a model called a **structure chart** (see Figure 3-7).

Two main principles of structured design are that program modules should be designed so they are (1) loosely coupled and (2) highly cohesive. Loosely coupled means each module is as independent of the other modules as possible, which allows each module to be designed and later modified without interfering with the performance of the other modules. Highly cohesive means that each module accomplishes one clear task. That way, it is easier to understand what each module does, and to ensure that if changes to the module are required, none will accidentally affect other tasks.

The structured design technique defines different degrees of coupling and cohesion and provides a way of evaluating the quality of the design before the programs are actually written. As with structured programming, quality is defined in terms of how easily the design can be understood and modified later when the need arises.

Structured design assumes the designer knows what the system needs to do: what the main system functions are, what the required data are, and what the needed outputs are. Designing the system is obviously much more than designing the organization of the program modules. Therefore, it is important to realize that the structured design technique helps the designer complete part of, but not the entire, systems design life cycle phase.

By the 1980s, file and database design techniques were developed to be used along with structured design. Newer versions of structured design assume database management systems are used in the system, and program modules are designed to interact with the database. Additionally, since more nontechnical people were becoming involved with information systems, user-interface design techniques were developed. For example, menus in an interactive system determine which program in the hierarchy gets called. Therefore, a key aspect of user-interface design is done in conjunction with structured design.

FIGURE 3-7
A structure chart created using the structured design technique.

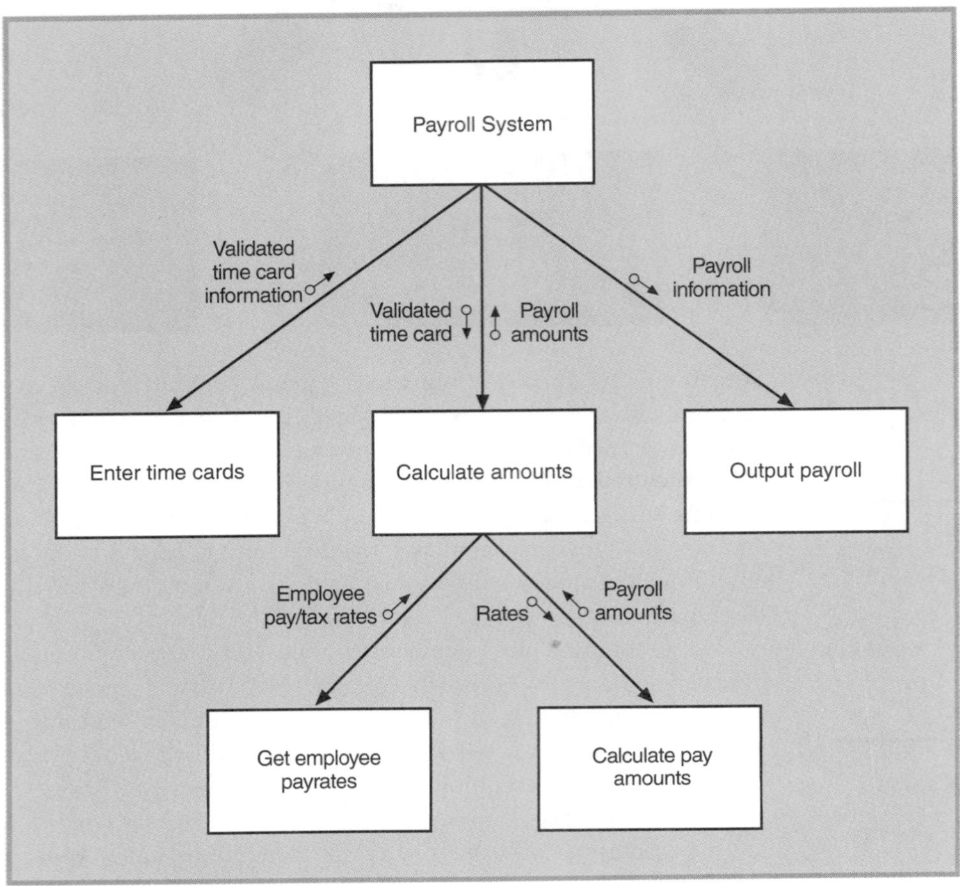

structured analysis
A technique that helps the developer define what the system needs to do (the processing requirements), what data the system needs to store and use (data requirements), what inputs and outputs are needed, and how the functions work together overall to accomplish tasks

data flow diagram (DFD)
A graphical model showing the inputs, processes, storage, and outputs of a system produced in structured analysis

Modern Structured Analysis Since the structured design technique requires the designer to know what the system should do, techniques for defining system requirements were developed. System requirements define what the system must do in great detail, but without committing to one specific technology. By deferring decisions about technology, the developers can sharply focus their efforts on what is needed, not on how to do it. If these requirements are not fully and clearly worked out in advance, the designers cannot possibly know what to design.

The **structured analysis** technique helps the developer define what the system needs to do (the processing requirements), what data the system needs to store and use (data requirements), what inputs and outputs are needed, and how the functions work together overall to accomplish tasks. The key graphical model of the system requirements used with structured analysis is called the **data flow diagram** (DFD), and it shows inputs, processes, storage, and outputs and how they function together (see Figure 3-8).

The most recent variation of structured analysis defines systems processing requirements by identifying all of the events that will cause the system to react in some way. For example, in an order-entry system, if a customer orders an item, the order-entry system must *process a new order* (a major system activity). Each event leads to a different system activity. The analyst takes each of these activities and creates a data flow diagram showing the processing details, including inputs and outputs.

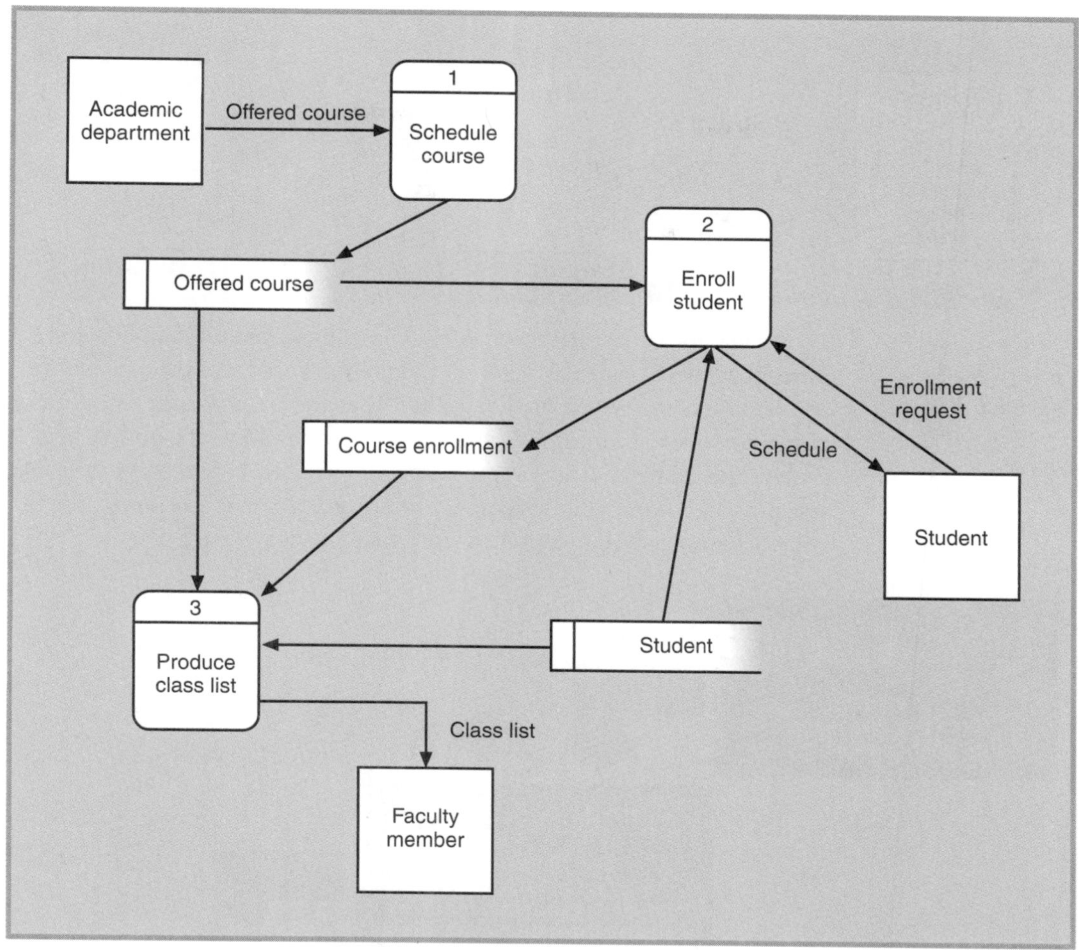

FIGURE 3-8
A data flow diagram (DFD) created using the structured analysis technique.

entity-relationship diagram (ERD)
A graphical model of the data needed by a system, including things about which information is stored and the relationships among them, produced in structured analysis and information engineering

A model of the needed data is also created based on the types of things about which the system needs to store information (data entities). For example, to process a new order, the system needs to know about the customer, the items wanted, and the details about the order. This model is called an **entity-relationship diagram** (ERD). The data entities from the entity-relationship diagram correspond to the data storage shown on data flow diagrams. Figure 3-9 shows an example of an entity-relationship diagram. Figure 3-10 illustrates the sequence from structured analysis to structured design to structured programming.

Weaknesses of the Structured Approach Because the structured approach to system development evolved over time, many variations can be found in practice. Some people are still following the original versions of structured analysis and structured design they learned years ago, ignoring many improvements. Others picked up bits and pieces of the techniques on the job and never formally studied the details.

Many people consider the structured approach to be weak in that the techniques only address some, but not all, of the activities of analysis and design. Critics desired a more comprehensive and rigorous set of techniques to make system development more like an engineering discipline and less like an art. In addition, many people thought the transition from the data flow diagram (in structured analysis) to the structure chart (in structured design) did not work well in practice. Many people thought that data modeling and the entity-relationship diagram were much more important than modeling processes with the data flow diagram. The structured approach, despite its inclusion of data modeling and database design, still made processes rather than data the central focus of the system.

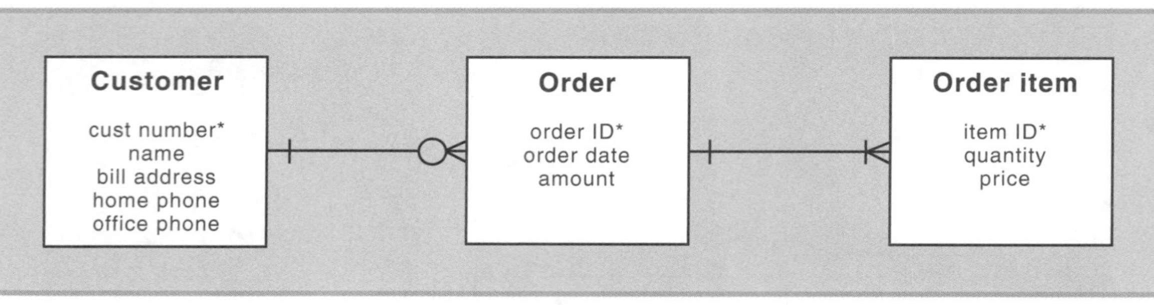

FIGURE 3-9
An entity-relationship diagram (ERD) created during structured analysis.

Finally, to ensure that systems are comprehensive and coordinated, many people thought the development of a system should begin only after the organization completed an overall strategic system planning effort. Therefore, they wanted a strategic system planning technique to be included in the approach to development, both to determine which systems should be built and to provide some initial requirements models that ensure all systems will be compatible. Because of these goals, some developers turned to another approach to systems development—information engineering.

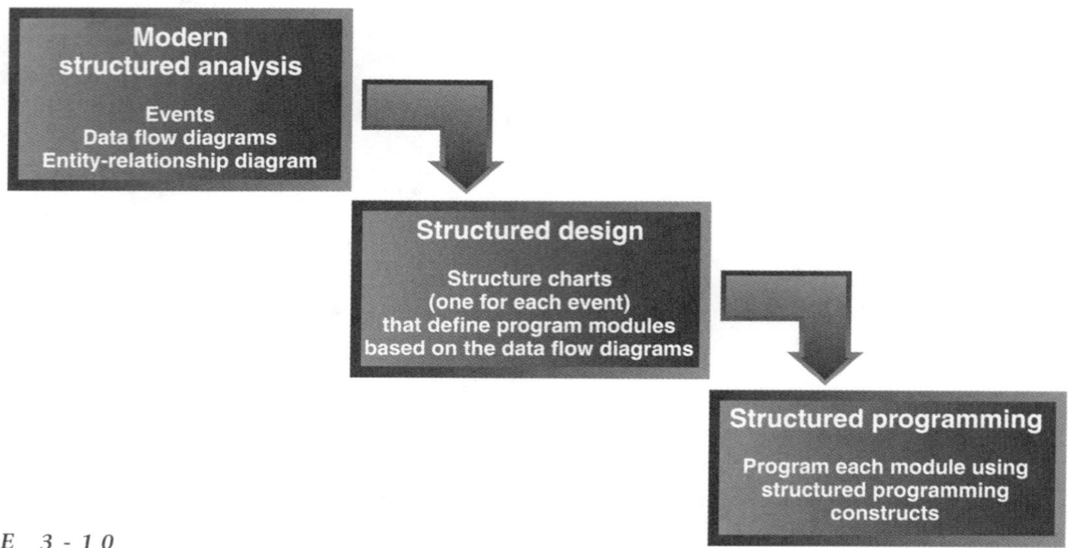

FIGURE 3-10
How structured analysis leads to structured design and to structured programming.

The Information Engineering Approach

The **information engineering** approach begins with overall strategic planning to define all of the information systems the organization needs to conduct its business (the application architecture plan). The plan also includes a definition of the business functions and activities the systems need to support, the data entities about which the systems need to store information, and the technological infrastructure the organization plans to use to support the information systems. This type of strategic information systems planning was described in Chapter 1.

information engineering
A system development methodology that focuses on strategic planning, data modeling, and automated tools, thought to be more rigorous and complete than the structured approach

Each new system project begins by using the defined activities and data entities created during strategic systems planning. Then the activities and data are refined as the project progresses. At each step, models are created of the processes and the data, and the ways they are integrated.

The type of data needed to conduct the business changes very little over time, but the processes followed to collect data change frequently. Therefore, the information engineering approach focuses much more on data than the structured approach. But just as the structured approach includes data requirements, information engineering includes processes, too. The processing model of information engineering, the *process dependency diagram,* is similar to a data flow diagram, but it focuses more on which processes are dependent on other processes and less on data inputs and outputs. Events trigger the processes, the same as with modern structured analysis.

A final major difference with information engineering is the more complete life cycle support it provides through the use of an integrated CASE tool. The CASE tool helps automate as much of the work as possible. It also forces the analyst to follow the information engineering approach faithfully, sometimes at the expense of flexibility. Final program code can be generated automatically by the CASE tool in many instances. CASE tools are also available for the structured approach, but they are often intentionally more flexible, and hence less rigorous.

Information engineering is mainly credited to James Martin, who wrote several books on information engineering and developed CASE tools to support it. By the late 1980s, information engineering was very popular for large mainframe systems. But because they lacked flexibility, the CASE tools that supported information engineering were less useful with smaller desktop applications and client-server applications. By the 1990s, fewer companies were using information engineering exclusively, although many of the concepts and techniques continue to be used, particularly the approach to planning and the emphasis on data modeling.

The information engineering approach takes many of the concepts of the structured approach and refines them into a rigorous and comprehensive methodology. Both approaches define information systems requirements, design information systems, and construct information systems by looking at processes, data, and the interaction of the two. This text merges key concepts from these two approaches into one—which we will refer to hereafter as the *traditional approach*. The traditional approach, in one version or another, is still widely used for information system development, although many information systems projects are also using object-oriented technology—a completely different approach.

object-oriented approach
An approach to systems development that views an information system as a collection of interacting objects that work together to accomplish tasks

object
A thing in the computer system that can respond to messages

object-oriented analysis (OOA)
Defining all of the types of objects that do the work in the system and showing how the objects interact to complete tasks

object-oriented design (OOD)
Defining all of the types of objects necessary to communicate with people and devices in the system and refining the definition of each type of object so it can be implemented with a specific language or environment

The Object-Oriented Approach

The objected-oriented approach is an entirely different approach to information systems. The **object-oriented approach** views an information system as a collection of interacting objects that work together to accomplish tasks (see Figure 3-11). There are no processes or programs; there are no data entities or files. The system consists of objects. An **object** is a thing in the computer system that is capable of responding to messages. This radically different view of what a computer system is requires a different approach to doing systems analysis, systems design, and programming.

The object-oriented approach began with the development of the Simula programming language in Norway in the 1960s. Simula was used to create computer simulations involving "objects" like ships, buoys, and tides in fiords. It is very difficult to write procedural programs that simulate ship movement, but a new way of programming simplified the problem. In the 1970s, the Smalltalk language was developed to solve the problem of creating graphical user interfaces that involved "objects" such as pull-down menus, buttons, check boxes, and dialog boxes. Other object-oriented languages include C++ and, more recently, Java. These languages focus on writing definitions of the types of objects needed in a system, and as a result, all parts of a system can be thought of as objects, not just graphical user interface objects.

Because the object-oriented approach views information systems as collections of interacting objects, **object-oriented analysis (OOA)** means defining all of the types of objects that do the work in the system and showing how the objects interact to complete tasks. **Object-oriented design (OOD)** means defining all of the additional types of objects necessary to communicate with people and devices in the system and refining the definition of each type of object so it can be implemented with a specific language or environment. **Object-oriented programming** (OOP) means writing statements in a programming language to define what each type of object does.

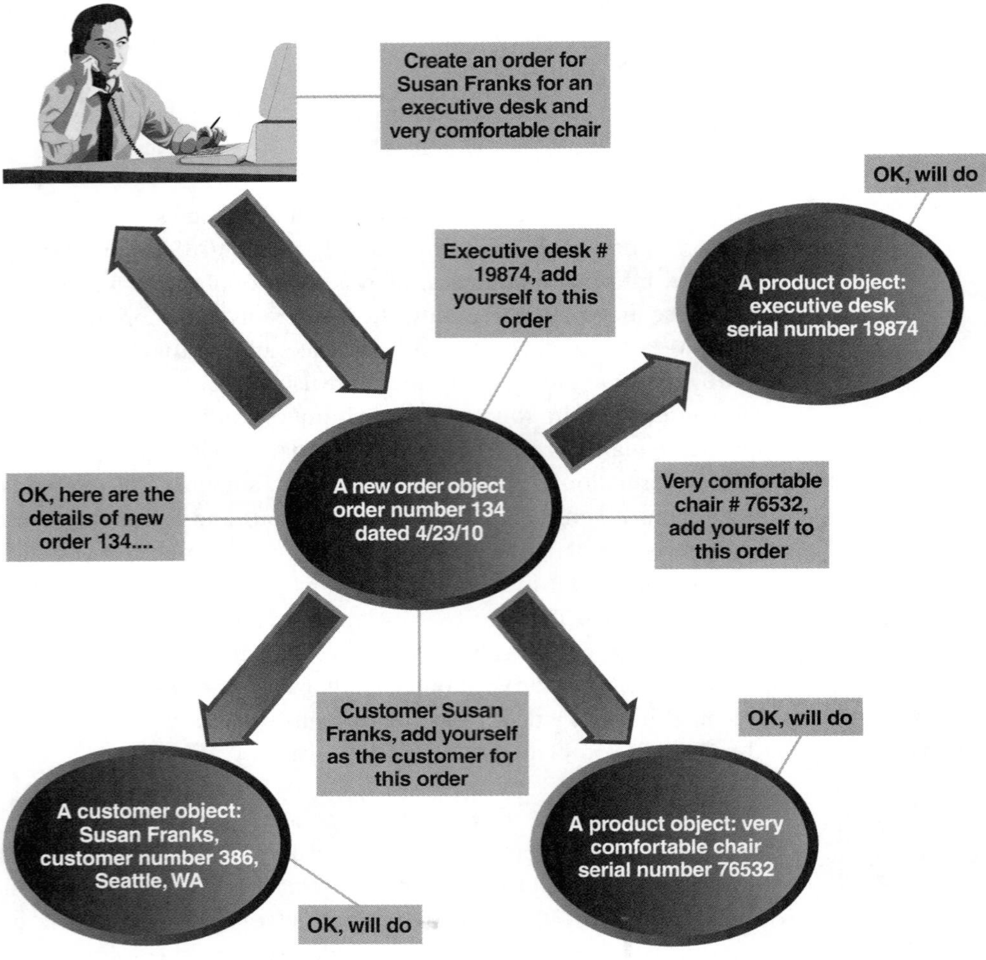

FIGURE 3-11
The object-oriented approach to systems (read clockwise starting with user).

object-oriented programming (OOP)
Writing statements in a programming language to define what each type of object does, including the messages the objects send to each other

class diagram
A graphical model that shows all of the classes of objects in the system in the object-oriented approach

An object is a type of thing—a customer or an employee, as well as a button or a menu. Identifying types of objects means classifying things. Some things, such as customers, exist outside the system (the real customer) and separately inside the system (a computer representation of a customer). A classification or "class" represents a collection of similar objects; therefore, object-oriented development uses a **class diagram** to show all of the classes of objects in the system (see Figure 3-12). For every class, there may be more specialized subclasses. For example, a savings account and a checking account are two special types of accounts (two subclasses of the class account). Similarly, a pull-down menu and a pop-up menu are two special types of menus. Subclasses exhibit or "inherit" characteristics of the class above it.

There are several key benefits of the object-oriented approach, among them naturalness and reuse. The object-oriented approach is natural—or intuitive—for people because people tend to think about the world in terms of tangible objects. It is less natural to think about complex procedures found in procedural programming languages.

Also, because the object-oriented approach involves classes of objects, and many systems in the organization use the same objects, these classes can be used over and over again whenever they are needed. For example, all systems use menus, dialog boxes, windows, and buttons, but many systems within the same company also use customer, product, and invoice classes that can be reused. There is less need to "reinvent the wheel" to create an object.

FIGURE 3-12
A class diagram created during object-oriented analysis.

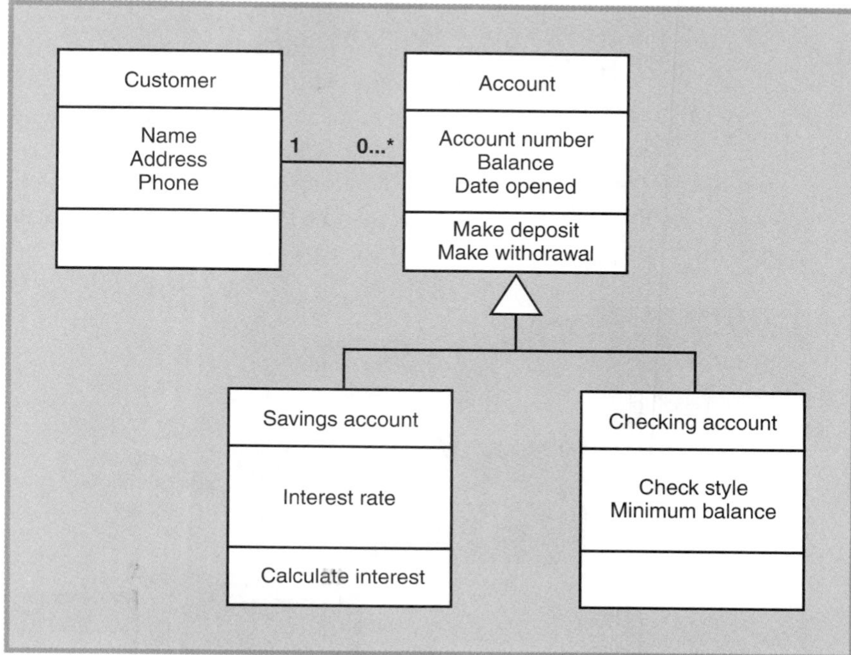

Clearly, the object-oriented approach is quite different from the traditional approach. But in other ways, quite a few traditional concepts are simply repackaged in the object-oriented (OO) approach. This makes the OO approach difficult to understand at first for some people. Parts 2 and 3 of this book discuss the similarities and differences in detail to help clarify each approach's strengths.

Many systems being developed today combine traditional and object-oriented technology in one system. Some IDEs also combine traditional and object-oriented technology in the same tool—for example, object-oriented programming is used for the user interface, and procedural programming for the rest. There are also many system projects that are exclusively traditional in analysis and design, and others that are exclusively object-oriented, even within the same information systems department. These are some of the reasons that it is important to cover both traditional approaches and newer object-oriented approaches in this text. Everyone should know the basic concepts of each, but your college's curriculum might emphasize one approach over the other.

System Development Life Cycle Variations

This text uses a generic system development life cycle throughout, but emphasizes that future systems analysts will encounter many variations in use in organizations. This section gives examples of the variations so that you will be familiar with and comfortable adapting to whatever variation you encounter. Some of the variations are based on the life cycle phases. Other variations are based on the degree of iteration that is allowed. Finally, some life cycles differ in their emphasis on the social, as well as technical, subsystems.

Variations of the Phases (Waterfall Models)

As we discussed in Chapter 2, the SDLC is often referred to as the "waterfall model." In a waterfall life cycle model, each life cycle phase is completed in sequence and then the results of the phase flow on to the next phase. Like a waterfall, there is no going back once a phase is completed (see Figure 3-13). This very rigid view of the SDLC insists on careful planning and control of the project. The decisions made at each phase are frozen, meaning they cannot be changed. Therefore, it is possible to define at any time exactly what has been decided. It is also possible to define at any point precisely how far along the project is. This waterfall model is based on tangible engineering projects—like building bridges, sky scrapers, and even space shuttles—that have very defined steps.

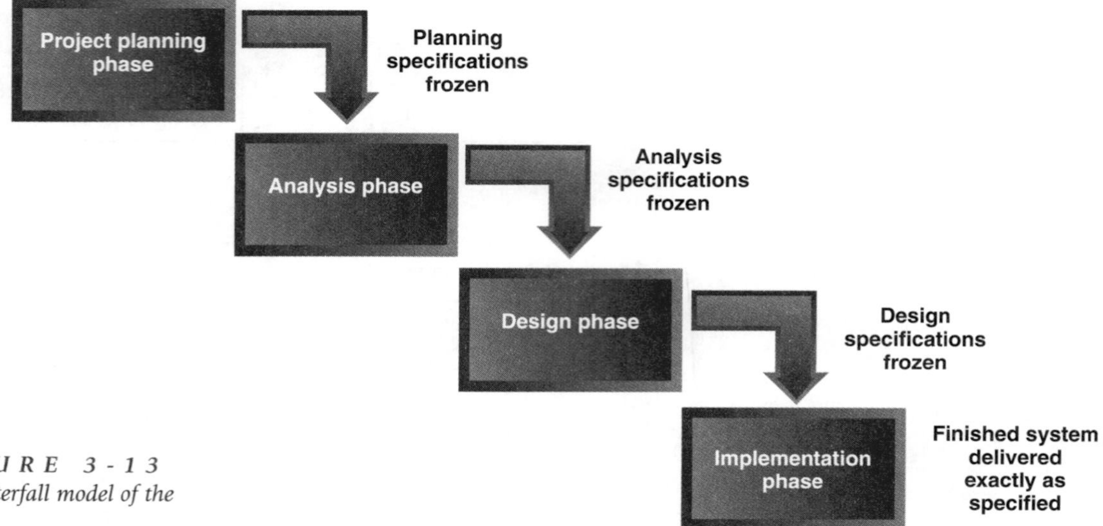

FIGURE 3-13
The waterfall model of the SDLC.

Figure 3-14 shows some examples of life cycles in use today that use different names for the phases. You may encounter one or several of these variations in your career. In a waterfall model, the more phases there are, the more rigid cutoffs there are. The first example (traditional) includes a feasibility study at the beginning, which usually assumes the project was unplanned rather than part of overall strategic system planning. System investigation and systems analysis together then make up the analysis phase in this SDLC. The second example is the information engineering life cycle. This life cycle includes the overall strategic planning effort as part of the life cycle (our life cycle includes planning for the project only). Two design phases are included—business system and technical design. Finally, construction and transition are two parts of the implementation phase.

The third life cycle model (objectory) includes only four phases, with less traditional names. The fourth life cycle model has eight phases, and it uses a different approach to naming the phases—they are named like activities (which we discuss below). The generic SDLC model used in this text uses five phases, named in a traditional way, and we name activities within the phases. No matter which of these life cycle models you use, the same overall tasks need to be carried out.

	Traditional SDLC	Information engineering	Objectory	SDLC with activity names for phases
Planning phase	Feasibility study	Information strategy planning		Organize the project and study feasibility
			Inception phase	Study and analyze the current system
Analysis phase	System investigation Systems analysis	Business area analysis		Model and prioritize the functional requirements
			Elaboration phase	Generate alternatives and propose the best solution
Design phase	Systems design	Business system design Technical design		Design the system
			Construction phase	Obtain needed hardware and software
Implementation phase	Implementation	Construction Transition		Build and test the new system
			Transition phase	Install and operate the new system
Support phase	Review and Maintenance	Production		

FIGURE 3-14
Life cycles with different names for phases.

The waterfall model has been criticized for information system projects because it can be too inflexible. Information systems are fairly intangible, compared to a bridge or a building. It is difficult to plan the information systems project until the requirements are fully understood. It is difficult to define all of the requirements until a specific design alternative has been chosen. In other words, a system development project cannot proceed sequentially through the waterfall in the same way a tangible building project can.

Variations Based on Iteration

A different view of the life cycle model is the notion of iteration. As we discussed in Chapter 2, iteration means that work tasks are done once, then again, and yet again (they are repeated). With each iteration, the result is refined so that it is closer to what is needed. Iteration assumes that no one gets the right result the first time. With an information system, you need to do some analysis and then some design before you really know whether the system will work and accomplish its goals. Then you do more analysis and design to make improvements (see Figure 3-15). In this view, it is not realistic to complete analysis (define the requirements) before starting work on the design. Similarly, it is very difficult to complete the design without knowing how the implementation will work (particularly with constantly changing technology). So you complete some design, then some implementation. And the iteration process continues—more analysis, more design, and more implementation. Naturally, the amount of iteration used depends on the complexity of the project.

How does iteration affect the SDLC? First, the idea of a waterfall no longer applies—the phases become blurred. One phase is not finished before another phase begins, so it is hard to know how far along the project is. The decisions are also not frozen at the end of each phase, so it is not certain at any point what has actually been decided. There is much uncertainty and ambiguity. For these reasons, it is much more difficult to manage a project when iteration is used.

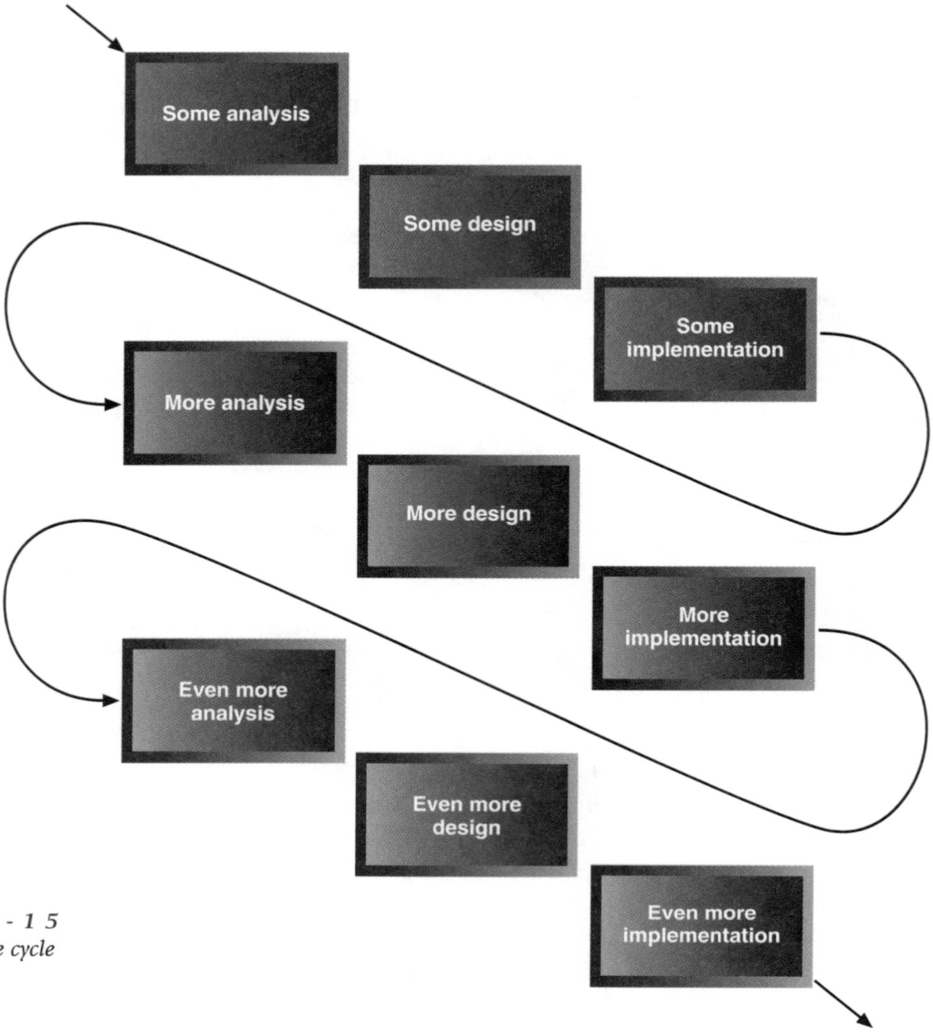

Some analysis
Some design
Some implementation
More analysis
More design
More implementation
Even more analysis
Even more design
Even more implementation

FIGURE 3-15
Iteration across life cycle phases.

It is important to recognize that any of the life cycle model variations discussed can use some amount of iteration. And most development is done with varying degrees of iteration. Even information engineering, which is sometimes considered one of the least flexible methodologies, can be and usually is done using iteration. The object-oriented approach is always described as highly iterative.

spiral model
Life cycle model with heavy iteration that breaks each project into smaller pieces, each with a different type of "risk"

The **spiral model** is a life cycle model with heavy iteration that is gaining recognition. The spiral model breaks each project into smaller pieces, each with a different type of risk. The risks could be anything from undefined requirements, to complex technology, to an uncertain competitive environment. The project starts out small, handling a few of the risks initially. Then the project expands in the next iteration to address more of the risks until eventually the system is completed (all risks are addressed). The spiral model, shown in Figure 3-16, begins the project in the center of the spiral where the project is still small, easy to manage, and low in risk. Then the project slowly expands. The iterative nature of the spiral model, coupled with the focus on reducing risk, has been very effective.

Variations Based on an Emphasis on People

Some life cycles in use reflect differences based on the underlying philosophies of the system development methodologies. Some methodologies recognize more explicitly

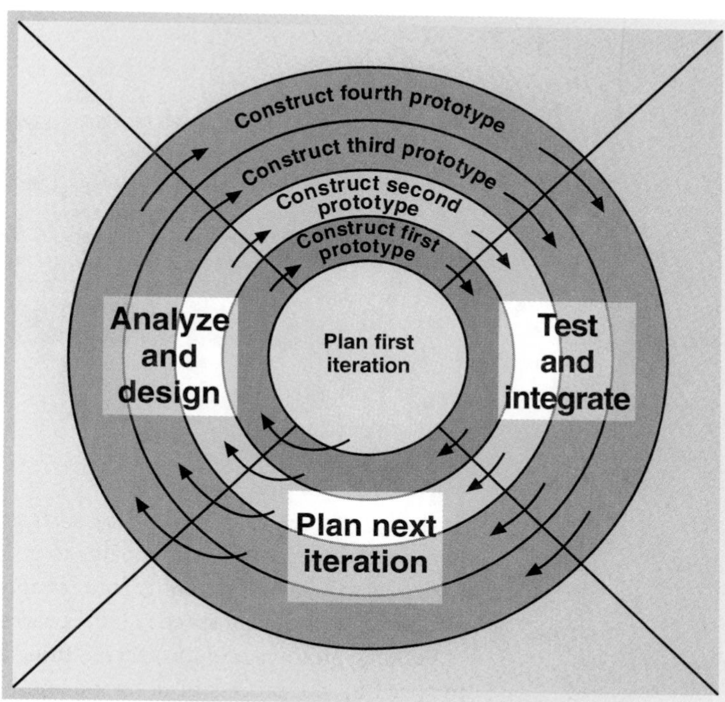

FIGURE 3-16
The spiral life cycle model.

sociotechnical systems
Information systems that include both social and technical subsystems designed to work well together

that information systems are **sociotechnical systems**—that is, they include both social and technical subsystems that must be considered and designed to work well together. Therefore, some life cycles explicitly emphasize these two aspects. Other terms for sociotechnical system development are user-centered design and participatory design. One example is called Multiview. The life cycle used with Multiview is shown in Figure 3-17. Note that human activity is studied in detail, and analysis and design of sociotechnical aspects of the system are considered in one phase. The design of the human-computer interface as a separate phase also emphasizes the importance of the users and user participation in the development of the system. End users often participate extensively in all aspects of system development.

The soft systems methodology is another approach that places more emphasis on people. Followers argue that it is relatively easy to model the data and processes in a system, but understanding the real world requires including people in the model. People have different and conflicting objectives, perceptions, and attitudes, making the system behavior unpredictable. A *rich picture diagram* that shows people and their attitudes can surface potential problems in a complex system.

Variations Based on Speed of Development

System developers have always looked for ways to speed up the development process. One reason to speed development is the continuing backlog of needed systems. To get more systems completed, it is necessary to speed up the process. Another reason for speed is the constantly changing technological and business environment. If it takes too long to get the system done, it might be too late to achieve the desired benefits.

rapid application development (RAD)
A variation of the system development life cycle that aims to speed up radically the development process using a variety of techniques

One variation of the systems development life cycle that aims to speed up the process radically is called **rapid application development** (RAD). Some variations of RAD try to

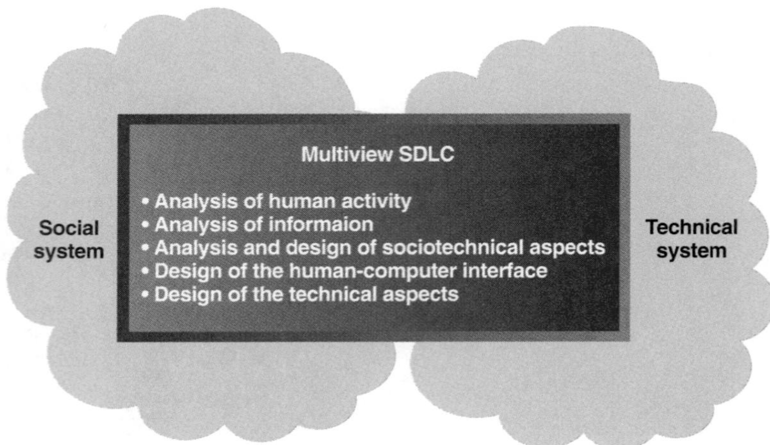

FIGURE 3-17
Phases of the Multiview SDLC.

speed up the activities in each phase, such as speeding the analysis phase by scheduling intensive meetings of key participants to get information gathered and decisions made rapidly. Iterative development is another approach often associated with RAD, as is the spiral life cycle model. It speeds up the process of getting to design and implementation.

Building prototypes of the system during analysis and design phases of the SDLC also can speed up the process of development, but prototyping is not always done just for RAD. Improved understanding of the system requirements is another objective of prototyping. Prototyping is discussed in Chapter 4 as an information-gathering technique.

Some approaches to RAD are even more radical. They involve creating working prototypes of the proposed system right away so that users can work with and critique them. Then the prototype is expanded into a finished system as soon as users agree on what the system is supposed to do. If not managed carefully, these approaches to RAD can be risky. RAD is discussed in more detail in Chapter 13.

Computer-Aided System Engineering (CASE) Tools

No matter which methodology is being used, it is important to use automated tools to improve the speed and quality of the system development work whenever appropriate. One type of tool discussed earlier is the CASE tool. The acronym CASE stands for *computer-aided system engineering*. CASE tools are specifically designed to help the systems analyst complete system development tasks. The analyst uses the CASE tool to create models of the system, many of them graphical models. But a CASE tool is much more than a drawing tool.

repository
A database that stores information about the system in a CASE tool, including models, descriptions, and references that link the various models together

The CASE tool contains a database of information about the models, called a repository. The **repository** stores information about the system, including models, descriptions, and references that link the various models together. The CASE tool can check the models to make sure they are complete and follow the correct diagramming rules. The CASE tool also can check one model against another to make sure they are consistent. If you consider that the analyst spends much time creating models, checking the models, revising the models, and then making sure the models all fit together, it is apparent how much help a CASE tool can provide. Figure 3-18 shows CASE tool capabilities that surround the repository. If system information is stored in a repository, it can be used in a variety of ways by the development team. Every time a team member adds information about the system, it is immediately available for everyone else.

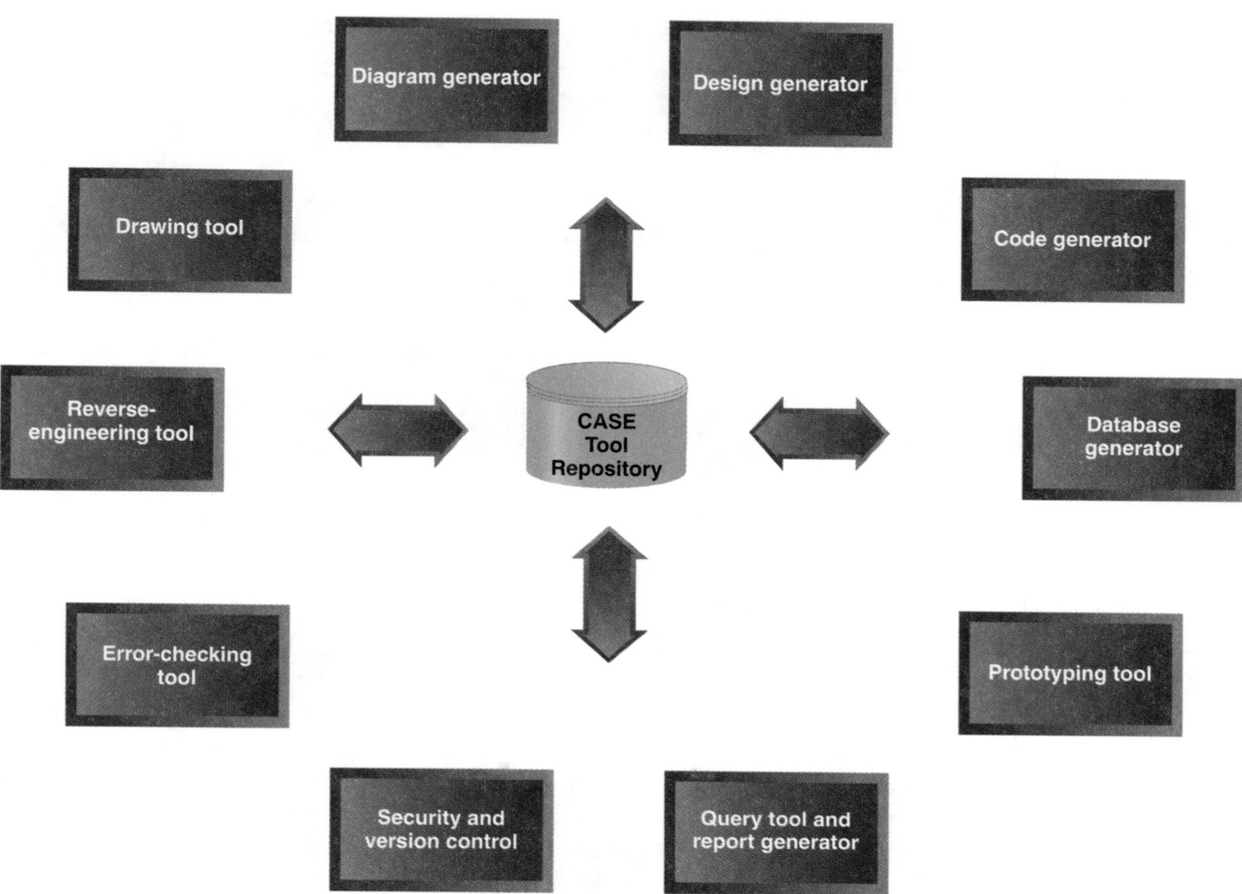

FIGURE 3-18
A CASE tool repository contains all information about the system.

CASE tools are often categorized as upper CASE or lower CASE tools. Upper CASE tools provide support for the analyst during the analysis and design phases, such as creating and checking models and storing information in the repository. Lower CASE tools provide support for implementation, primarily generating programs and database schemas based on specifications in the repository. CASE tools that combine support for the full life cycle are called integrated CASE, or ICASE tools.

Some CASE tools are designed to be as flexible as possible, allowing the analyst to use any approach to development desired. Other CASE tools are designed for very specific methodologies. Information engineering CASE tools have been fairly successful because they have been designed from the beginning for the information engineering approach. Two information engineering CASE tools are KnowledgeWare and Texas Instruments' Information Engineering Facility (IEF). Sterling Software (www.cool.sterling.com) now owns both of these tools, and has made both products into more flexible tools now called CoolTools.

Figure 3-19 shows a flexible CASE tool called Visual Analyst from Visible Systems Corporation (www.visible.com). Visual Analyst helps the analyst create traditional models, like data flow diagrams and entity-relationship diagrams, and supports object-oriented models. Like many vendors, Visible Systems Corporation has stopped using the term *CASE tool*. To many developers and managers, CASE tools failed to meet expectations and fell into disuse. Visual Analyst is now called an *integrated application development tool.*

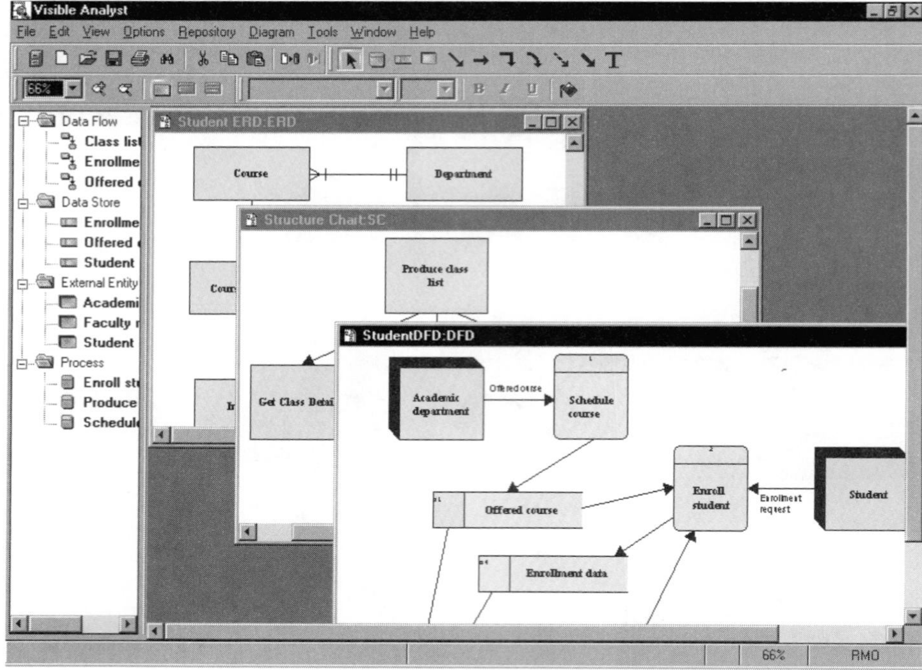

FIGURE 3-19
The CASE tool Visual Analyst showing a data flow diagram.

Rational Rose is a tool from Rational Software (www.rational.com) that specifically supports the object-oriented approach. Rational Rose is referred to as a *visual modeling tool*. The object-oriented models in this text conform to the unified modeling language (UML) standard developed primarily by researchers at Rational Software. Rational Rose provides reverse-engineering and code-generating capability and can be integrated with additional tools to provide a complete development environment. Figure 3-20 shows Rational Rose with a variety of UML models.

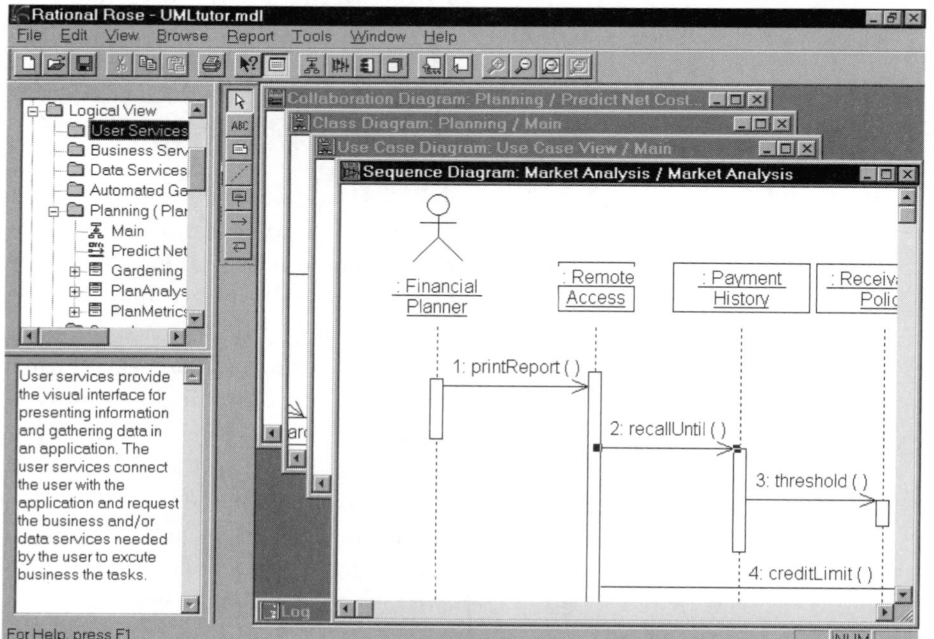

FIGURE 3-20
The visual modeling tool Rational Rose showing diagrams from the object-oriented approach.

A recent advance is the concept of *round-trip engineering*. Because system development can be so iterative, particularly in the object-oriented approach, it is important that the graphical models (such as class diagrams) be synchronized with generated program code. If the analyst changes the program code, the class diagram is updated. If the class diagram is changed, the program code is updated. Together J from Object International (www.oi.com) uses UML diagrams with the Java programming language to provide round-trip engineering support. Figure 3-21 shows Together J with a class diagram and synchronized Java code. Innovations such as these are renewing interest in CASE tools.

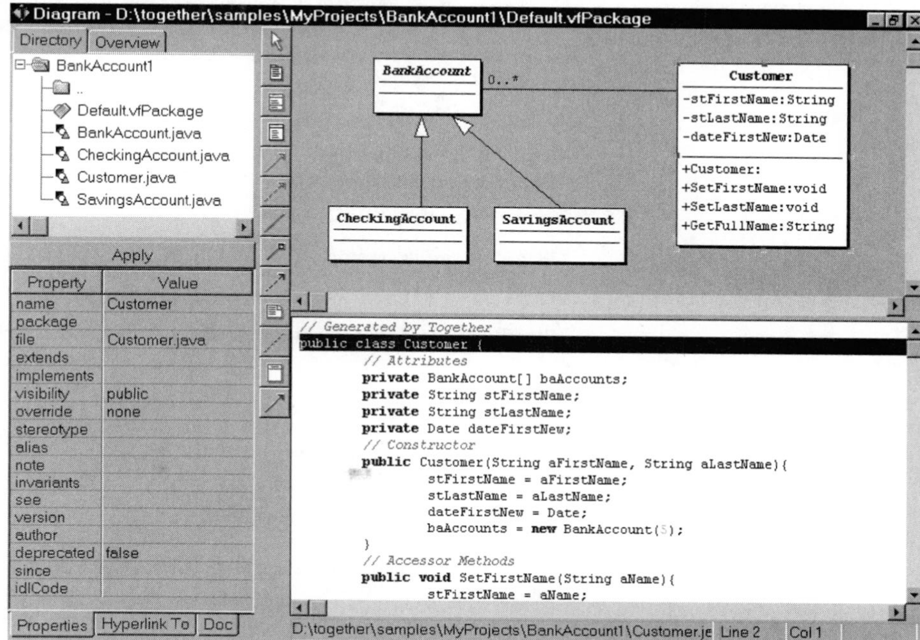

FIGURE 3-21
The round-trip engineering tool Together J showing a class diagram with synchronized Java source code.

Analysis and Design Phases in More Detail

All of the system development approaches described and virtually all of the specific methodologies you will encounter in organizations include similar activities in the analysis and design life cycle phases. This book is primarily about systems analysis and design, so these activities are covered in detail in the chapters that follow. Naturally, different system development methodologies recommend different techniques for completing these activities (and often have different names for them). In some cases, different types of models might be created to complete an activity. But the activities always involve answering the same key questions discussed in the following sections.

The Analysis Phase

The analysis phase involves defining in great detail what the information system needs to accomplish to provide the organization with the desired benefits. Many alternative design ideas should be proposed and the best design solution selected from among them. Later, during systems design, the selected alternative is designed in detail. There are six activities that must be completed during the analysis phase (see Figure 3-22). These activities are often completed in parallel. For example, the analyst gathers information continuously, not just at the beginning of the analysis phase. These activities are discussed in detail in Part 2; this section provides an overview.

Gather Information The analysis phase involves gathering a considerable amount of information. Systems analysts obtain some information from people who will be using the system, either by interviewing them or by watching them work. They obtain other information by reviewing planning documents and policy statements. Documentation from the existing system should be studied carefully. In short, analysts need to talk to nearly everyone who will use the new system and read nearly everything available about the existing system.

FIGURE 3-22
Analysis phase activities and key questions.

Analysis phase activities	Key questions
Gather information	*Do we have all of the information (and insight) we need to define what the system must do?*
Define system requirements	*What (in detail) do we need the system to do?*
Prioritize requirements	*What are the most important things the system must do?*
Prototype for feasibility and discovery	*Have we proven that the technology proposed can do what we think we need it to do?* *Have we built some prototypes to ensure the users fully understand the potential of what the new technology can do?*
Generate and evaluate alternatives	*What is the best way to do it?*
Review recommendations with management	*Should we continue and design and implement the system we propose?*

Beginning analysts often underestimate how much there is to learn about the work the user performs. The analyst really needs to become an expert in the business area the system will support. For example, if you are implementing an order-entry system, you need to become an expert on the way orders are processed (including accounting). If you are implementing a loan processing system, you need to become an expert on the rules used for approving credit. If you work for a bank, you need to think of yourself as a banker. The most successful analysts become very involved with their organization's main business. The analyst also needs to collect technical information.

The type of information gathered involves understanding the existing system, identifying and understanding all current and future users, identifying all present and future locations where work goes on, and identifying all system interfaces with other systems inside and outside the organization. Beyond that, the analyst needs to identify possible software package solutions that might be used to satisfy the system requirements. In Chapter 4, these specifics are discussed, including communication skills, interviewing skills, meeting skills, and other concepts related to gathering information.

The key question to be answered when completing the activity "Gather information" is, "Do we have all of the information (and insight) we need to define what the system must do?"

Define System Requirements As all of the necessary information is gathered, it is very important to record it. Some of this information describes technical requirements (for example, facts about needed system performance requirements or expected number of transactions). Other information involves functional requirements—what the system is required to do. It is very important to define carefully what the system is required to do.

Defining functional requirements is not just a matter of writing down facts and figures. Instead, many different types of models are created to help record and communicate what is required.

The modeling process is a learning process for an analyst. As the model is developed, the analyst learns more and more about the system. The process of modeling continues while information is gathered. The analyst continually reviews the models with the end users to verify that each model is complete and correct. The analyst studies each model, adds to it, rearranges it, and then checks how well it fits with other models being created. Just when the analyst is fairly sure the system requirements are fully specified, an additional piece of information surfaces requiring yet more changes, and the process of refinement begins again. It is important for the analyst to recognize that modeling can continue for quite some time, and it does not always have a defined end. The uncertainty involved makes some programmers uncomfortable.

logical model
Any model that shows what the system is required to do without committing to any one technology

physical model
Any model that shows how the system will actually be implemented

A requirements model (or collection of models) is a logical model. A **logical model** shows what the system is required to do in great detail, without committing to any one technology. By being neutral about technology, the development team can focus its efforts on what is needed no matter what technology is chosen. For example, a model might specify an output of the system as a list of data elements without committing to either paper or on-screen formats for the data. The focus of the model is what information the users need. A **physical model**, on the other hand, shows how the system will actually be implemented. A physical model of the output would include details about format.

The difference between logical and physical models is a key concept in systems analysis and systems design. Generally, systems analysis involves creating detailed logical models, and systems design involves detailed physical models. The design alternatives created during the analysis phase are physical models, but not very detailed.

The specific models created depend upon the technique being used for systems analysis. The modern structured analysis technique uses data flow diagrams (DFDs) and entity-relationship diagrams (ERDs). Information engineering uses process dependency diagrams and entity-relationship diagrams. Object-oriented techniques produce class diagrams and use case diagrams. Specific examples of these models are described in detail in Chapters 5, 6, and 7.

The key question to be answered when completing the "Define system requirements" activity is, "What (in detail) do we need the system to do?"

Prioritize Requirements Once the system requirements are well understood and detailed models of the requirements are completed, it is important to establish which of the functional and technical requirements are most crucial to the objectives of the system. Sometimes users suggest additional system functions that are desirable but inessential. However, users and analysts need to ask themselves which functions are truly important and which are fairly important, but not absolutely required. Again, the analyst who understands the organization and the work done by the users will have more insight for answering these questions.

Why ask these questions? Resources are always limited, and the analyst must always be prepared to justify the feasibility of the system. Therefore, it is important to know what is absolutely required. Unless the analyst carefully evaluates priorities, system requirements tend to expand as users make more suggestions (called *scope creep*).

The key question to be answered when completing the "Prioritize requirements" activity is, "What are the most important things the system must do?"

Prototype for Feasibility and Discovery Creating prototypes of part of the new system can be very valuable during the systems analysis phase. If the system involves new technology, it is important for the team to get firsthand experience working with it. Only then can the team be sure the technology is feasible. It is surprising how often new technology fails to live up to marketing hype. Prototypes can prove that the technology will do what it is supposed to do. Prototyping for feasibility is increasingly important when working with new or recently upgraded technology to assess the possibilities and limitations of the technology early in the process.

Additionally, if the new system will include new or innovative technology, it might be necessary to develop some prototypes to help the users visualize the possibilities when defining what they require. Prototypes can be built to help the users discover requirements they might not have thought about otherwise. These prototypes help get the users (and the analysts) thinking creatively "outside of the box."

The prototyping activity during the analysis phase helps answer two key questions: "Have we proven that the technology proposed can do what we think we need it to do?", and equally important, "Have we built some prototypes to ensure the users fully understand the potential of what the new technology can do?"

Generate and Evaluate Alternatives Many alternatives exist for the final design and implementation of a system. It is very important to carefully define and then evaluate all of the possibilities. When requirements are prioritized, the analyst can always generate several alternatives just by including some of the highest priority requirements rather than all of the requirements in the final system. But technology also raises several alternatives for implementing the system. Beyond that, the analyst can consider more than one method to build the system, perhaps using in-house development staff or a consulting firm. Further, one or more off-the-shelf software packages could possibly satisfy all of the requirements.

Clearly, lots of alternatives are open to the project team. Each alternative needs to be described or modeled at a high (summary) level. Each alternative also has its own costs, benefits, and other characteristics that need to be carefully measured and compared (as in the feasibility study described in Chapter 2). The best alternative is then chosen. This is not as easy as it sounds, because costs and benefits are very difficult to measure. And many design details are still uncertain. The analyst is concerned about feasibility of the project overall during the project planning phase, then the feasibility of each alternative during the analysis phase.

The key question to be answered when completing the "Generate and evaluate alternatives" activity is, "What is the best way to do it?"

Review Recommendations with Management All of the preceding activities are done in parallel—gather information, define requirements, prioritize requirements, prototype for feasibility and discovery, and generate and evaluate alternatives. The final analysis phase activity—review recommendations with management—is usually done when all of the other analysis activities are complete. Although management should be kept informed through regular project reporting, the project manager must eventually recommend a solution and obtain a decision from management. Questions the analyst must consider are the following: Should the project continue at all? If the project continues, which alternative is the best choice? Given the recommended alternative, what are the revised budget and schedule for completing the project?

Making a recommendation to senior executives is a major management checkpoint in the project. Every alternative—including cancellation—should be explored. Even though quite a bit of work might have been invested in the project, it is still possible that the best choice is to cancel the project. Perhaps the benefits are not as great as originally thought. Perhaps the costs are much greater than originally thought. Or, because of the rapidly changing business environment, perhaps the organization's objectives have changed since the project was originally proposed, making the project less important to the organization. For any of these reasons, it might be best to recommend the project be canceled.

If the project is worthwhile, the project team has detailed documentation of the system requirements and a proposed design alternative, so the project manager should be in a good position to produce a more accurate estimate of the budget and schedule necessary to complete the project. If a case can be made for continuing the project, then management will probably provide the requested resources and the project will continue. The key point to remember is that continuing the project to the design phase is never automatic. Good project management techniques always require continual reassessment of the feasibility of the project and formal management reviews.

The key question to be answered when completing the "Review recommendations with management activity" is, "Should we continue and design and implement the system we propose?"

The Design Phase

The design phase involves specifying in detail how a system will work using particular technology. Some of the design details will have been developed during systems analysis, when the alternatives were described. But much more detail is required. Sometimes systems design work is done in parallel with the analysis phase, and usually the activities of systems design are done in parallel. For example, the database is designed at the same time the user interface is designed. These activities are described in detail in Part 3. Figure 3-23 provides an overview of the activities for the design phase.

FIGURE 3-23
Design phase activities and key questions.

Design phase activity	Key question
Prototype for design details	*Have we created prototypes to ensure all detailed design decisions have been fully understood?*
Design the user interface(s)	*Have we specified in detail how all users will interact with the system?*
Design the system interface(s)	*Have we specified in detail how the system will work with all other systems inside and outside our organization?*
Design the application architecture	*Have we specified in detail how each system activity is actually carried out by the people and computers?*
Design and integrate the database	*Have we specified in detail how and where the system will store all of the information needed by the organization?*
Design and integrate the network	*Have we specified in detail how the various parts of the system will communicate with each other throughout the organization?*
Design and integrate the system controls	*Have we specified in detail how we can be sure that the system operates correctly and the data maintained by the system is safe and secure.*

Prototype for Design Details During the design phase, it is important to continue creating and evaluating prototypes. Often associated with interface design, prototyping can also be used to confirm design choices about database, network architecture, controls, or even programming environments being used. Therefore, when considering all of the design activities, think about how prototypes might be used to help understand a variety of design decisions. It is also important to recognize that rapid application development (RAD) approaches develop prototypes during design that evolve into the finished system. In those cases, the prototype is the system.

The key question to be answered when completing the "Prototype for design details" activity is, "Have we created prototypes to ensure all detailed design decisions have been fully understood?"

Design the User Interfaces A critical aspect of the information system is the quality of the user interface. The design of the user interface defines how the user will interact with the system. To most users, the interface is a graphical user interface with windows, dialog boxes, and mouse interaction. Increasingly, it can include sound, video, and voice commands. Users' capabilities and needs differ widely; each user interacts with the system in different ways. Additionally, different approaches to the interface might be needed for different parts of the system. Therefore, there are many user interfaces to consider. And as information systems become increasingly interactive and accessible, the user interface is becoming a larger part of the system.

Analysts should remember that to the user of the system, the user interface *is* the system. The user interface is more than just the screens—it is everything the user comes into contact with while using the system, conceptually, perceptually, and physically. Therefore, the user interface is not just an add-on component to the system. It is something that needs to be considered throughout the development process.

The nature of the user interface begins to emerge very early in the development process, when requirements are being defined. The specification of the kinds of tasks the users will complete begins to define the user interface. Then when alternatives are being defined, a key aspect of each alternative is its type of user interface. The activity of designing the user interface in detail, however, occurs during systems design.

Sometimes specialists in user interface design are brought in to help with the project. These specialists might be called interface designers, usability consultants, or human factors engineers. The visual programming environments now available make it easy for developers to create graphical user interfaces for applications. But it is still very difficult to make a graphical user interface friendly or intuitive.

The key question to be answered when completing the "Design the user interfaces" activity is, "Have we specified in detail how all users will interact with the system?"

Design the System Interfaces No system exists in a vacuum. Many other information systems will be affected by the new information system. Sometimes one system provides information that is later used by another system. Other times systems exchange information continuously as they run. The component that enables systems to share information is the system interface, and each system interface needs to be designed in detail.

Some system interfaces are with internal organizational systems, so there may be information available about the other systems. All of these systems should be designed to work together from the beginning. In some cases, the new system needs to interface with a system outside the organization—for example, at a supplier or customer site. In other cases, the new system needs to interface with a package application that the organization has purchased and installed. These system interfaces can become quite complex, particularly with so many types of technology available. Often, people with very specialized technical skills are required to work on these interfaces.

The key question to be answered when completing the "Design the system interfaces" activity is, "Have we specified in detail how the system will work with all other systems inside and outside our organization?"

Design the Application Architecture Designing the application architecture involves specifying in detail how all system activities will actually be carried out. These activities are described during systems analysis in great detail as logical models, without indicating what specific technology would be used. Once a specific design alternative is chosen, the detailed computer processing—the physical models—can be designed. A key decision is to define the automation boundary, discussed in Chapter 1, which separates the manual work done by people from the automated work done by computer. Models created include physical data flow diagrams, structure charts, object-interaction diagrams, or other physical models.

The approach to application design and the design models created will vary depending upon the technology that will be used. If the programming language used is Visual Basic, for example, the type and nature of the models developed will be different than if the language were COBOL. If client-server architecture is used, the models used will be different than with a centralized architecture. If object-oriented technology is used, the models will definitely be quite different. Additionally, some of the activities will be carried out by people rather than computers, so manual procedures need to be designed.

The key question to be answered when completing the "Design the application architecture" activity is, "Have we specified in detail how each system activity is actually carried out by people and computers?"

Design and Integrate the Database Designing the database for the system is another key design activity. The data model (a logical model) created during systems analysis is used to create the physical model of the database. Sometimes the database is a collection of traditional computer files. More often, it is a relational database consisting of dozens or even hundreds of tables. Sometimes files and relational databases are used in the same system. Sometimes object-oriented databases might be used instead of relational databases.

Analysts must consider many important technical issues when designing the database. Many of the technical (as opposed to functional) requirements defined during systems analysis have to do with database performance needs (such as response times). Much of the design work might involve performance tuning to make sure the system actually works fast enough. Another key aspect of designing the database is making sure that new databases are properly integrated with existing databases.

The key question to be answered when completing the "Design and integrate the database" activity is, "Have we specified in detail how and where the system will store all of the information needed by the organization?"

Design and Integrate the Network Sometimes a new system is implemented along with a new network. If this is the case, then the network needs to be designed. More often, though, the network has been established by network specialists based on an overall strategic plan. The systems design alternative chosen was one that fit the existing network plan. So rather than designing a network, the project team typically must integrate the system into an existing network.

Important technical issues arise when making the system operate over a network, such as reliability, security, and throughput. Again, specialists are often brought in to help with the technical details. The requirements developed during systems analysis specify what work goes on at what locations, so these locations need to be connected. Technical requirements (as opposed to functional requirements) often have to do with communication via networks.

The key question to be answered when completing the "Design and integrate the network" activity is, "Have we specified in detail how the various parts of the system will communicate with each other throughout the organization?"

Design and Integrate the System Controls A final design activity involves ensuring that the system has adequate safeguards to protect organizational assets. These safeguards are referred to as system controls. This activity is not listed last because it is less important than the others. On the contrary, it is a crucial activity. It is listed last because controls have to be considered for all other design activities—user interface, system interfaces, application architecture, database, and network design.

User-interface controls limit access to the system to authorized users. System interface controls ensure that other systems cause no harm to this system. Application architecture controls ensure that transactions recorded and other work done by the system are done correctly. Database controls ensure that data are protected from unauthorized access and from accidental loss due to hardware failure. Finally, and of increasing importance, network controls ensure that communication through networks is protected. All of these controls need to be designed into the system, based upon the specific technology in place. Specialists are often brought in to work on some controls, and all system controls need to be thoroughly tested.

The key question to be answered when completing the "Design and integrate the systems controls" activity is, "Have we specified in detail how we can be sure that the system operates correctly and the data maintained by the system is safe and secure?"

Summary

There are lots of ways to develop an information system. All development approaches use a system development life cycle (SDLC) to manage the project, plus models, techniques, and tools that make up a system development methodology. A system development methodology provides guidelines to follow for completing every activity in the SDLC, and there are many different methodologies in use. Most methodologies are based on one of three approaches to information system development—the structured approach, the information engineering approach, or the object-oriented approach.

There are quite a few variations of the SDLC. The original life cycles were waterfall models with rigid starting and ending points for each phase. Most variations now use iteration across phases, where some analysis, then some design, and then some implementation are completed and then the process is repeated to refine the system. Some variations focus on people who use the system, called participatory or user-centered design. Other variations have the goal of speeding up development, called rapid application development (RAD). Computer-aided system engineering (CASE) tools are special tools designed to help the analyst complete development tasks, including modeling and generating program statements directly from the models.

Most of this book addresses the analysis and design life cycle phases. Each phase involves completing specific activities, often in parallel. The analysis phase determines in detail what the system is required to accomplish with a preliminary description of the design alternative that should be implemented. The activities of the analysis phases are to gather information, define system requirements, prioritize requirements, prototype for feasibility and discovery, generate and evaluate alternatives, and review recommendations with management. The design phase determines in detail how the system will be implemented, including the activities to prototype for design details, design the user interfaces, design the system interfaces, design the application architecture, design and integrate the database, design and integrate the network, and design and integrate the system controls.

Key Terms

CASE tool, p. 66
class diagram, p. 74
data flow diagram (DFD), p. 70
entity-relationship diagram (ERD), p. 71
information engineering, p. 72
logical model, p. 85
model, p. 64
object, p. 73
object-oriented analysis (OOA), p. 73
object-oriented approach, p. 73
object-oriented design (OOD), p. 73
object-oriented programming (OOP), p. 74
physical model, p. 85
rapid application development (RAD), p. 79
repository, p. 80
sociotechnical systems, p. 79
spiral model, p. 78
structure chart, p. 69
structured analysis, p. 70
structured approach, p. 67
structured design, p. 69
structured program, p. 68
system development methodology, p. 64
technique, p. 66
tool, p. 65
top-down programming, p. 68

Review Questions

1. What is the difference between a model and a tool?
2. What is the difference between a technique and a methodology?
3. Which of the three approaches to system development was the earliest?
4. Which of the three approaches to system development is the most recent?
5. Which of the three approaches focuses on overall strategic systems planning?
6. Which of the three approaches is fundamentally different from the others?
7. What are the three constructs used in structured programming?
8. What graphical model is used with the structured design technique?
9. What graphical model is used with the modern structured analysis technique?
10. What model is the central focus of the information engineering approach?
11. Explain what is meant by a waterfall life cycle model.
12. What concept suggests repeating activities over and over until you achieve your objective?
13. What is user-centered and/or participatory design?
14. What is meant by rapid application development (RAD)?
15. What are CASE tools? Why are they used?
16. What information is contained in a CASE tool repository?
17. What are the main activities of the analysis phase?
18. What are the main activities of the design phase?

Thinking Critically

1. List some of the models architects create to show different aspects of a house they are designing. Explain why several models are needed.
2. What models might an automotive designer use to show different aspects of a car?
3. Sketch the layout of your room at home. Now write a description of the layout of your room. Are these both models of your room? Which is more accurate? More detailed? Easier to follow for someone unfamiliar with your room?
4. Describe a "technique" you use to help you complete the activity "Get to class on time." What are some "tools" you use with the technique?
5. Describe a "technique" you use to make sure you get assignments done on time. What are some "tools" you use with the technique?
6. What are some other techniques you use to help you complete activities in your life?
7. Think of completing college as a project. As with any other project, you should follow some sort of "college education completion" methodology. What might be the phases of your personal college education completion life cycle? What are some of the activities of each phase? What are some techniques you use to help complete the activities? What models might you create during the process of completing college? Differentiate between models you create

that get you through college compared to those that help you plan and control the process of completing college. What are some of the tools you use to help you complete the models?

8. There are at least three approaches to system development, a variety of life cycles, and a long list of techniques and models that are used in some approaches but not in others. Consider why this is so. Discuss these possible reasons: the field is so young, the technology changes so fast, different organizations have such different needs, there are so many different types of systems, and there are people with great differences in background who are developing systems.

Experiential Exercises

1. Go to the campus placement office and gather some information on companies that recruit information system graduates on your campus. Can you find any information about the approach they use to develop systems? Is their SDLC described? Do any mention a CASE tool? Visit the company web sites and see whether you can find any more information.
2. Visit the web sites for a few leading information system consulting firms. Can you find any information about the approach they use to develop systems? Are their SDLCs described? Is there any mention of a CASE tool?
3. Take a look at the web page *The CASE Tools Homepage*, listed in the "Further Resources" section at the end of this chapter. This site provides links to vendors that sell and support CASE tools. Follow some of the links and see whether you can determine which approach to system development each CASE tool emphasizes: the structured approach, information engineering, or the object-oriented approach. Try to find at least one tool that fits each approach and write a summary of its features.

Case Study

Factory System Development Project

Sally Jones is assigned to manage a new system development project that will automate some of the work being done in her company's factory. It is fairly clear what is needed: to automate the tracking of the work in progress and the finished goods inventory. What is less clear is the impact of any automated system on the factory workers. Sally has several concerns: How might a new system affect the workers? Will they need a lot of training? Will working with a new system slow down their work or interfere with the way they now work? How receptive will the workers be to the changes the new system will surely bring to the shop floor?

At the same time, Sally recognizes that the factory workers themselves might have some good ideas about what will work and what won't, especially concerning the technology that is more likely to survive in the factory environment. And what sort of user interface will work best for the workers? Sally doesn't know much about factory operations, although she does understand inventory accounting.

1. Is the proposed system an accounting system? A factory operations system? Or both?
2. Which life cycle variations might be appropriate for Sally to consider using?
3. Which activities of analysis and of design should involve factory workers as well as factory management?

Further Resources

Some classic and more recent texts include the following:

D. E. Avison and G. Fitzgerald, *Information Systems Development: Methodologies, Techniques and Tools* (2nd ed.). McGraw-Hill, 1995.

Tom DeMarco, *Structured Analysis and System Specification*. Prentice Hall, 1978.

C. Gane and T. Sarson, *Structured Systems Analysis: Tools and Techniques*. Prentice Hall, 1979.

Ivar Jacobson et al., *Object-Oriented Software Engineering: A Use Case Driven Approach*. Addison-Wesley, 1992.

Ivar Jacobson, Grady Booch, and James Rumbaugh, *The Unified Software Development Process*. Addison-Wesley, 1999.

James Martin, *Information Engineering: A Trilogy* (books 1, 2, and 3). Prentice Hall, 1990.

Meilir Page-Jones, *The Practical Guide to Structured System Design* (2nd ed.). Prentice Hall, 1988.

John Satzinger and Tore Orvik, *The Object-Oriented Approach: Concepts, Modeling and System Development*. Course Technology, 1996.

Ed Yourdon, *Modern Structured Analysis*. Prentice Hall, 1989.

Some web sites of interest include the following:

CASE Tools Homepage	http://osiris.sunderland.ac.uk/sst/casehome.html
Sterling Software	http://www.sterling.com/cool/
Visible Systems Corporation	http://www.visible.com
Rational Software Corporation	http://www.rational.com
Object International	http://www.togethersoft.com

Part 2

Systems Analysis Tasks

Investigating System Requirements

Learning Objectives

After reading this chapter, you should be able to:

- Describe the difference between functional and technical system requirements
- Identify and understand the different types of users who will be involved in investigating system requirements
- Describe the kind of information that is required to develop system requirements
- Determine system requirements through questionnaires, review of documentation, interviews, observation, prototypes, and joint application design sessions
- Explain the steps involved in conducting a structured walkthrough
- Understand the contribution of business process reengineering to the definition of requirements

Chapter Outline

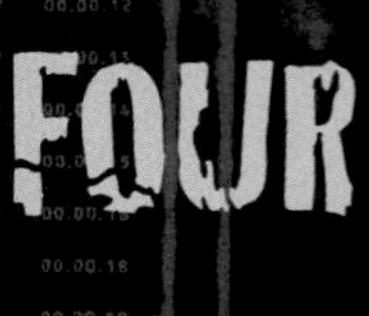

PROVIDING CUSTOMER SERVICE—AND GASOLINE—AT CONVENIENCE AMERICA

Convenience America had finally put its project team together. Hal Turley, the project leader, realized that his team of five systems analysts would change and grow as the project progressed. In fact, five people was a pretty large team to begin the project. The project schedule indicated that the activities composing the analysis phase of the system development life cycle were slated to begin. Hal had been through two other projects as project leader, so he had a good idea how to get started.

The new system was going to be entirely new for his company, not an added feature to an existing system. Convenience America owns a national chain of gas stations and convenience stores. The company had just concluded a six-month marketing study on a new service to offer its customers. To compete against the large oil company chains, Convenience America wants to connect all of its stores to a central database to keep track of gasoline and oil purchases for each customer. At the end of each month, customers would receive a statement of their gasoline purchases. This information could even be provided by different automobiles if desired. With this unified system, customers could access information about their purchases no matter which station they used throughout the United States. Payment could be made with cash or any credit card and would not require a special gasoline credit card.

Convenience America saw this as a way to improve customer service and loyalty. It had determined that many of its customers were fairly consistent repeat customers who would benefit from gasoline purchase information. In addition, once a person or family was enrolled, additional benefits could be provided such as rebates for volume purchases and information about special promotions. Convenience America planned to structure the program so that there were benefits both for individual families and for business customers. The primary benefit for Convenience America was to encourage customers to buy all of their gasoline from Convenience America.

Hal's first task was to identify which users he needed to solicit to get their requirements. On this type of a system, it would be easy to leave out some important users, and consequently some important requirements. He needed to be sure to get the information and database requirements from those in the marketing department who directed the original market trial. It was important to get suggestions from some of the salespeople who worked in the stores to make sure the user interface was easy to use—a lot of information would need to be collected at the time of the sale. The quality of this interface could make or break the success of this system.

Middle and upper management would certainly want tracking information on this new venture. Preliminary estimates indicated that family purchases should increase by 30 to 50 percent. Business purchases should also increase by a similar amount. The president would want hard facts to determine the success of the new service.

Hal needed to resolve some unique technical issues to get all of the stores connected to the central database. Fortunately, one of the members of his team was from the technical staff. However, Hal wanted to be sure to schedule discussion sessions with the other appropriate technical staff.

It was going to be a fun and challenging project. In his first project team meeting, Hal had two major objectives on the agenda:

- Construct a list of all the important users and technical staff who need to be involved in the definition of systems requirements.
- Begin building a list of questions and issues that need to be discussed to define the system requirements.

OVERVIEW

In the previous chapters, you learned that system development consists of four major phases: planning, analysis, design, and implementation. In the last chapter, you learned about the activities of the analysis and the design phases and the objective of each of the activities in the phases. This chapter focuses on the skills and the associated tasks that you will undertake in the analysis phase. As discussed in Chapter 1, many skills are involved in system development. Two of the skills that are needed to perform systems analysis are (1) fact-finding for the investigation of system requirements and (2) modeling of business processes based on the system requirements. Even though the analysis phase includes many related activities, these two specific skills are fundamental to analysis. In this chapter, you will develop your fact-finding and investigation skills. Later chapters will teach you about modeling.

During the fact-finding and investigation activities, you learn details of the business processes and daily operations. In fact, your objective during these activities is to try to become as knowledgeable of how the business operates as the users you interview. Why become an expert? Because only then can you ensure that the system meets the needs of the business. And only then can you use your unique combination of skills to make a difference. You bring a fresh perspective to the problem, and you can make a great difference in the organization by identifying new and better ways to accomplish business objectives using information technology. Many current users are so used to the way they have been performing their tasks that they cannot envision better, more advanced ways to an end result. Your technical knowledge combined with your newly acquired problem domain knowledge can bring unique solutions to implementing business processes.

An additional benefit to becoming an expert in the problem domain is that you build credibility with the users. Your domain knowledge increases the credibility of your suggestions and ensures that your suggestions meet users' specific needs. During the development of the new system, you will have many suggestions and recommendations about the daily procedures for running the business. The installation of new information systems usually requires major changes in business operating procedures. If you can "walk the walk and talk the talk" of the users' business operations, then they are much more likely to accept your recommendations. Otherwise, you may be viewed as an outsider who really does not understand the problems.

The sections that follow define system requirements. Then they explain several techniques for learning about the business processes and gathering information using both traditional and newer accelerated methods. Finally, the chapter explores business processing reengineering as a tool for business change and your involvement in it.

Functional and Technical Requirements

system requirements
The definition of specifications for functions to be provided by a system

System requirements are the functions that the new system must perform. Recall that one of the activities during the planning phase is to identify the system scope. In that activity, the analyst identifies a set of system capabilities. During analysis, the analyst defines and describes those capabilities in greater detail. In other words, the analyst expands those high-level capabilities into detailed system requirements. Generally we divide system requirements into two categories: functional requirements and technical requirements.

functional requirement
A system requirement that describes a function or process that the system must support

Functional requirements are the activities that the system must perform—that is, the business uses to which the system will be put. They derive directly from the capabilities identified in the planning phase. For example, if you are developing a payroll system, the required business uses might include functions such as "write paychecks," "calculate commission amounts," "calculate payroll taxes," "maintain employee-dependent information," and "report year-end tax deductions to the IRS." These are the functions of the new system. The process of identifying and describing *all* of these business uses requires a substantial amount of time and effort. The list of functions and their interrelated nature can become very complex.

Functional requirements are based on the procedures and business rules that the organization uses to run its business. Sometimes these procedures are well documented and easy to identify and describe. An example might be, "All new employees must fill out a W-4 form to enter information about their dependents in the payroll system." Other business rules might be more obtuse or difficult to find. An example of this kind of rule from Rocky Mountain Outfitters might be that, "An additional 2 percent commission rate is paid to order takers on telephone sales for 'special promotions' that are added to the order." These special promotions are unadvertised specials that are sold by the telephone order clerk—thus the special commission rate. This rule might be a new rule that the sales manager decided to institute. It might not be documented anywhere, and its inclusion in the functional requirements would depend on the manager remembering to describe it to you. Discovering this rule can have major impacts on the final design of the system, however. Without knowing this rule, you may design a system that allows only fixed commission rates. Including this rule would require the system to be able to handle variable commission rates by type of order. It is critical that you develop good skills to ferret out all of the applicable business rules that affect the new system.

technical requirement
A system requirement that describes an operating environment or performance objective

Technical requirements are all the operational objectives related to the environment, hardware, and software of the organization. Examples are "must run in a client-server environment with Windows NT," "must have one-half second response time on all screens," and "must be able to support 100 terminals at once (with the same response time)." These technical requirements are often expressed as specific objectives that the system must attain.

Both types of system requirements are needed for a complete definition of the new system. Both are included in the investigation of system requirements. Functional requirements are most often documented in the analysis models that are built. Technical requirements are usually documented in a narrative description of technical requirements.

Stakeholders—the Source of System Requirements

stakeholders
All the people who have an interest in the success of a new system

Your primary source of information for functional system requirements is from the various stakeholders of the new system. **Stakeholders** are all people who have an interest in the successful implementation of the system. Generally, we categorize stakeholders into one of three groups: (1) the users, those who actually use the system on a daily basis, (2) the clients, those who pay for and own the system, and (3) the technical staff, the people who must ensure that the system operates in the computing environment of the organization. Figure 4-1 illustrates the various kinds of stakeholders who have an interest in a new system. We have discussed earlier the difference between the users and the clients. During analysis, the analyst needs to consider the technical staff as well. One of the most important first steps in determining systems requirements is to identify these various system stakeholders. In the past, problems arose with new systems because only some of the stakeholders were included in the project and the system was built exclusively for them. One of your first tasks is to identify every type of stakeholder who has an interest in the new system. Your second task is to ensure that you identify the critical persons from each stakeholder type to be available as the business expert.

FIGURE 4-1
Stakeholders with an interest in new systems development.

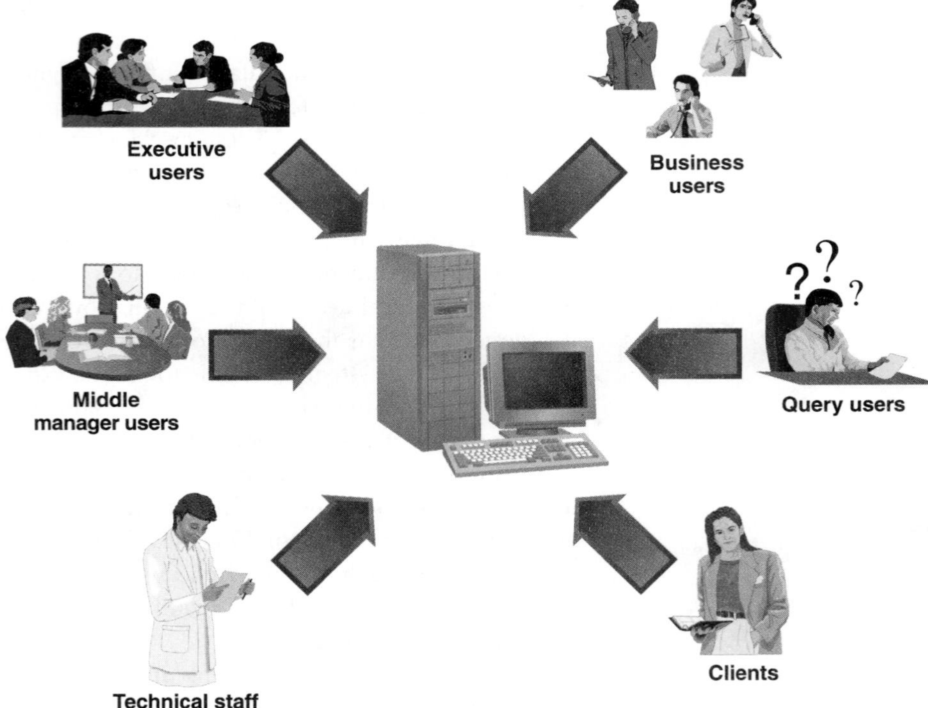

User Stakeholders

User roles—that is, types of system users—should be identified in two dimensions: horizontally and vertically. By horizontally, we mean that the analyst must look for information flow across business departments or functions. For example, a new inventory system may affect receiving, warehousing, sales, and manufacturing. You must be sure that individuals from each of these departments get to describe their requirements. The sales department may provide requirements to determine when and how to update inventory quantities or maybe to commit a certain quantity of a product at the time of the sale but before it is shipped. Manufacturing may have informational needs from the inventory system to assist in the scheduling of production. The horizontal dimension is an indication that many different departments, even those that may appear unrelated to the new system, should be included in the definition of requirements.

By vertically, we mean the information needs of clerical staff, of middle management, and of senior executives. Each of these stakeholders will have different information requests for the system that must be included in the design. The following sections describe the characteristics and information needs of the various users on the vertical dimension. These same characteristics also apply to each department across the horizontal dimension.

transaction
A single occurrence of a piece of work or an activity done in an organization

query
A request for information from a system or from a database

Business Operations Users Business operations users are the people who use the system to perform the day-to-day operations of an organization. We often call these operations *transactions*. A **transaction** is a piece of work done in an organization, such as "enter an order." In Chapter 1, you learned that a transaction processing system is one that processes these types of business operations. Business users will provide information about the daily operations of the business and how the system must support them.

Query Users A **query** is a request for information. A query user is a person who needs current information from the system. This person may be the same person as a business operations user or someone else. In some cases, a business may want to make information directly available to customers. However, a customer is not allowed to enter information on business transactions, just to view specific information. A query user, then, provides an analyst with insight about what kinds of information should be available daily, weekly, monthly, and annually, and about what format is most convenient.

Management Users Managers are responsible for seeing that the company is performing its daily procedures efficiently and effectively. Consequently, they need statistics and summary information from a system. Management will help an analyst answer the following types of questions:

- What kinds of reports must the system produce?
- What kind of performance statistics must the system maintain?
- What kind of volume information must be kept, and what volumes of transactions must the new system support?
- Are the controls in the system adequate to prevent errors and fraud?
- How many requests for information will be made and how often?

Executive Users The top executives of an organization are interested in strategic issues, as well as the daily issues just described. They typically want information from a system so that they can compare overall improvements in resource utilization. They may want the system to interface with other systems to provide strategic information on trends and directions of the industry and the business.

Client Stakeholders

Although the project team must meet the information processing needs of the users, it also has a major responsibility to satisfy the client. Chapter 2 defined the client as the person or group who is providing the funding for the project. In many cases, the client is the same group as the executive users. However, clients may also be a separate group of people, such as a board of trustees or executives in a parent company. We include the client in our list of important stakeholders because the project team must provide project status reviews to the client from time to time throughout the development project. The client or a direct representative on a steering or oversight committee also usually maintains ongoing approval and release of funds.

Technical Stakeholders

Although the technical staff is not a true user group, it is the source of many of the technical requirements. The technical staff includes the people who establish and maintain the computing environment of the organization. These people provide guidance in such areas as programming language, computer platforms, and other equipment. For some projects, the project team includes a member of the technical staff. For other projects, technical personnel are available on an as-needed basis.

The Stakeholders for Rocky Mountain Outfitters

To demonstrate the different perspectives of stakeholders, let's look at the proposed customer support system for Rocky Mountain Outfitters.

An important part of investigating system requirements is to identify all of the stakeholders. The set of requirements will be incomplete if there are users, clients, or important technical staff who are not consulted as information is being gathered. At RMO, operational users of the new order-processing system include inside sales representatives who take orders over the phone, as well as clerks who process mail orders. They all have different views about what the system should do for them. Sales representatives talk about looking up product information for customers and confirming availability and shipping dates. Mail-order clerks talk about scanning order information into the system to eliminate typing. The warehouse workers who put the shipments together after the orders are taken need information about orders that have been shipped, orders to be shipped, and back orders, as well as their normal operational screens that allow them to put orders together into shipments with bills of lading printed.

John and Liz Blankens, as owners, have special interests in reports of the products that have been ordered and shipped. They are interested in watching seasonal trends within and across products. In the sports equipment business, it is critically important to push the trendy items quickly and move on when the trend is past.

The development of the customer support system has been funded in part from internal cash flows. Funds have also been obtained, however, through a special line of credit at the bank. RMO normally has a short-term line of credit for seasonal needs. Since this project is a longer-term investment for a capital good, the Blankenses obtained a different line of financing for the new system. Their banker is extremely interested in the success of the project, so in this case, the project team even met with bank staff to see what special formats of financial information the bank would like the system to maintain.

Finally, since this system will involve new technology—the Internet and distributed systems—very heavy involvement is required by the technical staff. Figure 4-2 is a partial organization chart of RMO, indicating some of the people who need to be involved in the requirements definition. This organization chart does not show all of the business operations staff, but it does identify departments in which these stakeholders work. The organization chart will later be supplemented with a detailed list of all stakeholders who need to be involved with the requirements definition.

FIGURE 4-2
Organization chart of RMO indicating stakeholders for the new system.

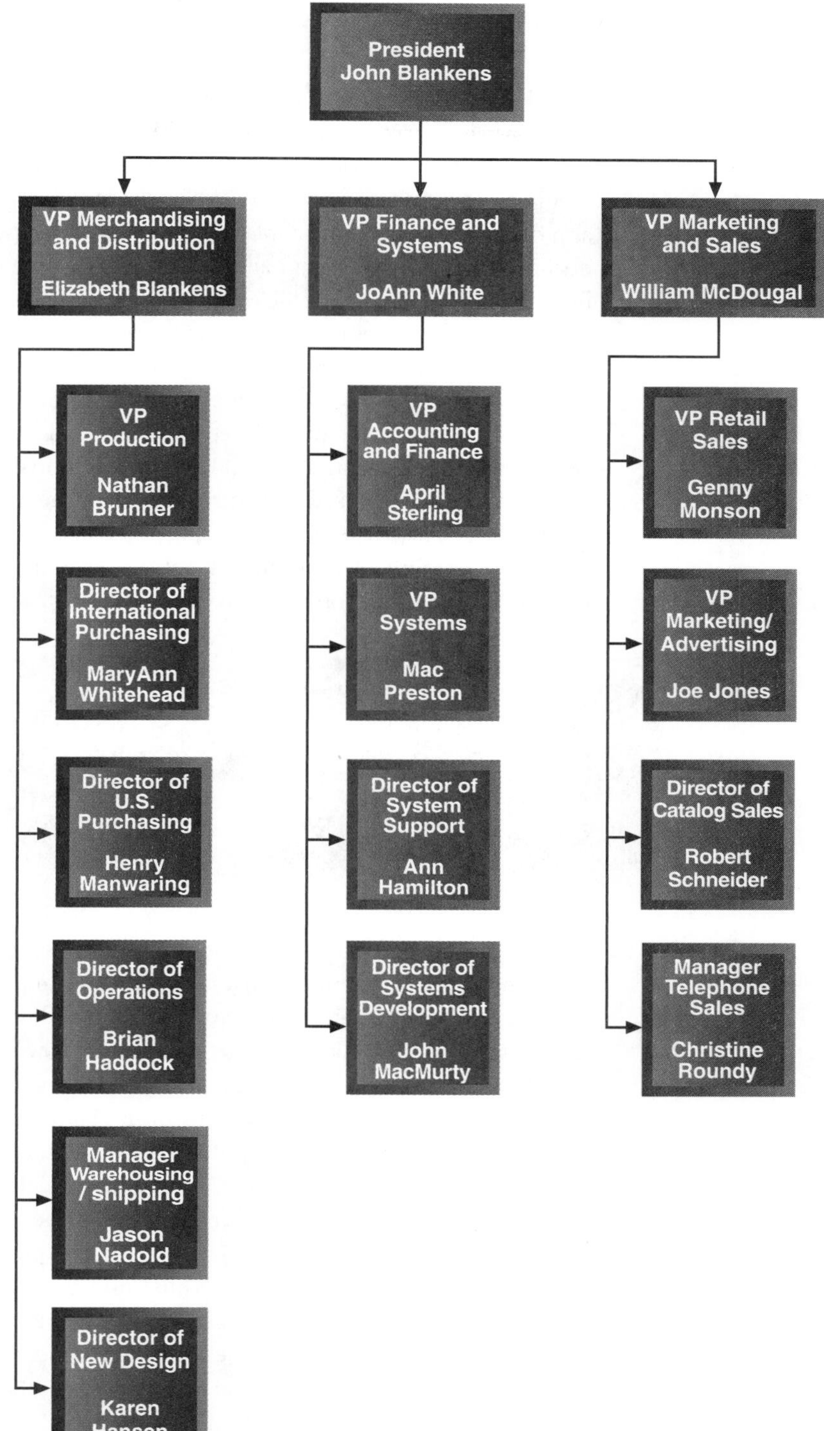

Identifying System Requirements

The objective of the analysis phase of system development is to understand the business functions and develop the system requirements. The question that always arises is whether to study and document the existing system, or whether to document only the requirements of the new system. When the structured approach, as well as the other approaches explained in Chapter 3, were first developed, systems analysts would first document the existing system and then extrapolate the requirements of the new system from the documentation of the existing system. In those days, the development of system requirements was a four-step process: (1) identify the physical processes and activities of the existing system, (2) extract the logical business function that was inherent in each existing physical process, (3) develop the logical business functions for the approach to be used in the new system, and (4) define the physical processing requirements of the new system. This approach, as shown in Figure 4-3, required a lot of time and effort. In Figure 4-3, we call each set of documents a model of the system. We use the term *model* as described in Chapter 3. The disadvantage of this approach was that it consumed an inordinate amount of time. It was not uncommon to spend a large portion of the budgeted funds simply on the analysis and development of the requirements.

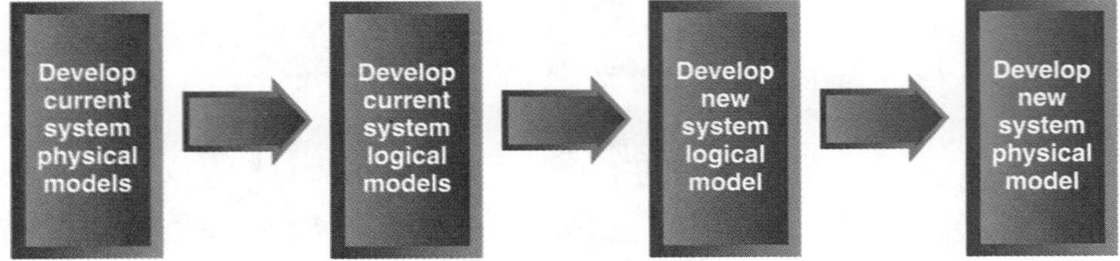

FIGURE 4-3
The previous approach to identifying system requirements.

The objective of analysis has not changed; however, the approach to developing system requirements has improved. It is still critically important to have a complete, correct set of system requirements, but in today's fast-paced world, there is not the time nor the money to develop all four sets of documents as shown in Figure 4-3. Today analysts use a more accelerated approach by balancing the review of current business functions with the new system requirements. As shown in Figure 4-4, the focus of analysis activities today is to develop a set of logical system requirements for the new system. In other words, the objective is to develop the logical model of the new system immediately. Review of the current system is done as much as necessary to understand the business needs, but not to define the specific processes. The physical model of the new system is developed as part of systems design. Thus, instead of developing four sets of documents, the project team develops only one set of documents. One of the important skills that a systems analyst must develop is the ability to balance the investigation of current procedures with the need to focus on the requirements of the new system.

FIGURE 4-4
The current approach to requirements development.

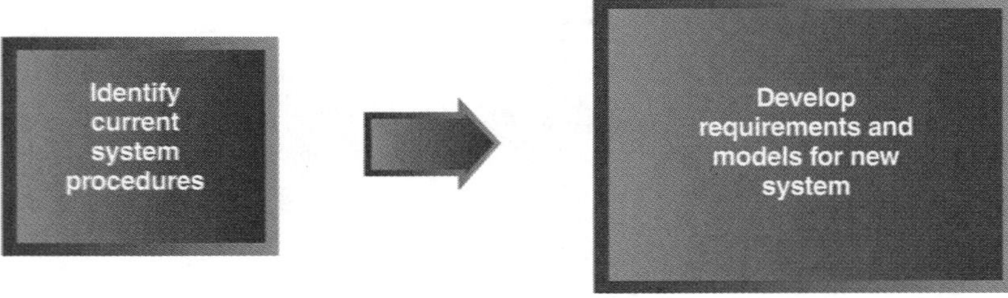

The first questions that new systems analysts ask are, "What kind of information do I need to collect? What is a requirement?" Basically, you want to obtain information that will enable you to build the logical model of the new business system; modeling is explained later in Chapters 5 and 6. Three major themes should guide you as you pursue your investigation:

- What are the business processes and operations? (that is, ask users, "What do you do?")
- How should the business processes be performed? (that is, ask, "How can it be done?" or "What steps are followed?")
- What are the information requirements? (that is, ask, "What information do you need in order to do it?")

In the first question—what do you do?—the focus is on understanding the business functions. In most cases, the users will provide answers in terms of the current system. As an analyst, you must carefully discern which of those functions are fundamental business functions, which will remain, and which may possibly be eliminated with an improved system. For example, sales clerks may indicate that the first thing they do when a customer places an order is to check the customer's credit history. In the new system, sales clerks may never need to perform that function; it might be performed by the system automatically. In this instance, the function remains a system requirement, but the method of carrying out the function moves from the clerks to the computer system.

The second question—how can it be done?—is the mechanism that moves the discussion from the current system to the new system. The focus is on how the new system should support the function rather than on how it is performed under the existing system. Thus, the first two questions go hand in hand to discover the need and begin to define the system requirement in terms of the new system.

The final question—what information is needed?—elaborates the second question by defining specific information that the new system must provide. The answers to the second and third questions form the basis for the definition of the system requirements.

If you focus your investigation around these three themes, you will be able to ask intelligent, meaningful questions in your investigation. Later, as you learn about models, you will be able to formulate additional meaningful detailed questions to ask.

One of the most critical skills that you should develop as a systems analyst is the ability to understand users' needs. As you develop skill in asking questions and building models, your problem-solving and analytical skills will increase. Remember, your value as a systems analyst is not that you know how to build a specific model or how to program in a specific language. Your value is in your ability to analyze and solve business information problems.

Fundamental to that skill is how effectively and efficiently you can identify and capture business rules that determine the system requirements. Effective requirements are complete, comprehensive, and correct. An efficient analyst is one who moves the project ahead rapidly with minimal intrusion on users' time and use of other resources. Typically, the development of new systems is a very expensive endeavor. One of the reasons is the tremendous investment of time by project members and users. Analysts need to control the amount of time spent by stakeholders and team members in identifying requirements.

The next sections present the various methods of information gathering to identify requirements. All these methods have been proven to be effective, although some are more efficient than others. In most cases, analysts use a combination of methods to increase both their effectiveness and efficiency. The combined methods provide a comprehensive fact-finding approach that is widely used for the development of large, complex systems. The six methods are:

- Distribute and collect questionnaires to stakeholders
- Review existing reports, forms, and procedure descriptions
- Conduct interviews and discussion with users
- Observe business processes and workflows
- Build prototypes
- Conduct joint application design (JAD) sessions

The last two methods—building prototypes and conducting JAD sessions—are used as part of a conventional SDLC method. In an expanded form, they also function as a complete development approach by themselves. Chapter 13 presents alternative methods of system development and explains how a prototyping method and a JAD method, as well as other accelerated methods, can be used to develop new systems. This chapter, however, explains how these techniques are used as part of a conventional SDLC method. This will provide you with a knowledge of the fundamentals of system development that will enable you to understand and use other methods when it is appropriate.

Distribute and Collect Questionnaires

Questionnaires have a limited and specific use in information gathering. The benefit of a questionnaire is that it enables the project team to collect information from a large number of stakeholders. Even if the stakeholders are widely distributed geographically, they can still help define requirements through questionnaires.

Frequently a questionnaire can be used to obtain preliminary insight on the information needs of the various stakeholders. This preliminary information can then be used to help determine the areas that need further research with document reviews, interviews, and observation. Questionnaires are also helpful to answer quantitative questions such as, "How many orders do you enter a day?" and, "How long does it take to enter one order?" Finally, questionnaires can be used to determine the users' opinions about various aspects of the system. Such questions as "On a scale of 1 to 7, how important is it to be able to access a customer's past purchase history?" These kinds of questions are often called **closed-ended questions**, because they direct the person answering the question to provide a direct response to only that question. They do not invite discussion or elaboration. The strength of close-ended questions, however, is that the answers are always limited to the set of choices. Tabulations of the answers can be made to determine averages or trends.

closed-ended questions
Questions that have a simple definitive answer

Figure 4-5 is a sample questionnaire showing three types of questions. The first part has closed-ended questions to determine quantitative information. The second part consists of opinion questions where respondents are asked whether they agree or disagree with the statement. Both of these kinds of questions are useful for tabulating and determining quantitative averages. The final part requests an explanation of a procedure or problem. This kind of question is good as a preliminary investigation to help direct further fact-finding activities.

RMO Questionnaire

This questionnaire is being sent to all telephone-order sales personnel. As you know, RMO is developing a new customer support system for order taking and customer service.

The purpose of this questionnaire is to obtain preliminary information to assist in defining the requirements for the new system. Follow-up discussions will be held to permit everybody to elaborate on the system requirements.

Part I. Answer these questions based on a typical four-hour shift.

1. How many phone calls do you receive?______________________
2. How many phone calls are necessary to place an order for a product?______________________
3. How many phone calls are for information about RMO products, that is, questions only?______________
4. Estimate how many times during a shift customers request items that are out of stock?______________
5. Of those out-of-stock requests, what percentage of the time does the customer desire to put the item on back order?______________%
6. How many times does a customer try to order from an expired catalog?______________________
7. How many times does a customer cancel an order in the middle of the conversation?______________
8. How many times does an order get denied due to bad credit?______________________

Part II. Circle the appropriate number on the scale from 1 to 7 based on how strongly you agree or disagree with the statement.

Question	Strongly Agree						Strongly Disagree
It would help me do my job better to have longer descriptions of products available while talking to a customer.	1	2	3	4	5	6	7
It would help me do my job better if I had the past purchase history of the customer available.	1	2	3	4	5	6	7
I could provide better service to the customer if I had information about accessories that were appropriate for the items ordered.	1	2	3	4	5	6	7
The computer response time is slow and causes difficulties in responding to customer requests.	1	2	3	4	5	6	7

Part III. Please enter your opinions and comments.

Please briefly identify the problems with the current system that you would like to see resolved in a new system.

____
____
____

FIGURE 4-5
A sample questionnaire.

open-ended questions
Questions that require discussion and do not necessarily have a simple short answer

Questionnaires are not well suited to helping you learn about processes, workflows, or techniques. The questions identified earlier, such as, "How do you do this process?" are best answered using interviews or observation. Questions that encourage discussion and elaboration are called **open-ended questions**. Although a questionnaire can contain a very limited number of open-ended questions, those that contain many open-ended questions are frequently not returned.

Review Existing Reports, Forms, and Procedure Descriptions

The review of existing documents and documentation serves two purposes. First, it is a good way to get a preliminary understanding of the processes. Often new systems analysts will not know much about the industry or the specific application that they are studying. A preliminary review of existing documentation will bring them up to speed fairly rapidly.

To begin the process, the analysts ask users to provide copies of the forms and reports that they currently are using. They also request copies of procedural manuals and work descriptions. The review of these materials provides an understanding of the business functions. They also form the basis for the development of detailed interview questions.

The second way to use documents and reports is in the interviews themselves. Forms and reports can serve as visual aids for the interview, and the working documents for discussion (see Figure 4-6). Discussion can center on the use of each form, its objective, distribution, and information content. It is always better to have forms that have been filled out with real information to ensure that a correct understanding of the fields and data content is obtained.

Rocky Mountain Outfitters—Customer Order Form

Rocky Mountain Outfitters

Name and address of person placing order.
(Please verify your mailing address and make correction below.)
Order Date __/__/__

Name
Address Apt. No
City State Zip

Phone: Day ( ) ________ Evening ( ) ________

Gift Order or Ship To: (Use only if different from adress at left.)
Name
Address Apt. No
City State Zip
Gift ☐ Address for this Shipment Only ☐ Permanent Change of Address ☐
Gift Card Message
Delivery Phone ( ) ________

Item No.		Description	Style	Color	Size	Sleeve Length	Qty	Monogram Style	Price Each	Total

Method of Payment
Check/Money Order ☐ Gift Certificate(s) ☐ AMOUNT ENCLOSED $ ______
American Express ☐ MasterCard ☐ VISA ☐ Other ☐
Account Number
MO YR __/__ Expiration Date
Signature

MERCHANDISE TOTAL ______
Regular FedEx shipping $4.50 per U.S. delivery address (Items are sent within 24 hours for delivery in 2 to 4 days) ______
Please add $4.50 per each additional U.S. delivery address ______
FedEx Standard Overnight Service ______
Any additional freight charges ______
International Shipping (see shipping information on back) ______

FIGURE 4-6
A sample order form for Rocky Mountain Outfitters.

Reviewing documentation of existing procedures will help you identify business rules that may not come up in the interviews. Written procedures also help discover discrepancies and redundancies in the business processes. However, be aware that procedure manuals frequently are not kept up to date. It is not uncommon for them to have errors. The assumptions and business rules that derive from the existing documentation should be reviewed with the users to ensure that they are correct.

Conduct Interviews and Discussions with Users

Interviewing stakeholders is by far the most effective way to understand business functions and business rules. It is also the most time-consuming and resource-expensive. In this method, members of the project team (systems analysts) meet with individuals or groups of users. A list of detailed questions is prepared and discussion continues until all the processing requirements are understood and documented by the project team. Obviously, this may take some time, so it usually requires multiple sessions with each of the users or user groups.

To conduct effective interviews, analysts need to organize in three areas: (1) preparing for the interview, (2) conducting the interview, and (3) following up the interview. Figure 4-7 is a sample checklist that summarizes the major points and is useful for preparing for and conducting an interview.

FIGURE 4-7
A sample checklist to prepare for user interviews.

Checklist for Conducting an Interview

Before

- Establish the objective for the interview
- Determine correct user(s) to be involved
- Determine project team members to participate
- Build a list of questions and issues to be discussed
- Review related documents and materials
- Set the time and location
- Inform all participants of objective, time, and locations

During

- Dress appropriately
- Arrive on time
- Look for exception and error conditions
- Probe for details
- Take thorough notes
- Identify and document unanswered items or open questions

After

- Review notes for accuracy, completeness, and understanding
- Transfer information to appropriate models and documents
- Identify areas needing further clarification
- Send thank-you notes if appropriate

Preparing for the Interview Every successful interview requires preparation. The very first, and most important, step in preparing for an interview is to establish the objective of the interview. In other words, what do you want to accomplish with this interview? Write down the objective so that it is firmly established in your mind. The second step is to determine which users should be involved in the interview. Frequently the first two steps are so intertwined that both are done together. Even if you don't do anything else to prepare for your interviews, you must at least complete these two steps. The objective and the participants drive everything else in the interview.

The interview participants include both users and project members. Generally, at least two project members are involved in every interview. The two project members help each other during the interview as well as compare notes afterward to ensure accuracy. The number of users varies depending on the objective of the interview. Usually it is better to limit the number of people in the interview. Interviewing more than three users at a time tends to cause long discussions, which sometimes are counterproductive. In many instances, only one user at a time is interviewed. This is especially true for small- to medium-sized projects.

The third step is to prepare detailed questions to be used in the interview. Write down a list of specific questions and prepare notes based on the forms or reports received earlier. It is generally recommended to prepare a list of questions that are consistent with the objective of the interview. Both open-ended questions and closed-ended questions are appropriate. Open-ended questions, such as, "How do you do this function?" encourage discussion and explanation. Close-ended questions, such as, "How many forms a day do you process?" are used to get specific facts. Generally, open-ended questions help get the discussion started and encourage the user to explain all the details of the business process and the rules.

The last step is to make the final interview arrangements and to communicate those arrangements to all participants. A specific time and location should be established, especially with the users. If possible, a quiet location should be chosen to avoid interruptions. Each participant should know the objective of the meeting and, where appropriate, should have a chance to preview the questions or materials to be reviewed. Interviews consume substantial time, and they can be made more efficient if each participant knows beforehand what is to be accomplished.

Conducting the Interview New systems analysts are usually quite nervous about conducting interviews. However, remember that, in most cases, the users are excited about getting a better system to help them do their jobs. Practicing good manners usually ensures that the interview will go well. Here are a few guidelines.

Dress appropriately. Dress at least as well as the best-dressed user. In many corporate settings, such as banks or insurance companies with managers present, business suits are appropriate. In factory or manufacturing settings, work dress may be appropriate. The objective in dressing is to project competence and professionalism without intimidating the user.

Arrive on time. If anything, be a little early. If the session is in a conference room, ensure that it is set up appropriately. For a large group or a long session, plan for refreshment breaks.

Limit the time of the interview. Both the preparation and the interview itself affect the time required for the interview. As you set the objective and develop questions, plan for about an hour and a half. If the interview will require more time to cover the questions, it is usually better to break off the discussion and schedule another session. (Other techniques that we discuss later have all-day sessions.) The users have other responsibilities and the systems analysts can absorb only so much information at one time. It is better to have several shorter interviews than one long marathon. A series of interviews provides an opportunity to absorb the material and go back to get clarification later. Both the analysts and the users will have better attitudes with several shorter interviews.

Look for exception and error conditions. Look for opportunities to ask "what if" questions. "What if it doesn't arrive? What if the signature is missing? What if the balance is incorrect? What if two order forms are exactly the same?" The essence of good systems analysis is understanding all of the "what ifs." Make a conscious effort to identify all of the exception conditions and ask about them. More than any other skill, the ability to think of the exceptions will strengthen the skill of discovering the detailed business rules. It is a hard skill to teach from a textbook; experience will hone this skill. You will teach yourself this skill by conscientiously practicing it.

Probe for details. In addition to looking for exception conditions, the analyst must probe to ensure a complete understanding of all procedures and rules. One of the most difficult skills to learn as a new systems analyst is to get enough details. Frequently, it is easy to get a general overview of how a process works. But do not be afraid to ask detailed questions until you thoroughly understand how it works and what information is used. You cannot do effective systems analysis by glossing over the details.

Take careful notes. It is a good idea to take handwritten notes. Usually tape recorders make users nervous. Note-taking, however, signals that you think the information you are obtaining is important, and the user is complimented. If two analysts conduct each interview, they can compare notes later. Identify and document in your notes any unanswered questions or outstanding issues that were not resolved. A good set of notes provides the basis for building the analysis models as well as establishing a basis for the next interview session.

Figure 4-8 is a sample agenda for an interview session. Obviously, you do not need to conform exactly to a particular agenda. However, as with the interview checklist shown in Figure 4-7, this figure will help prod your memory on issues and items that should be discussed in an interview. Make a copy and use it. As you develop your own style, you can modify it for the way you like to work.

Discussion and Interview Agenda

Setting

Objective of Interview
Determine processing rules for sales commission rates

Date, Time, and Location
July 1, 1999, at 9:00 a.m. in William McDougal's office

User Participants (names and titles/positions)
William McDougal, vice president of marketing and sales, and several of his staff

Project Team Participants
Mary Ellen Green and Jim Williams

Interview/Discussion

1. Who is eligible for sales commissions?
2. What is the basis for commissions? What rates are paid?
3. How is commission for returns handled?
4. Are there special incentives? Contests? Programs based on time?
5. Is there a variable scale for commissions? Are there quotas?
6. What are the exceptions?

Follow-Up

Important decisions or answers to questions
See attached write-up on commission policies

Open items not resolved with assignments for solution
See Item numbers 2 and 3 on open items list

Date and time of next meeting or follow-up session
July 7, 2000, 9:00 a.m.

FIGURE 4-8
Sample interview session agenda.

Following Up the Interview Follow-up is an important part of each interview. The first task is to absorb, understand, and document the information that was obtained. Generally analysts document the details of the interview by constructing models of the business processes. Review your findings with the other project members in the interview and document the results (that is, build the models) within a day or at most two.

During the interview, you probably asked some "what if" questions that the users could not answer. These are usually policy questions raised by the new system but which management has not considered before. It is extremely important that these questions not get lost or forgotten. As explained in Appendix A and shown in Figure A-1, an open-items log is an effective method to track these items. Figure 4-9 is a sample form with representative types of questions from Rocky Mountain Outfitters. If several teams are working, a combined list can be maintained.

OUTSTANDING ISSUES CONTROL TABLE

ID	Issue Title	Date Identified	Target End Date	Project Person	User Contact	Comments
1	Partial Shipments	6-12-2000	7-15-2000	Jim Williams	Jason Nadold	Ship partials or wait for full shipment
2	Returns and Commissions	7-01-2000	9-01-2000	Jim Williams	William McDougal	Are commissions recouped on returns
3	Extra Commissions	7-01-2000	8-01-2000	Mary Ellen Green	William McDougal	How to handle special promotions

FIGURE 4-9
A sample open-items list.

Make a list of new questions based on areas that need more elaboration or that are missing information. This will prepare you for the next interview.

Where appropriate, send a thank-you note or e-mail message to the users who participated in the interview.

Observe Business Processes and Workflows

The old adage that a picture is worth a thousand words is also true with systems analysis. Early in the investigation activities, time should be scheduled to observe the business procedures that the new system will support. There is no better way to learn how users actually use a system and what information they need than to observe first-hand the work being done.

You can observe the work in several ways, from a quick walkthrough of the office or plant to doing the work yourself. A quick walkthrough gives a general understanding of the layout of the office, the need and use of computer equipment, and the general workflow. Scheduling several hours to observe a user actually doing his or her job provides an understanding of the details of actually using the computer system and carrying out business functions. By being trained as a user and actually doing the job, you can discover the difficulties of learning new procedures, the importance of a system that is easy to use, and the stumbling blocks and bottlenecks of existing procedures and information sources.

It is not necessary to observe all processes at the same level of detail. A quick walkthrough may be sufficient for one process, while another process that is critical or more difficult to understand might require an extended observation period. If you remember that the objective is a complete understanding of the business processes and rules, then you can assess where to spend your time to gain that thorough understanding. As with interviewing, it is usually better if two analysts combine their efforts in observing procedures.

Observation often makes the users nervous, so you need to be as unobtrusive as possible. There are several ways to do this, such as working with a user or observing several users at once. Common sense and sensitivity to the needs and feelings of the users will usually result in a positive experience.

Build Prototypes

prototype
A preliminary working model of a larger system

The fundamental idea of a **prototype** is an initial, working model of a larger, more complex entity. Prototypes are used for many different purposes, and there are many names to describe these uses: throwaway prototypes, discovery prototypes, design prototypes, and evolving prototypes. Each can be used during different phases of the project to test and validate ideas that apply to that phase. As explained in Chapter 3, during analysis, prototypes are used to test feasibility and help identify processing requirements. These prototypes may be in the form of simple screens or report programs. During design, prototypes may be built to test various design and interface alternatives. Even during implementation, prototypes may be built to test the effectiveness and efficiency of different programming techniques. Prototyping is such a strong tool that you will find it used in almost every development project in some way.

As its name indicates, a discovery prototype is used for a single discovery objective and then discarded once the concept has been tested. For example, if you use a prototype to determine screen formats and processing sequences, once that definition is done, the prototype is thrown away. Evolving prototypes, on the other hand, are prototypes that grow and evolve and may eventually even be used as the final, live system. As we will see later in Chapter 13, one approach to prototyping is to keep modifying the prototype and adding to it until it actually becomes the system to be installed.

As discussed in Chapter 3, the purpose of a discovery prototype used during analysis is to have a working model to test a concept or verify an approach. Keeping in mind the following characteristics of prototypes will help project members develop effective prototypes.

mock-up
An example of a final product that is for viewing only, not executable

- *Operative.* Generally, a prototype should be a working model, with the emphasis on *working*. A simple start to a prototype, called a **mock-up**, is an electronic form (such as a screen) that only shows what it looks like, but does not provide execution capability. A prototype will actually execute and provide both "look and feel" characteristics, but may lack some functionality.
- *Focused.* To test a specific concept or verify an approach, a prototype should be focused on a single objective. Extraneous execution capability, not part of the specific objective, should be excluded. Although it may be possible to combine several simple prototypes into a larger prototype, the focused objective still applies. Later, prototypes can be combined to test the integration of several components.
- *Quick.* Tools, such as CASE tools, are needed so that the prototype can be built and modified quickly. Since a prototype's purpose is to validate an approach, if the approach is wrong, then the tools must be available to modify and test quickly to determine the correct approach.

Integrating prototyping activities in the project is fairly simple. The important point to keep in mind is to have an overall philosophy and purpose for building prototypes, and to maintain a consistent focus across all the prototypes that are built.

Conduct Joint Application Design Sessions

joint application design
A technique to define requirements or design a system in a single session by having all necessary people participate together

Joint application design (JAD) is a technique used to expedite the investigation of system requirements. The normal interview and discussion approach, as explained earlier, requires a substantial amount of time. The analysts first meet with the users, then document the discussion by writing notes and building models, then the models are reviewed and revised. Unresolved issues are placed on an open-items list and may require several additional meetings and reviews to be finalized. This process can extend from several weeks to months depending on the size of the system and the availability of user and project team resources.

The objective of JAD is to compress all of these activities into a shorter series of JAD sessions with users and project team members. An individual JAD session may last from a single day to a week. During the session, all of the fact-finding, model-building, policy decisions, and verification activities are completed for a particular aspect of the system. If the system is small, the entire analysis may be completed during the JAD session. The critical factor in a successful JAD session is to have all of the important stakeholders present and available to contribute and make decisions. The actual participants vary depending on the objective of the specific JAD session. Those who are involved may include the following:

- *The JAD session leader.* One of the more important members of the group, the session leader is a person who is experienced or has been trained in group dynamics and in facilitating group discussion. Normally, a JAD session involves quite a few people. Each session has a detailed agenda with specific objectives that must be met, and the discussion must progress toward those objectives. Maintaining focus requires someone with skills and experience to tactfully keep people on task. Often it is tempting to appoint one of the systems analysts as the session leader. However, experience indicates that successful JAD sessions are conducted by someone who is trained to lead group decision-making.
- *Users.* Earlier in this chapter, we identified various classes of users. It is important to have all of the appropriate users in the JAD sessions. Frequently, as requirements are discovered, policy decisions need to be made. If managers are not available in the sessions to make those decisions, then progress is halted. Due to business pressures, it may be difficult for top executives to be present during the entire session. In that case, arrangements should be made for executives to visit the session once or twice a day to become involved in those policy discussions.
- *Technical staff.* A representative from the technical support staff should also be present in the JAD session. There are always questions and decisions about technical issues that need to be answered. For example, participants may need details of computer and network configurations, operating environments, and security issues.
- *Project team members.* Both systems analysts and user experts from the project team should be involved in JAD sessions. These members assist in the discussion, clarify points, control the level of detail needed, build models, document the results, and generally see that the system requirements are defined to the necessary level of detail. The session leader is a facilitator, but often the leader is not the expert on how much detail and definition is required. Members of the project team are the experts on ensuring that the objectives are completely satisfied.

JAD sessions are usually conducted in special rooms with supporting facilities. First, since the process is so intense, it is important to be away from the normal day-to-day interruptions. Sometimes an off-site location may be necessary, or notification that interruptions are not welcome may be posted. On the other hand, it is usually helpful to have telephone access to executives and technical staff who are not involved in the meetings but who may be invited from time to time to finalize policy or technical decisions.

Resources in the JAD session room should include an overhead projector, a black or white board, flip charts, and adequate workspace for the participants. JAD sessions are work sessions, and all the necessary work paraphernalia should be provided.

Recently, JAD sessions have been taking advantage of electronic support to increase their efficiency. Analysis and documentation can be enhanced if participants have personal or laptop computers connected in a network. Then as requirements are documented with narrative descriptions or models, or even as some simple discovery prototypes are built, they can be made available to everybody. Often a CASE tool is provided on the computers to assist in visualization of screen and report layouts and file design. The CASE tool also provides a central repository for all the requirements developed during the session.

group support system
A computer system that enables multiple people to participate with comments at the same time, each on his or her own computer

Group support systems (GSSs), which also run on the network of computers, allow all participants to post comments (anonymously if desired) on a common working chat room. This approach helps participants who may be shy in group discussions to become more active and contribute to the group decisions. GSSs also provide the capability to store final requirements as decisions are made. Normally JAD sessions are conducted with everyone in the same room. However, GSSs on wide networks provide the opportunity for virtual meetings with participants at geographically distinct locations.

Figures 4-10 and 4-11 show two examples of conference rooms that have been set up to support JAD sessions. Figure 4-10 shows a layout that may be typical of a company that has a smaller development staff and fairly infrequent JAD sessions. The room is equipped with manual support tools such as white boards, marking pads, and so forth. Figure 4-11 shows a room with electronic support. Such a room might be available in larger companies that have development projects in progress more frequently. The room shown in Figure 4-11 has workstations available to develop model diagrams and prototypes during the JAD sessions. This room could even be quite sophisticated by having computer support for collaborative work (CSCW) software on the computers to facilitate comment and discussion. With CSCW software available, certain executives could even participate from remote locations, if necessary.

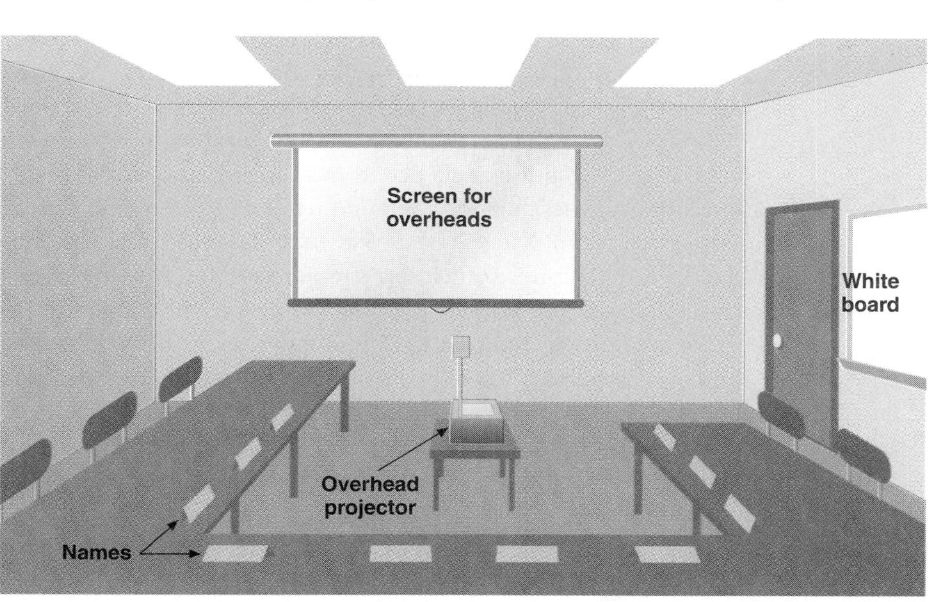

FIGURE 4-10
A standard JAD facility.

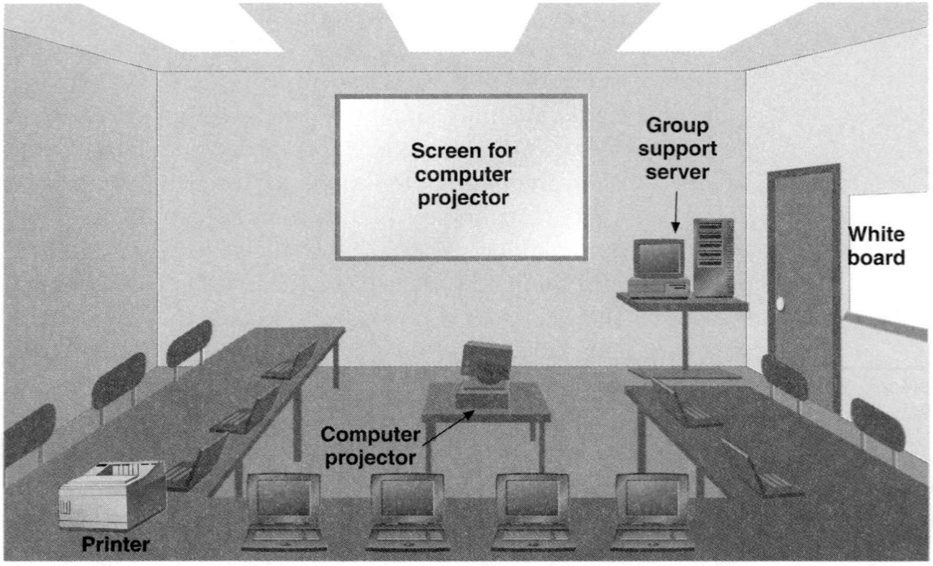

FIGURE 4-11
A high-tech JAD facility.

As stated earlier, one of the dangers of JAD is the risk involved in expediting decisions. Since the objective of JAD is to come to a conclusion quickly on policy decisions and requirements details, sometimes decisions are not optimal. At times, details are inappropriately defined or missed altogether. However, JAD sessions have been largely successful in reducing project development efforts and shortening the schedule.

The next section discusses a quality-control technique that is used to ensure that the system requirements are correct and comprehensive.

Structured Walkthroughs

One of the responsibilities of the project manager is to ensure the quality of the final system. However, it is insufficient to wait until the end of the project and the beginning of program testing to implement quality-control measures. One effective technique to implement quality control early in the project is with structured walkthroughs.

If we compare the development of a new information system to the construction of a house, the requirements determined during analysis are like a house's foundation. All of the design and programming of an information system is based on the accuracy and completeness of the system requirements. However, how do we know that the foundation is sufficiently strong to support the house? It would be helpful if some type of test could be devised to ensure the foundation is strong enough before the house is built on it. In building construction, there are rules and guidelines that have been developed over the years to indicate the requirements of the foundation based on the size and weight of the house. In software development, since each project is unique from every other project, these types of rules are not available. Thus, each set of system requirements should be reviewed and tested as much as is feasible.

In system development jargon, analysts often say they must verify and validate (sometimes called V&V) the system requirements. As explained in Appendix A, verification means determining that the requirements are internally consistent, while validation means ensuring that the requirements correctly express the needs of the users. Verification tests such things as whether the field definitions are consistent throughout all of the subsystems of a system. Validation tests such things as whether the requirements are complete—whether they have captured all of the information needs of the users. But how can we perform verification and validation during analysis?

When writing a computer program, a programmer must test the accuracy of the code by conducting various tests. Executing the program on a computer—by entering appropriate input data and observing the resultant output—tests a computer program. We cannot test the requirements that way, so we have to use a different approach. A technique called structured walkthroughs has been developed as a way to verify and validate the system requirements.

structured walkthrough
A review of the findings from your investigation and of the models built based on those findings

A **structured walkthrough**, sometimes just called a walkthrough, is a review of the findings from your investigation and of the models built based on those findings. A walkthrough is considered structured because analysts have formalized the review process into a set procedure. The objective of a structured walkthrough is to find errors and problems. The fundamental concept is one of documenting the requirements as you understand them and then reviewing them for any errors, omissions, inconsistencies, or problems with the description. A review of the findings can be done informally with colleagues on the project team, but a structured walkthrough must be more formal.

It is important to note one critical point: A structured walkthrough is *not* a performance review. Managers should be involved only if they were involved in the original fact finding and thus are required for verification or validation. The review is of an analyst's work and not the person. To understand the more structured approach, this section reviews the what, when, who, and how of a structured walkthrough.

What and When

The first item to be reviewed during a structured walkthrough is the documentation that was developed as part of the analysis phase. It can be a narrative describing a process, a flow chart showing a workflow, or a model diagram documenting an entire procedure. Normally, it is better to conduct several smaller walkthroughs that review three to six pages of documentation than to cover 30 pages of details. Any written work that is a fairly independent package can be reviewed in a walkthrough. It is not uncommon to hold smaller walkthroughs every week or two with members of the project team. The frequency of the walkthroughs is not as critical as the timing—that a walkthrough be scheduled as soon as possible after the documents have been created.

Who

The two main parties involved in walkthroughs are the person or persons who need their work reviewed and the group who reviews it. For verification—that is, internal consistency and correctness—it is best to have other experienced analysts involved in the walkthrough. They are used to looking for inconsistencies and problems. For validation—that is, ensuring that the system satisfies all the needs of the various stakeholders—the appropriate stakeholders should be involved. The nature of the work to be reviewed dictates who the reviewers should be. If it is a diagram showing a business process, then the users who supplied the original definition should be involved. If it is a technical specification of design details, then the technical staff should be involved in the review. At times, the reviewers may be members of the project team. In other instances, they are external users or technical staff. Those who can validate the correctness of the work are the ones who should be invited.

How

As with an interview, preparation, execution, and follow-up are required for a structured walkthrough.

Preparation The analyst whose work is being reviewed gets the material ready for review. Next he identifies the appropriate participants and provides them copies of the material. Finally, he schedules a time and place for the walkthrough and notifies all participants.

Execution During the walkthrough, the analyst presents the material point by point. If it is a diagram or flow chart, he walks through the flow explaining each component. One effective technique is to define a sample test case and process it through the defined flow. The reviewers look for inconsistencies or problems and point them out. A librarian, a helper for the presenter, documents the comments made by the reviewers. Presenters should never be their own librarians because they should not be distracted from explaining the documentation. Someone else should record the errors, comments, and suggestions to ensure accuracy.

Corrections and solutions to problems are not made during the walkthrough. At most, some suggested solutions may be provided, but the documentation should not be corrected during the walkthrough. Since it is not uncommon for presenters to be a little nervous, it is unfair to ask them to make wise decisions on the spur of the moment. If a misunderstanding of the user requirements is uncovered, a brief review may be in order. However, if an error is fairly complex, it is better to schedule an additional interview to clarify the misunderstanding. The walkthrough should not get bogged down into a fact-finding session. The reviewer should only provide feedback, and the presenter can integrate it into the material later, when he or she has the entire set of comments and can give the corrections his or her undivided best effort without interruptions or further criticism.

Follow-Up Follow-up consists of making the required corrections. If the reviewed material has major errors and problems, an additional walkthrough may be necessary. Otherwise, the corrections are made, and the project continues to the next activities.

Figure 4-12 is a sample review form that was used in one of the review sessions at Rocky Mountain Outfitters, for the sales commission rates and rules. Not shown are several attached sheets including a couple of flow charts of procedures. The material reviewed in this case is simply a list of business rules for commission rates. The reviewers are senior managers from the user community. Since sales commission business rules are critical, and these managers make the policy decisions on commissions, they are the obvious choices to review the rules as uncovered in discussions and interview sessions.

Walkthrough Control Sheet

Project Control Information

Project: *On-line Catalog System, Customer Support Subsystem*

Segment of project being reviewed: *Review of business rules for sales commission rates*

Team leader: *Mary Ellen Green*

Author of work: *Jim Williams*

Walkthrough Details

Date, time, and location
July 20, 2000. 10:00 a.m. MIS conference room.

Description of materials being reviewed:
This is a review of the business rules before they are integrated into the diagrams and models. There is a short flow chart attached showing the flow of the commission process. There is another flow chart showing the process to set commission rates. We will also review outstanding issues to ensure that all understand the policy decisions that must be made.

Participating reviewers:
William McDougal, Genny Monson, Robert Schneider

Results of Walkthrough

________XX__Accept. Sign-Offs:______________________________
___________Minor revisions. Description of revisions:

___________Rework and schedule new walkthrough. Description of required rework:
Excellent and thorough. No rework required.

FIGURE 4-12
A structured walkthrough evaluation form.

Business Process Reengineering

Business process reengineering (BPR) is a movement of the last 10 years that has been the source of many new information systems. This section provides a brief explanation of BPR and how it has influenced the definition of system requirements.

In many systems development projects, organizations add new levels of automation for support of existing business procedures. In other projects, the new system is to support a new service or line of business. For example, in RMO, the new system is required to support the new catalog sales area of the business. In addition, however, new levels of automation for all order processing and inventory management are provided.

Over the last decade, as global competition has become more intense, many companies have found that it is necessary to completely rethink their most fundamental assumptions about how they do business. The old rule of business procedures was, "If it isn't broken, don't fix it." A newer way of thinking is, "There is always a better way to do it. Let's improve it." Business process reengineering extends this newer approach even further with a more revolutionary idea: "Let's question basic assumptions to find a completely new way to do it that will bring dramatic and profound improvement." The only way such radical improvements are possible is through information technology; thus, IT is the enabler of BPR.

One of the classic stories of dramatic improvement is the accounts payable function in the North American division of Ford Motor Company. In the mid-1980s the accounts payable department employed over 500 people. The original project was to develop an information system to achieve a 20 percent improvement in productivity. However, as the project team and the executives began looking at ways to utilize automation to improve performance, they found that Mazda, Inc., of which Ford owned 20 percent, employed only five people in accounts payable. Even though Ford was a much larger organization, it still had over 100 times as many people to perform basically the same function. With a better vision of what was possible, the project team completely redesigned the payables function utilizing a much higher level of automation. In essence, the accounts payable function was subsumed into a larger purchasing function so that tracking of payables became automated from the time the purchase was made. At the conclusion of the project, Ford was able to carry out its accounts payable functions with a little more than 100 people. Instead of a 20 percent improvement, the company achieved a 400 percent improvement.

Since BPR and IT are so closely aligned, many systems analysts are finding that they must also become business analysts and assist in this process of rebuilding all internal business procedures based on new and innovative information systems. In addition, many of the modeling techniques and skills that you learn about in the next chapters are used to model both information systems processes and business procedures.

One important consideration that you should keep in mind during the investigation activities is that there may be opportunities for improvements in business procedures during the project. There may also be the need for a complete reengineering of some areas of the business. You, as a systems analyst, are frequently in a position to identify and recommend these types of improvements during the project. You can have a great impact on the continued success and even survival of your organization.

Summary

Generally we divide system requirements into two categories: functional requirements and technical requirements. The functional requirements are those that explain the basic business functions that the new system must support. Technical requirements involve the objectives of the system for performance, operating environment, and other nonfunctional issues.

To ferret out the requirements, you must work with various stakeholders in the new system. We categorize stakeholders into three groups: (1) the users, those who are the ones actually using the system day to day, (2) the clients, those who pay for and own the system, and (3) the technical staff, the people who must ensure that the system operates within the computing environment of the organization. One of the most important first steps in determining systems requirements is to identify these various system stakeholders.

A fundamental question to investigate system requirements is, "What kinds of information do I need?" This chapter provides you with some general guidelines. As you learn more about modeling, you should also understand better what information you need. Generally you want to discover information in the following three areas:

- What are the business processes and operations?
- How are the business processes performed?
- What are the information requirements?

We use six primary techniques to gather this information, and one technique to ensure its correctness. The six fact-finding techniques are:

- Distribute and collect questionnaires to stakeholders
- Review existing reports, forms, and procedure descriptions
- Conduct interviews and discussions with users
- Observe business processes and workflows
- Build prototypes
- Conduct JAD sessions

The review technique to ensure that analysis is accurate and complete is called a structured walkthrough. Remember that a structured walkthrough has the objective of reviewing and improving the *work*. It is not a performance review.

The fundamental idea of a prototype is an initial, working model of a larger, more complex entity. The primary purpose of a prototype is to have a working model that will test a concept or verify an approach. Discovery prototypes are built to define requirements but are then usually discarded, or at least not used for the final programming.

Joint application design is a technique used to expedite the investigation of system requirements by holding several marathon sessions with all the critical participants. Discussion results in requirements definition and policy decisions immediately, without the delays of interviewing separate groups and trying to reconcile differences. When done correctly, it is a powerful and effective technique.

Business process reengineering is becoming a widespread method of improving business processes. It entails a complete redesign of the business processes. Under BPR, new system development is not done merely to automate existing procedures, but the entire process is completely rethought. The objective is to use IT in novel ways to achieve dramatic improvements in efficiencies and levels of service. Due to the special problem-solving, analytical, and modeling skills of systems analysts, they frequently play an important role in all BPR efforts.

Key Terms

closed-ended questions, p. 106
functional requirement, p. 98
group support system, p. 115
joint application design, p. 113
mock-up, p. 113
open-ended questions, p. 107
prototype, p. 113
query, p. 101
stakeholders, p. 99
structured walkthrough, p. 117
system requirements, p. 98
technical requirement, p. 99
transaction, p. 101

Review Questions

1. What is the difference between functional requirements and technical requirements?
2. Explain the use of a discovery prototype and an evolutionary prototype.
3. List and describe the three fact-finding techniques.
4. What is the objective of a structured walkthrough?
5. Explain the steps in preparing for an interview session.
6. What kinds of information do you need to obtain during your fact-finding sessions?

7. What categories of stakeholders should you include in fact finding?
8. What is meant by vertical and horizontal dimensions when determining users to involve?
9. What is JAD? When is it used?
10. What is BPR? What does it have to do with systems analysis?
11. What is the difference between verification and validation?
12. Describe the open-items list and explain why it is important.
13. What do *correct*, *complete*, and *comprehensive* mean with regard to systems analysis?

Thinking Critically

1. One of the toughest problems in investigating system requirements is to make sure they are complete and comprehensive. What kinds of things would you do to ensure that you get all of the right information during an interview session?
2. What kinds of things can you do to ensure that you have included all of the right stakeholders on your list of people to interview? How can you double-check your list?
3. One of the problems you will encounter during your investigation is "scope creep,"—that is, user requests for additional features and functions. This happens because sometimes users have many unsolved problems and this systems investigation may be the first time that anybody has listened to their needs. How do you keep the system from growing and including new functions that should not be part of the system?
4. It is always difficult to observe users in their jobs. It frequently makes both you and them uncomfortable. What things could you do to ensure their behavior is not changing due to your visit? How could you make observation more natural?
5. What would you do if you get opposite answers for the same procedure from two different people you interviewed? What would you do if one was a clerical person and the other was the department manager?
6. You are a team leader of four systems analysts. You have one analyst who never does structured walkthroughs of his work. How would you help the analyst to get started? How would you ensure that the walkthrough was effective?
7. You have been assigned several issues to resolve on the open-items list, and you are having a hard time getting policy decisions from the user contact. How can you gently prod the user to finalize these policies?
8. You are going on your first consulting assignment to do systems analysis. Your client does not like to pay to train new, inexperienced analysts. What should you do to appear competent and well prepared? How should you approach the client?
9. In the running case of Rocky Mountain Outfitters, you have set up an interview with Jason Nadold in the shipping department. Your objective is to determine how shipping works and what the information requirements for the new system will be. Make a list of questions, open-ended and closed-ended, that you would use. Include any questions or techniques you would use to ensure you find out about the exceptions.

Experiential Exercises

1. Conduct a fact-finding interview with someone involved in a procedure that is used in a business or organization. This person could be someone at the university, in a small business in your neighborhood, in the student volunteer office at the university, in a doctor's or dentist's office, in a volunteer organization, or at your local church. Identify a process that is done, such as keeping student records, customer records, or member records. Make a list of questions, and conduct the interview. Remember, your objective is to understand that procedure thoroughly—that is, to become an expert on that single procedure.
2. Follow the same instructions as for exercise 1, except make this exercise an observation experience. Either observe the other person do the work, or ask to carry out the procedure yourself. Write down the details of the process you observe.
3. Get a group of your fellow students together and conduct a structured walkthrough of your results from exercise 1 or 2. Take the results of your interview or observation and document the procedure in a flow chart with some narrative. Then conduct a walkthrough with several colleagues. Or take another assignment, such as application question 9, and walk through your preparation for that assignment. Follow the steps outlined in the text.

4. Research and write a one- to two-page research paper using at least three separate library sources on one of the following topics:
   a. Joint application design
   b. Prototyping as a discovery mechanism
   c. Business process reengineering
   d. Computer support for collaborative work (CSCW)
   e. Workflow systems
   f. Structured walkthrough
5. Using Rocky Mountain Outfitters as your guide, and the customer support subsystem, develop a list of all the procedures that may need to be researched. You may want to think about the exercise in the context of your experience with retailers such as L.L. Bean, Lands' End, or Amazon.com. Get some catalogs, check out the Internet marketing done on the retailers' web sites, and then think about the underlying business procedures that are required to support those sales activities. List the procedures and describe your understanding of each.

Case Study

John and Jacob, Inc. On-Line Trading System

John and Jacob, Inc., is a regional brokerage firm that has been successful over the last several years. Competition for customers is intense in this industry. The large national firms have very deep pockets with many services to offer to clients. Severe competition also comes from discount and Internet trading companies. However, Jacob and Jacob has been able to cultivate a substantial customer base from upper middle income people in the Northeast United States. To maintain a competitive edge with its customers, Jacob and Jacob is in the process of developing a new on-line trading system. The plan for the system identifies many new capabilities that would provide new services to its clients.

Edward Finnigan, the project manager, is in the process of identifying all the groups of people who should be included in the development of the system requirements. He isn't quite sure exactly who should be included. Here are the issues he's considering:

Users. The trading system is to be on-line to each of the company's 30 trading offices. Obviously, the brokers who are going to use the system need to have input, but how should this be done? He also isn't sure what approach would be best to ensure that the requirements are complete, yet not require tremendous amounts of time. Including all of the offices would increase enthusiasm and support for the system, but would take a lot of time. Yet, involving more brokers would bring divergent opinions that would have to be reconciled.

Customers. The trading system will also include confirmations, reports of trades, and customer statements. Web access is also planned, which will enable customers to effect trades and to check their accounts. Consequently, another question Edward has is about involving Jacob and Jacob customers in the development of system requirements. Normally customers are not asked to participate in the development of systems. However, it would be nice to know how best to serve Jacob and Jacob's customers. He is sensitive to this issue because some of the brokers have indicated to him that many customers do not like the format of their statements coming from the current system. He would like to involve customers, but he does not know how.

Other stakeholders. Edward knows he should involve other stakeholders to help define system requirements. He isn't quite sure whom he should contact. Should he go to senior executives; should he contact middle management; should he include back-office functions like accounting and investing? He isn't quite sure how to get organized or how to decide who should be involved.

1. What is the best method for Edward to involve the brokers (users) in development of the new on-line trading system? Should he use a questionnaire? Should he

interview the brokers in each of the company's 30 offices, or would one or two brokers representing the entire group be better? How can Edward ensure that the information about requirements is complete, yet not lose too much time doing so?

2. Concerning customer input for the new system, how can Edward involve customers in the process? How can he interest them in participating? What are some ways that Edward can be sure that the customers he does involve are representative of Jacob and Jacob's entire customer group?

3. As Edward considers what other stakeholders he should include, what are some criteria he should use? Develop some guidelines to help him build a list of people to include.

Further Resources

Vangalur S. Alagar. *Specification of Software Systems.* Springer-Verlag, 1998.

Lowell Jay Arthur. *Rapid Evolutionary Development: Requirements, Prototyping and Software Creation.* Wiley & Sons, 1991.

Alan M. Davis. *Software Requirements: Analysis and Specification.* Prentice Hall, 1990.

Stan Magee. *Guide to Software Engineering Standards and Specifications.* Artech House, 1997.

Mike F. Smith. *Software Prototyping: Adoption, Practice, and Management.* McGraw-Hill, 1991.

Jane Wood. *Joint Application Development.* Wiley & Sons, 1995.

Modeling System Requirements: Events and Things

Learning Objectives

After reading this chapter, you should be able to:

- Explain the many reasons for creating information system models
- Describe three types of models and list some specific models used for analysis and design
- Explain how events can be used to define system requirements
- Identify and analyze events to which a system responds
- Explain how the concept of things in the system also defines requirements
- Explain the similarities and the differences between data entities and objects
- Read and interpret an entity-relationship diagram
- Read and interpret a class diagram

Chapter Outline

Models and Modeling

Events and System Requirements

Things and System Requirements

The Entity-Relationship Diagram

The Class Diagram

Where You Are Headed

FIVE

WAITERS ON WHEELS: COMPUTERIZED DELIVERY TRACKING

Waiters on Wheels is a restaurant meal-delivery service started in 1997 by Sue and Tom Bickford. The Bickfords both worked for restaurants while in college and always dreamed of opening their own restaurant. But unfortunately, the initial investment was always out of reach. The Bickfords noticed that many restaurants offer take out food, and some restaurants, primarily pizzerias, offer home delivery service. Many people they met, however, seemed to want home delivery service but with a more complete food selection.

Waiters on Wheels was conceived as the best of both worlds for Sue and Tom—a restaurant service but without the high initial investment. The Bickfords contracted with a variety of well-known restaurants in town to accept orders from customers and to deliver the complete meals. After the restaurant prepares the meal to order, it charges Waiters on Wheels a wholesale price, and the customer pays retail plus a service charge and tip. Waiters on Wheels started modestly, with only two restaurants and one delivery driver working the dinner shift. Business rapidly expanded, and the Bickfords realized they needed a custom computer system to support their operations. They hired a consultant, Sam Wells, to help them define what sort of a system they needed.

"What sort of events happen when you are running your business that make you want to reach for a computer?" asked Sam. "Tell me about what usually goes on."

"Well," answered Sue, "when a customer calls in wanting to order, I need to record it and get the information to the right restaurant. I need to know which driver to ask to pick up the order, so I need drivers to call in and tell me when they are free. Sometimes customers call back wanting to change their orders, so I need to get my hands on the original order and notify the restaurant to make the change."

"Okay, how do you handle the money?" queried Sam.

Tom jumped in, "The drivers get a copy of the bill directly from the restaurant when they pick up the meal, which should agree with our calculations. The drivers collect that amount plus a service charge. When drivers report in at closing, we add up the money they have and compare it to the records we have. After all drivers report in, we need to create a deposit slip for the bank for the day's total receipts. At the end of each week, we calculate what we owe each restaurant at the agreed-to wholesale price and send them a statement and check."

"What other information do you need to get from the system?" continued Sam.

"It would be great to have some information at the end of each week about orders by restaurant and orders by area of town—things like that," added Sue. "That would help us decide about advertising and contracts with restaurants. Then we need monthly statements for our accountant."

Sam made some notes and sketched some diagrams as Sue and Tom talked. Then after spending some time thinking about it, he summarized the situation for Waiters on Wheels. "It sounds to me like you need a system that does some processing when these events occur:

- A customer calls in to place an order
- A driver is finished with a delivery
- A customer calls back to change an order
- A driver reports for work
- A driver submits the day's receipts

"Then you need the system to produce information at specific points in time—for example, when it is:

- Time to produce an end-of-day deposit slip
- Time to produce end-of-week restaurant payments
- Time to produce weekly sales reports
- Time to produce monthly financial reports

"Based on the way you have described your business operations, I am assuming you will need to store information in a database about these types of things:

- Restaurants
- Menu items
- Customers
- Orders
- Order payments
- Drivers

"Then I suppose you are going to need to maintain information in a database about restaurants and drivers. You'll need to do some processing when:

- A new restaurant is added
- A restaurant changes the menu
- A restaurant is dropped
- A new driver is hired
- A driver leaves

"Am I on the right track?"

Sue and Tom quickly agreed that Sam was talking about the system in a way they could understand. They were confident they had found the right consultant for the job.

OVERVIEW

The last chapter introduced the many tasks and techniques involved when gathering information about the system and its functional and technical requirements. This chapter, along with Chapters 6 and 7, presents techniques for documenting the functional requirements by creating a variety of models. These models are created during the analysis phase activity we have named *Define system requirements*. After discussing the types and roles of models, we focus on two key concepts that help define system requirements in both the traditional and the object-oriented approach: events and things. Specific models for both the traditional approach and the object-oriented approach are covered in this chapter, including those based on the Rocky Mountain Outfitters (RMO) customer support system. Keep in mind, though, that in any given system development project, either the traditional approach or the object-oriented approach will be used. However, the two key concepts—events and things—are common to both approaches. Chapter 6 continues the discussion of requirements models for the traditional approach, and Chapter 7 continues the object-oriented approach.

Models and Modeling

An analyst can best describe the requirements for an information system using a collection of models (as we discussed in Chapter 3). Recall that a model is a representation of some aspect of the system being built. Because a system is so complex, a variety of models are created to encompass the detailed information collected and digested by an analyst during the analysis phase (see Figure 5-1). Also, many different types of models are used to show the system at different levels of detail (or levels of abstraction), including a high-level overview as well as detailed views of certain aspects of the system. Some models show different parts of the problem and solution; for example, one model might show inputs, another the data stored. Some models show the same problem and solution from different perspectives; one model might show how objects interact from the perspective of outside actors, and another how objects interact in terms of sequencing.

FIGURE 5-1
An analyst needs a collection of models to understand system requirements.

The Purpose of Models

Some developers think of a model as documentation produced after the analysis and design work is done. But actually, the process of creating a model helps an analyst clarify and refine the design. The analyst learns as he or she completes and then studies parts of the model. Analysts also raise questions while creating a model and answer them as the modeling process continues. New pieces are added; the consequences of changes are evaluated and again questioned. In this respect, the modeling process itself provides direct benefits to the analyst. The technique used to create the model is valuable in itself even if the analyst never shows a particular model to anyone else. But usually models are shared with others as analysis and design progresses.

Another key reason that modeling is important in system development is the complexity of describing information systems. Information systems are very complex, and the parts of the information system are intangible. Models of the various parts of the system help simplify and focus the analyst's efforts on a few aspects of the system at a time. The reason an analyst uses so many different models is that each focuses on different aspects of the system. In fact, some of the models created by the analyst may serve only to integrate these aspects—showing how the other models fit together.

Because of the amount of information gathered and digested and the length of time each analyst spends, the analyst needs to review the models frequently to help recall details of work previously completed. People can retain only a limited amount of information, so we all need memory aids. Models provide a way of storing information for later use in a form that can be readily digested.

The support for communication is one of the most often cited reasons for creating the models. Given that the analyst learns while working through the modeling process and that the collection of models reduces the complexity of the information system, the models also serve a critical role in supporting communication among project team members and with system users. If one team member is working on models of inputs and outputs, and another team member is working on models of the processes that convert the inputs to outputs, then communication is necessary to make sure these models fit together. The second team member needs to see what outputs are desired before modeling the process that creates them. At the same time, both of these team members need to know what data are stored (the data model) so they know what inputs are needed and what processes are needed to access the required data. Models support communication among the project team members.

Models also assist in communication with the system users. Typically, an analyst reviews the models with a variety of users to get feedback on the analyst's understanding of the system requirements. Users need to see clear and complete models to understand what the analyst is proposing. Additionally, the analyst sometimes works with users to develop the models, so the modeling process helps users better understand the possibilities that the new system can offer. Users also need to communicate among themselves using the models. And the analyst and the users together can use models to communicate system capabilities to managers who are responsible for approving the system.

Finally, the requirements models produced by the analyst are used as documentation for future development teams when they maintain or enhance the system. Considering the amount of resources invested in a new system, it is critical for the development team to leave behind a clear record of what was created. An important activity during implementation is to package the documentation accurately, completely, and in a form that future developers can use. Much of the documentation consists of the models created throughout the project. Figure 5-2 summarizes the reasons modeling is important to system development.

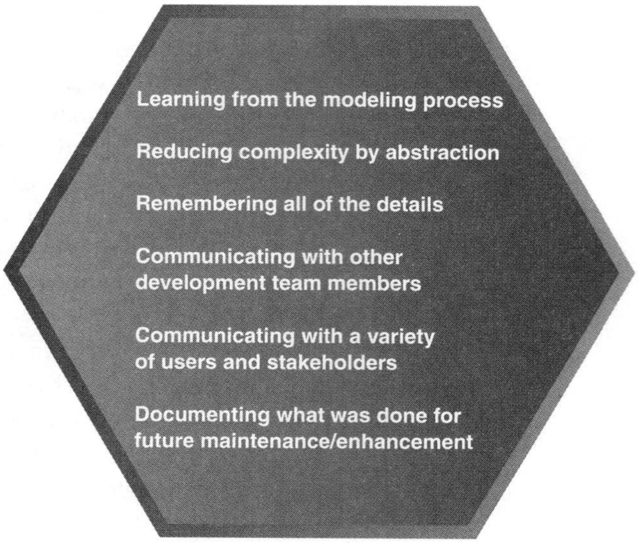

FIGURE 5-2
Reasons for modeling.

Learning from the modeling process
Reducing complexity by abstraction
Remembering all of the details
Communicating with other development team members
Communicating with a variety of users and stakeholders
Documenting what was done for future maintenance/enhancement

Types of Models

Many different types of models are used when developing information systems. The type of model used is based on the nature of the information being represented. Types of models include mathematical models, descriptive models, and graphical models.

mathematical model
A series of formulas that describe technical aspects of a system

Mathematical Models A **mathematical model** is a series of formulas that describe technical aspects of a system. Mathematical models are used to represent precise aspects of the system that can be best represented using formulas or mathematical notation, such as equations that represent network throughput requirements or a function expressing the response time required for a query. Additionally, scientific and engineering applications tend to compute results using elaborate mathematical algorithms. The mathematical notation is the most appropriate way to represent these functional requirements, and it is also the most natural way for scientific and engineering users to express those requirements. An analyst working on scientific and engineering applications had better be comfortable with math.

But mathematical notation is also sometimes efficient for simpler requirements for business systems. For example, in a payroll application, it is reasonable to model gross pay as regular pay plus overtime pay. A reorder point for inventory, a discount price for a product, or a salary adjustment for a promotion might be modeled with a simple formula.

descriptive model
Narrative memos, reports, or lists that describe some aspect of a system

Descriptive Models Not all requirements can be precisely defined with mathematics. For these requirements, analysts use **descriptive models**, which can be narrative memos, reports, or lists. Figure 5-3 provides examples of descriptive models. Initial interviews with users might require the analyst to jot down notes in a narrative form. Sometimes users describe what they do in reports or memos to the analysts. The analyst can then convert these narrative descriptions to a modeling notation while compiling all of the information.

FIGURE 5-3
Some descriptive models.

A narrative description of processing requirements as verbalized by an RMO phone-order representative:

"When customers call in, I first ask if they have ordered by phone with us before, and I try to get them to tell me their customer ID number that they can find on the mailing label on the catalog. Or, if they seem puzzled about the customer number, I need to look them up by name and go through a process of elimination, looking at all of the Smiths in Dayton, for example, until I get the right one. Next I ask what catalog they are looking at, which sometimes is out of date. If that is the case, then I explain that many items are still offered, but that the prices might be different. Naturally, they point to a page number, which doesn't help me because of the different catalogs, but I get them to tell me the product ID somehow..."

List of inputs for the RMO customer support system:

Item inquiry
New order
Order change request
Order status inquiry
Order fulfillment notice
Back order notice
Order return notice
Catalog request
Customer account update notice
Promotion package details
Customer charge adjustment
Catalog update details
Special promotion details
New catalog details

Sometimes a narrative description is the best form to use for recording information. Many useful models of information systems involve simple lists, such as lists of features, inputs, outputs, events, or users. Lists are a form of descriptive or narrative models that are concise, specific, and useful.

A final example of a descriptive model involves writing a process or procedure in a very precise way, referred to as structured English or pseudocode. Programmers are familiar with structured English or pseudocode for modeling algorithms that, when followed, always obtain the same result. Therefore, such algorithms are very precise models of processing.

graphical model
Diagrams and schematic representations of some aspect of a system

Graphical Models Probably the most useful models created by the analyst are graphical models. **Graphical models** include diagrams and schematic representations of some aspect of a system. Graphical models make it easy to understand complex relationships that are too difficult to follow when described verbally. Recall the old saying that a picture is worth a thousand words. In system development, a carefully constructed graphical model might be worth a million words!

Some graphical models actually look similar to a real-world part of the system, such as a screen design or a report layout design. But for most of the analyst's work, the graphical models use symbols to represent more abstract things, such as external agents, processes, data, objects, messages, and connections. The key graphical models used for the analysis phase tend to represent the more abstract aspects of a system, since the analysis phase focuses on fairly abstract questions about system requirements without indicating the details of how they will be implemented. The more concrete models of screen designs and report layouts are completed during the system design phase.

A variety of graphical models are used. Each model highlights (or abstracts) important details of some aspect of the information system. Each type of model should ideally use unique and standardized symbols to represent pieces of information. That way, whoever looks at a model can understand it. However, the number of available symbols is limited—circle, square, rectangle, line, and so on—so be careful when you are first learning the symbols of each model. You will also find variations in the notation used for each type of model in practice.

Overview of Models Used in Analysis and Design

The analysis phase activity named *Define system requirements* involves creating a variety of models. These are referred to as *logical models* (as discussed in Chapter 3) because they define in great detail what is required without committing to one specific technology. Many types of logical models are created to define system requirements. Figure 5-4 lists some of the more commonly used models.

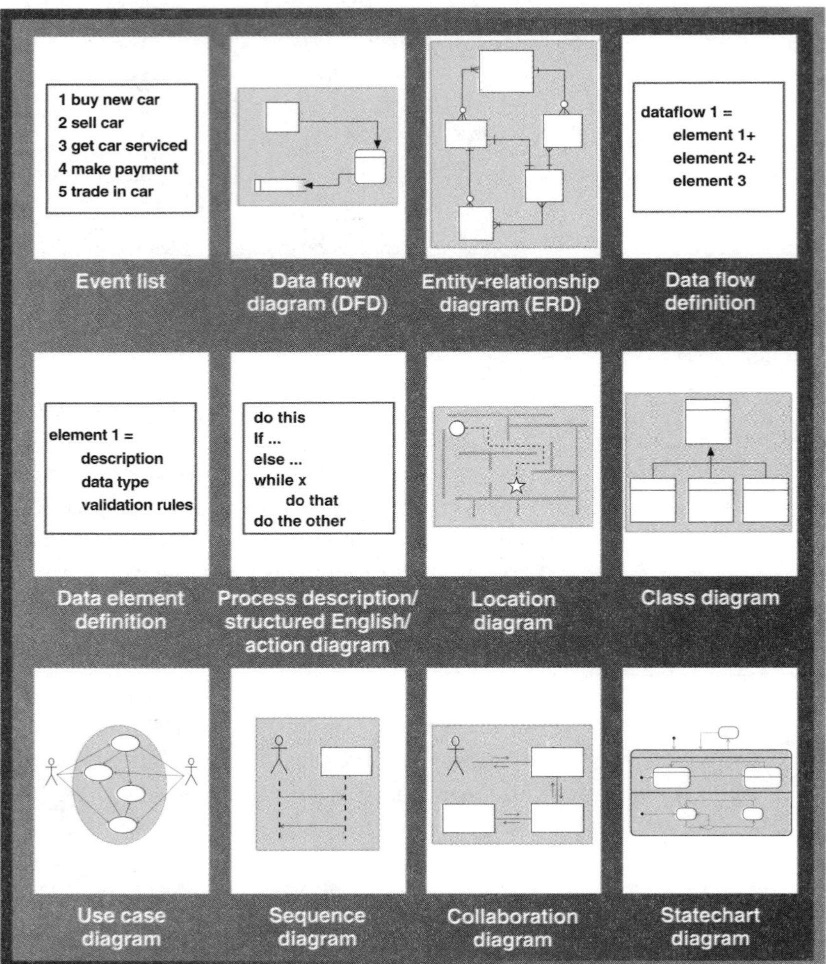

FIGURE 5-4
Models created during the analysis phase.

Many models are also created during the design phase (see Figure 5-5). These are physical models because they show how some aspect of the system will be implemented with specific technology. Some of these models are extensions of requirements models created during systems analysis or derive directly from the requirements models. Some models (for example, a class diagram) are used during analysis and during design.

FIGURE 5-5
Some models created during the design phase.

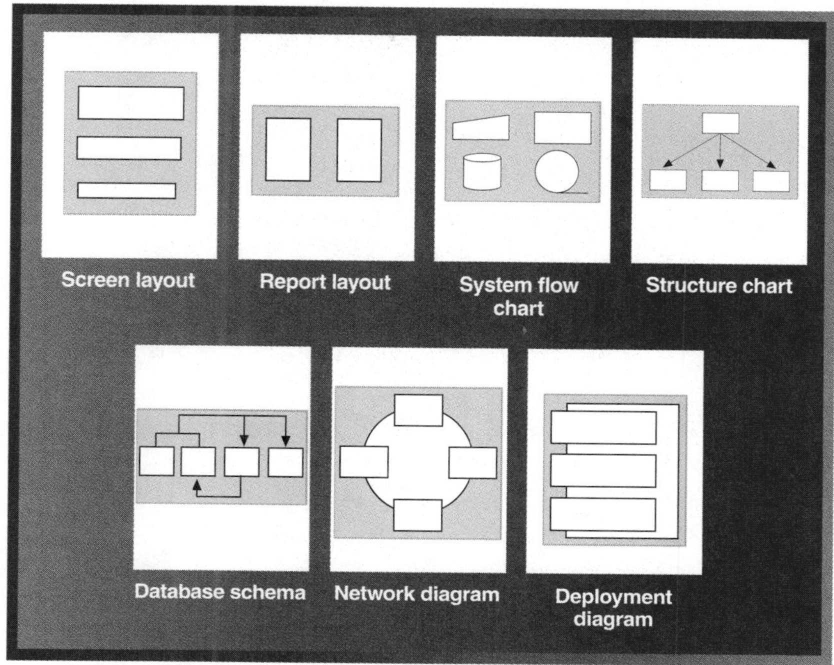

This chapter and Chapters 6 and 7 describe some of the requirements models in detail. Here we describe events and things, two key concepts common to requirements models in all approaches to development. The models discussed in this chapter include event lists, event tables, entity-relationship diagrams, and class diagrams.

Events and System Requirements

event
An occurrence at a specific time and place, that can be described and is worth remembering

Now that you know the basic types of models, we turn to the specifics for modeling the functional requirements for an information system. Virtually all approaches to system development begin the modeling process with the concept of an event. An **event** occurs at a specific time and place, can be described, and should be remembered by the system. Events drive or trigger all processing that a system does, so listing them and then analyzing them makes sense when you need to define system requirements.

When defining the requirements for a system, it is useful to begin by asking what events occur that will affect the system being studied. More specifically, what events occur that will require the system to respond? By asking about the events that occur that affect the system, you direct your attention to the external environment and look at the system as a black box. This initial view helps keep your focus on a high-level view of the system (looking at the scope) rather than on the inner workings of the system. It also focuses your attention on the interfaces of the system to outside people and other systems. End users, the ones who will actually use the system, can readily describe system needs in terms of events that affect their work. So, the external focus on events is appropriate when working with users. Finally, focusing on events gives you a way to divide (or decompose) the system requirements so you can study each separately. Complex systems need to be broken into manageable units to be understood, and decomposing the system based on events is one way to accomplish this, as shown in Figure 5-6.

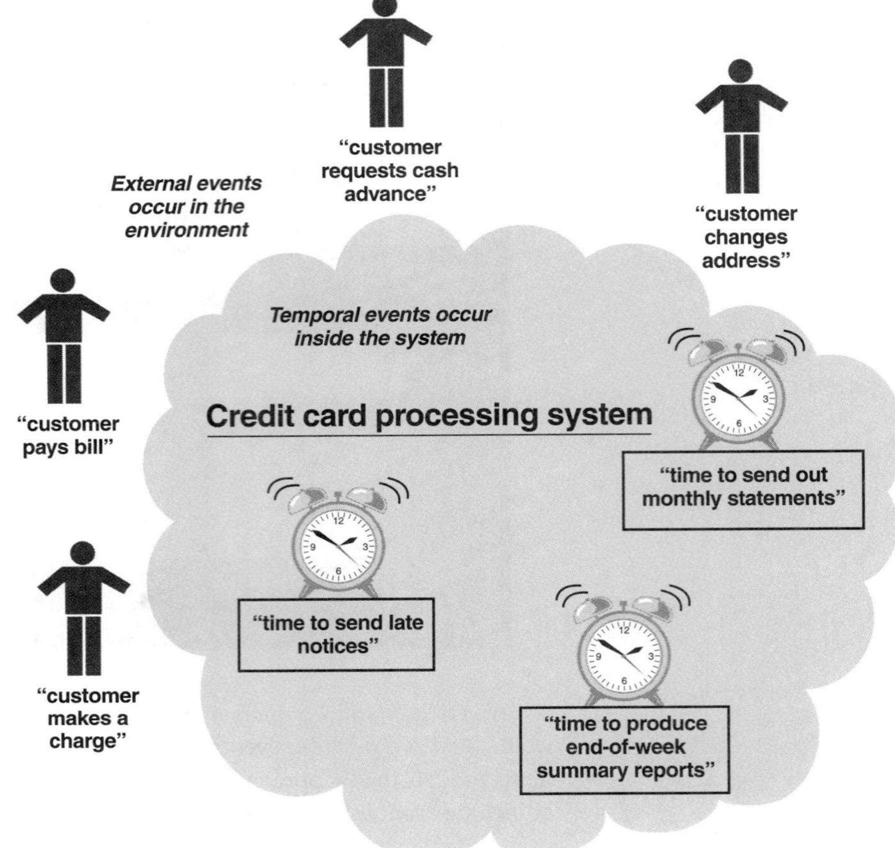

FIGURE 5-6
Events affecting a system.

The Background of the Event Concept

The importance of the concept of events for defining system requirements was first emphasized for modern structured analysis as it was adapted to real-time systems in the early 1980s. Real-time systems require the system to react immediately to events in the environment. Early examples of real-time systems included control systems such as manufacturing process control or avionics guidance systems. For example, in process control, if a vat of chemicals is full, then the system needs to turn off the fill valve. The relevant event is "vat is full," and the system needs to respond to that event immediately. In an airplane guidance system, if the plane's altitude drops below 5,000 feet, then the system needs to turn on the low-altitude alarm. Business systems analysts did not have to create similar types of applications, so real-time system extensions to structured analysis did not receive much emphasis in business systems analysis and design at the time.

However, information systems have become much more interactive over the last decade. Now most information systems being developed are so interactive that they can be thought of as real-time systems (indeed, people expect a real-time response to almost everything). The information engineering approach now uses the event concept (although not as the first step), and more importantly, the object-oriented approach has also embraced the event concept. No matter which approach you use, a key analysis step is listing and analyzing the events to which the system must respond.

Types of Events

There are three types of events to consider—external events, temporal events, and state events. The analyst begins by trying to identify and list as many of these events as possible, refining the list while talking with system users.

external event
An event that occurs outside the system, usually initiated by an external agent or actor

External Events External events are events that occur outside the system, usually initiated by an external agent or actor. An external agent (or actor) is a person or organizational unit that supplies or receives data from the system. To identify the key external events, the analyst first tries to identify all of the external agents that might want something from the system. A classic example of an external agent is a customer. The customer may want to place an order for one or more products. This event is of fundamental importance to an order-processing system like the one needed by Rocky Mountain Outfitters. But other events are associated with a customer. Sometimes a customer wants to return an ordered product, and a customer needs to pay the invoice for an order. External events such as these are the types the analyst looks for because they begin to define what the system needs to be able to do. They are events that lead to important transactions the system must process.

When describing external events, it is important to name the event so that the external agent is clear. The action the external agent wants to pursue is included, too. So, the event *Customer places an order* describes the external agent (a customer) and what the customer wants to do (to place an order for some products) that directly affects the system. Again, if the system is an order-processing system, the system needs to process the order for the customer.

Important external events can also result from the wants and needs of people or organizational units inside the company—for example, management requests for information. A typical event in an order-processing system might be *Management checks order status*. Perhaps managers want to follow up on an order for a key customer, and the system must routinely provide that information.

Another type of external event occurs when external entities provide new information that the system simply needs to store for later use. For example, a regular customer reports a change in address, phone, or employer. Usually one event for each type of external agent can be described to handle updates to data, such as *Customer updates account information*. Figure 5-7 provides a checklist to help in identifying external events.

FIGURE 5-7
External event checklist.

External Events to Look for Include:
√ External agent wants something resulting in a transaction
√ External agent wants some information
√ Data changed needs to be updated
√ Management wants some information

temporal event
An event that occurs as a result of reaching a point in time

Temporal Events A second type of event is a **temporal event**, an event that occurs as a result of reaching a point in time. Many information systems produce outputs at defined intervals, such as payroll systems that produce a paycheck every two weeks (or monthly). Sometimes the outputs are reports that management wants to receive regularly, such as performance reports or exception reports. These events are different from external events in that the system should automatically produce the required output without being told to do so. In other words, no external agent or actor is making demands, but the system is supposed to generate needed information or other outputs when they are needed.

The analyst begins identifying temporal events by asking about the specific deadlines that the system must accommodate. What outputs are produced at that deadline? What other processing might be required at that deadline? These events are usually named by defining what needs to be produced at that time. The payroll example discussed previously might be named *Time to produce biweekly payroll.* The event defining the need for a monthly summary report might be named *Time to produce monthly sales summary report.* Figure 5-8 provides a checklist to use in identifying temporal events.

FIGURE 5-8
Temporal event checklist.

Temporal Events to Look for Include:
- √ Internal outputs needed
  - √ Management reports (summary or exception)
  - √ Operational reports (detailed transactions)
  - √ Statements, status reports (including payroll)
- √ External outputs needed
  - √ Statements, status reports, bills, reminders

Temporal events do not have to occur on a fixed date. They can occur after a defined period of time has elapsed. For example, a bill might be given to a customer when a sale has occurred. If the bill has not been paid within 15 days, a late notice might be sent. The temporal event, *Time to send late notice,* might be defined as a point 15 days after the billing date.

state event
An event that occurs when something happens inside the system that triggers the need for processing

State Events A third type of event is a **state event**, an event that occurs when something happens inside the system that triggers the need for processing. For example, if the sale of a product results in an adjustment to an inventory record and the inventory in stock drops below a reorder point, it is necessary to reorder. The state event might be named *Reorder point reached.* Often state events occur as a consequence of external events. Sometimes they are similar to temporal events, except the point in time cannot be defined. The reorder event might be named *Time to reorder inventory,* which sounds like a temporal event.

Identifying Events

It is not always easy to define the events that affect a system. But some guidelines can help an analyst think through the process.

Events versus Conditions and Responses It is sometimes difficult to distinguish between an event and part of a sequence of conditions that leads up to the event. Consider an example of a customer buying a shirt from a retail store (see Figure 5-9). From the customer's perspective, this purchase involves a long sequence of events. The first event might be that a customer wants to get dressed. Then the customer wants to wear a red shirt. Next his red shirt appears to be worn out. Then the customer decides to drive to the mall. Then he decides to go in Sears. Then he tries on a red shirt. Then the customer decides to leave Sears and go to Wal-Mart to try on a shirt. Finally, the customer wants to purchase the shirt. The analyst has to think through a sequence such as this to arrive at the point where an event directly affects the system. In this case, the system is not affected until the customer is in the store, has a shirt in hand ready to purchase, and says, "I want to buy this shirt."

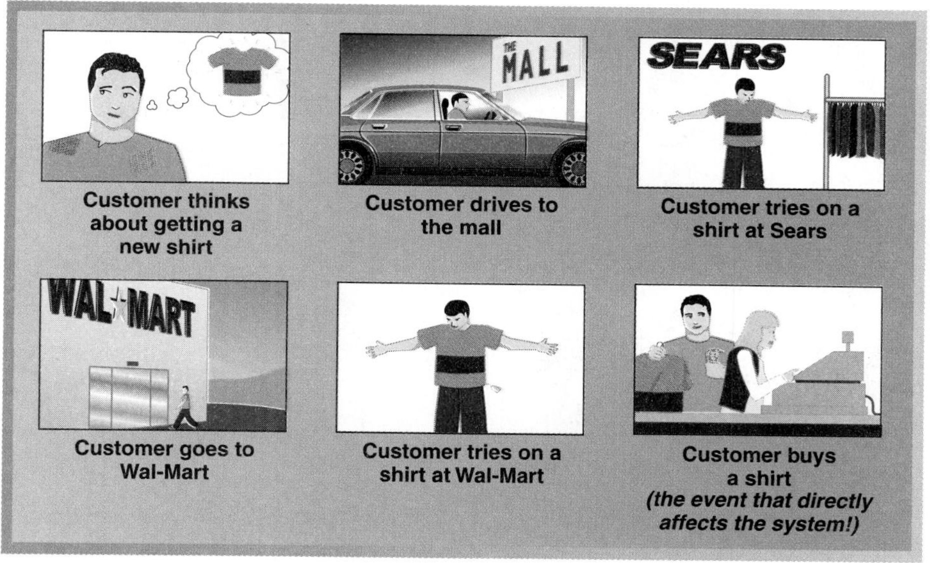

FIGURE 5-9
Sequence of actions that lead up to an event affecting the system.

In other situations, it is not easy to distinguish between an external event and the system's response. For example, when the customer buys the shirt, the system requests a credit card number, and the customer supplies the credit card. Is the act of supplying the credit card an event? In this case, no. It is part of the interaction that occurs while completing the original transaction.

The way to determine whether an occurrence is an event or part of the interaction following the event is by asking whether any long pauses or intervals occur—that is, can the system transaction be completed without interruption? Or is the system at rest again waiting for the next transaction? Once the customer wants to buy the shirt, the process continues until the transaction is complete. There are no significant stops once the transaction begins. Once the transaction is complete, the system is at rest, waiting for the next transaction to begin.

On the other hand, an example of separate events occurs when the customer buys the shirt using his store credit card account. Is the processing when the customer later pays the bill at the end of the month included as part of the interaction involving the purchase? In this case, no. The system records the transaction and then does other things. It does not halt all processes to wait for the payment. A separate event occurs later that results in sending the customer a bill (this is a temporal event: *Time to send monthly bills*). Eventually, another external event occurs (*Customer pays the bill*).

The Sequence of Events: Tracing a Transaction's Life Cycle It is often useful in identifying events to trace the sequence of events that might occur for a specific external agent or actor. In the case of Rocky Mountain Outfitters' new order-processing system, the analyst might think through all of the possible transactions that might result from one new customer (see Figure 5-10). First, the customer wants a catalog or asks for some information about item availability, resulting in a name and address being added to the database. Next, the customer might want to place an order. Perhaps she will want to change the order, correcting the size of the shirt, for example, or buying another shirt. Next, the customer might want to check the status of an order to find out the shipping date. Perhaps the customer has moved and wants an address change recorded for future catalog mailings. Finally, the customer might want to return an item. Thinking through this type of sequence can help identify events.

FIGURE 5-10
The sequence of "transactions" for one specific customer.

Technology-Dependent Events and System Controls Sometimes the analyst is concerned about events that are important to the system but do not directly concern users or transactions. Such events typically involve design choices or system controls. During analysis, the analyst should temporarily ignore these events. They are important for the design phase, however.

Some examples of events that affect design issues include external events that involve actually using the physical system, such as logging on. Although important to the final operation of the system, such a detail of implementation should be deferred. At this stage, the analyst should focus only on the functional requirements—the work that the system needs to complete. A logical model does not need to indicate how the system is actually implemented, so the implementation details should be left out.

system controls
Checks or safety procedures put in place to protect the integrity of the system

Most of these events involve **system controls**, which are checks or safety procedures put in place to protect the integrity of the system. Logging on to a system is required because of system security controls, for example. Other controls protect the integrity of the database, such as backing up the data every day. Both of these controls are important to the system, and they will certainly be added to the system during design. But spending time on these controls during analysis only adds to the requirements model details that the users are not typically very concerned about (they trust information services to take care of such details).

perfect technology assumption
The assumption that events should be included during analysis only if the system would be required to respond under perfect conditions

One technique used to help decide which events apply to controls is to assume that technology is perfect. The **perfect technology assumption** states that events should be included during analysis only if the system would be required to respond under perfect conditions—that is, with equipment never breaking down, capacity for processing and storage being unlimited, and people operating the system being completely honest and never making mistakes. By pretending that technology is perfect, analysts can eliminate events like *Time to back up the database* because they can assume the disk will never crash. Again, during design, the project team adds these controls because technology is obviously not perfect. Figure 5-11 lists some examples of events that can be deferred until the design phase.

FIGURE 5-11
Events deferred until the design phase.

Don't worry much about these until the design phase

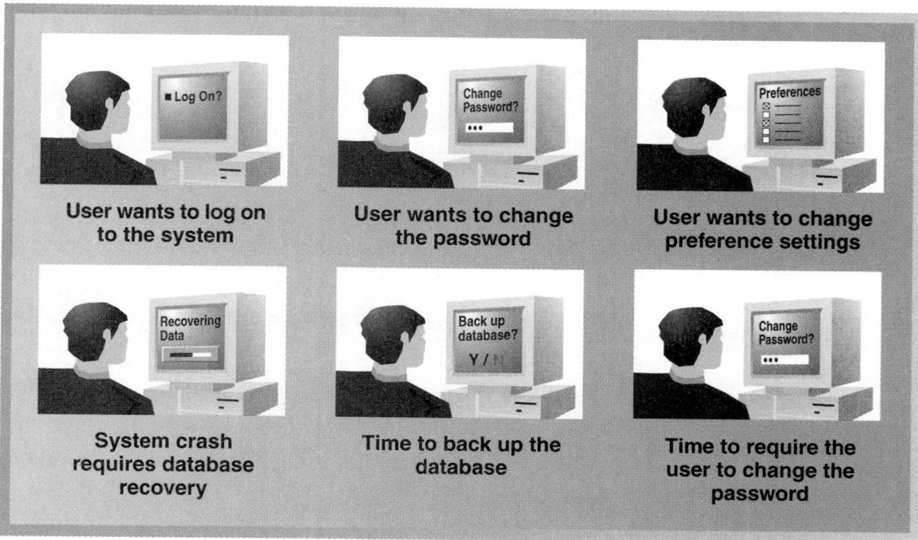
Log On?
User wants to log on to the system
Change Password?
User wants to change the password
Preferences
User wants to change preference settings
Recovering Data
System crash requires database recovery
Back up database?
Y / N
Time to back up the database
Change Password?
Time to require the user to change the password

Events in the Rocky Mountain Outfitters Case

The Rocky Mountain Outfitters customer support system involves a variety of events, many of them similar to those just discussed. A list of the external events is shown in Figure 5-12. Some of the most important external events involve customers: *Customer wants to check item availability, Customer places an order, Customer changes or cancels an order*. Other external events involve RMO departments: *Shipping fulfills order, Marketing wants to send promotional material to customers, Merchandising updates catalog.* The analyst can develop this list of external events by looking at all of the people and organizational units that want the system to do something for them.

FIGURE 5-12
External events for the RMO customer support system.

Customer wants to check item availability
Customer places an order
Customer changes or cancels order
Customer or management wants to check order status
Shipping fulfills order
Shipping identifies back order
Customer returns item (defective, changed mind, full or partial returns)
Prospective customer requests catalog
Customer updates account information
Marketing wants to send promotional materials to customers
Management adjusts customer charges (correct errors, make concessions)
Merchandising updates catalog (add, change, delete, change prices)
Merchandising creates special product promotion
Merchandising creates new catalog

There are also quite a few temporal events for the customer support system, shown in Figure 5-13. Many of these produce periodic reports for organizational units: *Time to produce order summary reports, Time to produce fulfillment summary reports, Time to produce catalog activity reports*. The analyst can develop the list of temporal events by looking for all of the regular reports and statements that the system must produce at certain times.

FIGURE 5-13
Temporal events for the RMO customer support system.

Time to produce order summary reports
Time to produce transaction summary reports
Time to produce fulfillment summary reports
Time to produce prospective customer activity reports
Time to produce customer adjustment/concession reports
Time to produce catalog activity reports

Looking at Each Event

event table
A table that lists events in rows and key pieces of information about each event in columns

While developing the list of events, the analyst should note additional information about each event for later use. This information is entered in an event table. An **event table** includes rows and columns, representing events and their details respectively. Each row in the event table records information about one event. Each column in the table represents a key piece of information about that event. The information about an event *Customer wants to check item availability* is shown in Figure 5-14.

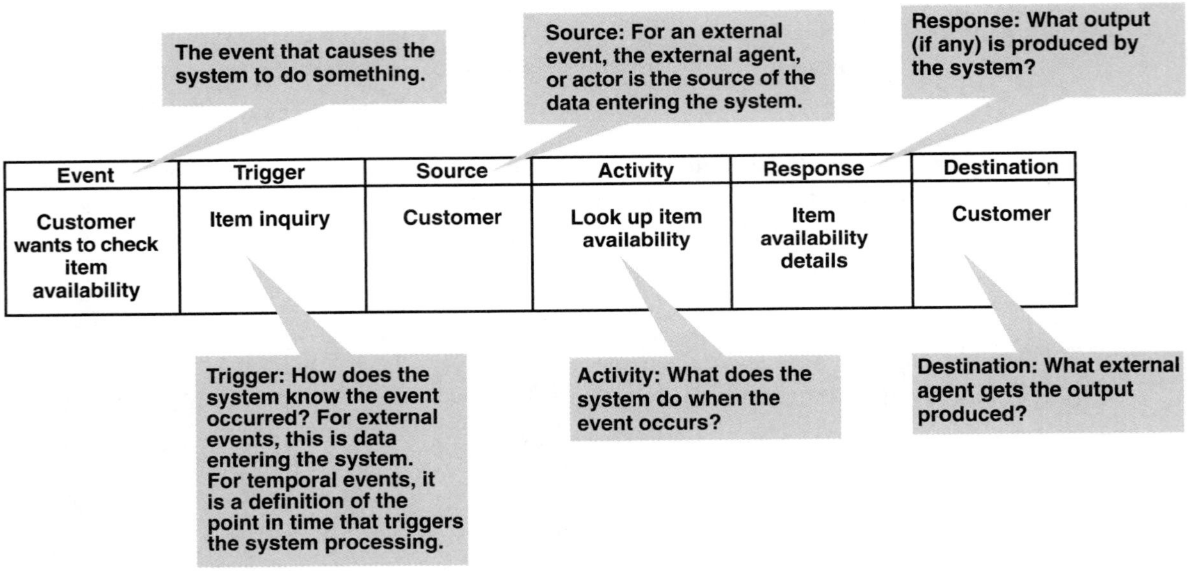

The event that causes the system to do something.

Source: For an external event, the external agent, or actor is the source of the data entering the system.

Response: What output (if any) is produced by the system?

Event	Trigger	Source	Activity	Response	Destination
Customer wants to check item availability	Item inquiry	Customer	Look up item availability	Item availability details	Customer

Trigger: How does the system know the event occurred? For external events, this is data entering the system. For temporal events, it is a definition of the point in time that triggers the system processing.

Activity: What does the system do when the event occurs?

Destination: What external agent gets the output produced?

FIGURE 5-14
Information about each event in an event table.

trigger
An occurrence that tells the system that an event has occurred, either the arrival of data needing processing or of a point in time

First, for each event, how does the system know the event has occurred? An occurrence that tells the system an event has occurred is called the **trigger**. For an external event, the trigger is the arrival of data that the system must process. For example, when a customer places an order, the new order details are provided as input. The source of the data is also important to know. In this case, the source of the new order details is the customer—an external agent or actor. For a temporal event, the trigger is a point in time. For example, at the end of each business day, the system knows it is time to produce transaction summary reports.

Next, what does the system do when the event occurs? The system's reaction to the event is called an **activity**. When a customer places an order, the system carries out the

source
An external agent or actor that supplies data to the system

activity
Behavior that the system performs when an event occurs

response
An output, produced by the system, that goes to a destination

destination
An external agent or actor that receives data from the system

activity *Create a new order*. When it is time to produce transaction summary reports, the system carries out the activity *Produce order summary reports*.

Finally, what response from the system does the activity produce? A **response** is an output from the system. There are often several responses from one activity. For example, when a customer places an order, an order confirmation goes to the customer, order details go to shipping, and a record of the transaction goes to the bank. When it is time to produce transaction summary reports, a transaction summary report is produced. The **destination** is where any response (output) is sent, again an external agent or actor. Sometimes there is no response at all. For example, if the customer wants to update account information, the information is recorded in the database but no output needs to be produced.

The list of events, together with the trigger, source, activity, response(s), and destination(s) for each event, can be placed in an event table to keep track of them for later use. An event table is a convenient way to record key information about the requirements for the information system. The event table for the RMO customer support system is shown in Figure 5-15. The event table will later be used in Chapters 6 and 7 to develop requirements models for both the traditional approach and the object-oriented approach.

Customer Support System Event Table

Event	Trigger	Source	Activity	Response	Destination
Customer wants to check item availability	Item inquiry	Customer	Look up item availability	Item availability details	Customer
Customer places an order	New order	Customer	Create new order	Real-time link Order confirmation Order details Transaction	Credit bureau Customer Shipping Bank
Customer changes or cancels order	Order change request	Customer	Update order	Change Confirmation Order change details Transaction	Customer Shipping Bank
Time to produce order summary reports	"End of week, month, quarter, and year"		Produce order summary reports	Order summary reports	Management
Time to produce transaction summary reports	"End of day"		Produce transaction summary reports	Transaction summary reports	Accounting
Customer or management wants to check order status	Order status inquiry	Customer or management	Look up order status	Order status details	Customer or management
Shipping fulfills order	Order fulfillment notice	Shipping	Record order fulfillment		
Shipping identifies back order	Back order notice	Shipping	Record back order	Back order notification	Customer
Customer returns item	Order return notice	Customer	Create order return	Return confirmation Transaction	Customer Bank

(continued)

Event	Trigger	Source	Activity	Response	Destination
Time to produce fulfillment summary reports	"End of week, month, quarter, and year"		Produce fulfillment summary reports	Fulfillment summary reports	Management
Prospective customer requests catalog	Catalog request	Prospective customer	Provide catalog info	Catalog	Prospective customer
Time to produce prospective customer activity reports	"End of month"		Produce prospective customer activity reports	Prospective customer activity reports	Marketing
Customer updates account information	Customer account update notice	Customer	Update customer account		
Marketing wants to send promotional materials to customers	Promotion package details	Marketing	Distribute promotional package	Promotional package	Customer and prospective customer
Management adjusts customer charges	Customer charge adjustment	Management	Create customer charge adjustment	Charge adjustment notification Transaction	Customer Bank
Time to produce customer adjustment/ concession reports	"End of month"		Produce customer adjustment reports	Customer adjustment reports	Management
Merchandising updates catalog	Catalog update details	Merchandising	Update catalog		
Merchandising creates special product promotion	Special promotion details	Merchandising	Create special promotion		
Merchandising creates new catalog	New catalog details	Merchandising	Create new catalog	Catalog	Customer and prospective customer
Time to produce catalog activity reports	"End of month"		Produce catalog activity reports	Catalog activity reports	Merchandising

FIGURE 5-15
The complete event table for the RMO customer support system.

Things
and System Requirements

Another key concept used to define system requirements involves understanding and modeling things that the system needs to store information about. To the users, these are the things they deal with when they do their work—products, orders, invoices, and customers—that need to be part of the system. For example, an information system needs to store information about customers and products, so it is important for the analyst to identify lots of information about them. Often these things are similar to the external agents or actors that interact with the system. For example, a customer external agent places an order, but the system also needs to store information about the customer. In other cases, these things are distinct from external agents. For example, there is no product external agent, but the system needs to store information about products.

In the traditional approach to development, these things make up the data about which the system stores information. The type of data that need to be stored is definitely a key aspect of the requirements for any information system. In the object-oriented approach, these things are the objects that interact in the system. No matter which approach to developing an information system you use, identifying and understanding these things are both key initial steps.

Types of Things

As with events, an analyst should ask the users to discuss the types of things that they work with routinely. The analyst can ask about several types of things to help identify them. Many things are tangible and therefore more easily identified, but others are intangible. Different types of things are important to different users, so it is important to include information from all types of users.

Figure 5-16 shows some types of things to consider. Tangible things are often the most obvious, such as an airplane, book, or vehicle. In the Rocky Mountain Outfitters case, a catalog and an item in the catalog are tangible things of importance. Another common type of thing in an information system is a role played by a person, such as employee, customer, doctor, or patient. A customer is obviously a very important role a person plays in the Rocky Mountain Outfitters case.

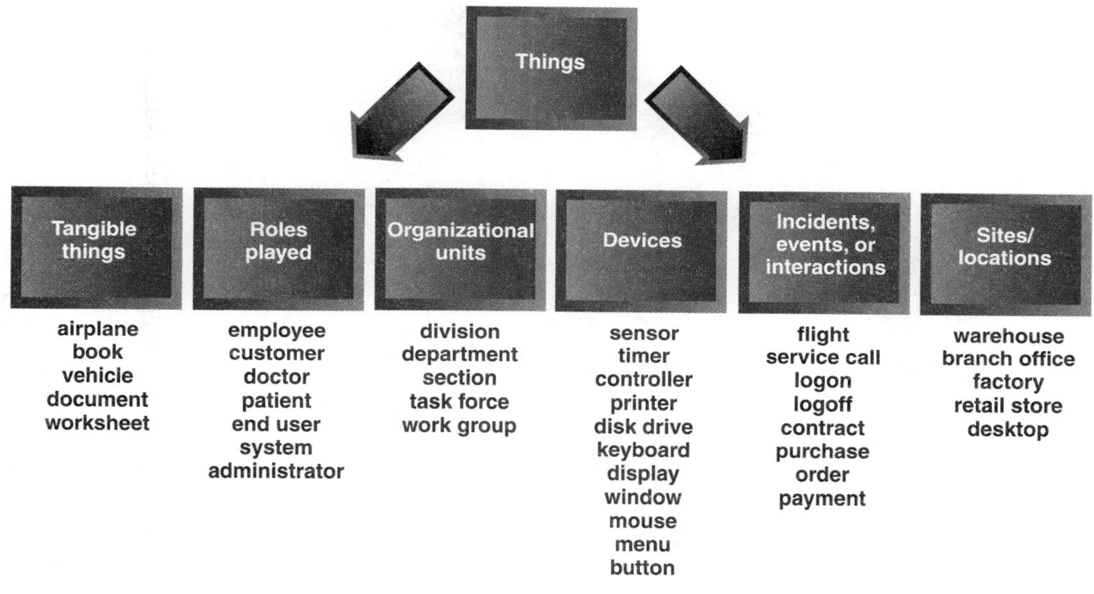

FIGURE 5-16
Types of things.

Other types of things can include organizational units, such as a division, department, or work group. Similarly, sites or locations might be important in a particular system, such as a warehouse, a store, or a branch office. Finally, information about an incident or interaction of importance can be considered a thing—information about an order, a service call, a contract, or an airplane flight. An order, a shipment, and a return are important incidents in the Rocky Mountain Outfitters case. Sometimes these incidents are thought of as relationships between things. For example, an order is a relationship between a customer and an item of inventory. Initially the analyst might simply list all of these as things and then make adjustments that might be required by different approaches to analysis and design.

The analyst identifies these types of things by thinking about each event in the event list and asking questions. For example, for each event, what types of things are affected that the system needs to know about and store information about? When a customer places an order, the system needs to store information about the customer, the items ordered, and the details about the order itself, such as the date and payment terms.

Relationships among Things

relationship
A naturally occurring association among specific things, such as *an order is placed by a customer* and *an employee works in a department*

There are many important relationships among things of importance in the system. A **relationship** is a naturally occurring association among specific things, such as an order is placed by a customer and an employee works in a department (see Figure 5-17). *Is placed by* and *works in* are two relationships that naturally occur between specific things. Information systems need to store information about employees and about departments, but equally important is storing information about the specific relationships—John works in the accounting department, and Mary works in the marketing department. Similarly, it is quite important to store the fact that Order 1043 for a shirt was placed by John Smith.

FIGURE 5-17
Relationships naturally occur between things.

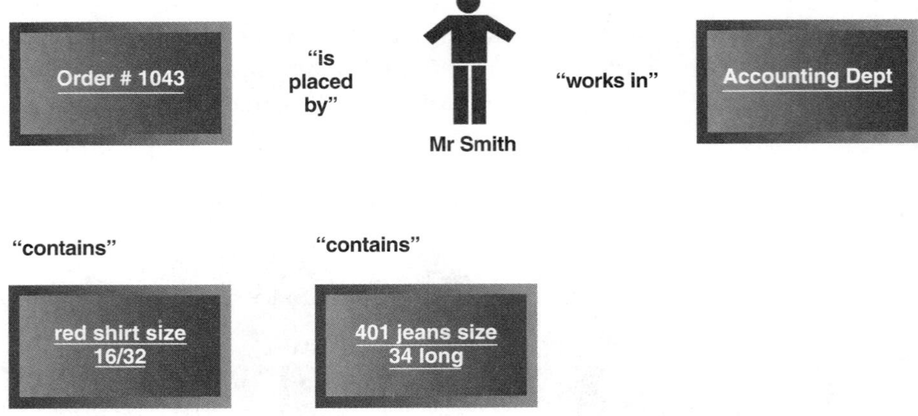

Relationships between things apply in two directions. For example, a customer places an order describes the relationship in one direction. Similarly, an order is placed by a customer describes the relationship in the other direction. It is important to understand the relationship in both directions because sometimes it might seem more important for the system to record the relationship in one direction than in the other. For example, Rocky Mountain Outfitters definitely needs to know what items a customer ordered so the shipment can be prepared. However, it might not be initially apparent that the company needs to know all of the customers who have ordered a particular item. What if the company needs to notify all customers who ordered a defective or recalled product? Knowing this information would be very important, but the operational users might not immediately recognize it.

cardinality
The number of associations that occur between specific things, such as *a customer places many orders* and *an employee works in one department*

multiplicity
A synonym for *cardinality* (often used with the object-oriented approach)

It is also important to understand the nature of each relationship in terms of the number of associations for each thing. For example, a customer might place many different orders, but an order is placed by only one customer. The number of associations that occur is referred to as the **cardinality** of the relationship. Cardinality can be one to one or one to many. Again, cardinality is established for each direction of the relationship. Sometimes the term **multiplicity** is used to refer to the number of associations. Figure 5-18 lists examples of cardinality associated with an order.

Sometimes it is important to describe not just the cardinality, but also the range of possible values of the cardinality (the minimum and maximum cardinality). For example, a particular customer might not ever place an order. In this case, there are zero associations. Alternatively, the customer may place one order, meaning one association exists. Finally, the customer might place two, three, or even more orders. The relationship for a customer placing an order can have a range of zero, one, or more, usually indicated as zero or more. The zero is the minimum cardinality and "more" is the maximum cardinality. These are referred to as "cardinality constraints."

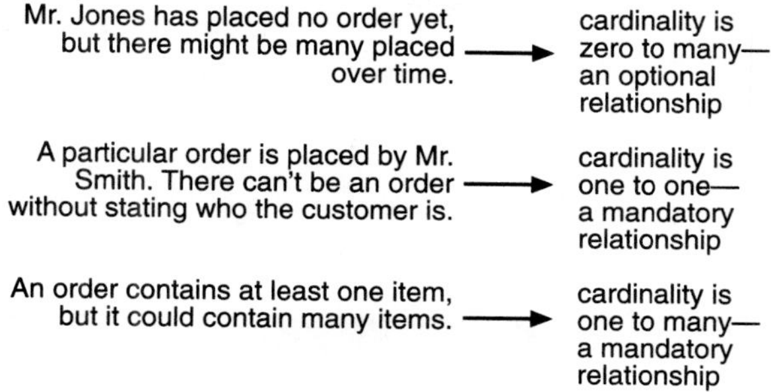

FIGURE 5-18
Cardinality of relationships.

In some cases, at least one association is required (a mandatory as opposed to optional relationship). For example, the system might not record any information about a customer until the customer places an order. Therefore, the cardinality would read "customer places one or more orders."

A one-to-one relationship can also be refined to include minimum and maximum cardinality. For example, the order is placed by one customer—it is impossible to have an order if there is no customer. Therefore, one is the minimum cardinality, making the relationship mandatory. Since there cannot be more than one customer for each order, one is also the maximum cardinality. Sometimes such a relationship is read as "an order must be placed by one and only one customer."

The relationships described here are between two different types of things—for example, a customer and an order. These are called **binary relationships**. Sometimes a relationship is between two things of the same type. For example, the relationship *is married to* is between two different people. This type of relationship is called a **unary relationship** (sometimes called a *recursive relationship*). Another example of a unary relationship is an organizational hierarchy where one organizational unit reports to another organizational unit—the packing department reports to shipping, which reports to distribution, which reports to marketing.

binary relationships
Relationships between two different types of things, such as a customer and an order

unary (recursive) relationship
A relationship between two things of the same type, such as one person being married to another person

A relationship can also be between three different types of things, called a **ternary relationship**, or any number of different types of things, called an ***n*-ary relationship**. One particular order, for example, might be associated with a specific customer plus a specific sales representative, requiring a ternary relationship.

ternary relationship
A relationship between three different types of things

***n*-ary relationship**
A relationship between *n* (any number) different types of things

Storing information about the relationships is just as important as storing information about the specific things. It is important to have information on the name and address of each customer, but it is equally important (or perhaps more so) to know what items each customer has ordered.

Attributes of Things

Most information systems store and use specific pieces of information about each thing. The specific pieces of information are called **attributes**. For example, a customer has a name, a phone number, a credit limit, and so on. Each of these is an attribute. The analyst needs to identify the attributes of each thing that the system needs to store. One attribute may be used to identify a specific thing, such as a social security number for an employee or an order number for a purchase. The attribute that uniquely identifies the thing is called an **identifier**, or **key**. Sometimes the identifier is already established (a social security number, vehicle ID number, or product ID number). Sometimes a specific identifier needs to be assigned by the system (an invoice number or transaction number).

attribute
One piece of specific information about a thing

identifier (key)
An attribute that uniquely identifies a thing

A system may need to remember many similar attributes. For example, a customer has several names—a first name, a middle name, a last name, and possibly a nickname. A **compound attribute** is an attribute that contains a collection of related attributes, so an analyst may choose one compound attribute to represent all of these names, perhaps naming it *Customer full name*. A customer might also have several phone numbers—a home phone number, office phone number, fax phone number, and cellular phone number. The analyst might start out by describing the most important attributes but later add to the list. Attribute lists can get quite long. Some examples of attributes of a customer and the values of attributes for specific customers are shown in Figure 5-19.

compound attribute
An attribute that contains a collection of related attributes

FIGURE 5-19
Attributes and values.

All customers have these attributes:	Each customer has a value for each attribute:		
Customer ID	101	102	103
First name	John	Mary	Bill
Last name	Smith	Jones	Casper
Home phone	555-9182	423-1298	874-1297
Work phone	555-3425	423-3419	874-8546

Data Entities and Objects

When describing things important to a system, so far we have mainly used examples of things the system needs to store information about. In the traditional approach to system development, these things are called **data entities**. The data entities, the relationship between data entities, and the attributes of data entities are modeled using an entity-relationship diagram (ERD). Computer processes interact with the data entities, creating them, updating attribute values, and associating one with another. The entity-relationship diagram (described in detail later) is used to create the database design, usually a relational database. The entity-relationship diagram is used with the structured approach and the information engineering approach to system development.

data entities
The things the system needs to store information about in the traditional approach to information systems

The other way to think about things is as objects that interact in the system. Objects in the work environment of the user (sometimes called objects in the problem domain) in the object-oriented approach are often similar to the data entities in the traditional approach. The main difference is that objects do the work in the system, they do not just store information. In other words, objects have behaviors as well as attributes. This simple distinction has profound effects on the way the system is viewed and constructed. But at the early stages of modeling requirements, there are few differences compared to the traditional approaches. Figure 5-20 compares the traditional and object-oriented views of data entities and objects.

With the object-oriented approach, each specific thing is an object (John, Mary, Bill) and the type of thing is called a **class** (in this case, customer). The word *class* is used because all of the objects are classified as one type of thing. The classes, associations among classes, and the attributes of classes are modeled using a class diagram. Additionally, the class diagram shows some of the behaviors of objects of the class, called *methods*.

class
The type or classification to which all similar objects belong

methods
The behaviors all objects of the class are capable of

encapsulation
Covering or protecting each object so that it contains values for attributes and methods for operating on those attributes, making the object a self-contained (and protected) unit

Methods of a class are the behaviors all objects in the class are capable of. A behavior is an action that the object processes itself. Instead of an outside process updating data values, the object updates its own values when asked to do so. To ask the object to do something, another object sends it a message. One object can send another object a message, or an object can send the user a message. The information system overall becomes a collection of interacting objects, as described in Chapter 3.

Since each object contains values for attributes and methods for operating on those attributes (plus other behaviors), an object is said to be **encapsulated** (covered or protected)—a self-contained unit. We will return to the discussion of the class diagram later in this chapter.

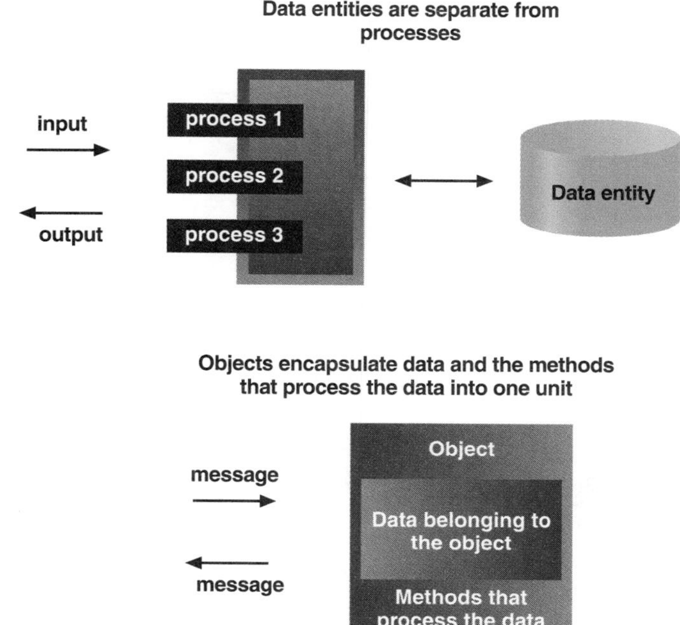

FIGURE 5-20
Data entities compared to objects.

The Entity-Relationship Diagram

The traditional approach to system development (the structured techniques and information engineering as described in Chapter 3) places a great deal of emphasis on data storage requirements for the new system. Data storage requirements include the data entities, their attributes, and the relationships among the data entities. As just discussed, the model used to define the data storage requirements is called the entity-relationship diagram (ERD).

Examples of ERD Notation

On the entity-relationship diagram, rectangles represent data entities, and the lines connecting the rectangles show the relationships among data entities. Figure 5-21 shows an example of a simplified entity-relationship diagram with two data entities, Customer and Order. Each Customer can place many Orders, and each Order is placed by one Customer. The cardinality is one to many in one direction and one to one in the other direction. The "crows feet" symbol on the line next to the Order data entity indicates "many" orders. But other symbols on the relationship line also represent the minimum and maximum cardinality constraints. See Figure 5-22 for an explanation of relationship symbols. The model in Figure 5-21 actually says that a Customer places a minimum of zero and a maximum of many Orders. Reading the other direction, the model says an Order is placed by at least one and only one Customer. This notation can express precise details about the system. The business policies that management has defined are reflected in the constraints, and the analyst must discover what these policies are. The analyst does not determine that two customers cannot share one order; management does.

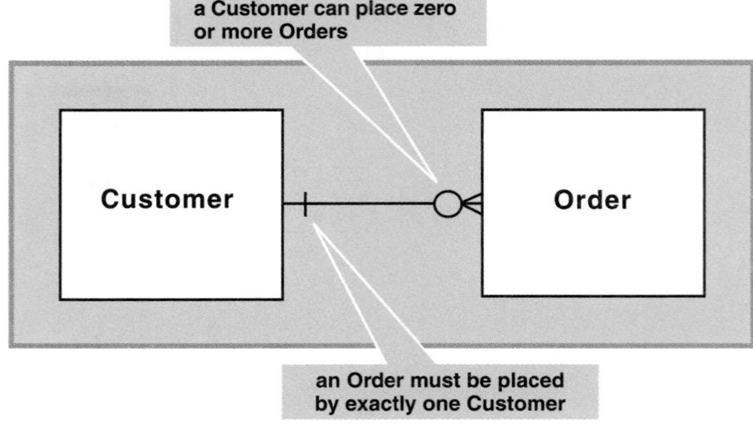

FIGURE 5-21
A simple entity-relationship diagram.

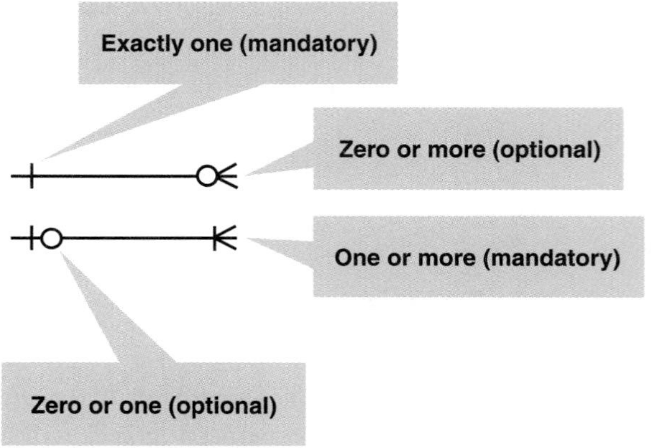

FIGURE 5-22
Cardinality symbols of relationships.

Figure 5-23 shows the model expanded to include the order items (one or more specific items included on the order). Each order contains a minimum of one and a maximum of many items. For example, an order might include a shirt, a pair of shoes, and a belt, and each of these items is associated with the order. This example also shows some of the attributes of each data entity: A customer has a customer number, a name, a billing address, and several phone numbers. Each order has an order ID, order date, and so on. Each order item has a quantity, price, and back-order status. The attributes of the data entity are listed below the name, with the key identifier listed first.

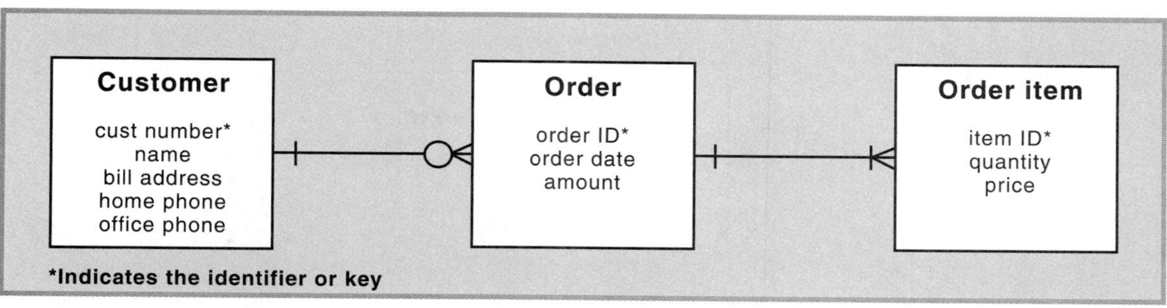

FIGURE 5-23
An expanded ERD with attributes shown.

Figure 5-24 shows how the actual data in some transactions might look. John is a customer who has placed two orders. The first order, placed on February 4, was for two shirts and one belt. The second order, placed on March 29, was for one pair of boots and two pairs of sandals. Mary is a customer who has not yet placed an order. Recall that a customer might place zero or more orders. Therefore, Mary is not associated with any orders. Finally, Sara placed an order on March 30 for three pairs of sandals.

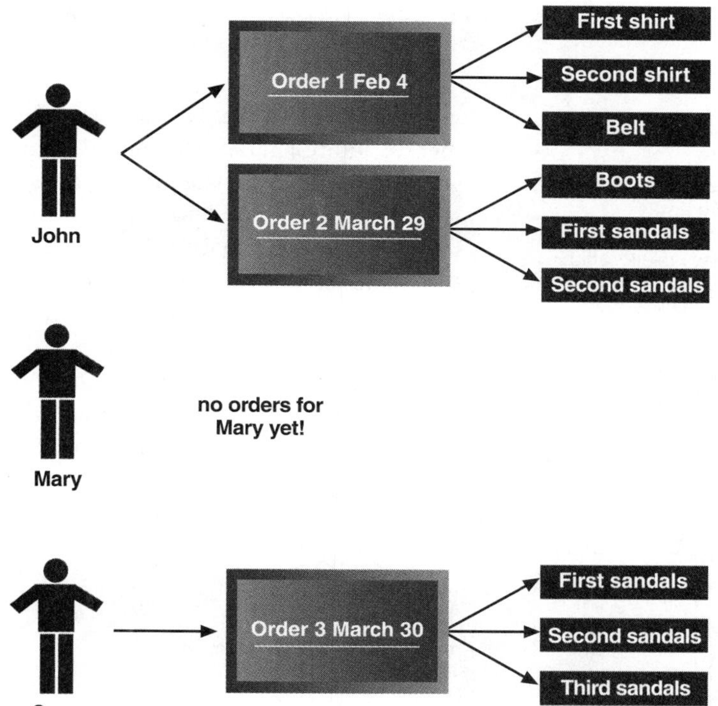

FIGURE 5-24
Customers, orders, and order items.

As an analyst works on the model, the ERD is often refined. One example of refinement is analyzing many-to-many relationships. An example of a many-to-many relationship is shown in Figure 5-25. At a university, courses are offered as course sections, and a student enrolls in many course sections. Each course section contains many students. Therefore, the relationship between course section and student is many to many. There are situations where many-to-many relationships occur naturally, and they can be modeled as shown, with "crows feet" on both ends of the relationship. If a relational database is designed from an ERD with a many-to-many relationship, a separate table containing keys from both sides of the relationship is created because relational databases cannot directly implement many-to-many relationships. Chapter 10 discusses relational databases in detail.

FIGURE 5-25
A university course enrollment ERD with a many-to-many relationship.

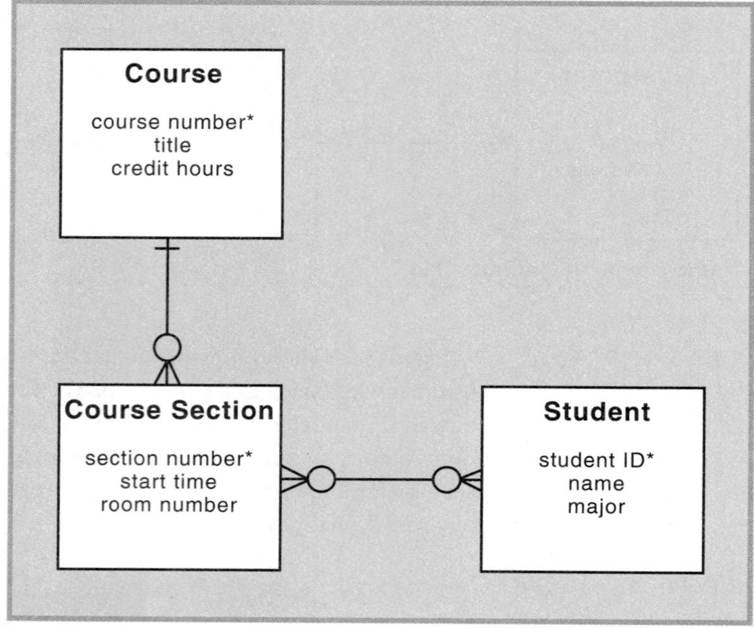

On closer analysis, however, we often discover that many-to-many relationships involve additional data that must be stored. For example, in the ERD in Figure 5-25, where is the grade each student receives for the course stored? This is important data, and although the model indicates which course section a student took, the model does not have a place for the grade. The solution is to add a data entity to represent the relationship between student and course section, sometimes called an **associative entity**. The associative entity is given the missing attribute. Figure 5-26 shows the expanded ERD with an associative entity named Course Enrollment, which has an attribute for the student's grade. Other refinements are made to the ERD during the modeling process. One major refinement process that applies to designing relational databases, called *normalization,* is discussed in Chapter 10.

associative entity
A data entity that represents a many-to-many relationship between two other data entities

FIGURE 5-26
A refined university course enrollment ERD with an associative entity.

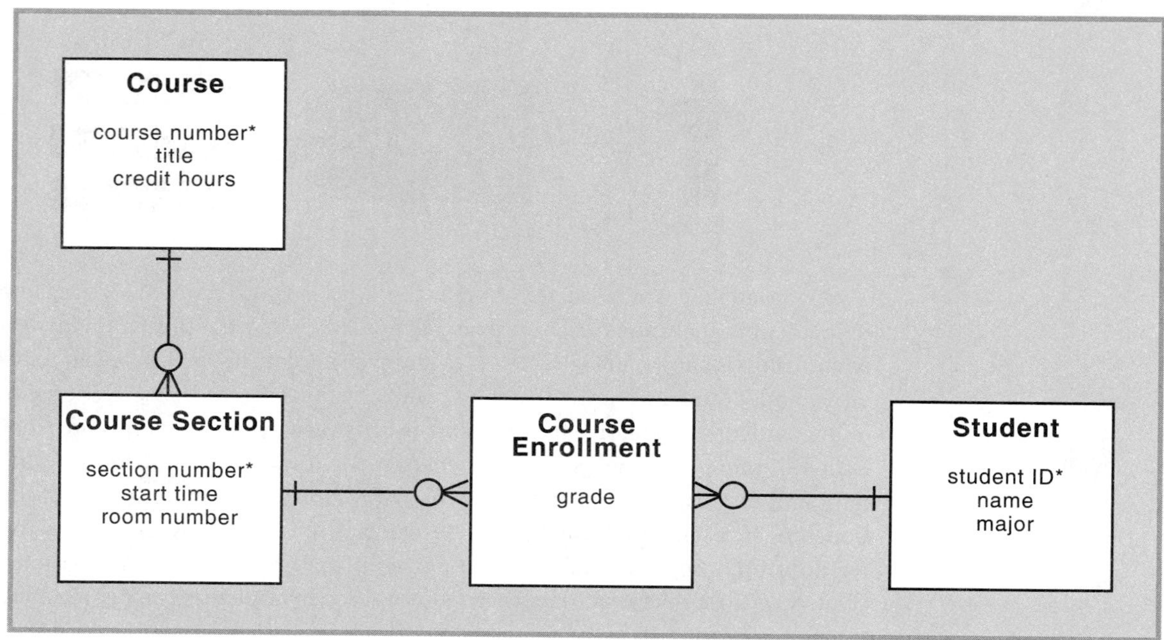

The Rocky Mountain Outfitters Case ERD

The Rocky Mountain Outfitters entity-relationship diagram is a variation of the customer and order example already described. Figure 5-27 shows a fairly complete version of the model but without the attributes.

Each customer can place zero or more orders. Each order can have one or more order items, meaning the order might be for one shirt and two sweaters. Each order item is for a specific inventory item, meaning a specific size and color of shirt. Although such an attribute is not shown on the diagram, an inventory item should have an attribute for quantity on hand of that size and color. Since there are many colors and sizes (each with its own quantity), each inventory item is associated with a product item that describes the item generically (vendor, description, season, normal price, special price). Each product item is contained in one or more catalogs.

The catalog also offers packages, such as a package that contains a shirt, pants, and belt combination at a reduced price. A package contains product items. Note that a catalog contains zero or more packages (an optional relationship), but a package must include at least one product item (a mandatory relationship). The minimum and maximum cardinality constraints can provide very specific information about the requirements.

The entity-relationship diagram for Rocky Mountain Outfitters also has information about shipments. Since this diagram includes requirements for orders, not retail sales, each order item is part of a shipment. A shipment may contain many order items. Each shipment is shipped by one shipper.

The ERD shown in Figure 5-27 contains a lot of very specific information about the requirements for the system. Be sure you can trace through all of the relationships shown and try to describe a specific example of each data entity involved in one specific order. Once it is developed, a model like this entity-relationship diagram needs to be walked through carefully, as you would walk through the logic of a program. Being able to walk through and "debug" any model is a very important skill in system development.

To test your understanding of the diagram, consider whether one order might have items shipped by different shippers. Is it possible, given the requirements shown in this diagram? The answer is yes. Operationally, some order items might be back ordered, so when they are finally shipped, they are part of a different shipment. A different shipper could handle this shipment.

Other requirements shown in the model include one or more order transactions for each order. An order transaction is a record of a payment or a refund for the order. One order transaction is created when the customer initially pays for the order. Later, though, the customer might add another item to the order, generating an additional charge. This involves a second order transaction. Finally, the customer might return an item, requiring a refund and a third order transaction.

FIGURE 5-27
Rocky Mountain Outfitters customer support system entity-relationship diagram (ERD) without attributes shown.

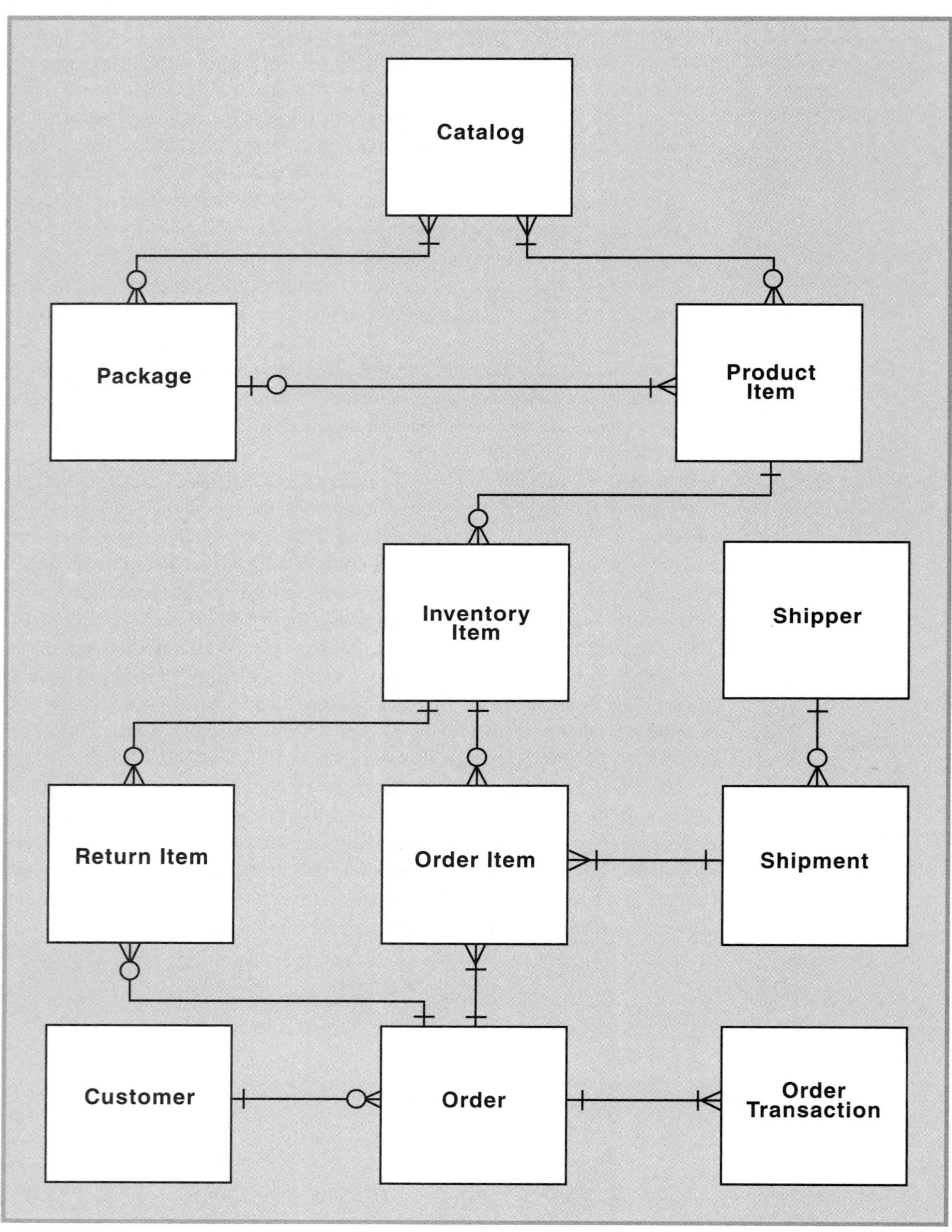

The Class Diagram

The object-oriented approach also emphasizes understanding the things involved in a system. As discussed previously, this approach models classes of objects instead of data entities. The classes of objects have attributes and associations, just like the data entities. Cardinality also applies among classes. The main difference, as already discussed, is that the objects do the actual processing in the system as well as storing information. The processing done (the behavior of the object) is possible because objects have methods as well as attributes. The sets of requirements models for the traditional approaches and object-oriented approaches eventually become quite different because of the object behavior. The design models are definitely very different.

More Complex Issues about Classes of Objects

Some other issues about things come up more frequently with the object-oriented approach than with the traditional approach, although the issues are not exclusively object-oriented. These issues are two additional ways that people structure their understanding of things in the real world: generalization/specialization and aggregation. This section discusses these concepts first before discussing the class diagram used to model things with the object-oriented approach.

generalization/specialization hierarchies
Hierarchies that structure or rank classes from the more general superclass to the more specialized subclasses; sometimes called inheritance hierarchies

Generalization/Specialization Hierarchies are based on the idea that people classify things in terms of similarities and differences. Generalizations are judgments that group similar types of things; for example, there are many types of motor vehicles—cars, trucks, and tanks. All motor vehicles share certain general characteristics, so a motor vehicle is a more general class. Specializations are judgments that categorize different types of things—for example, special types of cars include sports cars, sedans, and sport utility vehicles. These types of cars are similar in some ways, yet different in other ways. Therefore, a sports car is a special type of car.

A generalization/specialization hierarchy is used to structure or rank these things from the more general to the more special. As discussed previously, classification refers to defining classes of things. Each class of thing in the hierarchy might have a more general class above it, called a *superclass*. At the same time, a class might have a more specialized class below it, called a *subclass*. In Figure 5-28, a car has three subclasses and one superclass (Motor Vehicle).

We mentioned that people structure their understanding by using generalization/specialization hierarchies. That is, people learn by refining the classifications they can make about some field of knowledge. A knowledgeable banker can talk at length about special types of loans and types of deposit accounts. A knowledgeable merchandiser like John Blankens at Rocky Mountain Outfitters can talk at length about special types of outdoor activities and types of clothes. Therefore, when asking users about their work, the analyst is trying to understand the knowledge the user has about the work, which the analyst can represent by constructing generalization/specialization hierarchies. At some level, the motivation for the new customer support system at RMO started with John's recognition that Rocky Mountain Outfitters might handle many special types of orders with a new system (web orders, telephone orders, and mail orders). These special types of orders are shown in Figure 5-29.

inheritance
A concept that allows subclasses to share characteristics of their superclasses

Inheritance allows subclasses to share characteristics of their superclasses. Returning to Figure 5-28, a car is everything any other motor vehicle is, but something special. A sports car is everything any other car is, plus something special. In this way, the subclass "inherits" characteristics. In the object-oriented approach, inheritance is a key concept that is possible because of generalization/specialization hierarchies. Sometimes these hierarchies are referred to as *inheritance hierarchies*.

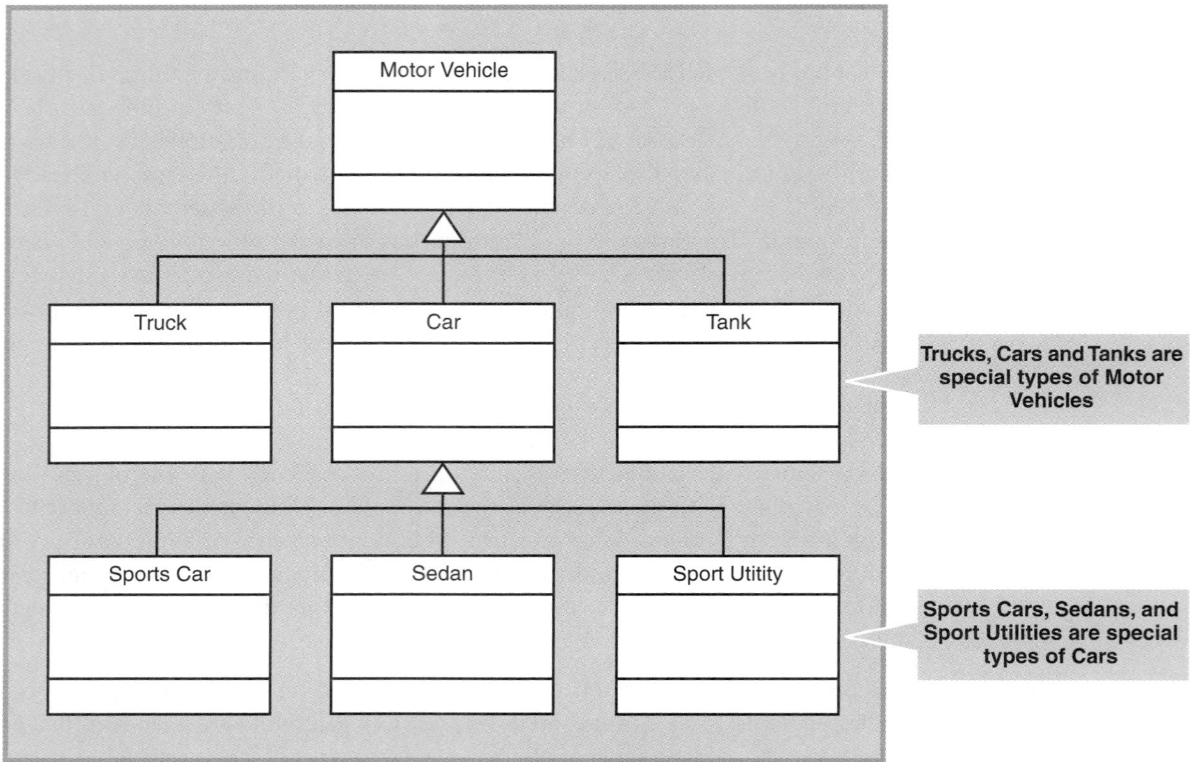

Motor Vehicle
Truck
Car
Tank
Trucks, Cars and Tanks are special types of Motor Vehicles
Sports Car
Sedan
Sport Utitity
Sports Cars, Sedans, and Sport Utilities are special types of Cars

FIGURE 5-28
A generalization/specialization hierarchy for motor vehicles.

aggregation
A relationship between an object and its parts

Aggregation (Whole-Part Hierarchies) Another way that people structure information about things is by defining them in terms of their parts. For example, learning about a computer system might involve recognizing that the computer is composed of different parts—processor, main memory, keyboard, disk storage, monitor. The term **aggregation** is used to describe a relationship of parts. A keyboard is not a special type of computer; it is part of a computer. Yet, it is also something entirely separate in its own right. Aggregation is much like any other relationship among types of things, except it is a much stronger relationship. Figure 5-30 demonstrates the concept of aggregation in a computer system. Figure 5-30 demonstrates the concept of aggregation in a computer system, showing the diamond symbol to represent aggregation.

Examples of Class Diagram Notation

As mentioned previously, the class diagram is the name for the model used to show classes of objects for a system. The notation is based on the Unified Modeling Language (UML), which has become the *de facto* standard for models used with object-oriented system development. This section introduces the notation for the class diagram. This notation has been used previously in Figures 5-28, 5-29, and 5-30.

Figure 5-31 shows an example of a symbol for one class: Customer. The class symbol is a rectangle with three sections. The top section contains the name of the class, the middle section lists the attributes of the class, and the bottom section lists the important methods of the class. Methods are not always shown in the class symbol if they are fairly standard. It can usually be assumed that a new customer can be added, deleted, updated, and connected to an account.

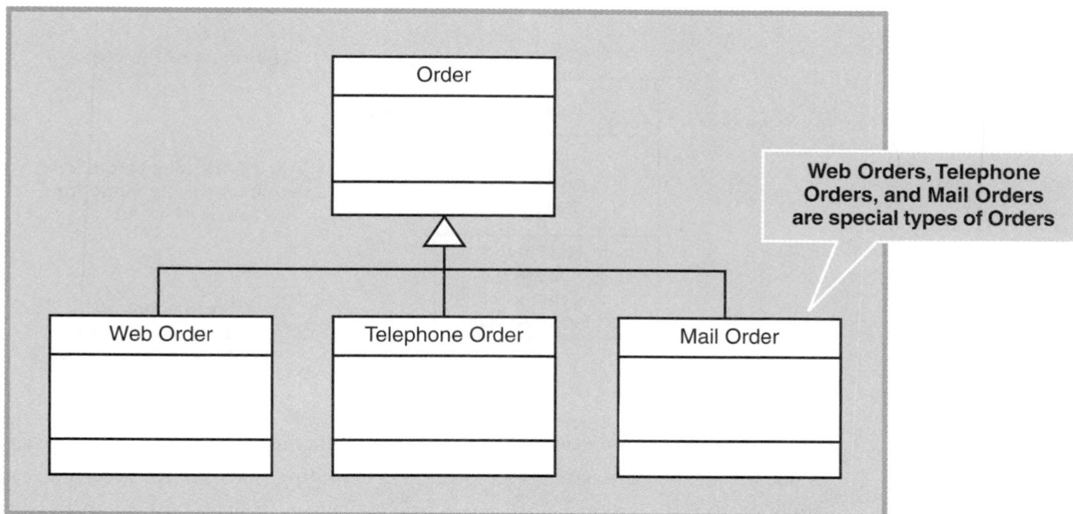
Order
Web Orders, Telephone Orders, and Mail Orders are special types of Orders
Web Order
Telephone Order
Mail Order

FIGURE 5-29
A generalization/specialization hierarchy for orders.

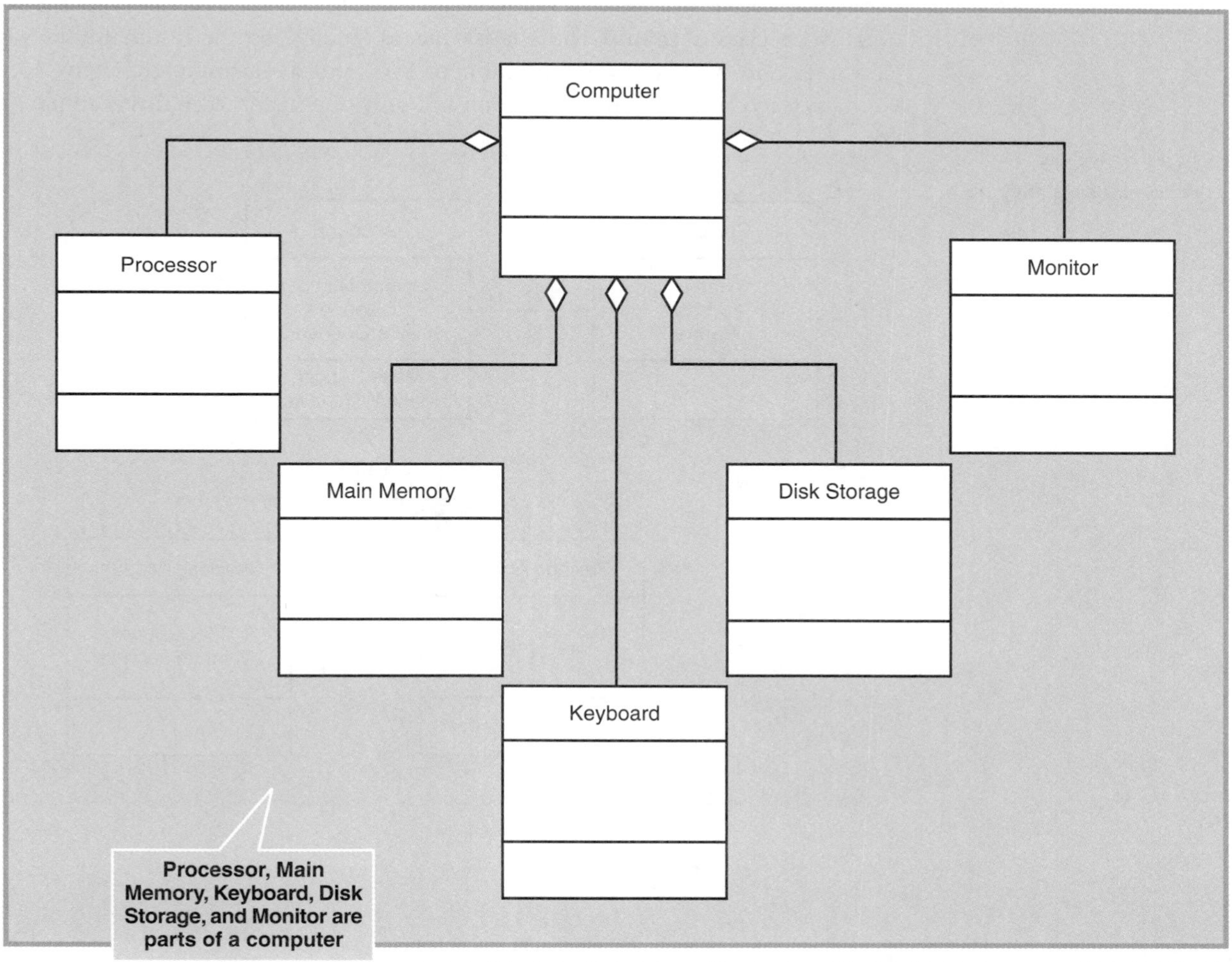

FIGURE 5-30
Aggregation or whole-part relationships.

Figure 5-32 shows an expanded example of a class diagram for a system that maintains bank accounts. Here the methods of the customer class are not shown. The account class has a few methods listed because they are unique to bank accounts and are central to the processing of the system. They include Make Deposit and Make Withdrawal. Sometimes even these methods would not be shown on the class diagram initially. Many analysts wait before worrying about object behavior (we will wait until Chapter 7).

Computer
Processor
Monitor
Main Memory
Disk Storage
Keyboard
Processor, Main Memory, Keyboard, Disk Storage, and Monitor are parts of a computer

F I G U R E 5 - 3 1
The class symbol for the class diagram.

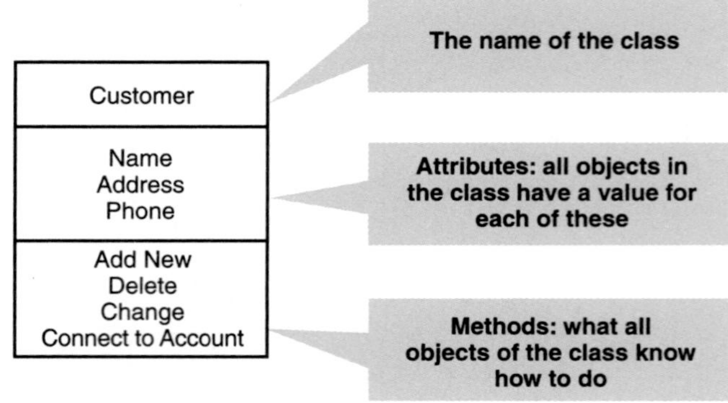

The bank account system includes a generalization/specialization hierarchy: Account is the superclass, and Savings Account and Checking Account are two subclasses. The triangle symbol drawn on a line connecting the classes indicates inheritance. The subclasses inherit attributes and behaviors from the superclass. Therefore, a Checking Account inherits the two methods plus all of the attributes from Account. Similarly, Savings Account inherits the same methods and attributes. But Savings Account also knows how to calculate interest; Checking Account does not. The result is that some attributes and methods are common to both types of accounts, but other attributes and methods are not.

The Customer class and the Account class can be associated, just as they would be in an entity-relationship diagram. Each customer can have zero or more accounts. Note that the diagram indicates the minimum and maximum cardinality on the line connecting the classes. (Recall that the term *multiplicity* is used as a synonym for cardinality on class diagram.) The asterisk means "many," so the cardinality between Customer and Account is a minimum of zero and a maximum of many (0..*). Similarly, each account is owned by one and only one customer in this example.

F I G U R E 5 - 3 2
A bank account class diagram.

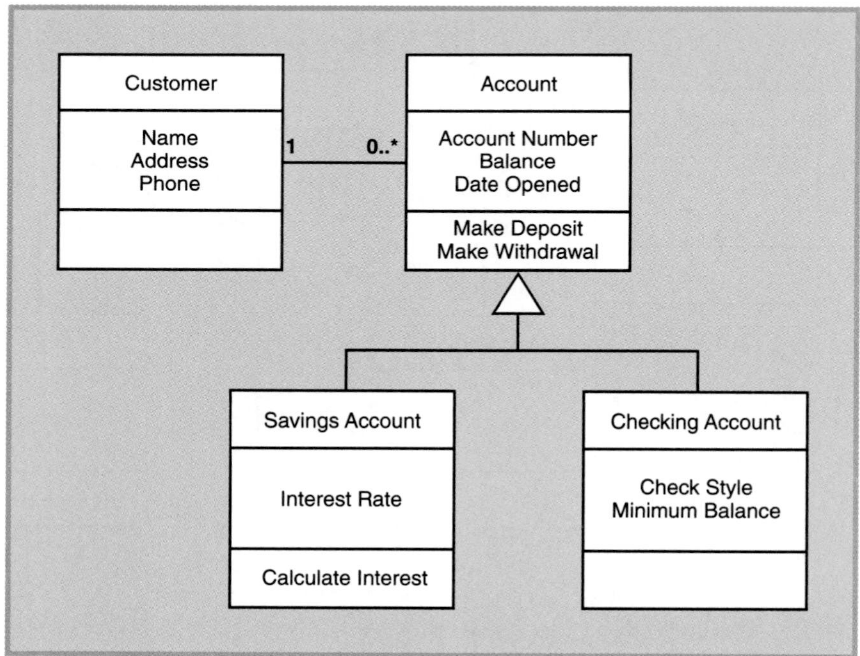

The Rocky Mountain Outfitters Case Class Diagram

The class diagram for Rocky Mountain Outfitters is shown in Figure 5-33. As you can see, it is very similar to the entity-relationship diagram shown previously in Figure 5-27.

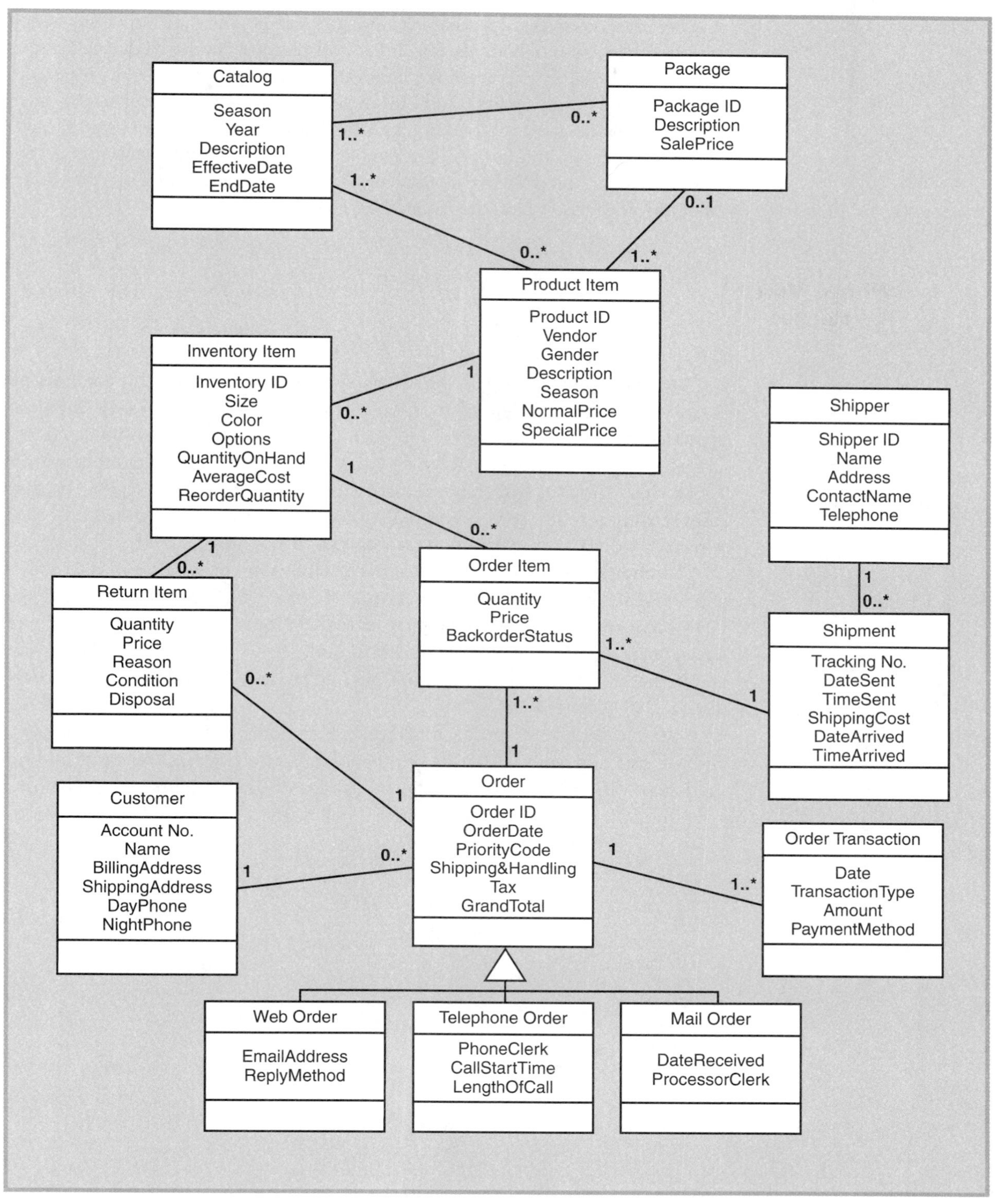

FIGURE 5-33
Rocky Mountain Outfitters class diagram.

Note that the main attributes are shown here, and they are often shown in the entity-relationship diagram also. A generalization/specialization hierarchy is included to show that an order can be any one of three types—web order, telephone order, and mail order—as discussed previously. Note that all types of orders share the attributes listed for Order, but each special type of order has some additional attributes.

The other classes and associations among classes are the same as before. Note that this diagram shows no methods. The initial class diagram developed during the systems analysis phase of the project would not include methods. As behaviors of objects are further developed during analysis and during design, methods are added to the diagram. For now, remember that objects of each class have behaviors as well as attributes. For example, remember that the Shipper class knows how to create a new shipper object, delete a shipper object, change a name or address, and connect to a specific shipment. All of the other classes have these standard capabilities also.

Where You Are Headed

The requirements models for a new system created during the analysis phase become quite different depending upon whether the project team uses the traditional approach or the object-oriented approach. The two key concepts discussed in this chapter—events and things—are the starting places in the modeling process for both approaches. The next two chapters discuss these two approaches separately, in both cases starting with the same preliminary information. Figure 5-34 shows how the two approaches diverge once the events and things are identified.

The traditional approach takes the event table and creates a set of data flow diagrams (DFDs) based on the information in the table, including the context diagram, DFD fragments, and detailed DFDs. The entity-relationship diagram (ERD) defines the data storage requirements that are included in the DFDs. Other information about the requirements includes data flow definitions and process descriptions. These models are discussed in Chapter 6.

The object-oriented approach takes the event table and creates a set of use cases and use case diagrams. Use cases and the class diagram are used to create additional models of object behavior, including sequence diagrams, statechart diagrams, and other models. These models are discussed in Chapter 7.

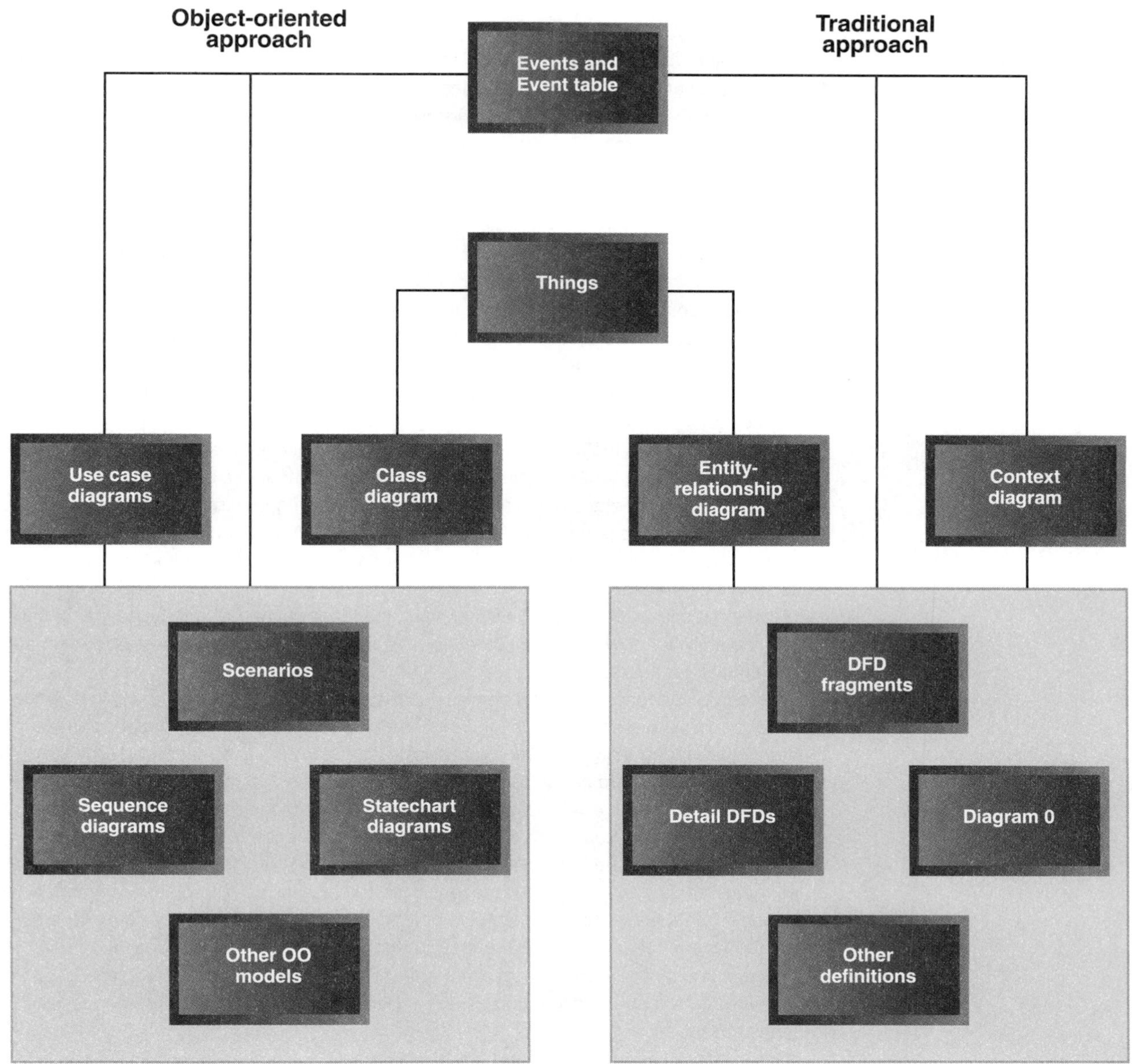

FIGURE 5-34
Requirements models for the traditional approach and the object-oriented approach.

Summary

This chapter is the first of three chapters that present techniques for modeling a system's functional requirements, highlighting the tasks that are completed during the analysis phase activity named *Define system requirements*. Models of various aspects of the system are created to document the requirements. Events and things in the user's work environment are key concepts common to all approaches to system development. The traditional approach uses entity-relationship diagrams (ERD), and the object-oriented approach uses class diagrams as key models.

Models are useful for many reasons, including the learning process that occurs while creating the models, reducing complexity, documenting the many details that need to be remembered, communicating with team members and users, and documenting system requirements for future use during system support. Many types of models are used, including mathematical models, descriptive models, and graphical models. Many of the models described in this text are graphical models, some used during analysis and some used during design.

A key early step in the modeling process is to identify and list the events that occur that require a response from the system. An event is something that can be described that occurs at a specific time and place and is worth remembering. External events occur outside the system, usually triggered by someone who interacts with the system. Temporal events occur at a defined point in time, such as the end of a work day or the end of every month. State events occur based on an internal system change. Information about each event is recorded in an event table, which lists the event, the trigger for the event, the source of the trigger, the activity the system must carry out, the response produced as system output, and the destination for the response.

The other key concept involves the things the users deal with when doing their work, such as products, orders, invoices, and customers. There are many naturally occurring relationships among things the user works with: A customer places an order, and an order requires an invoice. Cardinality (or multiplicity) of a relationship refers to the number of associations involved in a relationship: A customer might place many orders, and each order is placed by one customer. Attributes are specific pieces of information about a thing, such as a name and an address for a customer. The traditional approach models these things as data entities that represent data that are stored. The object-oriented approach models these things as objects belonging to a class, which have attributes as well as behaviors (called methods). Although the two approaches to development use the same basic concept to identify these data entities or objects, the way objects are used in the object-oriented approach is quite different from the traditional approach because of the object behavior.

The traditional approach uses the entity-relationship diagram to show data entities, attributes of data entities, and relationships. The object-oriented approach uses the class diagram to show the classes, attributes, and methods of the class, and associations among classes. Two additional concepts are used in class diagrams (although they are sometimes used in entity-relationship diagrams, too): generalization/specialization hierarchies, which allow inheritance from a superclass to a subclass, and aggregation, which allows a collection of objects to be associated as a whole and its parts.

The next two chapters discuss requirements models produced by the traditional approach and the object-oriented approach separately.

Key Terms

activity, p. 141
aggregation, p. 154
associative entity, p. 150
attribute, p. 145
binary relationships, p. 145
cardinality, p. 144
class, p. 146
compound attribute, p. 146
data entities, p. 146
descriptive model, p. 130
destination, p. 141
encapsulation, p. 147
event, p. 133
event table, p. 140
external event, p. 135
generalization/specialization hierarchies, p. 153
graphical model, p. 131
identifier (key), p. 145
inheritance, p. 153
mathematical model, p. 130
methods, p. 147
multiplicity, p. 144
***n*-ary relationship,** p. 145
perfect technology assumption, p. 138
relationship, p. 144
response, p. 141
source, p. 141
state event, p. 136
system controls, p. 138
temporal event, p. 135
ternary relationship, p. 145
trigger, p. 140
unary (recursive) relationship, p. 145

Review Questions

1. What are some of the reasons for creating models during system development?
2. What are three types of models?
3. What are the two key concepts used to begin defining system requirements?
4. What is an event?
5. What are the three types of events?
6. Which type of event results in data entering the system?
7. Which type of event occurs at a defined point in time?
8. Which type of event does not result in data entering the system but always results in an output?
9. What type of event would be named *Employee quits job*?
10. What type of event would be named *Time to produce paychecks*?
11. What are some examples of system controls?
12. What does the perfect technology assumption state?
13. What are the columns in an event table?
14. What is a trigger? A source? An activity? A response? A destination?
15. What is a "thing" called in models used in the traditional approach?
16. What is a "thing" called in the object-oriented approach?
17. What is a relationship?
18. What is cardinality of a relationship (sometimes called multiplicity)?
19. Describe how the minimum and maximum cardinality are shown in an entity-relationship diagram.
20. What are unary, binary, and *n*-ary relationships?
21. What are attributes and compound attributes?
22. What is an associative entity?
23. What are the symbols shown in an entity-relationship diagram?
24. What are the symbols shown in a class diagram?
25. What is encapsulated along with the values of attributes in an object?
26. What is a generalization/specialization hierarchy?
27. From what type of class do subclasses inherit?
28. What type of relationship shows an object and its parts?
29. What three pieces of information about a class are put in the three parts of the class symbol?
30. What does the triangle symbol indicate on a line connecting classes on the class diagram?

Thinking Critically

1. Provide an example of each of the three types of models that might apply to designing a car, to designing a house, and to designing an information system.
2. Explain why requirements models are logical models rather than physical models.
3. Explain why the event concept did not initially receive much attention in business systems analysis.
4. Review the external event checklist in Figure 5-7 and think about a university course registration system. What is an example of an event of each type in the checklist? Name each event using the guidelines for naming an external event.
5. Review the temporal event checklist in Figure 5-8. Would a student grade report be an internal or external output? Would a class list for the instructor be an internal or external output? What are some other internal and external outputs for a course registration system? Using the guidelines for naming temporal events, what would you name the events that trigger these outputs?
6. In a course registration system, for the event *Student registers for classes*, create an event table entry listing the event, trigger, source, activity, response(s), and destination(s). For the event *Time to produce grade reports*, create another event table entry.
7. Consider the following sequence of actions taken by a customer at a bank. Which action is the event the analyst should define for a bank account transaction-processing system? (1) Kevin gets a check from grandma for his birthday. (2) Kevin wants a car. (3) Kevin decides to save his money. (4) Kevin goes to the bank. (5) Kevin waits in line. (6) Kevin makes a deposit in his savings account. (7) Kevin grabs the deposit receipt. (8) Kevin asks for a brochure on auto loans.
8. Consider the perfect technology assumption, which states that events should be included during analysis only if the system would be required to respond under perfect conditions. Could any of the events in the event table for Rocky Mountain Outfitters be eliminated based on this assumption? Explain. Why are events such as *User logs on to system* and *Time to back up the data* required only under imperfect conditions?
9. Draw an entity-relationship diagram including minimum and maximum cardinality for the following: The system stores information about two things—cars and owners. A car has attributes for make, model, and year. The owner has attributes for name and address. Assume a car must be owned by one owner, and an owner can own many cars, but an owner might not own any cars (perhaps she just sold them all, but you still want a record of her in the system).
10. Draw a class diagram for the cars and owners described in exercise 9 above but include subclasses for sports car, sedan, and minivan with appropriate attributes.
11. Consider the entity-relationship diagram shown in Figure 5-26, the refined ERD showing course enrollment with an associative entity. Does this model allow a student to enroll in more than one course section at a time? Does the model allow a course section to contain more than one student? Does the model allow a student to enroll in several sections of the same course and get a grade for each enrollment? Does the model store information about all grades earned by all students in all sections?
12. Again consider the entity-relationship diagram shown in Figure 5-26. Add the following to the diagram and list any assumptions you had to make. A faculty member usually teaches many course sections, but some semesters a faculty member may not teach any. Each course section must have at least one faculty member teaching it, but sometimes course sections are taught by teams. Further, to make sure all course sections are similar, one faculty member is assigned as course coordinator to oversee the course, and each faculty member can be coordinator of many courses.
13. If the entity-relationship diagram you drew in exercise 12 showed a many-to-many relationship between faculty member and course section, a further look at the relationship might reveal the need to store some additional information. What might this information include? (Hint: Does the instructor have specific office hours for each course section? Do you give an instructor some sort of evaluation for each course section?) Expand the ERD to allow this additional information to be stored.
14. Draw a class diagram for the course enrollment system completed in exercise 13.
15. Consider a system that needs to store information about computers in a computer lab at a university, such as the features and location of each computer. What are the things that might be included in a model? What are some of the relationships among these things? What are some of the attributes of these things? Draw an entity-relationship model for this system.
16. Draw a class diagram for the computer lab system described in exercise 15.
17. Refer to the class diagram for bank accounts in Figure 5-32. Expand the model to show that there

are special types of customers—personal and commercial. All customers have a name and mailing address. Commercial customers have additional attributes for credit rating, contact person, and contact person phone. Personal customers have attributes for home phone and work phone. Additionally, expand the model to show that the bank has multiple branches, and each account is serviced by one branch. Naturally, each branch has many accounts.

18. Consider the class diagram for Rocky Mountain Outfitters shown in Figure 5-33. If a web order is created, how many attributes does it have? If a telephone order is created, how many attributes does it have? If an existing customer places a phone order for one item, how many new objects are created overall for this transaction?
19. A product item for RMO is not the same as an inventory item. A product item is something like a men's leather hunting jacket supplied by Leather 'R' Us. An inventory item is a specific size and color of the jacket—like a size medium brown leather hunting jacket. If RMO adds a new jacket to its catalog, and there are six sizes and three colors available in inventory, how many objects need to be added overall?
20. Consider the following class diagram showing college, department, and faculty member.

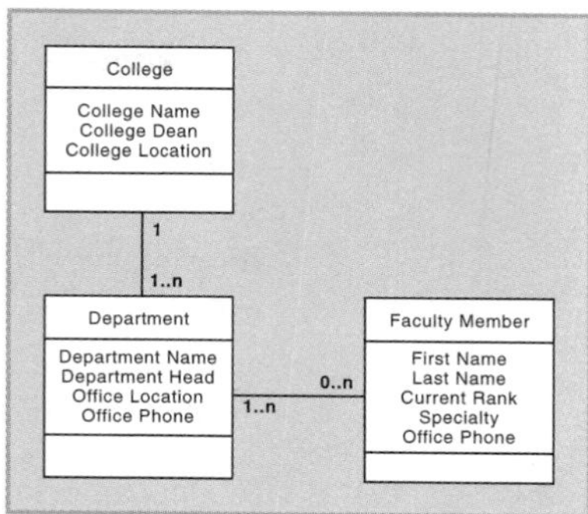

   a. What kind of relationships are shown in the model?
   b. How many attributes does a "faculty member" have? Which (if any) have been inherited from another class?
   c. If you add information about one college, one department, and four faculty members, how many objects are added to the system?
   d. Can a faculty member work in more than one department at the same time? Explain.
   e. Can a faculty member work in two departments at the same time, where one department is in the college of business and the other department is in the college of arts and sciences? Explain.

Experiential Exercises

1. Set up a meeting with a librarian. During your meeting, ask the librarian to describe the situations that come up in the library to which the book checkout system needs to respond. List these external events. Now ask about points in time, or deadlines, that require the system to produce a statement, notice, report, or other output. List these temporal events. Does it seem natural for the librarian to describe the system in this way? Similarly, ask the librarian to describe the things the system needs to store information about. See whether you can get the librarian to list the important attributes and describe relationships among things. Does it seem natural for the librarian to describe these things? Create either an ERD or a class diagram based on what you learn.
2. Visit a restaurant or the college foodservice and talk to a server (or talk with a friend who is a food server). Ask about the external events, temporal events, and data entities or objects, as you did in exercise 1. What are the events for order processing at a restaurant? Complete an event table and either an ERD or class diagram.
3. Review the procedures for course registration at your university and talk with the staff in advising, in registration, and in your major department. Think about the sequence that goes on over an entire semester. What are the events that students trigger? What are the events that your major department triggers? What are the temporal events that result in information going to students? What are the temporal events that result in information going to instructors or departments?
4. Again review information about your own university. Create generalization/specialization hierarchies using the class diagram notation for (1) types of faculty, (2) types of students, (3) types of courses, (4) types of financial aid, and (5) types of housing. Include attributes for the superclass and the subclasses in each case.

Case Studies

The Spring Breaks 'R' Us Travel Service Booking System

Spring Breaks 'R' Us Travel Service (SBRU) books spring break trips at resorts for college students. During the fall, resorts submit availability information to SBRU indicating rooms, room capacity, and room rates for each week of the spring break season. Each resort offers bookings for a different number of weeks each season, and rooms have different rates depending on the week. A variety of rooms with different capacities are usually made available by the resorts so students can book the right size room. Couples can book a two-person room, for example, and four people can book a room for four.

In December, SBRU generates a list of resorts, available weeks, and room rates that is distributed to college campus representatives all over the country. When a group of students submits a reservation request for a week at a particular resort, SBRU assigns the students to a room with sufficient capacity and sends each student a confirmation notice. When the cutoff date for a spring break week arrives, SBRU sends each resort a list of students booked in each room for the following week. When the students arrive at the resort, they pay the resort directly for the room. Resorts send commission checks directly to the SBRU accounting system, which is separate from the booking system. When spring break is over, students return to their schools and hit the books.

1. What events must the SBRU booking system respond to? Create a complete event table listing the event, trigger, source, activity, response, and destination for each event. Be sure to consider only the events that trigger processing in the booking system, not the SBRU accounting system or the systems operated by the resorts.

2. List the data entities (or classes) that are mentioned. List the attributes of each data entity (or class). List the relationships among data entities (or classes).
3. Which classes might be refined into a generalization/specialization hierarchy? List the superclass and any subclasses for each of them.

The Real Estate Multiple Listing Service System

The Real Estate Multiple Listing Service system supplies information that local real estate agents use to help them sell houses to their customers. During the month, agents list houses for sale (listings) by contracting with homeowners. The agent works for a real estate office, which sends information on the listing to the multiple listing service. Therefore, any agent in the community can get information on the listing.

Information on a listing includes the address, year built, square feet, number of bedrooms, number of bathrooms, owner name, owner phone number, asking price, and status code. At any time during the month, an agent might directly request information on listings that match customer requirements, so the agent contacts the multiple listing service with the request. Information on the house, on the agent who listed the house, and on the real estate office the agent works for is provided. For example, an agent might want to call the listing agent to ask additional questions or call the homeowner directly to make an appointment to show the house. Twice each month (on the 15th and 30th), the multiple listing service produces a listing book that contains information on all listings. These books are sent to all of the real estate agents. Many real estate agents want the books (which are easier to flip through), so they are provided even though the information is often out of date. Sometimes agents and owners decide to change information about a listing, such as reducing the price, correcting previous information on the house, or indicating that the house is sold. The real estate office sends in these change requests to the multiple listing service when the agent asks the office to do so.

1. To what events must the multiple listing service system respond? Create a complete event table listing the event, trigger, source, activity, response, and destination for each event.

2. Draw an entity-relationship diagram to represent the data storage requirements for the multiple listing service system, including the attributes mentioned. Does your model include data entities for offer, buyer, and closing? If so, reconsider. Include information the multiple listing service needs to store, which might be different from information the real estate office needs to store.

3. Draw a class diagram that corresponds to the ERD but shows that different types of listings have different attributes. The description in the case assumes all listings are for single-family houses. What about multifamily listings or commercial property listings?

The State Patrol Ticket Processing System

The purpose of the State Patrol ticket processing system is to record driver violations, keep records of the fines paid by drivers when they plead guilty or are found guilty of moving violations by the courts, and to notify the court that a warrant for arrest should be issued when such fines are not paid in a timely manner. A separate State Patrol system records accidents and verification of financial responsibility (insurance). Yet a third system produces driving record reports from the ticket and accident records for insurance companies. Finally, a fourth system issues, renews, or suspends driver's licenses. These four systems are obviously integrated in that they share access to the same database, but otherwise, they are operated separately by different departments of the State Patrol. State Patrol operations (what the officers do) are entirely separate.

The portion of the database used with the ticket processing system involves driver data, ticket data, officer data, and court data. Driver data, officer data, and court data are used by the system. The system creates and maintains ticket data. Driver attributes include license number, name, address, date of birth, date licensed, and so on. Ticket attributes include ticket number (each is unique and preprinted on each sheet of the officer's ticket book), location, ticket type, ticket date, ticket time, plea, trial date, verdict, fine amount, and date paid. Court and officer data include the name and address of each respectively. Each driver may have zero or more tickets, and each ticket applies to only one driver. Officers write quite a few tickets.

When an officer gives a ticket to a driver, a copy of the ticket is turned in and entered into the system. A new ticket record is created and relationships to the correct driver, officer, and court are established in the database. If the driver pleads guilty, he or she mails in the fine in a preprinted envelope with the ticket number on it. In some cases, the driver claims innocence and wants a court date. When the envelope is returned without a check and the trial request box has an "X" in it, the system notes the plea on the ticket record, looks up driver, ticket, and officer information, and sends a ticket details report to the appropriate court. A trial date questionnaire form is also produced at the same time and is mailed to the driver. The instructions on the questionnaire tell the driver to fill in convenient dates and mail the questionnaire directly to the court. When the court receives this information, it schedules a trial date and notifies the driver of the date and time.

When the trial is completed, the court sends the verdict to the ticketing system. The verdict and trial date are recorded for the ticket. If the verdict is innocent, the system that produces driving record reports for insurance companies will ignore the ticket. If the verdict is guilty, the court gives the driver another envelope with the ticket number on it for mailing in the fine.

If the driver fails to pay the fine within the required period, the ticket processing system produces a warrant request notice and sends it to the court. This happens if the original envelope is not returned within two weeks or if the court-supplied envelope is not returned within two weeks of the trial date. What happens then is in the hands of the court. Sometimes the court requests that the driver's license be suspended, and the system that processes driver's licenses handles the suspension.

1. To what events must the ticket processing system respond? Create a complete event table listing the event, trigger, source, activity, response, and destination for each event.
2. Draw an entity-relationship diagram to represent the data storage requirements for the ticket processing system, including the attributes mentioned. Explain why it is important to understand how the system is integrated with other State Patrol systems.
3. Draw a class diagram that corresponds to the ERD but assumes there are different types of drivers. Classifications of types of drivers vary by state. Some states have restricted licenses for minors, for example, and special licenses for commercial vehicle operators. Research your state requirements and create a generalization/specialization hierarchy for the class Driver, showing the different attributes each special type of driver might have. Consider the same issues for types of tickets. Include some special types of tickets in a generalization/specialization hierarchy in the class diagram.

Further Resources

Some classic and more recent texts include the following:

Events and event analysis:

Stephen McMenamin and John Palmer. *Essential Systems Analysis*. Prentice Hall, 1984.

Ed Yourdon. *Modern Structured Analysis*. Prentice Hall, 1989.

Data modeling, entity-relationship diagrams, and database management:

Peter Rob and Carlos Coronel. *Database Systems: Design Implementation, and Management* (4th ed.). Course Technology, 1999.

Objects, object behavior, and the class diagram:

Grady Booch, Ivar Jacobson, and James Rumbaugh. *The Unified Modeling Language User Guide*. Addison-Wesley, 1999.

Peter Coad, David North, and Mark Mayfield. *Object Models, Strategies, Patterns, and Applications* (2nd ed.). Prentice Hall, 1997.

SIX

The Traditional Approach to Requirements

Learning Objectives

After reading this chapter, you should be able to:

- Explain how the traditional approach and the object-oriented approach differ when an event occurs
- List the components of a traditional system and the symbols representing them on a data flow diagram
- Describe how data flow diagrams can show the system at various levels of abstraction
- Develop data flow diagrams, data element definitions, data store definitions, and process descriptions
- Develop tables to show the distribution of processing and data access across system locations
- Read and interpret additional models that can be incorporated within traditional structured analysis, including process decomposition and workflow diagrams

Chapter Outline

PERFORMANCE AUTO ELECTRIC: FOLLOWING THE DATA FLOW

Sarah Conrad and Jim Robles were meeting to review a first draft of several data flow diagrams for Performance Auto Electric's new billing system. Sarah was the analyst assigned to define the new system requirements. Jim was the manager in charge of accounts payable, and he knew virtually all there was to know about how the current system operated. The two had met several times before. Their most recent meeting (just last week) reviewed details of the current system events, transaction processing, and process participants. Sarah left that meeting with many pages of notes and a large number of current system forms and reports. Sarah had telephoned Jim several times since that meeting to ask additional questions.

Sarah began the review by saying, "The materials and information that you gave me last week took me quite a while to absorb, but I think I was able to understand and document all of the system functions that we discussed. The purpose of this meeting is to review the processing requirements that I've written down and to ensure that they're complete and accurate. So let's start with a few of the diagrams that I've created."

Sarah laid out three of the diagrams on the table. Jim looked at them briefly and said, "I've never seen diagrams like this before—they look like strategic plans for playing a game of marbles. And I thought those entity relationship diagrams are strange!" Sarah replied, "I expect this meeting to be a little rough since this is your first look at this style of documentation. I'll explain how to interpret the diagrams as we go along. Please feel free to stop me with a question at any time. The quality of our work depends on your understanding the diagrams, so don't be shy."

Sarah continued, "These are called data flow diagrams. They divide your system into a set of processing functions. The processes are the rectangles with the rounded-off corners. The arrows show data flowing among processes and between processes and files." Jim pointed to a square on one of the diagrams and said, "I assume that this represents an auto repair shop?" Sarah replied, "Yes, the squares represent people or organizations that supply inputs or expect output data from the system."

Jim said, "I think I can get the hang of this. I recognize most of the names that you've used for the processes and data. I'm not sure what these other symbols are—they're named for things that we store in our manual files and database, but they don't seem to correspond exactly to our system." "They don't," Sarah replied. "They're entities from the entity-relationship diagram that we developed a couple of weeks ago. But let's skip over those for the moment. Why don't we walk through the processing sequence for generating a billing statement, and we'll discuss the entities as we get to them?"

Sarah and Jim continued reviewing the DFDs, and the next thing they knew, over an hour had passed. Several pages of Sarah's notepad had been filled, and there were 25 corrections and comments noted in red on the data flow diagrams. Jim said, "My brain feels completely drained. I don't think that I can do any more of this today." Sarah replied, "You've given me plenty of things to work on, so let's call it quits for now. Can we meet for two hours at nine o'clock on Thursday?" Jim replied, "Yes, I'm free then. So will you be bringing more data flow diagrams or do you have something even weirder up your sleeve?" Sarah smiled and said, "The toughest stuff is behind us. But I do have a few more surprises for you."

OVERVIEW

Chapter 5 described two key concepts associated with modeling system requirements in both the traditional and the object-oriented approaches to information systems development: events and things. In this chapter, the focus turns to what the system does when an event occurs: activities and interactions.

This chapter describes the traditional structured approach to representing activities and interactions. We describe and present the diagrams and other models that are used in the traditional approach. We also provide examples of these models from the Rocky Mountain Outfitters customer support system to show how each model is related in the traditional approach. Also, we briefly discuss how the structured and Information Engineering approaches and their models can be used together to describe a system. Details of the object-oriented approach to representing activities and interactions are described in Chapter 7.

Traditional and Object-Oriented Views of Activities

The traditional and object-oriented approaches differ in what happens when an event occurs. A key question about system requirements is, What does the system do in response to an event? The traditional approach differs from the object-oriented approach in the way the system is modeled and implemented. The traditional approach views a system as a collection of processes, some done by people and some done by computers. Computer processes are much like conventional computer programs—they contain instructions that execute in a sequence. When the process executes, it interacts with data, reading data values and then writing data values back to the data file. The process might also interact with people, such as when an instruction asks the user to input a value or it displays information to the user on the computer screen. The traditional approach to systems, then, involves processes, data, inputs, and outputs. When modeling what the system does in response to an event, the traditional approach includes processing models that emphasize these components.

In contrast, the object-oriented (OO) approach views a system as a collection of interacting objects. The objects are the things discussed in Chapter 5. Objects are capable of behaviors (called methods) that allow the objects to interact with each other and with people using the system. One object asks another object to do something by sending it a message. There are no conventional computer processes or data files *per se*. Objects carry out the activities and remember the data values. When modeling what the system does in response to an event, the object-oriented approach includes models that show objects, their behavior, and their interactions with other objects. Figure 6-1 summarizes the differences between traditional and object-oriented approaches to systems.

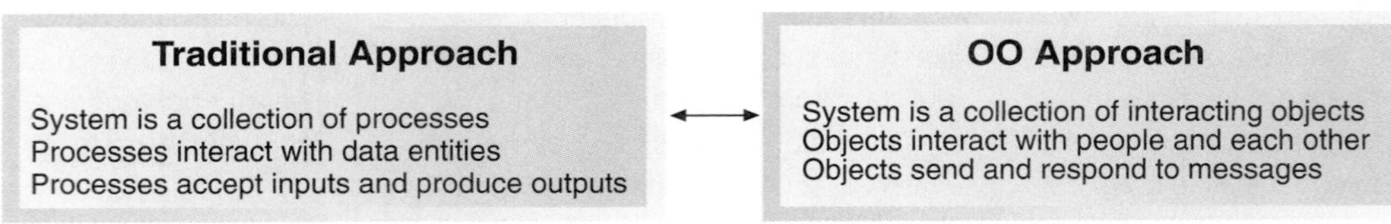

Traditional Approach
System is a collection of processes
Processes interact with data entities
Processes accept inputs and produce outputs
OO Approach
System is a collection of interacting objects
Objects interact with people and each other
Objects send and respond to messages

FIGURE 6-1
Traditional versus OO approaches.

Data Flow Diagrams

The traditional approach to information systems development describes activities as processes carried out by people or computer. A graphical model that has proven to be quite valuable for modeling processes is the data flow diagram. There are other process models, such as the process dependency diagram used in the Information Engineering approach and the workflow diagrams used with business process reengineering, but the data flow diagram is the most commonly used process model.

A data flow diagram (DFD) is a graphical system model that shows all of the main requirements for an information system in one diagram: inputs and outputs, processes, and data storage. Everyone working on a development project can see all aspects of the system working together at once with the DFD. That is one reason for its popularity. The DFD is also easy to read because it is a graphical model and because there are only five symbols to learn (see Figure 6-2). End users, management, and all information systems workers typically can read and interpret the DFD with minimal training.

FIGURE 6-2
Data flow diagram symbols.

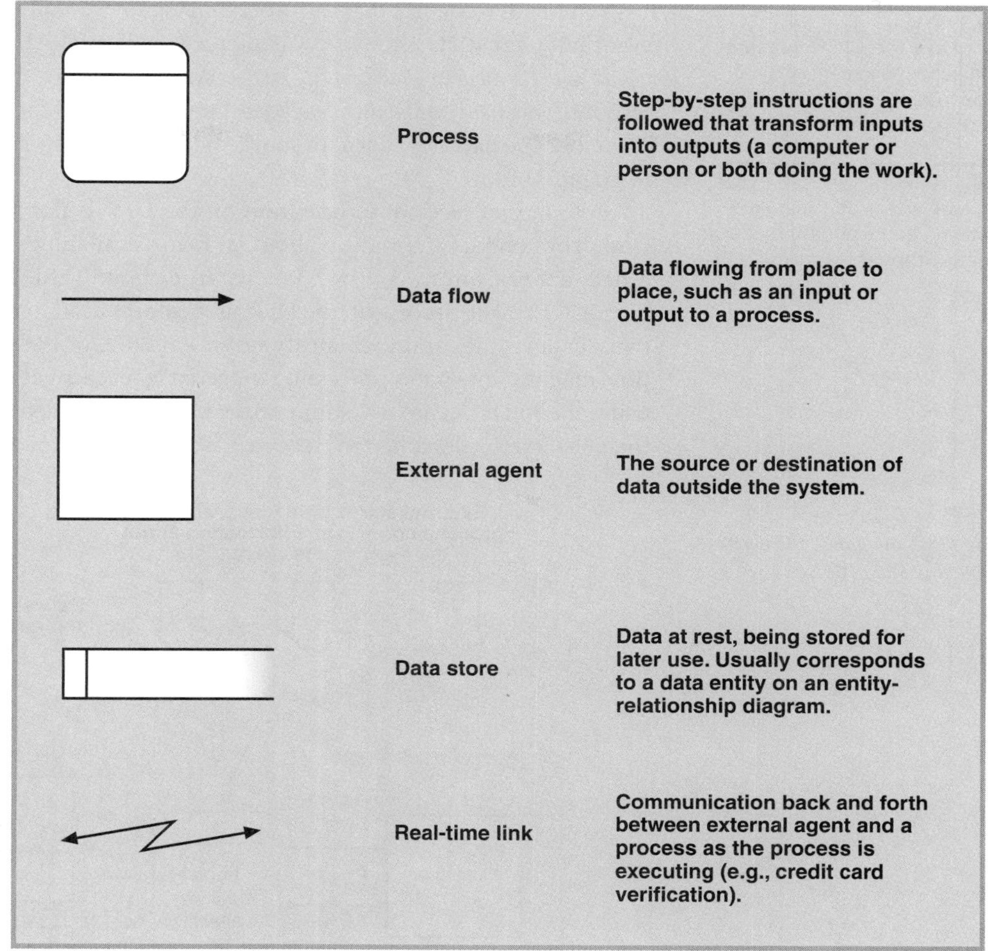
Process
Step-by-step instructions are followed that transform inputs into outputs (a computer or person or both doing the work).
Data flow
Data flowing from place to place, such as an input or output to a process.
External agent
The source or destination of data outside the system.
Data store
Data at rest, being stored for later use. Usually corresponds to a data entity on an entity-relationship diagram.
Real-time link
Communication back and forth between external agent and a process as the process is executing (e.g., credit card verification).

external agent
A person or organization, outside the boundary of a system, that provides data inputs or accepts data outputs

process
A symbol on a DFD that represents an algorithm or procedure by which data inputs are transformed into data outputs

Figure 6-3 shows an example of a data flow diagram representing a portion of the Rocky Mountain Outfitters (RMO) customer service system. The square is an **external agent**, Customer, the source and destination for some data outside the system. The rectangle with rounded corners is a **process** named "Look up item availability" that can also be referred to by its number, 1. A process defines rules for transforming inputs to outputs. The lines with arrows are **data flows**. Figure 6-3 shows two data flows between Customer and process 1: a process input named "Item inquiry" and a process output named "Item availability details." The final symbol, the flat three-sided rectangle, is a **data store**. Each data store represents a file or part of a database that stores information about a data entity. In this example, data flows (lines with arrows) point from the data stores to the process, meaning the process looks up information in the data stores named "Catalog," "Product item," and "Inventory item."

FIGURE 6-3
A DFD showing the process "Look up item availablility," (a DFD fragment from the RMO case).

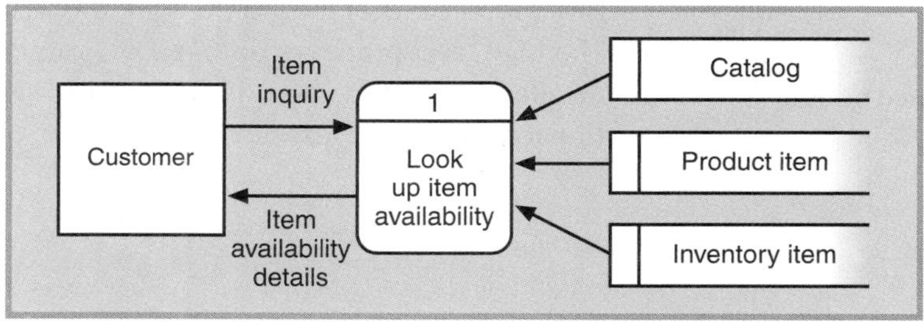
Item inquiry
Customer
Item availability details
1
Look up item availability
Catalog
Product item
Inventory item

data flow
An arrow on a DFD that represents the movement of data among processes, data stores, and external agents

data store
A place where data is held pending future access by one or more processes

You might recognize that the process in Figure 6-3 corresponds to an activity in the event table for RMO shown in Chapter 5 (see Figure 5-15). The event was *Customer wants to check item availability*, the trigger was *Item inquiry*, the source was *Customer*, the response was *Item availability details*, and the destination for the response was *Customer*. Therefore, the data flow diagram shows the system activity in response to this one event in graphical form.

But another piece of information on the DFD is not in the event table: the data stores containing information about an item's availability. Each data store represents a data entity from the entity-relationship diagram (ERD) shown in Chapter 5 (see Figure 5-27). The process on the DFD uses information that we provided by including these data entities and their attributes in the ERD for the system. Therefore, the data flow diagram integrates processing triggered by events with the data entities modeled using the ERD. Figure 6-4 summarizes the correspondence between components of the DFD, events described in the event table, and entities defined in the ERD.

FIGURE 6-4
The DFD integrates the event table and the ERD.

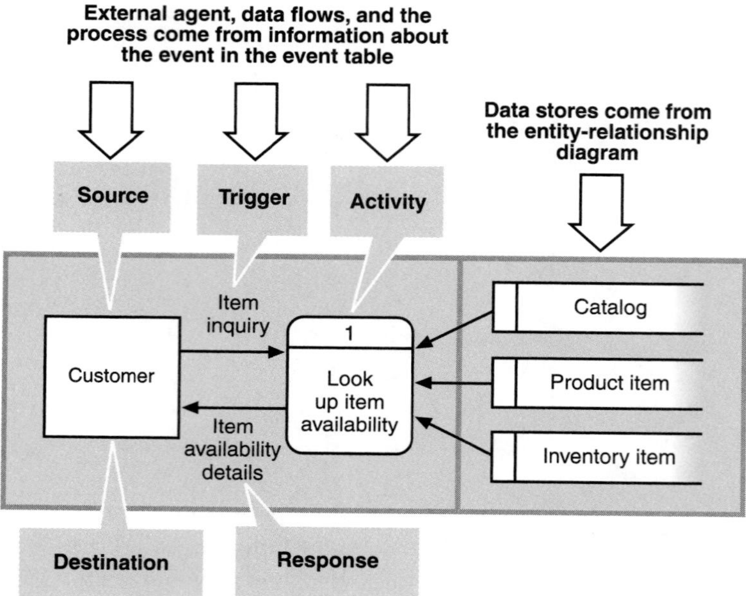

Data Flow Diagrams and Levels of Abstraction

levels of abstraction
Any modeling technique that breaks the system into a hierarchical set of increasingly more detailed models

Many different types of data flow diagrams are produced to show system requirements. The example just described is a DFD fragment, showing one process in response to one event. Other data flow diagrams show the processing at either a higher level (a more general view of the system) or at a lower level (a more detailed view of one process). These differing views of the system (high level versus low level) are referred to as levels of abstraction. Being able to show either higher-level or lower-level views of the system is another very useful feature of the data flow diagram. The high-level processes on one DFD can be decomposed into separate lower level, detailed DFDs. Processes on the detailed DFDs can also be decomposed into additional diagrams to provide multiple layers, or levels of abstraction.

Context Diagrams

context diagram
A DFD that summarizes all processing activity within the system in a single process symbol

A **context diagram** is a DFD that describes the highest-level view of a system. All external agents and all data flows into and out of the system are shown in one diagram, with the whole system represented as one process. Figure 6-5 shows the context diagram for a simple university course registration system that interacts with three external agents: academic department, student, and faculty member. Academic departments supply information on offered courses, students request enrollment in offered courses, and faculty members receive class lists when the registration period is complete.

FIGURE 6-5
A context diagram for a course registration system.

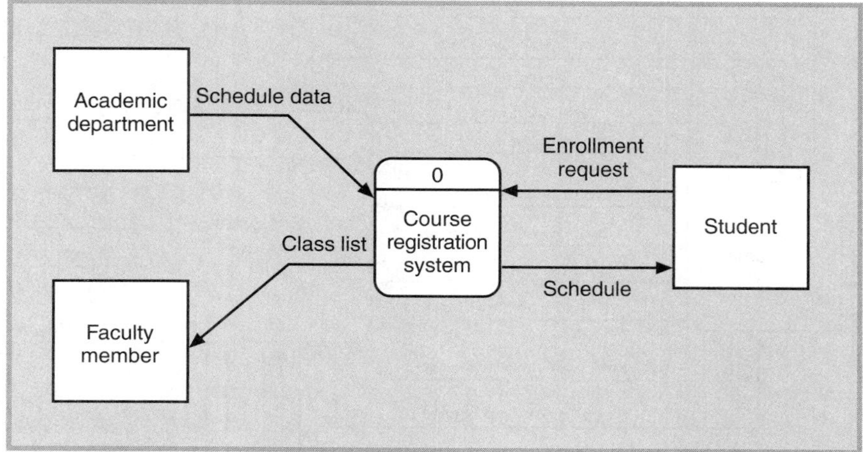

The context diagram is useful for showing system boundaries. The system scope is defined by what is represented within the single process and what is represented as an external agent. External agents that supply or receive data from the system are outside of the system scope. Everything else is inside the system scope. Data stores are not usually shown on the context diagram because all of the system's data stores are considered to be within the system scope (that is, part of the internal implementation of the process that represents the system). We first introduced the idea of a context diagram in Chapter 2 (Figure 2-8) during the planning phase as a tool to identify the scope of the new system. Now that you know what a DFD is, you should realize that the context diagram is simply the highest-level DFD.

Figure 6-6 shows the context diagram for the Rocky Mountain Outfitters customer support system. More than 30 data flows are shown, involving nine different external agents. The data flows come from the event table; they are triggers and responses for all of the events. The context diagram provides a good overview of the scope of the system, showing the system in "context." But it does not show any detail about the processing that takes place inside the system.

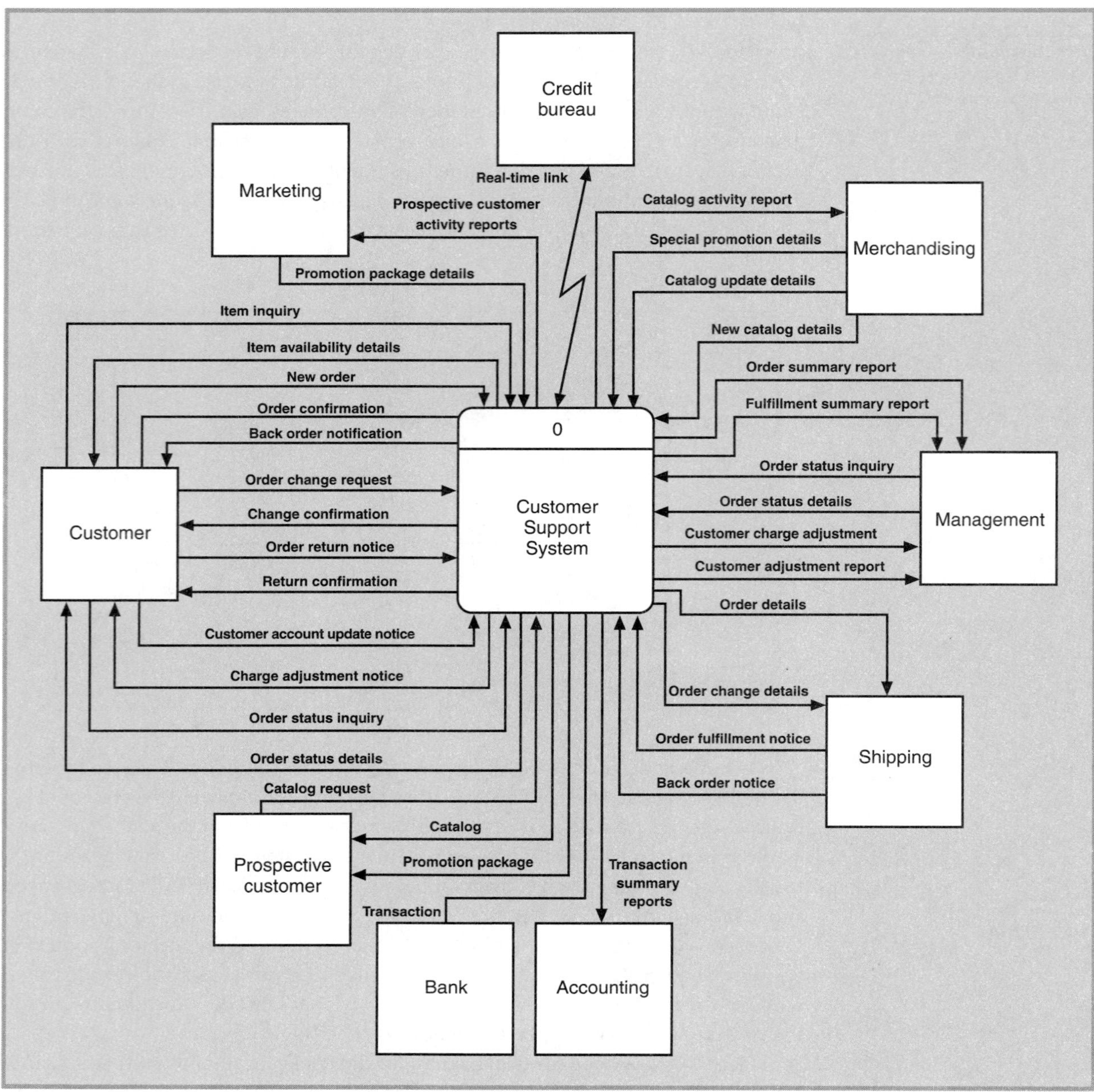

FIGURE 6-6
A context diagram for the Rocky Mountain Outfitters customer support system.

DFD fragment
A DFD that represents the system response to one event within a single process symbol

DFD Fragments

A **DFD fragment** is created for each event in the event list (expanded into the event table). Each DFD fragment is a self-contained model showing how the system responds to a single event. The analyst usually creates DFD fragments one at a time, focusing attention on just part of the system.

Figure 6-7 shows the three DFD fragments for the simple course registration system. Each DFD fragment represents all processing for an event within a single process symbol. But the fragments show details of interactions between the process, external agents, and internal data stores. The data stores used on a DFD fragment represent entities on the ERD. Each DFD fragment shows only those data stores that are actually needed to respond to the event.

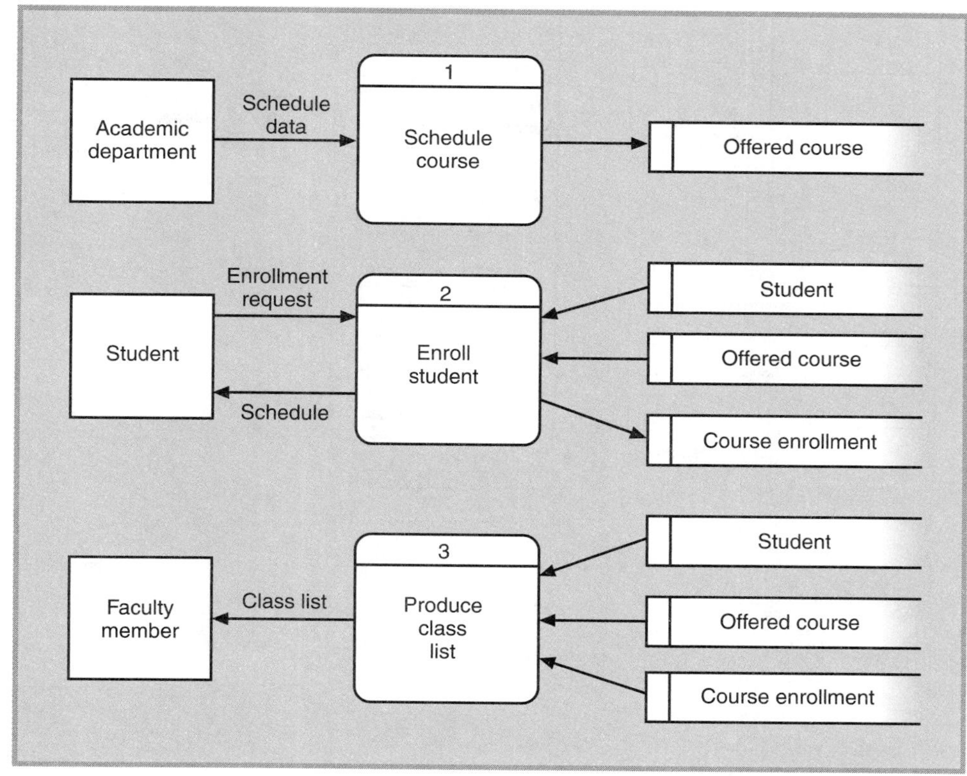

FIGURE 6-7
DFD fragments for the course registration system.

Figures 6-8 and 6-9 show the DFD fragments from the Rocky Mountain Outfitters case. Note that there are 20 DFD fragments, the same number of events in the RMO event table developed in Chapter 5 (Figure 5-15).

The Event-Partitioned System Model

event-partitioned system model
A DFD that models system requirements using a single process for each event in a system or subsystem

diagram 0
Synonym of the event-partitioned system model

An entire set of DFD fragments can be combined on a single DFD called the **event-partitioned system model** or **diagram 0**. Diagram 0 is used to show the entire system on a single DFD in greater detail than on the context diagram. Figure 6-10 shows a set of four related DFDs to show how each level provides additional information about the level above it. The topmost DFD is the context diagram for the course registration system (the same as Figure 6-5). The DFD directly below the context diagram is the event-partitioned system model (that is, diagram 0). Note that diagram 0 is actually a decomposition of the single process from the context diagram. It is also a combined version of the three DFD fragments shown in Figure 6-7. Each process on diagram 0 represents processing for a single event.

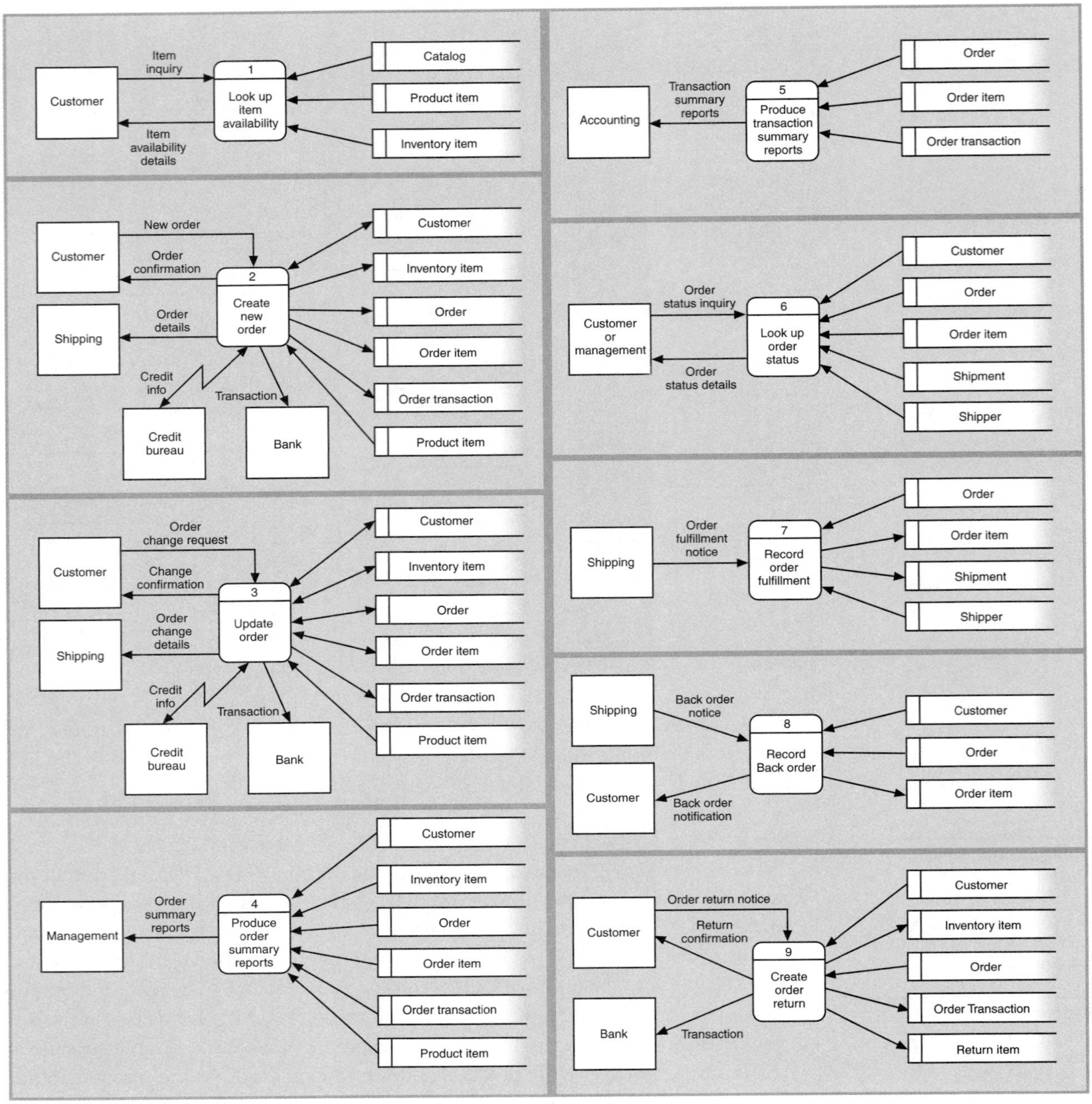

Customer
Item inquiry
1
Look up item availability
Item availability details
Catalog
Product item
Inventory item

Accounting
Transaction summary reports
5
Produce transaction summary reports
Order
Order item
Order transaction

Customer
New order
Order confirmation
Shipping
Order details
2
Create new order
Credit info
Credit bureau
Transaction
Bank
Customer
Inventory item
Order
Order item
Order transaction
Product item

Customer or management
Order status inquiry
Order status details
6
Look up order status
Customer
Order
Order item
Shipment
Shipper

Customer
Order change request
Change confirmation
Shipping
Order change details
3
Update order
Credit info
Credit bureau
Transaction
Bank
Customer
Inventory item
Order
Order item
Order transaction
Product item

Shipping
Order fulfillment notice
7
Record order fulfillment
Order
Order item
Shipment
Shipper

Shipping
Back order notice
Customer
Back order notification
8
Record Back order
Customer
Order
Order item

Management
Order summary reports
4
Produce order summary reports
Customer
Inventory item
Order
Order item
Order transaction
Product item

Customer
Order return notice
Return confirmation
Bank
Transaction
9
Create order return
Customer
Inventory item
Order
Order Transaction
Return item

FIGURE 6-8
DFD fragments for RMO customer support system (part 1).

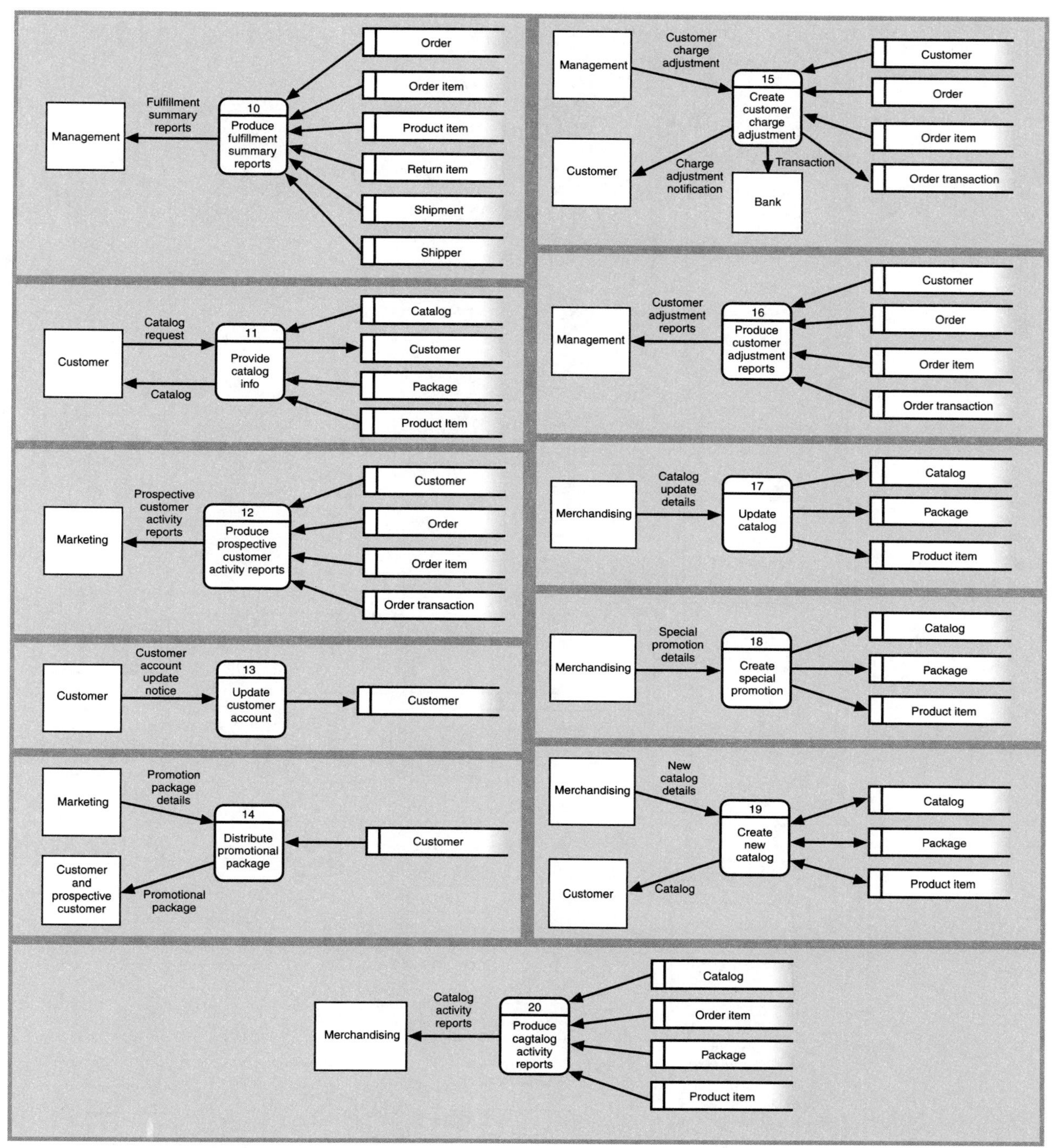

FIGURE 6-9
DFD fragments for RMO customer support system (part 2).

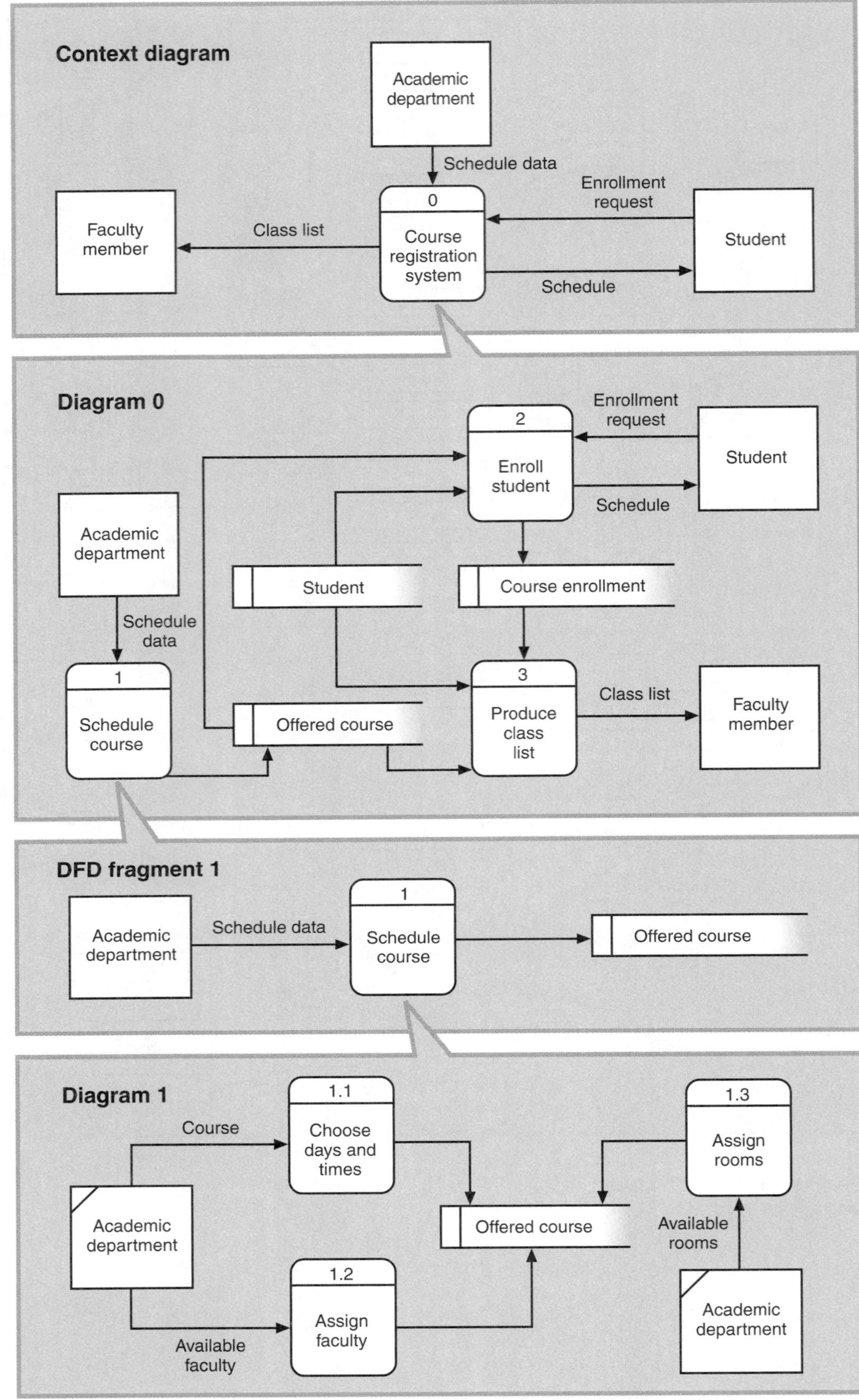

FIGURE 6-10
Layers of DFD abstraction for the course registration system.

The third DFD in Figure 6-10 shows a single DFD fragment corresponding to process 1 on diagram 0. Since there are three processes on diagram 0, there should be three separate DFD fragments, one for each process or event. But only one DFD fragment is shown in the figure. The DFD at the bottom of the figure is a process decomposition of process 1. Process decompositions are discussed fully in the next section.

The RMO customer support system involves 20 events; therefore, the event-partitioned system model (diagram 0) should contain 20 processes. But such a diagram would be crowded and difficult to read. A natural solution to this problem is to divide the system into subsystems. The analyst looks at the events and defines groups of events that make up a subsystem. Figure 6-11 lists the RMO customer support system subsystems and their events. The events have been grouped into subsystems based on similarities, including interactions with external agents, interactions with data stores, and similarities in required processing.

Order-entry subsystem

Customer wants to check item availability
Customer places an order
Customer changes or cancels an order
Time to produce order summary reports
Time to produce transaction summary reports

Order fulfillment subsystem

Customer or management wants to check order status
Shipping fulfills an order
Shipping identifies a back order
Customer returns the item (the item is defective, the customer has changed his mind, full or partial returns)
Time to produce fulfillment summary reports

Customer maintenance subsystem

Prospective customer requests a catalog
Time to produce prospective customer activity reports
Customer updates account information
Marketing wants to send promotional materials to customers
Management adjusts customer charges (correct errors, make concessions)
Time to produce customer adjustment/concession reports

Catalog maintenance subsystem

Merchandising updates the catalog (add, change, delete, change prices)
Merchandising creates a special product promotion
Merchandising creates a new catalog
Time to produce catalog activity reports

FIGURE 6-11
RMO subsystems and events for each subsystem.

A DFD is then created to represent the division of the system into subsystems. Once the subsystem DFD is created, the analyst draws a separate event-partitioned DFD for each subsystem. When subsystems are defined, the set of DFDs are related to one another, as shown in Figure 6-12. The context diagram is decomposed into a subsystem DFD, which is in turn decomposed into a set of event-partitioned DFDs. There is no single diagram 0. Instead there is an event-partitioned DFD for each of the subsystems. In essence, each event-partitioned DFD is a diagram 0 for a single subsystem.

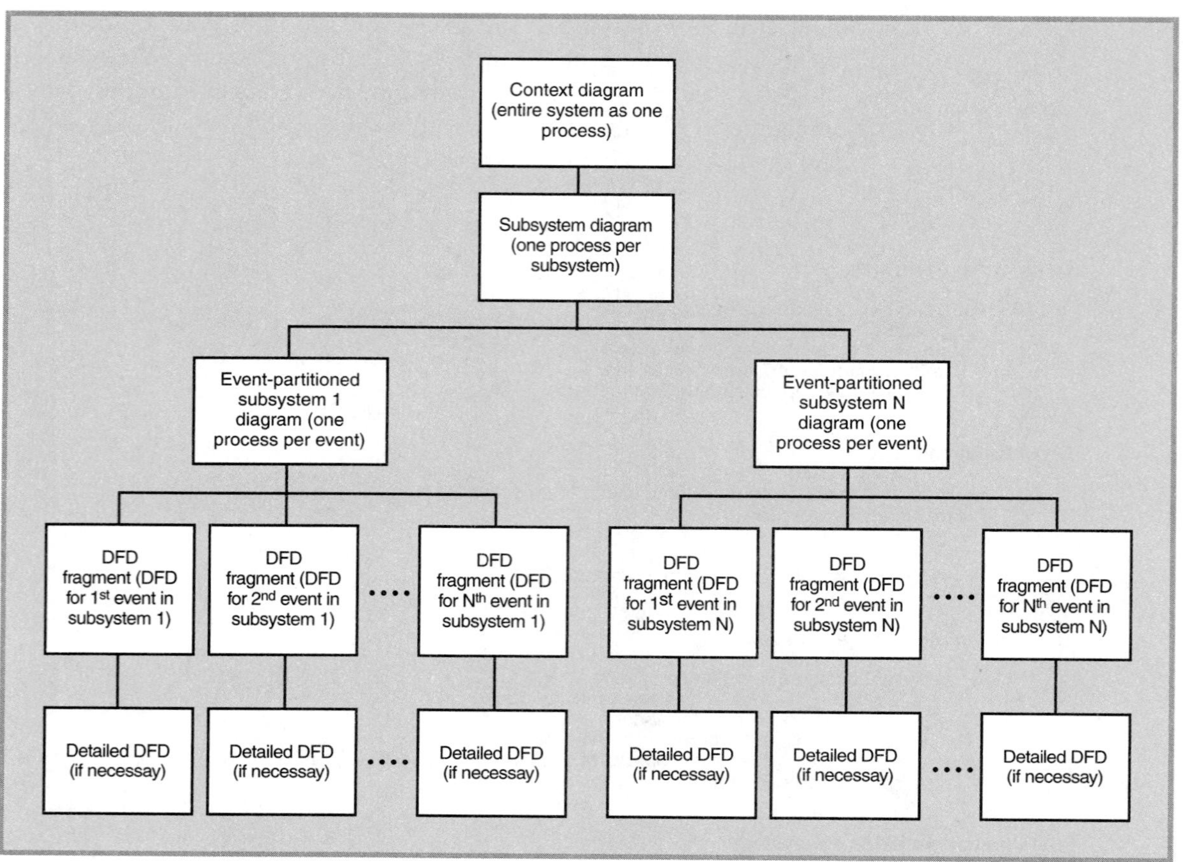

FIGURE 6-12
The relationship among DFD abstraction levels when subsystems are defined.

Figure 6-13 shows the RMO subsystem DFD from Figure 6-11 in graphic form. Note the complexity of the diagram and imagine how much more complex it would be if all 20 events had been included instead of just the four subsystems. Figure 6-14 shows an event-partitioned view of the order-entry subsystem. The analyst would also need to create event-partitioned DFDs for the other three subsystems.

Note that several external agents (such as Shipping, Customer, Management, and Bank) have been represented in multiple places in Figure 6-13 to minimize crossing data flows and improve readability. By convention, a line is drawn at a 45-degree angle in one corner of any external agent symbol that appears multiple times in a DFD (for example, the short line in the upper-left corner of the Customer symbol in Figure 6-13). When a reader is focusing on an external agent, the line is a visual cue that additional copies of the external agent (and data flows to and from the agent) appear elsewhere on the diagram.

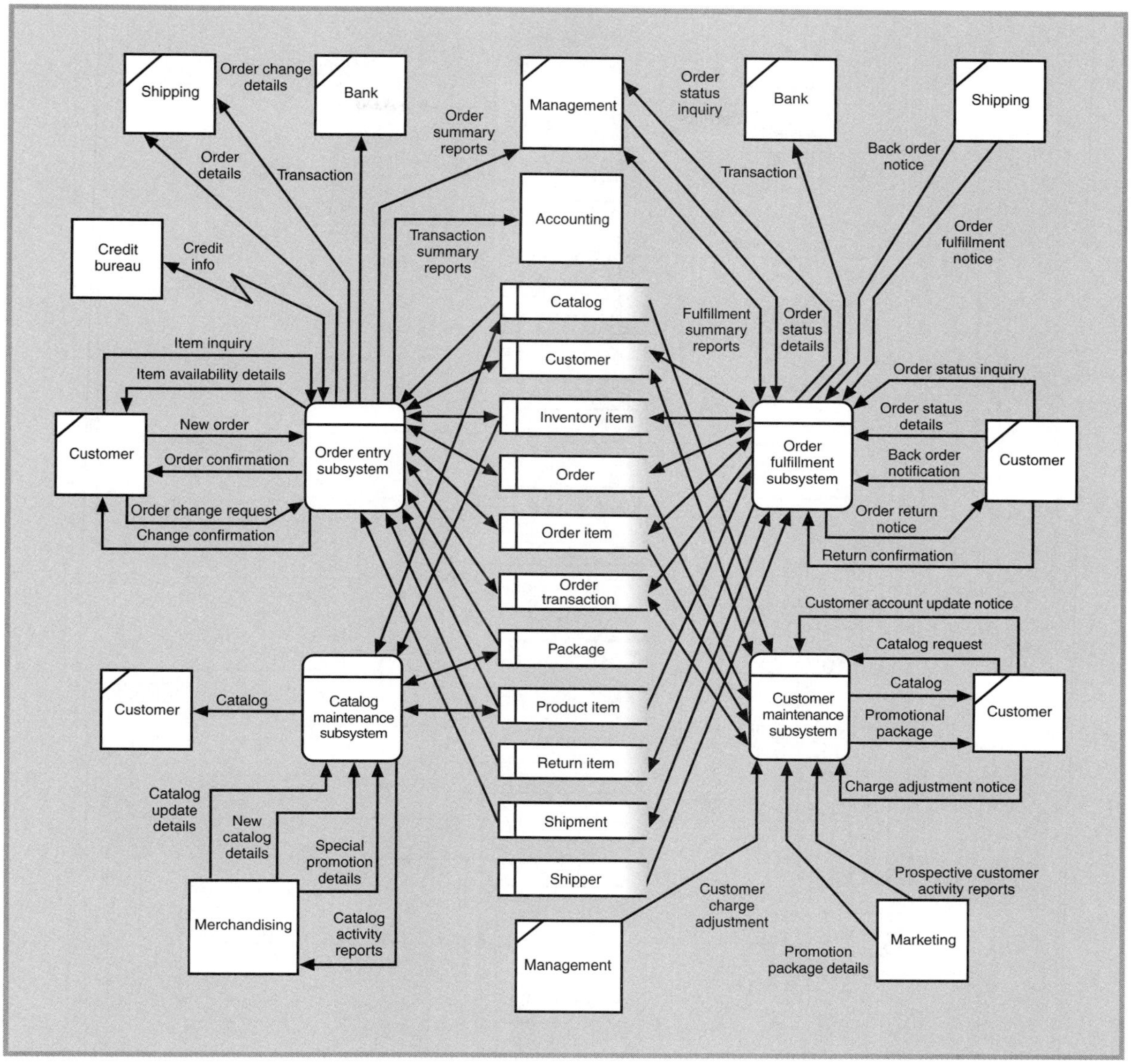

FIGURE 6-13
The RMO subsystem DFD.

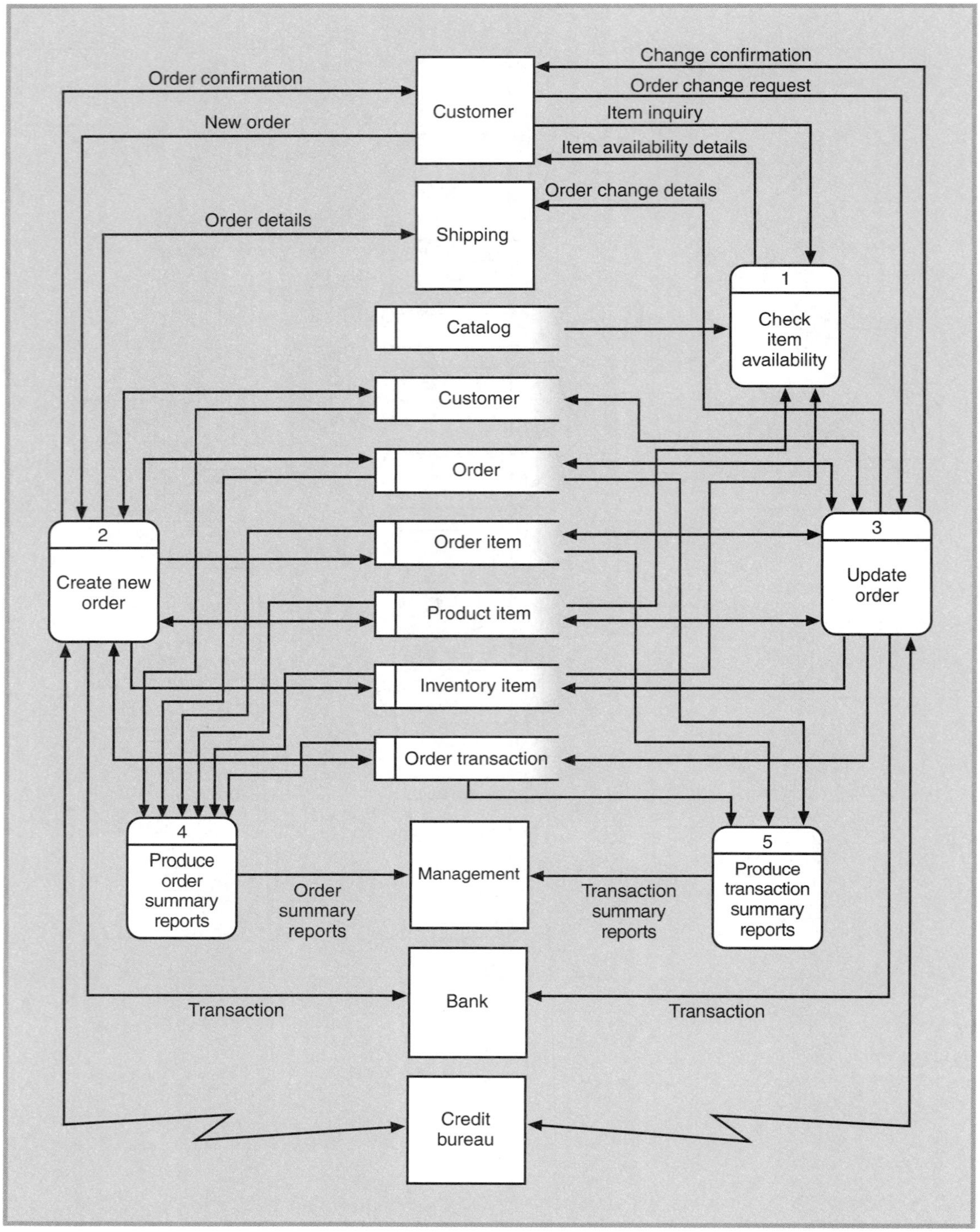

FIGURE 6-14
An event-partitioned model of the order-entry subsystem.

Decomposing Processes to See Detail of One Activity

Sometimes certain DFD fragments involve a lot of processing that the analyst needs to explore in more detail. A DFD fragment can be decomposed into subprocesses just like the context diagram is decomposed into diagram 0. As with any modeling step, further decomposition helps the analyst learn more about the requirements while also producing needed documentation.

Figure 6-15 shows an example of a more detailed diagram for DFD fragment 2, "Create new order." It is named diagram 2 since it shows the "insides" of process 2. The subprocesses are numbered 2.1, 2.2, 2.3, and 2.4. The numbering system does not necessarily imply sequence of subprocess execution, though.

FIGURE 6-15
A detailed diagram for Create new order (diagram 2).

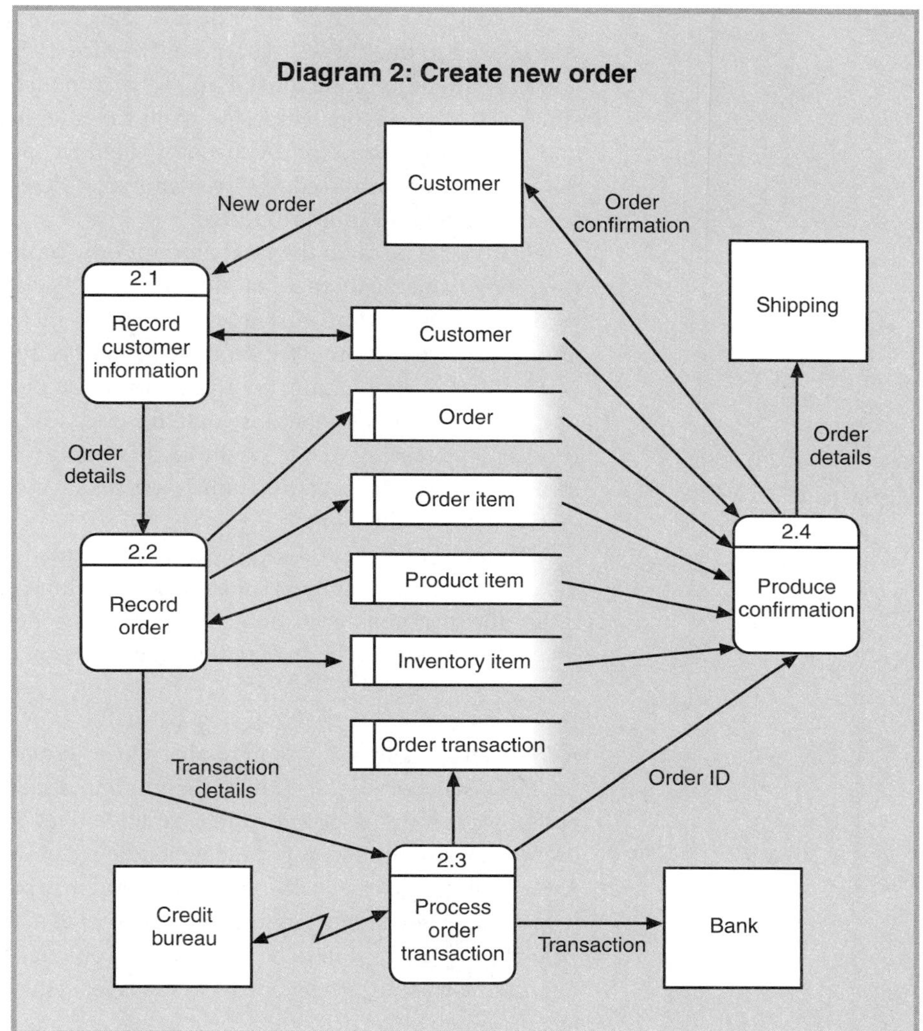

The diagram decomposes process 2 into four subprocesses: "Record customer information," "Record order," "Process order transaction," and "Produce confirmation." These subprocesses are viewed as the four major steps required to complete the activity. This is just one way to divide up the work. Another analyst might arrive at a different solution.

The first step begins when the customer provides the information making up the "New order" data flow. The "New order" data flow contains all of the information about the customer and the items the customer wants to order. Process 2.1 stores the customer information in the data store named "Customer" (creating a new customer record or updating existing customer information as required). Remember that the data store represents the customer data entity on the ERD developed in Chapter 5 (see Figure 5-27). The data put in the data store correspond to the list of attributes for the customer data entity. If the customer is an existing customer, the data are already in the data store, so the process reads data. Process 2.1 then sends the rest of the information about the order, a data flow named "Order details," on to process 2.2.

Process 2.2 takes the "Order details" data flow and creates a new order record by adding data to the "Order" data store. Then for each item ordered, the stock on hand and the current price are looked up in the "Product item" and "Inventory item" data stores. If there is adequate stock on hand, an order item record is created for that item, and the stock on hand for the inventory item record is changed. This repeats until all items have been processed. If three items are ordered, one order record is created and three order item records are created.

Process 2.2 adds up the total amount due for the order (price times quantity for each item) and sends the data flow named "Transaction details" to process 2.3 to record the transaction. "Transaction details" include the order number, amount, and credit card information. Process 2.3 needs a real-time link to a credit bureau to get a credit authorization for the customer's credit card. This is a real-time link rather than a data flow because data need to flow back and forth rapidly while the process is executing. If the credit card is approved, a record of the transaction is created in the "Order transaction" data store, and a data flow for the transaction goes directly to the bank.

The final process produces the order confirmation for the customer and the order details that go to shipping. Using the order number, process 2.4 looks up data on the order, the customer, each order item (plus the item description from the product item), and produces the required outputs.

Physical and Logical DFDs

A DFD can be a physical system model, a logical system model, or a blend of the two. If the DFD is a logical model, then it assumes that the system will be implemented with perfect technology, as described in Chapter 5. If the DFD is a physical model, then one or more assumptions about implementation technology will be embedded in the DFD. These assumptions may take many forms and may be very difficult to spot.

Consider whether diagram 2 in Figure 6-15 is a logical model. First, is it clear what type of computer is doing the processing? Could it be a desktop system? A centralized mainframe system? A networked client-server system? Or, could the entire process as just described be carried out by people without any computer at all? Similarly, are the data stores sequential computer files? Are they tables in a relational database? Or are they files of paper in a file cabinet? How does the system get the data flow "New order" from the customer so it can be processed? By clicking check boxes and list boxes in a Windows application? Or on a web page? Or by manually filling out a form that a clerk types into the system? Or by talking to the clerk over the phone? Or by talking to a speech recognition program over the phone?

All of the alternatives described are possible, and if the model is a logical model, you should not be able to tell how the system is implemented. At the same time, the processing requirements (what must go on) should be fairly detailed, down to indicating what attribute values are needed. The model could be even more detailed and still be a logical model.

Now consider whether the DFD in Figure 6-16 is a physical system model by comparing it to diagram 1 in Figure 6-10. A number of elements indicate assumptions about implementation technology, including:

- Technology-specific processes
- Actor-specific process names
- Technology- or actor-specific process orders
- Redundant processes, data flows, and files

The most obvious technology assumption is embedded in the name of process 1.1. Making copies is an inherently manual task, which implies that the data store "Old schedules" and the data flows into and out of process 1.1 are paper. It is possible that the data store and flows are electronic, but if so, the question arises why an explicit process would be needed to make electronic copies.

Many of the process names specifically reference actors in the system. References to "Chair," "Assoc. Dean," and "University scheduling" all indicate that a process is performed by a particular individual or department. The sequential flow of data among the processes is a by-product of the person or department that carries out each process. One can imagine alternative implementations with fewer processes, different process orders, or different assignment of processes to individuals and/or departments. The DFD clearly models one very specific set of decisions about process ordering and responsibility.

The DFD also includes processes with similar or redundant processing logic. For example, faculty input is accepted early but faculty members later perform error checking twice (the data flows from processes 1.4 and 1.7). Also, rooms are assigned at two different times from two different data stores ("Reserved room" for process 1.5, and "General room" for process 1.6). As before, these features indicate very specific assumptions about technology and division of responsibility. The redundant error checking indicates that it is possible for previous processes to make mistakes. A system implemented with perfect technology needs no internal error checking. The partitioning of room assignment among two files and processes may be related to technology (for example, no one process could successfully assign all rooms at once) or it could indicate a historic division of responsibility for room assignment.

Inexperienced analysts often develop DFDs such as the one in Figure 6-16. The path to developing such a model is simple—model everything the current system does exactly the way it does it. The problem with this approach is that design assumptions and technology limitations of the old system can inadvertently become embedded in the new system. This can easily occur when analysis and design are performed by different persons or teams. The new designer(s) may not realize that some of the "requirements" embedded in the DFDs are simply reflections of the way things are now, not the way they necessarily should be in the future.

Physical DFDs are sometimes developed and used during the last stages of analysis or early stages of design. They are useful models for describing alternative implementations of a system prior to developing more detailed design models. But physical DFDs should be avoided during all analysis phase activities except for generating alternatives. Even during that activity, physical DFDs should be clearly labeled as such so that readers know that the model represents one possible implementation of the logical system requirements.

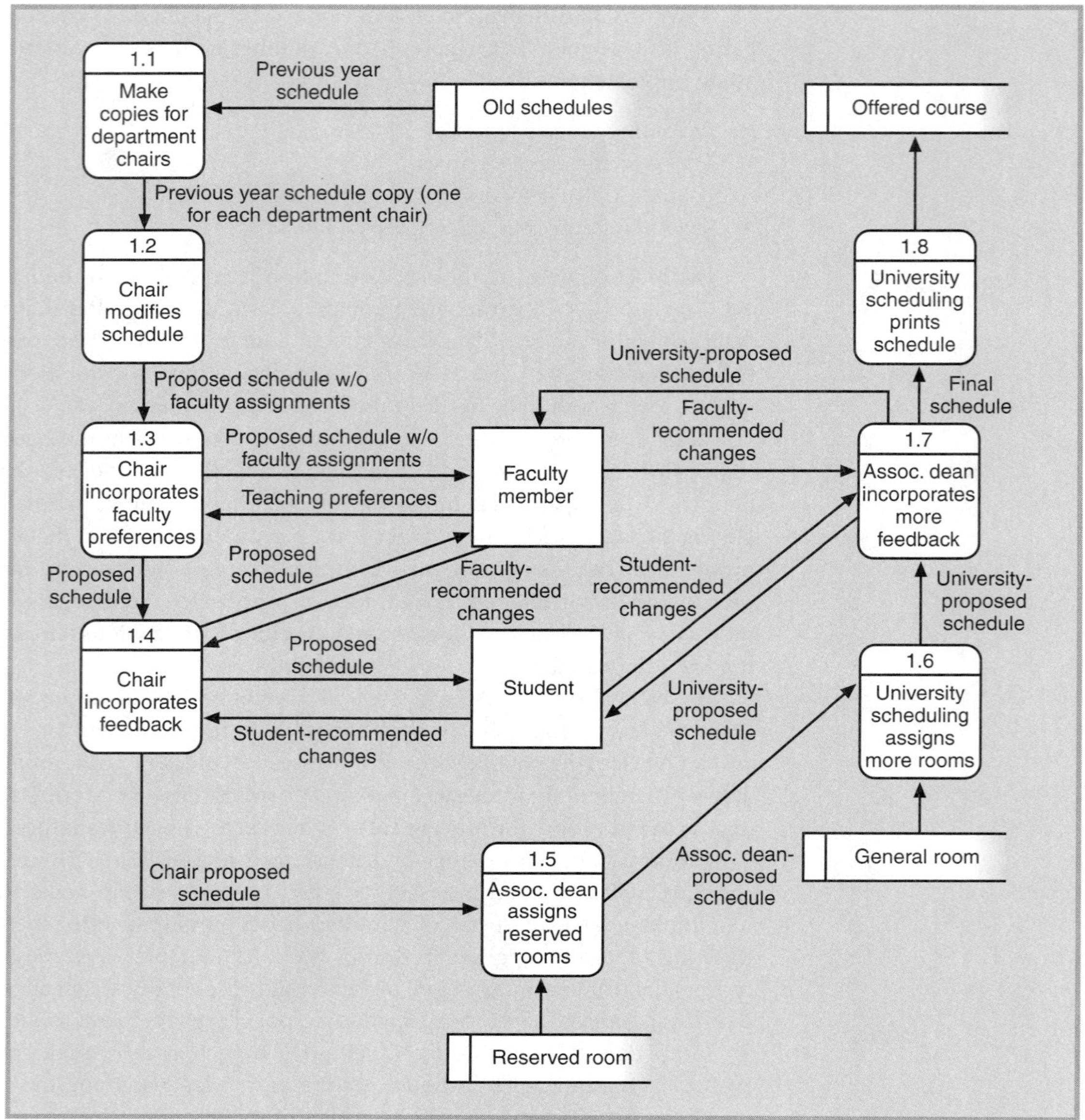

186

FIGURE 6-16
A physical DFD for scheduling courses (resist the temptation to create physical DFDs during analysis).

Evaluating DFD Quality

A high-quality set of DFDs is readable, is internally consistent, and accurately represents system requirements. Accuracy of representation is determined primarily by consulting users and other knowledgeable stakeholders. A project team can ensure readability and internal consistency by applying a few simple rules to DFD construction. Analysts can apply these rules while developing the DFDs or during a separate quality check after preparing DFD drafts.

information overload
An undesirable condition that occurs when too much information is presented to a reader at one time

Minimizing Complexity People have a limited ability to manipulate complex information. If too much information is presented at once, people experience a phenomenon called **information overload**. When information overload occurs, a person has difficulty understanding the information presented. The key to avoiding information overload is to divide information into small and relatively independent subsets. Each subset should contain a comprehensible amount of information that can be examined and understood in isolation.

A layered set of DFDs is an example of dividing a large set of information into small, independent subsets. Each DFD can be examined in isolation. The reader can find additional detail about a specific process by moving down to the next level, or find information about how a DFD relates to other DFDs by examining the next higher-level DFD.

An analyst can avoid information overload within any single DFD by following two simple rules of DFD construction:

- 7 ± 2
- Interface minimization

rule of 7 ± 2
The rule of model design that limits the number of model components or connections among components to no more than nine

The **rule of 7 ± 2** (also known as Miller's Number) derives from psychology research, which shows that the number of information "chunks" that a person can remember and manipulate at one time varies between five and nine. A larger number of chunks causes information overload. Information chunks may be many things, including names, words in a list, digits, or components of a picture.

Some applications of the rule of 7 ± 2 to DFDs include:

- A single DFD should have no more than 7 ± 2 processes.
- No more than 7 ± 2 data flows should enter or leave a process, data store, or data element on a single DFD.

Of course, these rules are general guidelines, not unbreakable laws. DFDs that violate these rules may still be readable, but violations should be considered a warning of potential problems.

minimization of interfaces
A principle of model design that seeks simplicity by minimizing the number of interfaces or connections among model components

Minimization of interfaces is directly related to the rule of 7 ± 2. An interface is a connection to some other part of a problem or description. As with information chunks, the number of connections that a person can remember and manipulate is limited, so the number of connections should be kept to a minimum. Processes on a DFD represent chunks of business or processing logic. They are related to other processes, entities, and data stores by data flows. A single process with a large number of interfaces (data flows) may be too complex to understand. This may show up directly on a process decomposition as a violation of the rule of 7 ± 2. An analyst can usually correct the problem by dividing the process into two or more processes, each of which should have fewer interfaces.

Pairs or groups of processes with a large number of data flows between them are another violation of the interface minimization rule. Such a condition usually indicates a poor partitioning of processing tasks among the processes. The way to fix the problem is to reallocate the processing tasks so that fewer interfaces are required. The best division of work among processes is the simplest, and the simplest division is one that requires the fewest interfaces among processes.

Data Flow Consistency An analyst can often detect errors and omissions in a set of DFDs by looking for specific types of inconsistency. Three common and easily identifiable consistency errors are:

- Differences in data flow content between a process and its process decomposition
- Data outflows without corresponding data inflows
- Data inflows without corresponding outflows

A process decomposition shows the internal details of a higher-level process in a more detailed form. In most cases, the data content of flows to and from a process at one DFD level should be equivalent to the content of data flows to and from all processes in a decomposition. This equivalency is called balancing, and the higher-level DFD and the process decomposition DFD are said to be "in balance."

balancing
Equivalence of data content between data flows entering and leaving a process and data flows entering and leaving a process decomposition DFD

Note the use of the term *data content* in the previous paragraph. Data flow names can vary among DFD levels for a number of reasons, including decomposition of one combined data flow into several smaller flows. Thus, the analyst must be careful to look at the components of data flows, not just data flow names. For this reason, detailed analysis of balancing should not be undertaken until data flows have been fully defined.

Unbalanced DFDs may be acceptable when the imbalance is due to data flows that were ignored at the higher levels. For example, diagram 0 for a large system usually ignores details of error handling, such as when an item is ordered but is later determined to be out of stock and discontinued by its manufacturer. A process called "Fulfill order" on diagram 0 would not have any data flows associated with this condition. In the process decomposition of "Fulfill order," the analyst might add a process and data flows to handle discontinued items.

Another type of DFD inconsistency can occur between the data inflows and outflows of a single process or data store. By definition, a process transforms data inflows into data outflows. In a logical DFD, data should not be needlessly passed into a process. The following consistency rules can derive from these facts:

- All data that flows into a process must flow out of the process or be used to generate data that flows out of the process.
- All data that flows out of a process must have flowed into the process or have been generated from data that flowed into the process.

Figure 6-17 shows an example where the first rule is violated. Data elements A, B, and C flow into the process but do not flow out. Data element A is used to determine what formula is applied to recompute the value of X, so that data element is a necessary input. However, data elements B and C play no role in generating process output and should thus be eliminated as unnecessary inflows. A process such as the one shown in Figure 6-17 is sometimes called a black hole (data enter but never escape).

black hole
A process with a data input that is never used to produce a data output

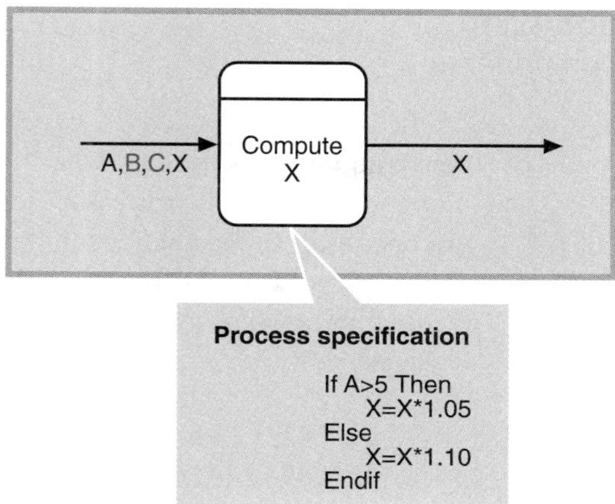

FIGURE 6-17
A process with unnecessary data input, a black hole.

Figure 6-18 shows an example where the second rule is violated. Data elements A, B, and Y flow out of the process. Data element A flows into the process. Data element Y is computed by an algorithm based on data element A. However, data element B does not flow into the process and is not computed by internal processing logic. Thus, data element B indicates either an error in the data outflow (B should be eliminated) or an omission in the internal processing logic (the rule that determines B is missing). A data element such as B is sometimes called a **miracle** (it miraculously "appears" without any apparent source).

miracle
A process with a data output that is created out of nothing

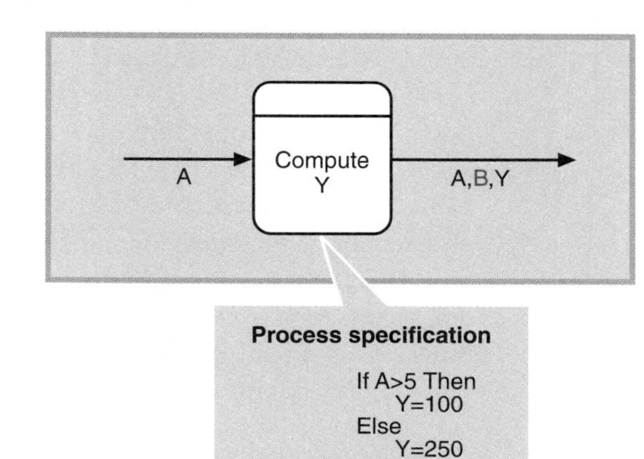

FIGURE 6-18
A process with an impossible data output, a miracle.

Note that both consistency rules apply to data stores as well as processes. Any data element that is read from a data store must have previously been written to that data store. Similarly, any data element that is written to a data store must eventually be read from the data store. Examining the consistency of data flows to and from a data store is complicated by the fact that a data element may flow into and out of a data store on completely different DFDs.

Evaluating data flow consistency is a straightforward but tedious process. Fortunately, most CASE tools automatically perform data flow consistency checking. But those CASE tools place rigorous requirements on the analyst to specify the internal logic of processes precisely. Without precise process descriptions, it is impossible for the CASE tool (or a human) to know what data elements are used as input or generated as output by internal processing logic.

Documenting
DFD Components

In the traditional approach, data flow diagrams show all three types of internal system components—processes, data flows, and data stores—on one diagram, but additional details about each component need to be described. First, each lowest-level process needs to be described in detail. Additionally, the analyst needs to define each data flow in terms of the data elements it contains. Data stores also need to be defined in terms of the data elements. Finally, the analyst also needs to define each data element.

Process Descriptions

Each process on a DFD must be formally defined. There are several options for process definition, including one that has already been discussed—process decomposition. In a process decomposition, a higher-level process is formally defined by a DFD that contains lower-level processes. These lower-level processes may in turn be further decomposed into even lower-level DFDs.

Eventually a point will be reached at which a process doesn't need to be defined by a DFD. This point occurs when a process becomes so simple that it can adequately be described by another process description method. These methods include structured English, decision tables, and decision trees. In each of these cases, the process is described as an algorithm. The most appropriate method is the one that is most compact, readable, and unambiguous. In most cases, structured English is the preferred method.

structured English
A method of writing process specifications that combines structured programming techniques with narrative English

Structured English uses brief statements to describe a process very carefully. Structured English looks a bit like programming statements, but without references to computer concepts. Rules of structured programming are followed, and indentation is used for clarity. For example, a simple set of instructions for processing ballots after a vote is shown in Figure 6-19. Some statements are simply instructions. Other statements include repetition of instructions. Other statements involve doing one set of instructions or the other. The procedure always starts at the top and ends at the bottom. Therefore, the rules of structured programming apply. Note, though, that this is not necessarily a computer program—it might be done by a person—so it is a logical model. It is unambiguous, so anyone following the instructions will arrive at the same result. That makes it an algorithm.

FIGURE 6-19
A structured English example.

```
Process Ballots Procedure

Collect all ballots
Place all ballots in a stack
Set Yes count and No count to zero
Repeat for each ballot in the stack
   If Yes is checked then
       Add one to Yes count
   Else
      Add one to No count
   End If
   Place ballot on counted ballot stack
End repeat
If Yes count is greater than No count then
   Declare Yes the winner
Else
   Declare No the winner
End If
Store the counted ballot stack in a safe place
End Process Ballots Procedure
```

An example of a process description for Rocky Mountain Outfitters is shown in Figure 6-20. Note how the process description provides more specific details about what the process does. If the process description becomes too complex, then the analyst should choose another method of process description. Excess length (for example, more than 20 lines) and/or multiple levels of indentation (indicating complex decision logic) indicate that a structured English description may be too complex. Excess indentation can sometimes be solved by converting the description to an equivalent decision table or decision tree. In other cases, a process decomposition may be required.

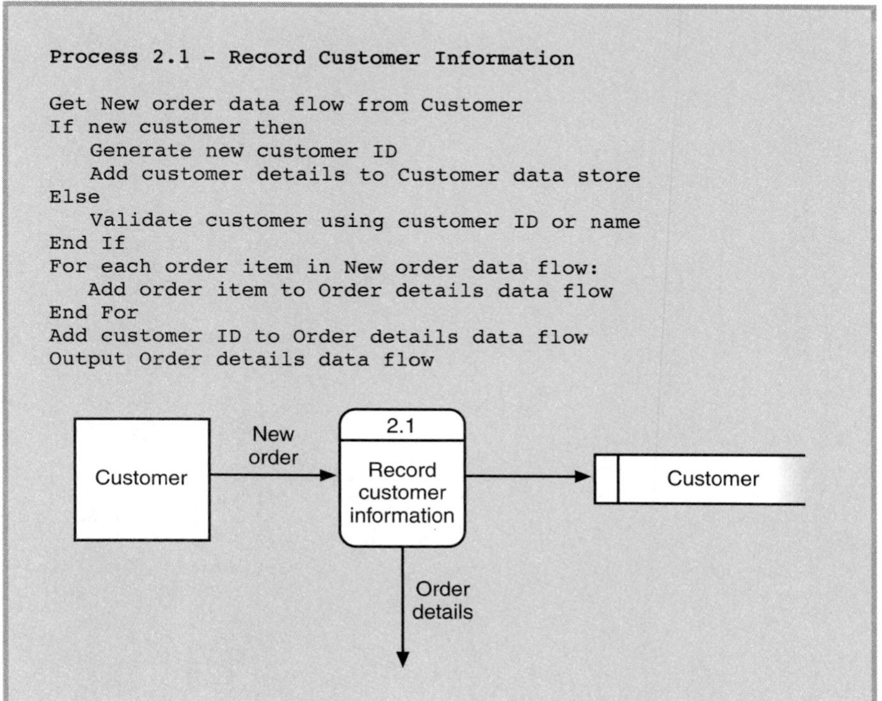

Process 2.1 - Record Customer Information
Get New order data flow from Customer
If new customer then
Generate new customer ID
Add customer details to Customer data store
Else
Validate customer using customer ID or name
End If
For each order item in New order data flow:
Add order item to Order details data flow
End For
Add customer ID to Order details data flow
Output Order details data flow
Customer
New order
2.1
Record customer information
Customer
Order details

FIGURE 6-20
RMO process 2.1 (Record Customer Information) and its structured English process description.

Structured English is well suited to describing processes with many sequential processing steps and relatively simple control logic (such as a single loop or an if-then-else statement). Structured English is not well suited for describing processes with the following characteristics:

- Complex decision logic
- Few (or no) sequential processing steps

Decision logic is complex when multiple decision variables and a large number of possible combinations of those decision variables need to be considered. When a process with complex decision logic is described with structured English, the result is typically a long and/or difficult-to-read description. For example, consider the structured English description for calculating shipping cost shown in Figure 6-21. Note that the description is relatively long and consists mostly of control structures (if, else, and endif statements).

FIGURE 6-21
A structured English process description for determining delivery charges.

```
If YTD purchases > $250 then
      If number of items ordered < 4 then
         If delivery date is next day then
               delivery charge is $25
         Endif
         If delivery date is second day then
               delivery charge is $10
         Endif
         If delivery date is seventh day then
              delivery charge is $1.50 per item
         Endif
      Else
         If delivery date is next day then
              delivery charge is $6 per item
         Endif
         If delivery date is second day then
              delivery charge is $2.50 per item
         Endif
         If delivery date is seventh day then
              delivery charge is zero (free)
         Endif
      Endif
Else
      If number of items ordered < 4 then
        If delivery date is next day then
              delivery charge is $35
        Endif
        If delivery date is second day then
              delivery charge is $15
        Endif
        If delivery date is seventh day then
              delivery charge is $10
        Endif
      Else
        If delivery date is next day then
              delivery charge is $7.50 per item
        Endif
        If delivery date is second day then
              delivery charge is $3.50 per item ordered
        Endif
        If delivery date is seventh day then
              delivery charge is $2.50 per item
        Endif
      Endif
Endif
```


decision table
A tabular representation of processing logic containing decision variables, decision variable values, and actions or formulas

decision tree
A graphical description of process logic that uses lines organized like branches of tree

Decision Tables and **decision trees** can summarize complex decision logic more concisely than structured English. Figures 6-22 and 6-23 show a decision table and decision tree that represent the same logic as the structured English example in Figure 6-21. Both incorporate decision logic into the structure of the table or tree to make the descriptions more readable than their structured English equivalent. The decision table is more compact, but the decision tree is easier to read. Sometimes an analyst needs to describe a process all three ways before deciding which approach describes a particular process best.

The following steps are used to construct a descision table:

1. Identify each decision variable and its allowable values (or value ranges).
2. Compute the number of decision variable combinations as the product of the number of values (or value ranges) of each decision variable.
3. Construct a table with one more column than the number of decision variable combinations computed in step 2 (the extra column is for decision variable names and process action or computation descriptions). The table should have a row for each decision variable and a row for each process action or computation.

4. Assign the decision variable with the fewest values (or value ranges) to the first row of the table. Put the decision variable name in the first column. Divide the remaining columns into sets of columns for each decision variable value (or value range).
5. Choose the next decision variable with the fewest values (or value ranges) for the second row. Put its name in the first column. Compute the number of column groups as the product of the number of values (or value ranges) of this variable and all the variables above it in the table. Divide the remaining columns into the computed number of groups and insert values (or value ranges) in a regular pattern.
6. Continue inserting rows as instructed in step 5 until all decision variables have been included in the table.
7. Add a row for each calculation or action. For each calculation cell, insert the appropriate constant value or formula for the combination of decision variable values that appear above the cell in the same column. For each action cell, place a check mark in the cell if that action is performed when the decision variables have the values shown in the column above the cell.

FIGURE 6-22
A decision table for calculating shipping charges.

YTD Purchases > $250	Yes						No					
Number of Items (N)	N ≤ 3			N ≥ 4			N ≤ 3			N ≥ 4		
Delivery Day	Next	2nd	7th	Next	2nd	7th	Next	2nd	7th	Next	2nd	7th
Shipping Charge ($)	25	10	N × 1.50	N × 6.00	N × 2.50	Free	35	15	10	N × 7.50	N × 3.50	N × 2.50

FIGURE 6-23
A decision table for calculating shipping charges.

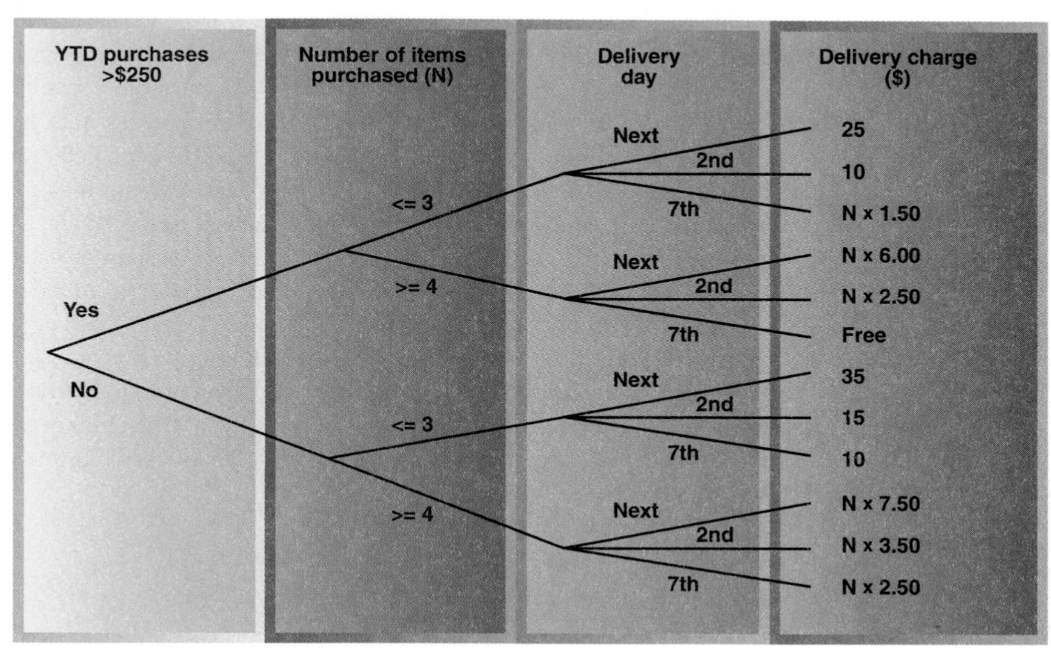

Now let's follow these steps to show how the decision table in Figure 6-22 was constructed. There are three decision variables—year to date (YTD) purchases, number of items ordered, and delivery date. YTD purchases have two relevant ranges: less than $250, and greater than or equal to $250. Note that decision variable ranges must be mutually exclusive and collectively exhaustive. Number of items ordered also has two relevant ranges: less than or equal to three, and greater than or equal to four. Delivery date has three possible values: next day, second day, and seventh day. There are 2 × 2 × 3 = 12 combinations of values, so there are 13 columns in the table to allow for a decision variable name, the formula, and action names.

Both YTD purchases and number of items have two relevant value ranges, so either can occupy the first row. We chose YTD purchases. It has two value ranges, so we created two groups of 12 ÷ 2 = 6 columns, and label one for each possible value. The next row is for number of items. It has two ranges, so we need four groups of three columns—that is, 12 ÷ 2 value ranges for number of items ÷ 2 value ranges for YTD purchases. The value ranges for number of items are inserted into the column groups in a regular pattern as shown in the sample figure. The delivery date is now inserted into the table. Since it is the last decision variable, we don't need to group any columns beneath it. We simply insert the values of the delivery date into individual columns in a regular pattern as we did for the other decision variables.

The final step is to insert the row containing formulas and values for the shipping charges. Each cell contains a value or formula for the combination of decision variable values in the columns above. For example, shipping is free for customers with YTD purchases greater than $250, orders of more than three items, and seventh-day delivery. The shipping charge is $35 for customers with YTD purchases less than $250, an order of three items or fewer, and next-day delivery.

If the decision table is used to represent a process that implements one or more actions, instead of value calculations as in the previous example, then the table must contain a row for each action. Cells in these rows are checked to indicate which actions are performed under which conditions. Figure 6-24 shows a simple example of this type of table. Two action rows are included and the action is performed if a check mark appears in the cell immediately below the decision variable values. For example, if the customer is new and the shipment contains an item back-ordered more than 25 days, then the shipment is expedited and the detailed return instructions are included in the container. If the customer isn't new and the order contains no items back-ordered more than 25 days, then neither action is taken.

You can construct a decision tree using almost the same steps as listed above for constructing a decision table. The primary difference is that rows in a decision table are columns in a decision tree and vice versa. To see this for yourself, draw an imaginary line through the table in Figure 6-22 from the top-left to bottom-right. Then flip the table along the imaginary line and compare the structure of the flipped table to the decision tree in Figure 6-23. The only other significant difference between a table and a tree is that a tree uses labeled branches instead of grouped columns to represent decision variable values.

FIGURE 6-24
A simple decision table with multiple action rows.

New Customer	Yes		No	
Item back order ≥ 25 days	Yes	No	Yes	No
Include detailed return instructions	✓	✓		
Expedite delivery	✓		✓	

Data Flow Definitions

data flow definition
A textual description of a data flow's content and internal structure

A data flow is a collection of data elements, so **data flow definitions** list all the elements. For example, a simplified "New order" data flow consists of a customer name, credit card number, and list of catalog item numbers and quantities. Some of these elements are actually structures of other elements, such as a customer name consisting of first name, middle initial, and last name. Most of these data elements are stored by the system, so they coincide with the attributes of data entities included in the ERD.

Sometimes data flow definitions contain a more complex structure. In the "New Order" example, each data flow consists of many catalog items and quantities (a repeating group). It is important to document this structure. The notations for data flow definitions vary. One approach is simply to list the data elements, as shown in Figure 6-25. The elements that can have many values are indicated. Another approach uses an algebraic notation such as that shown in Figure 6-26. The data flow "equals" or "consists of" one element plus another element, and so on. Groups of elements that can have many values are enclosed in curly braces. This example shows "New Order" "equals" the customer name "plus" credit card information "plus" "one or more" inventory item number and quantity. In this example, the customer name can be defined separately as a structure of elements.

FIGURE 6-25
Data flow definitions simply listing elements.

```
Customer-Name
Customer-Address
Credit-Card-Information
Item-Number
Quantity
```

FIGURE 6-26
Algebraic notion for data flow definition (New Order).

```
New-Order = Customer-Name + Customer-Address +
Credit-Card-Information + 1{ Item-Number + Quantity }N
```

Figures 6-27 and 6-28 show a complex report and its corresponding data flow definition. The structure of the report is a repeating group over products with an embedded repeating group over inventory items. The data flow definition captures this structure by embedding the item repeating group within one set of curly braces and the product repeating group as the outermost set of curly braces.

FIGURE 6-27
A sample report produced by the RMO customer support system.

Products and Items

ID	Name	Season	Category	Supplier	Unit Price	Special	Special Price	Discontinued
RMO210	HikeGood Pant	Spr/Sum	Clothes	8201	$35.00	☑	$0.00	No

Description Full Length Cotton Outdoor Pants

Size	Color	Style	Units in Stock	Reorder Level	Units on Order
Large	Blue	Pleated	3000	200	
Large	Charcoal	Plain	2000	200	
Large	Charcoal	Pleated	3000	200	
Small	Blue	Plain	3000	200	0
Small	Blue	Pleated	5000	200	
Small	Charcoal	Plain	3000	200	
Small	Charcoal	Pleated	3000	200	

ID	Name	Season	Category	Supplier	Unit Price	Special	Special Price	Discontinued
RMO212	CottonTees	Spr/Sum	Clothes	8201	$28.00	☑	$0.00	No

Description Lightweight Cotton T Shirts

Size	Color	Style	Units in Stock	Reorder Level	Units on Order
Large	Blue	Round	3000	300	
Large	Blue	V-neck	3000	300	
Large	Yellow	Round	3000	300	
Large	Yellow	V-neck	3000	300	
Small	Blue	V-neck	3000	300	
Small	Yellow	Round	2000	300	
Small	Yellow	V-neck	3000	300	

FIGURE 6-28
A data flow definition for the RMO products and items report.

```
products-and-items-report =

 N
 1{ product-id + product-name + season + category +
   supplier + unit-price + special + special-price +
   discontinued + description +
    N
    1{ size + color + style + units-in-stock +
      reorder-level + units-on-order
    }
 }
```

Data Element Definitions

Data element definitions describe a data type, such as string, integer, floating point, or Boolean. Each element should also be described to indicate specifically what it represents. Sometimes these descriptions are very specific. A date of sale might be defined as the date the payment for the order was received. Alternatively, the date of sale might be the date an order is placed. Sometimes different departments in the same company have different definitions for the same element, so it is very important for the analyst to confirm exactly what the element means to users.

Other parts of a data element definition vary depending on the type of data. A length is usually defined for a string. For example, a middle initial might be one character maximum, but how long should a first name be? Numeric values usually have a minimum and maximum value that can be defined as a valid range. Sometimes specific values are allowed for the element, such as valid codes. If the element is a code, it is important to define the valid codes and their meaning. For example, code A might mean ship immediately, code B might mean hold for one day, and code C might mean hold shipment pending confirmation. Some sample data element definitions are shown in Figure 6-29.

FIGURE 6-29
Data element definitions.

```
units-in-stock =
  a positive integer

supplier =
  a four digit numeric code

unit-price =
  a positive real number accurate to two decimal places,
  always in U.S. dollars

description =
  a text field containing a maximum of 50 printable characters

special =
  a coded field with one of the following values
  0: item is not "on special"
  1: item is "on special"
```


Data Store Definitions

Since a data store on the DFD represents a data entity on the ERD, no separate definition is typically needed (except perhaps a note referring the reader to the ERD). If data stores are not linked to an ERD, then the analyst simply defines the data store as a collection of elements (possibly with a structure) in the same way data flows are defined.

DFD Summary

Figure 6-30 shows each of the components of a traditional analysis model—entity-relationship diagram, data flow diagrams, process definitions, and data definitions. The four components form an interlocking set of specifications for most system requirements. The data flow diagram provides the highest-level view of the system, summarizing processes, external agents, data stores, and the flow of data among them. Each of the other components describes some aspect of the data flow diagram in greater detail.

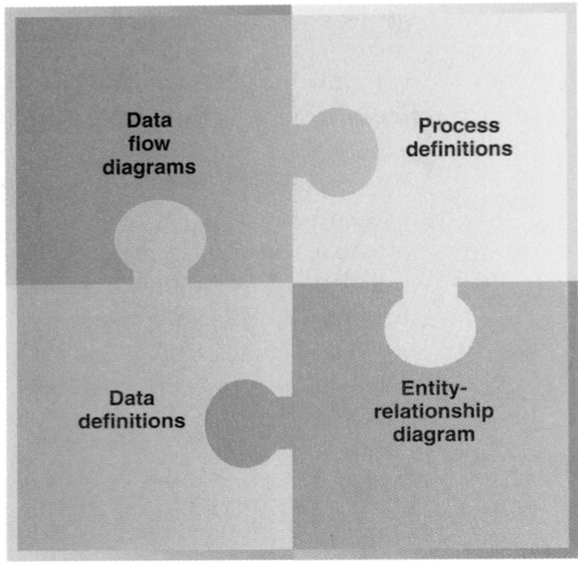

FIGURE 6-30
The components of a traditional analysis mode.

The models described thus far were developed in the 1970s and 1980s as part of the modern structured analysis methodology (see Yourdon 1989 in the "Further Resources" section). They were designed to document completely the logical requirements of a system. However, some analysts choose to augment the structured models with other models borrowed from other methodologies. Such models may be used to describe information not captured by the structured models. Or they may be used to present similar information in a slightly different form. The remainder of this chapter describes some of these "borrowed" models and how they can be used to augment the traditional structured analysis models.

Information Engineering Models

As we discussed in Chapter 3, the Information Engineering approach to development focuses on strategic systems planning and the data requirements of a system. **Information Engineering (IE)** was developed by James Martin in the early 1980s. At that time, IE and the structured system development methodology were considered competitors, although they have some features in common. As both methodologies matured, many analysts sought to combine the best features of both. This section provides an overview of the IE methodology and describes some IE analysis models and how they can be used with the structured analysis models covered earlier in this chapter.

The IE System Development Life Cycle

Figure 6-31 shows the phases of the IE system development life cycle (SDLC). The first phase of the IE SDLC concentrates on modeling the entire enterprise and producing coordinated information systems plans. The enterprise is modeled in terms of its strategic goals and plans, information technology infrastructure, stored data requirements, and major functional areas. Overlaps between the data and functional models of the enterprise are used to identify specific business areas for further analysis in the second IE SDLC phase.

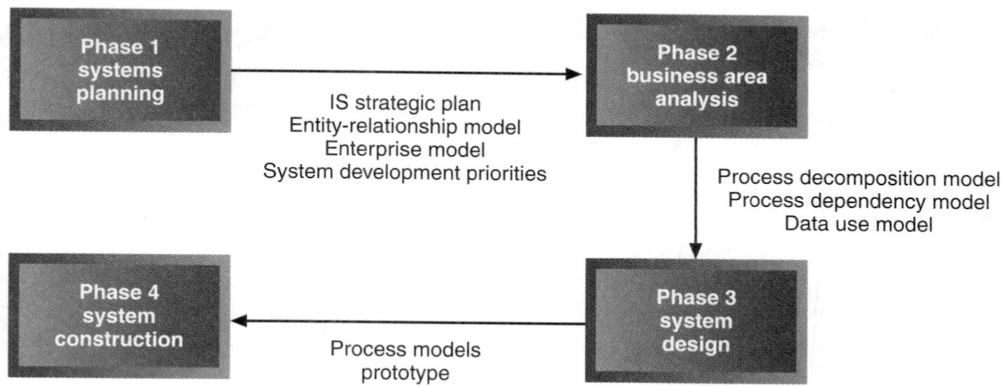

FIGURE 6-31
Information Engineering system development life cycle phases

Two key features of IE that are apparent in the first phase of its SDLC are its focus on the business as a whole and its focus on stored data. IE concentrates first on describing the enterprise and its environment in terms of strategies, plans, goals, objectives, and organization structure. Only then does it define requirements for information technology infrastructure and information processing applications. The focus on stored data follows from a key assumption that business data is an organizational resource that is much less subject to change than business processes. Internal and external data needs are assumed to drive processing requirements and process structure, not vice versa.

The second phase of the IE SDLC describes the processing requirements of each business area in the enterprise. Each business area is decomposed into a hierarchy of related processes. A process decomposition model describes the hierarchy, and a process dependency model describes interrelationships among processes. A set of data use models specifically describes relationships between the process and data models.

The third phase of the IE SDLC develops the detailed design of business processes. IE uses a number of different techniques and models to describe processes. The methodology stresses user input to process design and process model validation. Developers develop graphical process models. Users then validate those models.

The final phase of the IE SDLC is system construction. The IE methodology assumes that CASE tools and code generators are used to implement the construction phase. Process models and prototypes developed in the previous phase are used to generate system components automatically. Code generators take data and process models developed with CASE tools as inputs and generate application programs and database creation commands as outputs. Historically, IE has targeted procedural programming languages such as COBOL and C and relational database management systems. Later generations of IE CASE tools and code generators have targeted object-oriented environments.

IE and Structured Development Compared

A brief comparison of IE and structured methodologies was provided in Chapter 3. That comparison is summarized in Table 6-1. The most significant difference between IE and the structured methodologies is their scope. IE is designed to address enterprisewide information processing requirements. Its first phase looks at the enterprise as a whole and describes data and information processing within the context of organizational structure, goals, and strategies.

	Information engineering	Structured methodology
For what system size is it best suited?	enterprise	any size
Does it explicitly incorporate strategic planning?	yes	no
How tightly integrated are its techniques and models?	highly integrated	moderately integrated
What is the central focus of analysis?	data	processes
Can techniques from other methodologies be easily incorporated?	no	yes

TABLE 6-1
A comparison of the information engineering and structured methodologies.

In comparison, the structured methodology makes no assumptions about either the scope of its application or about planning activities that precede structured analysis. Structured development can be applied to entire enterprises, relatively small systems within an enterprise, or anything in between. Structured analysis does not explicitly require enterprise-level planning and description activities. Thus, its models do not directly incorporate information about organizational structure, enterprise strategies and goals, or existing or proposed information technology infrastructure.

Another significant difference between the methodologies is how tightly integrated their tools and techniques are. IE is a tightly integrated methodology in which the outputs of each phase completely specify inputs required in the next phase. The methodology is complete and all-encompassing. Thus, it is difficult (and generally unnecessary) to incorporate techniques and models from other methodologies within IE. In contrast, the structured methodology is a general set of techniques and models that are only moderately integrated. There are some gaps in the methodology; thus, it is possible (and usually desirable) to incorporate techniques and models from other methodologies such as IE.

An ERD for the entire enterprise is developed during the first IE phase. This model provides information that is used extensively in later IE modeling and design activities. Structured analysis also develops and uses an ERD, though it plays a less central role in the structured methodology than in the IE methodology. Because both methods develop an ERD, it is possible to incorporate into the structured methodology other IE models that depend on the ERD. The following sections describe some of the IE models that are useful additions to the structured analysis models described earlier in the chapter.

Process Decomposition and Dependency Models

IE takes a different approach to modeling processes and data flows than does structured analysis. IE process models show three types of information:

- Decomposition of processes into other processes
- Dependency relationships among processes
- Internal processing logic

process decomposition diagram
A model that represents the hierarchical relationship among processes at different levels of abstraction

Analysts first model business areas by decomposing them into subsidiary processes. Figure 6-32 shows an example of a **process decomposition diagram** for the RMO customer support system. This diagram shows the decomposition of the system (or business area) into major subsystems. The order-entry subsystem is further decomposed into processes that are in turn decomposed further. Processes that have further decomposition shown on a different diagram or page are indicated by three dots in their upper-left corner.

FIGURE 6-32
The RMO customer support system process decomposition diagram.

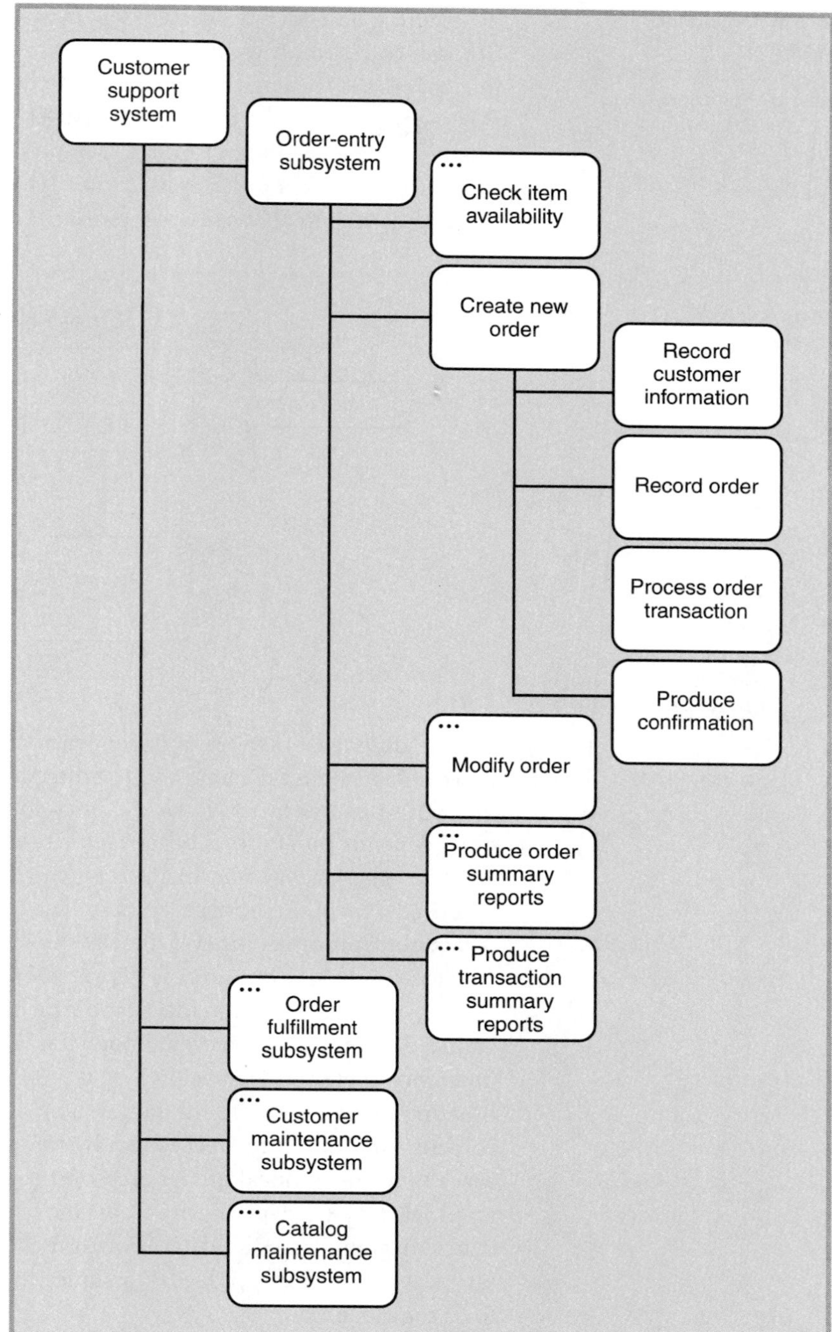

Compare the information content of Figure 6-32 to earlier examples of RMO DFDs (Figures 6-6, 6-13, and 6-15). There are several similarities between process representations used with IE and structured development. One obvious similarity is the symbol used to represent a process. Both methodologies use an oval, though it is oriented differently on the page.

Another similarity is the hierarchical relationship among processes at different levels of abstraction. Structured DFDs show these relationships indirectly by placing process decompositions in different DFDs and using a hierarchical numbering system to tie the various DFDs together. The process dependency diagram directly represents hierarchical relationships among process levels on a single page. Both methods show the same process decomposition information in different ways.

process dependency diagram
A model that describes the ordering of processes and their interaction with stored entities

Figure 6-33 shows a **process dependency diagram** for "Create new order." This diagram corresponds to the "Create new order" process decomposition DFD shown in Figure 6-15. The large arrow labeled "Order received" represents the event that triggers the first process on the diagram. The lines and arrows connecting processes indicate dependency among processes. The process "Produce confirmation" depends on the process "Process order transaction," which implies that "Produce confirmation" cannot begin until "Process order transaction" is completed.

FIGURE 6-33
A process dependency diagram.

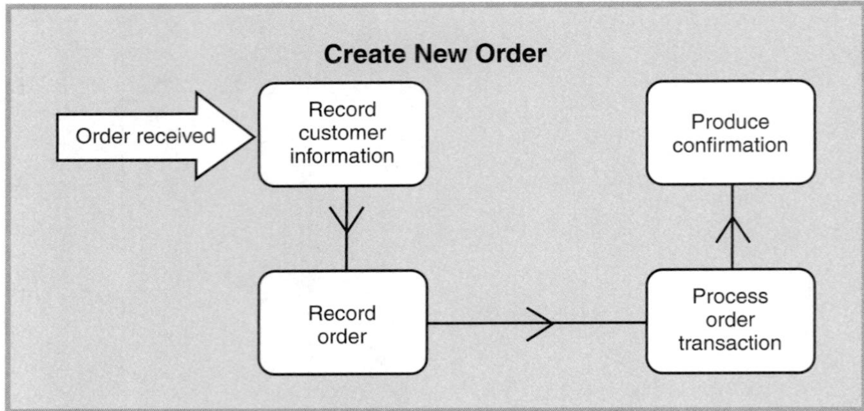

The difference between a dependency and a data flow is subtle but important. Process dependency diagrams do not directly show data flows among processes (note the lack of names or labels on the lines connecting processes). "Produce confirmation" depends on "Process order transaction" because it makes no sense to confirm an order that hasn't been finalized (such as an order for which credit has not been approved). Even if one process creates data that another process uses, the flow of that data among processes is never directly shown on a process dependency diagram.

Figure 6-34 shows a process dependency diagram with added data flows. Process dependencies are shown in black, and data flows are shown in red. IE calls such a diagram a data flow diagram even though it is quite different from a structured DFD. The differences between Figures 6-15 and 6-34 are striking. Although the number and ordering of processes are the same, there are significant differences in information content. The structured DFD explicitly shows data flows to and from processes. These flows (and the entities) are ignored on the process dependency diagram. The structured DFD routes data directly from one process to another, but the process dependency diagram routes all data through the data stores. The reason is that during the business area analysis phase, IE assumes that processes read and write data to and from the database.

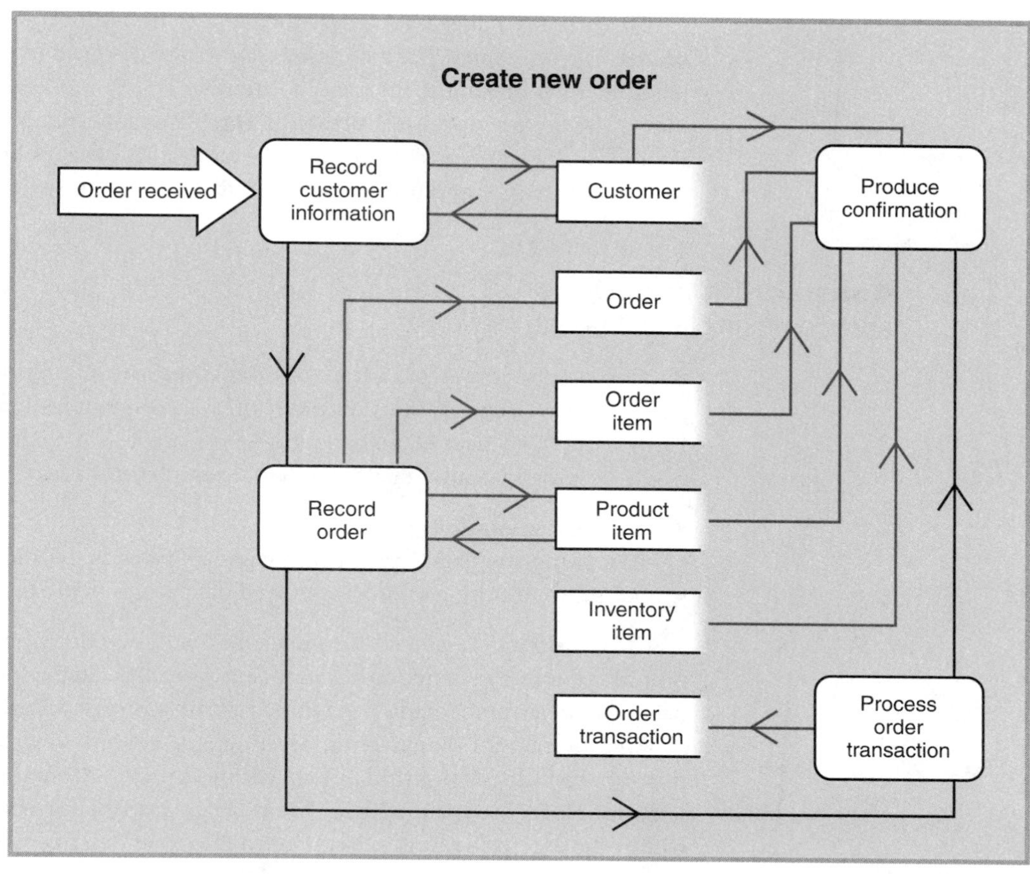

FIGURE 6-34
A process dependency diagram with data flows.

The differences between the diagrams demonstrate a different approach to representing a logical (rather than physical) model of processing. IE purposely avoids showing data flows among entities and processes because they might introduce unnecessary physical assumptions about process implementation. IE assumes that the "when" and "where" of data flows are design decisions. Data sources, destinations, and routing are not explicitly modeled until the system design phase.

In contrast, structured analysis requires representing flows to and from entities and processes. Data flow among processes can be forced through data stores, but this is neither required nor encouraged by structured analysis rules. The IE approach makes it difficult to create a model that is not logical. The structured approach provides a general set of modeling tools that can be applied to logical or physical models. It is up to the analyst to decide what restrictions (if any) are needed when developing logical DFDs.

Both process decomposition and process dependency diagrams can be useful additions to the structured methodology. Process decomposition diagrams are useful for graphically showing the hierarchical relationships among a large number of processes. Process dependency diagrams can be a useful alternative for examining and describing the logical essence of process relationships and data flows.

The primary danger of using either model in addition to traditional structured models is redundancy. The structured models are designed to be as nonredundant as possible, thus making them easier to modify. The IE models represent information that is also represented on structured DFDs. Thus, any change to a structured model requires a change to a corresponding IE model. This makes implementing a change more cumbersome, and opens up the possibility of inconsistency among the models.

Considering Locations and Communication through Networks

Because structured systems analysis concentrates on logical modeling, physical issues such as processing locations and networks are sometimes ignored during analysis. However, a great deal of information about process, data, and user distribution is needed during the early stages of design. Examples include:

- Number of locations of users
- Processing and data access requirements of users at specific locations
- Volume and timing of processing and data access requests

This information is needed to make initial design decisions such as the distribution of computer systems, application software, and database components. It is also needed to determine required network capacity among user and processing locations.

The Information Engineering development methodology specifies several tables that describe the relationship among entities on the ERD and other information system and enterprise components. These tables describe the sources and uses of data within the enterprise. Examples include tables that describe which processes initially capture information about an entity and which processes later use that data. IE develops these tables early in the SDLC and uses them to guide later decisions about process design and structure.

The structured development methodology does not explicitly recognize the need for gathering such information during the analysis phase. However, many analysts do gather and summarize distribution information during structured analysis. The information is usually readily available during the analysis phase, and gathering that information during analysis instead of design is typically more efficient. Gathering the information during analysis also provides a more informed basis for performing the last two analysis phase activities—*Generate and evaluate alternatives* and *Review recommendations with management*.

The first step in gathering distribution data is to identify and describe the locations where work is being or will be performed. These locations include business offices, warehouses, and manufacturing facilities, but they also include less obvious locations that need to be supported. Users might conduct work in customer or supplier offices, at home, in hotel rooms while traveling, and even in their cars. All of these locations should be listed, and a location diagram should be drawn to summarize the locations graphically. A location diagram for Rocky Mountain Outfitters is shown in Figure 6-35. The location diagram shows the analyst what network connections might be required, but also has the added benefit of reminding everyone that users at all locations should be consulted about the system.

location diagram
A diagram or map that identifies all of the processing locations of a system

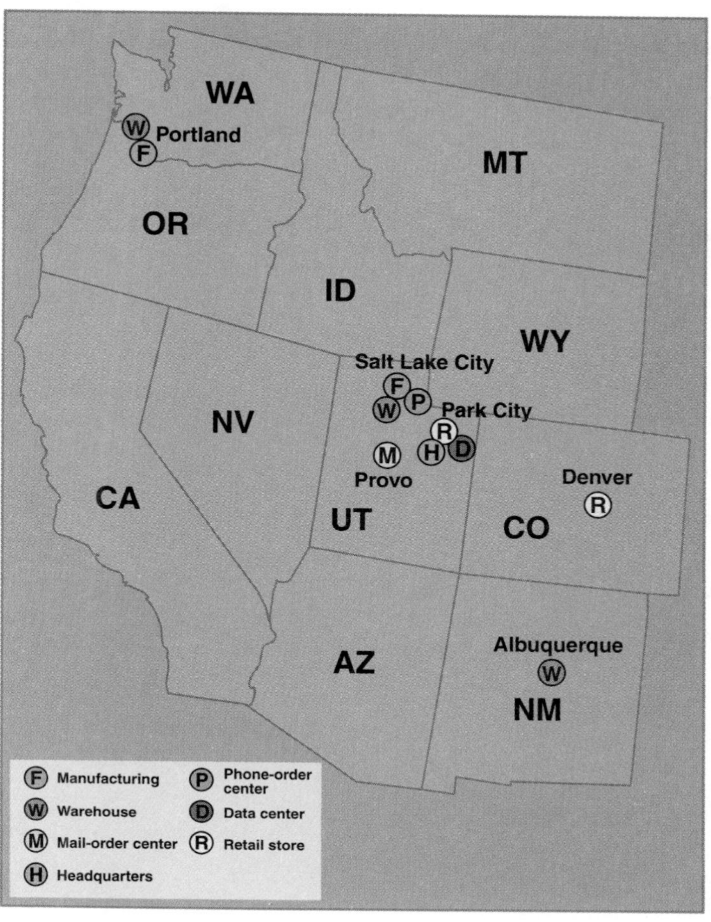

FIGURE 6-35
The Rocky Mountain Outfitters location diagram.

activity-location matrix
A table that describes the relationship between processes and the locations in which they are performed

activity-data matrix
A table that describes stored data entities, the locations from which they are accessed, and the nature of the accesses

CRUD
An acronym for create, read, update, and delete

The next step is to list the functions that are performed by users at each location. Using the event table, the analyst can list where each activity is performed. Figure 6-36 shows an **activity-location matrix** that summarizes this information. Each row is a system activity, and each column represents a location. Many activities are performed at multiple locations.

Recall that Rocky Mountain Outfitters also has a system project under way for the inventory management system. The inventory management system will involve many activities at the manufacturing facilities, but the customer support system will not. Additionally, there is a plan for integrating the system at the retail stores with the inventory management system, but not with the customer support system. Therefore, these locations are not shown on the activity-location matrix.

Other matrices can be created to highlight access requirements. One approach is to list activities and data entities (or classes of objects) in an **activity-data matrix**. This matrix shows which activities require access to the data or objects. This information can be found on the DFD fragments for the traditional approach and on the sequence diagrams for the object-oriented approach. In either approach, creating a matrix to summarize this information can be useful.

Figure 6-37 shows an activity-data matrix for Rocky Mountain Outfitters. In the cells of the matrix, additional information is shown to clarify what the activity does to the data. The letter *C* means the activity creates new data, *R* means the activity reads data, *U* means the activity updates data, and *D* means the activity might delete data. The acronym **CRUD** (create, read, update, and delete) is often used to describe this type of matrix.

ACTIVITY	LOCATION				
	Corporate offices (Park City)	**Distribution warehouses** (Salt Lake, Albuquerque, Portland)	**Mail-order** (Provo)	**Phone sales** (Salt Lake)	**Customer direct interaction** Anticipated
Look up item availability	X	X	X	X	X
Create new order			X	X	X
Update order				X	X
Look up order status	X	X		X	X
Record order fulfillment		X			
Record back order		X			
Create order return		X			
Provide catalog info			X	X	X
Update customer account			X	X	X
Distribute promotional package	X				
Create customer charge adjustment	X				
Update catalog	X				
Create special promotion	X				
Create new catalog	X				

FIGURE 6-36
Activity-location matrix for the Rocky Mountain Outfitters customer support system.

FIGURE 6-37
Rocky Mountain Outfitters activity-data matrix.

ACTIVITIES	DATA ENTITIES										
	Catalog	Customer	Inventory item	Order	Order item	Order transaction	Package	Product item	Return item	Shipment	Shipper
Look up item availability			R								
Create new order		CRU	RU	C	C	C	R	R		C	R
Update order		RU	RU	RUD	RUD	RUD	R	R		CRUD	R
Lookup order status		R		R	R	R				R	R
Record order fulfillment					RU					RU	
Record back order					RU					CRU	
Create order return		CRU		RU		C			C		
Provide catalog info	R		R				R	R			
Update customer account		CRUD									
Distribute promotional package	R	R	R				R	R			
Create customer charge adjustment		RU				CRUD					
Update catalog	RU		R				RU	R			
Create special promotion	R		R				R	R			
Create new catalog	C		R				CRU	R			

C = Creates new data, R = Reads existing data, U = Updates existing data, D = Deletes existing data

Workflow Modeling

workflow
The flow of control through a processing activity as it moves among people, organizations, computer programs, and specific processing steps

A **workflow** is the sequence of processing steps that completely handles one business transaction or customer request. A workflow encompasses a trigger, the processing steps implemented to respond to the trigger, participants (people and machines), and the flow of data and control among participants. Workflows may be simple or complex. Complex workflows may be composed of dozens or hundreds of processing steps and may include participants from different parts of an organization.

Analysts sometimes model workflows to gain a better understanding of the current system. A workflow model can be easily validated by a user or system participant because it directly models the sequence of processing activities. Workflow models may also be developed during the transition between analysis and structured design. They provide a way of describing the complex interactions among system components and participants. Workflow models can be used to describe alternative approaches to system organization and human-computer interaction. Analysts also commonly use workflow modeling when performing business process reengineering (as discussed in Chapter 4).

No single method is commonly used to model workflows. Modeling methodologies commonly employed include flow charts, data flow diagrams, and activity diagrams. Data flow diagrams do a good job of capturing the flow of data within a workflow, but aren't designed to represent control flows. Flow charts and activity charts are specifically designed to represent control flow among processing steps, but they don't represent data flow.

Figure 6-38 shows a workflow model for a university scheduling process. An activity diagram is used to represent the workflow. Rounded rectangles represent processing steps, arrows represent control flows, diamonds represent decisions, and horizontal bars represent synchronization points. The model shows each processing step within a vertical box that represents a participant. Thus, the model shows not only the flow of control among processing steps but also among participants.

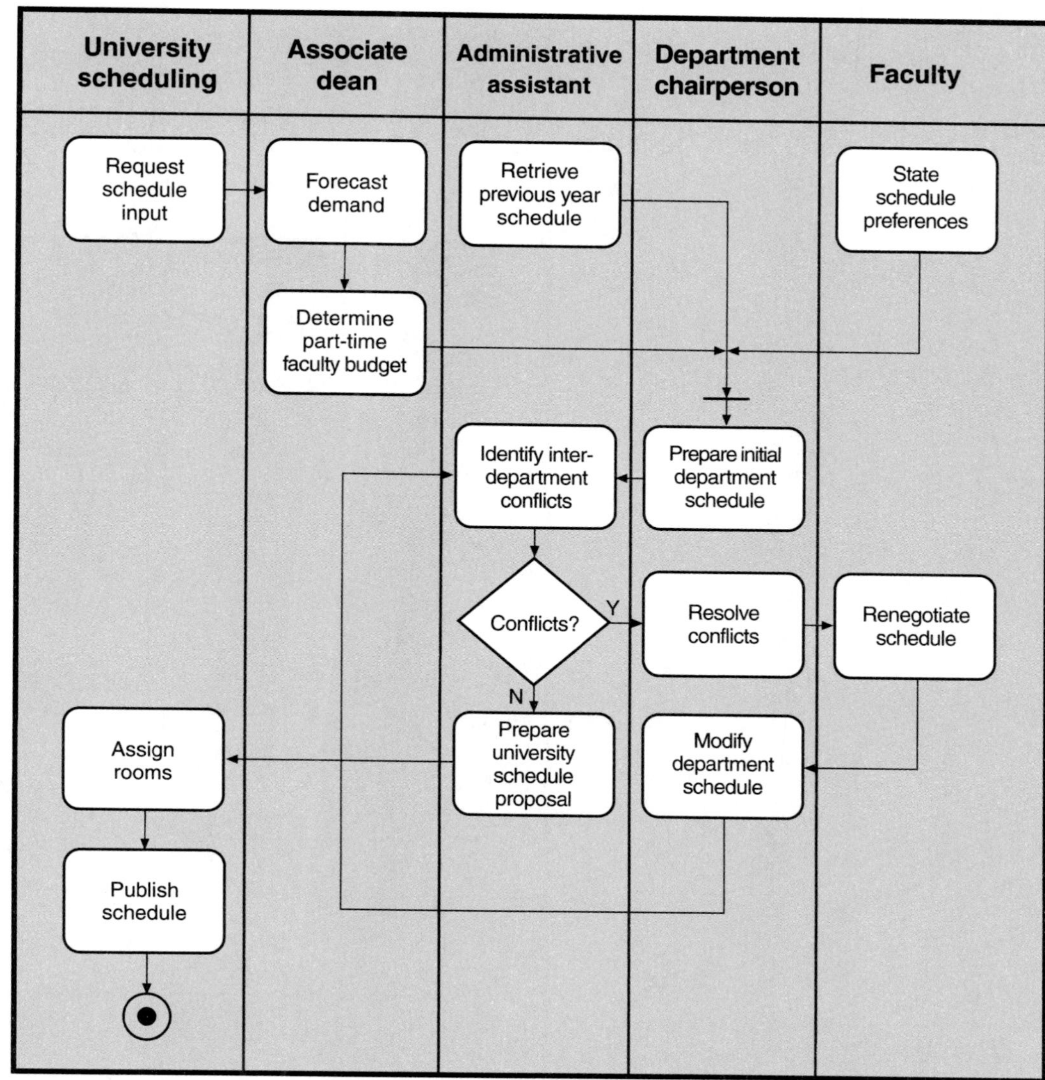

FIGURE 6-38
A workflow model of a university scheduling system.

Figure 6-39 shows one possible workflow model for the RMO telephone order entry process. This model shows the interactions between the customer, an order clerk, and the automated system. An analyst can use this type of model to document the flow of control through a process prior to constructing software to implement the automated system. Alternative workflows can be documented and analyzed to determine the best basis for designing and implementing the man-machine interface.

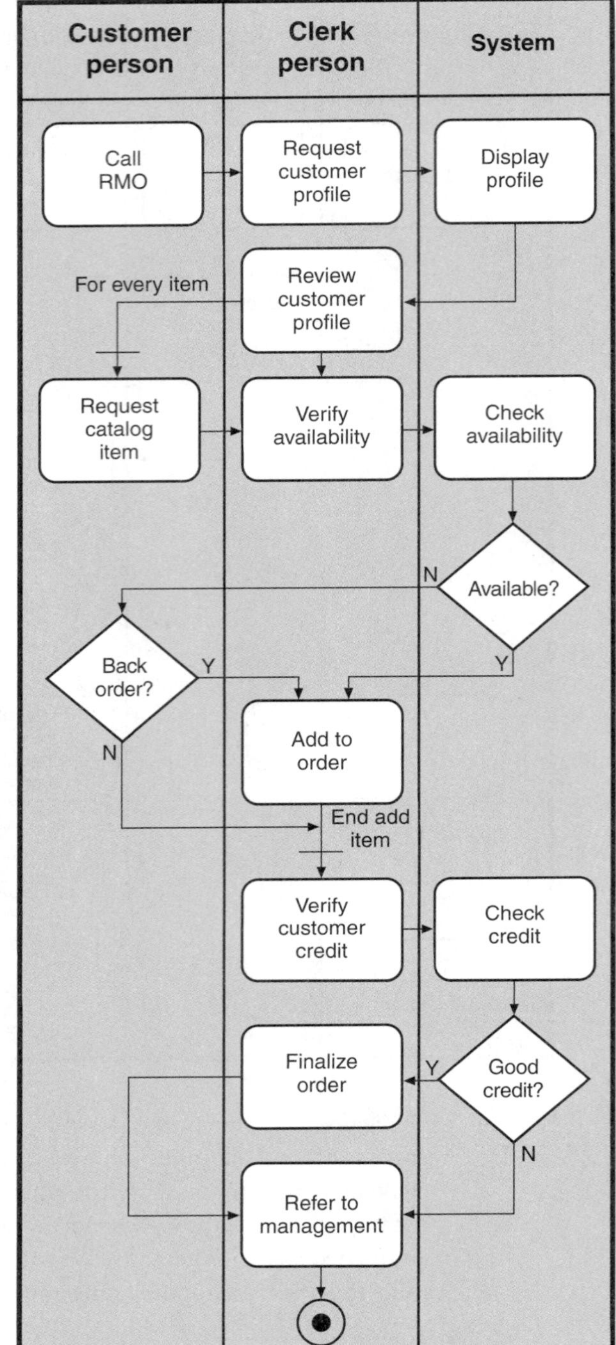

FIGURE 6-39
A workflow model of the RMO telephone order entry system.

Summary

Data flow diagrams (DFDs) are used in combination with the event table and entity-relationship diagram (ERD) to model system requirements. DFDs model a system as a set of processes, data flows, external agents, and data stores. DFDs are relatively easy to read because they graphically represent key features of the system using a small set of symbols. Because there are many features to be represented, many types of DFDs are developed including context diagrams, DFD fragments, subsystem DFDs, event-partitioned DFDs, and process decomposition DFDs.

Each process, data flow, and data store requires a detailed definition. Processes may be defined by a number of methods, including a structured English process specification, a decision table, a decision tree, or a process decomposition DFD. Process decomposition DFDs are used when internal process complexity is too great to create a readable, one-page definition by any other means. Data flows are defined in terms of their component data elements and their internal structure. Data elements may be further defined in terms of their type and allowable content. Data stores correspond to entities on the ERD, and thus require no additional definition.

Data flow diagrams may be supplemented by models borrowed from other methodologies to provide additional information or an alternative view of requirements. Information Engineering (IE) models are a useful complement to traditional analysis models. Particularly useful models include the process decomposition diagram, process dependency diagram, location diagram, activity-location matrix, and activity-data (or CRUD) matrix. The process decomposition model provides a useful summary of how processes on multiple DFD levels are related to one another. The process dependency diagram provides a view of processing details that emphasizes interaction with stored entities instead of data flow among processes.

The location diagram, activity-location matrix, and activity-data matrix describe important information about system locations. The location diagram summarizes geographic locations where the system is to be used. The activity-location matrix describes which processes are implemented at which locations. The activity-data matrix summarizes where and how each data store is used.

We've now covered all of the models that are used to document system requirements in the traditional approach to systems analysis. Chapter 7 will cover the models used to document system requirements in the object-oriented approach to systems analysis. Chapter 8 will cover the transition from systems analysis to systems design. Chapter 9 will show how traditional models developed in this chapter and the object-oriented models developed in Chapter 7 are extended or supplemented to design a new system.

Key Terms

activity-data matrix, p. 205
activity-location matrix, p. 205
balancing, p. 188
black hole, p. 188
context diagram, p. 173
CRUD, p. 205
data flow, p. 172
data flow definition, p. 195
data store, p. 172
decision table, p. 192
decision tree, p. 192
DFD fragment, p. 174
diagram 0, p. 175
event-partitioned system model, p. 175
external agent, p. 171
information overload, p. 187
levels of abstraction, p. 172
location diagram, p. 204
minimization of interfaces, p. 187
miracle, p. 189
process, p. 171
process decomposition diagram, p. 200
process dependency diagram, p. 202
rule of 7 ± 2, p. 187
structured English, p. 190
workflow, p. 208

Review Questions

1. List at least three different types of DFDs. What is each diagram type used to represent?
2. List the five component parts (symbols) of a DFD. Briefly describe what each symbol represents.
3. How does an analyst determine whether a person or organization should be represented on a DFD as an external agent or one or more processes?
4. Processes on an event-partitioned DFD can be described by a detailed DFD or a process specification. How does an analyst determine which is the most appropriate form of description?
5. Describe how each column of an event table is represented on a DFD (that is, what symbols are used?).
6. How are entities from the ERD represented on a DFD? How are relationships from the ERD represented on a DFD?
7. What features may be present on a physical DFD that should never be present on a logical DFD?
8. What DFD characteristics does an analyst examine when evaluating DFD quality?
9. What is a black hole? What is a miracle? How can each be detected?
10. Why might an analyst describe a process with a decision table or tree instead of structured English?
11. List IE process models that are useful supplements to the traditional analysis-phase modeling tools (such as DFDs, data definitions, and process specifications). Describe the advantages and disadvantages of each IE model compared to a similar traditional model.
12. What are the disadvantages of using both traditional and IE process models to describe the same system?
13. List and briefly describe the life cycle phases of Information Engineering.
14. What is an activity-location matrix? How is it related to DFDs?
15. What is an activity-data matrix? How is it related to DFDs and the ERD?
16. What is a workflow model? What information does it describe that isn't (or can't be) described on a DFD?

Thinking Critically

1. Assume that you are preparing a DFD to describe the process of creating, approving, and closing a mortgage loan by a mortgage broker. Should the broker be represented as an external agent or by one or more processes? Why? What about the closing agent, the credit bureau, and the bank that issues the mortgage note?
2. Examine the course registration system described in Figure 6-10. Are there any other processes that would be required to implement a fully functioning system? Hint: Black holes and miracles may indicate processing steps that were left out of the DFD.
3. Develop a process decomposition DFD for process 3 ("Modify existing order") in Figure 6-14. What similarities are there between your DFD and the one shown in Figure 6-15? Try to redraw all three DFDs to eliminate redundancies. Are your revised DFDs more or less readable than the original DFDs?
4. Assume that the transaction summary report for the RMO order-entry subsystem (see process 5 in Figure 6-8) contains a listing of every order that was created during a date range entered by the user. The report title page contains the report name, the date range, and the date and time the report was prepared. For each order, the report lists the order number, order date, order total, and form of payment. Within each order, the report lists all order items and returns, including item number, quantity ordered (or returned), and price. Report totals include the sum of all order totals, average order total, average item price, and average return price. Write a data flow definition entry for the report and write a process specification for the process that produces the report.
5. Create a process dependency diagram for all the processes shown in Figure 6-10.
6. Create a process decomposition diagram (with data flows) for process 1 in Figure 6-10.
7. Create an activity-data (CRUD) matrix for the course registration system in Figure 6-10.

Experiential Exercises

1. Develop a physical DFD that models the process of grocery shopping from the time you write down a shopping list until the time you store purchased groceries in your home. Construct your DFD as a linear sequence of processes. Now develop a logical DFD to describe the same scenario. Try to develop a diagram that is equally valid as a logical description of the way you currently buy groceries and as a logical description of ways you might buy groceries without ever leaving your home.
2. Consider the admissions requirements for a degree program, major, or concentration at your school. Look up the requirements in the school catalog and rewrite them in structured English. Develop an equivalent decision table and/or decision tree. Which is easier to understand? Why?
3. Get a copy of your school transcript. Write a data definition that describes its contents. Write data element definitions for the fields Grade, Credits, and Degree.
4. Define process 2 in Figure 6-7 as it is implemented at your school. Use whatever combination of process decomposition and process specification is appropriate. If you develop any process decomposition DFDs, then be sure to define all data flows.
5. Create a workflow diagram to represent the process of checking out books at your school library.

Case Studies

The Real Estate Multiple Listing Service System

Refer to the description of the Real Estate Multiple Listing Service system in the Chapter 5 case studies. Use the event list and ERD for that system as a starting point for the following exercises:

1. Draw a context DFD.
2. Draw an event-partitioned DFD.
3. Draw any required process decomposition DFDs.

The State Patrol Ticket Processing System

Refer to the description of the State Patrol ticket processing system in the Chapter 5 case studies. Use the event list and ERD for that system as a starting point for the following exercises:

1. Draw a context DFD.
2. Draw an event-partitioned DFD.
3. Draw any required process decomposition DFDs.
4. Create data flow definitions for any data flows that are fully described in the written system description.

The Reliable Pharmaceutical Service

The Reliable Pharmaceutical Service (RPS) provides pharmaceutical preparation and delivery services to 12 different nursing homes in Albuquerque, NM. RPS accepts orders for nursing home patients and delivers the orders in locked cases to each nursing home. The service employs approximately 12 delivery personnel, 20 pharmacist's assistants (PAs), six licensed pharmacists, and various office and clerical staff.

Orders are normally phoned in by nurses or administrative personnel within each unit of a client nursing home. A nursing home unit is either an entire floor or a wing within a floor (for example, third floor east wing). Orders identify the unit, individual patients by name and room number, drugs prescribed, dosage, and quantity. Some prescriptions are specified as a standing order, which is filled for each delivery until the prescription is specifically canceled. Orders are entered into a computer system as they are received, and a case manifest is immediately prepared. Orders are phoned or faxed in by 3:00 A.M. and P.M. for deliveries starting at 6:00 A.M. and P.M.

A licensed pharmacist reviews each case manifest as it is received and allocates the work needed to fill each portion of the order to PAs, another licensed pharmacist, or to himself or herself. All drugs for a single patient are collected in one plastic drawer of a locking case. Each case is marked with the institution name, floor number, and wing number (if applicable). Each drawer is marked with the patient's name and room number. Dividers are inserted within a drawer to separate multiple prescriptions for the same patient.

When all of the individual components of an order have been assembled, a pharmacist makes a final check of the contents to the case manifest, signs each page of the manifest, and places in the case the bottom two copies of the manifest (the manifest is a four-part carbonless form). The case is then locked and moved to the loading area. One of the remaining copies is placed in a mail basket for later processing and the other stored in file cabinet in the assembly area. When all of the cases have been assembled, they are loaded onto a truck and taken to the nursing homes.

Each Monday morning a billing report is prepared for each nursing home based on orders for the previous week ending Sunday evening. Each billing report is organized by patient and lists the total quantity of each dosage delivered during the previous week, the drug price, total charge per drug, and total charge per patient. The report totals include the total charge for each drug (for all patients prescribed that drug) and the total charge for all drugs. Billing reports are mailed to each institution.

When payments are received from nursing homes, they are checked against outstanding billing reports. If the payment doesn't match any outstanding billing reports, an inquiry is immediately made of the nursing home by telephone. Depending on the instructions of the customer, the payment is either returned or processed as a partial payment, or the billing report is adjusted. Once a payment is matched (either fully or partially), the check number and date of the payment are recorded. A deposit slip is then prepared for any payments just processed and the deposit is hand carried to the bank. A record of the deposit (individual check numbers, check amounts, and customer names) is also made and filed.

Prepare a set of DFDs, process specifications, and data flow definitions based on the preceding system description and the RPS event table (Table 6-2) and ERD (Figure 6-40).

Event	Trigger	Source	Activity	Response	Destination
Nursing home sends order	Order (phone or fax)	Nursing home	Fill order	Filled order	Nursing home
Time to create billing report	"Monday morning"		Create billing reports	Billing reports	Nursing home
Payment arrives	Payment	Nursing home	Record payment	Deposit	Bank

TABLE 6-2
The Reliable Pharmaceutical Service event table.

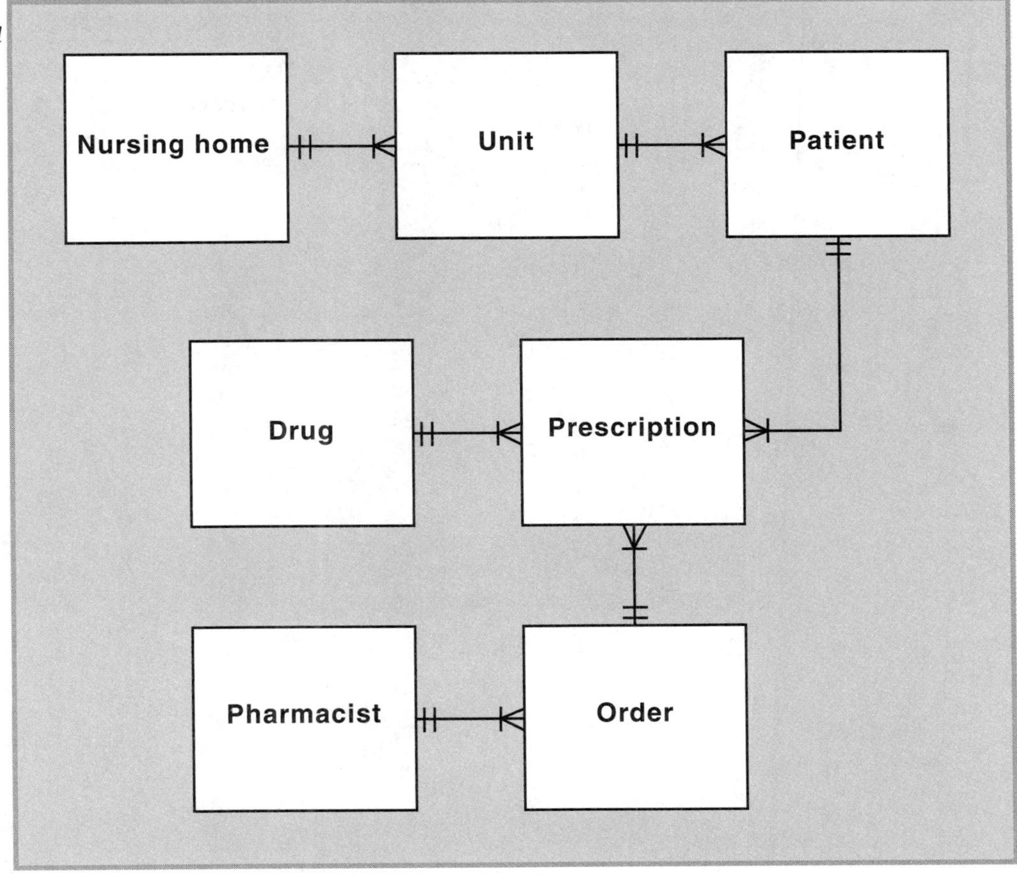

FIGURE 6-40
The Reliable Pharmaceutical Service entity-relationship diagram.

Further Resources

Stefan Jablonski and Christoph Bussler. *Workflow Management: Modeling Concepts, Architecture, and Implementation*. International Thompson Computer Press, 1996. ISBN 1-85032-222-8.

J. Martin. *Information Engineering: Book I Introduction*. Prentice Hall, 1988. ISBN 0-13-464462-X.

J. Martin. *Information Engineering: Book II Planning and Analysis*. Prentice Hall, 1989. ISBN 0-13-464885-4.

Stephen M. McMenamin and John F. Palmer. *Essential Systems Analysis*. Yourdon Press, 1984. ISBN 0-13-287905-0.

G. A. Miller. "The magical number seven, plus or minus two: Some limits on our capacity for processing information." *Psychological Review*, volume 63 (1956), pp. 81–97.

Edward Yourdon. *Modern Structured Analysis*. Yourdon Press, 1989. ISBN 0-13-598624-9.

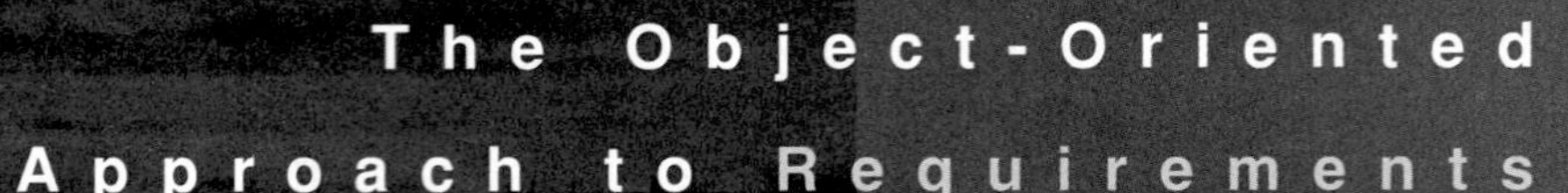

The Object-Oriented Approach to Requirements

Learning Objectives

After reading this chapter, you should be able to:

- Develop use case diagrams
- Develop use case scenarios
- Develop sequence diagrams
- Develop collaboration diagrams
- Develop statechart diagrams
- Explain the way that object-oriented diagrams work together to provide functional requirements

Chapter Outline

SEVEN

NORTHWOODS KAYAKS: IDENTIFYING THE METHODS WITHIN THE OBJECTS

This is an effective way to do system development for these new event-driven systems. I think you will learn a lot using the object-oriented techniques for this project," Bill Hinson said. Bill, who was a team leader on the project for the new payroll system at Northwood's Kayaks, was talking to Roger Tomason, who had been assigned to Bill's team for the summer. Roger was a summer intern from the university and was asking how UML (Unified Modeling Language) worked, since it was being used on their project.

Bill continued, "We're developing most of our new systems using object-oriented principles. The complexity of the new system, along with the fact that it is interactive, makes the object-oriented approach a natural way to develop requirements. It takes a little different thought process than you may be used to, but the object-oriented models track very closely with the new object-oriented programming languages."

Bill continued, "This new way of thinking about a system in terms of objects is very interesting. It also is consistent with the object-oriented programming techniques you learned in your programming classes. You probably first learned to think about objects when you developed screens for the user interface. All of the controls on the screen, such as buttons, text boxes, and drop-down boxes, are objects. Each has its own set of trigger events that activate its program functions.

"Now you just extend that same thought process so that you think of things like time cards and employees as objects. We can call them business objects to differentiate them from screen objects such as windows and buttons. During analysis, we have to find out all of the trigger events and methods associated with each business object."

"I don't think it will be hard for me to identify the business objects," Roger said. "I learned how to do that in my database and data modeling class."

"That's right," said Bill. "The classes you build in your data model are the business objects that we will work with. The next thing you need to learn is how to identify the trigger events associated with each object class and what the methods must do. In other words, we must determine the object's behavior."

"How do we do that?" Roger asked.

"You continue with your fact-finding activities and build a scenario for each business process. The way the business objects interact with each other in the scenario determines how you identify the trigger events. We refer to those trigger events as the messages between objects. The tricky part is that you need to think in terms of objects instead of just processes. Sometimes it helps me to pretend I am an object. I will say, 'I am a time card object. What functions and services are other objects going to ask me to do?' Once you get the hang of it, it works very well, and it is very enlightening to see how the system requirements unfold as you develop the diagrams."

"Ahaa! I think I get it. But how do we write the requirements? Do we just write them down?"

Bill responded, "No, of course not. As with the other parts of the analysis, we build models to document what we learn. In this case, we build an interaction diagram, which shows how the business objects interact. We also build other diagrams, called statechart diagrams, to show the internal logic for each business object."

Roger replied, "I'm looking forward to learning how to do object-oriented analysis. I enjoyed the concepts I learned in my object-oriented programming class. I think this summer will be a good learning experience for me."

OVERVIEW

As you saw in Chapter 5, systems analysis begins by defining business events and things. Business events are documented in the event table, which provides the basis for discovering the data flows and processes that are included in the new system. The things are documented in an entity-relationship diagram (ERD) for the traditional approach and in a class diagram for the object-oriented approach.

In Chapter 6, you learned how the event table and the entity-relationship diagram were used to drive the development of the application requirements for the new system based on the structured approach to development. In this chapter, you will learn how the event table and the class diagram are used to develop the requirements for the new system based on an object-oriented approach. We demonstrate object-oriented diagrams with the Rocky Mountain Outfitters (RMO) customer support system example.

From the object-oriented viewpoint, the world consists of objects. Any given business function is supported by a set of objects that work together, or collaborate, to carry out the function. To do object-oriented analysis, we first define the set of objects that will make up the system. Then we describe how those objects work together. There are two parts to this description. First, we must understand and describe the interactions or communication among the various objects. These interactions take the form of messages between objects. Second, we must describe the internal processing that goes on within each object to respond to messages from other objects, and to initiate messages to other objects.

The Unified Modeling Language and the Object Management Group

The object-oriented (OO) modeling notation that we present in this textbook is the Unified Modeling Language (UML). UML is the successor to the modeling techniques found in Grady Booch's Object Technology, James Rumbaugh's Object Modeling Technique (OMT), Ivar Jacobson's Object Oriented Software Engineering, and several other methods. In 1995, a preliminary version of UML was presented to the OOPSLA (Object-Oriented Programming, Systems, Languages, and Applications) conference. By January 1997, UML had gone through several iterations and reviews incorporating public feedback and revision by the primary authors. In January 1997, UML was presented to the Object Management Group (OMG) in response to its request for a standard modeling technique.

OMG is a consortium of over 800 software vendors, developers, and organizations that have combined their efforts to develop and propagate uniformity in object-oriented systems. Established in 1989, OMG's mission is to promote the theory and practice of object technology for the development of distributed computing systems. The goal is to provide a common architectural framework for object-oriented applications based on widely available interface specifications. Since January 1997, other revisions of UML have been developed and submitted to OMG for consideration as a widely accepted standard for OO modeling.

Object-Oriented Requirements

Five separate, but interrelated, object-oriented models, or diagrams, are used to define the application requirements from the object-oriented perspective. In most cases, analysts use all five diagrams to get a complete definition of business requirements. However, in some situations, only three or four diagrams may be required to specify the requirements accurately. The five diagrams are (1) the class diagram, (2) the use case diagram, (3) the collaboration diagram, (4) the sequence diagram, and (5) the statechart diagram.

You learned about the notation and form of a class diagram in Chapter 5. The purpose of the class diagram is to identify, and classify, the objects that will make up the new system. In a class diagram, the properties, or attributes, of each object that need to be recorded are also identified. For example, the RMO system will contain customer objects. Therefore, we defined a class called the *Customer class* along with the attributes for each customer. Generally, a single comprehensive diagram is used to show all classes and relationships for the entire system.

use case diagram
A diagram to show the various user roles and how those roles use the system

collaboration diagram
A diagram showing the objects that collaborate together to carry out a use case

message
The communication between objects within a use case

sequence diagram
A diagram showing the sequence of messages between objects during a use case

interaction diagram
Either a collaboration diagram or a sequence diagram that shows the interactions between objects

statechart diagram
A diagram showing the life of an object in states and transitions

The purpose of the **use case diagram** is to identify the "uses," or use cases, of the new system—in other words, to identify how the system will be used. The use case diagram is essentially an extension of the event table. A use case diagram is a convenient way to document the functions that the system must support. Sometimes a single, comprehensive use case diagram is used to describe the entire system. At other times, a set of smaller use case diagrams make up the use case model.

The purpose of a **collaboration diagram** is to identify which objects "collaborate" to carry out a given business function. For example, if one of the business uses of the RMO system is to "record a customer order," then the collaboration diagram identifies all of the objects involved. To record a customer order requires a customer object, some inventory objects, a new order object, and so forth. A single collaboration diagram is used to identify these objects and show the interactions, or **messages**, that are sent between them to carry out this function. Generally, many collaboration diagrams are needed.

A **sequence diagram** is just another view of the same information contained in a collaboration diagram, but it provides a little different perspective. It is a graphical view that emphasizes the sequence of the messages rather than the collaborating objects. A sequence diagram is drawn so that the sequence of the messages is graphically depicted by their position on the page. The sequence proceeds from top to bottom. Both the collaboration diagrams and the sequence diagrams are referred to as **interaction diagrams**.

Finally, the last diagram used for the application requirements is the statechart diagram. A **statechart diagram**, or more simply just a statechart, describes the states and the behavior of each individual object. Each object class has a statechart. Within the statechart are action statements that eventually become the logic in the final system. These logic components within each class are called methods.

Rather than try to remember how all of these diagrams relate to each other at this point just remember the following:

OO Requirements = Event Table + Class Diagram + Use Case Diagrams + Interaction Diagrams (collaboration and/or sequence diagrams) + Statechart Diagrams

Throughout this chapter, we will show how these various diagrams support each other to give a complete requirements statement. At the end of the chapter, we will tie them all together again.

Again, the difference between the traditional structured approach and the object-oriented approach to systems development is not in the development method, but in the set of models that is used. The SDLC for each approach has the same phases: planning, analysis, design, and implementation. The activities within the phases are also the same. The set of models used is the difference. Of course, since the individual tasks within the activities are focused on building the models, they will be different. For example, instead of building DFDs, the analysts build interaction diagrams and statecharts. The amount and level of prototyping for the object-oriented approach may be different, but the activity of prototyping is part of the SDLC for either approach. Figure 7-1, which repeats the information in Figure 5-34, shows the various diagrams that make up the requirements diagrams for both the structured approach and the object-oriented approach. Note that both approaches begin with the event table, and both approaches require the identification of things. As you saw in Chapter 5, the ERD and the class diagram use different notation, but share many common concepts. The primary difference between the two methods is that the diagrams, or models, used to represent the processes or system activities are quite different. You learned how to build DFDs in the last chapter. Now you will learn how to build use cases, interaction diagrams, and statecharts. Let's begin with a refresher on the contribution made by a class diagram for the development of the new system.

FIGURE 7-1
Requirements diagrams for traditional and object-oriented models.

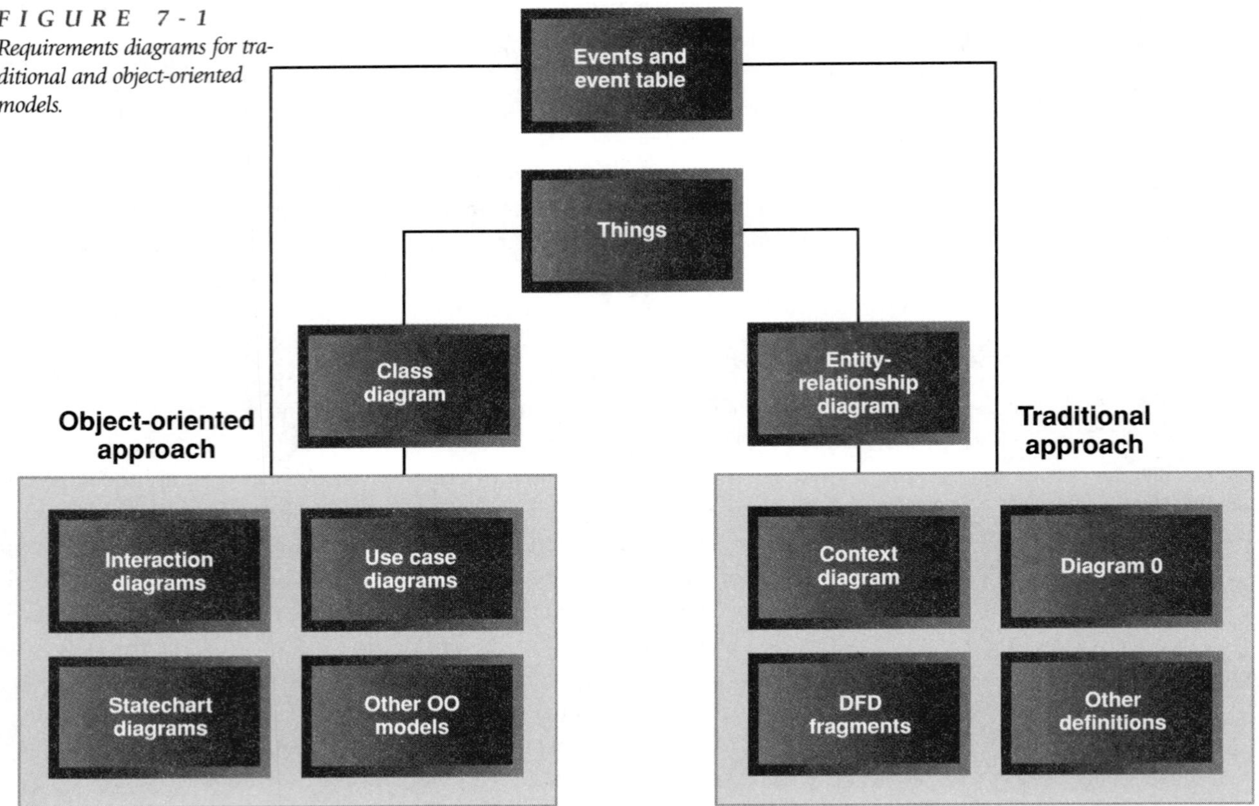

The Class Diagram

Often when we discuss system development, we divide the description of the new system into two categories: structural information and behavioral information. By structure we mean the component parts of the system, and by behavior we mean the logic performed by the components. The class diagram provides the definition of the components of the system. The other diagrams that you learn about in this chapter—namely, use case diagrams, interaction diagrams, and statechart diagrams—focus on the activities that the system performs. In other words, they describe the behavioral aspects of the new system. Thus the class diagram tells what the components of the system are, and the other diagrams tell what these components do or the actions they perform.

In Chapter 5, you learned about classes and objects and how to build a class diagram showing the classes of things in the problem domain of the system. The classes identified in the class diagram provide the structure for the database. Since in the object-oriented approach a program consists of object classes, this same set of classes provides the definition of the classes that will become the object-oriented program. So not only do the classes define the structure of the database, they also define the structure of the computer program. You will learn about object-oriented program structure in Chapter 9 when we discuss program design.

As Chapter 5 explains, the identified classes are those associated with the problem domain. Additional classes, such as user-interface classes, are also specified in the class diagram. User-interface classes include such classes as windows for input screens and output report classes. For example, the customer input form window is a user interface class that is designed during the development of the user interface. Those types of classes are called **implementation classes**, and you will learn how they are added to the class diagram in Chapter 11.

implementation classes
Classes in a class diagram needed for physical implementation, such as user-interface classes

In summary, the class diagram provides important structural information for the new system. The diagram includes information about the system database as well as the object-oriented program. The class diagram for the program consists of both problem domain classes and implementation classes. Next, we move to the diagrams that describe the system activities. These models are used as the basis for developing the algorithms and program routines for the new system.

The System Activities: An Object-Oriented Use Case/Scenario View

Use Cases and Actors

use case
A single use or function performed by the system for those who use the system

actor
A role played by a user of the system

The object-oriented approach uses the term **use case** to describe an activity the system carries out in response to an event. You can think of a use case as a case or situation where the system is used for some purpose. For example, consider a word processing system. A use case analysis of the word processor might identify various use cases such as "write a letter" or "print a newsletter." There are two important concepts in this analysis: (1) a person is involved, and (2) the person uses the system. In UML, the person involved is called an **actor**. An actor is always outside of the automation boundary of the system. Another way to think of an actor is as a role. In the word processing example, there may be a student actor who uses the word processor to write an essay and an editor actor who uses the word processor to edit articles. The same person may play several roles, such as a student, an editor, or a publisher. For example, in the RMO case, the use case "Create new order" might involve a sales representative talking to the customer on the phone. On the other hand, the customer might be the actor if the customer places the order directly over the Internet.

Figure 7-2 shows how a use case is documented. A simple stick figure is used to represent an actor. The stick figure is given a name that characterizes the role being played by the actor. The use case itself is symbolized by an oval with the name of the use case inside. Lines between actors and use cases indicate which actors participate with which use cases.

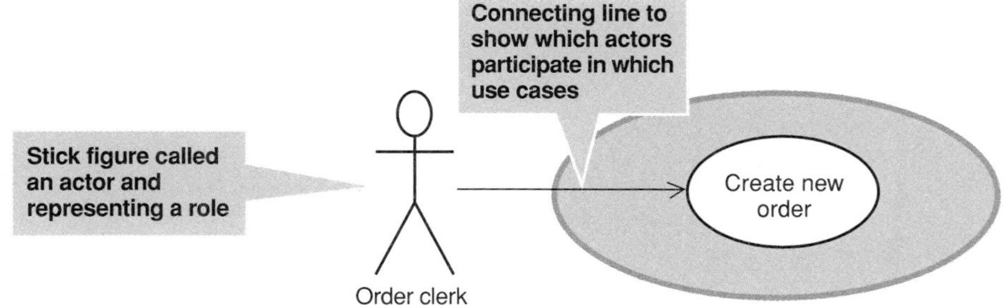

FIGURE 7-2
A simple use case with an actor.

Scenarios

A use case only shows that an actor interacts with the computer system to carry out a business activity. A use case is a high-level description and may include a whole sequence of individual steps to accomplish the use case. We describe these individual steps with a narrative that is called a *flow of activities*. A flow of activities describes the internal steps or activities within a use case. For the "Create new order" use case, the main steps might be similar to the subprocesses used in the data-flow diagram (diagram 2 in Figure 6-15): "Record customer information," "Record order," "Process order transaction," and "Produce confirmations." Each of these steps can be further defined.

The flow of activities is a general description of the steps within a use case. In most cases, we need to extend the description to even more detail. Frequently, a use case may have several alternative sequences of internal activities. For example, the exact sequence of tasks will probably be different for the "Create new order" use case, depending on whether it is a customer or a clerk interacting with the system. It is the same use case, but a different sequence. We describe these different sequences as **scenarios**. Thus, a scenario is the identification and description of a unique set of internal activities within a use case. It represents a unique path through the use case. In the RMO example, there are at least two scenarios for the use case "Create new order." One scenario might be named "Customer creates telephone order" and the other might be called "Customer creates web order." Figure 7-3 details these two scenarios.

scenario
A particular sequence of activities within a use case; a use case may have several different scenarios

Flow of Activities for Customer Creates Telephone Order

Main Flow: The customer calls RMO and talks to an order clerk. The order clerk verifies the customer information. If this is a first-time customer, a new customer record is created. The customer requests the first item for an order; this triggers the creation of a new order. The clerk verifies that the item is in stock and adds the requested quantity to the order. The customer then requests each item, and the clerk verifies it and then adds it to the order. Finally, the customer provides payment information, and the clerk verifies it. The order is marked as ready, and the scenario ends.

Exception Conditions: If an item is not in stock, then the customer can choose not to order it or to put it on the order as a backordered item. If the customer credit is no good, then the order is put on hold until a check is received and cleared.

Flow of Activities for Customer Creates Web Order

Main Flow: A customer connects to RMO's home page and then links to the order page. If this is a new customer, he/she requests the new customer page and enters the required information, including a secret password. The system then links to the catalog pages. The customer searches the catalog by category and item. When she finds something to order, she clicks on the "Add to shopping cart" button and the item is added to the cart. After all things have been added, a review of the items ordered is displayed. The customer can change any item. The system then displays the payment form, and the customer enters her payment information. The system verifies it and the customer has one last chance to cancel the order. After the order is accepted, a summary web page is displayed and a summary e-mail is sent to the customer. The system also builds an order with the appropriate line items.

Exception Conditions: If it is a repeat customer and she has forgotten her password, a new customer account is created with new confidential billing information. If the credit card does not pass the validation, the order is put on hold until a check is received and cleared. The customer has 24 hours to go back into the order and make any changes.

FIGURE 7-3
Two scenarios for "Create new order."

The Use Case Diagram

A use case diagram is a graphical model that summarizes the information about the actors and use cases. To do use case analysis, we look at the system as a whole and try to identify all of the major uses of the system. These uses normally derive from the business events identified in the event table. From the event table for RMO in Chapter 5, twenty primary use cases have been identified. Another way to think about a use case diagram is that it is a functional description of the entire system. It identifies the functions that must be supported by the new system.

Automation Boundary and Organization Figure 7-4 expands the use case given in Figure 7-2 to include an additional actor and additional use cases. As indicated by the relationship lines, each actor can use every use case. A boundary line is also drawn around the entire set of use cases. This boundary is the automated system boundary, or the automation boundary. It denotes the boundary between the environment, where the actors reside, and the internal functions of the automated system.

FIGURE 7-4
A use case diagram with a system boundary.

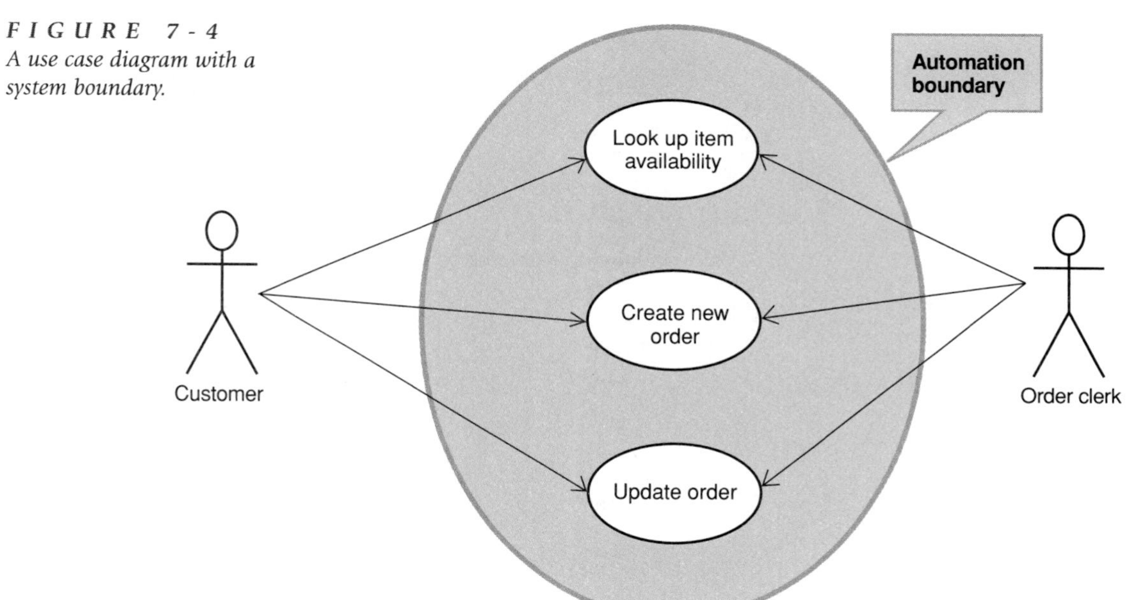

There are various ways to organize the use cases to depict different points of view. One approach is to organize the use cases by subsystem. Figure 7-4 is one such subsystem, namely, the "Order-entry subsystem." Figure 7-5 expands that figure to show all the subsystems for the customer support system of RMO with their associated use cases. Comparing Figure 7-5 with the event table in Chapter 5, we see that the 14 use cases correspond to events triggered by external actors. The internally triggered events are often not represented on the use case diagram. Another approach is to include all use cases that involve a specific actor. Figure 7-6 shows all the use cases involving the customer actor. This diagram is useful for showing all of the activities that are accessible through the Internet.

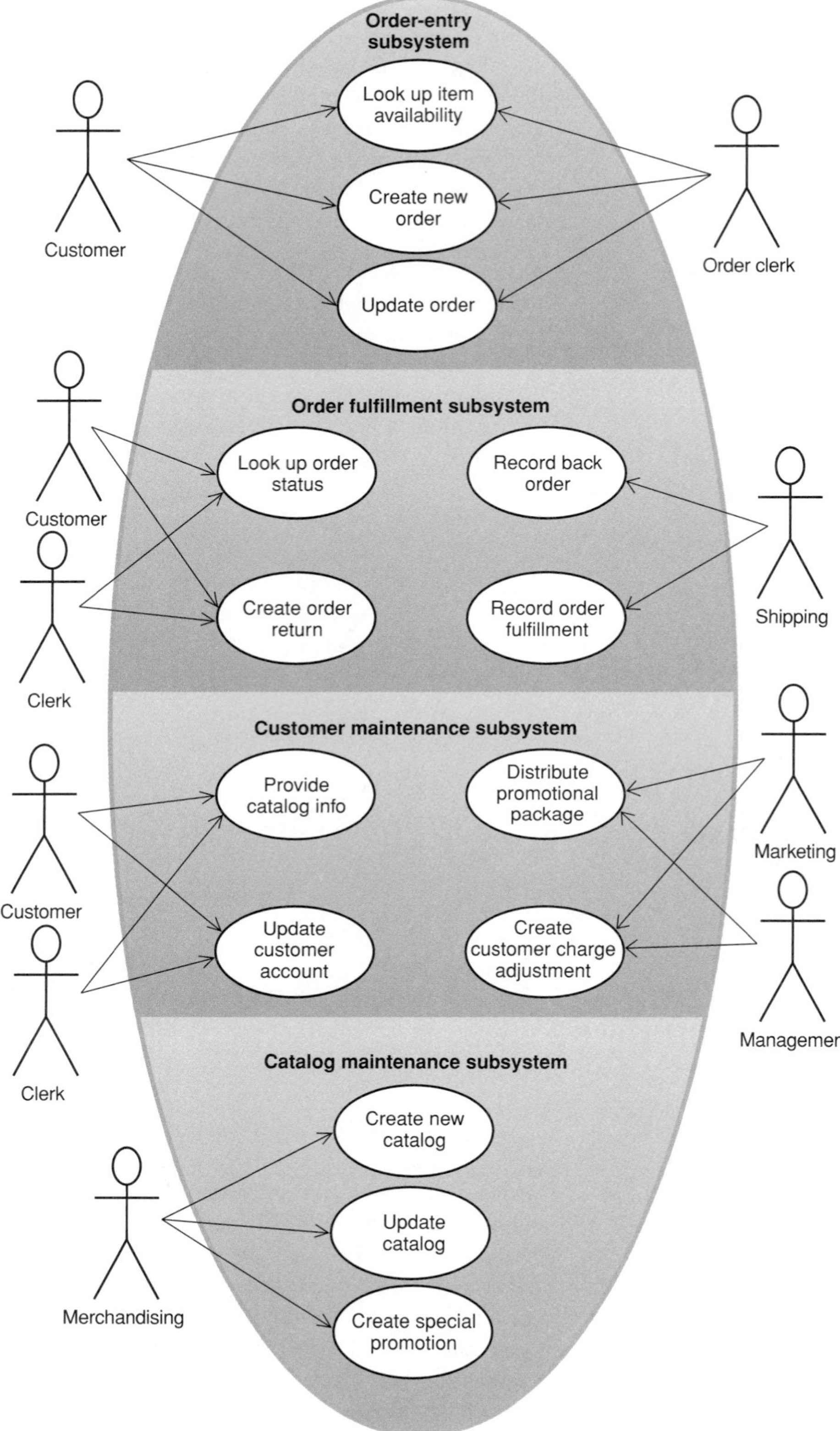

FIGURE 7-5
A use case diagram example of the customer support system (by subsystem).

FIGURE 7-6
All use cases involving customers.

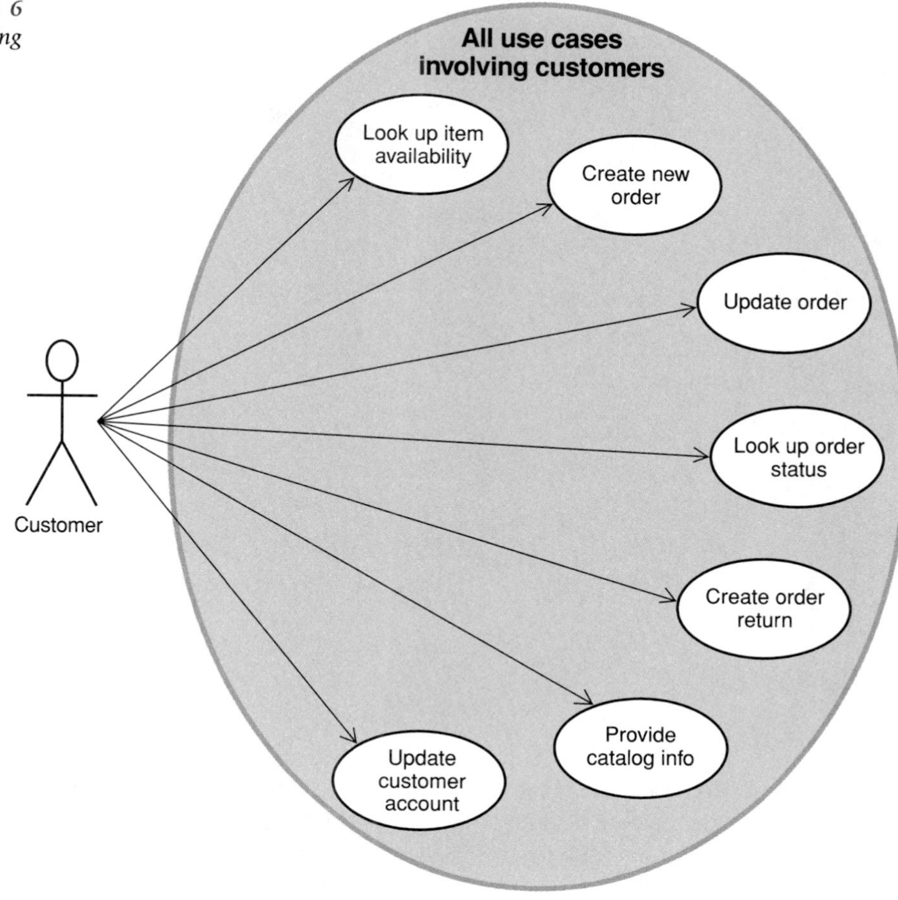

«Includes» Relationships Frequently during the development of a use case diagram, it becomes necessary for one use case to use the services of another use case. For example, two of the "Order-entry subsystem" use cases are "Create new order" and "Update order." Each of these use cases may need to validate the customer account. A common use case may be defined to carry out this function. Figure 7-7 shows an additional use case, named "Validate customer account," which is used by both of the other use cases. The relationship between these use cases is denoted by the connecting line with the arrow. The relationship is read *Create new order «includes» Validate customer account.* Sometimes this relationship is referred to as the «includes» relationship, or sometimes as the «uses» relationship. In programming jargon, such a relationship is referred to as a *common* subroutine, meaning that it is common to both use cases.

Figure 7-7 also shows that "Look up item availability" can be part of an «includes» relationship. Thus, you can define a common use case that is entirely an internal use case, such as "Validate customer account," or have both external actors and other use cases reference a use case. "Look up item availability" is an example of the latter.

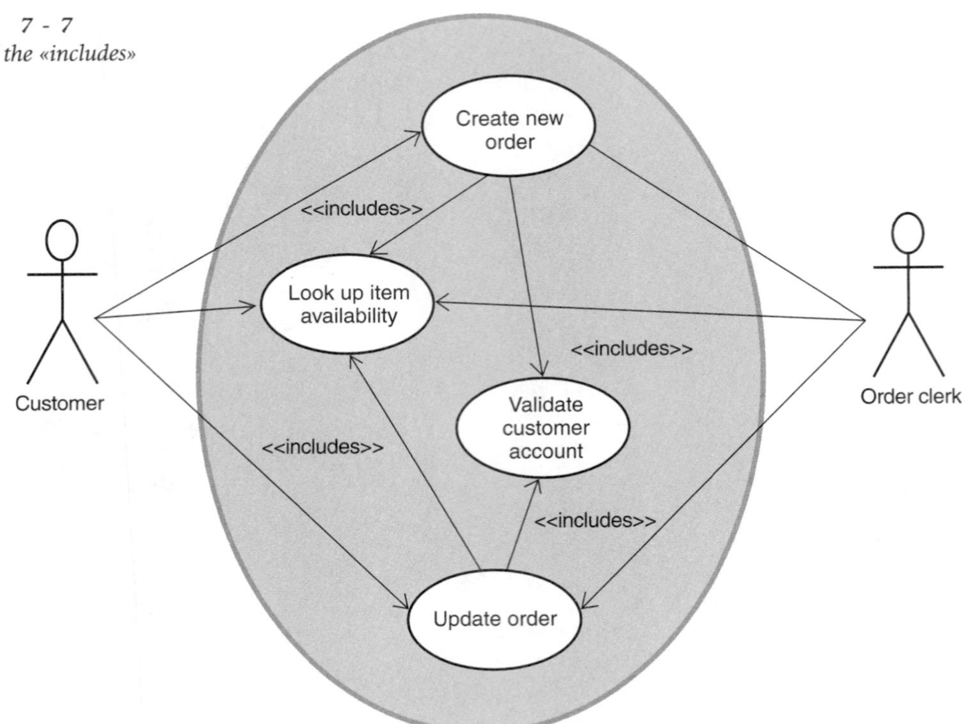

FIGURE 7-7
An example of the «includes» use case.

The Use Case Diagram Compared to Structured Techniques The objective of a use case diagram is to provide an overview of the system, including the actors who use the system and the functions they perform with the system. In the sense that it helps define the scope of the system, a use case diagram is like a context diagram. However, the individual use cases appear more like a DFD fragment in that they identify an individual function that the system must support.

One of the primary differences between structured and object-oriented modeling is that the thought process for developing a use case diagram begins by defining the automation boundary. In the development of a DFD, the automation boundary often is not defined until the entire process has been detailed. Thus, in a DFD an external agent is always the original source or destination of the information and may not necessarily be the one interacting with the system. In a use case diagram, an actor is the one who actually interacts with the system, whether that actor is the original source of the information or not.

For example, in the RMO case, a customer may call an RMO clerk and make an order. The DFD would identify the external entity as the customer. The clerk's activity would be embedded in a process described as "Enter customer order." In the use case diagram, the clerk would be identified as the actor who used the system to enter or create a customer order.

Another major difference is that the use case diagram does not indicate data flows. The information flowing into and out of the system is not identified until the next level of modeling, in the interaction diagrams.

Development of a Use Case Diagram There are two possible starting points to develop a use case diagram. The most common method is to use the event table. We analyze each event in detail to determine the way the system is used to support that event, the actors who initiate the event, and other use cases that may be invoked because of the event. Generally, each event will become a use case. In some instances, an event may spawn more than one use case.

One other method to start the development of a use case diagram is to identify all of the actors who use the system. You can build a list of actors based on the event table by looking at the Trigger and Source columns in the event table. An analysis of those columns helps identify who or what is using the system and what its use is. Sometimes the actors are the sources, at other times the actor may be someone other than the source. Instead of listing the actors as Bob, Mary, or Mr. Hendricks, you should identify the specific roles that these people play. Remember that the same person may play various roles as he or she uses the system. Those roles become such things as order clerk, department manager, auditor, and so forth. It is important to be comprehensive and to identify every possible role that will use the system.

Once the actors and use cases have been identified, the next step is to develop a flow of activities as the starting point for identifying the various scenarios. As the different flows of activities are developed, any common internal use cases can be identified and separated into different use cases.

Object Interactions:
Collaboration and Sequence Diagrams

As stated earlier, two types of interaction diagrams can be used to describe the flow of information and interactions between objects. A collaboration diagram and a sequence diagram contain the same information but each has a slightly different focus. A collaboration diagram emphasizes the objects that interact together to support a use case while a sequence diagram focuses more on the details of the messages themselves. During system development, you may use both types or either type of diagram. Those who prefer a top-down approach tend to draw a collaboration diagram first to get an overview of the objects that collaborate to carry out the use case. Those who prefer to use a bottom-up approach frequently develop a sequence diagram first and, in fact, may never even draw a collaboration diagram. However, both are useful models that you should understand and be able to draw. Even though a sequence diagram is slightly more complex than a collaboration diagram, it is used more frequently in industry. Thus let's begin with a sequence diagram first and explain a collaboration diagram second.

A use case diagram does not show the flow of information into, out of, or within the system. To define the flow of information in object-oriented modeling, we go to the next level of diagram, which is the interaction diagram. Figure 7-8 shows the relationship among a use case diagram, a class diagram, and an interaction diagram such as a sequence diagram. The interaction diagram includes objects from the class diagram and actors from the use case diagram. A particular sequence diagram documents the information flows within a single use case, or a single scenario.

FIGURE 7-8
The relationship of use case, class, and sequence diagrams.

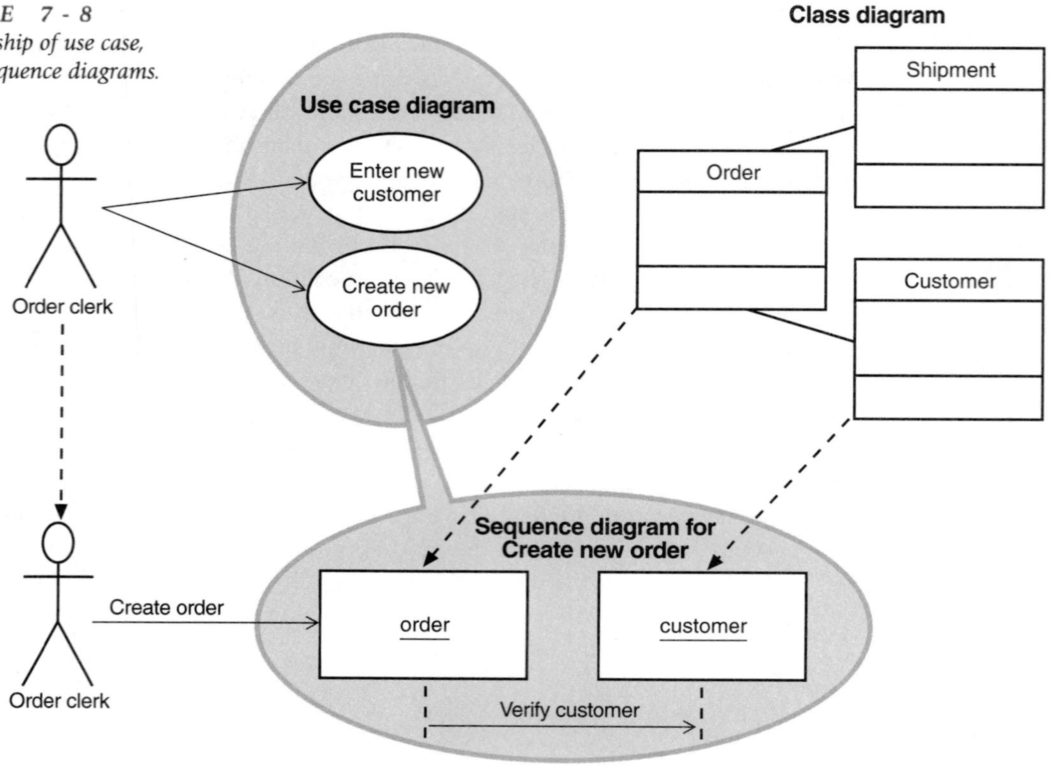

The use cases, which we developed as an initial step in defining OO requirements, do not identify separate objects. As we begin to develop the sequence diagram, we relate the objects identified in the class diagram with the use cases. To develop a sequence diagram, we use both the class diagram and the flow of events for each scenario. We identify which objects collaborate and what interactions between those objects are necessary to carry out the steps in the flow of events for the scenario.

Sequence Diagrams

A sequence diagram shows the sequence of the interactions between objects that occurs during the flow of events of a single scenario or use case. Four basic symbols are used on a sequence diagram: (1) the actor symbol, represented by a stick person, (2) the object symbol, represented by a rectangle with the name underlined, (3) the lifeline symbol, represented by a dashed line or a narrow vertical rectangle, and (4) the message symbol, represented by a directional arrow with a message descriptor. Figure 7-9 illustrates a generic sequence diagram showing the general format and the symbols used.

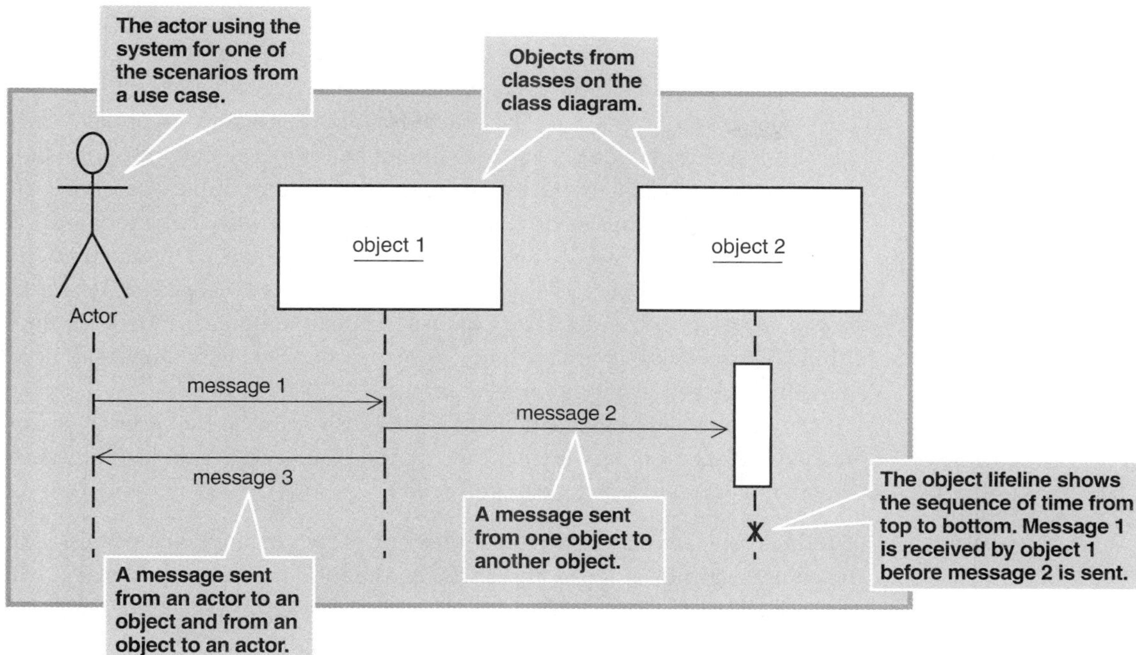

FIGURE 7-9
The symbols of a sequence diagram.

Actors and Objects The stick figures represent the actors of the use case being described. The objects are the internal objects involved in the use case and are depicted with the rectangles.

Chapter 5 discussed "things" and then identified object classes based on those things. The object classes are object sets—that is, sets of objects that have the same characteristics. An object class is represented by a rectangle with the name of the class inside. To identify a particular object within the object class, a rectangle is also used, but the text inside the rectangle is underlined. Figure 7-10 shows the different ways to name an object. An object may have a specific identifier or may just be any random object within the class. A sequence diagram represents a specific set of messages and interchanges between objects. Thus, objects are indicated on the diagram instead of classes.

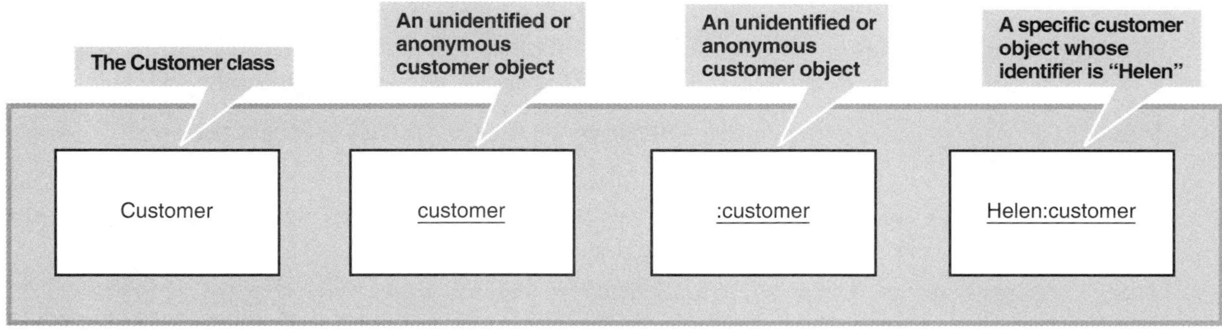

FIGURE 7-10
Object and class names.

Sometimes students get actors and objects mixed up, particularly when they have the same name. For example, in the RMO order-entry use case, there may be a customer actor and a customer object. What is the difference? The customer actor is the external, physical person who is playing the role of a customer. This actor represents the physical person. The customer object is the computer object that maintains information about the RMO customer. It is purely an internal, computer artifact—a virtual customer. Obviously, we want to synchronize the internal object with the desires and actions of the external actor. In fact, that is the purpose of the system, to maintain internal computer information about the ordering activities of the external person. When describing how to develop a class diagram, Chapter 5 noted the difference between the external "things" (actors) and internal "things" (objects). It is even more important now to note the difference between these views of things.

During the period that a scenario is active, the actors and objects are also active. A sequence diagram represents the time period of this active life with a dashed line—the lifeline—beneath each actor and object.

lifeline
The vertical line under an object on a sequence diagram to show the passage of time for the object

Lifelines Each actor and object has a **lifeline**, which is simply the vertical dashed line below the actor or object that depicts the duration of the actor or object within the scenario. Time flows from top to bottom; the lifeline shows the sequence of the messages. The messages higher up on the lifeline occur before those below.

A dashed line is used when the creation and destruction of the actor or object is not important for the scenario. When it is important to note the creation or destruction of an object, then a long narrow rectangle can be used (see Figure 7-9). The narrow rectangle is used when a particular object comes into existence and then later may be destroyed during the sequence. The narrow box is often referred to as the **activation lifeline**. The object is active during the period depicted by the narrow rectangle. For example, suppose you are describing a scenario that has a web page as an object. That web page may exist as an object only while it is on the screen. Once the browser window is closed, or another URL is requested, then that page is no longer active—that is, it no longer exists as an object. The point at which the object ceases to exist in the system is depicted by a big "X" at the end of the activation lifeline.

activation lifeline
The narrow vertical rectangle to emphasize that an object is active only during part of a scenario for a sequence diagram

Messages Since an object-oriented system gets work done by a cooperative effort of individual objects that send messages to each other, the internal events identified by the flow of events within a scenario become the messages between objects and actors or other objects. A message symbol has two parts: the directional arrow and the message descriptor. The syntax for the message descriptor is the following:

[true/false condition] return-value := message-name (parameter-list)

The true/false condition is a test to see whether the message can be sent. It is like the decision point or if statement in a programming language. If the condition evaluates to true, the message is sent. If not, it is not sent.

A message is a request from one actor or object to another to do some action. Chapter 5 discussed class methods as the instrument that an object uses to perform a function. When one object or actor sends a message to another object, it is a message to invoke a particular method. The receiving object responds by executing the invoked method. In a sequence diagram, actors are like objects because they also perform some function when they receive a message. The message-name is usually just some simple descriptor to identify the action of the message. Names such as "ring telephone," "update account," or "create customer" are used. However, an extended syntax for the name can also be used, as indicated in the previous example. When we want to indicate more precisely that the message is invoking a method, we use the extended name, which includes a possible return value from the method and the list of input variables needed by the method. In fact, the message-name often has the same, or very similar, syntax as the name of the method it is invoking.

Figure 7-11 shows several alternatives for message-names. As can be seen, different combinations of components may be used. The first example has ony a message-name. The second has a message-name with parentheses but no parameters. The first two examples are essentially the same, since they provide no additional information. The third example includes parameters. The parameters are data items that are passed to the destination object as input parameters to the receiving method.

FIGURE 7-11
Example message-name formats.

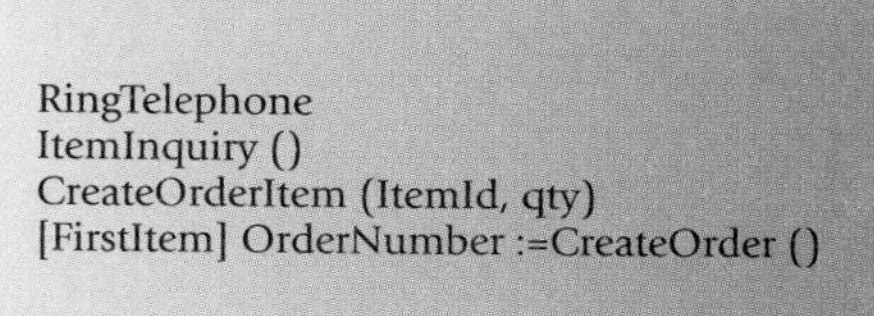

RingTelephone
ItemInquiry ()
CreateOrderItem (ItemId, qty)
[FirstItem] OrderNumber :=CreateOrder ()

The last example contains both a condition statement and a return value. In this example, the condition state, [FirstItem], is a test to see whether this is the first item to be added to an order. If it is the first item, then send the CreateOrder message. For all subsequent items, a new order does not need to be created. At this point in the analysis, we do not force the syntax of the condition statement to be formally correct in programming language. The primary purpose is to write it in such a way that the meaning is easily understood.

The other addition to the last message is a return value. A return message indicates that the sending object sends the message and then waits for the return value. Another, more flexible method to indicate messages and returning messages with a return arrow is shown in Figure 7-12, which is discussed next. Either approach is acceptable and are used interchangeably. Figure 7-12 shows an example of each technique.

RMO Sequence Diagrams Figure 7-12 is the sequence diagram from RMO for the "Look up item availability" use case. The order of the messages on the lifelines provides information about the sequence of the messages. In this case, since the customer is the actor, we assume that the customer actually interacts with the system, probably through the Internet. The objects that are involved are the catalog object, a specific product item object, and a specific inventory item object. This example is considered an "informal" example because the format of the message descriptors is fairly nonspecific. In other words, there are no parameter lists or return values. The message descriptors generally describe the meaning of the message, but not specific details about them. It is perfectly acceptable to develop sequence diagrams at this level of detail during this initial fact-finding activity. Later on in the SDLC, when we start the design activities, we will want more detailed and precise message descriptors. For now, however, this level of detail provides a good overview of the types of messages and of the sequence of events that must occur to look up an item.

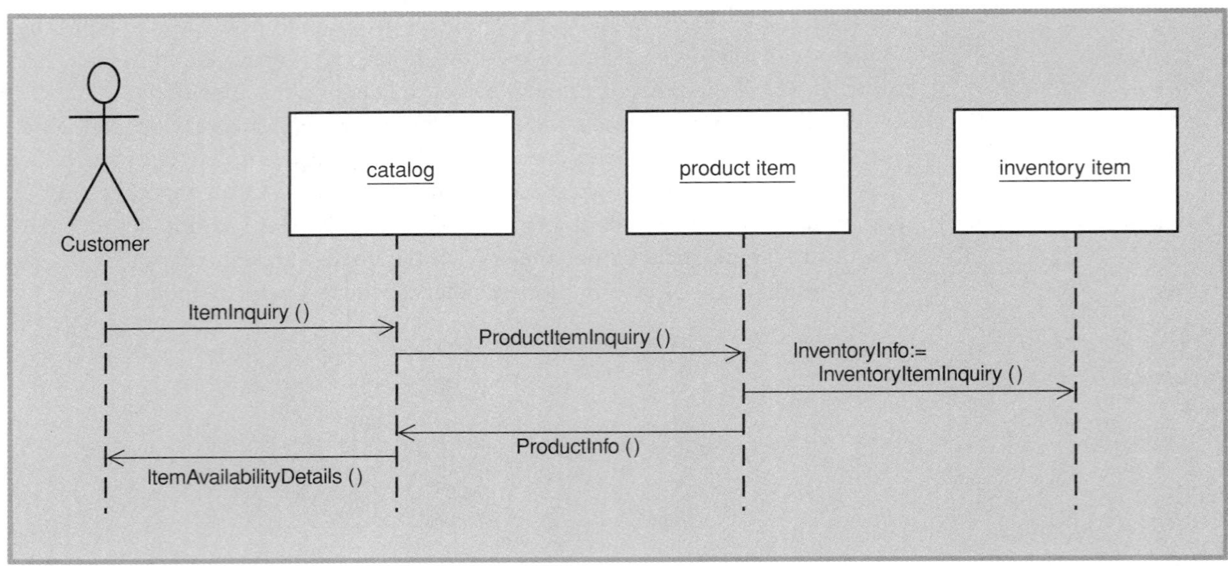

FIGURE 7-12
A sequence diagram for "Look up item availability."

The first message comes from the customer to a catalog object, such as "Spring 2000 catalog." Specific products are identified in the catalog, so a message is sent to a product item. Individual inventory items can be found from products, so the product item forwards the message to the specific inventory item, which sends back the requested information and which the customer eventually receives. The next section describes how to develop the messages in this flow of events when describing how to develop sequence diagrams.

The example in Figure 7-13 shows a more comprehensive scenario illustrating several of the more elaborate capabilities of a sequence diagram. Figure 7-13 describes the scenario for the "Create new order" use case. The scenario is for a telephone order from a customer to a customer service clerk. We will point out several additional features of sequence diagrams, and of the messages themselves.

The customer initiates the scenario with a phone call to the clerk. The diagram shows two messages going from the clerk to a customer object. If it is a new customer, then a CreateCustomer message is sent directly to the customer object rectangle to create a new customer object. If the customer already exists in the system, then the status of the customer is checked. This either/or situation is indicated by the true/false condition shown within the square brackets. These two messages also begin at the same location. Messages that begin at the same location are either (1) concurrent messages that are sent at the same time or (2) mutually exclusive messages (as in this case).

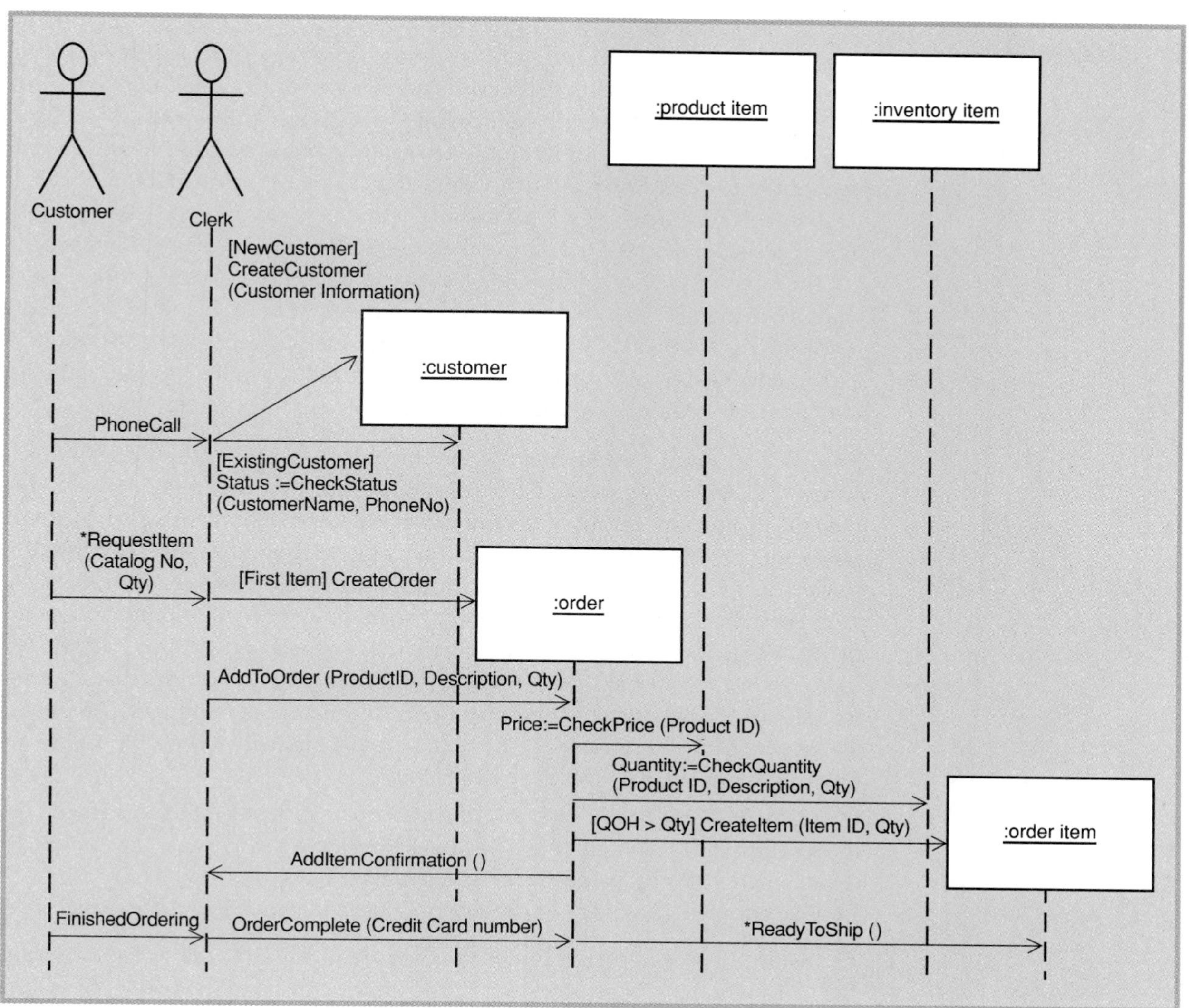

FIGURE 7-13
A sequence diagram for the telephone-order scenario of the "Create new order" use case.

Both of these messages also contain parameter lists of input information. The CreateCustomer message is a little less precise in that it only describes the parameters as customer information, whatever that might be. The CheckStatus message, however, does identify a specific input parameter as the customer name and telephone number. We can assume that these two parameters are sufficient to uniquely identify the customer. These two examples illustrate that, at least during early analysis, model components may range from being very precise to only identifying a general type of parameter.

Reading down the diagram, the next message is a RequestItem message from the customer to the clerk. At this point, the customer is requesting individual items to purchase. The input parameter is the catalog number. This message includes an asterisk, which indicates that this message may be sent multiple times.

The RequestItem message triggers the rest of the process. If it is the first item requested, as shown by the true/false condition, then a new order is created. Then the clerk sends more messages to the order object to add each requested item to the order. When the order object receives the AddToOrder message, it then initiates three other messages: first to the product item object to get the correct price, then to the inventory item object to verify the quantity on hand and to update the quantity on hand, and finally to create a new order item—that is, a line item on the order. The new order item is created only if the quantity on hand is greater than the quantity ordered.

The focus in the figure is on the problem domain classes found on the class diagram. User interface classes and components have not been included. For example, how does the clerk actually initiate a CreateCustomer message? The clerk will have a user-interface object, such as an input screen, that will accept the message and send it to the system. Additionally, data management objects are also added later.

Sequence Diagram Development To develop the proper thought process to develop sequence diagrams, we must first understand a little more about the object-oriented method of program execution. An object-oriented program consists of a group of interacting objects, and the attributes of each object are usually hidden or private. So, to view or update any attributes, we must send the object a message requesting a service. The program logic within an object works only on the attributes and components within the object itself. For example, to update an account balance for a customer, we would send a message to a particular customer object that in essence would say, "Please update your account balance with the amount parameter that we are sending with this message." The syntax of this message may be "updateAccount (amount)" or perhaps "updateAccount (50.00)."

To create a new object, such as a line item on an order, we send a Create message to the class along with all of the necessary parameters to create a new object. The Create method then is invoked and creates a new object.

One effective way to develop a sequence diagram is to follow these steps:

1. Identify all of the objects and actors that are involved in the scenario. Use only actors who have been identified in the use case diagram. Use only objects from classes that are identified in the class diagram. If objects or actors are needed that have not been previously identified, then update those other diagrams.

2. Based on the flow of activities, identify each message that will be required to carry out the scenario. Identify both the source object or actor for the message and the destination object or actor. Sometimes this task is difficult, especially if you are not used to thinking in object-oriented terms. Remember that objects can only do things to themselves. For example, if you want to check the quantity of an inventory item, only the inventory object itself can do that, not the order object, a clerk actor, or any other object or actor. Thus, a message with a destination of the inventory object is needed. A tougher problem is to identify the source of the message. A few guidelines to help identify the source of a message are (1) identify the object that needs the service, (2) identify the object that has access to the required input parameters, and (3) remember that if there is a one-to-many association relationship from the class diagram, then frequently the object on the one side will create and send messages to the object on the many side. This third condition is always true when an independent/dependent relationship exists between classes. For example, order items cannot exist without orders. Thus, order items are dependent on orders. For this situation, always send the messages to the dependent class, i.e., order items, through the independent class, i.e., orders.

For example, in the telephone order, who should send messages to check the inventory? Who should create the order line item—the clerk or the order? The order must control and ensure that line items are connected to it correctly, so let the order create line items. The order therefore also needs to receive a product description (hence the message to the product item) and to verify inventory counts (hence the message to the inventory item). In every system development project, frequently there are several possible solutions.

3. Next determine whether each message is always sent or whether it is sent only under certain conditions. To keep things simple, you may not want to try to identify every single response that returns from a message. Eventually, you will want to do so, but in the first iteration, focus on getting the major messages identified to support the flow of events.
4. Sequence the messages correctly and attach them to the appropriate lifelines of the actors and objects.
5. Add the formal syntax on the messages to describe conditions, message names, and passed parameters.
6. If desired, add the response messages and communications to make the sequence diagram complete.

Let's go through the thought process to build the telephone-order sequence diagram given in Figure 7-13:

1. For a telephone order, an order clerk actor is necessary. How does the order clerk know what to do? We include the customer actor to give directions to the clerk and initiate the telephone order. Obviously, we need an order object. An order is for a customer, so we need a customer object. The ordered items are line items on the order, so we need order item objects. We will also want product descriptions and quantities. To get those in the system, we must have a product item object and an inventory item object provide that information. Let's consider only the building of the order at this point. In a more expanded analysis, we might want to consider payments, back-orders, and shipments. Reviewing the class diagram, we do not see any other objects that will be needed to build the order.
2. The customer first calls the order clerk (a message). The order clerk gets the information from the customer and either (1) verifies customer information (a Verify message to a customer object) or (2) creates a new customer object (a Create message to the customer class). The customer will tell the clerk what item she wants to order (a message). If it is the first item, the order clerk will first create a new order (a "Create a new order" message to the order class), then try to add that item to the order (an AddToOrder message to the order). The program logic within the order will say to itself, "I need to know the price and to commit inventory." This is an example of the first guideline in step 1—that is, to determine who needs the service. The order contacts the product object (a message) and requests the name and price for the item. The order will also contact the inventory object (a message) to verify that there is stock on hand and to commit the ordered quantity from inventory. If those two messages are completed successfully, then the order says, "I better create a line item object and connect it to me" (a Createorder item message). This last sequence of adding items repeats until the customer indicates that she is finished. The sequence continues with the payment and shipping information.
3. As we worked through the sequence in step 2, we automatically determined the sequence of messages. However, some of those messages may be concurrent, some may be conditional, and the order may not be exactly right. In this step, we ensure the correct sequence, sources, and destinations of the messages.

4. To ensure that the correct parameters are sent in the messages, we need to look at the attributes for each object from the class diagram. For example, to create a new customer, look at the customer class and determine which attributes need input information. To create a new order item, consider what information comes from the customer (via the order clerk) and what information can be obtained from the product object and the inventory object. Pass the appropriate parameters in the request message or the return message.

At this point, you should begin to build the relationships indicated in Figure 7-8. The use case diagram and flow of events are used to build the sequence diagrams. The details of the sequence diagram require the use of the class diagram. As you build the sequence diagram, you may discover that the use cases, the flow of events, or the class diagram are incomplete and may require updating. One of the strengths of object-oriented development is that this interconnectedness helps to ensure that the final system is complete, comprehensive, and correct. Take the time as you develop the various models to ensure that you conduct your analysis correctly.

Collaboration Diagrams

The primary use of a collaboration diagram is to quickly get an overview of all of the objects that collaborate to support a given scenario. A collaboration diagram uses the same symbols for actors, objects, and messages found in a sequence diagram. The lifeline symbol is not used. However, a different symbol, the link symbol, is used. Figure 7-14 shows the four symbols in a generic collaboration diagram.

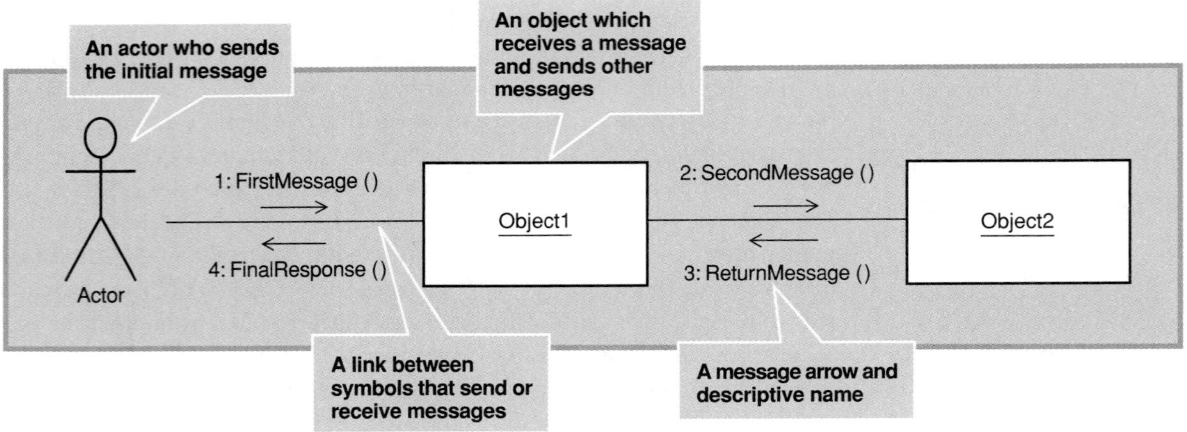

FIGURE 7-14
The symbols of a collaboration diagram.

The format of the message descriptor for a collaboration diagram differs slightly from the one for a sequence diagram. Since no lifeline shows the passage of time during a scenario, each message is numbered sequentially to indicate the order of the messages. The syntax of the message descriptor in a collaboration diagram is the following:

[true/false condition] sequence-number: return-value := message-name (parameter-list)

As can be seen in Figure 7-14, a colon always directly follows the sequence number.

The connecting lines between the objects or between actors and objects represent links. In a collaboration diagram, a link shows that two items share a message—that one sends a message and the other receives it. The connecting lines are essentially used only to carry the messages. You can think of them as the wires used to transmit the messages.

link
A relationship between two objects rather than between object classes

Figures 7-15 and 7-16 present collaboration diagrams for the same two RMO scenarios as for the sequence diagrams: "Look up item availability" and "Create new order."

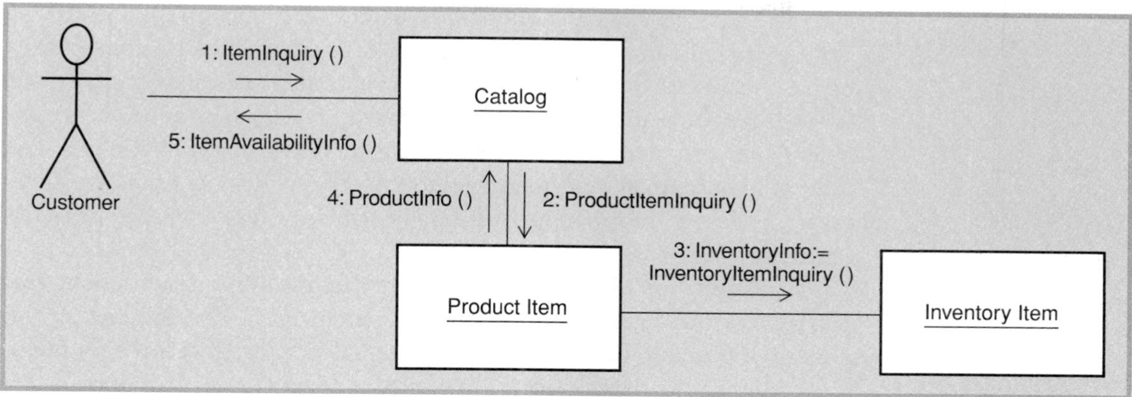

FIGURE 7-15
A collaboration diagram for "Look up item availability."

When comparing Figure 7-15 to the sequence diagram for the same scenario, Figure 7-12, it should be clear that the emphasis of Figure 7-15 is the collaborating actors and objects. Message information is included, but clearly the focus of the diagram is on the collaboration itself. The sequence numbers on the messages are used to present information about the order of the messages.

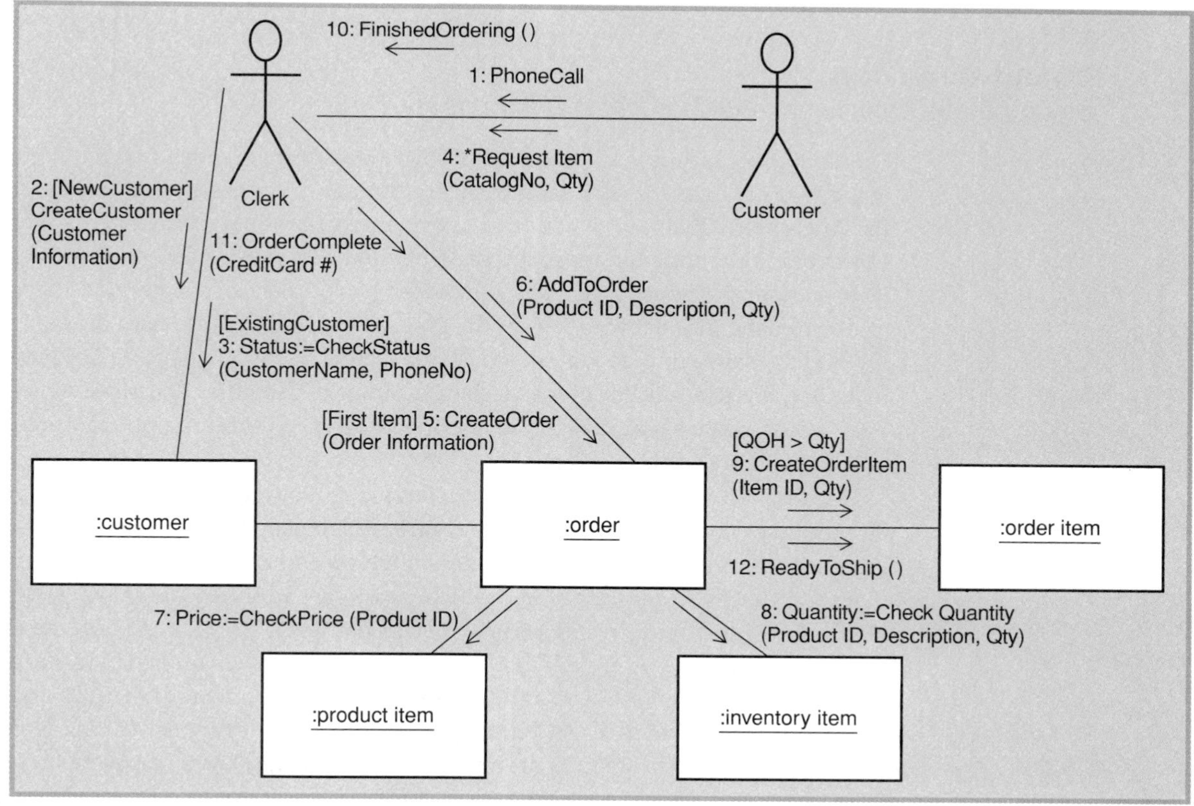

FIGURE 7-16
A collaboration diagram for the telephone-order scenario of "Create new order."

As Figure 7-16 reveals, the collaboration diagram for a telephone order is more detailed. Message 1, Phone Call, is not detailed at this point in the analysis. Messages 2 and 3, however, are a little more detailed. Each has a true/false condition to control the sending of the message. In this case, either message 2 or 3 is sent, but not both. Each also has a general description of the parameter list but no specifics of which variables to include. In message 4, the customer requests items to add to the order. As before, the asterisk on the sequence-number indicates a repeating message. Message 5, which is sent only after the first item is ordered, creates a new order. Message 6, which is sent after every repeating message 4, tells the order object to add this item to the order. The order then sends messages to the product item and the inventory item to gather necessary information, and finally sends a Create message to create a new order item.

A collaboration diagram cannot easily describe information about concurrent messages or messages that are initiated simultaneously. The diagram also does not indicate information about the creation or deletion of objects within the scenario. Sometimes in the development of interaction diagrams, a collaboration diagram is drawn without the messages simply to identify the collaborating actors and objects. Then a sequence diagram is developed to describe the detailed information about the interactions and messages within the scenario.

Once the scenarios have been described in detail with the interaction diagrams, the next task is to define the internals of each object class. The message interactions that have been defined in the interaction diagrams are used to help describe the behavior of each object class. We next turn to this activity.

Object Behavior: States, State Transitions, and Statechart Diagrams

The final piece of information that is needed in developing functional requirements is the internal logic for each object. This information is a description of the actions that the objects themselves perform. A sequence diagram gives an external view of object behavior. It identifies the messages that the objects send and receive. But what does an object do when it receives a message? The purpose of a statechart diagram is to describe the internal workings of the objects. Each class in the class diagram for a system has its own unique statechart diagram. Figure 7-17 is an extension of Figure 7-8, including the relationship of a statechart diagram to the other diagrams. As seen in Figure 7-17, a statechart diagram is developed based on information from the class diagram and the sequence diagrams.

Figure 7-18 is another view of the relationship between the sequence diagrams and the statecharts. In the figure, each column is labeled with the name of a class so that the columns identify all of the classes in the system. Each row is for one sequence diagram. The Xs in the cells show which classes are involved in which classes are involved in which sequence diagrams. A statechart for a class is based on the entire column for that class. Every cell that has an X in it will provide information about the messages to and from the class and will impact the development of the statechart.

In the object-oriented approach, how an object performs its actions is called object behavior. Each object is an instance of a class, and has a complete life cycle, from creation to destruction. An object comes into existence within the system in some manner. During the period of its existence, it is "in" certain states, and makes transitions from state to state. These states and the changes an object makes from state to state are shown in a statechart diagram. Two main symbols on the statechart diagram include states and transitions.

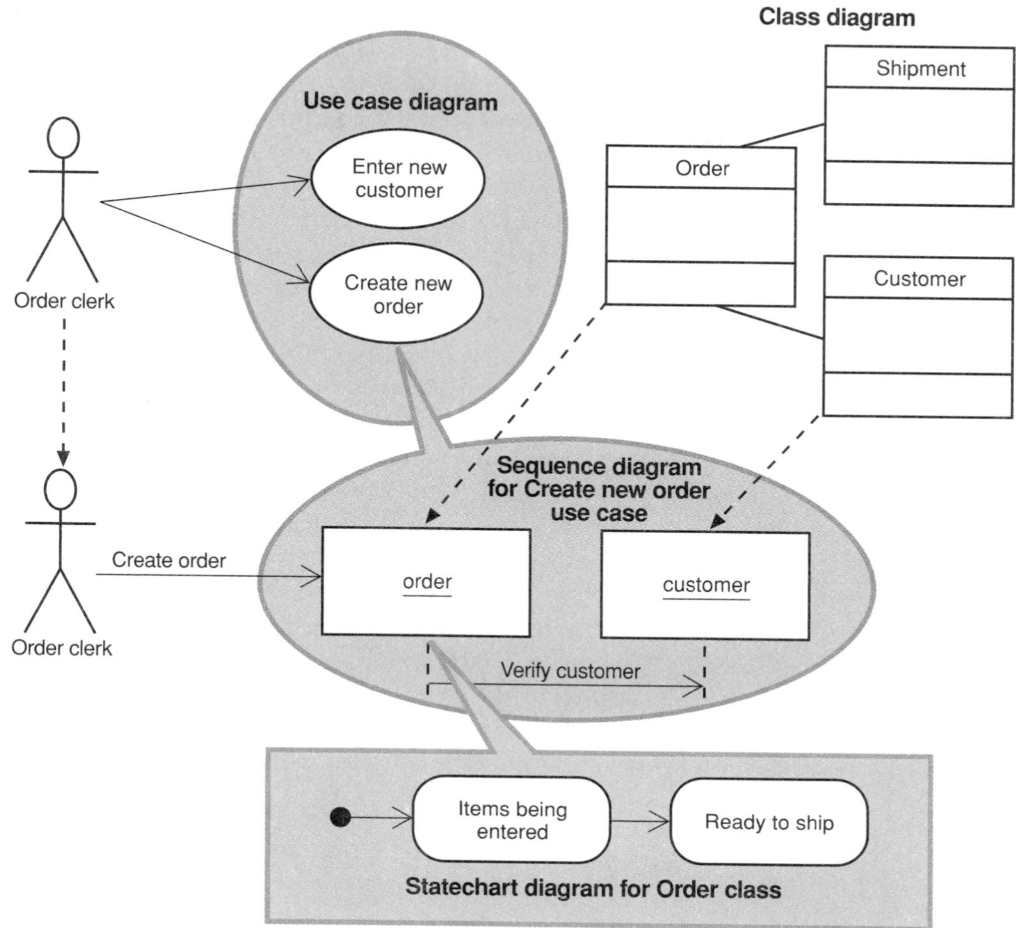

FIGURE 7-17
Relationships among OO models.

Object States

state
A condition of being for an object; part of a statechart

A **state** for an object is a condition during its life when it satisfies some criteria, performs some action, or waits for an event. Each unique state has a unique name. A state is a semipermanent condition of an object, such as "on," "working," or "loading equipment." States are described as semipermanent conditions because external events can interrupt them. An object remains in a state until some event causes it to move to another state.

	Customer class	Order class	Order item class	Product class
Create new customer	X			
Create new customer	X	X	X	X
Check item availability				X

FIGURE 7-18
Relationships between class statecharts and sequence diagrams.

The naming convention we use helps identify valid states. A state may have a name of a simple condition such as "on" or "in repair." Other states are more active, with names consisting of gerunds or verb phrases such as "being shipped" or "working." For example, a specific order object comes into existence when a customer orders something. Right after it is created, the object is in a state such as "adding new order items," then a state of "waiting for items to be shipped," and finally a state of "order completed" when all items have been shipped. If you find yourself trying to use a noun to name a state, you probably have an incorrect idea about states or object classes. Rethink your analysis and describe only states of being of the object itself. Another way to help identify states is to think about status conditions that may need to be reported to management or to customers. Such a status condition as "shipped" for an order is something a customer may want to know.

While the object is in a particular state, it is quiescent with nothing happening or it may be active and performing some action. For example, a computer monitor in the "on" state is displaying items on the screen. In the "off" state, however, the monitor is quiescent, with no action whatsoever. An **action**, then, is an activity that the object performs while it is in a state. (Later we will see that an action can also be performed during a transition.)

action
An activity performed by an object in a particular state

A state is represented by a rectangle with rounded corners with the name of the state placed inside. Any actions that must be performed during the period the state is active are placed beneath the state name within the rounded rectangle. Figure 7-19 is an example of two states in a portion of a statechart for a milling machine. The arrows are the transitions, which will be discussed shortly. Notice that the first state, Idle, has no actions. The machine is just "idle"—that is, it is on but doing nothing. In the second state, the machine is working. The working state is an active state, and the machine first loads a part, then mills it, and then unloads it.

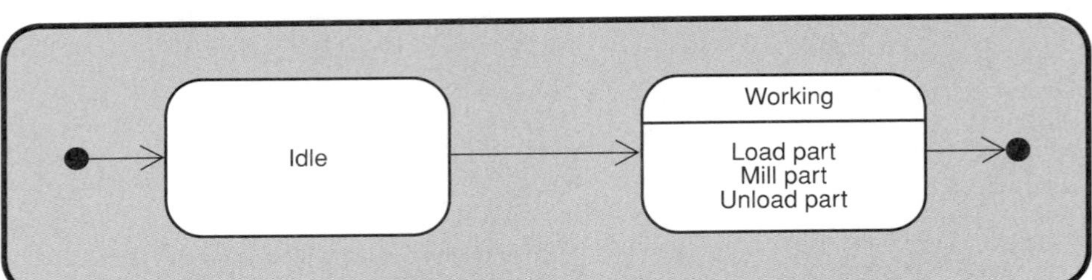

FIGURE 7-19
States for a milling machine.

As shown in Figure 7-19, we use two special states to indicate the beginning and the ending of a statechart. A darkened circle denotes an initial state and simply indicates the entry point into a statechart. This initial state is called a pseudostate, because the entry point may be prior to the creation of the object itself. Concentric circles with the inside circle darkened denote the final state. This state indicates the exit from the statechart, frequently indicating the removal of the object from the system. The purpose of these pseudostates is to make the statechart more understandable. They have no other particular importance.

Objects in the world, both system objects and physical objects, behave in more complex ways than being in one state at a time. The condition of being in more than one state at a time is called **concurrency** or **concurrent states**. We will describe some of this more complex behavior later. For now, let's just look at a state that represents concurrency.

concurrency or concurrent states
The condition of being in more than one state at a time within a statechart

To capture an object's complex behavior, we use **composite states**. A composite state represents a higher level of abstraction and can contain nested states and paths. For example, with the previous piece of machinery, we may want to identify an "on"

composite state
A high-level state that has other states nested within it

state. While it is on, it may also be working or idle. Thus, we may draw lower-level statecharts within the "on" state. Figure 7-20 represents this idea. Notice the notation—that each concurrent path has its own little compartment within the composite state.

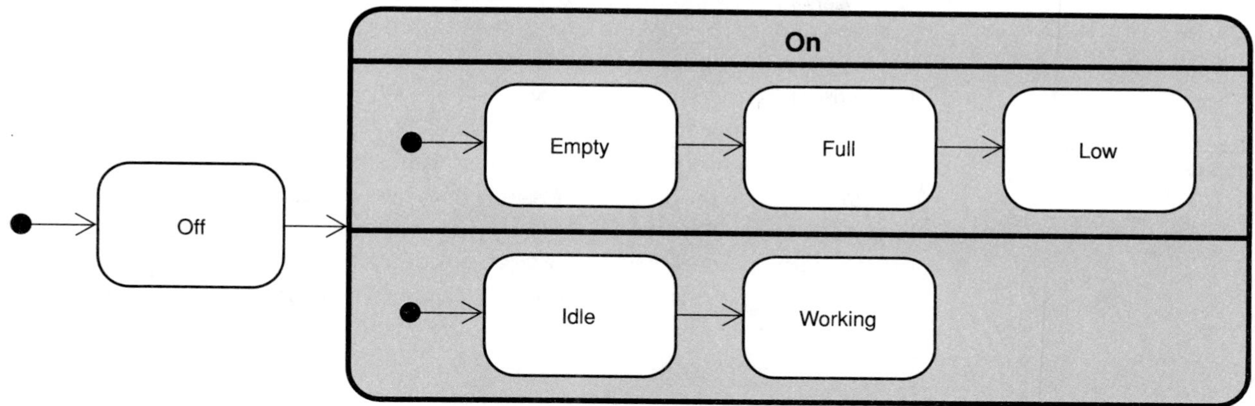

FIGURE 7-20
Sample composite states for the milling machine object.

To document concurrent behavior of a single object, we draw a composite state with the lower portion divided into compartments, one for each concurrent path of behavior. For example, imagine a piece of machinery that has an input bin to hold pieces of raw material. This machinery also cycles between two states in its work cycle of "working" and "idle." We may want to describe two separate paths, one representing the states of the input bin and the other the states of the machining portion. The first path will have states of "empty," "full," and "low." The second path will contain the two states "working" and "idle." These two paths are independent so that the movement between states in one compartment is completely independent of the other compartment. When an object enters the composite state, it begins at the black dot for each path.

Object Transitions

transition
A component of a statechart that signifies the movement from one state to the next

destination state
The destination of a transition connected to the arrowhead of the transition symbol

origin state
The origin of the transition connected to the tail of a transition arrow symbol

message event
The trigger for a transition consisting of a message that has the properties of an event

guard-condition
A true/false test to see whether a transition can be taken

A **transition** is the mechanism that causes an object to leave a state and change to a new state. Remember that states are semipermanent conditions. They are semipermanent because transitions interrupt them and cause them to end. Generally transitions are considered to be short in duration compared to states, and cannot be interrupted. In other words, once a transition begins, it runs to completion by taking the object to the new state, called the **destination state**. A transition is represented by an arrow from an **origin state**—the state prior to the transition—to a destination state and is labeled with a string to describe the components of the transition.

The transition label consists of three components:

transition-name (parameters, ...) [guard-condition] / action-expression

The transition-name is the name of a **message event** that triggers the transition and causes the object to leave the origin state. Notice the similarity to the format of the messages described in the previous section. They are called message events because (1) they are messages from other objects and (2) they happen instantaneously like an event. The parameter list identifies any parameters that the message is to pass to the object, such as a customer name or inventory item identifier. These parameters may be required by the actions identified in the transition or by the actions in the destination state.

The **guard-condition** is a qualifier or test on the transition, and it is simply a true/false condition that must be satisfied before the transition can fire. For the transition to fire, first the trigger event must occur and then the guard must also be true. Sometimes a transition has only a guard-condition and no trigger event. In that case, the trigger event is constantly firing, and whenever the guard becomes true, the transition fires.

Recall from the discussion on sequence diagrams that messages have a similar test, which is called a true/false condition. This true/false condition is a test on the sending side of the message. Thus before a message can be sent, the true/false condition must be true. The guard-condition is on the receiving side of the message. The message may be received, but the transition fires only if the guard-condition is true. This combination of tests, messages, and transitions provides tremendous flexibility in defining requirements. This level of flexibility is also needed to represent accurately how things happen in the real world.

action-expression
The statement on a transition to describe the action to be performed

The **action-expression** is a procedural expression that executes when the transition fires. Just like the action-expressions in a state, this expression is the activity that occurs during the transition. However, since a transition cannot be interrupted, an action within the transition must execute to completion once it begins. It may be any type of procedural expression and may be written in descriptive English or programming language to describe operations, methods, or functions. Parameter lists may also be included as appropriate.

Any of the three components—transition-name, guard-condition, or action-expression—may be empty. If either the transition-name or the guard-condition is empty, then they automatically evaluate to true. Either of them may also be complex, with AND and OR connectives. The action-expression may be a compound set of steps with a series of actions that are delimited with commas.

Figure 7-21 provides an example of a transition for the piece of machinery between the "Off" and "On" states. In this case, the message event that triggers the transition is "On-button-pushed." For the transition to occur, however, the guard must also be true. In this instance, the guard is a compound condition. Both "Bin not empty" and "Safety cover closed" must be true. Before entering the "On" state, the machine runs its self-test and places the result in a field called "status." The action expression in this figure is similar to a programming statement from a language such as C++.

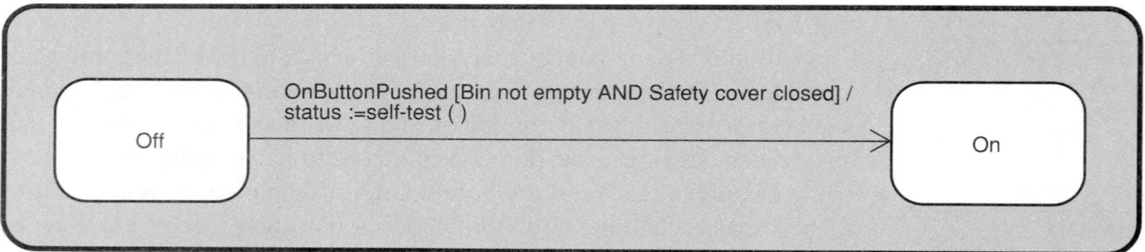

FIGURE 7-21
A sample transition.

internal transition
A transition within a state that does not remove the object from the state

Normally, a transition causes an object to leave a state and move to a new state or to return to the same state. An **internal transition** is a special kind of transition that does not cause the object to leave the state. Three versions of an internal transition are important to us. These three, identified by their transition names, are the *entry/*, the *exit/*, and the *do/* transitions. These three transitions are used to define when action statements are executed within a state. In other words, the notation is:

entry/ action expression
do/ action expression
exit/ action expression

The *entry/* internal transition, along with its action expression, is activated at the moment that the object enters the state. The *do/* action executes while the object is within the state. The *exit/* internal transition is activated just before the object leaves the state. Figure 7-22 shows how the milling machine states in Figure 7-19 should be modified to use these internal transitions. The part is loaded upon entry to the state, milled while the object is in the "working" state, and uploaded just before exiting. The benefit of using these internal transitions instead of just writing action statements as in Figure 7-19 is that we can now distinguish when and how these action statements are controlled. In this example, the action expressions within the "working" state are simple English statements.

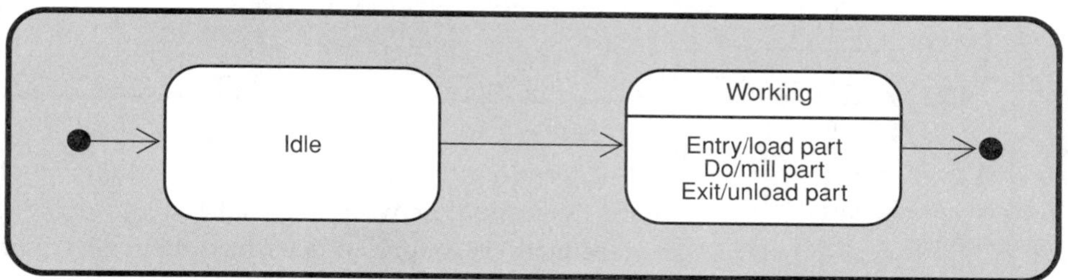

FIGURE 7-22
States for the milling machine object with internal transitions.

Messages, Transitions, and Actions

Two types of transition-names are observed. Figure 7-20 shows a simple trigger that occurs and fires the transition. The other type of transition-name corresponds more to the formal syntax presented in the last section—*transition-name (parameters,...)*—and corresponds to the syntax of a message-name. It is common practice to use the same name and parameter list for a message and a transition. This is one way that sequence diagrams and statecharts are related, as indicated in Figure 7-17. Figure 7-23 is a simple statechart for a line item object (order item class) on an order. Note that the transition name *CreateItem (ItemId, Qty)* corresponds to a message name in Figure 7-13. Additional detailed analysis will show that every transition name in Figure 7-23 corresponds to a message in some sequence diagram.

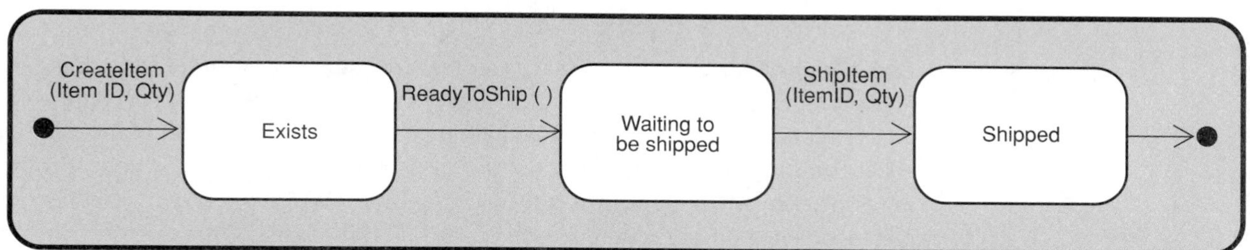

FIGURE 7-23
Transition names and message names for a statechart.

dot notation
A programming notation to indicate the class to which the method refers

Objects not only receive messages but send messages. Messages are normally sent as part of an action statement. Remember that action statements are part of transitions or states. They are placed wherever activity occurs within the object. The syntax that we normally use is called **dot notation**, and it comes from programming languages. Using dot notation, we simply name the object or class to which the message will be directed, then add a dot or period, then add the name of the message, as in the following example:

ControlPanel.UpdateStatus (status)

Figure 7-24 is an updated version of Figure 7-21, which shows the addition of an action statement that will send an output message. Two statements within the action expression are separated by a comma. The control panel object is the destination of an output message named UpdateStatus ( ). The variable *status* is sent as a parameter in the message. Someplace on a sequence diagram, there will be a control panel object and a milling machine object with a message going from the milling machine to the control panel.

FIGURE 7-24
An action expression with an output message.

The focus of the analysis during the development of the interaction diagrams was on the interactions between objects. During the development of the statechart, more detailed analysis is done to determine the logic required to send the messages and what happens when an object receives a message. You should now understand how object-oriented analysis is done and how these various drawings support and strengthen each other to provide a comprehensive requirements definition. So far, we have discussed only states and transitions as independent ideas. The next section discusses how to combine them to develop a statechart to show the entire life span of an object in the new system. Statechart diagrams are the most complex elements of object-oriented development. You should read this next section very carefully, making sure you understand each figure and its explanation.

Statechart Diagrams

The previous section, while explaining the concepts related to states and transitions, presented several simple statechart diagrams. This section continues with that discussion and pursues some of the more complex issues related to statecharts and their development. First we focus on understanding how to read a statechart, then we explain how to develop statecharts.

Order Items Figure 7-23 showed a statechart diagram for an order item object for Rocky Mountain Outfitters. An order item is a passive object with a very simple life cycle. An order item object is the line item on an order. It is an artifact of the ordering system and is purely internal information. It is, however, a genuine object. It comes into existence when an item is added to the order. While the order is being completed, it is in a do-nothing state, which we shall call "exists." Once the order is complete, then the order item moves to a state of "waiting to be shipped." This means that the inventory that is described by the order item can be shipped. Once the items have been shipped, the order item changes to a state of "shipped." This is the final state or final disposition of the order item. The final pseudostate is optional. As indicated previously, it simply represents the end of the path. Note that the names of the states are names that apply to the order item object itself. A state name should describe a state of being or a condition of the object. As indicated earlier, one good way to help think of states is to think about status conditions of the object. If the customer calls up and asks, "What is the status of my order?" what kind of status conditions, i.e., states, will enable you to answer that question.

The first step in the development of this statechart is to look at all the sequence diagrams that contain order item objects and list all of the input and output messages associated with the statechart. Figure 7-13 lists two messages, *CreateItem ( )* and *ReadyToShip ( )*. If we had developed all of the sequence diagrams, we would also have found another message called *ShipItem ( )*. Once the messages are identified, then the states and transitions are developed. More explanation is given later on developing statecharts.

Orders The example in Figure 7-23 illustrates how an object moves from state to state based on transitions and message events. Since an order item object is such a simple object, that is the extent of its life. As we saw in the sequence diagram in Figure 7-13, an order object is more complex because it not only receives messages, but also sends messages to request information and to initiate other behavior. It must have internal logic to accomplish these ends. It is still a passive object because it initiates behavior only when it receives a message. A truly active object initiates activity on its own accord.

Figure 7-25 contains a statechart diagram for an order object. This statechart is considerably more complex, with several new constructs. We first explain how to read and understand the statechart and then discuss how to develop one. Even though this statechart is more complex than the previous one, we have simplified it by leaving out exception conditions like back orders, lost shipments, and cancellations.

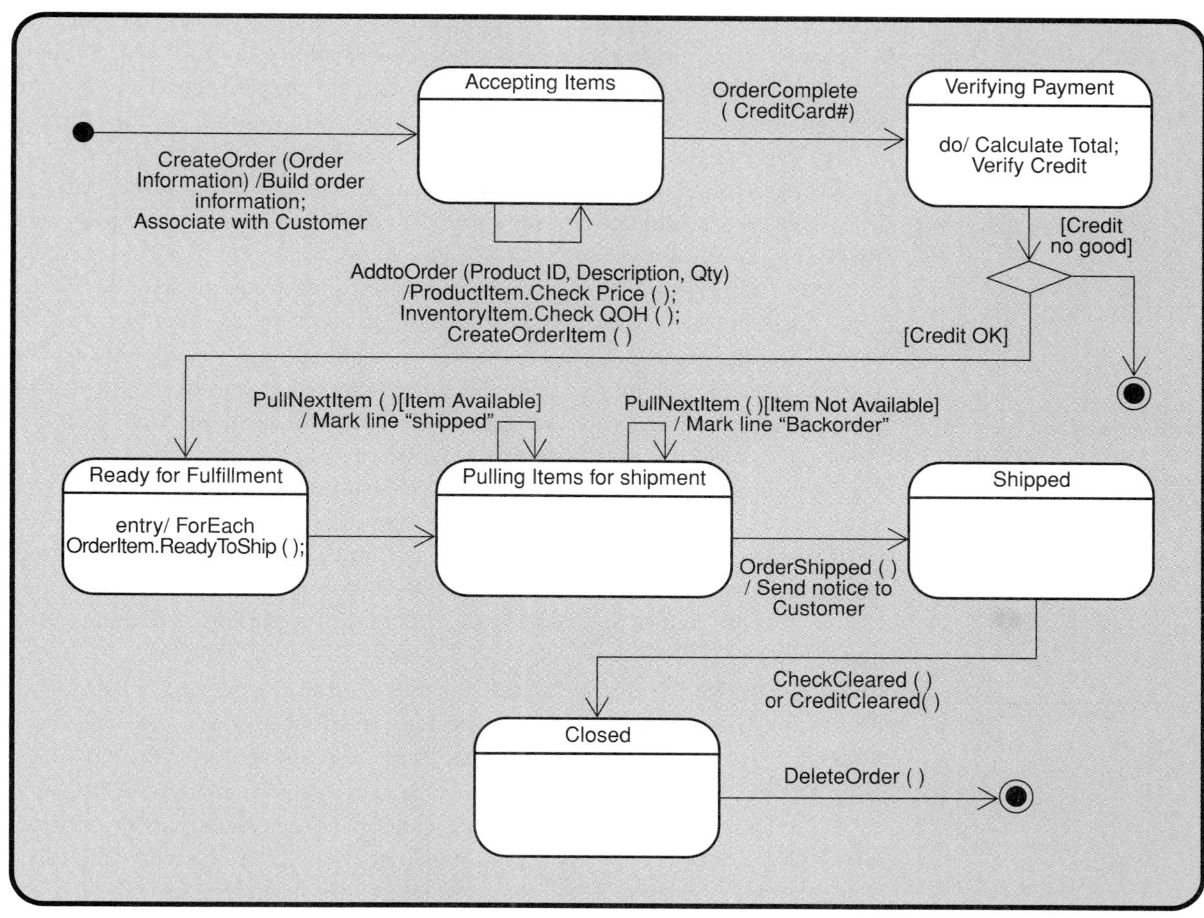

FIGURE 7-25
A statechart diagram for an order.

The initial transition that causes the order item to be created is *CreateOrder ( )*. Almost all statecharts begin with a create message to instantiate a new object in the class. Note that it is the same as the message on the sequence diagram in Figure 7-16. In this initial transition, the action expression is to build order information and associate this order with the correct customer. The action-expression is quite informal. It could be expressed more precisely using structured English. If you know the target programming language, you can even write the expression using programming language syntax. Note that an action-expression is an activity that is performed *by* the object itself. It does not represent an action done *to* the object. These action-expressions will later become the code that is contained within the methods of the object class.

An order first comes into existence in the system as an empty shell. Its first state is waiting for all of the order items to be added to it. Once an order is complete—that is, it has all the items listed as line items—then the customer credit and payment method is verified. The diamond is called a decision pseudostate. Its purpose is simply to allow the transition to split based on the guard conditions. Once payment is ensured, the order is ready to be shipped. The shipping department then pulls items from stock and packs them up. Sometimes not all items are in stock, so only a partial shipment can be made. In that case, items that are not shipped are back-ordered, and the order stays open until all items have been shipped. Finally, when all items have been shipped, the order is closed.

The first state, "Accepting items," describes the order object while it is still being built. We include a state called *Accepting Items* because in our system we want the customer to be able to add items to the order for a full 24 hours. Thus we purposely added this state to keep the order open for additions. While in this state, all the order items are added to the order by a transition that fires every time a new item is added. The action-expression on this transition corresponds to the output messages sent by the order object, as noted in Figure 7-16. The action-expressions use the dot notation to indicate which object class is the destination for the messages. These messages get the price, check the quantity on hand, and create a new order line item. Note that the transition fires when it receives a message, executes the action-expressions, and then returns to the same state.

The "Accepting items" state has two output transitions, *AddToOrder ( )* and *OrderComplete ( )*. Only one of the output transitions is taken at any point in time. The output transition that is taken is the one whose trigger (message event) fires.

The states in Figure 7-25 also contain action-expressions. In the state "Verifying payment" the *do* internal transition has action-expressions for the order object to calculate the total amount due and verifies the customer credit. The diagram does not explain the details of how this is done, but a message is probably sent to another object. Out of this state there is a transition with no trigger. This is called a **completion transition** because it fires automatically when the state finishes its actions.

completion transition
A transition with no trigger event so that it is taken when the origin state completes its actions

The completion transition then enters a **decision pseudostate**, indicated by a diamond shape. The decision pseudostate is a temporary condition that exists while the guard-conditions are being checked. The two transitions out of the decision pseudostate are also completion transitions, but each has a true/false guard-condition statement. If the guard-condition is true, the transition can fire; if not, the transition cannot fire.

decision pseudostate
A diamond on a statechart to represent a decision point on a path

The "Ready for fulfillment" state is the next state along the path. A **path** in a statechart is simply a valid sequence of states and transitions in which the states are interconnected by the transitions. In this state, the entry internal transition has action-expressions that send *ReadyToShip* messages to all the order items on the order.

path
A valid sequence of transitions and states through a statechart

The remaining states and transitions contain no new statechart constructs. The order object continues with its states and transitions as explained previously. You should carefully review the statechart to make sure you understand how it describes the life of an order.

The approach used to develop this statechart is similar to the one used to develop the order item statechart. First, all of incoming message events are identified across all sequence diagrams involving orders. Then all output messages are identified. Figure 7-13 shows that there are three input messages and four output messages. These are not all the messages, because this chapter does not include all the sequence diagrams. Other messages come from the sequence diagrams describing the shipment use cases and scenarios.

Once the messages are identified, then the analyst brainstorms the possible states that might exist for the object. The incoming message events help to identify transitions, and transitions are always connected to states. So, we think of possible states based on ideas from the transitions. In this case, we identified two states, "Accepting items" and "Ready for fulfillment," which seem to correspond to the three input message events.

After a little more thought, we decided to insert an intermediate state for "Verifying payment." The rest of the statechart is based on the other messages from the shipment use case. More explanation on developing statecharts is provided later.

Concurrent Behavior

In today's event-driven systems, the ability to express concurrent behavior is becoming more and more important. Concurrent behavior means that an object can do multiple things in parallel. Most of us are familiar with Microsoft Windows operating systems that allow multiple programs to be open and executing at the same time. For example, you may be writing with a word processor while the Internet browser is downloading some files and Print Manager is running some print jobs in the background. This is concurrent behavior. All three programs are active at once and the operating system alternates among each of the three—giving each a few microseconds of CPU time so that all can remain active and perform their necessary functions.

Earlier in the chapter, we discussed a composite state, which is one way to represent concurrent behavior. This section reviews that approach and introduces other ways to model concurrent behavior.

thread
An active path, particularly when multiple paths can be active concurrently

synchronization pseudostate
A vertical bar denoting a point where transitions can split or combine for multiple threads

A statechart diagram represents concurrent behavior with multiple paths. Each separate path is a called a **thread**. Figure 7-26 shows the two types of notation that are used in statechart diagrams to denote multiple threads of concurrent behavior. Part (a), which has been illustrated earlier, is a composite state with multiple compartments. Part (b) illustrates how a single transition is split into multiple concurrent paths. The solid vertical bar is called a **synchronization pseudostate**, which denotes a point where the paths split into multiple threads. In part (b), both output arrows are followed and multiple threads are active. This same solid bar is used to merge multiple threads back into a single thread.

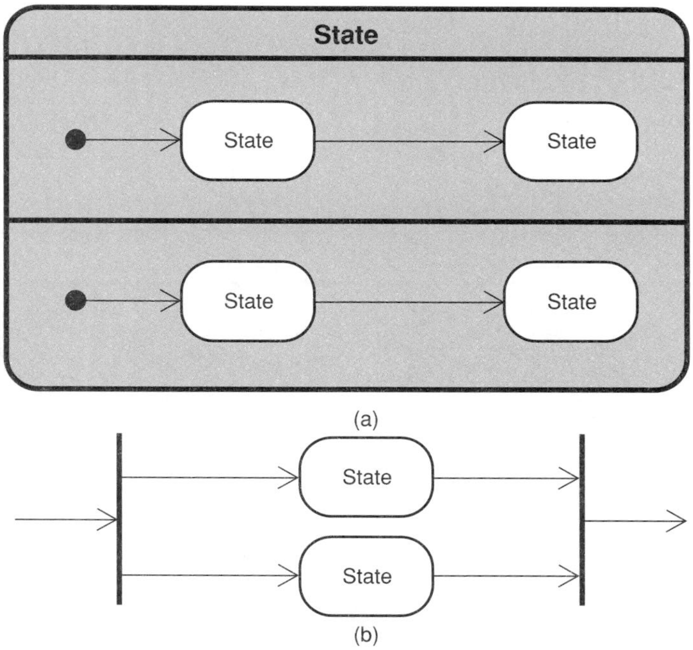

FIGURE 7-26
Concurrent threads of execution in statechart diagrams.

Figure 7-27 illustrates the difference between the two types of output transitions. Two transitions coming out of a state represent an OR condition. One or the other path is taken, but not both. Two transitions out of a synchronization bar represent an AND condition. Both path one and path two are taken simultaneously.

FIGURE 7-27
Transitions out of state or synchronization bar.

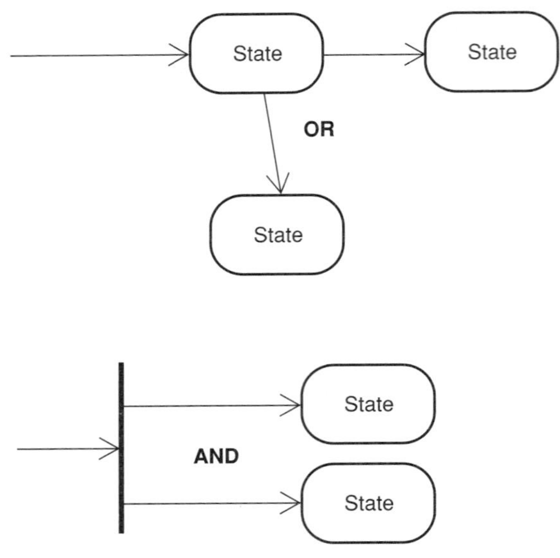

Figure 7-28 shows an example of concurrent behavior with a composite state for an RMO customer. Once a customer object is created within the RMO system, the object is either active or inactive. When a customer first contacts RMO, a new customer object is created and placed in an active state. Within the composite active state, two compartments represent two concurrent threads. The top compartment is a thread that denotes the purchasing status of a customer. High-volume customers become special customers and receive additional benefits. The bottom compartment contains a thread that monitors whether a customer has any outstanding orders. These two threads are completely independent, and movement between states in one compartment has no effect on the other compartment.

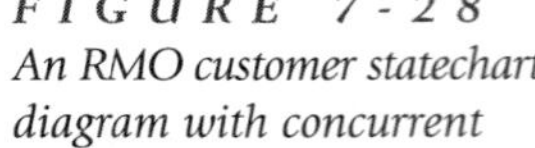

FIGURE 7-28
An RMO customer statechart diagram with concurrent threads.

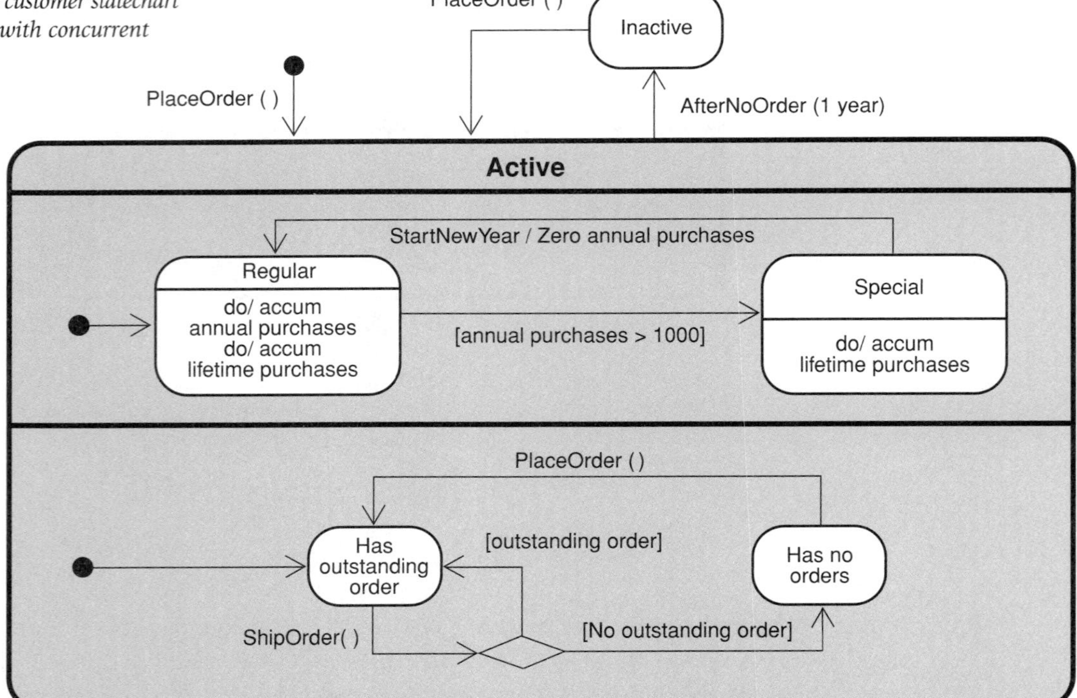

Note that each path begins with an initial pseudostate. The initial pseudostate indicates where the path starts within each compartment. When a customer object enters the active state, it enters every compartment concurrently. When an object leaves the composite state, it leaves every compartment and, of course, every embedded state. In Figure 7-28, if the customer has not purchased anything for a year, then he or she goes to the inactive state. A return to the active state occurs when an order is placed.

Figure 7-29 is an alternative statechart showing concurrent paths by utilizing synchronization pseudostates. The difference between the two statecharts is that Figure 7-27 does not provide a composite active state. If you a need to refer to the customer by a high-level condition, such as active, then Figure 7-28 is a better approach. Otherwise, either figure is acceptable, based on the preferences of the analyst.

Remember, a customer is a real-world entity that exists separately and apart from any computer artifacts. There are really two customers: the customer person and the internal customer object that represents the customer. The computer customer is an artifact that is created to maintain information about the real-world customer that is important to the system. The states and transitions of the artifact essentially attempt to reflect accurately the actions and states of the real-world customer. Several of the transitions within this statechart diagram are triggered by the actions of the external customer, such as placing an order. Other transitions are used to maintain purely internal information that RMO wants to keep about the customer.

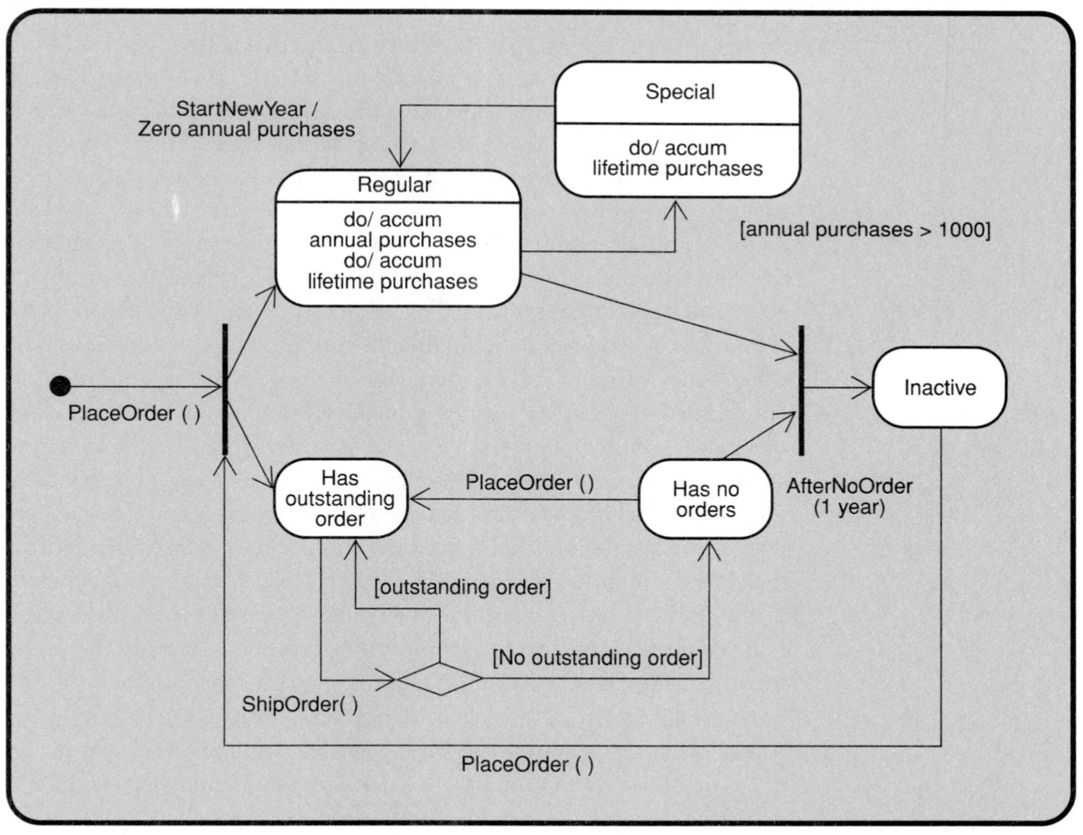

FIGURE 7-29
A customer statechart with synchronization pseudostates.

This statechart represents the life of a customer from the viewpoint of the system. The states are the important "states of being" that RMO needs to track. States like "making a phone call" and "has returned an item" are not states that are important to track. Development of a statechart requires a careful analysis to determine important states for the system to track.

Statechart Development

Statecharts are some of the more difficult diagrams for new systems analysts to develop. So don't be discouraged if at first you have a hard time developing them. Usually the primary problem is identifying the right states for the object. It may be helpful to pretend you are the object itself. It is easy to pretend to be a customer. It is a little more difficult to say, "I am an order," or, "I am a shipment. How do I come into existence? What states am I in?" However, if you can begin to think this way, it will help you develop statechart diagrams.

The other major area of difficulty for new analysts is to identify and handle composite states with nested threads. Usually the primary cause of this difficulty is a lack of experience in thinking about concurrent behavior. The best solution is to remember that developing statecharts is an iterative behavior, more so than developing any other type of diagram. Analysts never get a statechart right the first time. They always draw it, then refine it again and again. They add new states, delete some states, make other states composite states, add new transitions, develop new paths, and so forth. This happens over and over until they finally think the statechart is right. Then they find out they forgot something and refine it some more.

Finally, don't forget to ask about the exception condition—especially where you see the word "verify" or "check." You normally will have two transitions out of states that are verifying: one for acceptance and one for rejection.

Here is a list of steps that will help you get started in developing state charts.

1. Review the class diagram and select the classes that will require statecharts. In most instances, it is a good idea to assume that all classes will need statecharts. Begin with the classes that appear to have the simplest statecharts, such as *Order Item* class. Sometimes it is helpful to develop several statecharts at once—for example, for several classes that work together in a scenario or use case.
2. Identify all the input messages across all sequence diagrams for the selected class. As shown earlier in the development of the *OrderItem* and *Order* statecharts, the use cases and scenarios provide the basis for identifying messages into and out of a class. One good assumption is that each message will be the trigger for a transition. With this step, you begin identifying the minimum set of transitions. There may be other transitions that you will add later on, but this set is a good starting point.
3. For each selected class in the group, make a list of all the states you can identify. At this point, simply brainstorm. If you are working in a team, have a brainstorming session with the whole team. One effective way to start is to note that each transition identified in step 2 requires at least one origin state and one destination state. Remember a state is a semipermanent condition or state of being of an object in the class from the perspective of the new system. Sometimes it is helpful to think of the physical object, identify states of the physical object, then translate those that are appropriate into corresponding system states. It is also helpful to think of the life of the object. How does it come into existence in the system? When and how is it deleted from the system? Does it have active states? Does it have inactive states? Does it have states where it is waiting? Think of activities done *to* the object or *by* the object. Often the object will be in a particular state as these actions are occurring. Identify these states. Begin associating these activities with the states.
4. Build statechart fragments and sequence the fragments in the correct order. From steps 2 and 3, you should have a fairly complete set of transitions and states. Next build these individual states and transitions into statechart fragments. A **statechart fragment** consists of either a state-transition-state sequence or a transition-state-transition sequence. Either is acceptable. This process is similar to what was done to develop DFD fragments. However, in this process

statechart fragment
A portion of a path in a statechart

you are identifying statechart fragments. Review the matrix in Figure 7-18 to ensure you have handled all the input messages from every sequence diagram.

As the fragments are being built, it is natural to begin to look for a natural life cycle or order of the fragments. Continue to build longer paths in the statechart by combining the fragments. At times, this process is very straightforward with obvious connections among fragments. At other times, there may be missing pieces or complex paths that are not easily identified.

5. Review the paths and look for independent, concurrent paths. Take different pairings of states and ask the question, "Can the object be in both of these states at the same time?" or, "Does one state naturally follow another state?" For example, for a customer order, as seen in Figure 7-24, the states "Accepting items," "Ready for fulfillment," and "Shipped" are obviously dependent states. If an order is accepting order items, it certainly cannot be in the shipped state. Thus, these states would not be on concurrent paths, but would be on the same path with one state followed by the next state.

   When an item can be in two states concurrently, there are two possibilities: Either the two states are on independent paths, or one state in reality should be a composite state and the two states should be nested, one inside the other. One way to identify a candidate as a composite state is if it is concurrent with several other states. For example, in Figure 7-26, a customer will obviously be active and have outstanding orders. But how about a customer whose order was just shipped and who has no outstanding orders? RMO also considers that customer an active customer. Since the "active" state is concurrent with multiple other states, it is a composite state.

6. Expand each transition with the appropriate message event, guard-condition, and action-expression. Include with each state appropriate internal transitions and action-expressions. Much of this may have been done as the statechart fragments were being built. Ensure that every output message identified from the sequence diagrams is somewhere within an action-expression for a state or transition. Double-check message event names and passed parameters for accuracy and completeness.

7. Review and test each statechart. When we write computer programs, we test them by executing them. Since we cannot execute a statechart, we test it by "desk-checking" it. Review each of your statecharts by doing the following:

   - Make sure your states are really states of the object in the class. Ensure that the names of states truly describe "states of being" of the object.

   - Follow the life cycle of an object from coming into existence to being deleted from the system. Be sure all possible combinations are covered and that the paths on the statechart are accurate.

   - Be sure your diagram covers all exception conditions as well as the normal expected flow of behavior.

   - Make sure you have included all the relevant transitions. One way to do this is to do a pair-wise comparison of all states by asking, "Is it permissible for the object to go from state A to state B?"

   - Review all possible messages and make sure they are handled somewhere in the statechart.

- Review all possible actions and activities (processes) and make sure they are included correctly in the statechart.
- Look again for concurrent behavior (multiple threads) and the possibility of nested paths (complex states).

Finally, don't be surprised if your statechart goes through several more iterations. Object behavior can be very complex, and it is difficult to model it correctly the first time. As you refine your statecharts, you will probably have to update the sequence diagrams. It is normal to make several iterations between sequence diagrams and statechart diagrams as you understand more and more about the behavior of individual objects and their interaction with other objects. For example, analysts frequently change their minds about which class should send a message. In Figure 7-13, we might ask, "Which class should send the messages to check the price and the quantity on hand? Should it be the clerk actor or the order object?" There are arguments for both, and the answer may not be obvious until you begin to develop the statechart.

One of the main strengths of object-oriented development is this interrelatedness among the various diagrams. This close association among diagrams can be used to advantage to help ensure the completeness and correctness of the analysis. You should take a minute and review Figure 7-17 to make sure you understand how these diagrams are related.

Summary

The set of systems requirements within the object-oriented approach is documented using the events table and the following four diagrams:

- The use case diagram
- The class diagram
- The interaction diagram (collaboration and/or sequence diagram)
- The statechart diagram

A use case diagram documents the various ways that the system can be used. It is developed based on the information included within the events table. A use case consists of actors, use cases, and connecting lines. A use case is an identification of a single function that the system supports. An actor represents a role of someone or something that uses the system. The connecting lines are used to indicate which actors invoke which use cases. Use cases can also invoke other use cases as a common subroutine. This type of connection between use cases is called the << includes>> relationship.

The internal activities of a use case are first described by an internal flow of events. It is possible to have several different internal flows, which represent different scenarios of the same use case. Thus, a use case may have several scenarios.

A diagram provides more detail of the scenarios by identifying the components that make up a scenario or an entire use case if there is only one scenario. This more detailed description—which documents the actors, the objects, and the interactions among them—is called a collaboration diagram. The purpose of a collaboration diagram is to describe what objects work together to carry out the scenario. The actors involved are those who are associated with the use case. The objects come from the object classes and are those entities that work together during the use case. The interactions are the messages that the objects send to each other to enable the collaboration. A collaboration is a high-level view of the activity within a scenario. A more detailed view is provided with another type of interaction diagram, called a sequence diagram. The objective of a sequence diagram is to show the detailed sequence of messages that are exchanged between the objects during a use case.

When objects receive and send messages, they do so based on their behavior. A statechart diagram is used to describe the complete life of an object, including all of its actions associated with sending and receiving messages. A statechart thus represents the internal logic associated with each object. A statechart diagram consists of states, transitions, and action-expressions. The states are the possible conditions or states of being of an object while it is carrying out an action or waiting for a message. The transitions are the moves that an object makes among states. Transitions are usually triggered by the receipt of a message. Action-expressions are the activities that an object carries out. Action-expressions may be associated with either states or transitions.

Key Terms

action, p. 240
action-expression, p. 242
activation lifeline, p. 230
actor, p. 221
collaboration diagram, p. 219
completion transition, p. 246
composite state, p. 240
concurrency or concurrent states, p. 240
decision pseudostate, p. 246
destination state, p. 241
dot notation, p. 243
guard condition, p. 241
interaction diagram, p. 219
internal transition, p. 242
implementation classes, p. 221
lifeline, p. 230
link, p. 236
message, p. 219
message event, p. 241
origin state, p. 241
path, p. 246
scenario, p. 222
sequence diagram, p. 219
state, p. 239
statechart diagram, p. 219
statechart fragment, p. 250
synchronization pseudostate, p. 247
thread, p. 247
transition, p. 241
use case, p. 221
use case diagram, p. 219

Review Questions

1. What is the purpose of a use case diagram?
2. What is the difference between a sequence diagram and a collaboration diagram?
3. What does an actor represent in a use case diagram?
4. What is a <<includes>> relationship?
5. Explain how a collaboration diagram is used to document a use case.
6. What is the relationship between a use case and a scenario?
7. Describe the syntax of a message descriptor text and explain the meaning of each component.
8. Describe what an object lifeline is and how it is used in a sequence diagram. Contrast an object lifeline with an activation lifeline.
9. What are the three guidelines to help determine the source object of a message in a sequence diagram?
10. What is the purpose of a statechart diagram? What does a statechart diagram document?
11. Define and explain the concept of a state.
12. List and describe three different types of pseudostates. Explain the use of each.
13. What is dot notation? How is it used?
14. What is a completion transition? How is it different from a normal transition?
15. Discuss the two ways that concurrency is documented in a statechart diagram. What conditions would cause one to be used instead of the other?
16. How does an internal transition differ from an external transition? Give some examples of internal transitions.
17. Describe the relationships among a statechart, interaction diagram, class diagram, and use case diagram.
18. What function does an action-expression serve in a statechart? Where are action-expressions located?

Thinking Critically

1. Given the following narrative, identify the actors and the use cases. Draw a use case diagram.
   A patient record and scheduling system in a doctor's office is used by the receptionists, the nurses, and the doctors. The receptionists use the system to enter new patient information when first-time patients visit the doctor. They also schedule all appointments. The nurses use the system to keep track of the results of each visit and to enter information about patient care, such as medications and diagnoses. The nurses can also access the information to print out patient reports or the history of patient visits. The doctors primarily use the system to view patient history. The doctors may enter some patient treatment information occasionally, but most frequently they let the nurses enter this information.
2. Interpret and explain the use case diagram in Figure 7-30. Explain the various roles of those using the system, and what functions each role requires. Explain the relationships and how the use cases are related to each other.

FIGURE 7-30
A use case diagram for the inventory system.

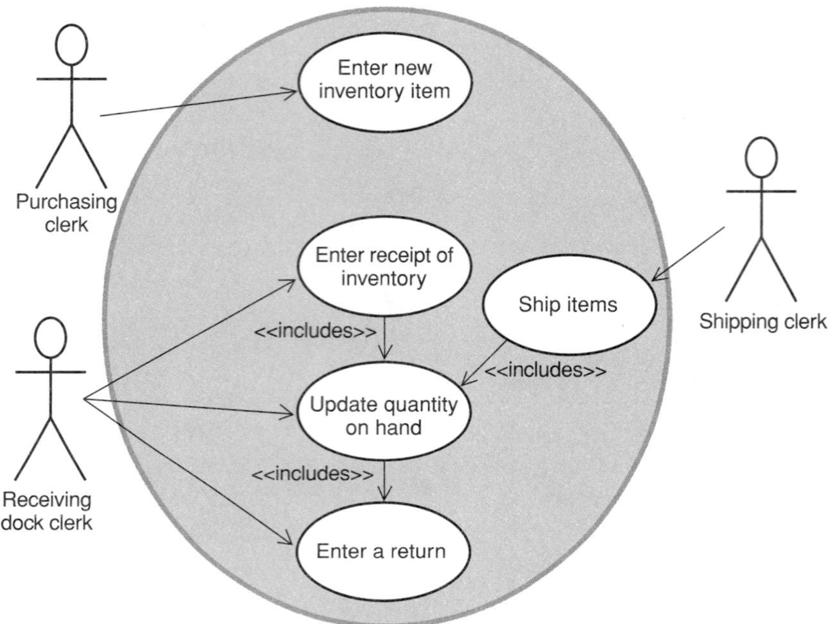

3. Given the following narrative, identify two possible scenarios and develop a flow of events for each.
   Quality Building Supply has two kinds of customers: contractors and the general public. Sales to each are slightly different. When a contractor buys materials, he or she takes them to the contractor checkout desk. The clerk scans in the items to be purchased. The system creates an electronic ticket for the items. The system compares the total amount against the contractor's current credit limit, and if it is acceptable, finalizes the sale. The contractor's credit limit is reduced by the amount of the sales. Once a month, the system sends an invoice to the contractor and, when the payment comes in, increases the credit limit back to the original value. The electronic ticket is maintained in the system and can be printed out at any time for the following 60 days. Some contractors like to keep a record of their purchases, so they request them to be printed (usually at the time of the sale). Others aren't interested in a printout.
   A sale to the general public is simply entered into the cash register and a paper ticket is printed as the items are identified. Payment can be by cash, check, or credit card. The clerk must enter the type of payment to ensure that the cash register balances at the end of the shift. For credit card payments, the system prints out a credit card voucher that the customer must sign.
4. Draw a collaboration diagram for each scenario identified in exercise 3.
5. Draw a sequence diagram for each scenario identified in exercise 3.
6. Given the following narrative, develop a statechart diagram for the Loan object class.

A library system must keep track of books, patrons, and book loans. The object classes for books and patrons represent physical items. A book loan is a genuine object class; however, it is a conceptual item that is not represented by a physical object.
A book loan first comes into existence when a patron takes some books and other lendable items to the checkout desk. The library clerk scans in the patron library card to create a new loan. The loan is then ready to record items that the patron is checking out. Each time the clerk scans a book barcode, the system associates the checkout item with the loan. Once all of the items have been scanned, the system considers the loan active and prints out a list of items on the loan if desired. While the loan is active, the items may be returned. The patron does not have to return them at the same time. The loan will remain active until the patron returns all the books on that loan. Once they are returned, the loan becomes inactive.

7. Given the following narrative, extend and change the statechart for the Loan object class to handle concurrent paths to keep track of the active status of the loan as well as the overdue status.
   If while the loan is active some items are not returned by the loan due date, the loan goes into a grace-period overdue status of seven days. A letter is mailed to the patron, but no fines are accrued. If after seven days there are still items that have not been returned, another letter is sent and the loan status becomes overdue with fines. A patron can return the items by dropping them in the return slot and still not pay the fines. If that happens, then the loan is inactive, but it can still have outstanding fines due. Once all of the books have been returned and all of the fines paid, the loan is closed.
8. Given the collaboration diagrams in Figure 7-31, develop a statechart diagram. Use the information from the collaboration diagrams and your own judgment to determine the life of the object.

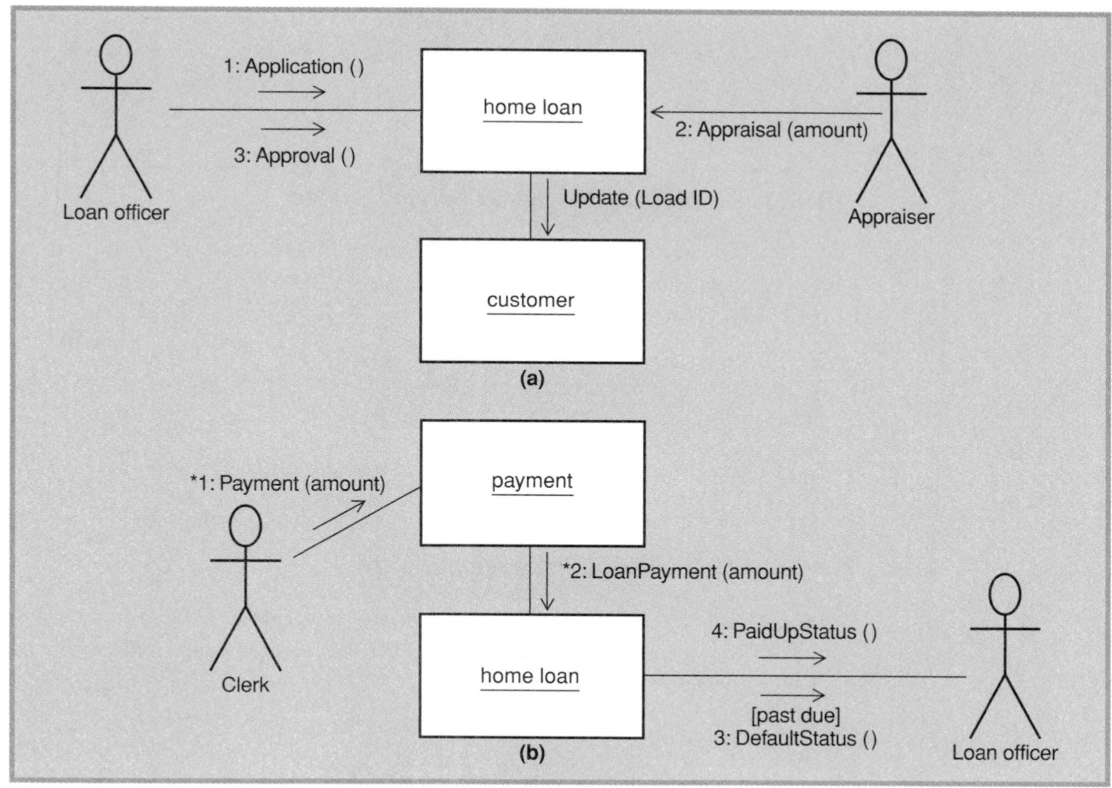

FIGURE 7-31
Collaboration diagrams for the HomeLoan class.

9. In this chapter, the first half of the statechart for the *Order* class was explained (see Figure 7-25). In your own words, explain the last part of the statechart, beginning with the state *ReadyForFulfillment*. Explain how the flow of activity occurs based on the transitions and states in the diagram.
10. The object-oriented approach is very useful to develop real-time, event-driven systems. One such system is the Minesweeper game that is part of the Windows 98 package. Your assignment is to play the game and analyze it. Develop the following diagrams: a class diagram, a use case diagram, a collaboration or sequence diagram for the "select a tile" use case, and a statechart for a tile. Hint: A tile object sends messages to other tile objects. In other words, an object sends messages to the same class, but to different objects in the class. You can show this on the collaboration diagram by using different objects with unique object names. Consider all the different views of a tile as the various states, such as blank, marked with a flag, and so forth. Consider the icons both on top of the tile and hidden underneath the tile as attributes of the tile object.

Experiential Exercises

1. Based on what you know about Rocky Mountain Outfitters, work with several other students to develop a sample flow of events for each use case in the RMO use case diagram.
2. Continue developing the RMO requirements by developing the rest of the collaboration and sequence diagrams for each use case in the RMO use case diagram.
3. Next, based on the information that you have developed in the first two exercises, develop statechart diagrams for all classes in the RMO class diagram.
4. An activity diagram is a type of statechart diagram. It describes workflows and business processes. Learn about activity diagrams and explain how they differ from statecharts. Develop an activity diagram for the telephone-order process.
5. Download from the Internet an evaluation copy of Rational Rose's UML CASE tool and use the tool to draw the RMO diagrams.

Case Studies

The Real Estate Multiple Listing Service System

Refer to the description of the Real Estate Multiple Listing Service system in the Chapter 5 case studies. Using the event list and ERD for that system as a starting point, develop the following object-oriented models:

1. Convert your ERD to a class diagram.
2. Develop a use case diagram.
3. Develop a sequence diagram for each use case identified in the use case diagram.
4. Develop a statechart diagram for each class in the class diagram.

The State Patrol Ticket Processing System

Refer to the description of the State Patrol ticket processing system in the Chapter 5 case studies. Using the event list and ERD for that system as a starting point, develop the following object-oriented models:

1. Convert your ERD to a class diagram.
2. Develop a use case diagram.
3. Develop a sequence diagram for each use case identified in the use case diagram.
4. Develop a statechart diagram for each class in the class diagram.

The DownTown Videos Rental System

DownTown Videos is a chain of 11 video stores scattered throughout a major metropolitan area in the Midwest. The chain started with a single store several years ago and has grown to its present size. Paul Lowes, the owner of the chain, knows that to compete with the national chains will require a state-of-the-art movie rental system. You have been asked to develop the system requirements for the new system.

Each store has a stock of movies and video games for rent. It is important to keep track of each movie title to know and to identify its category (classical, drama, comedy, and so on), its rental type (new release, standard), movie rating, and other general information such as movie producer, release date, cost, and so forth. In addition to tracking each title, the business must track each individual copy to note its purchase date, its condition, and its rental status. User functions must be provided to maintain this inventory information.

Customers, the lifeblood of the business, are also tracked. DownTown considers each family to be a customer, so special mailings and promotions are offered to each household. For any given customer, several people may be authorized to rent videos and games. The primary contact for each customer can also establish rental parameters for other members of the household. For example, if a parent wants to limit a child's rental authorization to only PG and PG-13 movies, the system will track that.

Each time a movie is rented, the system must keep track of which copies of which movies and games are rented; the rental date and time and the return date and time; and the household and person renting the movie. Each rental is considered to be open until all of the movies and games have been returned. Customers pay for rentals when checking out videos at the store.

For this case, develop the following diagrams:

1. A class diagram
2. A use case diagram
3. A sequence diagram for each use case
4. A statechart diagram for the customer and rental object classes

Further Resources

Grady Booch, James Rumbaugh, and Ivar Jacobson. *The Unified Modeling Language User Guide*. Addison-Wesley, 1999.

Hans-Erik Eriksson and Magnus Penker. UML Toolkit. John Wiley & Sons, 1998.

Martin Fowler. *UML Distilled: Applying the Standard Object Modeling Language*. Addison-Wesley, 1997.

Ivar Jacobson, Grady Booch, and James Rumbaugh. *The Unified Software Development Process*. Addison-Wesley, 1999.

Environments, Alternatives and Decisions

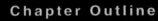

Learning Objectives

After reading the chapter, you should be able to:

- Describe the strategic decisions that integrate the target processing environment and the design approach for the new system
- Prioritize the system requirements based on the desired scope and level of automation for the new system
- Determine alternative approaches for system development
- Evaluate and select a development approach based on the needs and resources of the organization
- Describe key elements of a request for proposal and evaluate vendors' proposals for outsourced alternatives
- Develop a professional presentation of findings to management

Chapter Outline

Assessing the Target Processing Environment

Deciding on Scope and Level of Automation

Generating Alternatives for Implementation

Choosing an Alternative for Implementation

Considering Outsourced Solutions

Presenting the Results and Making the Decisions

EIGHT

TROPIC FISH TALES: NETTING THE RIGHT SYSTEM

Robert Holmes wasn't exactly sure how to proceed with his project. He had six proposals from software vendors to develop an Internet-based ordering system for his company, Tropic Fish Tales. He and his project team had to figure out some way to make a meaningful comparison among the proposals to determine which alternative best fit the needs of the company. Then he had to make a presentation of his analysis and recommendations.

The problem was that none of the six proposals was the same. He and his team had spent a tremendous amount of time developing a request for proposal (RFP) that they had sent to several firms providing custom solutions. They had worked hard on the RFP to make sure it contained very precise definition of business requirements. Even with this well-designed RFP, none of the six proposals looked the same. He was going to have to devise a method to do a fair comparison between the proposals. Otherwise, how would he know which solution was the best for Tropic Fish Tales?

His company had made an early decision to develop an RFP and obtain outside assistance with the development. The project appeared to be pretty large, and the information system staff was quite small and inexperienced. The least expensive solution was from a company that had a standard off-the-shelf ordering system. The advantage was that it would be quick and fairly inexpensive to install it and get it working. However, the disadvantage was that it did not quite fit all of the requirements. Robert wasn't sure how important the missing functionality was to his company. The system could be made to work with some modifications to work procedures and forms.

At the other end of the spectrum was a proposal for a completely new state-of-the-art system for Internet sales and with electronic interfaces to suppliers and shippers. This system was a complete electronic commerce solution with fully automated support. The proposal also indicated that substantial transaction, customer, and order history information would be retained and available in real time. The system also contained automated inventory management functions. Although the system had more capability than the company really needed, it would certainly bring Tropic Fish Tales to the forefront of high-technology solutions. He wondered if the company could afford the price, however, which was about three times the cost of the low-cost solution.

The other proposals ranged in between the two extremes. One company proposed to develop a system from the ground up, working very closely with his firm to ensure that the system fit the requirements perfectly. Another company had a base system that it proposed to modify. The base system was for a different industry and was not currently Internet-based, so substantial modifications would be necessary. One solution ran only on UNIX machines. Even though the system appeared to have most of the desired functionality, it would take some work to modify it for a Novell network, which is the current environment for the company.

Robert was scheduled to meet Bill Williams, the director of information systems, later in the day. He hoped Bill would have some suggestions about how to address this problem.

OVERVIEW

As we discussed in previous chapters, the six major activities of the analysis phase of system development are the following:

- Gather information
- Define system requirements
- Prototype for feasibility and discovery
- Prioritize requirements
- Generate and evaluate alternatives
- Review recommendations with management

You have already learned about fact finding, defining system requirements, and prototyping activities of the analysis phase. You learned how to develop requirements using either a traditional approach or an object-oriented approach. This chapter explains the last three activities of the analysis phase—that is, the transitional activities that refocus the project from discovery and analysis to solutions and design. These final activities are pivotal in the project; they set the direction for the design and implementation of the solution system.

One important consideration that affects the tasks associated with prioritizing requirements and considering alternatives is the environment that exists or will exist for the new system—that is, what computers and operating systems will be used for the new system. This chapter begins with a broad review of computer environments prior to discussing the details of requirements for the system functionality. Even though these topics do not relate to specific tasks of the SDLC, they are important elements in the overall strategic planning and system deployment. We also demonstrate the important points to consider in the environment by discussing them in the context of Rocky Mountain Outfitters and its new system.

Following the section on the computer environment, we discuss the tasks required to prioritize and evaluate requirements. The primary objectives of the final stages of analysis are (1) to decide what to include within the scope of the system and (2) to determine what method to use to develop the system.

It is normal during analysis to uncover many more requirements and needs than can reasonably be included within the system, so one of these activities is to categorize and prioritize the requirements to determine what to include. Frequently, two or three alternative packages of requirements will be developed, along with the required resources for development, and then a steering committee of executives and users will decide which approach is most viable. This chapter discusses various strategies for prioritizing and selecting a scope and level of automation.

Once the system scope is determined, then several alternative methods of development are reviewed. These alternatives can range from building the new system from scratch to buying a system from someone else to outsourcing the entire development and daily operation. We review the most popular of these alternatives and provide some direction on making a selection. Included within this discussion are instructions on how to develop and use a request for proposal (RFP).

Assessing the Target Processing Environment

target processing environment
The configuration of computer equipment, operating systems, and networks for the new system

In selecting an appropriate solution, analysts need to consider first the target processing environment. The **target processing environment** is the configuration of computer equipment, operating systems, and networks that will exist when the new application system is deployed. The client and users of the new system are obviously most interested in the functions of the application itself because they need it to carry out the business of the organization. However, the application does not function in a vacuum. There must be a stable environment of supporting components to enable the application to execute successfully. If the environment is not suitable and stable, then the application will not function. An important part of any project is defining and ensuring that the target processing environment is defined, developed, and deployed so that it is stable.

Generally, most projects do not require a complete redeployment of the computer hardware and operating system software. The hardware and operating environment are major decisions made as part of the IT strategic plan, which we discussed in Chapter 1. However, it is not uncommon for an organization's strategic IT plan to include deploying additional equipment with each new project. Thus, the design and implementation of the processing environment is one of the important early activities of moving the project from the analysis of the requirements to the design of the solution.

It is not our intent to teach you how to design networks, configure hardware, and implement operating systems. That is a course in and of itself. The intent of this section is to identify issues related to the target processing environment that will impact the development of the application system. To start, we describe the various processing environment alternatives and recent trends in system development.

Centralized Systems

Prior to the early 1970s, there was generally only one technological environment for an information system—a mainframe computer system operated from a central location (such as a regional or national headquarters). The only significant options were the form of input (such as keypunch, key-to-tape, or interactive input using video display terminals) and whether input/output devices would be placed in remote locations. Although centralized mainframes are no longer the preferred platform for deploying information systems, they are still widely used for batch processing applications and as one part of larger distributed systems.

Centralized mainframes are generally used for large-scale batch processing applications. Such applications are common in industries such as banking, insurance, and catalog sales. Information systems in such industries often have the following characteristics:

- Some input transactions do not need to be processed in real time (for example, out-of-state checks delivered in large nightly batches from central bank clearinghouses).
- On-line data-entry personnel can be centrally located (for example, a centrally located group of telephone order takers can serve geographically dispersed customers).
- Large numbers of periodic outputs are produced by the system (for example, monthly credit card statements mailed to customers).

Any application that has two or three of these characteristics is a viable candidate for implementation on a centralized mainframe.

Centralized computer systems are seldom used as the sole hardware platform for an information system. Most systems have some transaction inputs that must be accepted from geographically dispersed locations and processed in real time (an example of such an input is a cash withdrawal from an ATM). Most systems also have some outputs that are requested from and delivered to remote locations in real time (an example of such an output is an insurance policy inquiry by a state motor vehicle department). Thus, centralized computer systems are typically used to implement one or more subsystems within a larger information system that includes on-line, batch, and geographically dispersed components.

Single-Computer Architecture As its name implies, single-computer architecture places all information system resources on a single computer system and its directly attached peripheral devices. Users interact with the system via simple input/output devices that are directly connected to the computer. Single-computer architecture requires that all system users be located near the computer. The primary advantage of single-computer architecture is its simplicity. Information systems deployed on a single-computer system are relatively easy to design, build, operate, and maintain.

The capacity limits of a single computer may make single-computer architecture impractical or unusable for large information systems. Many systems are so large that even the largest mainframe computer cannot perform all the required processing, data storage, and data retrieval tasks. For such systems, a clustered or multicomputer architecture is required.

clustered architecture
A group of computers of the same type that have the same operating environment and share resources

multicomputer architecture
A group of dissimilar computers that are linked together

Clustered and Multicomputer Architectures A **clustered architecture** employs a group (or **cluster**) of computers to provide needed processing or data storage and retrieval capacity. Computers from the same manufacturer and model family are networked together. Similar hardware and operating systems allow application programs to execute on any machine in the cluster without modification. Application programs can be allocated to an underutilized computer when they execute so that the processing load can be balanced across all machines. Program movement and access to resources on other machines occur quickly and efficiently due to rapid and direct communication at the operating system level. In effect, a cluster acts as a single large computer system. Often one computer acts as the entry point to the system. The other computers in the system function as slave computers and are assigned tasks by the controlling computer.

A **multicomputer architecture** also employs multiple computer systems, but hardware and operating systems are not required to be as similar as in a clustered architecture. Hardware and software differences make it impractical to move application programs from one machine to another. Instead, a suite of application programs and data resources is exclusively assigned to each computer system. Resource access among machines is not implemented at the operating system level, thus reducing simplicity and efficiency. Even though this architecture is similar to a distributed configuration (discussed in the next section), we classify it as a centralized system since it functions as a single large computer. As with a clustered architecture, one computer is the central computer, and the other computers are slaved to it. One frequent example of a multicomputer architecture is a central computer with a front-end communication computer. The main computer executes the programs and holds the database, while the front-end computer handles all the communication lines with other computers or simple terminals.

If all parts of an information system have similar hardware and operating system requirements, then a clustered architecture is typically simpler to implement, operate, and maintain. Clustered architectures are generally more cost efficient and provide greater total capacity due to their streamlined methods of load balancing and resource sharing. Multicomputer architectures are most useful when a centralized information system can be decomposed into relatively independent subsystems, each of which is naturally suited to a different hardware platform and/or operating system. It may also be used to graft new capabilities or subsystems onto an existing single-computer information system. Figure 8-1 illustrates the three types of centralized systems. Part (a) shows a single, central mainframe type of system. Part (b) shows a clustered architecture. Note that the computers are all of the same type and are very closely connected. Part (c) shows a multicomputer architecture. In part (c), note that the computers can consist of different makes and models.

FIGURE 8-1
Single, clustered, and multicomputer architectures.

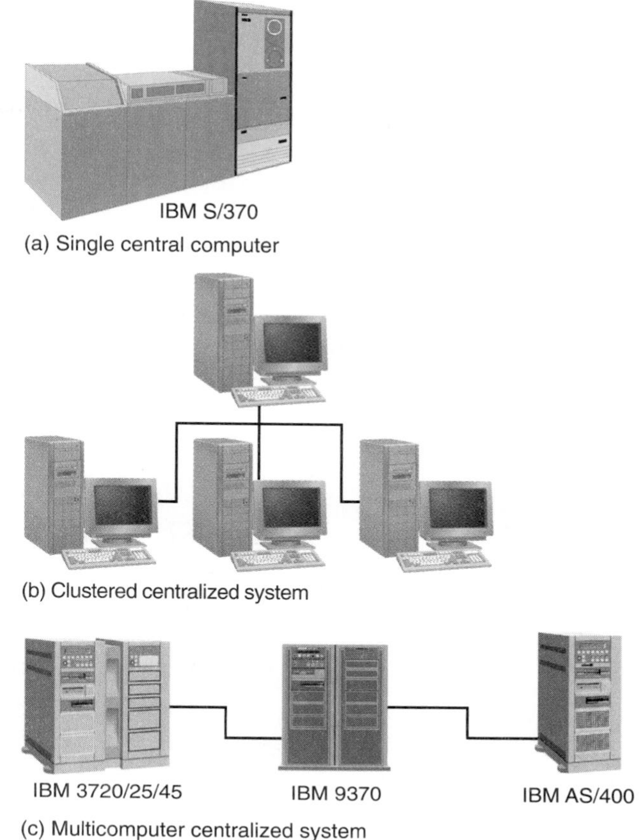

Distributed Computing

Components of a modern information system are typically distributed across many computer systems and geographic locations. For example, corporate financial data might be stored on a centralized mainframe computer. Minicomputers in regional offices might be used to periodically generate accounting and other reports based on data stored on the mainframe. Personal computers in many locations might be used to access and view periodic reports as well as to directly update the central database. Such an approach to distributing components across computer systems and locations is generically called **distributed computing** or **distributed processing**.

distributed computing
The approach to distributing a system across several computers and locations

Distributed computing relies on communication networks to connect geographically dispersed computer hardware components. Networking technology has undergone rapid change over the last three decades. The changes include rapid increases in transmission capacity, significant reductions in cost, and standardized methods of constructing and interacting with networks. These improvements have allowed distributed computing to become the preferred method of deploying the vast majority of business applications.

computer network
A set of transmission lines, equipment, and communication protocols to permit sharing of information and resources

local area network (LAN)
A computer network where the distances are local, such as within the same building

Computer Networks A **computer network** is a set of transmission lines, specialized hardware, and communication protocols that allow communication among different users and computer systems. Computer networks are divided into two classes depending on the distance they span. A **local area network (LAN)** is typically less than one kilometer long and connects computers within a single building or floor. The term **wide area network (WAN)** can describe any network over one kilometer, though much greater distances spanning cities, countries, continents, or the entire globe are typically implied.

wide area network (WAN)
A computer network across large distances such as a city, state, or nation

Figure 8-2 shows a possible computer network for RMO. Each geographic location is served by a single LAN, and all LANs are connected by a WAN. Communication among users and computers in a single location uses a LAN. Communication among geographically dispersed sites uses the LANs at both sites and the WAN. A **router** connects each LAN to the WAN. A router scans messages on the LAN and copies them to the WAN if they are addressed to a user or computer on another LAN. The router also scans messages on the WAN and copies them to the LAN if they are addressed to a local user or computer.

router
A piece of equipment that is used to direct information within the network

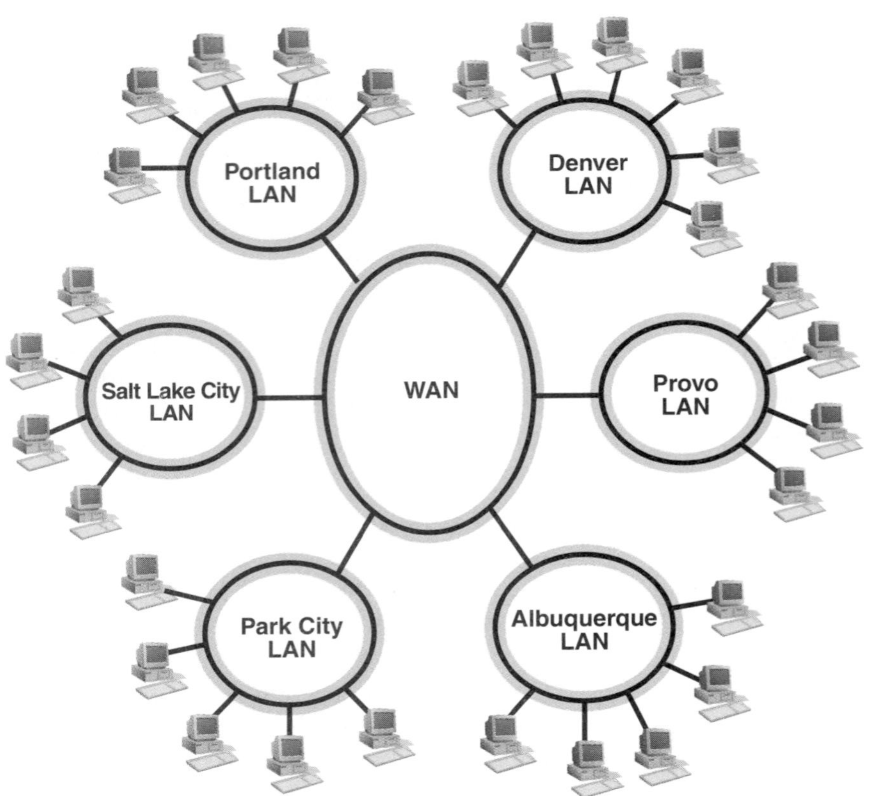

FIGURE 8-2
Network configuration for RMO.

LANs and WANs can be built using many technologies. Technologies such as Ethernet and token ring are typically used to implement LANs. They provide low to moderate amounts of message-carrying capacity at relatively low cost. WAN technologies such as asynchronous transmission mode and frame relay are more complex and expensive, though they typically provide higher message-carrying capacity and greater reliability. WANs may be constructed using purchased equipment and leased long-distance transmission lines. WAN setup and operation may also be subcontracted from a long-distance telecommunications vendor (such as AT&T, Sprint, and MCI).

Computer networks provide a generic communication capability among computer systems and users. This generic capability can be used to implement many services including direct communications (such as telephone service and video conferencing), message-based communications (such as e-mail), and resource sharing (such as access to electronic documents, application programs, and databases). A single network can simultaneously support all types of service with appropriate hardware and sufficient transmission capacity.

There are many ways to distribute information system resources across a computer network. Users, application programs, and databases can be placed on the same computer system, on different computer systems on the same LAN, or different computer systems on different LANs. Application programs and databases can also be subdivided and each distributed separately. The variety of distribution possibilities may seem daunting. But extensive experience with distributed systems over the last two decades has led to the emergence of a few standard approaches to distributing resources.

server computer, or server
A computer that provides services to other computers on the network

client computer
A computer that requests services from other computers on the network

middleware
Computer software that implements communication protocols on the network and helps different systems communicate

data layer
The layer on a client-server configuration that contains the database

Client-Server Architecture Client-server architecture is currently the dominant architectural model for distributing information system resources. Client-server architecture divides information system processes into two classes: client and server. A **server computer** manages one or more system resources and provides access to those resources through a well-defined communication interface. A **client computer** uses the communication interface to request resources, and the server responds to those requests. Software that implements the communication interface is usually called **middleware**.

Client-server architecture is a general architectural model that can be implemented in many different ways. A typical example is the architecture used to support a shared printer on a local area network (see Figure 8-3). An application program on a personal computer (PC) sends a document to a server computer on the LAN. The LAN server computer receives the request via its network interface card and dispatches the request to a management process for the specified printer (the server computer operating system may manage multiple printers). When the printer has successfully printed the document, a message is sent back to the PC to notify the user that the printed document is ready.

FIGURE 8-3
Client-server architecture with a shared printer.

business logic layer
The part of a client-server configuration that contains the programs that implement the program logic of the application

view layer
The part of the client-server configuration that contains the user interface and other components to access the system

three-layer architecture
A client-server architecture that contains the three layers of view layer, business logic layer, and data layer

N-Layer Client-Server Architecture An information system application program can be divided into the following set of client and server processes or layers:

- **The data layer** manages stored data, usually implemented as one or more databases.
- **The business logic layer** implements the rules and procedures of business processing.
- **The view layer** accepts user input, and formats and displays processing results.

This approach to client-server architecture is sometimes called **three-layer architecture**. Figure 8-4 illustrates the interaction of the three layers. The view layer acts as a client of the business logic layer, which, in turn, acts as a client of the data layer.

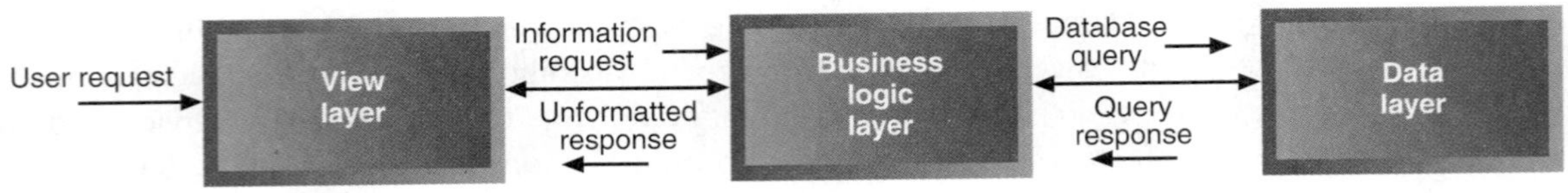

User request → View layer; Information request → / ← Unformatted response; Business logic layer; Database query → / ← Query response; Data layer

FIGURE 8-4
Three-layer architecture.

Application software that is divided into three layers is relatively easy to distribute and replicate across a network. Interactions among the layers are always requests or responses. This makes the layers relatively independent of one another. Thus, they can be placed on different computer systems, with network connections and middleware serving as the glue that binds them together into a single application.

Three-layer architecture can be expanded into a larger number of layers when processing requirements or data resources are complex. Architectures that employ more than three layers are called **n-layer architectures** or **n-tiered architectures**. For example, the database layer as illustrated in Figure 8-4 can be split into multiple data layers. Figure 8-5 shows an example in which the data layer is split into two separate layers. The business logic layer interacts with a combined database server that provides a unified view of the data stored in several different databases. The combined database server determines where requested data is located and sends access requests to the servers that control the other databases. The responses from the individual database servers are then combined to create a single response to send to the business logic layer.

n-layer architectures, or n-tiered architectures
A client-server architecture that contains n layers

FIGURE 8-5
N-layer architecture.

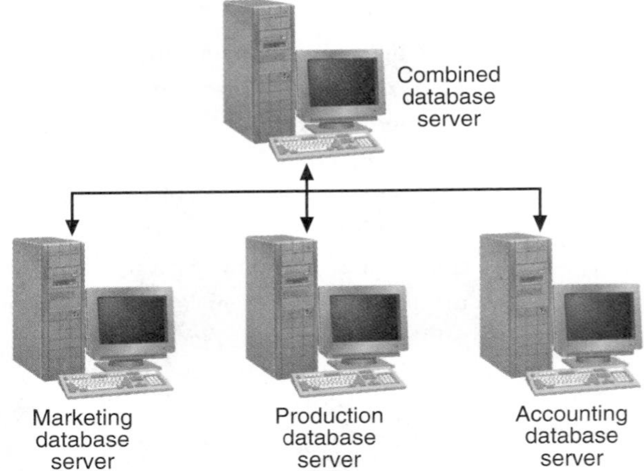

The Internet and Intranets

The Internet and World Wide Web are becoming an increasingly popular framework for implementing and delivering information system applications. However, there is a great deal of confusion regarding what is or is not an Internet or web application. This confusion results from many factors, including lack of agreement about the exact meaning of the terms *Internet* and *web*. We will now provide a set of definitions for these terms.

The **Internet** is a global collection of networks that are interconnected using a common low-level networking standard—TCP/IP (Transmission Control Protocol/Internet Protocol). Many services are provided by the Internet, including the following:

Internet
A global collection of networks that use the same networking protocol—TCP/IP

- E-mail protocols such as Simple Mail Transfer Protocol (SMTP), Post Office Protocol (POP), and Internet Message Access Protocol (IMAP)
- File transfer protocols such as File Transfer Protocol (FTP)
- Remote login and process execution protocols such as Telnet, tn3270, and remote procedure call (RPC)

World Wide Web (WWW) or web
A collection of resources such as files and programs that can be accessed over the Internet using standard protocols

The **World Wide Web (WWW)**, also called simply the **web**, is a collection of resources (programs, files, and services) that can be accessed over the Internet by a number of standard protocols, including the following:

- Formatted and linked document protocols such as HyperText Markup Language (HTML), Extensible Markup Language (XML), and HyperText Transfer Protocol (HTTP)
- Executable program standards including Java, JavaScript, and Visual Basic Script (VBScript)

The Internet is the infrastructure upon which the web is based. In other words, resources of the web are delivered to users over the Internet.

intranet
A private network that is accessible to a limited number of users, but which uses the same TCP/IP protocol as the Internet

An **intranet** is a private network that uses Internet protocols but is accessible only by a limited set of internal users (usually members of the same organization or workgroup). The term also describes a set of privately accessible resources that are organized and delivered via one or more web protocols over a network that supports TCP/IP. An intranet uses the same protocols as the Internet and web but restricts resource access to a limited set of users. Restricted access can be accomplished by a number of means including unadvertised resource names, firewalls, and user/group account names and passwords.

extranet
An intranet that has been extended outside of the organization to facilitate the flow of information

virtual organization
A loosely coupled group of people and resources that work together as though they were an organization

An **extranet** is an intranet that has been extended to include directly related business users outside the organization (such as suppliers, large customers, and strategic partners). An extranet allows separate organizations to exchange information and coordinate their activities, thus forming a **virtual organization**.

The web is organized using a client-server architecture. Web resources are managed by server processes that can execute on dedicated server computers or on multipurpose computer systems. Clients are programs that send requests to servers using one or more of the standard web resource request protocols. Web protocols define valid resource formats and a standard means of requesting resources. Note that any program (not just a web browser) can be designed to use web protocols. Thus, Weblike capabilities can be embedded in ordinary application programs.

The Internet as an Application Platform Internet and web technologies present an attractive alternative for implementing information systems. For example, consider the problem of data entry and access by a buyer who works for RMO to purchase items from RMO suppliers. Buyers are typically on the road for several months a year, often for weeks at a time. A traveling buyer needs some means of remotely interacting with RMO's inventory management system to record purchasing agreements and query inventory status.

One way of providing these capabilities is to design custom application software and a private means of connecting to that software. The client portion of the application is installed on the buyers' laptop computers. A modem is attached to a computer that executes the server portion of the application. A buyer then dials in to the system from remote locations to gain access to the application server.

Another alternative for implementing remote access for buyers is to construct an application that uses a web browser interface. Such an application executes on a web server and is accessible from any computer with an Internet connection. Buyers can use a web browser on their laptop computer and connect to the application via an Internet service provider wherever they're currently located. Buyers could also access the application from any other computer with Internet access (for example, a computer in a vendor office, hotel business suite, or copy center such as Kinko's).

Implementing the application via the Internet greatly expands the application's accessibility. It also eliminates the need to install custom client software on buyers' laptop computers. Client software can be updated by simply updating the version stored on the web server. The application is relatively cheap to develop and deploy because existing web standards and networking resources are employed. Custom software and private access via modems require a more complex development effort and the maintenance of a greater number of customized resources.

Implementing an application using the web, an intranet, or an extranet has a number of advantages over traditional client-server approaches to application architecture, including the following:

- *Accessibility.* Web browsers and Internet connections are nearly ubiquitous. Internet, intranet, and extranet applications are accessible to a large number of potential users (including customers, suppliers, and off-site employees).
- ***Low-cost communication.*** The high-capacity WANs that form the Internet backbone are funded primarily by governments. Traffic on the backbone networks travels free of charge to the user, at least for the present. Connections between private LANs and the Internet can be purchased from a variety of private Internet service providers at relatively low cost. In essence, a company can use the Internet as a low-cost WAN.
- ***Widely implemented standards.*** Web standards are well known, and many computing professionals are already trained in their use. Server, client, and application development software is widely available and relatively cheap.

Information resource delivery via an intranet or extranet enjoys all of the advantages of web delivery since intranets and extranets use web standards. In many ways, intranets, extranets, and the web represent the logical evolution of client-server computing into an off-the-shelf technology. Organizations that had shied away from client-server computing due to the costs and required learning curve can now enjoy client-server benefits at substantially reduced complexity and cost.

Of course, there are negative aspects of application delivery via the Internet and web technologies, including the following:

- ***Security.*** Web servers are a well-defined target for security breaches because web standards are open and widely known. Wide-scale interconnection of networks and the use of Internet and web standards have created a common and easily accessible target.
- ***Reliability.*** Internet protocols do not guarantee a minimum level of network throughput per user or even that a message will ever be received by its intended recipient. Standards have been proposed to address these shortcomings, but they have yet to be widely adopted.
- ***Throughput.*** The data transfer capacity of most home users and many business users is limited by analog modems to under 56 kilobits per second. Internet service providers and backbone WANs can become overloaded during high-traffic periods, resulting in slow response time for all users and long delays when accessing large resources.
- ***Volatile standards.*** Web standards change rapidly. Client software is updated every few months. Developers of widely used applications are faced with a dilemma: use the latest standards to increase functionality or use older standards to ensure greater compatibility with user software.

The primary disadvantages to RMO of implementing the customer order application via the Internet are security, performance, and reliability. If a buyer can access the system via the web, then so can anyone else. Access to sensitive parts of the system can be restricted by a number of means including user accounts and passwords. But the risk of a security breach will always be present. Performance and reliability are limited by the buyer's Internet connection point and the available Internet capacity between that connection point and the application server. Unreliable or overloaded local Internet connections can render the application unusable. RMO has no control over these factors.

Development and System Software Environments

The development environment consists of the standards and tools that are in use in the organization. For example, specific languages, CASE tools, and programming standards may be required. The system software environment includes operating systems, network protocols, database management systems, and so forth. In some projects, the development and system software environment may be open to choice. In other situations, they must conform to the existing environment. In either case, an important activity of the analysis phase is to determine the components of the environment that will control the development of the new application system.

The important components of the development and system software environment that will affect the project are the language environment and expertise, existing CASE tools and methodologies, required interfaces to other systems, the operating system, and the database system.

- **Language environment and expertise.** Normally companies have a preferred language for system development. However, as technology changes, newer languages frequently provide additional levels of productivity. Development in today's environment can be extremely complex due to the numerous options and alternatives. There are numerous development languages, from structured languages such as COBOL, to object-oriented languages such as Smalltalk, C++, and Visual Basic, to web-based languages such as Java and Perl Script, to other development environments such as PowerBuilder. Using a new language does require additional commitment and funding to provide the necessary training to the development team.
- **Existing CASE tools and methodologies.** If a company has invested heavily in a CASE tool, then all new development may have to conform to the tool's methodology. Using a CASE tool also frequently dictates the implementation language and methodology. Some CASE tools also generate programming data structures or code components, which will constrain the development environment.
- **Required interfaces to other systems.** Rarely does an organization replace all of its systems at the same time. Typically a new system must provide information to and receive data from existing systems. This becomes especially important when information must be shared across different hardware platforms, operating systems, and databases and at various locations. As the development plan is refined, specific miniprojects may be required to define interface requirements and write interface programs.
- **Operating system environment.** As with the hardware configuration, the available operating system has an important influence on the design and implementation approaches. Strategic goals may also exist to change operating systems, especially in client-server and multitier environments. In many companies, especially those that are migrating from one hardware architecture to another, multiple platforms (types of computers and system software) may exist. It is not uncommon today to have multiple platforms and multiple operating systems with complex requirements for interfaces and communication links. Legacy systems frequently still provide transaction support and must be linked to newer client-server applications and databases. The operating system environment, in conjunction with the hardware configuration, determines the need for a multiplatform system and possibly multiple versions of the new application system.

- **Database management system (DBMS) environment.** Many corporations have committed to a particular database vendor. When this is the case, all the corporate data is maintained under the auspices of a particular database engine. Obviously, the new system must conform to that engine. In other projects, the new application may be part of a total IT strategy to move the organization to a different DBMS.

Many alternatives exist depending on the platform, the operating system, and the application. Various database management systems are available, some of which are very complex and sophisticated. Even though communication capabilities continue to increase dramatically, heavy transactions may require a system that can be manipulated by the user or owner. In many instances, this necessitates a distributed database environment, with portions of the database spread over a country or even the world. Options exist to link to existing databases or to integrate the data into one large consolidated database. New data warehousing technology may require all new applications to connect to the data warehouse. In any case, the database is an important aspect of the processing environment that must be finalized during analysis and before detailed design is commenced.

The Environment at Rocky Mountain Outfitters

The systems environment at Rocky Mountain Outfitters (RMO) had been built piecemeal over the life of the company to support the business functions at the various locations. See Figure 6-35 to review the locations of RMO's business activities.

The Current Environment Table 8-1 illustrates the current computer environment at RMO. Existing technology consists of a mainframe computer located at the home office in Park City with dedicated lines connected to the three warehouse distribution sites in Salt Lake City, Portland, and Albuquerque. A dedicated line also connects the mainframe to the mail-order center in Provo, Utah. Both mail-order and distribution functions are connected directly to the mainframe to allow real-time connection of terminals. The communication technology is based on high-volume mainframe transaction technology. The merchandising and distribution system is a mainframe application written in COBOL and using a combination of DB2 database technology and VSAM files. The system was written by RMO staff with assistance from outside technical consultants.

Dial-up telephone lines are used to communicate with the manufacturing sites in Salt Lake City and Portland. Each manufacturing facility also has its own LAN system to support specific manufacturing information. Updates to the central inventory system are done in a batch mode daily via the dial-up connection.

The retail stores have local client-server retail systems that collect sales and financial information through the cash registers. This information is also forwarded to the central accounting and financial systems residing on the mainframe. The transmittal is done in batch mode daily.

The phone-order system, in Salt Lake City, is a fairly small Windows application running in a client-server environment. It was built by RMO staff as an independent application and is not well integrated with the rest of the inventory and distribution systems. Information is forwarded daily in batches to the system in Park City.

Location and facility	Equipment	Connection
Park City—Data Center	Mainframe	
Park City—Retail	Client-server	Daily dial-up
Salt Lake City—Manufacturing	Local LAN	Daily dial-up
Salt Lake City—Warehouse	Midrange computer	Dedicated line to Data Center
Salt Lake City—Phone Order	Client-server	Daily dial-up
Provo—Mail Order Center	Client-server	Dedicated line to Data Center
Portland—Warehouse	Midrange computer	Dedicated line to Data Center
Portland—Manufacturing	Local LAN	Daily dial-up
Denver—Retail	Client-server	Daily dial-up
Albuquerque—Warehouse	Midrange computer	Dedicated line to Data Center

TABLE 8-1
The existing processing environment at RMO.

Other applications, such as human resources and general accounting, are also mainframe systems running in Park City.

The Proposed Environment Many of the decisions associated with the target environment are made during strategic planning, which establishes long-term directions for the organization. In other situations, the strategic plan is modified as new systems are developed to use the latest technological advancements.

Rocky Mountain Outfitters is at a pivotal point with its new system development. The scope and impact of the new customer support system and the inventory management system were of such a magnitude that the project merited a review of the long-term strategic plan. In Table 8-2, the various target environments possible to RMO are listed. The alternatives are listed by type of technology and degree of centralization. The first three alternatives consider whether to

- move to Internet technology
- utilize internal LAN/WAN technology or
- use a mix of the two options.

The next two alternatives focus on the equipment—whether to

- use a mainframe central processor or
- distributed client-server processors.

Finally, the location and type of database are considered. The decision is whether to use more traditional relational database technology or to move to more advanced object-oriented databases.

TABLE 8-2
Processing environment alternatives.

Alternative	Description
1. Move all functionality to the Internet/intranet	Make both the internal applications and external customer support web-based technology. Will facilitate continual electronic commerce growth.
2. Use 100% internal LAN/WAN technology	Internal transactions can be faster. No need to interface the database with the web. Put only catalog on the web.
3. Mix of 1 and 2	Use the web for customer interaction but use the internal LAN/WAN for back-end processing such as shipping, inventory, etc.
4. Use the mainframe as the central database server	Support high-volume transactions. Will serve as a centralized database for control and consistency.
5. Use a distributed database on local servers	Distributed data provides more rapid response. Has the problem of keeping all the data up to date.
6. Use completely OO components such as Common Object Request Broker Architecture (CORBA) objects	Makes seamless applications with object brokers. Looks to the future for an OO database. May still be too "state-of-the-art."
7. Use OO for a user interface over relational database management system (RDMS)	Use PowerBuilder or Visual Basic to develop an application over a relational database such as Oracle. Supports high volume. Very efficient.
8. Use OO for a user interface plus some CORBA, with RDMS	Starts the move to a complete OO environment. Would require middleware to integrate solutions.

During the requirements gathering activities of the analysis phase, these alternatives were enumerated and investigated. As the application requirements were unraveled, the target environments were defined that would support both the immediate and the long-range requirements. RMO wanted to be state-of-the-art, but it also did not want to have a high-risk project and attempt new technology that was not yet proven.

Table 8-3 lists the major components of the strategic direction for Rocky Mountain Outfitters. The strategic plan for RMO is to move away from a COBOL mainframe environment to a combination client-server configuration. In this approach, the mainframe will remain as the central database server. The other two tiers are application servers, which will contain business logic as well as Internet server capabilities. The users will have individual client personal computers that are connected to the application servers.

Issue	Direction(s)
Required interfaces to other systems	1. Temporary interface for existing inventory management system 2. Permanent connection with the new inventory management system 3. Interface to feed to the accounting/general ledger 4. Interface that automatically feeds to external systems–credit check and shipper
Equipment configurations	1. Conform to the new strategic plan for the increased use of the client-server environment (see Figure 8-2) 2. Mainframe limited to database and telecommunications; All applications are on client-server machines.
Operating system environment	1. Move to the Windows NT environment
Network configurations	1. Windows NT network 2. Internet capabilities 3. Intranet capabilities
Language environment	1. Move away from COBOL 2. Use Visual Basic and Java
Database environment	1. Maintain DB2 database capability 2. May move to an OO database over the long term
CASE tools	1. Only drawing capabilities used, so no strong constraints

TABLE 8-3
Strategic directions for the processing environment at RMO.

Deciding on Scope and Level of Automation

Prioritizing requirements includes tasks to define both the scope and the level of automation for the new system. Scope and level of automation are two aspects of the new application system that are very closely related. Since they are so closely linked, at times systems analysts treat both of them together. But the scope of the system defines which business functions will be included in the system. For example, in the current RMO point-of-sale system, the scope includes handling mail and telephone sales, but not Internet sales. The level of automation is how much automated computer support exists for the functions that are included. In the new system, a very low level of automation for telephone sales may be to require the telephone clerks to have printed catalogs at their desks to verify customer requests. The system would then support only simple data entry of the order information. A higher level of automation for telephone sales would be to have the catalog and customer information on-line so that the telephone clerks get automated entry and verification of inventory items and customer name and address information.

Determining Scope

One common problem with development projects is "scope creep." As implied by the name, requests for the addition of new system functions continue after the requirements have been defined and decisions made. One way to help control this problem is by formalizing the process to identify, categorize, and prioritize the functions that will be included within the new system.

It is normal during analysis to uncover many more requests for business functions than are possible based on the schedule and budget constraints imposed on the project. Decisions need to be made on which functions are critical and must be included and which can be deferred until later. A common approach to determining the scope is to list each requested function and rate its importance, using such categories as "mandatory," "important," and "desirable."

scoping table
A tabular list of all the functions to be included within a system

A **scoping table** of requested business functions is built during analysis. Frequently, the table is an expanded version of the event table. The level of detail (called the level of granularity) in the event table is often increased to further describe the functions for scoping the system. The event table may only identify a customer sale, but the scoping table might need to distinguish among mail sales, telephone sales, and Internet sales. Table 8-4 is an example of the first draft of the scoping table for Rocky Mountain Outfitters. It includes the events from the event table, plus some other events that were identified during later discussions with the users. The additional functions are indicated by background shading.

Functions (expanded from the event table)	Priority (mandatory, important, desirable)	Description
Check item availability	Important	Check QOH and restock date. Clerk or via web.
Place order	Mandatory	Create a new order. Clerk or via web.
Change or cancel order	Important	Make changes or cancel the existing order. Clerk or via web.
Check order status	Important	Check shipment and back order status. Clerk or via web.
Fulfill order	Mandatory	Update QOH. Print label; send notice to shipper.
Create back order	Important	For out-of-stock items.
Return item	Important	Return to inventory; update customer account.
Mail catalog	Mandatory	Print customer address information.
Correct customer account	Important	Make changes to customer information.
Send promotional material	Important	Print address information; track promotions.
Adjust customer charges	Mandatory	Make corrections to customer accounts.
Update catalog	Mandatory	Change seasonal items, offerings, descriptions.
Create promotional materials	Important	Track contents and mailings of promotions.
Create new catalog	Mandatory	Computer-generated catalog.
Maintain customer purchase history	Important	Keep five-year history of customer purchase records.
Ongoing feed to manufacturing	Desirable	Daily order volumes and inventory levels provided to manufacturing plants.
EDI feed to suppliers from sales data	Desirable	Daily order volumes and inventory levels provided to suppliers.
Tie-in to shipper system	Desirable	RMO access to shipment status from shipper.
Data warehousing and analysis	Desirable	Daily extraction of order data. Analysis of trends, buying patterns, demographics, etc.

(*continued*)

Functions (expanded from the event table)	Priority (mandatory, important, desirable)	Description
Automated sales prompting	Desirable	The system makes suggestions for accessories or other related purchases based on selected items.
Expanded sales analysis with DSS	Desirable	Decision Support System (DSS) does detailed demographic and purchase patterns for promotions.
System Reports		
Produce order summary reports	Important	Order volumes by various categories.
Produce activity reports	Desirable	Orders, fulfillments, cancelations, back orders, etc.
Produce transaction summary reports	Desirable	Payments, defaults, credit cards, checks, return credits.
Produce customer adjustment report	Desirable	Returns, adjustments, trends, etc.
Produce fulfillment reports	Desirable	Timing, level of service, outages and back orders, etc.
Produce catalog activity reports	Desirable	Response rates of catalogs, promotions, web, phone, mail.

TABLE 8-4
Scoping list of potential functions for RMO (shading indicates new additions).

In Table 8-4, each business function has been prioritized as mandatory, important, or desirable. Users and clients prioritize the functions based on the needs of the business and the objectives of the new system. For example, if one of the objectives of the system is to increase customer support, then functions that allow RMO to respond to customer requests will be mandatory functions.

Defining the Level of Automation

The level of automation is a description of the kind of support the system will provide for each function. For most functions of an application system, at least three levels of automation can be defined: low, middle, and high. At the lowest level, the computer system only provides simple record keeping. Data input screens are provided to capture information and insert it into a database. Simple types of field edits and validations on input data are also included. For example, a low level of automation for an order-entry function will have a data-entry screen to enter customer and order information. The system date may be used for the order date. Line items for the order are each entered manually. The system may or may not automatically calculate the price. Usually stock on hand and anticipated shipment dates are not verified. At the end of entering the order, the information is stored in the database and the order-entry function is concluded.

We define a middle-range level of automation alternative for each function, which may be a single midrange point or various midrange alternatives. Usually the midrange alternative is a combination of features from the high-level and the low-level automation alternatives. We try to make our best guess of what is necessary and what is justified at the current stage of technology and budget.

A high level of automation occurs when the system takes over, as much as possible, the processing of the function. Usually it is more difficult for an analyst to define high-end automation than low-end automation. Low-end automation is basically an automated version of a current manual procedure. However, generating a high-level automation system requires brainstorming and thinking "outside the box" to create new processes and procedures. Chapter 6 discussed some of the ideas associated with business process reengineering (BPR). The idea of BPR is to rethink completely the way that business functions are performed to achieve radical improvements in processing speeds and levels of service. Successful reengineering of business functions depends on computer systems that provide high levels of automation support.

Let's take the order-entry function within RMO as an example. We start by identifying the best customer service possible. We ask the question, "Why does a customer need to be able to order exactly what she wants at the most convenient time?" Also, "What does a customer want to know about her order at the time she has finished ordering?" (Normally, we would be reengineering the entire process and would also ask questions like, "When, or how soon, does the customer want to receive the items she has ordered?" But, let's limit our discussion only to the order-entry component of the process.)

To answer the order-entry questions, RMO staff decide that they want to provide their customers all the benefits of catalog shopping: convenience, order any time of day or night, no crowds, order from home, wide selection, simplicity of ordering, and privacy. RMO also wants to provide, as much as is feasible, the benefits of store shopping: being able to inspect the items; to try them on and compare sizes, colors, and patterns; to have items and related accessories near each other; to examine several components together for match and compatibility; and so forth.

Given the stated desires, a high-end system might have the following characteristics:

- Customers can access the catalog on-line, with full-color, three-dimensional pictures. For more technical products, the catalog should include detailed descriptions and diagrams showing construction and specific characteristics. This service can be provided for Internet customers through the web. For telephone customers, the catalog can be provided through the phone line directly to a television set.
- The catalog is also interactive and allows the customer to combine several items, with appropriate graphical imaging to display them together (for example, showing a shirt, jacket, and shorts on a simulated person).
- The user interface to the catalog and order system is either voice-activated or keypad-activated.
- The system should make suggestions of related items that customers may need or desire to purchase at the same time.
- The system should verify that all items are in stock and establish a firm time when shipment will occur. (The fulfillment portion of the system should support shipment within 24 hours or less or, even better, guarantee same-day delivery.)
- Items not in stock should be immediately ordered from the manufacturer or other supply source (the system will immediately send the transaction to the other systems), enabling RMO to ensure delivery to the customer at a future date.
- Payment is verified on-line, just as in a store.
- The customer can see a history of all prior orders and can check the status of any individual order either with the telephone or on the web.

It is interesting to note that all of these capabilities can be supported with current technology. The question, of course, is whether RMO can justify the cost at this point in time. In any event, we have defined high-end automation for the order-entry portion of the new system.

Table 8-5 expands Table 8-4 by replacing the description column with three columns to show the various levels of automation that are possible for each function. The table does not attempt to describe every characteristic of each level of automation. Supporting descriptions will describe each cell in more detail. The objective of the table is to provide an overview of the functions, their priority, and the various methods to implement those functions at the different levels of automation.

Functions (expanded from the event table)	Priority (mandatory, important, desirable)	Low-end automation	Medium (most probable) automation	High-end automation
Check item availability	Important	Periodic listing of QOH.	Real-time internal QOH	+ Sales prompting web
Place order	Mandatory	Clerk data entry	Clerk real-time and customer via web	+ Promotion prompting and stock-out alternatives
Change or cancel order	Important	Clerk overnight	Clerk real-time and customer via web for 24 hours	Clerk real-time and customer via web up to shipment
Check order status	Important	Clerk overnight	Clerk real-time and customer via Web	+ Automatic notification
Fulfill order	Mandatory	Print pull list and shipping label	Pull list, shipping label, real-time update	Automated warehouse, real-time update
Create back order	Important	Clerk data entry	Real-time	+ System automatic + Notify supplier
Return item	Important	Clerk data entry	Real-time, clerk update restock and customer	Automatic inventory and account update
Mail catalog	Mandatory	Print labels	Personalize cover letter	+ Personalize throughout
Correct customer account	Important	Data entry	Real-time	+ Automatic from activity
Send promotional material	Important	Print labels	Personalized cover page	Personalized based on buying history
Adjust customer charges	Mandatory	Data entry	Real-time update	+ Automatic from activity
Update catalog	Mandatory	Data entry	Real-time	+ Automatic suggestions from sales history
Create promotional materials	Important	Data entry	Real-time	Recommendations based on sales history
Create new catalog	Mandatory	Record-keeping of products, prices, etc.	Record-keeping of products, prices, pictures, layouts	Digital scan and page layout
Maintain customer purchase history	Important	Archive files with summary reports	Archive, printed promotional notices	Automatic, real-time for sales prompting
Ongoing feed to manufacturing	Desirable	Printed reports	Daily update	Real-time and trend analysis

(*continued*)

Functions (expanded from the event table)	Priority (mandatory, important, desirable)	Low-end automation	Medium (most probable) automation	High-end automation
EDI feed to suppliers from sales data	Desirable	Printed reports and history	Daily update	Real-time and trend Analysis
Tie-in to shipper system	Desirable	No link	Daily update + email notification to customer	Automatic feed and shipment tracking via web link
Data warehousing and analysis	Desirable		Trend analysis	Trend analysis, data visualization tools
Automated sales prompting	Desirable		Based on promotions	Based on sales promotions and history
Expanded sales analysis with DSS	Desirable		Printed reports	Data visualization tools
System reports				
Produce order summary reports	Important	Printed on request	On-line view and real-time	Data visualization tools
Produce activity reports	Important	Printed on request	On-line view and real-time	Data visualization tools
Produce transaction summary reports	Important	Printed on request	On-line view and real-time	Data visualization tools
Produce customer adjustment report	Important	Printed on request	On-line view and real-time	Data visualization tools
Produce fulfillment reports	Important	Printed on request	On-line view and real-time	Data visualization tools
Produce catalog activity reports	Important	Printed on request	On-line view and real-time	Data visualization Tools

TABLE 8-5
System functions with three levels of automation.

Selecting Alternatives

During the 1980s and early 1990s, the primary purpose for building information systems was to automate manual tasks of business processes. Frequently these systems simply automated the existing tasks and were built at the low level of automation. During the late 1990s and into the 2000s, new systems are being used to provide competitive advantage in the marketplace. More and more new systems are being developed to provide high-level automation solutions. This activity of defining scope and level of automation is a critical and important aspect of new system development. Like RMO, more and more companies are developing strategic plans that move them from simply automating current procedures to focusing on high levels of customer service and support.

The criteria to decide which of the functions to support and at what level of automation are based on the strategic IT plan and the feasibility analysis, as was explained in Chapter 2. The feasibility study conducted at the beginning of the project should now be revisited with a better understanding of user functions that was developed during the fact-finding activities.

- **Economic feasibility.** Obviously, the higher levels of automation will require substantially more funds to implement. It may not be possible to determine the exact cost of each function and level of automation. However, as the most desirable alternatives are selected, a cost/benefit calculation is done to determine the expected benefits of the selected alternatives.
- **Operational, organizational, and cultural feasibility.** Broader scope and higher levels of automation usually require reengineering of the business functions and their manual processes. Benefits can be substantial and dramatic; however, support for the change must be offered to maintain morale and commitment to the new system. Typically, information systems staff underestimate the importance and difficulties associated with changing people's work procedures and job activities. It is usually a good practice to involve people who have been trained in organizational behavior to assist in managing changes.
- **Technological feasibility.** Not only is it necessary to review the technical feasibility of the desired alternatives, but careful consideration must be given to whether the organization has in-house expertise available to develop and implement the system. It is frequently necessary to hire or contract with outside resources to obtain the technical expertise for the more advanced levels of automation. Usually it is more prudent to select alternatives that do not require the organization to push the state-of-the-art. Yet, some companies, those with substantial funds and broad access to resources, do so. Careful consideration needs to be given for cutting-edge projects.
- **Schedule and resource feasibility.** The inclusion of more advanced features in the system will not only cost more but also expand the schedule. One effective method to minimize this impact is to develop a plan of future upgrades of the system. All commercial software developers work this way, and it is a viable alternative for in-house development. A new system often has less capability than the organization ultimately desires. Users can gain experience with the new system, and information systems staff can learn from past experience so that together they can enhance the system until they achieve the desired level of automation.

Evaluation of Alternatives for RMO The selection of which functions to include and at what level of automation is a preliminary decision at this point in time. The final decision is dependent on implementation alternatives, which are explained in the following sections.

Based on preliminary budget and resource availability, the project team at RMO determined that it is possible to include all functions that were categorized either as mandatory or important. For each of those functions, a detailed analysis is done for the desired level of automation. Fundamental decisions about level of automation affect several functions at the same time. For example, three levels of automation affect *Check item availability*, *Place order*, *Change order*, and so forth. The three basic alternatives listed in Table 8-5 are (1) data entry of information with overnight processing, (2) real-time entry for both employee clerks and for customers via the web, and (3) the same system as the medium level with added sales prompting based on promotions and even personal customer purchase history. In fact, these three alternatives could even be divided into more alternatives, such as web versus no web and sales prompting for promotions but not purchase history. A fundamental decision on the automation level will then need to be consistent across the listed functions.

Table 8-6 lists the functions and shows by shading which functions are to be included and at what level of automation. Since some basic decisions affect several functions, it was not necessary to do a detailed calculation of the cost/benefit, operational impacts, technical feasibility, and resource impacts for each function. The low level of automation was not acceptable to RMO management. Most of the current systems already provided that level of automation. As can be seen from the table, the medium or most probable level of automation was the one selected for most functions. The high-end automation would have required a substantial jump not only in software but also in hardware processing power. For example, the high-end support for placing an order would require both clerks and the web-based system to encourage the customer to purchase other items based on sales promotions and past purchases. This would require a very large on-line database with very fast access speeds. The cost for that capability is not within the budget for RMO at this point in time.

Functions (expanded from the event table)	Priority (mandatory, important, desirable)	Low-end automation	Medium (most probable) automation	High-end automation (+ means medium +)
Check item availability	Important	Periodic listing of QOH	Real-time internal and web	+ sales prompting
Place order	Mandatory	Clerk data entry	Clerk real-time and customer via web	+ Promotion prompting and stock-out alternatives
Change or cancel order	Important	Clerk overnight	Clerk real-time and customer via web for 24 hours	Clerk Real-time & customer via Web up to shipment
Check order status	Important	Clerk overnight	Clerk real-time and customer via web	+ Automatic notification
Fulfill order	Mandatory	Print pull list and shipping label	Pull list, shipping label, real-time update	Automated warehouse, real-time update
Create back order	Important	Clerk data entry	Real-time	+ System automatic + Notify supplier
Return item	Important	Clerk data entry	Real-time, clerk update restock and customer	Automatic inventory and account update
Mail catalog	Mandatory	Print labels	Personalize cover letter	+ Personalize throughout
Correct customer account	Important	Data entry	Real-time	+ Automatic from activity
Send promotional material	Important	Print labels	Personalized cover page	Personalized based on buying history
Adjust customer charges	Mandatory	Data entry	Real-time update	+ Automatic from activity
Update catalog	Mandatory	Data entry	Real-time	+ Automatic suggestions from sales history
Create promotional materials	Important	Data entry	Real-time	Recommendations based on sales history

(*continued*)

Functions (expanded from the event table)	Priority (mandatory, important, desirable)	Low-end automation	Medium (most probable) automation	High-end automation (+ means medium +)
Create new catalog	Mandatory	Record-keeping of products, prices, etc.	Record-keeping of products, prices, pictures, layouts	Digital scan and page layout
Maintain customer purchase history	Important	Archive files with summary reports	Archive, printed promotional notices	Automatic, real-time for sales prompting
Ongoing feed to manufacturing	Desirable	Printed reports	Daily update	Real-time and trend analysis
EDI feed to suppliers from sales data	Desirable	Printed reports and history	Daily update	Real-time and trend Analysis
Tie-in to shipper system	Desirable	No link	Daily update and e-mail notification to customer	Automatic feed and shipment tracking via web link
Data warehousing and analysis	Desirable		Trend analysis	Trend analysis, data visualization tools
Automated sales prompting	Desirable		Based on promotions	Based on sales promotions and history
Expanded sales analysis with DSS	Desirable		Printed reports	Data visualization tools
System Reports				
Produce order summary reports	Important	Printed on request	On-line view and real-time	Data visualization tools
Produce activity reports	Important	Printed on request	On-line view and real-time	Data visualization tools
Produce transaction summary reports	Important	Printed on request	On-line view and real-time	Data visualization tools
Produce customer adjustment report	Important	Printed on request	On-line view and real-time	Data visualization tools
Produce fulfillment reports	Important	Printed on request	On-line view and real-time	Data visualization tools
Produce catalog activity reports	Important	Printed on request	On-line view and real-time	Data visualization tools

TABLE 8-6
Preliminary selection of alternative functions and level of automation for RMO (selections are shaded).

Of the newly identified functions, three functions must be added to the project. The first addition is to prepare for the high-end support by including the function to maintain customer history and use it in the development of special promotions. Analysis of the feasibility factors indicates that this alternative does not require substantial increase in cost or project schedule.

The second addition is a more rapid update of inventory levels to the manufacturing facilities. The cost/benefit analysis of this alternative indicates an immediate return by a reduction of back orders and stock-outs.

The third addition is a subsystem to provide more sophisticated analysis of sales trends. This subsystem will utilize the database of sales orders and time series data based on customer histories. Thus, it builds on the data being included for individual customers. It was difficult for the project team to calculate a precise cost/benefit ratio for this new capability, but the sales manager convinced the steering committee that the capability is a critical factor in the future competitiveness of RMO. So, resources will be dedicated to add this capability. This subsystem is somewhat independent, so, to minimize its impact on the schedule, the project team will implement the subsystem several months after the rest of the system.

Generating Alternatives for Implementation

So far, we have described the analysis and fact-finding activities in the development project. The previous section discussed the issues of scoping the solution system, including the level of automation. The next logical question is, "Where do we go from here?" Obviously, a system needs to be designed and then programmed. But there are numerous ways to achieve that goal. For example, if an application is fairly standard, perhaps the organization could just buy a computer program or system to support it. Even for more sophisticated systems, companies may have already developed standard systems. Companies also develop systems from the ground up. If purchasing is not an option, then the organization may decide to build the system in-house. Even then there are various alternatives. Outside help is available as contract programmers for a range of services or specific technical expertise. The point is that the organization must now plan the rest of the project, and there are a multitude of options.

Figure 8-6 presents some variations on obtaining a system. The vertical column is the build-versus-buy axis. At the bottom of this axis, the entire system is built from the ground up. At the top of the axis, the entire system is purchased as a package. In between is a combination of build and buy. The horizontal axis represents computer software, in other words computer programs, on the left and development services, in other words the work of a development team, on the right. The purpose of this diagram is to illustrate that the various alternatives, such as pursuing in-house development or purchasing an ERP system, all contain aspects of building, buying, software, and services. In-house development may have components that are purchases. A turnkey project, although mostly a purchase, may include some customization done either by the vendor or even by the in-house staff. As the diagram illustrates, there is considerable overlap, and the only invalid combination is to build development services, which does not make sense.

FIGURE 8-6
Implementation alternatives.

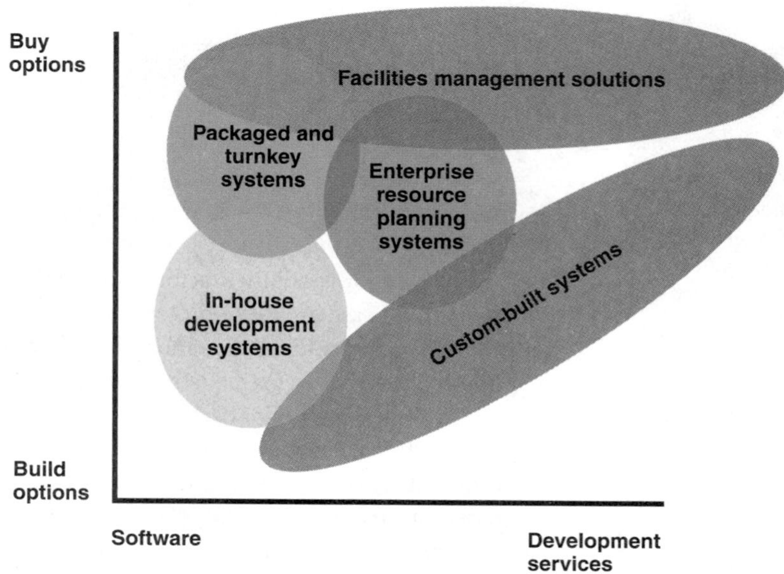

Facilities Management

Facilities management is not an actual development technique or implementation alternative. It is an organization's strategic decision to move all development and all processing to an outside provider. In other words, **facilities management** is the outsourcing of the entire data processing and information support capability. For example, a bank may hire a facilities management firm to provide all of the data processing capability. The computers, software systems, networks, even the technical staff all belong to the outside firm. The bank in essence has opted to let another firm become the information systems department.

facilities management
The outsourcing of all data processing and information technology to an outside vendor

This kind of solution is a long-term, strategic decision. It applies to the entire organization and not just a single development project. So, even though we discuss it as one of the alternatives for implementation, this decision is not typically made under the auspices of any project. It is usually a top executive decision. Normally a contract between an organization and a facilities management provider is a multimillion dollar contract that covers facilities management services for eight to 10 years. Electronic Data Systems (EDS), which is a multibillion dollar company, is one company that obtains the majority of its revenues by providing facilities management services to many industries. Banking, health-care (such as Blue Cross and Blue Shield) grocery, insurance, retailing, and government are industries that EDS supports. EDS can provide high-quality facilities management services in these various industries by maintaining an internal organization around highly experienced industry specialists.

Packaged and Turnkey Software Systems

Packaged software comprises those software systems that are purchased to support a particular application. A strict definition implies that the software is used as is, with no modifications. We all have packaged software on our personal computers, such as a word processor or an accounting/general ledger package. We buy the executing components without the source code but with documentation, install it, and use it. We don't modify it or try to add new capabilities. We use it exactly as it comes, with only the built-in options. The advantages of this software are that it works well and is inexpensive for the amount of capability provided. It is also usually well documented, relatively error free, and stable.

packaged software
Software that is already built and can be purchased as a package

There is a place for packaged software in the overall scheme of an organization's strategy. First, many packages can serve as a portion of a total project. For example, a standard reporting system package may provide reporting capabilities to users. Generally, whenever possible, you should try to find packaged software to perform those standard functions.

turnkey system
A complete system solution, including software and hardware, that can be turned over

A variation of this solution is called a **turnkey system**. A turnkey system is one in which an outside company provides a complete solution, including hardware and software, and the organization only needs to turn it on. In most cases, the outside vendor specializes in a particular industry and application software for that industry. Literally hundreds of firms, many of them small to medium-sized, specialize in systems for particular needs. These turnkey system firms advertise in trade journals for an industry. A few examples are legal systems for law firms, video systems for video stores, patient record systems for dentists and doctors, point-of-sale systems for small retail firms, construction management systems for construction firms, library systems for libraries, and so forth. The list is almost endless.

One critical problem with turnkey systems is that they often do not exactly meet the needs of an organization. In that case, the organization frequently has the onerous task of modifying the way it does business to conform to the computer system. Another variation on the turnkey system to alleviate this problem is to have the vendor customize part of the system. An organization normally purchases the base system, a certain number of customization changes, and a service agreement. The vendor firm does an analysis to determine the unique requirements of the organization and makes those changes to the code. The service agreement can range from simple bug fixes to more extensive modifications over a period of months or years. In some cases, only executable code is provided; in others, both executable and source code are provided so that the organization can also make its own modifications. Sometimes all modifications are done by the vendor firm; other times the purchasing organization may have programmers work with the vendor project team to reduce the cost of customization and to gain experience on the new system. Again, numerous combinations are possible. This is a very popular method of obtaining software for small and medium-sized applications that are somewhat, but not completely, standard.

In the past, turnkey systems were used only for specialized systems within an organization. However, in recent years, several large firms have introduced this approach for enterprisewide systems. These systems, called **enterprise resource planning (ERP)**, support all the operational functions of the entire organization. Companies such as SAP and PeopleSoft have had good success introducing ERP systems into organizations. Obviously, when the support is enterprisewide, the deployment is a major undertaking. Many of these projects take longer than a year to install and cost millions of dollars.

enterprise resource planning (ERP)
A turnkey system that includes all operational functions of an organization

The advantage of ERP systems is that a new system can usually be obtained at a much lower cost and risk than in-house development. The cost is lower because 60 to 80 percent of the application already exists in the base system. Risk is lower because the base system is usually well developed and tested. Other organizations are already using it, so it has a track record of success.

The disadvantage is that the ERP system may not do exactly what the organization needs even after the system has been customized. Frequently, there is a gap between the exact needs of the organization and the functionality provided. The company then must modify its internal processes and train its users to conform to the solution provided. We explain ERP systems and their implementation in detail in Chapter 13.

Custom Software Development

Custom software development is defined as a solution that is developed by an outside service provider. The new system is developed from scratch, based on the system development life cycle. The project team, however, consists primarily of outside consultants. In some cases, the project team is entirely staffed by the consulting firm; in others, the project team is a combination of in-house staff and outside consultants.

The advantage of custom development is that an organization purchases a tremendous amount of experience and expertise to build a new system. Usually the consulting firm has developed similar systems in the past and has extensive domain knowledge for a particular industry and application. It also will have a large pool of very experienced staff to solve complex technical problems. In addition, a large, experienced staff can be brought to the project rapidly to meet schedules and deadlines. Outsourcing and contract development are the fastest growing segments of the IS industry. The consulting arm of the major accounting firms, American Management Systems, and other companies all continue to experience substantial growth.

The major disadvantage of custom development, of course, is the cost. Not only is the organization paying for the development of a new system, but it is paying for it in consulting hourly wages. Typically, organizations opt for custom development when they do not have in-house expertise or have very aggressive schedules that must be met. Normally, the anticipated return on investment for the new system must be quite high to justify the cost of this approach. Custom systems are usually large systems with very high transaction volumes. One example is a health-care system with millions of claims to process. When the system can reduce the cost of processing a claim by one or two dollars, the total savings reaches millions of dollars quickly due to the high volume.

In-House Development

Most large and medium-sized companies have an in-house information systems development staff. In fact, you may find excellent employment opportunities as a member of the development staff in such companies. Depending on staff experience, a company will tailor in-house development efforts. One of the main problems with in-house development, particularly in medium-sized firms, is that special technical expertise beyond employees' experience may be required for a portion of the project. One alternative is to use company employees to manage and staff the project, but to hire special consultants to assist in those areas where extra expertise is required. Control and progress are still maintained by the organization, but assistance is obtained when needed.

The advantages of this approach are primarily control of the project and knowledge of the project team. Company staff also have a better understanding of the internal culture of the organization as well as the specific processing needs of the various business groups. One other major benefit is that the organization can build internal expertise by developing the system in-house.

The major disadvantage is that the in-house staff may not recognize when they need assistance. At times, the "not invented here" syndrome, which means, "If we did not think of it or develop it, it is no good," complicates development because perfectly good, reasonably priced solutions are not utilized. Sometimes the technical problems are more complex than anticipated and in-house people do not recognize the need to obtain expert assistance.

Choosing an Alternative for Implementation

At times, choosing an implementation alternative is straightforward. At other times, deciding between alternatives can be difficult, especially when outside providers are included. For example, one solution may have some of the required functions, but not all. Another solution may have the requisite functions but may only run on an undesirable platform and operating system. Some solutions may provide a quick, inexpensive solution for existing problems but be limited for future growth, others offer long-term capabilities but are very expensive and take a long time to develop. One vendor may propose a turnkey system, another custom development, another a turnkey with a particular database management system and platform, and yet another a joint development project. The problem in selecting is the proverbial comparison of apples and oranges. Frequently, there is very little commonality among the solutions proposed by outside vendors. Each vendor will propose to its own strength. Your objective as the systems analyst is to establish a set of common criteria that can be used to compare the alternatives with as much consistency as possible.

Identifying Criteria for Selection

To begin selection, you must identify the criteria that will be used to compare the various alternatives. These criteria will be used to compare all viable alternatives, both in-house development projects and outsourced solutions. Obviously, not all criteria will apply to both in-house and outsourced alternatives, but there are three major areas to consider in selecting an implementation alternative: (1) general requirements, (2) technical requirements, and (3) functional requirements. Some of the evaluation criteria for outside vendors do not necessarily apply if the project is developed in-house.

General requirements include considerations that are important but not directly associated with the computer system itself. The first major component of general requirements is the feasibility assessment, which was discussed earlier in the context of selecting the scope and level of automation desired. Each of the implementation alternatives under consideration must meet the requirements for cost, technology, operations, and schedule defined in the feasibility analysis. For outsourced alternatives, general requirements may be a set of criteria that evaluates the provider, its stability, and performance record. For in-house alternatives, these criteria will include the risks associated with a development project, such as the length of the schedule and availability of in-house expertise. Both alternatives will be evaluated for total cost and impact on the organization. The following list identifies several criteria that can be included in this section:

- The performance record of the provider
- Level of technical support from the provider
- Availability of experienced staff
- Development cost
- Expected value of benefits
- Length of time (schedule) until deployment
- Impact on internal resources
- Requirements for internal expertise
- Organizational impacts (retraining, skill levels)
- Expected cost of data conversion
- Warranties and support services (from outside vendors)

Obviously some criteria are more important to the organization than others. For example, in the preceding list, we may want to purchase only from a very reputable, stable, and experienced provider. So, the performance record of the provider is extremely important. On the other hand, we may have some leeway in the schedule, so a very short deployment schedule may not be critical. The relative importance of each item in the list can be weighted with a numbering scale. Table 8-7 provides a sample table of general criteria and weighting factors for RMO. That table uses a five-point weighting scale. Criteria that are more important are given a higher number, such as a five or maybe a four. Those that are less important are assigned lower numbers. The extended score is the weight times the raw score for each category.

The four alternatives along the top of the table represent various implementation options. The first alternative is to develop the system in house. The second and third alternatives are different turnkey systems which start with a basic package and modify it. The last option is to contract with a development, consulting firm to develop a completely new system from the ground up. The four alternatives are for illustration only to show the various weighting values possible.

General requirements criteria	Weight (5=high, 1=low)	Alternative 1 In-house		Alternative 2 Package #1 + modify		Alternative 3 Package #2 + modify		Alternative 4 Custom development	
		Raw	Extended	Raw	Extended	Raw	Extended	Raw	Extended
Availability of experienced staff	4	3	12	3	12	3	12	5	20
Development cost	3	5	15	5	15	3	9	1	3
Expected value of benefits	5	5	25	3	15	4	20	3	15
Length of time until deployment	4	2	8	5	20	4	16	2	8
Low impact on internal resources	2	2	4	4	8	5	10	4	8
Requirements for internal expertise	2	2	4	4	8	5	10	4	8
Minimize organizational impacts	3	4	12	3	9	4	12	4	12
Performance record of the provider	5	5	25	4	20	4	20	4	20
Level of technical support provided	4	5	20	3	12	3	12	3	12
Warranties and support services provided	4	5	20	4	16	4	16	4	16
Total			**145**		**135**		**137**		**117**

*This table uses a five-valued scale for raw scores (5 through 1).

TABLE 8-7
Matrix and weighting factors for general requirements for RMO.

As discussed in Chapter 4, every new system has a unique set of technical and functional requirements based on the environment, the needs of the organization, and the system to be developed. In addition to the technical requirements unique to a project, the project team should include other general criteria to evaluate the quality of the software system. This is especially true when packaged software is available for purchase. To test the quality of the software, we would use the following criteria:

- Robustness (the software does not crash)
- Programming errors (the software calculates correctly)
- Quality of code (maintainability)
- Documentation (user and system, on-line and written)
- Easy installation
- Flexibility (the software makes it easy to adjust to new functionality and new environments)
- Structure (maintainable, easy to understand)
- User-friendliness (naturalness and intuitiveness of use)

Probably the most difficult part of this exercise is establishing the weighting factors. The client, system users, and the project team should all have a voice in establishing the weighting factors. In addition, some criteria may be identified as mandatory. In that case, we first eliminate alternatives that do not meet the mandatory requirements. Tables 8-8 and 8-9 show matrices for evaluating technical and functional requirements for RMO.

Technical requirements criteria	Weight (5=high, 1=low)	Alternative 1 In-house		Alternative 2 Package #1 + modify		Alternative 3 Package #2 + modify		Alternative 4 Custom development	
		Raw	Extended	Raw	Extended	Raw	Extended	Raw	Extended
Robustness	5	?	*18	3	15	4	20	?	*18
Programming errors	4	?	*16	4	16	4	16	?	*16
Quality of code	4	?	*18	4	16	5	20	?	*18
Documentation	3	5	15	3	9	4	12	4	12
Easy installation	3	5	15	5	15	4	12	4	12
Flexibility	3	4	12	3	9	4	12	5	15
Structure	3	4	12	4	12	4	12	4	12
User-friendliness	4	5	20	3	12	4	16	5	20
Total			**126**		**104**		**120**		**123**

*This table uses a five-valued scale for raw scores (5 through 1).

*Note: Since alternatives 1 and 4 have not yet been built, values cannot be assigned to the technical requirements concerning quality of code. Extended values are the averages of other values.

TABLE 8-8
Matrix of technical requirements for RMO.

288

Functional requirements criteria	Weight (5=high, 1=low)	Alternative 1 In-house		Alternative 2 Package #1 + modify		Alternative 3 Package #2 + modify		Alternative 4 Custom development	
		Raw	Extended	Raw	Extended	Raw	Extended	Raw	Extended
Inquiry on items	4	5	20	4	16	5	20	5	20
Customer order	5	5	25	5	25	5	25	5	25
Change order	4	5	20	5	20	5	20	5	20
Inquiry on orders	4	5	20	5	20	4	16	5	20
Packaging of order	5	5	25	5	25	5	25	5	25
Shipment of order	5	5	25	5	25	5	25	5	25
Create back order	4	5	20	5	20	5	20	5	20
Accept return	4	5	20	5	20	4	16	5	20
Correct customer account	4	5	20	3	12	4	16	5	20
Update catalog	5	5	25	2	10	3	15	5	25
Create special promotions	3	5	15	0	0	2	6	5	15

(*continued*)

Functional requirements criteria	Weight (5=high, 1=low)	Alternative 1 In-house		Alternative 2 Package #1 + modify		Alternative 3 Package #2 + modify		Alternative 4 Custom development	
		Raw	Extended	Raw	Extended	Raw	Extended	Raw	Extended
Initiate a promotion mailing	3	5	15	0	0	2	6	5	15
Sales summaries	3	5	15	3	9	3	9	5	15
Order summaries	2	5	10	3	6	3	6	5	10
Shipment summaries	2	5	10	2	4	5	10	5	10
Total			**285**		**209**		**234**		**285**

*This table uses a five-valued scale for raw scores (5 through 1).

TABLE 8-9
Matrix of functional requirements for RMO.

Making the Selection

Once requirements have been considered and rated, each alternative can then be evaluated with a raw score based on how well it meets the criteria. Ranges for raw score generally vary from a simple three-point scale to a more finely ratcheted six-point scale. For example, a three-point scale could be Fully Satisfy (2), Partially Satisfy (1), and Not Satisfy (0). A six-point scale would represent Superior (5), Excellent (4), Good (3), Fair (2), Poor (1), and Disqualify (0). To calculate a weighted score in each criteria for each alternative, you would multiply the raw score by the weighting factor. An overall score, which is the sum of the individual criteria scores, determines a ranking among the various alternatives for this category.

RMO decided to undertake most of the development with in-house staff. As seen in Table 8-7, the first three alternatives rank very close together on general requirements, with in-house development having a slight advantage. In Table 8-8, which shows technical requirements, alternatives 1, 3, and 4 are very close. In-house development does provide a slight advantage. In Table 8-9, alternatives 1 and 4 are equal and much better than alternatives 2 and 3. The in-house systems analysts and technical staff had proven several times in the past that they could handle complex systems. In addition, this approach would enable RMO to continue to build in-house expertise. Although the approach was to do development in-house, the information systems group was not averse to hiring specialists when needed.

RMO has sufficient in-house expertise to develop the networks and the database portions of the system. The web-based development will also be done in-house, but RMO will probably have to hire several specialists. RMO wanted to have the expertise within the company, and hiring some new specialists seemed like a viable method.

Some of the integration issues, such as integrating new client-server architecture with mainframe systems, could possibly become quite complicated. Some very experienced consultants were available, and RMO anticipated that it would need to retain a couple to oversee this portion of the project.

RMO staff is now ready to proceed with the project. A review of the feasibility constraints identified no serious problems. The project is still on schedule and within budget. The attitude within the company is very positive. With the availability of these additional resources, the project also appears to be technically feasible.

Considering Outsourced Solutions

In the RMO example, in-house development was the alternative chosen for implementation. To get the necessary information needed to evaluate the other alternatives, the customer support system team sent out a formal request for proposal (RF) to each prospective vendor. Use of RFPs is almost universal in government contracts, and even fairly common in industry. An RFP is a notice to development firms to send in a proposal to develop and install a solution system. This section discusses the mechanics of developing and evaluating an RFP.

Generating a Request for Proposal

request for proposal (RFP)
A formal document, containing details on the system requirements, sent to vendors to request that they bid on supplying hardware, software, and/or support services

A **request for proposal (RFP)** is a formal document sent to vendors. Its basic purpose is to state requirements and solicit proposals to meet those requirements. It is often, particularly in governmental purchasing, a legal document as well. Vendors rely on information and procedures specified in the RFP. That is, they invest resources in responding with the expectation that stated procedures will be followed consistently and completely. Thus, an RFP is generally considered to be a contract offer, and a vendor's response represents an acceptance of that offer.

A good RFP includes a detailed explanation of the information needs of an organization and the processing requirements that must be fulfilled. Chapter 4 defined system requirements as consisting of functional and technical requirements. A good RFP will provide detailed explanations of both these types of system requirements. If the early project assumptions indicated that the most viable option for the new system is to purchase a turnkey solution or outsource a custom development, then the analysis activities are oriented toward developing an RFP. When the outside firm is selected, it will then ensure that in-depth domain knowledge is obtained before detailed design or customization begins.

To develop and distribute a good RFP, the purchasing organization must do a fair amount of analysis. This is not usually a problem for firms that have in-house information systems staff. However, for firms that do not have information systems staff, determining processing requirements is a very common problem. It is also difficult to generate a meaningful RFP or evaluate the various purchase alternatives. Smaller, unsophisticated firms tend to ignore this problem and simply try to make the best decision they can, often from a position of ignorance. A wiser approach is to hire an independent consultant, one who will not be involved in the development, to help establish the selection criteria and decide on a solution vendor. The same criteria for choosing a final vendor should be used in selecting this independent consultant.

The general outline of an RFP is as follows:

1. Identification of requestor
2. Format, content, and timing requirements for responses
3. Requirements
4. Evaluation criteria

The RFP should clearly state the procedural requirements for submitting a valid proposal. Where possible, an outline of a valid proposal should be given with a statement of the contents of each section. In addition, the RFP should clearly state deadlines for questions, proposal delivery, and other important events.

The requirements statement constitutes the majority of the RFP. The body of the RFP can formalize and state the guidelines previously described. Requirements should be separated into those that are absolute (essential) and those that are optional or subject to negotiation. This categorization is a more formal version of the prioritization discussed earlier. The RFP should also explicitly state the evaluation criteria. This might include the categories that are to be evaluated as well as the weighting factors. An example table of contents for an RFP is illustrated in Figure 8-7. As can be seen from the table of contents, many of the items identified in the Tables 8-7, 8-8, and 8-9 are the same issues that are evaluated for an RFP.

FIGURE 8-7
A sample RFP table of contents.

Request for Proposal
Table of Contents

I. Description and background of the requesting organization

II. Format and method of submitting the proposal

III. Overview of information system needs

IV. Expected business benefits

V. Description of technical requirements
- A. Operating environment
- B. Performance requirements
- C. Integration, interfaces, and compatibility
- D. Hardware specifications
- E. Expansion and growth requirements

VI. Description of functional requirements
- A. Specification of primary functions
- B. Description of outputs and information requirements
- C. User-interface requirements

VII. General requirements
- A. Maintenance and support
- B. Documentation and training
- C. Future releases
- D. Project schedule
- E. Project resources

VIII. Evaluation criteria
- A. Technical requirements
- B. Functional requirements
- C. General requirements
- D. Evaluation schedule

IX. Request for provider information
- A. Statement of experience of provider
- B. References
- C. Performance bond
- D. Resumés of project personnel

Benchmarking and Choosing a Vendor

One method to evaluate the quality of a system provided by an outside vendor is either to observe it in use or to install it on a trial basis and test it out. However productive this approach may be, it is also expensive and difficult. The format of the data, the forms, and even the platform may be alien to existing configurations. If the system is complex, then people must be trained to use it. Resources must also be available to do testing. Although this approach can be quite expensive, it may be less expensive than a bad decision.

benchmark
An evaluation of a system against some standard

Some applications are amenable to a more rigorous evaluation called a **benchmark**. A benchmark is a performance evaluation of application software (or test programs) using actual hardware and systems software under realistic processing conditions. In years past, benchmarking was often difficult to perform due to the expense of the hardware and software configurations and the length and cost of installation. Currently, these problems are less severe due to cheaper hardware, streamlined installation procedures, and fierce competition among vendors.

Another way to observe a system in use is to visit another company. Sometimes the vendor will have demonstration versions already installed in its own facilities. As a potential purchaser, you will be permitted to go to the vendor's plant and test the system. Vendors also have previous clients who are using the system. It is almost always a good idea to visit these previous clients. You will not be permitted to test the system yourself with your own data, but you will be permitted to observe the other company using the system. You can also talk to previous clients about their experience with the system and the vendor. It is always a good idea to get references and to talk with other companies that have done business with the vendor.

Frequently you will want to be able to add capabilities to the software as the business environment changes. To do so, you must make the enhancements yourself or have your vendor supply upgrades and fixes. You should ascertain the level of ongoing research and development being done by the vendor. How compatible are these new capabilities to your particular system after it has been customized? Is there an active users group that can suggest new enhancements and modifications to the vendor? Your company is making a major investment, and it is important that the investment be considered in the long term. The largest investment in any new system is the long-term cost of maintenance. A good vendor will help you leverage your maintenance dollars by providing upgrade support and new capabilities based on feedback from existing clients.

Presenting the Results and Making the Decisions

The results of the investigation and analysis that were conducted as part of the activities described in this chapter are normally summarized in a written report and presented in an oral executive-level presentation. The intended audience is the executive steering committee, which has decision-making and funding responsibility for the project. The objective of the documentation and presentation is to provide the necessary background so that informed decisions can be made.

The responsibility of the project team, including both technical and user members, is to do the detailed investigation and calculations to enable an informed analysis of all of the alternatives to be made. However, the final decision of which alternative, or mix of alternatives, is chosen rests with the executive steering committee. This committee not only controls the budget and provides the funding but is responsible for the overall strategic direction of the company.

Probably one of the more difficult tasks for the project team is to compile, organize, and present the alternatives and critical issues in a way that is easy to understand yet accurate and complete. The executive steering committee usually consists of people who are, first and foremost, business executives. They generally are not technical experts, yet they need to make decisions that affect the entire organization. We emphasize again that the presentation of findings is one of the most difficult tasks that the project team will have. It requires careful consideration to find the right balance of detail. At one extreme is so much technical detail that the steering committee cannot understand or follow the logic and becomes lost or bored. At the other end are recommendations without sufficient supporting detail or logic.

The formality of the presentation varies from organization to organization. Some companies require very formal written reports and oral presentations. Other organizations require nothing written and only informal discussion between the client (the person funding the project) and the project team leader. Smaller organizations tend to be less formal, and large corporations typically have standard policies and procedures for approval.

The format of the document and report varies considerably depending on the desires of the audience. A detailed description of how to develop this presentation is provided in Appendix D. Generally, the written documentation will follow the same format as the presentation.

Summary

The target processing environment is the configuration of computer equipment, operating systems, and networks in which the new system must operate. It determines the constraints that are imposed on the system development alternatives. The two primary hardware environments that are in use today consist of a centralized computing configuration, which could be a central mainframe or a cluster of similar computers, and a distributed computing configuration, which could be a local area network, a wide area network, or even an Internet, intranet, or extranet environment. Many new systems are being developed within a client-server network on either a two-tier or three-tier network. The number of tiers indicates the division of functionality among various server computers.

One of the important activities in the analysis phase is to prioritize the system requirements based on the scope and level of automation desired. The scope of the new system determines which functions will be supported by the system. The level of automation is a measure of how automated the selected functions will be. Highly automated functions have sophisticated computer systems such as expert systems to help carry out the business functions.

Another activity that is done in conjunction with prioritizing the requirements is to determine what alternatives are possible for developing the solution and then to select one of those alternatives. Implementation alternatives include such things as building the system in-house, buying a package or turnkey solution, or contracting with a developer to build it (outsourcing). When outsourcing is anticipated, a request for proposal (RFP) is developed and sent out. The RFPs are then evaluated for match to the desired requirements. Making the selection of the various alternatives should be a careful process. The evaluation includes consideration of such factors as the match of the proposed system to the functional and technical requirements and the reputation and performance record of the submitting vendor.

One of the final tasks in the analysis phase is the development of the recommendations and the presentation to management for a decision. After the analysis phase, a more knowledgeable decision can be made about the direction, cost, feasibility, and approach of the rest of the project. It is the responsibility of the systems analyst to document the results of the analysis phase and present them in a logical fashion that is focused toward the executives who make funding decisions.

Key Terms

benchmark, p. 287
business logic layer, p. 265
client computer, p. 265
clustered architecture, p. 262
computer network, p. 263
data layer, p. 265
distributed computing, p. 263
enterprise resource planning (ERP), p. 284
extranet, p. 267
facilities management, p. 283
Internet, p. 266
intranet, p. 267
local area network (LAN), p. 263
middleware, p. 265
multicomputer architecture, p. 262
n-layer architectures, or n-tiered architectures, p. 266

packaged software, p. 283
request for proposal (RFP), p. 286
router, p. 264
scoping table, p. 274
server computer, or server, p. 265
target processing environment, p. 260
three-layer architecture, p. 265
turnkey system, p. 283
view layer, p. 265
virtual organization, p. 267
wide area network (WAN), p. 264
World Wide Web (WWW) or web, p. 267

Review Questions

1. What is meant by the *target processing environment*? Why is it important in the consideration of a development approach?
2. Explain the difference between a centralized computer system and a distributed computer system.
3. Explain the difference between a clustered architecture and a multicomputer architecture on a centralized processing system.
4. What is a multitier network? What role does middleware play?
5. Explain the difference between the Internet, intranets, and extranets.
6. Explain the fundamentals of facilities management.
7. Explain the difference between scope and level of automation.
8. What is meant by the build-versus-buy decision?
9. Define a *packaged solution*. Explain what is entailed in the packaged solution approach.
10. What is meant by *ERP*? How does an ERP approach affect acquiring a new solution?
11. What does *outsourcing* mean? How does it impact a project?
12. Define *benchmark*. Why is it useful in selecting a new system?
13. What is an *RFP*? Why is it developed at the end of the analysis phase instead of at the beginning?
14. What is the difference between general requirements, technical requirements, and functional requirements?

Thinking Critically

1. How can the Internet Protocol (IP) be used to develop application programs?
2. What are the advantages of purchasing a packaged solution? What are the disadvantages or dangers?
3. What are the advantages of building a solution from the ground up? What are the disadvantages?
4. What are the advantages to outsourcing a development project? What are the disadvantages?
5. Explain the fundamental concepts of client-server computing. What is the difference between a LAN, a WAN, and client-server computing?
6. Discuss the importance of developing a formal technique and definite criteria for evaluation alternatives.
7. Given the following narrative, identify the functions to be included within the scope. Also identify several levels of automation for each function. The purpose of this question is to give you an opportunity to think creatively, especially to identify high-level automation alternatives for the various functions.

   Conference Coordinators (CC) assists organizations or corporations in coordinating and organizing conferences and meetings. It provides such services as designing and printing brochures, handling registration of attendees, fielding questions from attendees, securing meeting spaces and hotel rooms, and planning extracurricular activities. CC gets its business in two ways: by following up on leads that a company is going to be holding a conference and by having the company contact CC directly. When a contact is made, the client is asked for basic information about the desired event: city, dates, anticipated number of attendees, price range, and external activities desired. From this information, a bid is prepared. CC likes to keep its turnaround time on bids to under five working days. Each project is assigned to a project manager, who will gather information from the support staff to prepare the bid. If necessary, he or she may also request information from the desired city's visitors' center.
8. What are important points that determine weighting factors for functions listed in the system requirements for a proposed system?
9. List the important points that determine weighting factors in the general and technical requirements for a proposed system.

10. Given the following matrix of various technical requirements, develop your own weighting factors for an inventory management system at a small plumbing supplier. Justify your weights. Extend the raw scores to the Extended column and calculate the totals. Which would you choose? Justify your selection: Did you go strictly by the numbers or are there other factors you might consider? How do you handle a number that is not given—give it an average of the others, pick the best of the others, guess a value, or assign a zero?
(Raw numbers use a six-point scale.)

Category	Weight	Alternative 1 Build in-house		Alternative 2 Buy turnkey		Alternative 3 Buy package	
		Raw	Extended	Raw	Extended	Raw	Extended
Robustness		5		3		3	
Programming errors		?		4		4	
Quality of code		?		4		5	
Documentation		4		4		3	
Easy installation		5		5		4	
Flexibility		5		4		3	
User-friendliness		5		5		5	
TOTAL							

Experiential Exercises

1. Discuss the evolution of client-server computing from file server, to application servers, to multitier networks. What has been the driving force causing this evolution? Where do you think network computing will be in the next five years? Ten years?
2. Find an example of an application system that is browser-based and uses TCP/IP standards. Explain how it works, showing sample screens and reports.
3. Set up an interview with an organization that uses information systems. Ask for an example of an RFP for a software system. Identify the parts of the RFP. Compare them to the recommended components discussed in this chapter.
4. From a news article or Internet information, find an example of a company that is installing an ERP package (SAP, Peoplesoft, etc.). If possible, get a copy of the overall project plan and analyze the various activities. Compare them to a standard SDLC. Find out the total budget for the project.
5. Develop an RFP for RMO to be sent out to various vendors.
6. Develop a recommended development approach for RMO. Also develop a presentation of your recommendation to upper management.
7. Look through some trade magazines (*Software*, *CIO*, *Datamation*, *Infoweek*, etc.) to find examples of companies that have done an evaluation of vendors. Describe their methods and comment on their strengths and weaknesses.

Case Study

Tropic Fish Tales' RFPs

Now that you have read and studied the chapter, go back and review the opening case on Tropic Fish Tales. Your job is to provide specific advice for Robert Holmes or Bill Williams on how to evaluate the various RFPs.

Assuming that you can build some matrices that measure relative strengths between the proposals, comment on the applicability of doing an evaluation based strictly on the numbers. In other words, assume that Robert and Bill were able to create criteria and weights to measure the benefit to the company of the different alternatives.

1. Do you think it would be possible to sum up the resulting values and make a decision based only on the numbers? Support your answer.
2. What kinds of factors, other than those in the matrix of weighted criteria, might need to be considered in making a decision? Can these other factors influence the decision as strongly as the quantified criteria?
3. What if the values of several alternatives are very close? What other factors might need to be considered?

Further Resources

Scott E. Donaldson and Stanley G. Siegel. *Cultivating Successful Software Development: A Practitioner's View.* Prentice Hall, 1997.

Ralph L. Kliem and Irwin S. Ludin. *Project Management Practitioner's Handbook.* American Management Association, 1998.

Sanjiv Purba, David Sawh, and Bharat Shah. *How to Manage a Successful Software Project, Methodologies, Techniques, Tools.* John Wiley & Sons, 1995.

John J. Rakos. *Software Project Management for Small to Medium Sized Projects.* Prentice Hall, 1990.

Neal Whitten. *Managing Software Development Projects: Formula for Success.* John Wiley & Sons, 1995.

System Design Tasks

Moving to Design

Learning Objectives

After reading this you chapter, you should be able to:

- Develop a system flow chart
- Develop a structure chart using transaction analysis and transform analysis
- Write pseudocode for structured modules
- Develop a package diagram
- Develop a design class diagram
- Discuss the issues related to the management and coordination of the design phase of the SDLC
- Explain the major components and levels of design
- Describe the difference between models produced by structured design and object-oriented design

Chapter Outline

NINE

NEW CAPITAL BANK: OBJECT-ORIENTED DEVELOPMENT OF A PAYROLL SYSTEM

The development project for New Capital Bank was running smoothly. Bill Santora, the project manager, was thinking, "The walk-through of the design of the new payroll system went very well. There are only a half-dozen minor changes that I need to make on the design class diagrams. This object-oriented design technique is very natural."

Just then, Mary Garcia, Bill's supervisor, walked into the room. "Your walk-through went very well, Bill. Evidently your training in UML has been well worth the effort. Not only is your design solid, but your presentation was easy to follow. I think we have a solid design, and I am requesting that we begin programming immediately on the parts that you have finished."

"That's great! I really appreciate your encouraging me to learn object-oriented development. It took me awhile to understand how to do analysis using UML. Once I understood that, however, it was easy to learn how to do object-oriented design."

"What do you mean?" asked Mary. (Even though Mary was the supervisor, she had not done object-oriented design before.)

"Well, the basic object-oriented design model is the class diagram. We first develop the class diagram during the analysis phase, then we simply enhance it during design to create the design model. We don't have to draw a different kind of diagram; we use the same class diagram, just enhanced."

"What kind of enhancements do you do?" she continued.

"Basically, we add new attributes to keep track of the states of the objects. We call these *state variables*. We make sure we have correctly identified all of the methods for each class. To do that, we double-check the methods we have identified against the collaboration diagrams. Then we develop the code for each method based on the action-expressions in the statecharts. What we end up with is the class diagram, with the new state variables and with the methods that have pseudocode.

"The hardest part about design is that we found several errors in our analysis models. When we got down to the details, some of our assumptions about how things would work were wrong. In some cases, we had to go back and rethink the requirements.

"Thanks for your encouragement and help. I really like this approach to system development." Bill was thinking that he ought to do something nice for Mary. She had encouraged him to learn the new technique. Now he had two sets of tools to do system development. She obviously had given his career a boost.

OVERVIEW

The last chapter described the activities that complete the analysis phase and begin the transition from analysis to design. This chapter completes that transition and discusses issues related to the design of the new system. While analysis focuses on what the system should do—that is, the requirements—design is oriented toward how the system will be built—that is, defining structural components. Such activities as defining the scope and prioritizing the requirements are clearly analysis tasks. However, defining the target processing environment and determining levels of automation are tasks begun during analysis but completed in more detail during design. Thus, we began the transition to design in the last chapter, and we move completely into design considerations in this chapter.

This chapter is the first of four chapters discussing design. In this chapter, we discuss the design of the application software. In the next chapter (Chapter 10), you will learn how to design databases. Chapters 11 and 12 contain the details of designing the user interface, the necessary controls, and other system interfaces. You have already begun to learn about considerations in the design of the network and hardware components of the system in Chapter 8.

We begin the chapter with a general discussion of design to explain how it fits into the SDLC. We then describe the detailed process of application program design, both for a structured program and an object-oriented program. We also describe some alternative methods of conducting the design activities. Finally, we discuss issues of managing the project team during the design phase.

Understanding the Elements of Design

Systems design is the process of describing, organizing, and structuring the components of a system at both the architectural level and a detailed level that will allow the construction of the proposed system. The important idea is that design describes, organizes and structures with a focus toward the construction of a new system. Systems design is like a set of blueprints necessary to build a house. The blueprints are organized by the different components of the house and describe the rooms, the stories, the walls, the windows, the doors, the wiring, the plumbing, and all other details. We do the same thing in systems design, although the components we are describing are the components of the new system. We design and specify various components of the solution.

To get an understanding of the components of design, we must answer three questions:

- What is used for input to the design?
- How is design done?
- What are the final design documents?

Inputs: Moving from Analysis to Design

During the activities of the analysis phase, we built documents and models. For traditional analysis, models such as the event table, data flow diagrams, and entity-relationship diagrams were built. For object-oriented analysis, we also used the event table and developed other models such as class diagrams, interaction diagrams, and statechart diagrams. Regardless of the approach, the input to the design phase activities is the set of documents and models that was built.

During analysis, we built models to represent the real world and to understand the desired business processes and the information used in those processes. Basically, analysis involves decomposition—taking a complex problem with complicated information requirements and breaking it into smaller, more understandable components. We then organize, structure, and document the problem domain knowledge by building requirements models. Analysis and modeling require substantial user involvement to explain the requirements and to verify that the models are accurate.

Design is also a model-building activity. Design uses the information gleaned during analysis—the requirements models—and converts the information into models that represent the solution system. Design is much more oriented toward technical issues and therefore requires less user involvement and more involvement by systems analysts and other technicians. Figure 9-1 illustrates this flow from analysis to design, highlighting the distinct objectives of each phase.

Major Components and Levels of Design

To do design, we first partition the total system into its major components. An information system is much too complex to design all at once. Figure 9-2 depicts how these various components fit together. The icons on the figure are pieces of hardware, and inside

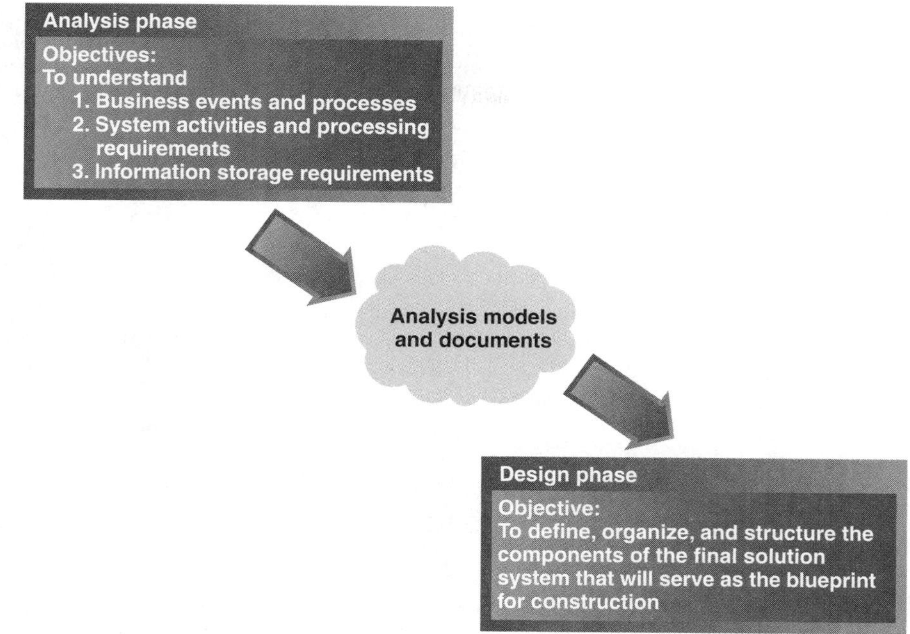

FIGURE 9-1
Analysis objectives to design objectives.

the hardware are the software components. The cloud represents the entire system, and the various icons show the parts of the system that must work together to make the system functional. Not only must all of the components of the design be integrated, but the design activities are also done simultaneously. This textbook describes the various activities separately, but the design of one component heavily influences the design of the other components. The design activities usually all proceed in parallel.

Chapter 3 introduced the various activities of the design phase:

- Design and integrate the network
- Design the application architecture
- Design and integrate the database
- Design the user interfaces
- Design the system interfaces
- Design and integrate system controls
- Prototype for design details

Each of the activities develops a specific portion of the final set of design documents. Just as a set of building blueprints has several different documents, a systems design package also consists of several different sets of documents that specify the entire system. In addition, just as the blueprints must all be consistent and integrated to describe the same physical building, the various systems design documents also must be integrated to provide a comprehensive set of specifications for the total system.

FIGURE 9-2
Designed components of the final system.

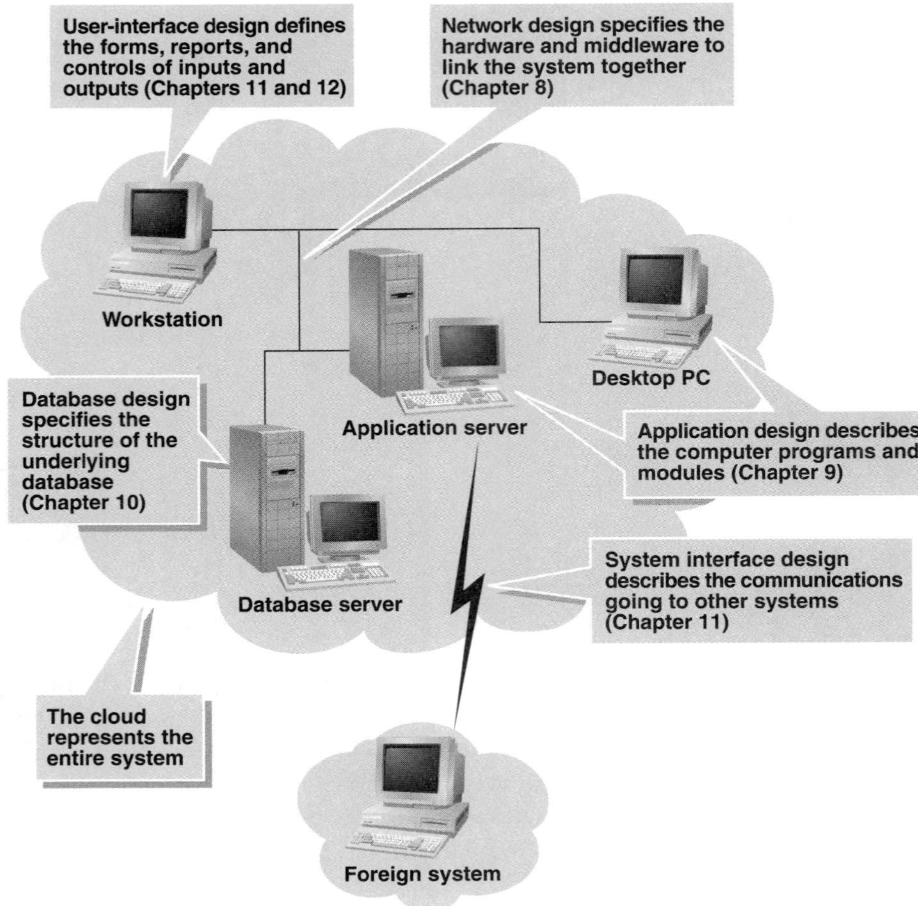
User-interface design defines the forms, reports, and controls of inputs and outputs (Chapters 11 and 12)
Network design specifies the hardware and middleware to link the system together (Chapter 8)
Workstation
Desktop PC
Database design specifies the structure of the underlying database (Chapter 10)
Application server
Application design describes the computer programs and modules (Chapter 9)
System interface design describes the communications going to other systems (Chapter 11)
Database server
The cloud represents the entire system
Foreign system

A second important idea is the concept of the different levels of design. During analysis, we first identified the scope of the problem before we tried to understand all of the details. We called this top-down analysis. Analysis, as it was presented, included both top-down activities (for example, scope first, then details) and bottom-up activities (for example, DFD fragments first, then middle-level diagram). The same ideas are applicable during design.

architectural design
High-level design; also called **general design**, or **conceptual design**

detail design
Low-level design that includes the design of specific program details

As you begin working in industry, you will find that various names are given to the design at the highest level, including **architectural design**, **general design**, and **conceptual design**. Architectural design first determines the overall structure and form of the solution before trying to design the details, which is usually called **detail design**. It is not so important to distinguish which activities are architectural design and which activities are detail design. Neither is it important to identify specific documents as belonging to architectural design and others to detail design. What is important is to recognize that design should begin in a top-down fashion. Let's review the implications of this approach for each of the identified design components from Figure 9-2.

For the entire system, the analysts first identify the overall target processing environment. The overall architectural requirements and structure of the network are determined before the details of the routers, firewalls, workstations, and so forth are specified. This was reviewed in Chapter 8.

For the application software, the first steps are to identify the various subsystems and their relationships to the network, the database, and the user-interface components. Part of that early design is the automation system boundary. The system boundary identifies which functions are included within the automated system and which are manual procedures. Notice that we began this process by identifying the level of automation, which was explained in Chapter 8.

For the database component, the first steps are to identify the type of database to be used and the database management system. Some details of the record structures and the data fields may have been identified, but the final design decisions will depend on the architecture.

For the user interface, the analysts first design the general form and structure of the user dialog and the inputs and outputs. The project team also describes the relationship of the user-interface elements with the application software and the hardware equipment. Afterward, the detail form and report layouts can be developed.

The approach to design is also determined by whether the project team is utilizing structured techniques and models or object-oriented techniques and models. The specific tasks within the activities of the SDLC vary depending on which approach is being used. Let's discuss those differences next.

Outputs: Structured Models and Object-Oriented Models

The original definition of design indicates that design involves describing, organizing, and structuring. The output of the design activities is a set of diagrams and documents that achieve that objective. These diagrams are the models and the documents of the various aspects of the solution system. As with the analysis models, some components are similar for structured and OO approaches, but other components are very different.

Figure 9-3 duplicates the information in Figure 7-1 and extends it with the design models for both structured design and object-oriented design. As noted in the figure, some areas are different and others are similar. In the area of database design, the structured technique usually uses a relational database model. The object-oriented technique can design to either a relational database model or to a newer object-oriented database model. In the area of the user interface, both techniques include the design of menus, forms, and reports as well as the human-computer dialog. These two areas share many of the same techniques whether a structured approach or an object-oriented approach is used.

In the area of the application architecture design, however, traditional structured techniques and object-oriented techniques do differ substantially.

Structured techniques, including analysis and design models, have been used for many years to describe the structure and organization of systems written using the input-process-output model of software. These models are well suited to describing business applications, most of which rely on databases or files and do not require sophisticated real-time processing. These models were originally developed to support application software design and programming using COBOL and BASIC programming languages. They are equally well suited to programming in other languages such as C, FORTRAN, Pascal, and other business-oriented programming languages.

Object-oriented techniques are newer techniques that have become widely used since the late 1980s. They are well suited to situations that are real-time, interactive, and event-driven, such as operating systems that require multitasking capabilities. Since many of the business applications today are interactive, event-driven applications, object-oriented development is rapidly becoming the standard for new development of the application software even for business applications.

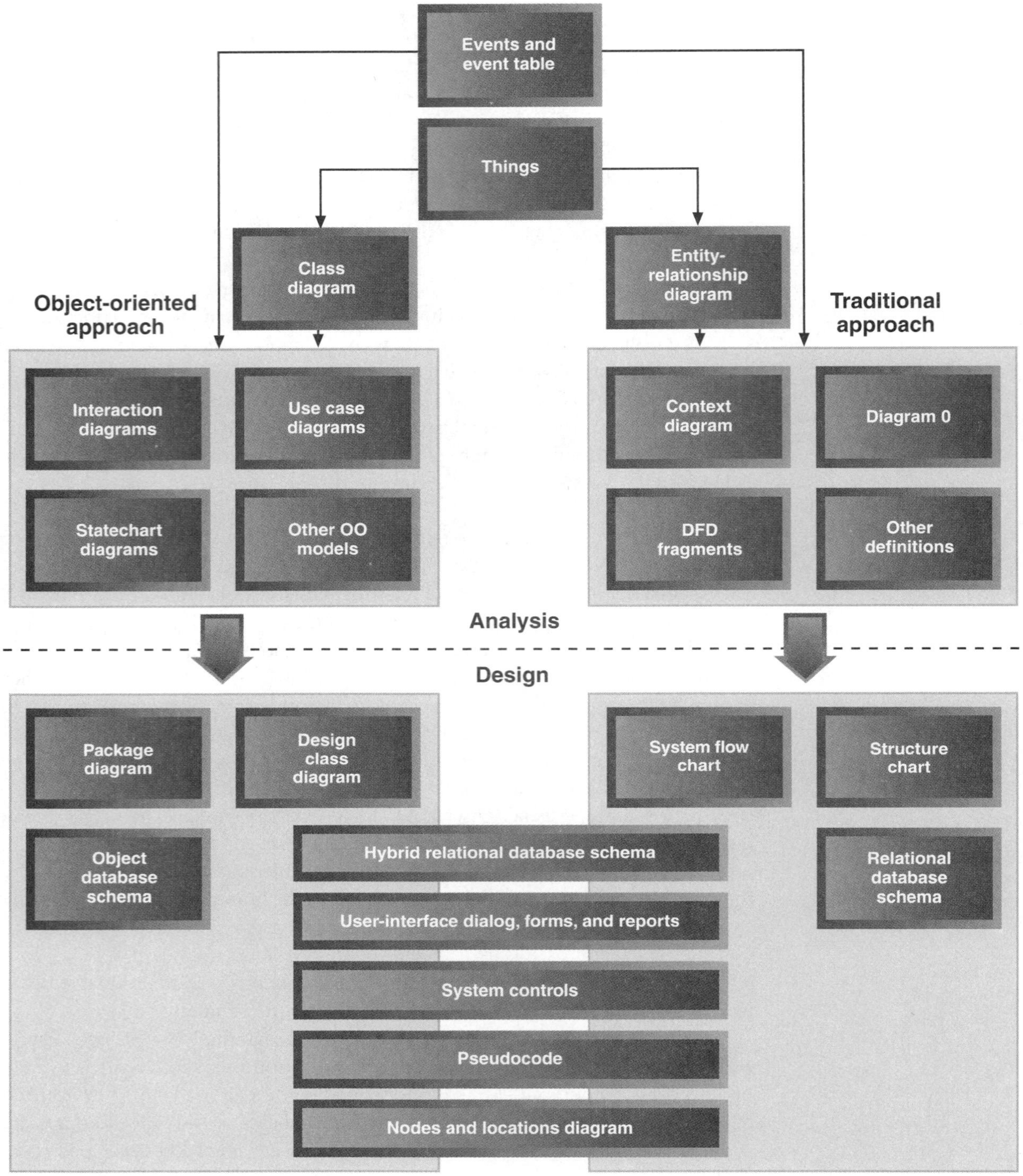

FIGURE 9-3
Structured and object-oriented models.

A frequently asked question is, can structured techniques and object-oriented techniques be mixed? In other words, is it possible to do structured analysis and then object-oriented design of the application or vice versa? In some situations it may be possible to mix and match, such as when designing and implementing the interface using OO after completing traditional structured analysis. But generally, such mixing and matching does not work well for application design, because the basic philosophies of the two techniques are so fundamentally different. The design of the application software using a traditional approach provides an architectural structure based on the functions of the system. A system designed using object-oriented techniques has an architectural structure based on a set of interacting objects.

Designing the Application Architecture: The Structured Approach

The application architecture consists of the application software programs that will carry out the functions of the system. Application design must be done in conjunction with the design of the database and the design of the user interface. However, to make this chapter easier to understand, we focus exclusively on the design of the computer software itself.

You may already have experience writing business programs using a third-generation language such as Visual Basic, C, COBOL, or Pascal. You know that the organization of one of these programs consists of **modules** that are arranged in a hierarchy like a tree. The top module is often called the boss module or the main module. The middle-level modules are control modules, and the leaf modules (those at the ends of the branches) are the detail modules that contain most of the algorithms and logic for the program. A module, then, is a small section of program code that carries out a single function. A **computer program** is the set of these modules that are compiled into a single executable entity. The design of a computer program is specified with a structure chart, which will be discussed in detail later in this section.

module
An identifiable component of a computer program that performs a defined function

computer program
An executable entity made up of a set of modules

In large systems, a single program usually cannot perform all of the required functions. Sometimes one program is written to perform on-line activities, and another program may carry out periodic functions executed once a day. Other programs may have specialized functions such as backing up the data or producing year-end financial reports. The composite of all these individually executable entities, or programs, makes up the entire system.

Both the structure of the overall system and any individual subsystems are documented using a **system flow chart**. The system flow chart identifies each program, along with the data it accesses. The system flow chart also shows the relationships among the various programs, subsystems, and their files or databases. It documents the architectural structure of the overall system. We describe how to design a system flow chart later in this chapter.

system flow chart
A diagram that describes the overall flow of control between computer programs in a system

Finally, the project team must also design the internal logic of individual modules. The internal algorithms that compose the logic of the module are usually documented using **pseudocode**. If you have taken programming classes, you probably had to write algorithms in pseudocode before you actually coded your programs. Pseudocode is very much like the structured English described in Chapter 6. Pseudocode is simply structured-programming-like statements that describe the logic of a module.

pseudocode
Structured-programming-like statements that describe the logic of a module

The inputs for the development of these design models and documents are the data flow diagrams and their detailed documentation, in structured English, and the detailed data flow definitions. In general, we use a top-down approach for doing design. Figure 9-4 illustrates the process. The source of information for all these models is the DFD along with its detailed documentation. There is an intermediate form of the DFD called the DFD with automation system boundary. The automation system boundary divides the computerized portions of the system from the manual portions and, as such, determines which portions need to be included in the design. This enhanced DFD is the one that is actually used as the source for the design models as shown in the figure.

FIGURE 9-4
Structured design models.

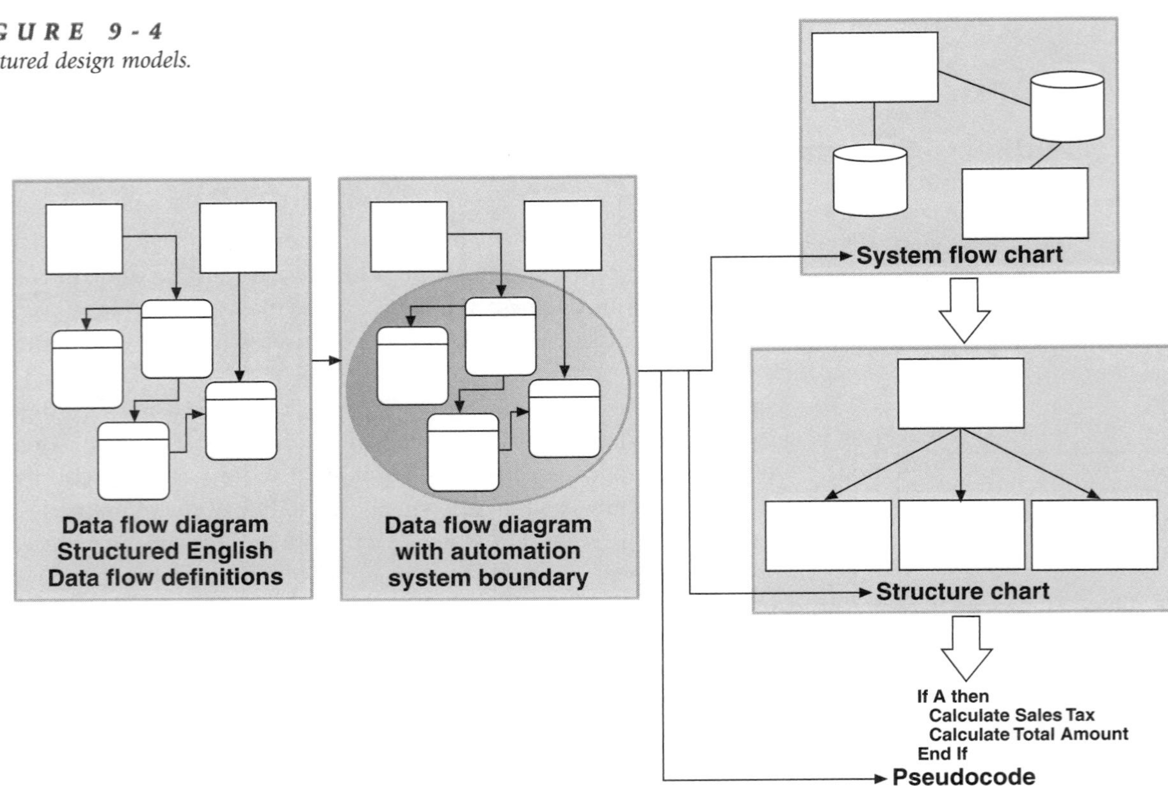

The following sections of the text follow the structure shown in Figure 9-4. First, the automation system boundary is discussed. Then the development of the system flow chart is explained. Next the approach to the design of the structure chart is elaborated. Finally, the form and method of writing pseudocode are given.

The Automation System Boundary

The automation system boundary partitions the data flow diagram processes into manual processes and those that are to be included in the computer system. During the analysis phase, we looked at the business events and all of the processes that described those events. At that time, we did not try to distinguish between which processes were manual and which were automated. To develop the computer system design, we must identify the processes that will be automated.

Figure 9-5 illustrates a typical data flow diagram with the automation system boundary added. This figure shows both the system boundary, which identifies the entire automated system, and program boundary lines, which partition the DFD into separate programs. This diagram is the first step in the design and determines what the programs are and what processes are included within those programs. In this context, we define a program as a separate executing entity.

Processes can either be inside or outside the system boundary. Processes that are outside are manual processes such as sorting and inspecting paper documents, entering customer orders, or visually inspecting incoming shipments. In some cases, the system boundary goes right through a process. This indicates that the process is a mid- or high-level process and is exploded on a more detailed diagram. Some of the processes in the exploded detail will be inside and some outside the system boundary.

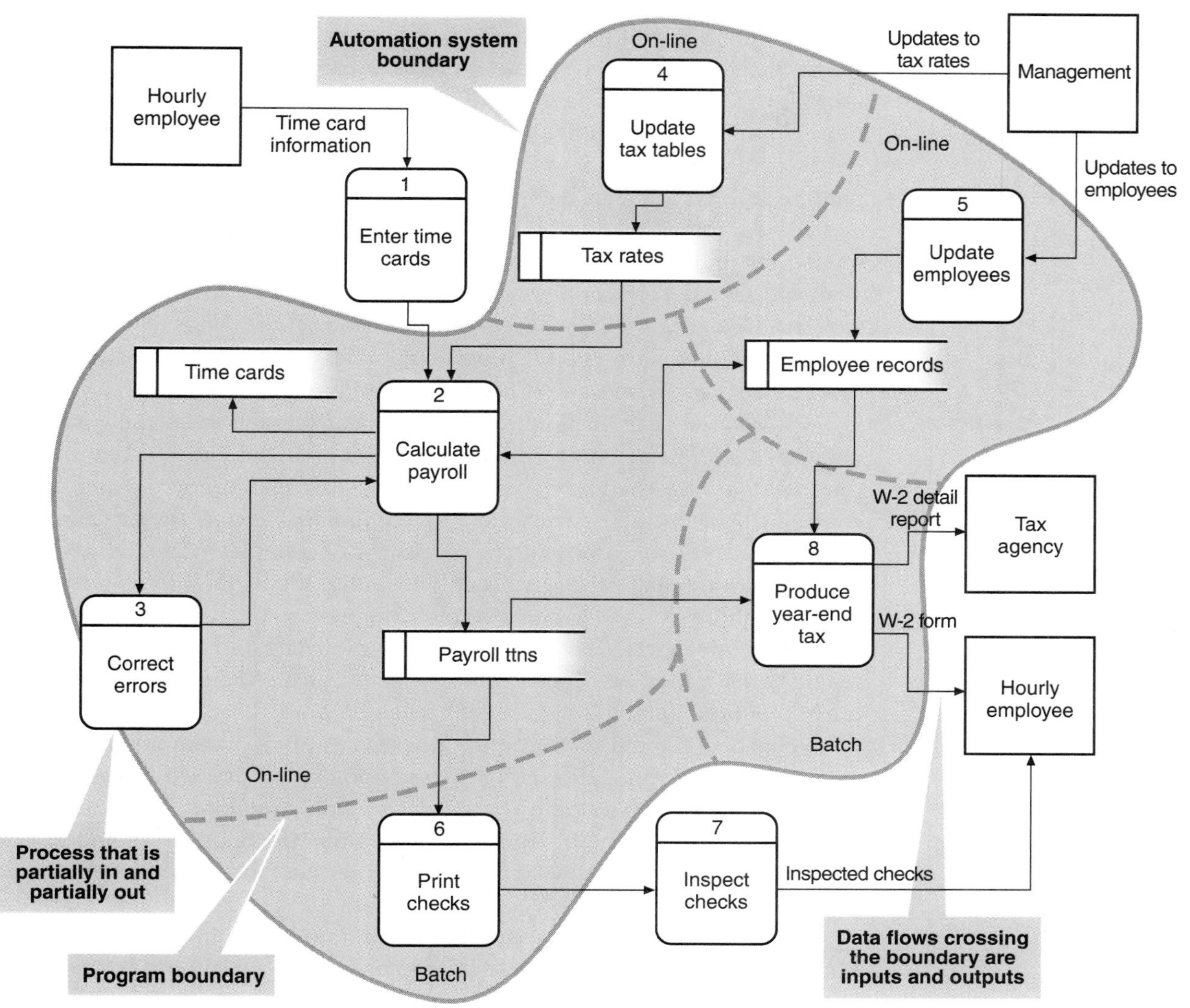

FIGURE 9-5
The data flow diagram with an automation system boundary.

Data flows are found inside, outside, or crossing the system boundary and the program boundaries. The data flows that cross the system boundary are particularly important—they represent inputs and outputs of the system. In other words, the design of the program interfaces, including both the user interface and transmittals to other systems, is defined by the boundary-crossing data flows. In the final system, these data flows will be forms or reports in the user interface, or files or telecommunication transmittals between systems. Data flows that cross the boundaries between programs represent program-to-program communication. In the final system, these data flows will also be files or telecommunication transmittals between programs.

Figure 9-5 is the high-level data flow diagram showing all of the major processes. The system boundary can also be drawn on each data flow fragment to show more detail about which processes are internal or external and which low-level data flows cross the boundary.

The System Flow Chart

The system flow chart is the computer system representation of various computer programs, files, databases, and associated manual procedures. Processes are grouped into programs and subsystems based on similarities such as shared timing (for example, a process performed monthly), shared accesses to stored data (for example, all processes that update employee data), and shared users (for example, processes that produce reports for the marketing department). The programs and subsystems thus created have complex interdependencies including flow of data, flow of control, and interaction with permanent data stores. This process is frequently done during the analysis activities. For example, the subsystems of the RMO customer support system were defined during the analysis phase (see Chapter 6), and the set of events allocated to each subsystem makes up the program modules. We also saw this division in the definition of various use cases (Chapter 7).

A system flow chart graphically describes the organization of the subsystems into automated and manual components showing the flow of data and control among them. System flow charts are primarily used to describe large information systems consisting of distinct subsystems and dozens or more programs. They are also used to describe systems that perform batch processing (for example, systems used to process bank transactions, payroll checks, and utility bills). A common characteristic of such systems is the division of processing into discrete steps (such as validation of input transactions, update of a master file with transaction data, and production of periodic reports) with a fixed execution sequence. Many batch systems also make extensive use of files in addition to or instead of databases.

System flow charts first came into widespread use to document the processing and data flow between programs that processed information through batch transaction files. Frequently, in these systems, one program would produce a file of all the daily transactions. Then another program would process the transactions and update a master file. Another program would be used to produce the various reports required from the system.

In today's newer systems, much processing is done in real time, as each transaction is entered. These systems also usually make updates to a relational database system instead of a master file. Centralized database management systems also include many of the processes that were previously done by individual programs. System flow charts may also be used for these newer systems, although the diagrams tend to be much simpler because the processing is more centralized to a single program or subsystem. But since the systems developed these days are generally much more complex overall, you will still see system flow charts that represent how all of the pieces fit together.

Many business systems today also have both real time and batch components. For example, today your credit card purchases are at a minimum verified realtime, and may even be posted real time. However, monthly account statements and customer payments are typically processed in batches. A system flow chart is used to describe the overall organization of this type of system and show the relationship between the real-time components and the batch processing.

Figure 9-6 illustrates the most common symbols that are used in system flow charts. These symbols are fairly common throughout the industry, although from time to time you will see variations.

FIGURE 9-6
Common system flow chart symbols

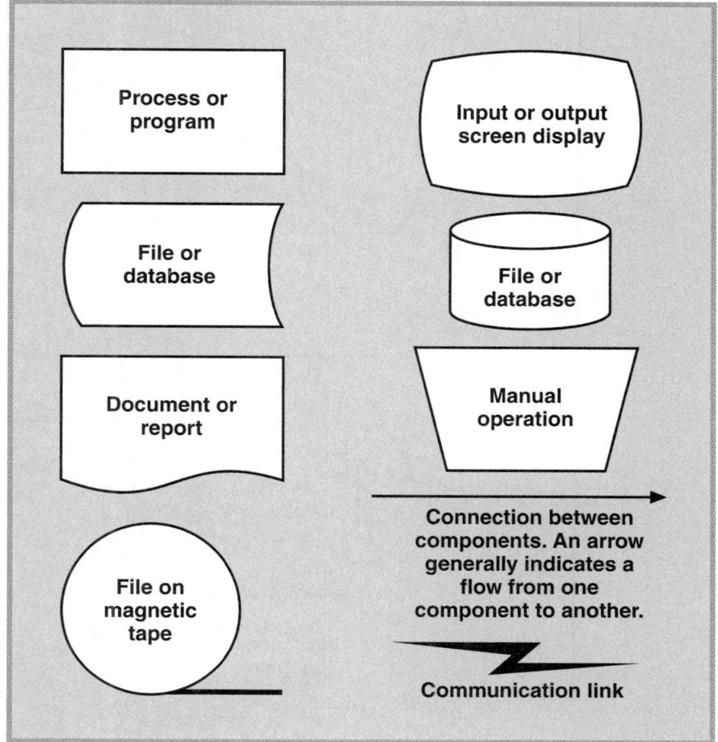

Figure 9-7 is an example system flow chart for the payroll system shown on the DFD in Figure 9-5. Note that the system flow chart identifies the files, programs, and manual processes of the total system. We have added physical implementation description by identifying the file mediums, disk or tape. Frequently, we also include additional system functions and files such as backups and history files. Even though the information between Figures 9-5 and 9-7 is very similar, the focus of the diagrams is different. The system flow chart is focused on physical artifacts and implementation such as executable programs, files, and documents.

Figure 9-7 shows that the payroll program has four inputs and produces three outputs. The inputs are the time cards, the tax rate table file, the employee database, and corrections. Outputs produced are an error report, a payroll transaction file, and an updated payroll history file. Other programs (that is, independent executables) are the two progams to maintain the tax rate tables and the employee database, and the other two programs to write checks and to produce year-end income tax reports.

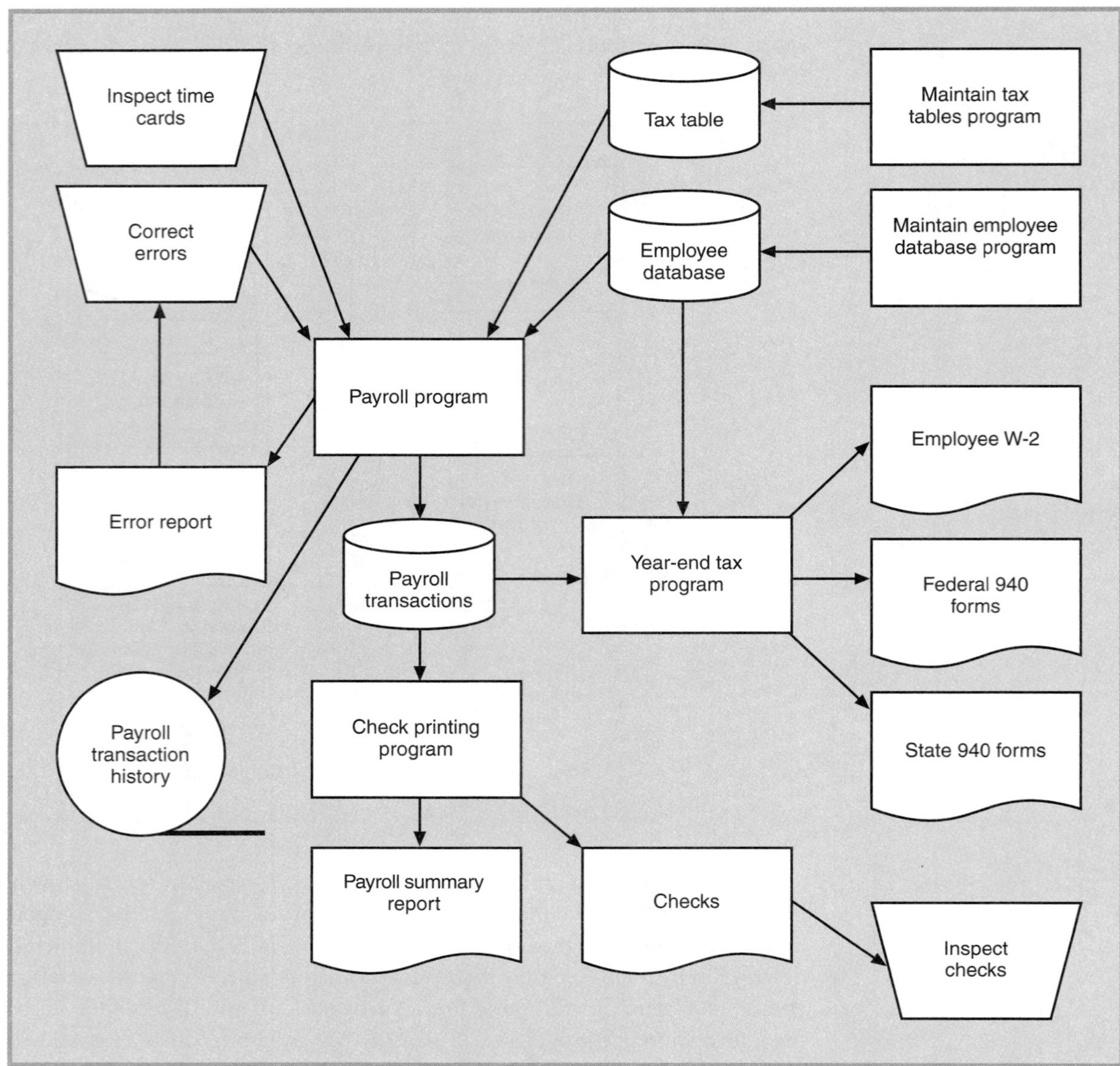

FIGURE 9-7
A sample system flow chart for a payroll system.

Figure 9-8 is an example of a system flow chart for Rocky Mountain Outfitters. The four main programs correspond to the subsystems defined in the subsystem DFD shown in Figure 6-13. In the system flow chart, the individual data stores have been converted into database files. To complete the system flow chart correctly, team members must also start the database design concurrently to define the database files correctly.

Figure 9-8 also shows one additional subsystem that is not found in Figure 6-13. During the scoping and level of automation review discussed in Chapter 8, RMO decided that it needed a higher level of automation for a couple of sales analysis reports. In this instance, instead of adding the reports to an existing subsystem, the project team defined a new subsystem. To maintain consistency across the diagrams, Figure 6-13 should also be updated to be consistent with the new additions.

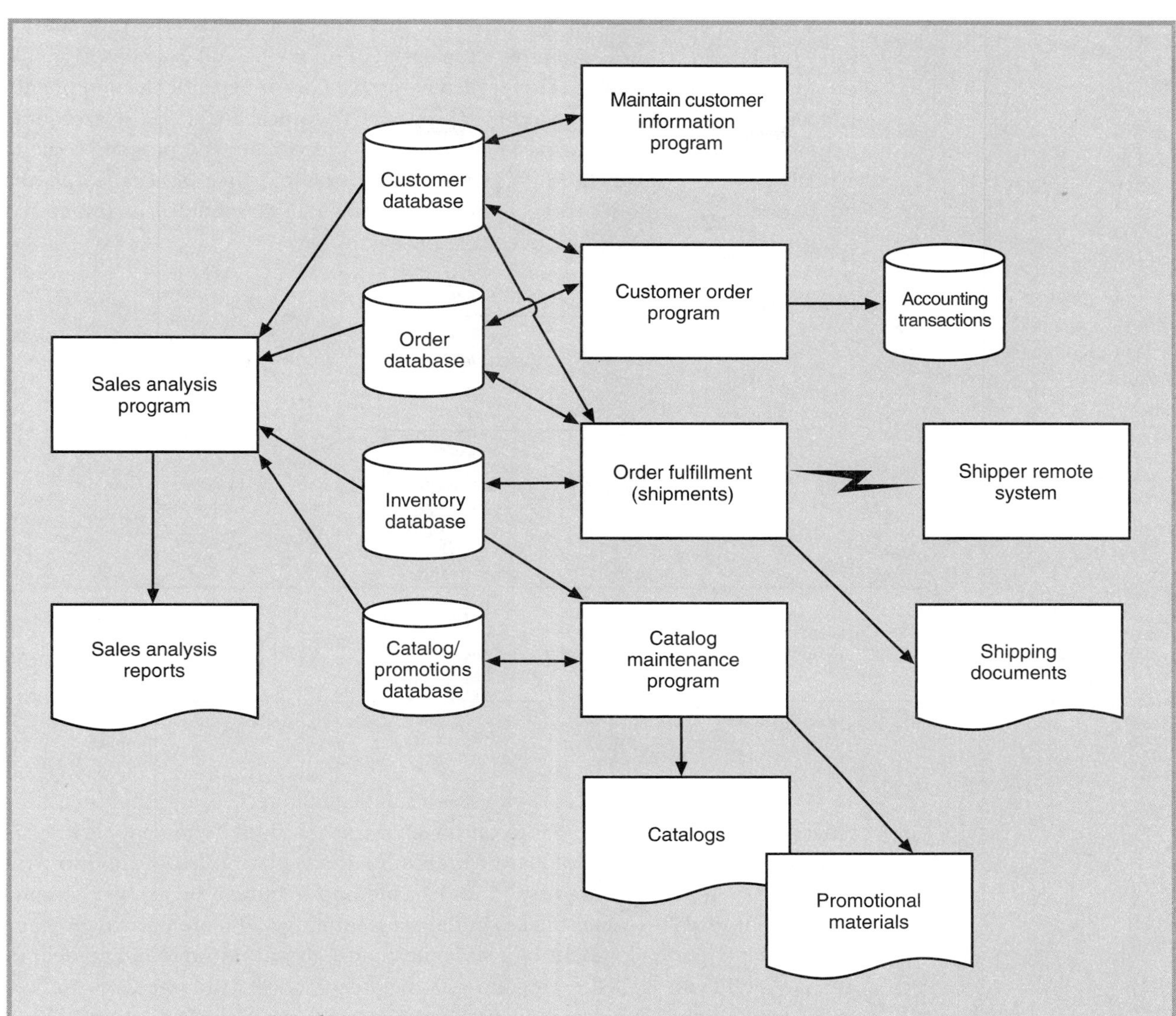

FIGURE 9-8
The system flow chart for Rocky Mountain Outfitters.

The Structure Chart

The primary objective of structured design is to create a top-down decomposition of the functions to be performed by a given program in a new system. Each independent program shown in the system flow chart performs a set of functions. Using structure chart techniques always provides a hierarchical organization of program functions. First this section explains what a structure chart is and how to interpret one. We explain how each structure chart is related to the DFDs created during systems analysis and then we show how a detailed data flow diagram is used to develop one.

structure chart
A hierarchical diagram showing the relationships between the modules of a computer program

A **structure chart** hierarchy describes the functions and the subfunctions of each part of a system. For example, the program may have a function called *Calculate pay amount.* Some subfunctions are *Calculate base amount, Calculate overtime amount,* and *Calculate taxes.* In a structure, these functions are drawn as a rectangle. Each rectangle represents a module.

A module is the basic component of a structure chart and is used to identify a function. Figure 9-9 shows a simple structure chart for the payroll example. Modules, as shown by the rectangles, are relatively simple and independent components. Higher-level modules are "control" modules that control the flow of execution. Lower-level modules are "worker bee" modules and contain the program logic to actually perform the functions. In Figure 9-9, for example, the *Calculate pay amount* module may do nothing more than call the lower-level modules in the correct sequence to carry out the logic of calculating payroll.

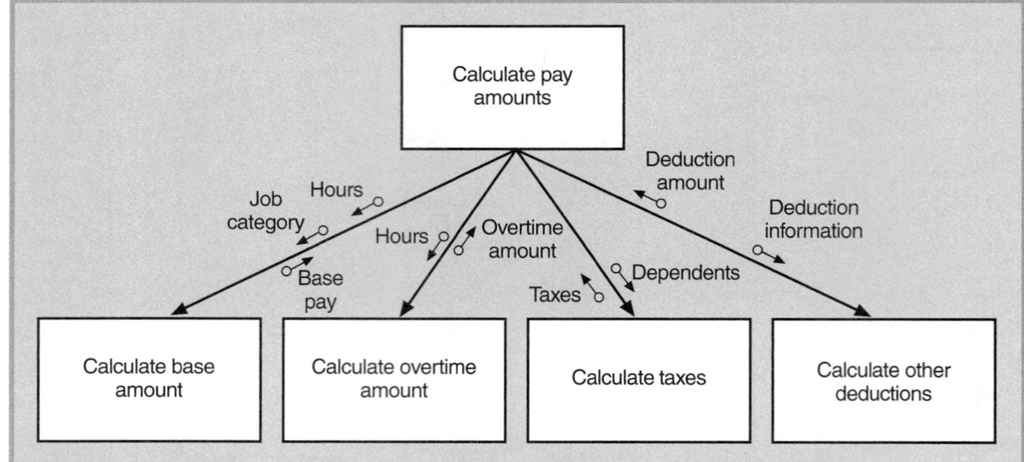

FIGURE 9-9
A simple structure chart for the Calculate pay amount.

Notice how a structure chart provides a simple and direct organization to a computer program whose purpose is to calculate payroll amounts. As you have learned in your programming class, modular programming is a well-accepted method to write computer programs that are easy to understand and maintain. Breaking a complex program into small modules makes initial programming and maintenance programming easier. The development of a structure chart is based on rules and guidelines. The key points are that the program is a hierarchy and that the modules are well formed with high internal cohesiveness and minimum data coupling. We describe in more detail the characteristics of good modules later.

The vertical lines connecting the modules indicate the calling structure from the high-level modules to the lower-level modules. The little arrows next to the lines show the data that are passed between modules and represent the inputs and outputs of each module. At the structure chart level, we are not yet concerned with what is happening inside the module. We only want to know that somehow the module does the function indicated by its name, using the input data and producing the output data.

Figure 9-10 shows the common symbols used to draw structure charts. The rectangle represents a module. In a structure chart, a module can represent something as simple as a block of code, such as a paragraph or section in a COBOL program. In other languages, a module typically represents a function, procedure, or subroutine. Examples of modules as program fragments include subroutines (as in FORTRAN and BASIC), paragraphs or subprograms (as in COBOL), procedures (as in Pascal), and functions (as in FORTRAN, C, and C++). A module can also be a separately compiled entity such as a complete C program. The rectangle with the double bars is simply an existing module or a module that is used in several places. Use of the double bar notation is optional.

FIGURE 9-10
Structure chart symbols.

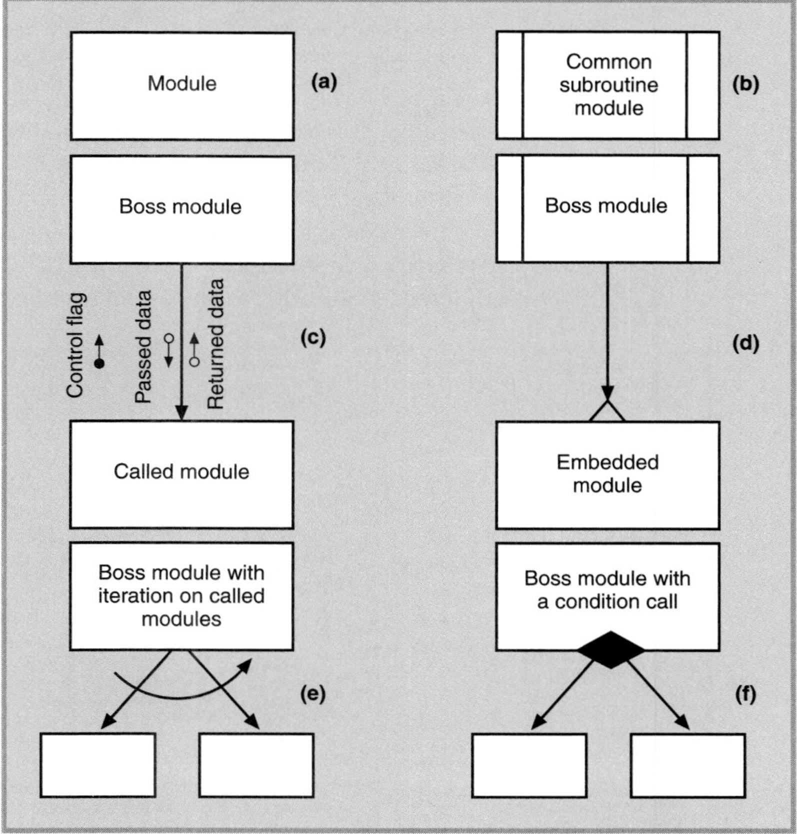

program call
The transfer of control from a module to a subordinate module to perform a requested service

data couples
The individual data items that are passed between modules in a program call

Figure 9-10 (c) shows a call from a higher-level module to the lower-level module. A **program call** occurs when one module invokes a lower-level module to perform a needed service or calculation. The implementation of a program call varies among programming languages. For example, a program call can be implemented as a function call in C or C++, a procedure call in Pascal, or a subroutine call in FORTRAN. In each case, control is passed to the called module, the called module executes a series of program statements, and control is then passed back to the calling module immediately after the program call was made.

Figure 9-10 (c) also indicates how data are passed between modules. The arrows with the open circle, called **data couples**, represent data being passed into and out of the module. A data couple can be an individual data item (such as a customer account number) or a higher-level data structure (such as an array, record, or other data structure). The type of coupling used at each level of the structure chart usually follows the principle of layering of detail. That is, coupling between modules near the top of a structure chart typically consists of data structures representing high-level aggregations of data. Coupling between modules at the bottom of the structure chart typically consists of single data items, flags, and relatively small data structures.

The arrow with the darkened circle is a control couple flag. A flag is purely internal information that is used between modules to indicate some result. Flags originating from lower-level modules often indicate a result such as a record passing a validation test. Another common use is to indicate that the end of file was reached.

Figure 9-10 (d) illustrates a lower-level module that is broken out on the structure chart, but that in all probability will be subsumed into the calling module for programming. It is primarily a documentation technique to ensure that the function performed by the module is highlighted. Figure 9-10 (e) and (f) show two alternatives for program calls. In (e) we show the notation used to indicate iteration through several modules. In (f) we show conditional calling of low-level modules—that is, the modules are called only when certain conditions exist.

Figure 9-11 is a more complete view of the payroll program, including the original *Calculate payroll* function from Figure 9-10. Notice that the entire structure chart shown in Figure 9-11 is based on the system activities following the temporal event *Time to produce payroll*. During systems analysis for the payroll system, the analyst would have identified this event as one that occurs at the end of every week for hourly employees. Many other events would have been identified, as well.

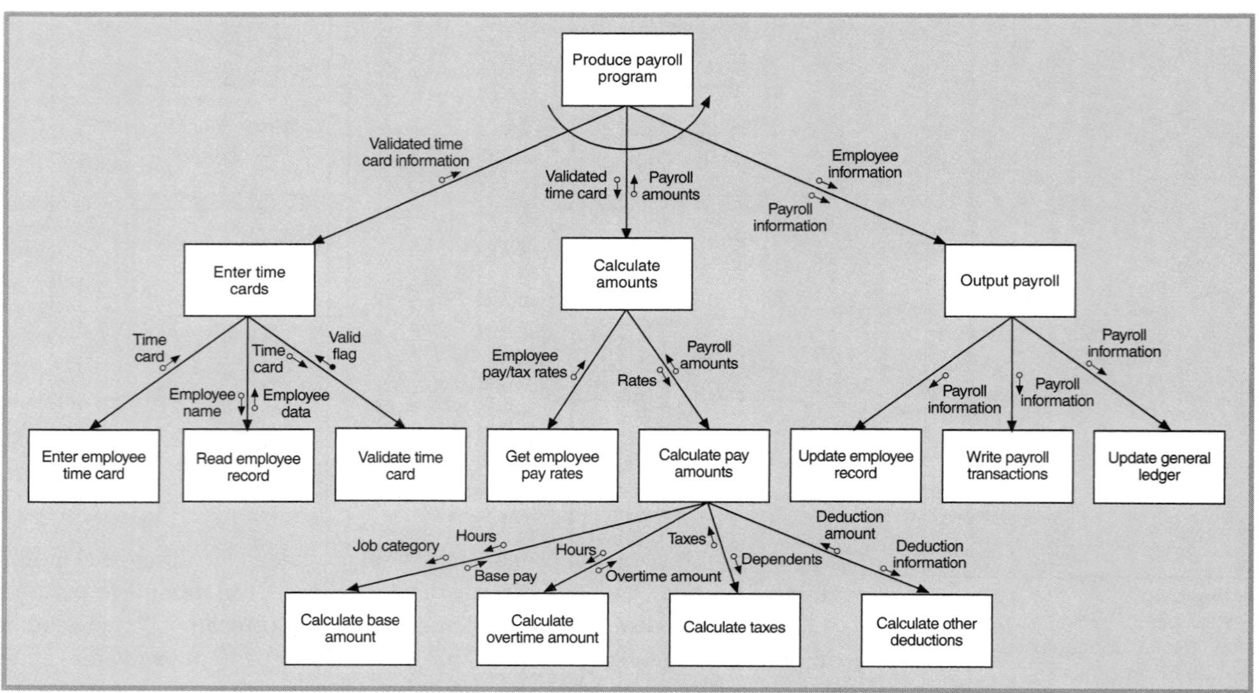

FIGURE 9-11
A structure chart for the entire Calculate payroll program.

A basic idea of structured programming is that each module only has to do a very specific function. The module at the very top of the tree is the boss module. Its functions will be to call the modules on the next tier, pass information to them, and receive information back. The function of each middle-level module is to control the processing of the modules below it. Each has control logic and any error-handling logic that is not handled by the lower-level module. The modules at the extremities, or the leaves, contain the actual algorithms to carry out the functions of the program. This approach to programming separates program control logic from business algorithm logic and makes programming much easier.

The arrows from the higher-level modules to the lower-level modules indicate the program call. The order of call is always from left to right. Notice that the structure chart maintains a strict hierarchy in the calling structure. A lower module never calls a higher module. The curved arrow immediately below the boss module indicates a loop across all three calls. In other words, the main module will have an internal loop that includes calls to all three lower-level modules within the same loop.

In the example, you can see the flow of information downward and back up. Usually a higher-level module will request a service of a lower-level module and pass the necessary input information. The lower-level module then returns the requested information or some control information, as a flag, to notify the higher-level module of the successful completion of the task. Looking at the *Enter time cards* subhierarchy, you can observe that the employee time card information is passed to the boss. Then the employee name is passed to the next module, which reads some employee information and passes it back up. Finally, employee data and time card information is passed to the rightmost module, which validates the time card. It passes up a control flag, indicating success or failure of the validation. If the validation fails, then error messages are written and the program goes into its error-handling routines. We have not shown all the complexities required in a real program, especially the error-handling modules.

The larger structure chart shows examples of various other symbols and illustrates how a structure chart can be used to write the computer code for the *Produce payroll program* portion of the system. By dividing the system into modules this way, the project team can program each module to do nothing more than call the lower-level modules in the correct sequence to carry out the logic of calculating payroll. Notice how a structure chart provides a simple and direct organization to a computer program. As you have learned in your programming class, modular programming is a well-accepted method to write computer programs that are easy to understand and maintain. A structure chart shows the hierarchy and calling relationship, but it leaves tremendous flexibility for the programmers to code the program.

Included within the structure chart will be the modules that access data from the outside world. It is important that the design of these modules be consistent with the design of the user interface, the interface to other systems, and the database design. The structure charts that have been developed should also be consistent with the system flow chart. If changes were made during this design activity, then the project team should update the system flow chart accordingly.

Developing a Structure Chart Structure charts are developed to design a hierarchy of modules for a program. A structure chart is in the form of a tree with a root module and branches. A subtree is simply a branch that has been separated from the overall tree. When the subtree is placed back in the larger tree, the root of the subtree becomes just another branch in the overall tree. Why is this important? The structure chart can be developed in pieces and combined for the final diagram.

Figure 9-8 showed the system flow chart for RMO's customer support system. Each major program corresponds to a subsystem in the event-partitioned diagram as illustrated in Figures 6-11 through 6-13. Each program will have its own structure chart. However, as can be seen in Figures 6-11 and 6-12, each program consists of several events. Each event corresponds to a process on the event-partitioned DFD, and each process will be detailed in a DFD fragment based on the event from the event table.

There are two methods to develop structure charts: transaction analysis and transform analysis. **Transaction analysis** uses as input the system flow chart and the event table to develop the top level of the tree—that is, the main program boss module and the first level of called modules. **Transform analysis** uses as input the DFD fragments to develop the subtrees, one for each event in the program. Each subtree root module

transaction analysis
The development of a structure chart based on a DFD that describes the processing for several types of transactions

transform analysis
The development of a structure chart based on a DFD that describes the input-process-output data flow

corresponds to the first-level branch of the main program structure chart. We discuss each method in turn.

In transaction analysis, the first step is to examine the system flow chart and identify each major program, such as the *Customer order* program in Figure 9-8. In Figure 9-12, we duplicate Figure 6-14, which is the event-partitioned DFD for the order-entry subsystem, and which shows the five processes derived from the five events of this subsystem. In this subsystem, the five primary processes are five different transactions that must be supported. These transactions are *Check item availability, Create new order, Modify existing order, Produce order summary reports*, and *Produce transaction summary reports.*

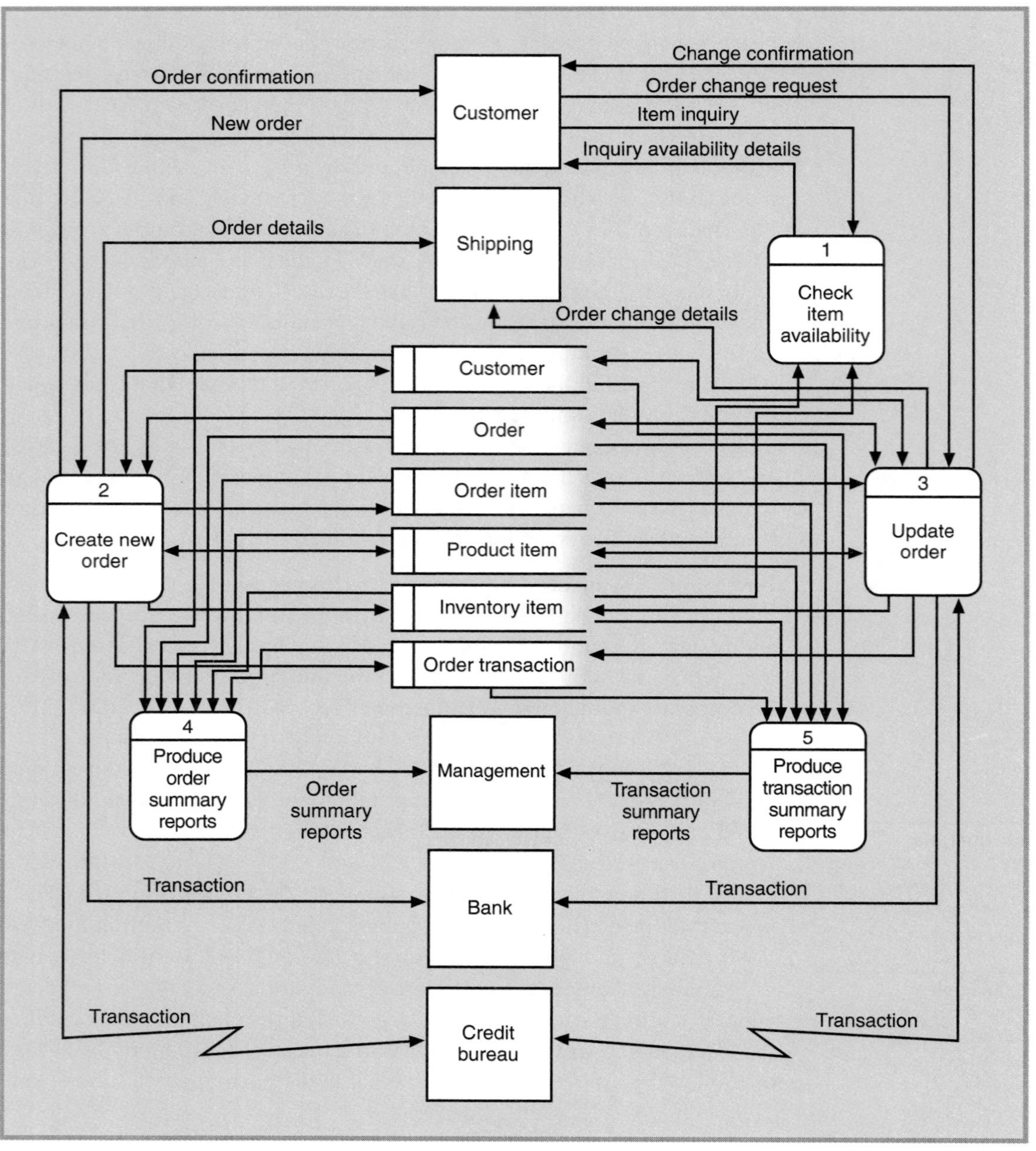

FIGURE 9-12
Event-partitioned DFD for the order-entry subsystem.

Figure 9-13 shows the structure chart based on transaction analysis for this program. Transaction analysis is the process of identifying each separate transaction that must be supported by the program and constructing a branch for each one. In essence, this program, at least at the highest level, is simply a module to display a screen and allow the user to enter the choice of transaction, then to invoke the appropriate module to process the transaction. This diagram does not show the additional detail below each of the transaction modules. Each of the transaction modules, which are named after the transactions, will be the main boss module for a subtree to process the transaction. Each subtree will be developed based on the DFD fragment for that event and will be developed utilizing transform analysis.

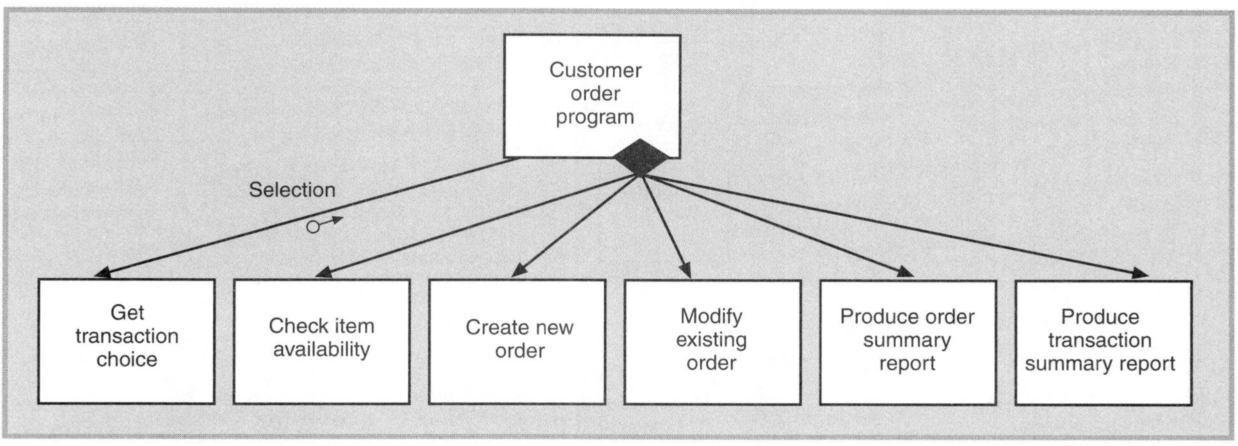

FIGURE 9-13
High-level structure chart for the customer order program.

This structure chart also has very few data couples. Essentially, the only information passed is the transaction selection from the *Get transaction choice* module. That information is used by the control module to select the correct processing module. The subtree beneath the processing module will display the appropriate screens to accept and pass the detailed information required.

Transform analysis is based on the idea that the computer program "transforms" input data into output information. Structure charts developed with transform analysis usually have three major subtrees: an input subtree to get the data, a calculate subtree to perform the logic, and an output subtree to write the results. Figure 9-11 is a good example of a structure chart that was developed using transform analysis, for the process of transforming time-card inputs into payroll outputs following one event. Note that a DFD fragment usually follows this pattern of input-process-output, and the structure chart converts the processing on the DFD fragment to a top-down structure of program modules.

Sometimes DFD fragments are decomposed into detailed diagrams. The detailed diagrams provide more detail than can be used for the structure chart. Figures 9-14 through 9-17 provide an example of transform analysis from Rocky Mountain Outfitters. Figure 9-14 shows the DFD fragment created for the *Create new order* event. Figure 9-15 contains the exploded view for that event, which is used for transform analysis. The structure chart is developed directly from the data flow diagram. The fundamental idea is that the detailed diagram processes from the data flow diagram become the leaf modules in the structure charts. The mid-level process processes—that is, the processes that were exploded to derive the low-level processes—become the intermediate-level boss modules on the structure chart. Thus, the hierarchy of the structure chart directly reflects the organization of the set of nested, or leveled, data flow diagrams. Additional boss modules may need to be developed to provide the correct structure to the structure chart.

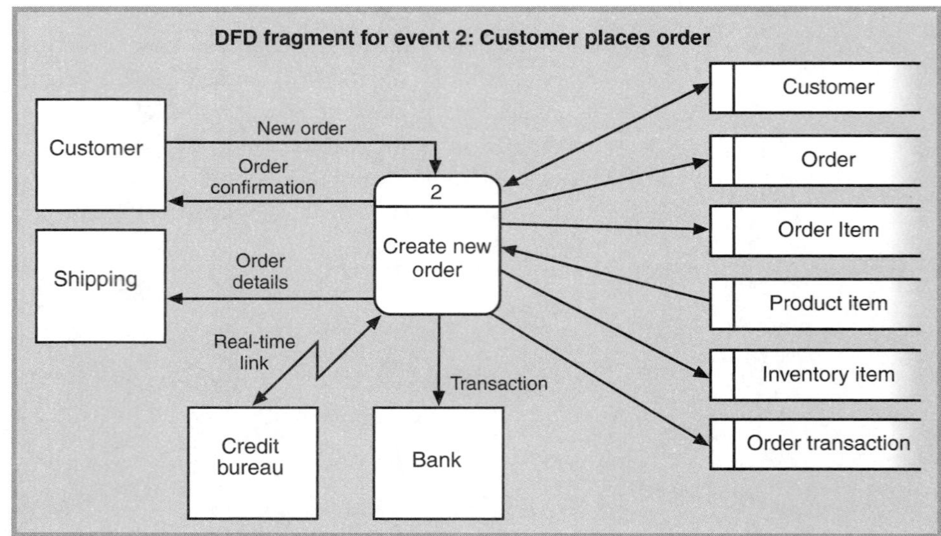

FIGURE 9-14
The Create new order *DFD fragment.*

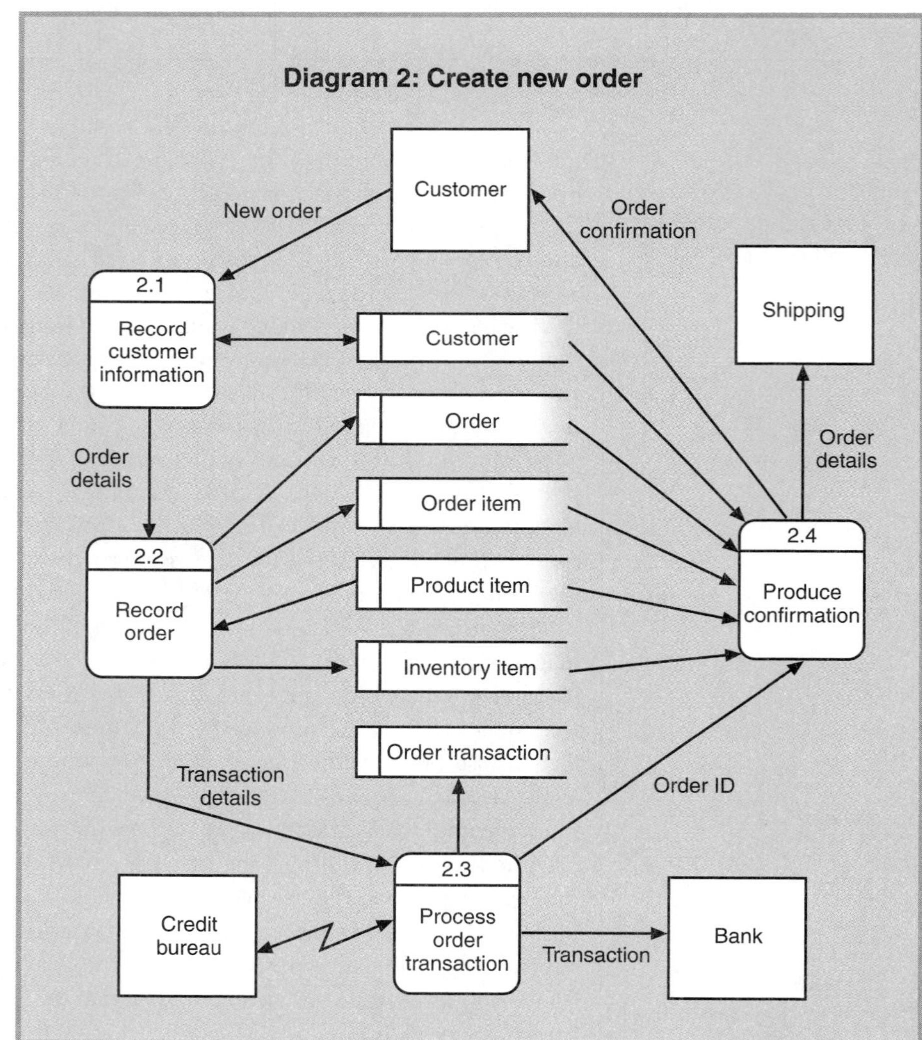

FIGURE 9-15
Exploded view of the Create new order *DFD.*

As stated above, the general form of a structure chart developed with transform analysis is input-process-output. The method to develop a structure chart from a data flow diagram fragment consists of the following steps:

1. Determine the primary information flow. This is the main stream of data that is transformed from some input form to the output form.
2. Find the process that represents the most fundamental change from an input stream to an output stream (see Figure 9-16). The input data stream is called the **afferent data flow**. The output data stream is called the **efferent data flow**. The center process is called the **central transform**.
3. Redraw the data flow diagram with the input to the left and the output to the right. The central transform process goes in the middle. If this is an exploded-view data flow diagram, add the parent process to the diagram. Nonprimary data flows can be omitted to simplify the drawing. An example of this redrawn data flow diagram is given in Figure 9-16.
4. Generate the first-draft structure chart, based on the redrawn data flow, with the calling hierarchy and the required data couples. An example of this is given in Figure 9-17.
5. Add other modules as necessary to (1) get input data via the user-interface screens, (2) read and write to the data stores, and (3) write output data or reports. Usually, these are lower-level modules, or utility modules. Add the appropriate data couples based on the data flows to and from these data stores.
6. Using any structured English or decision table documentation as a basis, add other required intermodule relationships such as looping and decision symbols.
7. Make the final refinements to the structure chart based on the quality control concepts discussed in the following section.

afferent data flow
The incoming data flow in a sequential set of process bubbles

efferent data flow
The outgoing data flow from a sequential set of process bubbles

central transform
The central processing bubbles in a transform analysis type of data flow

FIGURE 9-16
Rearranged view of the Create new order *DFD.*

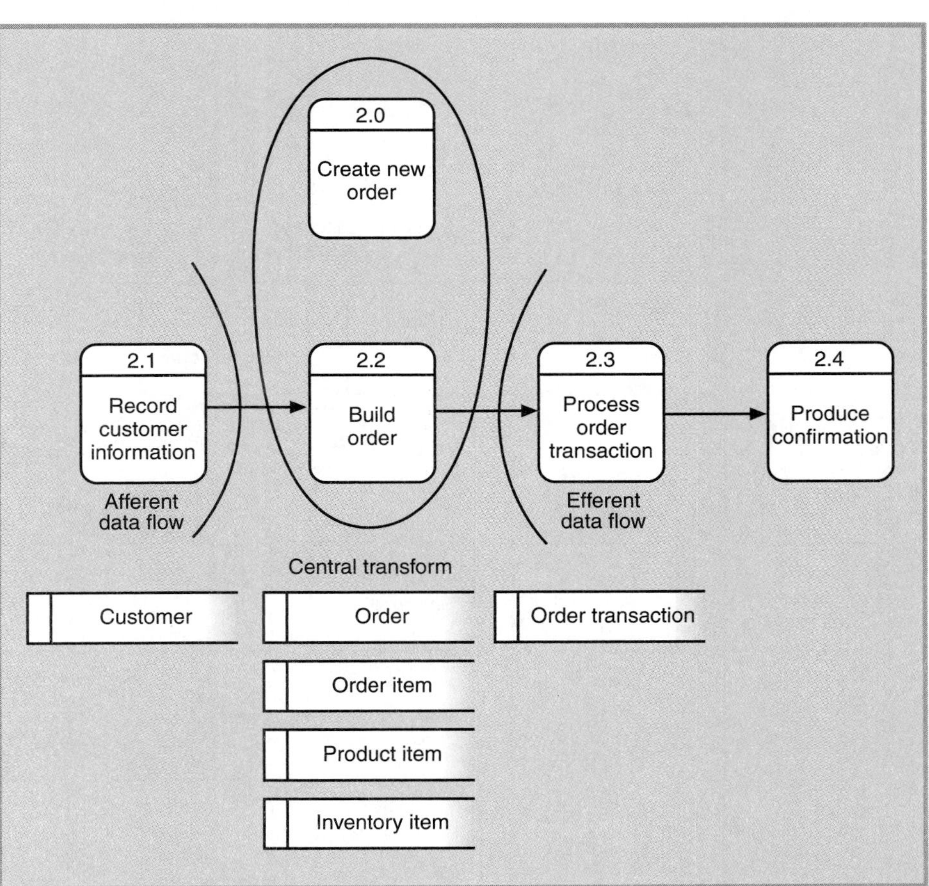

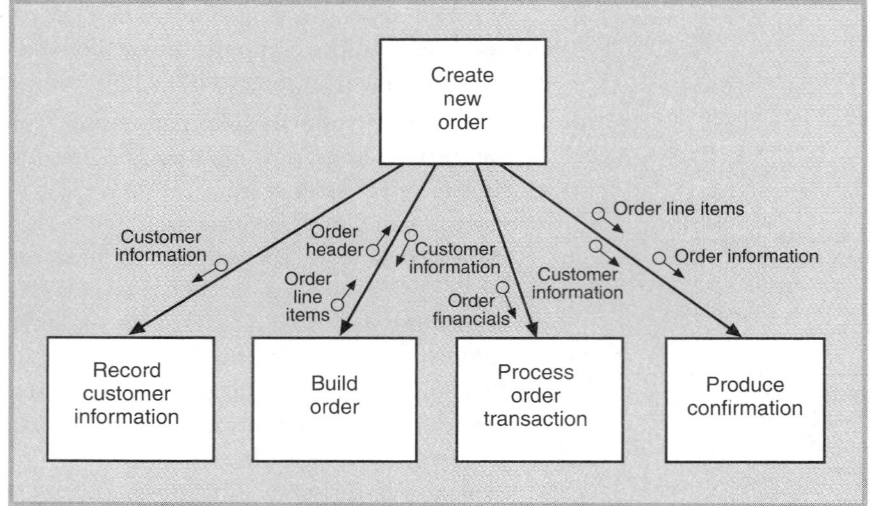

FIGURE 9-17
First draft of the structure chart.

Through step 4, as can be seen in Figure 9-17, the organization of the structure chart very closely mirrors the data flow diagram from which it derives. In step 5 we begin to enhance the first-draft structure chart with additional modules to provide modules to read and write data. Frequently there are no corresponding process on the data flow diagram. Thus, at this point we depend less on the data flow diagram information and more on the requirements of a good design. Figure 9-18 illustrates the structure chart for the next step—step 5.

FIGURE 9-18
The structure chart for the create new order program.

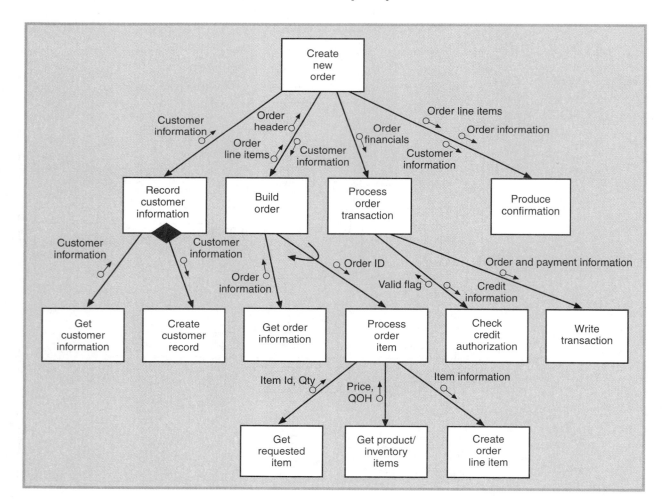

Comparing Figure 9-17 and 9-18, notice that Figure 9-17 indicates that all the input information comes from the far-left module, *Read customer information*. In Figure 9-18, observe that the accessing of information has been distributed across other branches of the structure chart. Customer information is retrieved through the far-left branch, but additional customer information about the order is retrieved in the second branch of the chart. Even though this organization is not exactly true to the data flow diagram, it is a more logical organization of the structure chart. The addition of these data access modules is truly a design process—the creation of new components based on systems design principles.

In addition to distributing the access of customer input data, the structure chart in Figure 9-18 has other data access modules to retrieve product and inventory information. This type of data retrieval corresponds to the data flows on the data flow diagram between the processes and the data stores. During design, we must explicitly identify the modules that actually read and write to the data stores. The module *Get product/inventory items* is added to the structure chart to provide the retrieval of the product information. As the additional modules are added to the structure chart, then the data couples are defined more precisely to reflect this more detailed design structure.

In Figure 9-18, we have also added the additional symbols concerning looping and optional calls. The black diamond indicates that the call to create a customer record is optional, and in fact is required only if the customer is a new customer. The general form of the structure chart has the inputs to the left and the outputs to the right.

The high-level boss module, *Create new order*, and its tree of modules can be plugged into the transaction structure chart in Figure 9-12. Figure 9-19 illustrates the process of combining the top-level structure chart, developed using transaction analysis, with the lower-level subtrees, developed with transform analysis.

Evaluating the Quality of a Structure Chart The process of developing structure charts from DFDs can become rather involved. There are rules and guidelines that can be used to test the quality of the final structure chart, however. Two measures of quality are **module coupling** and **module cohesion**. Generally, it is desirable to have highly cohesive and loosely coupled modules.

module coupling
The manner in which modules relate to each other; the preferred method is data coupling

module cohesion
A measure of the internal strength of a module

The principle of coupling is a measure of how a module is connected to the other modules in the program. The objective is to make modules as independent as possible. A module that is independent can execute in almost any environment. It has a well-defined interface of several predefined data fields. It passes back a well-defined result in predefined data fields. The module does not need to know who invoked it and, in fact, can be invoked by any module that conforms to the input and output data structure. The best coupling is through simple data coupling. In other words, the module is called, and a specific set of data items is passed. The module performs its function and returns the output data items. This kind of module can be reused in any structure chart that needs the specific function performed.

Cohesion refers to the degree to which all of the code within a module contributes to implementing one well-defined task. Modules with high cohesion implement a single function. All of the instructions within the module are part of that function, and all are required for the function. Modules with low (or poor) cohesion implement multiple or loosely related functions.

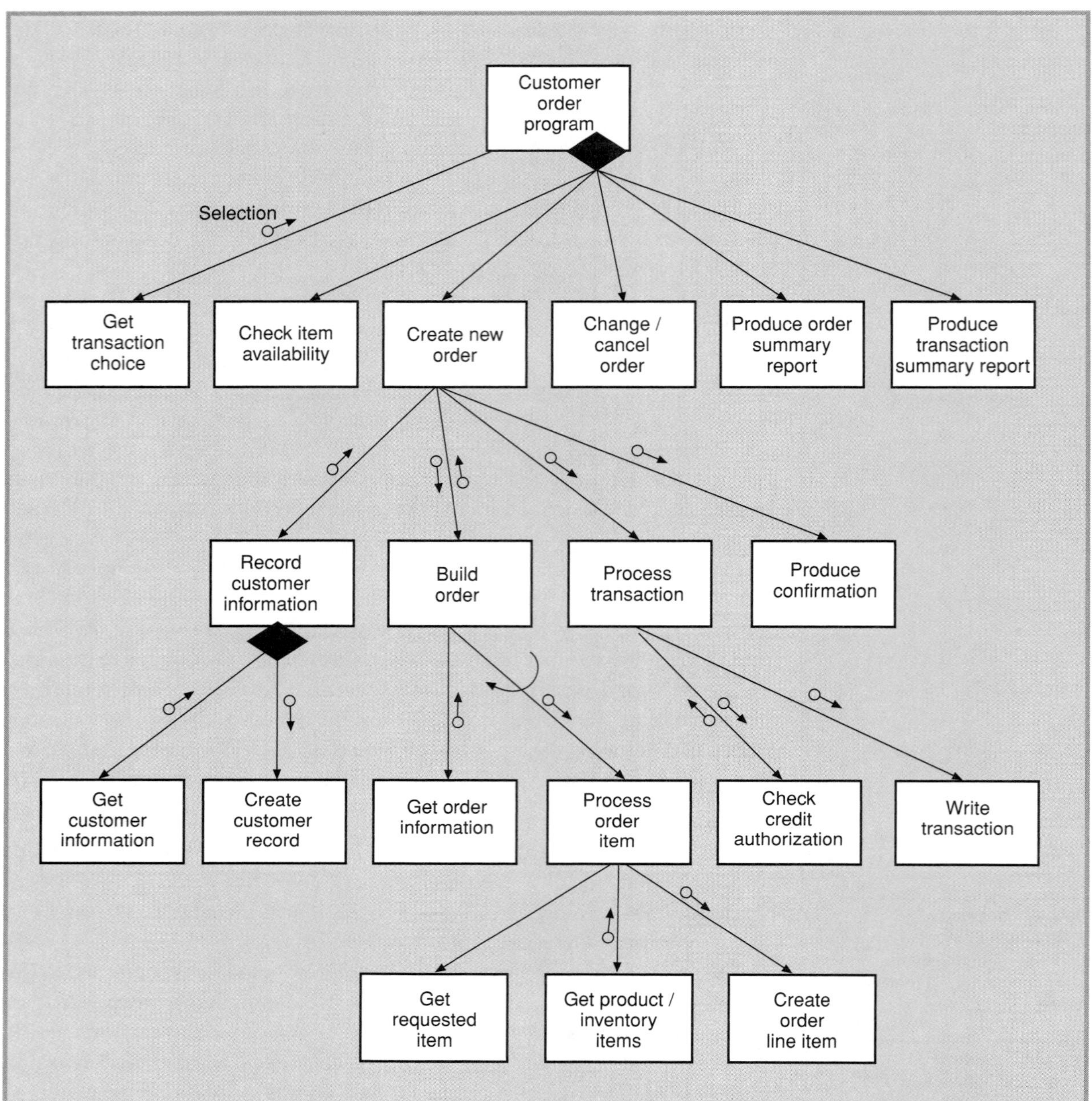

FIGURE 9-19
Combination of structure charts (data couple labels are not shown).

Note that the amount of coupling and the specific data items being passed are good indicators of the degree of module cohesion. Modules that implement a single task tend to have relatively low coupling because all of their internal code acts on the same data item(s). Modules with poor cohesion tend to have high coupling because loosely related tasks are typically performed on different data items. Thus, a module with low cohesion generally has several unrelated data items passed by its superior.

A flag passed down the structure chart is also an indicator of poor cohesion in the lower-level module. Flags passed into a module are typically used to select the part of the recipient module's code that will be executed. Part (a) of Figure 9-20 shows an example of poor cohesion. A project team can improve cohesion by partitioning the module into separate modules, one for each value of the flag as shown in part (b) of Figure 9-20. The code of the superior module is programmed to use the flag to decide which of the partitioned subordinate modules to call.

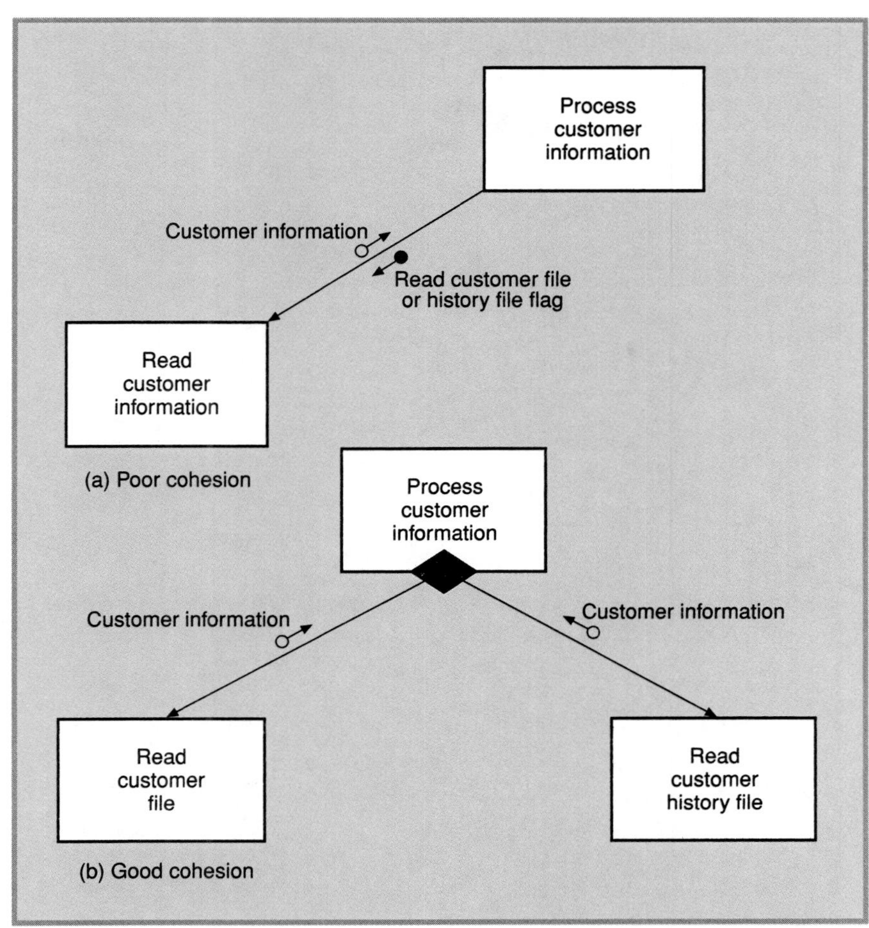

FIGURE 9-20
Examples of module cohesion.

Module Algorithm Design: Pseudocode

The previous two models, the system flow chart and structure chart, provide the overall structure of the system and the structure within each program. The next requirement of design is to describe the internal logic within each module. There are three common methods used to describe module logic: flow charts, structured English, and pseudocode. All three methods are equivalent in their ability to describe logic. Flow charting is a visual method that uses boxes and lines to describe the flow of logic in a program. In the early days of computing, flow charting was used almost exclusively. Today, however, versions of pseudocode and structured English have replaced flow charting. In Chapter 6, you learned about structured English. Pseudocode is a variation of structured English that is more similar to a programming language. Frequently, we write pseudocode using statements that are very similar to the target language. If we are writing to COBOL, we use COBOL-like syntax. If we are writing in Visual Basic or C, we use a syntax that mirrors those languages.

Figure 9-21 shows a simple example of the logic of the payroll system. Pseudocode statements for the Produce payroll program module, the Calculate amounts, Calculate pay amounts module and the Calculate taxes module are shown. This figure shows examples of each of the three types of control statements used in structured programming: (1) sequence, that is, a sequence of executable statements, (2) decision, that is, if-then-else logic, and (3) iteration, that is, do-until or do-while.

```
Produce payroll program
DoUntil No more time cards
     Call Enter time card
     Call Calculate amounts
     Call Output payroll
End Until

Calculate amounts
     Call Get employee payrates
     Call Calculate pay amounts

Calculate pay amounts

Call Calculate base amount
If (HoursWorked > 40) Then
     Call Calculate overtime amount
End If
Call Calculate taxes
If (SavingsDeduction=yes) or (MedicalDeduction=yes) or (UnitedWay=yes) Then
     Call Calculate other deductions
End if

Calculate taxes

Get Tax Rates based on Number Dependents, Payrate
Calculate Income Tax = PeriodPayAmount * IncomeTaxRate
If YTD Pay < FICA MaximumAmount Then
     Calculate EmpFICA = PeriodBasePay * FICAEmployeeRate
     Calculate CorpFICA = PeriodBasePay * FICACorpRate
End If
If StateTax Required Then
     Get StateTaxRate based on State, NumberDependents, Payrate
     CalculateStateTax = PeriodPayAmount * StateTaxRate
End If
If StateTaxRequired Then
     Get StateTaxRate based on State, NumberDependents, Payrate
     Calculate StateTax = PeriodPayAmount * StateTaxRate
End If
If OvertimePay > 0 Then
     Calculate OvertimeIncomeTax = PeriodOvertimePay * IncomeTaxRate
     Add to OvertimeIncomeTax to Incometax
     If YTDPay < FICAMaximum Amount Then
          Calculate EmpOvertimeFICA = PeriodOvertimePay * FICAEmployeeRate
          Calculate CorpOvertimeFICA = PeriodOvertimePay * FICACorpRate
     End If
     If StateTaxRequired Then
          Calculate StateOvertimeTax = PeriodOvertimePay * StateTaxRate
     End If
End If
```


FIGURE 9-21
Pseudocode for the Calculate pay amounts *hierarchy.*

Integration of the Structured Application Design with User-Interface Design, Database Design, and Network Design

In the previous sections, you learned how to develop a structure chart based on the information in a data flow diagram. The primary focus was on capturing the information in the structural relationship between the processes. The structure chart developed from either transaction analysis or transform analysis will be correct, but it may not be complete. Before the structure chart can be considered complete, it usually must be modified or enhanced to integrate the design of the user interface, the database, and the network. Since we have not covered those design topics yet, we will only briefly discuss the types of changes that need to be made.

The user interface will consist of a set of input forms, interactive input/output forms, and output forms or reports. Sometimes a given input, such as a customer order, requires an interactive dialog between the user and the system and includes a series of input forms. Every form must be displayed and the data retrieved somewhere in a module in the structure chart. As these forms are developed, the structure chart needs to be evaluated from three aspects:

- Do additional user-interface modules need to be added?
- Does the pseudocode in the interface modules need to be modified?
- Do additional data couples need to be added to pass data?

In earlier chapters, you learned that the entity-relationship diagram (ERD) must be consistent with the data stores found on the data flow diagrams. There is not necessarily a one-to-one correspondence between data stores and database tables, but the information on every data store must be somewhere in the database. In addition, every database table and field must be represented by a data store somewhere. During design, the project team performs this same type of analysis and makes appropriate changes to the structure chart.

The same three aspects—modules, pseudocode, and data couples—need to be evaluated for the database. If a database management system is being used, then a common interface is usually provided. The designer can make the database accessible either by calling a database interface module or by embedding SQL (Structured Query Language) statements within the pseudocode.

Finally, to ensure that the structure chart is integrated with the network, we evaluate which functions will be executing on which nodes of the network. As discussed earlier in the chapter, at the time that the system boundary is developed, each program can be assigned to a node. The activity at the end of design is to ensure that assignments were correctly made and that the system flow chart correctly reflects the distributed nature of the system.

Designing the Application Architecture: The Object-Oriented Approach

When the analysis phase is performed with object-oriented models, then the design is also completed with object-oriented models. Object-oriented design models provide the bridge between the object-oriented analysis models and the object-oriented program. Before going any further, let's review how an object-oriented program works. Then we will discuss the design models and how they must be structured to support object-oriented programming.

Object-Oriented Programs

To understand object-oriented design, we must first understand how an object-oriented program works. The basic concept of an object-oriented program is that the program consists of a set of program objects that cooperate to accomplish a result. Each program object has program logic and any necessary attributes encapsulated into a single unit. These objects work together by sending each other messages. They then work in concert to support the functions of the main program.

Let's compare structured programs to object-oriented programs through an analogy—a computer analogy, but also one with which you may be familiar. A mainframe computer is a large computer with a massive amount of computing capability. A mainframe computer may be connected to thousands of individual work terminals. It controls them all. No individual terminal does work unless the main computer directly instructs it to do so. The mainframe also does all the database access and execution. This is much like a structured program.

In contrast, a network of personal computers consists of many individual computers connected by a network cable. Each computer has its own computing capability. If you are working at one computer, you can communicate with another personal computer, with a message, and ask for assistance. For example, some of the individual computers on the network may be connected to printers. We call them print servers. Others may have large disk drives and special databases that we can access. They are called file servers. If you are working on a network and you want to print something, your computer sends a message to the computer that controls the printer and asks it to print something for you. Typically, in most networks with which you may be familiar, there is no single, coherent purpose to all of the individual computers. But suppose that there were a central purpose and that each computer had a specific set of functions that it performed to support that purpose. The computers could all work together, again by sending messages to each other to fulfill the overall computing purposes. This analogy is much like the way a system or program using object-oriented techniques is designed.

Figure 9-22 depicts how an object-oriented program works. The program includes a window object that displays a form in which to enter customer ID and other information. After the customer ID is entered, the window object sends a message (message number 2) to the customer class to tell it to create a new customer object (instance) in the program, and also to go to the database and get the customer information and put it in the object (message 3). Once that is done, the new customer object also sends that information back to the window object to display it on the screen. The clerk then enters the deposit amount (message 4), and another sequence of messages is sent to update the customer object in the program and the customer information in the database.

One common question about object-oriented programs is, who is in charge? In a structured program, it is obvious who the boss module is and who controls the computing. In an object-oriented program, it is not so obvious. In fact, no one may be in charge. In the network of personal computers analogy, let's assume that each personal computer has a specific function. Whenever that function needs to be done, then that personal computer is in charge, and can call on the other personal computers and devices in the network to provide the necessary support.

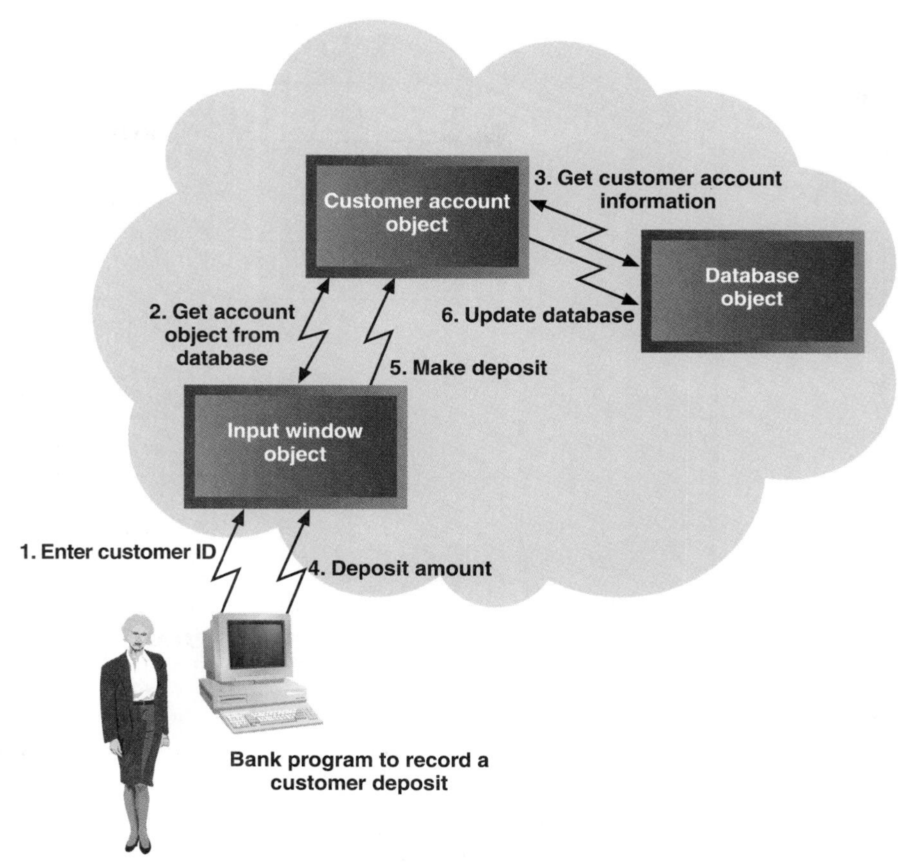

FIGURE 9-22
Object-oriented event-driven program flow.

An object-oriented system consists of sets of computing objects. Each object has encapsulated within itself data and program logic. We define the structure of the program logic and data fields by defining a class. The class definition describes the structure or a template of what an executing object looks like. The object itself does not come into existence until the program begins to execute. We call that instantiating the class, or making an instance (an object), based on the template provided by the class definition.

In Chapter 5 you saw a class diagram. Remember that each class definition included the attributes for the class and the methods, or logic, applicable to the class. For example, in Figure 5-32, which we have reproduced as Figure 9-23, there are some classes associated with a banking example. The *Account* class defines the template for an account object. The attributes are *Account number*, *Balance*, and *Date opened*. We also define the methods of the class. The methods define the operations and calculations that can be performed on the attributes. The methods for the *Account* class are *Make deposit* and *Make withdrawal*. Methods in an object-oriented program are where the programming logic is contained. There are as many instances of that class as necessary for all the accounts in the bank system. Each instance has its own data values. However, all instances share the method logic since it is common to all.

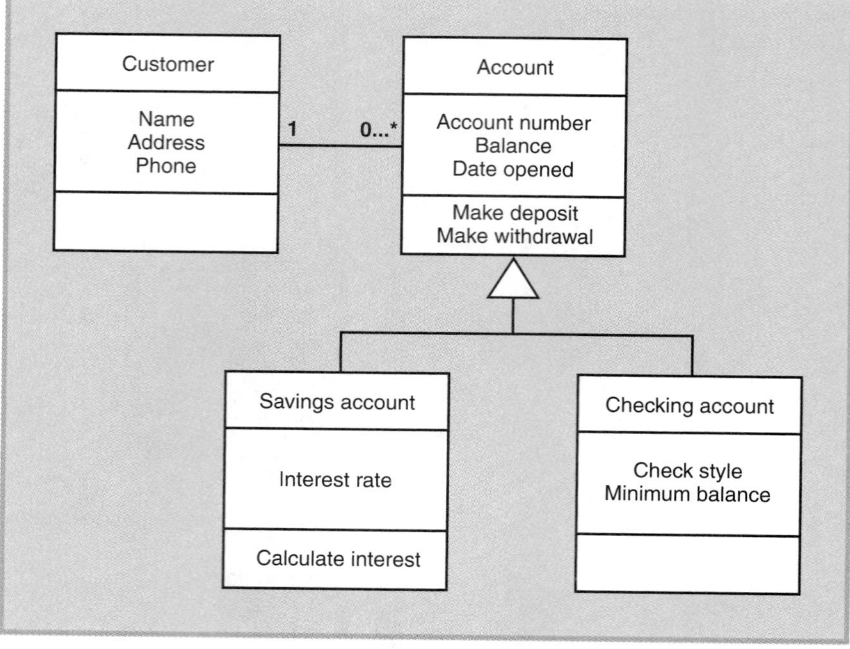

FIGURE 9-23
A simplified bank class diagram.

Two important principles of object-oriented programs make them robust and easy to maintain. The first principle is called *encapsulation*, which is the idea that we have been discussing: that all the data required by the instantiated object is encapsulated within the object and that the logic for the object is encapsulated within the class, which is the template for the object. Encapsulation simplifies debugging, since any changes to the data are caused by the code in the class.

information hiding
Making the data fields of an object class invisible to other classes

The second principle, called **information hiding**, is related to encapsulation and means that the data fields in an object are not visible to other objects in the system. If an object wants to get the value of data—for example, the name attribute—it must ask that object to provide the value. For example, if you are at a party where everybody has a name tag, then there is no information hiding of names. However, a party where the guests do not have name tags does have information hiding and requires you to ask another person, "What is your name?" In object-oriented parlance, the name attribute is hidden and you must invoke the object's "get-name" method to find out. You invoke the "get-name" method by sending a message to the other person—namely, the spoken words "what is your name?" Generally, in object-oriented programming, the attribute values are invisible to the outside world, but the methods are visible—that is, the methods can be invoked by an external message.

To write an object-oriented program for the customer deposit example, we need to know the classes (Customer Input Window, Customer Account, and Database). We also need to know the attributes for each class, the methods for each class, and the messages to be sent between the classes. All of this information is obtained from the object-oriented models we build during design.

Many of the objects in the program correspond to the classes you identified in the class diagram. The messages look a lot like the interactions you defined in the collaboration and sequence diagrams. In fact, one of the strengths of object-oriented development is that the design models are very similar to the analysis models. It is usually a straightforward process to build the design models from the analysis models.

Another strength is that the final program is very similar to the object-oriented design models. Programming flows directly from the design models.

Object-Oriented Models

Referring back to Figure 9-3, you can see the flow from OO analysis to OO design. Figure 9-24 summarizes the process, showing the input analysis models and the output design models. The package diagram is a high-level diagram that documents the subsystems by showing which classes are contained within which subsystems. Information for the package diagram is obtained primarily from the use case diagram and the class diagram.

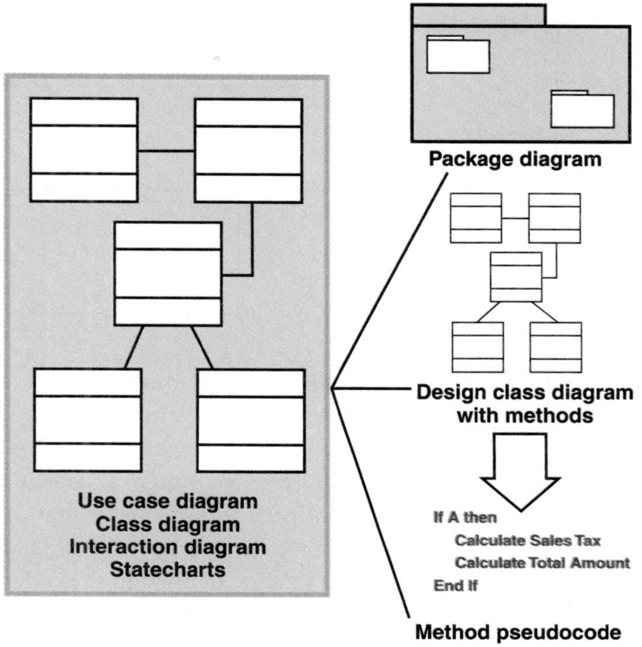

FIGURE 9-24
Object-oriented analysis models to design models.

The design class diagram is an extension of the class diagram but is extended to show more details of the attributes and the methods. Input information to create the design class diagram comes from the class diagram, the interaction diagrams, and the statecharts.

Package Diagrams

package diagram
A high-level diagram that identifies the major components of a system

The **package diagram** is a high-level diagram similar in concept to the traditional approach's system flow chart. The objective of a package diagram is to identify the major components of a complete system. In large systems, it is usually necessary to divide the system into various subsystems. Each subsystem functions independently from the other subsystems, although subsystems normally exchange information and frequently share the same database.

Figure 9-25 is an example of a simple package diagram from Rocky Mountain Outfitters. The four subsystems were shown on the use case diagram in Chapter 6. Only two symbols are used in a package diagram: a tabbed rectangle and a dashed arrow. The tabbed rectangle identifies the major system and the subsystems. Enclosure of a subsystem within the primary system indicates that it is part of the major system. Subsystems may be nested to any level of depth. However, no other overlap is allowed. In other words, a subsystem may not be part of two higher-level systems.

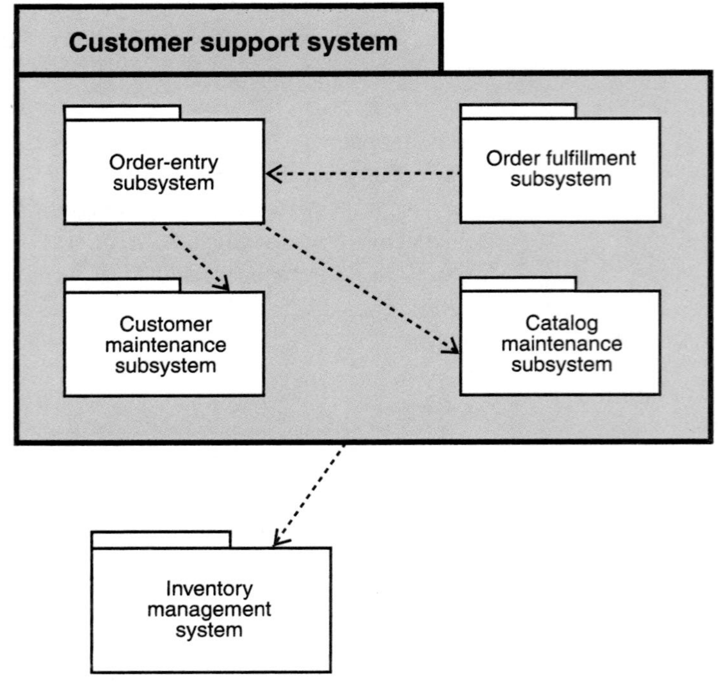

FIGURE 9-25
The RMO customer support system package diagram.

The arrow is a dependency arrow. The arrow tail is connected to the package that is dependent, and the arrowhead is connected to the independent package. The easiest way to read the diagram is to follow the direction of the arrow. For Figure 9-25, the order fulfillment subsystem is dependent on the order-entry subsystem.

Dependency simply means that the structure and code in one package is dependent on another package. The order fulfillment subsystem may use the order data structures that are defined in the order-entry subsystem. Or information may be sent from the order-entry subsystem. In any event, if anything changes in the order-entry subsystem, the order fulfillment subsystem may also require modification.

The package diagram is developed essentially from design decisions on how best to partition the final computer system. To make these design decisions, the designer considers the event list, the use case diagram, and the class diagram. The event list and use case diagram provide insight into which business functions are logically associated. The class diagram identifies the classes that are partitioned into the various packages or subsystems. Each element in a package belongs exclusively to that package. Figure 9-26 expands Figure 9-25 to show which classes are assigned to which packages. For example, the Customer class is assigned to customer maintenance package.

Note that the dependency arrows can also be determined once the classes are assigned to packages. The order-entry subsystem obviously needs to access the Customer class; thus, it is dependent on the customer maintenance package.

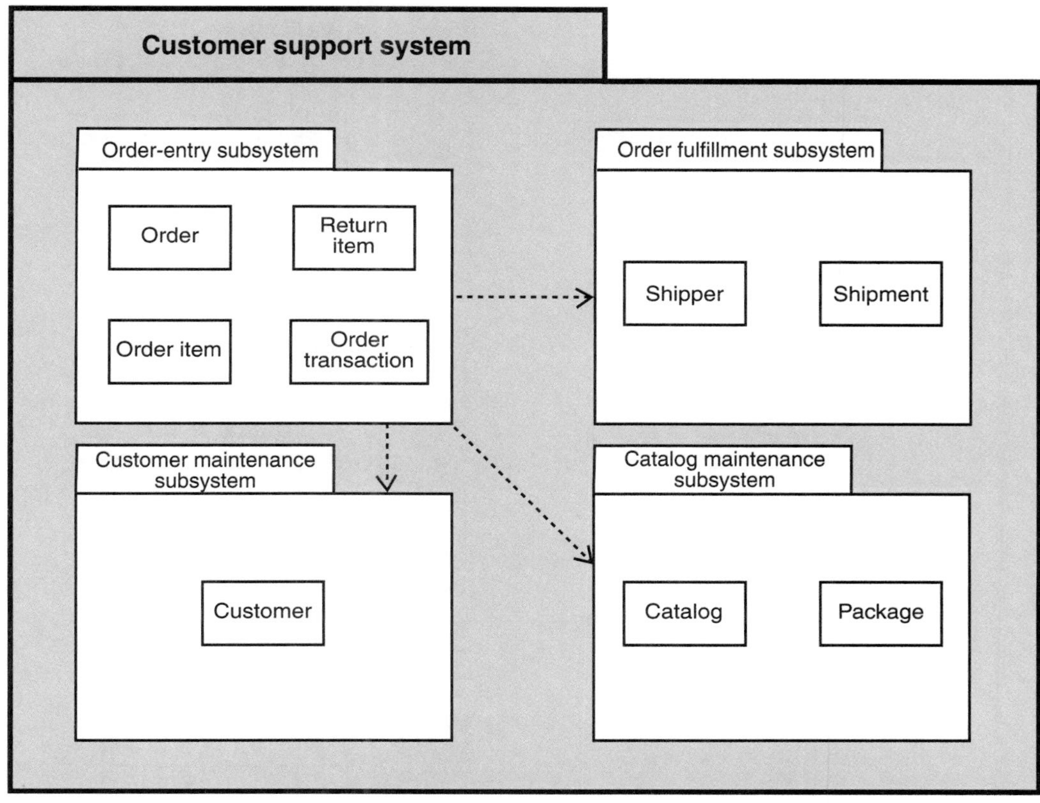

FIGURE 9-26
A package diagram including design classes from RMO.

Design Class Diagrams

design class diagram
A class diagram with notation to describe design components within the classes

The **design class diagram** is a variation of the class diagram. The class diagram shows a set of classes and their relationships. During analysis, since it is a discovery process, we generally do not worry too much about the details of the attributes or the methods. In object-oriented programming, the attributes of a class have a visibility characteristic to denote whether other classes can access that attribute. Each attribute also has a type definition, such as character or numeric. During design, we want to elaborate these items, as well as to define parameters that are passed to the methods, return values from methods, and the internal logic of each method. Thus, the class diagram for design is very similar to the class diagram for analysis, but it is just more complete. We complete it by integrating information from statecharts and interaction diagrams into the classes of the design class diagram.

Figure 9-27 shows the symbols used for a design class diagram. Figure 9-27 (a) is a simplified version showing only the attributes and method identifiers. Figure 9-27 (b) is an expanded version of a design class. The expanded version includes ovals that will contain the logic of each method.

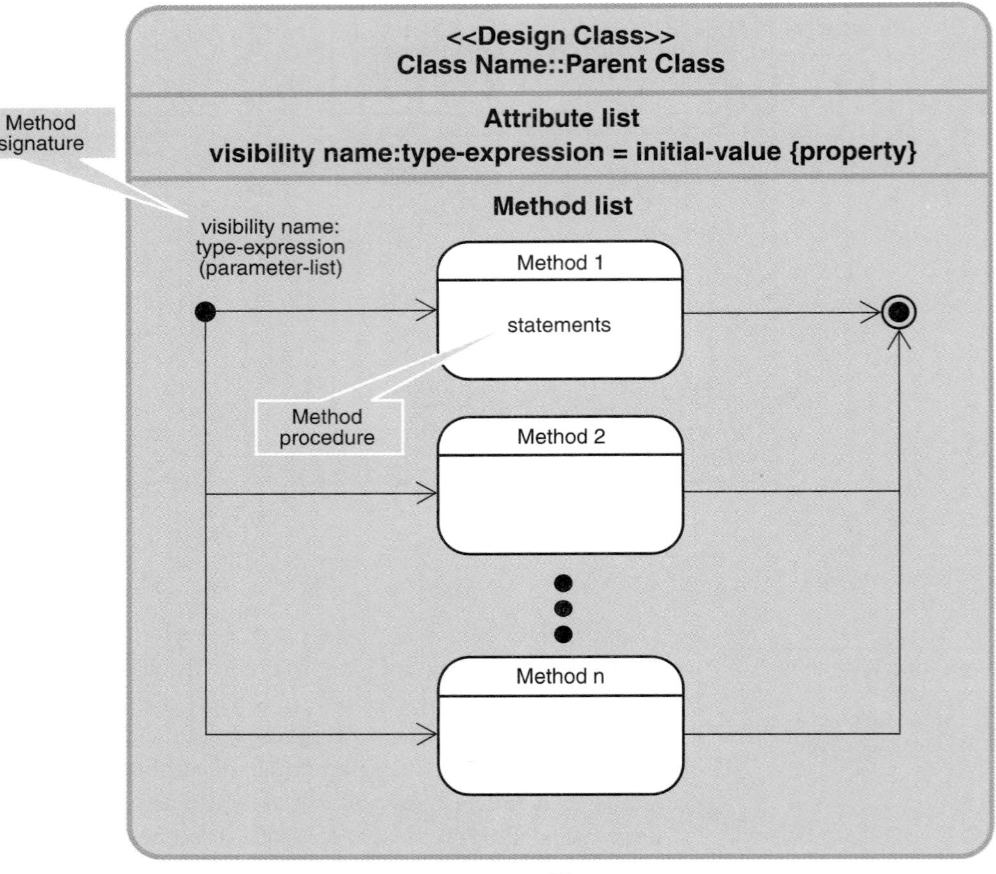

FIGURE 9-27
Internal symbols used to define a design class.

The large rectangle is a design class, as indicated by the <<Design Class>> indicator at the top. The name of the class is in the top compartment. If appropriate, the name of a superclass is also included. A superclass is a type of parent class, as explained later. The middle compartment contains the list of attributes. The format to define each attribute, however, includes the following:

- Visibility (+ is visibile, – is not visible. Visibility denotes whether other objects can directly access the attribute)
- Attribute name
- Type expression (such as character, string, integer, number, currency, or date)
- Initial value

The third compartment contains the list of methods, as did the basic class rectangle. However, we have added much more information about the details of the methods. The notation with the rounded rectangles and the arrows is simply an indication that this information derives primarily from the statechart, hence the similarity with statechart notation. Each method has two components: (1) a method signature and (2) a method procedure. The method signature is located on the incoming arrow. The method procedure is placed inside the rounded box.

method signature
The method name, return type, and parameter list used to provide a unique identifier for each method

A **method signature** shows all of the information needed to invoke (or call) the method. It shows the format of the message that must be sent, which consists of the following:

- Method name
- Type expression (the type of the return parameter from the method)
- Method parameter list (incoming arguments)

In object-oriented programming, we use the entire signature to identify a method. Some OO languages allow multiple methods to have the same name, so to distinguish between them, we also use the parameter list. In those languages, you need both the method name and the parameter list to invoke the correct method. For example, suppose that we want to be able to get to a customer record either by the customer ID number or by the customer name. We could identify two methods, each with the same name, such as "GetCustomer (CustomerID)" and "GetCustomer (CustomerName)." We say that GetCustomer is an **overloaded method**—that is, there are two methods with the same name. To know which method to invoke, the compiler must also note what parameters are included—whether a number (CustomerID) or a text field (CustomerName) was entered.

overloaded method
A class method that has several versions of the same method and is identified by distinct parameter lists

Figure 9-28 shows a design class for the account class from Figures 5-32 and 9-23. The figure shows both the abbreviated form in (a) and the expanded form in (b). We have also included several attributes and methods that did not appear in the analysis class diagram but that are added as part of the design process.

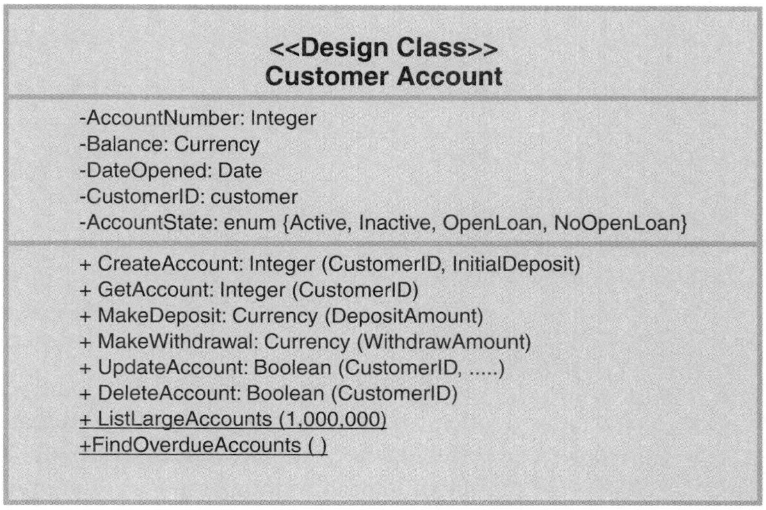

FIGURE 9-28(a)
The short form diagram for the customer account design class.

<<Design Class>>
Customer Account

-AccountNumber: Integer
-Balance: Currency
-DateOpened: Date
-CustomerID: customer
-AccountState: enum {Active, Inactive, OpenLoan, NoOpenLoan}

+ CreateAccount: Integer (CustomerID, InitialDeposit)
+ GetAccount: Integer (CustomerID)
+ MakeDeposit: Currency (DepositAmount)
+ MakeWithdrawal: Currency (WithdrawAmount)
+ UpdateAccount: Boolean (CustomerID, ...)
+ DeleteAccount: Boolean (CustomerID)
<u>+ ListLargeAccounts (1,000,000)</u>
<u>+FindOverdueAccounts ( )</u>

FIGURE 9-28(b)
The long form diagram for the customer account design class.

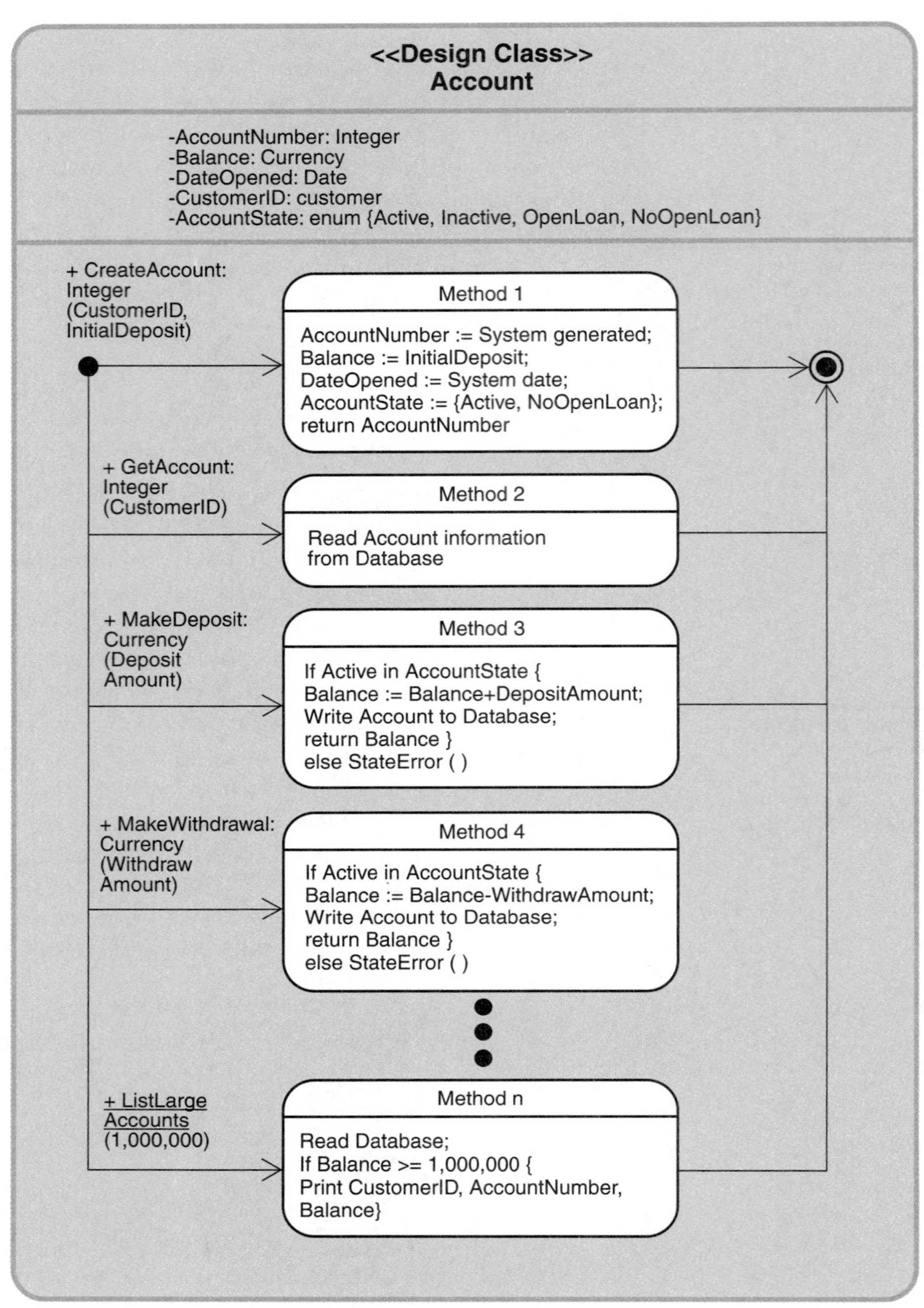

The attribute list contains all attributes discovered during analysis activities. In addition, we have included other attributes that represent object references for associations with objects of another class. For example, an account is associated with a customer, so the customer account class needs a way to point to the right customer. There are two ways to do this. One way is to have a field, which is a direct reference or connection to the appropriate customer object. That field is called an object reference or object pointer. The other method is to insert a foreign key for the customer class in the customer account class. For object-oriented databases, the first method is used. For relational databases, the second method is the correct one. This is an excellent example of the need to design the database concurrently with the class definitions for the application design. More about this relationship will be explained in Chapter 10's discussion of object-oriented databases.

One additional attribute is the state value attribute. This attribute is added to the class to maintain information about which state or states an object is in. State information is extracted from the statechart and added to the method logic. This variable is the repository of state information.

The method list contains the two methods identified in Figure 9-23, *Make deposit* and *make withdrawal*. We also show the appropriate input arguments that the methods need in order to execute properly. Part of design is identifying and specifying the input argument list as well as the output return type.

The method called ListLargeAccounts, as denoted with an underlined name, is a special kind of method. Remember in object-orientation, a class is a template to create individual objects or instances. Most of the methods apply to each instance of the class. However, frequently there is a need to look through all of the instances at once. That type of method is called a **class method**, which is denoted by an underline. In this example, the ListLargeAccounts method looks through all the instances of this class and returns those having balances larger than $1,000,000.

class method
A method that is associated with a class instead of with each object of the class

We have also included examples for the body of the method or the method procedure. As you look at the method, you will notice that there are statements to update and verify the state of the object, as well as to carry out the required processing of the method. Also notice that the method includes a statement to send a message to the database to update the database.

Design Class Diagram Development

Now that you understand the design class notation, we will briefly demonstrate the design process using the Rocky Mountain Outfitters case. There are several ways that this design could be done. The one we recommend, however, not only produces a correct design, but also helps to check the completeness of the analysis models. This method uses the analysis class diagram, the sequence diagrams, and the statechart diagram.

The first step is to decide on the class to be designed. In this case, let's select the Order class from the RMO class diagram in Figure 5-32. To complete this step, elaborate the attribute list as much as possible with the visibility and type information.

The second step is to find all methods that belong to this class. A method on a class is invoked by a message to that class. To identify all of the methods for this class, we simply look through all of the sequence diagrams and find all of the messages going to the *Order* class. We elaborate the design class with as much information as is available, including the message name, any parameters that are passed, and any return values. Figure 9-29 (a) shows the list of messages found from searching all the sequence diagrams for RMO. Part (b) shows the short form of the *Order design class.*

FIGURE 9-29
The list of messages from sequence diagrams and the abbreviated Order design class.

Messages
CreateNewOrder ( )
AddItemToOrder ( )
Complete ( )
RequestShipment ( )
ShippingOrderItem ( )
CloseShipment ( )
BackOrderItemArrives ( )
ReturnItem ( )
ListReadyOrders ( )
ListOrdersWithBackorder ( )

(a) List of messages to the Order class

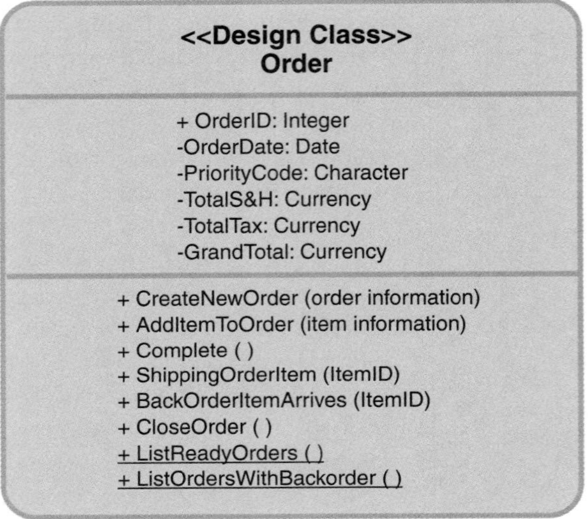

(b) Short form of the design class for Order

The final step is to elaborate the methods with the logic. To do this, we must get the information from the statechart for this class. Remember that a transition in a statechart is fired by a trigger. This trigger represents a message being sent to the object. Therefore, all the messages found from the sequence diagrams need to have a corresponding transition in the object's statechart. If there is no transition, then the statechart has not defined a way to handle the incoming message.

Figure 9-30 presents a variation of the statechart for the Order class. Table 9-1 shows the message-transition pairing. Note that the same names are used for message names and transition names. We recommended a similar type of verification during the analysis activities that developed the interaction diagrams and the statecharts. During analysis, it was only a strong recommendation, but now it must be done to ensure a correct design.

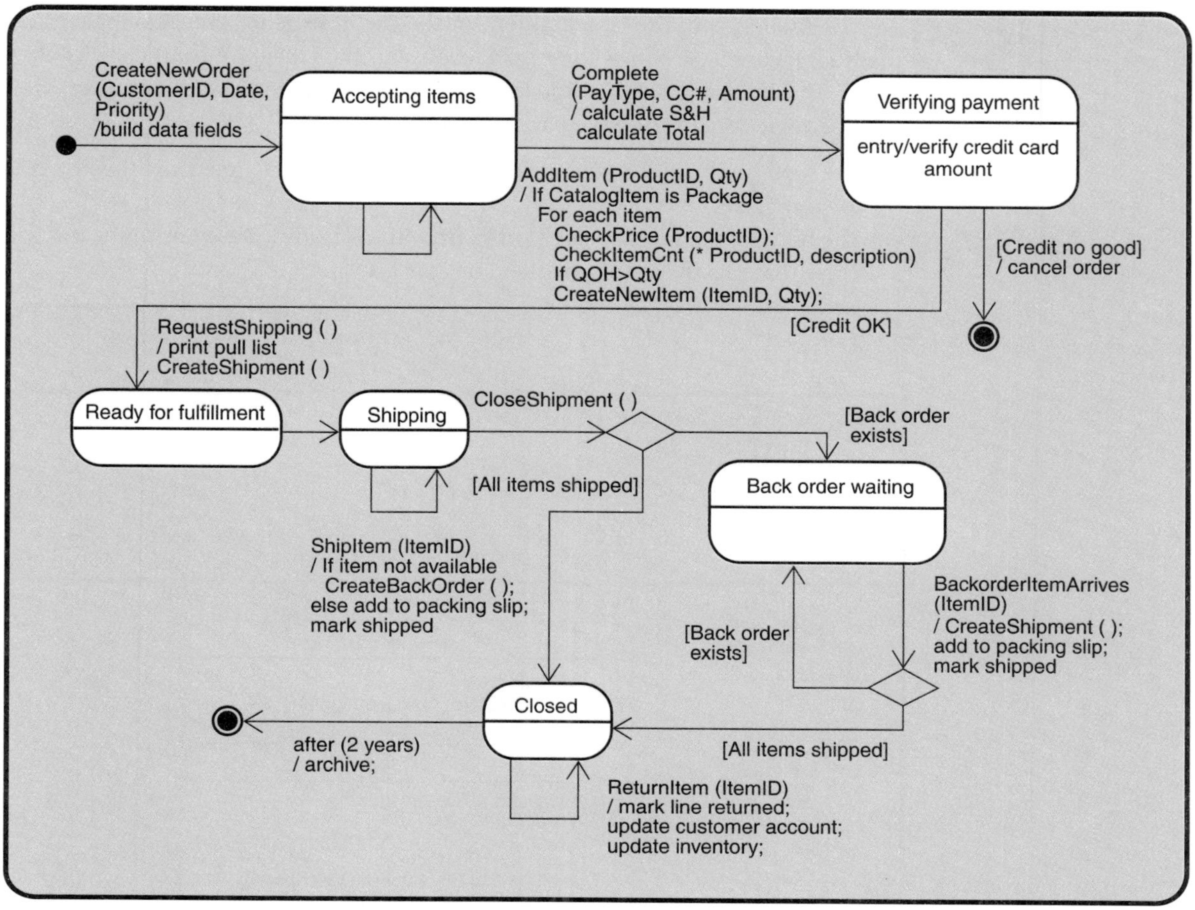

FIGURE 9-30
Statechart for the Order class.

TABLE 9-1
Message name and transition name pairings.

Message name	Transition name
CreateNewOrder ( )	CreateNewOrder ( )
AddItemToOrder ( )	AddItemToOrder ( )
Complete ( )	Complete ( )
RequestShipment ( )	RequestShipment ( )
ShippingOrderItem ( )	ShippingOrderItem ( )
CloseShipment ( )	CloseShipment ( )
BackOrderItemArrives ( )	BackOrderItemArrives ( )
ReturnItem ( )	ReturnItem ( )
ListReadyOrders ( )	
ListOrdersWithBackorder ( )	

The final step in the development of the methods is to write the logic for each method. Figure 9-32 shows the expanded design class diagram for the Order class in RMO. The logic for the method comes directly from the statechart diagram. The method logic consists of three parts:

- Decision logic (an if statement) to test the state variable to ensure the object is in the right state
- The body of the method that comes from the action-expressions in the statechart
- Logic that sets the state variable to the correct value

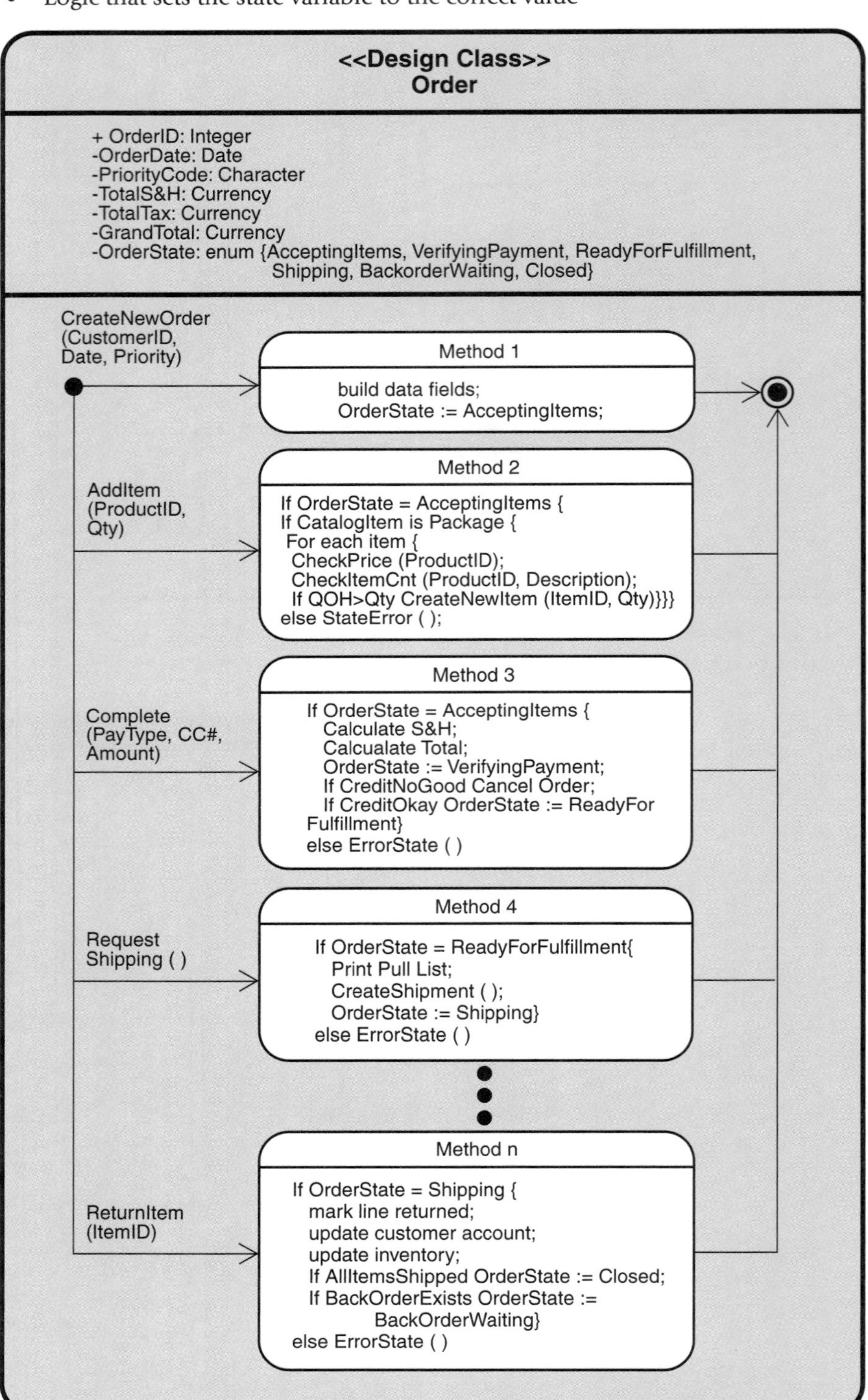

FIGURE 9-32
Expanded design class for RMO's Order class.

Comparing Figure 9-30 and 9-32, you can see how this method logic works. For the first example, look at method 1 in Figure 9-32. Since this is the constructor *CreateNewOrder*, there is no object, and hence no beginning state. The body of the method simply captures the information on the action-expression on the *CreateNewOrder ( )* transition in Figure 9-30. The final statement sets the state variable to be the *AcceptingItems* state.

For the second example, look at method 3, which derives from the *Complete ( )* transition. The initial statement verifies the state variable to be in the *AcceptingItems* state. The next two statements derive from the action-expression on the transition. In accordance with the statechart, the object enters the *VerifyingPayment* state. Therefore, an assignment statement to set the state variable is added. The *VerifyingPayment* state has an internal transition to verify the credit card information. This entry internal transition belongs to the *Complete ( )* transition. No other external message causes it to execute, and it immediately follows the incoming transition. Thus, it is also added to the body of the method. Going out of the state are two completion transitions—that is, they have no external trigger. Since they also automatically fire in the statechart, they also belong to the same transition. Hence the logic to test the guard-conditions on each outgoing transition is included in the body of the method. If the credit is no good, then the method cancels the order. If the customer's credit is okay, then the state variable is set to show that the object moves to the *ReadyforFulfillment* state. Since there are no completion transitions out of this state, that is the end of the method.

The development of method logic is this process of extracting the statements from the statechart. In this example, we have not tried to convert information action-expressions to more formal pseudocode. If the conversion is done by a CASE tool, then no conversion can be done. If the systems analysts are involved, then part of the design activity can be to decipher the semantics of the action-expressions and write them in pseudocode.

Method Development and Pseudocode

As noted previously, a method is invoked by an incoming message. Thus, each transition that is associated with a message identifies a separate method. The detailed logic for the method derives from the action statements associated with the transition and from other action statements following the transition. Part of the design activity is to analyze the statechart to determine the scope of the method being designed. The method scope is the determination of which action statements should be included within the method.

For example, for every transition that identifies a method, there is also a destination state, possibly with more action statements. Past the destination state there may be completion transitions (with an automatic "true" trigger), decision pseudostates, concurrent pseudostates, and constructs other than transitions with message triggers. Generally, all action statements up to the next identifiable message trigger are included within the message. However, the systems analyst must carefully analyze the action statements to ensure that the transformation from statechart to design class is done correctly. More detail on this process can be obtained in the "Further Resources" section.

Chapter 7 identified various forms for the action statements, including informal versus formal structured English statements, dot notation, and internal transitions such as *entry/*, *do/*, and *exit/*. During the development of class methods, the logic of each method is documented using pseudocode. In Figure 9-32 and other examples, the pseudocode is placed on the design class diagram itself. However, it should be easy to think of situations in which the diagram will become much too limiting for the pseudocode. In those instances, use the short form of the design class diagram and write the method pseudocode as supporting documentation in another document. Many CASE tools have facilities to record method or module pseudocode. You will notice in Figure 9-32 that the method pseudocode has the same structure as described earlier in the chapter.

Inheritance, Overriding, and Polymorphism

One of the strengths of OO development is that the object-oriented languages have built-in capabilities that facilitate programming and maintenance. During analysis, we learned that class inheritance represents generalization and specialization. Specialization classes inherit the attributes and relationships of the parent class. In OO programming, inheritance is used to extend both attributes and methods. In Figure 9-27, which specifies the symbols for a design class, the name of the parent class is identified next to the class name. Including the parent name indicates that the class is a specialization class that will inherit both the attributes and the methods of the parent class. For example, Figure 9-23 includes two specialized classes from the *Account* class,—namely, *Checking account* and *Savings account*. The notation for the checking account would be *Checking Account::Account* to indicate that a checking account is a specialization of an account. The account would inherit such attributes as Name and Address, and such methods as *MakeDeposit* and *MakeWithdrawal*. Thus, the code is reused. This inheritance happens automatically, and the checking account class automatically can refer to those methods. A good design practice is to utilize this inheritance capability whenever possible. It is used extensively in the design of the user interface. Where applicable, it is also a good practice for the design of the application classes.

override
To enable a method in a subclass to replace the logic from the method in the parent class

polymorphism
The reuse of methods for different classes that are distinct even though they may have the same name and parameter list

OO programming also provides the ability to **override** the inherited methods. The specialization class typically inherits all of the methods of the parent class. But say the application requires that a savings account be opened with a minimum deposit of $100. Then the *CreateAccount* method for a savings account would need to be a little different than the code for opening other accounts. By identifying (and later programming) a method with the same name in the specialization class, the inherited method is overridden. In other words, it is not used by the subclass.

Polymorphism is the reuse of methods for different classes. The idea of polymorphism stems from fact that every class is self-contained, and we can, therefore, use the same method name in completely distinct classes. For example, from Figure 9-23 let's say we have defined a method on the *Customer* class called *Get Balance*. This method looks at all the accounts for that customer and returns the sum of all savings and checking accounts. We can also define a *Get Balance* method for the *Account* class that would return only the balance in that single account. Polymorphism permits us to use the same name across methods without confusing them. Of course to invoke the method, it must be invoked on the correct class. But that is a normal activity of object-orientation, and is done by sending a message to the desired class.

Integrating the Object-oriented Application Design with User Interface Design, Database Design, and Network Design

As with the structured approach, we have focused primarily on application design without considering how the design classes are affected by the design of the user interface, the database, or the network. We provide a quick overview here to highlight the major concerns.

Since object-oriented design results in a set of independent, interacting classes, the overall structure of the system will not change due to the user interface. What does happen is that a set of user interface classes are designed and added to the overall design class diagram. There are many tools and libraries of components that can be used for object-oriented user interface design. Integration of the application design and the user interface design usually entails selecting a library of tools, designing the forms and reports using the provided components, and then inserting logic within the methods to access the methods of those interface classes. Design of the application classes is best done in conjunction with the design of the user interface so that messages to the correct user interface methods are known during design.

Database access is also usually provided through a set of database classes. Database access can be quite complex since some object-oriented languages automatically read and write to the database, while other languages require specific calls to database interface objects. The important point is that the target language and database system be identified so that appropriate integration is done during design.

Object-oriented applications frequently execute in a distributed environment. Individual objects (or classes) can be assigned to separate nodes. However, appropriate middleware will be necessary to ensure that there are no conflicts between objects, especially with accessing the database.

Coordinating the Project

The initiation of design activities is a pivotal point in the development project. The focus changes from discovery to solution development, and the whole tenor of the project changes. Coordinating all of the ongoing activities is challenging for even the best project managers. There are a myriad of details and tasks that must be coordinated to keep the project on track.

Even though analysis is essentially complete, some tasks will remain that need to be coordinated. Every new system has a multitude of business rules that must be integrated into the system. For example, a set of business rules concerning sales commissions will include when and how commissions are calculated, what happens to commissions on merchandise returns, when commissions are paid, how the commission schedule will vary to encourage sales of high-margin items and sale items, and so forth. All of these business rules must be defined to develop the commission programs properly. However, what if management is still making decisions on these business rules? You do not want to hold up the entire project for a few of these decisions. On the other hand, you want to make sure that these decisions do not fall through the cracks.

In addition, as the project team members, including the users, better understand the potential capabilities of the new system, they may want to modify the business definition to provide higher levels of automation and support. This request is good for the company, which will benefit from a better system. However, the request is bad for the project because it will increase the scope and delay the project. How does the project manager, in conjunction with the steering committee, decide which additional capabilities to include?

Design activities also require substantial coordination. The project begins to fragment based on the number of design issues to be addressed. Frequently, the system is divided into subsystems, and each subsystem has unique design requirements. The project team may be divided into smaller teams to focus on the various subsystems and on other design issues. Some technical issues—such as network configuration, database design, distributed processing needs, and communications capability—will be common to all subsystems. Other issues—such as response time, specialized input equipment, and so forth—may be limited to specific subsystems. Technical issues with the coordination and integration of all subsystems, along with any middleware software, are fundamental to the overall system. The point is that there are frequently many smaller teams of people working to carry out the activities of systems design. The activities and results of these teams must be coordinated to ensure a successful project.

Two other miniprojects may be initiated at this point: a data-conversion project and a test-case development project. We will explain more about the details of these projects in Chapter 15. We mention them here simply to note that they insert additional complexity to the management of the project.

Finally, activities of the implementation phase, such as programming, also begin around this time. In fact, design and programming frequently are conducted concurrently. As design decisions are made, then programming can begin immediately. So, in addition to the various groups working on the design issues, groups of programmers and programmer/analysts will probably be added to the project team.

The complexity of control and coordination of these various activities can be exacerbated by the fact that people involved in the project may work at different locations. Open communication, which is essential to successful coordination, becomes exponentially more complicated as more people are added to the team.

Given these complexities, let's discuss several project management tools and techniques to help in the coordination of the project.

Coordinating Project Teams

The fundamental tool in coordinating the activities of the various project teams is the project schedule. As the activities of the design phase begin, the project manager must update the project schedule by identifying and estimating all of the tasks associated with design and implementation, as well as the outstanding tasks associated with ongoing definition. The project schedule usually requires major rework to ensure that the project is organized during these remaining phases.

During the analysis phase, project management is often done by the project manager and an assistant. When the project expands and several teams are formed, the management of the project becomes larger and more complicated. Frequently, a committee, composed of the leaders of the key design and implementation teams, assumes more responsibility for carrying out the coordination and control aspects of project management. Weekly, and sometimes daily, status meetings are held. If this group includes people at remote locations, teleconferencing support may be required.

Coordinating Information

As you noticed from the previous design models, design begins to generate a tremendous amount of detail. Modules, classes, data fields, data structures, forms, reports, methods, subroutines, tables, and so forth are being defined in substantial detail. It becomes a tremendous coordination problem to keep track of all of these pieces of information. Two kinds of tools can help in this process.

The most common and widespread technique to record and track project information is with a CASE tool. Most CASE tools have a central repository to capture information. In Chapter 3 you learned about CASE tools. Figure 3-18 illustrated the various component tools that make up a comprehensive CASE system. A major element in a CASE tool system is the central repository of information. The central repository not only records all design information but is normally configured so that all teams can view project information to facilitate communication among the teams of a project. Figure 9-33 illustrates the various information components that may exist within a CASE data repository.

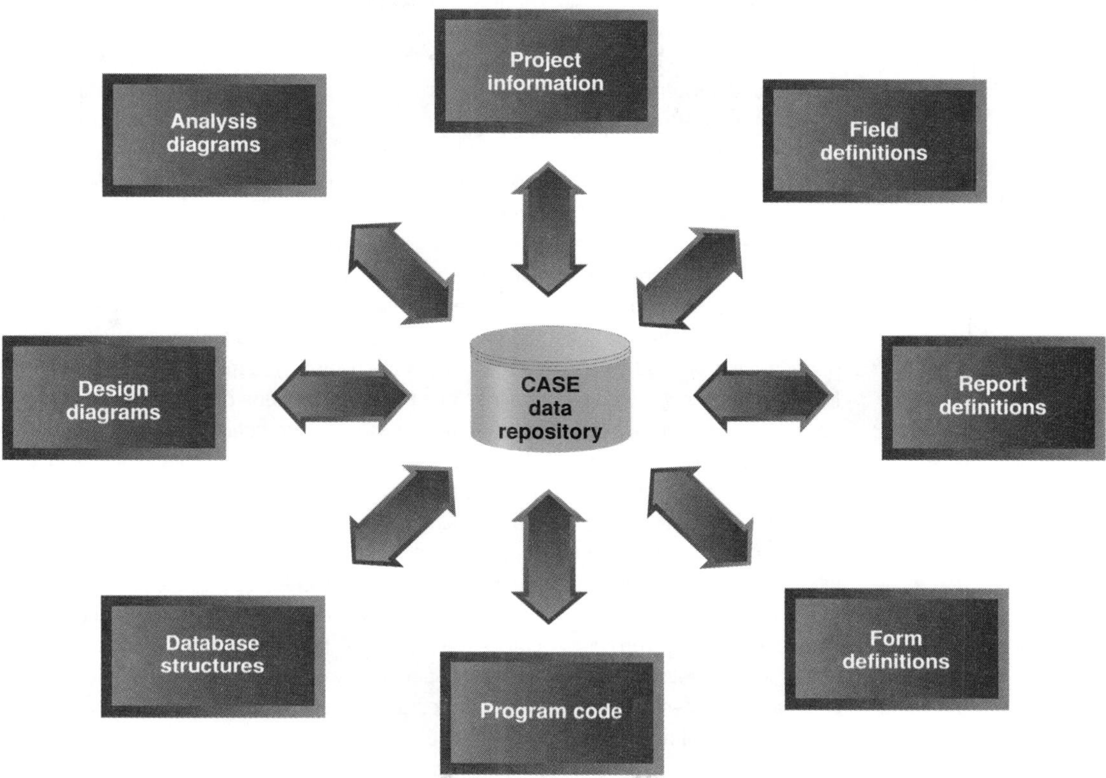

FIGURE 9-33
System development information stored in the CASE repository.

Other electronic tools are also available to help with team communication and information coordination. These tools and techniques, often referred to as computer support for collaborative work, not only record final design information but assist in collaborative work. Often during the development process, several people need to work together in the development of the design. Thus they need to discuss and dynamically update the working documents or diagrams. One system that is frequently used is Lotus Notes. Other systems allow figures and diagrams to be updated with tracking and version information so that there is a history of the evolution of the result.

One difficult part of development projects is keeping track of open items and unresolved issues. We add this here, not because it requires a new technique, but because the problem is pervasive and all good project managers develop techniques to track these items. One simple way is to have an "open items control log." Appendix A, which discusses project management concepts, describes the need for a control log. Figure A-1 is an example of a log. It is a sequential list of all open items with information to track responsibilities and resolution of the open item, as shown by the column headings. Frequently, closed items are shaded for easy identification.

Summary

Systems design is the process of organizing and structuring the components of a system to allow the construction (that is, programming) of the new system. Various components need to be designed during systems design, including the user interface, the application programs, the database, and the network. This chapter focused on application design.

The inputs to the design activities are the diagrams, or models, that were built during analysis. Ideally, the analysis models represent all of the requirements for the new system, and user involvement during design decreases dramatically. In reality, the design activities always raise questions about the requirements that necessitate additional detail and review.

The outputs of the design are also a set of diagrams, or models, that describe the architecture of the new system and the detailed logic within the various programming components. The inputs, design activities, and outputs are different depending on whether a structured approach or an object-oriented approach is used.

For the structured approach to systems design, the primary input is the data flow diagram. The data flow diagram is first enhanced by the addition of a system boundary. The system boundary is sketched on a diagram 0 to show the overall system. It is also sketched on the DFD fragments to show program boundaries at a lower level. On the data flow diagrams, system boundaries are converted to structure charts using either transaction analysis or transform analysis. Transaction analysis is used for higher-level DFDs and captures the information about the various transactions that are processed by the system. Transform analysis is used to develop a structure chart that is based on transforming a single transaction from its input form to an output.

Object-oriented design uses three analysis diagrams as its input: the class diagram, the collaboration diagrams, and the statechart diagrams. The class diagram is enhanced so that it becomes a design class diagram that describes the variables in each class and the methods of each class. The design process takes the messages from the collaboration diagram and makes them into the methods. The statechart is used to develop the logic internal to each method.

Finally, the tenor of the project changes at this point. During design, technical staff are involved. Sometimes programming is begun, which requires the addition of programming staff. Other side projects for data conversion and test-data development are also begun at this time.

Key Terms

afferent data flow, p. 319
architectural design, p. 302
central transform, p. 319
class method, p. 335
computer program, p. 305
conceptual design, p. 298
data couples, p. 313
design class diagram, p. 331
detail design, p. 302
efferent data flow, p. 319
general design, p. 298
information hiding, p. 328
method signature, p. 333
module, p. 305
module cohesion, p. 321
module coupling, p. 321
overloaded method, p. 333
override, p. 340
package diagram, p. 329
polymorphism, p. 340
program call, p. 313
pseudocode, p. 305
structure chart, p. 311
system flow chart, p. 305
transaction analysis, p. 315
transform analysis, p. 315

Review Questions

1. What is the primary objective of systems design?
2. What is the difference between analysis and design?
3. Explain the relationship and difference between a module and a program.
4. What is the purpose of the automation system boundary? How do you develop one?
5. What is a system flow chart used for? What symbols are used on a system flow chart?
6. What is the purpose of a structure chart? What are the symbols used on a structure chart?
7. Explain *transaction analysis*.
8. Explain *transform analysis*. What is meant by the term *central transform*?

9. Explain module coupling and module cohesion. Why are these concepts important?
10. What is the purpose of a package diagram?
11. What is meant by a method signature? What is it made up of and why is it important?
12. What is the difference between the analysis class diagram and the design class diagram?
13. Explain the concept of inheritance.
14. Explain the concept of polymorphism. How is it different from overloading?
15. What is the purpose of the problem control log? How is it used?

Thinking Critically

Note: Problems 1 through 5 contain data flow diagrams. Normally DFDs always have data flow names on the diagram, but since the focus of the problem is on the structure not the flow, the names have not been included.

1. Given the data flow diagram shown in Figure 9-34, do the following: (a) draw a system boundary, (b) divide the DFD into program components such as real-time, monthly, daily, periodic, and so forth, (c) draw a system flow chart based on the division into program components.

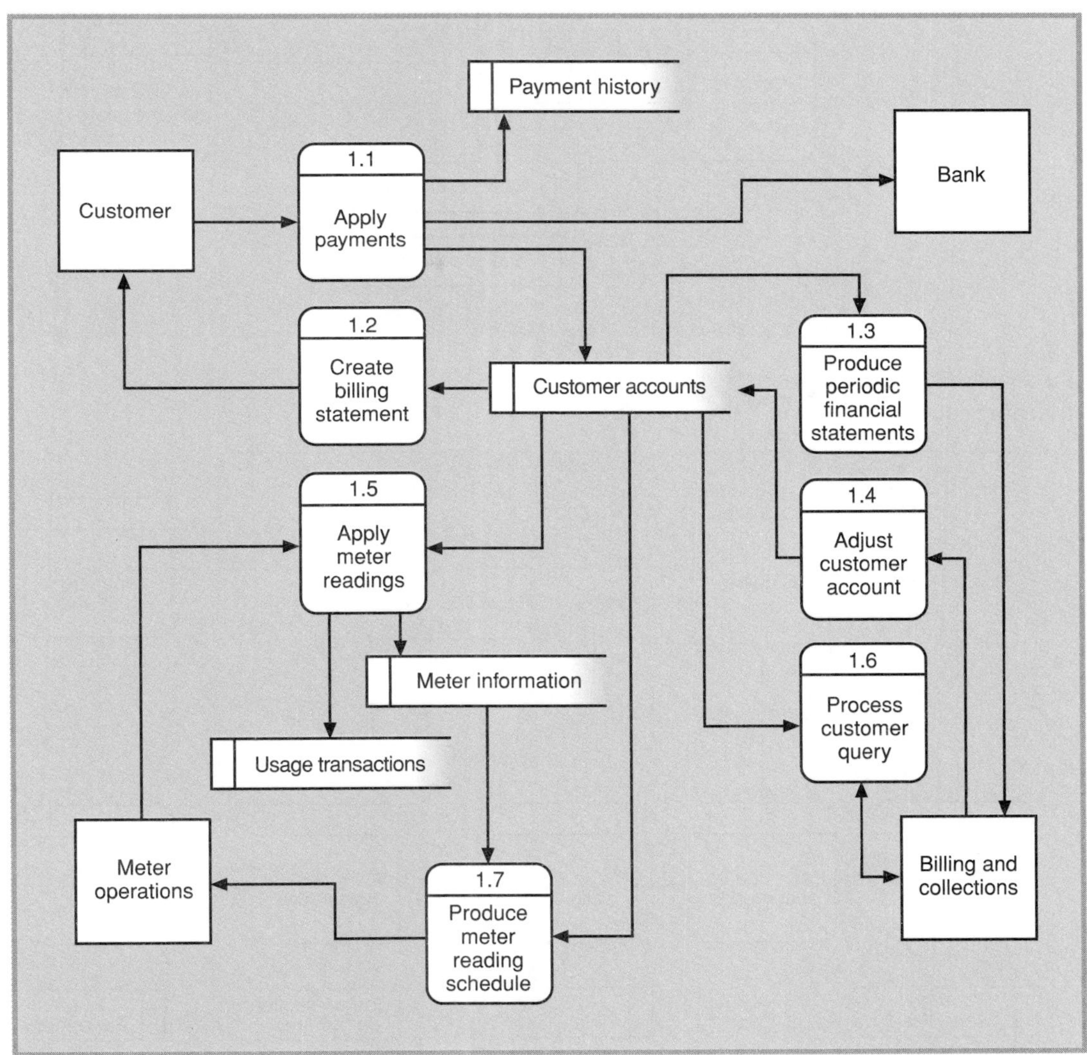

FIGURE 9-34
Electric company customer billing.

2. Given the data flow diagram shown in Figure 9-35, and using transaction analysis, develop a structure chart.

FIGURE 9-35
Student registration program.

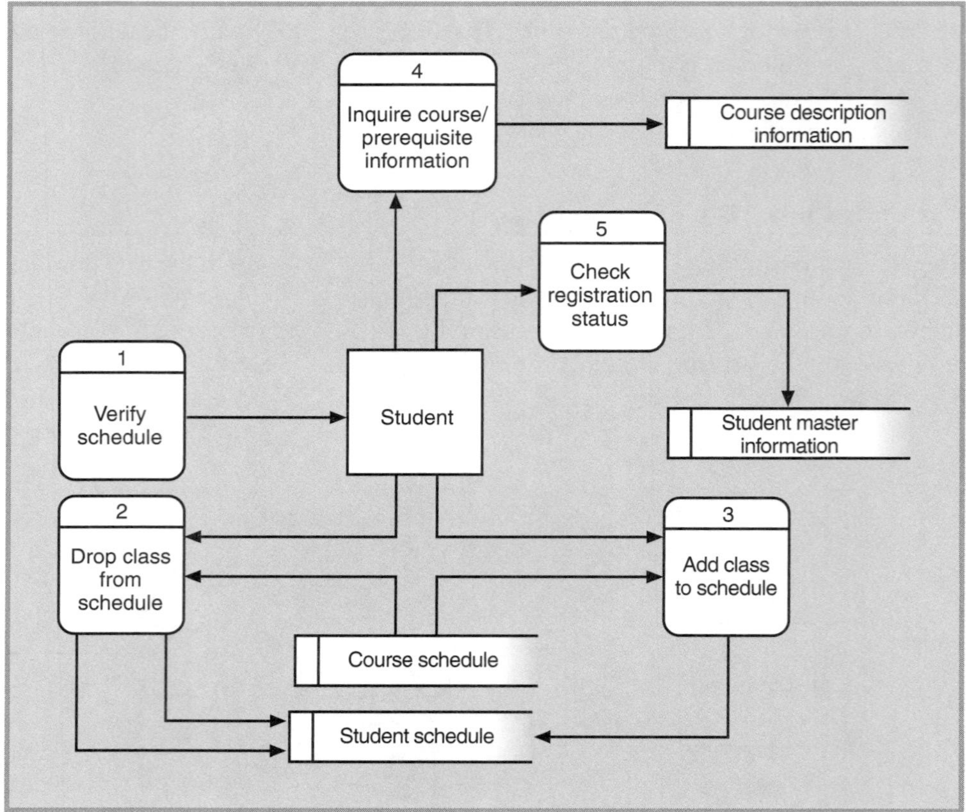

3. Given the data flow diagram shown in Figure 9-36, and using transform analysis, develop a structure chart.

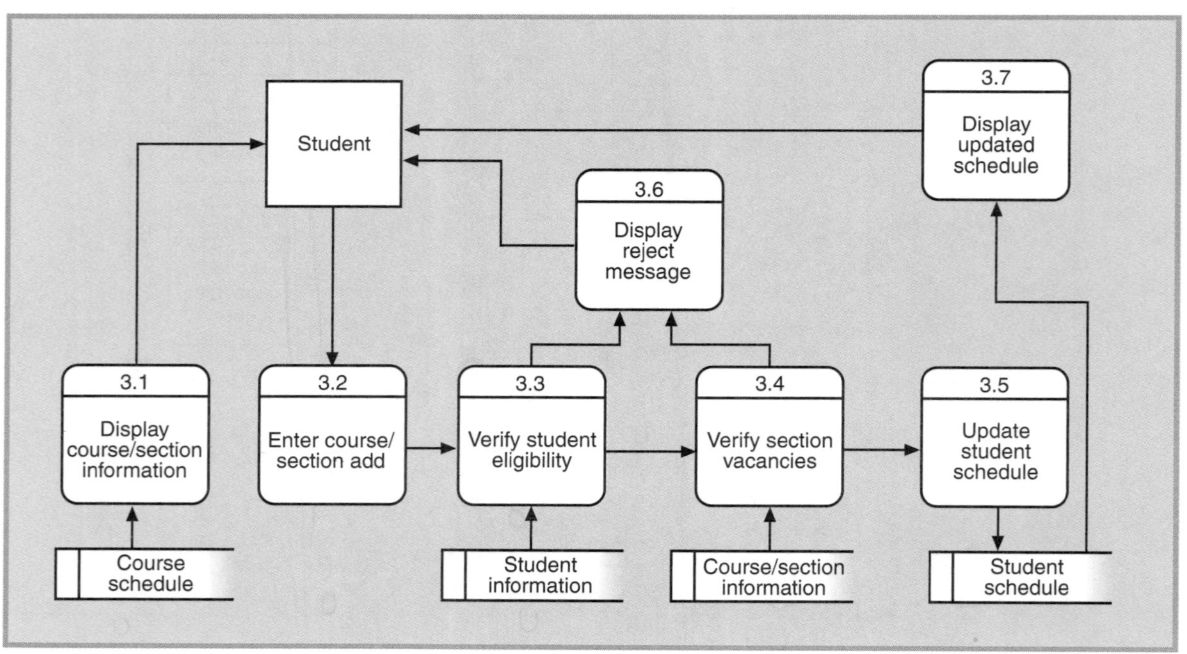

FIGURE 9-36
Expolosion of Add classes to schedule.

4. Integrate the structure charts from problems 2 and 3 into a single structure chart.

5. Given the data flow diagram shown in Figure 9-37, and using transform analysis, develop a structure chart.

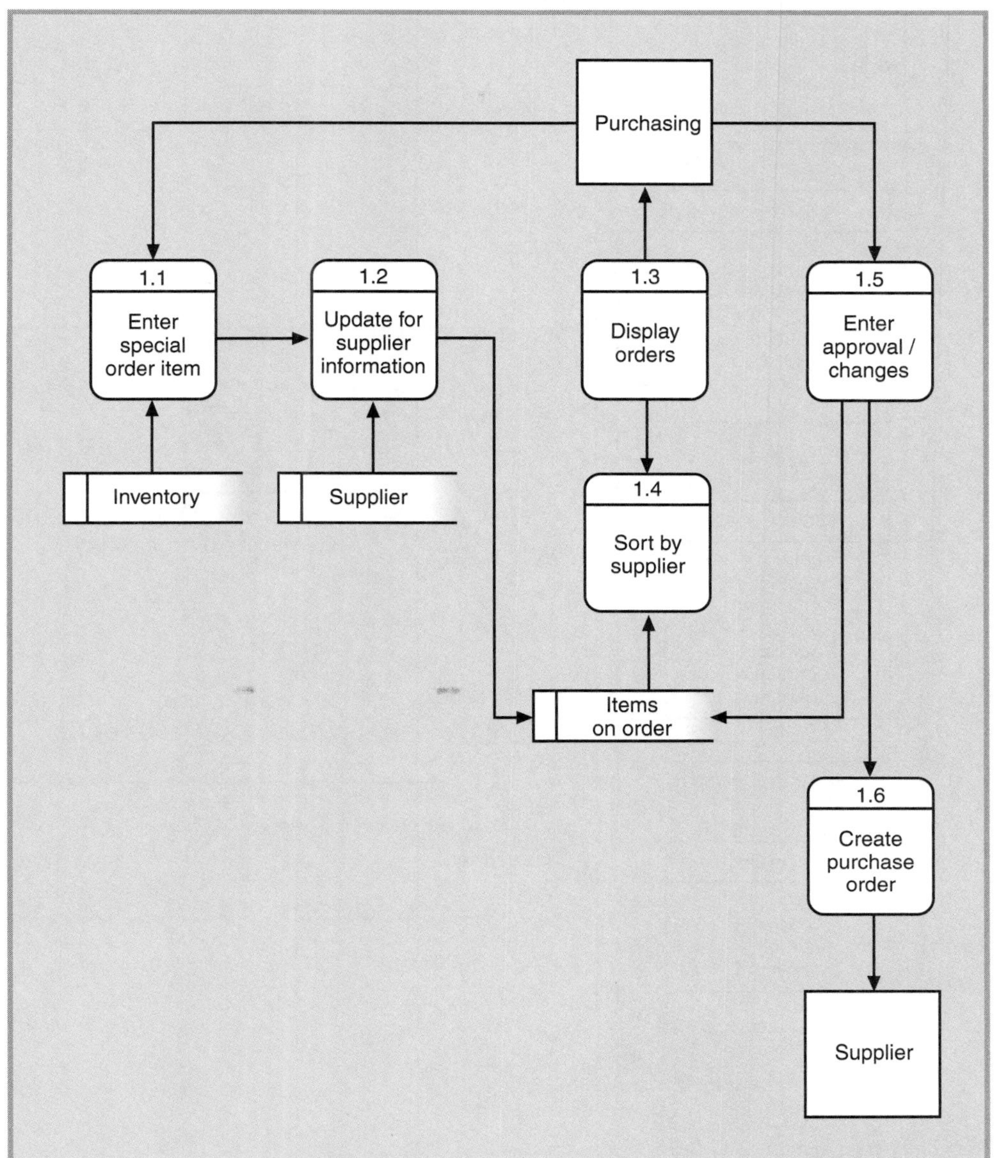

FIGURE 9-37
Special order purchasing.

6. Given the class and collaboration diagrams in Figure 9-38, add the methods to the *Book Copy* class including accurate method signatures.

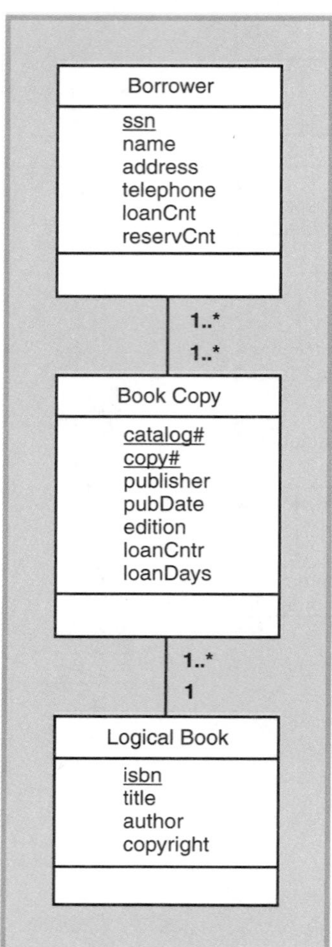

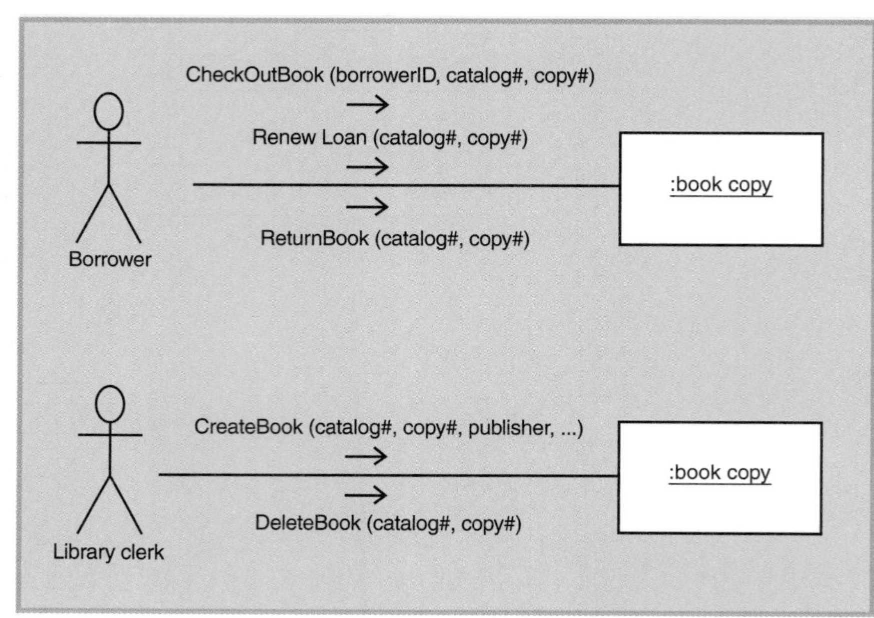

FIGURE 9-38
The library class diagram and Book Copy collaborations.

7. Given the statechart diagram in Figure 9-39, add the logic to each of the methods identified in problem 6.

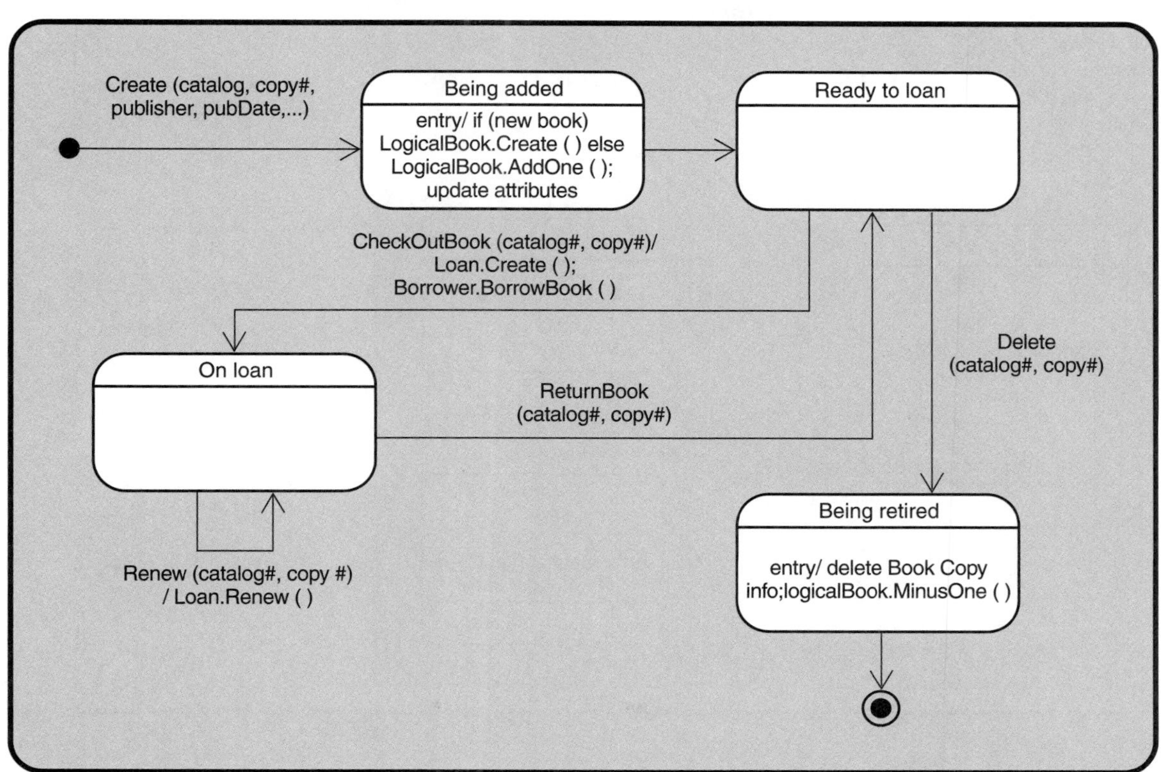

FIGURE 9-39
Statechart for Book Copy.

8. Given the class and statechart diagrams in Figure 9-40, add state variables, method names, and method logic to the class to develop the design class.

FIGURE 9-40
Human Resources class diagram and Employee statechart.

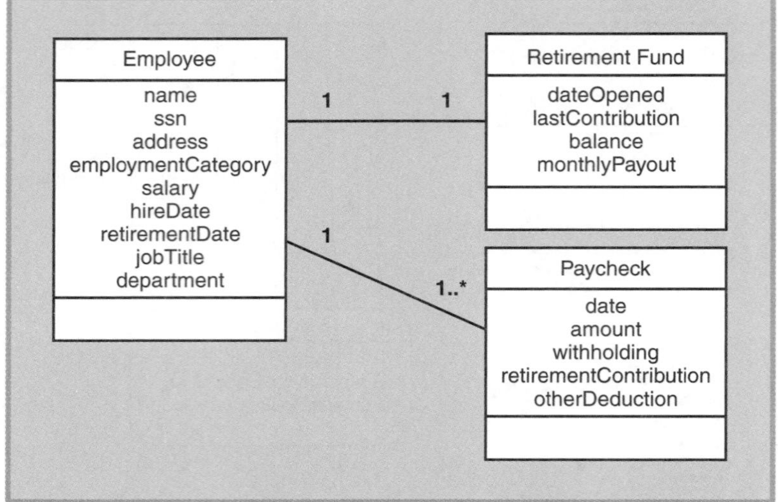

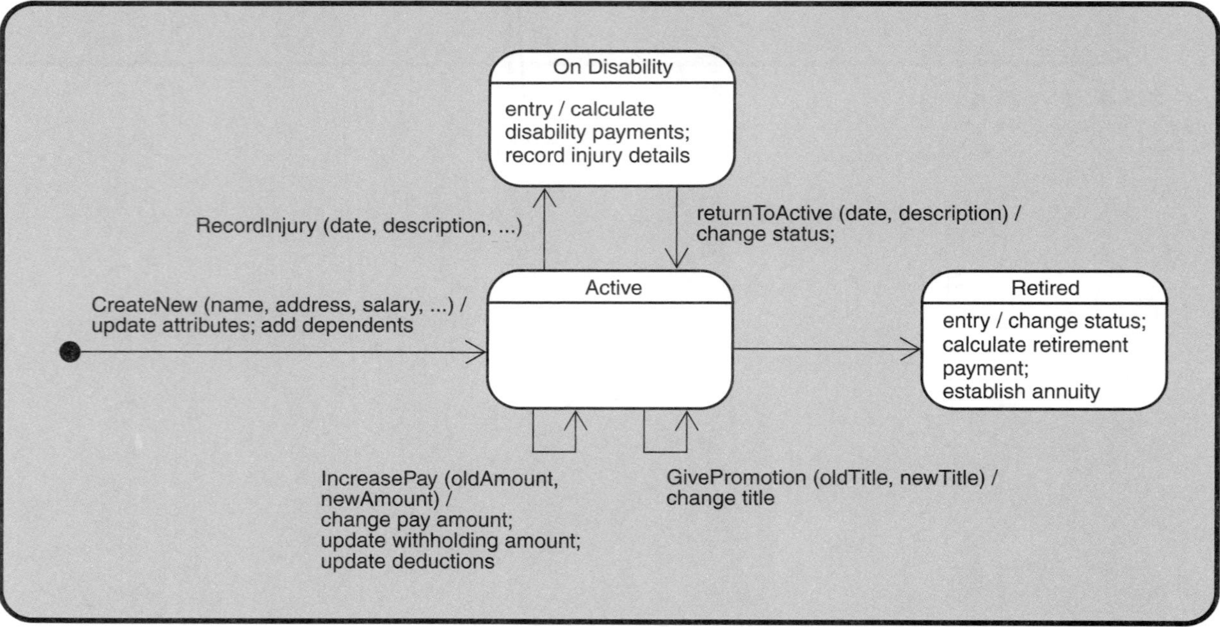

9. Compare and contrast the two designs for RMO—the structured and the object-oriented. Point out strengths and weaknesses of each. Also select your preference and justify your selection.

Experiential Exercises

1. Discuss the basic differences between structured design and object-oriented design. What kinds of systems are naturally more inclined to a hierarchy and which kinds of systems are more closely related to interacting objects?
2. Find an example of a business system written in COBOL or BASIC that has a hierarchical structure and was probably developed using traditional methods. Search the Internet for COBOL, Microfocus COBOL, or Visual Basic.
3. Find an example of a business system that is written in Java or C++ and is a group of interacting objects and consequently was developed using object-oriented techniques. You might search the web for Java, Visual J++, C++, or companies doing development.
4. Find a local company that is doing development using structured techniques. Set up an interview with an IS staff member. Gather as much information as you can about the company's systems. Review the company's techniques and SDLC methodology.
5. Find a local company that is using UML. Set up an interview with an IS staff member. Gather as much information as you can about the company's systems. Discuss the company's techniques and SDLC methodology. Also review how the company trained its systems personnel.

Case Studies

The Real Estate Multiple Listing Service System (Structured)

Refer to the description of the Real Estate Multiple Listing Service system in the case studies of Chapter 5, and the DFDs you developed in the case studies for Chapter 6. Develop a structure chart for the system. Follow the steps indicated in this chapter, including any additional modules required for accessing data.

The Reliable Pharmaceutical Service System

Based on the description of the Reliable Pharmaceutical Service system, Table 6-2, Figure 6-40, and the DFDs you developed for Chapter 6, develop a structure chart for the system. Be sure to include modules for input, output, and data access.

The Real Estate Multiple Listing Service System (Object-Oriented)

Refer to the description of the Real Estate Multiple Listing Service system in the case studies of Chapter 5, and the object-oriented diagrams you developed for Chapter 7. Develop a design class diagram for the system. You may use the notation illustrated in the chapter, or you may simply write the method logic as a text document.

The DownTown Video Rental System

Using the description and the object-oriented modules you developed in Chapter 7 for the DownTown Video Rental system, develop a design class diagram for the system. You may use the design class notation illustrated in the chapter, or you may simply write the method logic as a text document.

Further Resources

Grady Booch, James Rumbaugh, and Ivar Jacobson. *The Unified Modeling Language User Guide*. Addison-Wesley, 1999.

Tom DeMarco. *Structured Analysis and System Specification.* Yourdon Press, 1979.

Ivar Jacobson, Grady Booch, and James Rumbaugh. *The Unified Software Development Process.* Addison-Wesley, 1999.

Edward Yourdon. *Modern Structured Analysis.* Yourdon Press, 1989.

Designing Databases

Learning Objectives

After reading this chapter, you should be able to:

- Describe the differences and similarities between relational and object-oriented database management systems
- Design a relational database schema based on an entity-relationship diagram
- Design an object database schema based on a class diagram
- Design a relational schema to implement a hybrid object-relational database
- Describe the different architectural models for distributed databases

Chapter Outline

Databases and Database Management Systems

Relational Databases

Object-oriented Databases

Hybrid Object-Relational Database Design

Data Types

Distributed Databases

TEN

NATIONWIDE BOOKS: DESIGNING A NEW DATABASE

The project leaders for Nationwide Book's (NB) new web-based ordering system were meeting with NB's database administrator to discuss how they were going to tackle the database design. Present at the meeting were Sharon Thomas (who had led the project since its inception), Vince Pirelli (a contractor who had done most of the analysis), and Bill Anderson (NB's database administrator, who hadn't directly participated in earlier phases of the project). Sharon started the meeting by saying, "When the project began, we planned to use the existing Oracle database. Maria Peña [the chief information officer] also wanted us to use this project to try out newer development methods and tools that we knew we'd need in the coming years. So we hired Vince to do the systems analysis using object-oriented [OO] methods. We also purchased a Java development environment and sent two of our programmers to a three-week training course."

Vince added, "I developed a traditional entity-relationship diagram as a basis for designing the interface between the new programs and the existing database. I checked the documentation for the current database schema. Most of the information needed by the new system is already there, though we'd need to add some new fields and we might have to change some table definitions. But now that we're looking at design and implementation, I'm concerned that we may be handicapping our new system with an outdated database management system."

Sharon added, "We decided to use this project as a pilot for OO development and implementation to speed up development. But I'm not sure that we'll actually achieve that if we use the existing Oracle database."

Bill said, "I understand your concerns about interfacing OO programs with a relational database. It sounds like trying to mix oil and water, but it's really not that difficult. We just hired Anna Jorgensen, a database developer who has experience writing C++ programs that interface with relational databases. I had her look over the class diagrams and use cases for the new system, and she assures me that the interfaces to Oracle will be straightforward and simple. I can assign her to your project for a few months if you need the help."

Vince responded, "There's no question that an interface can be written, but it may not be as easy as Anna thinks. There are also problems with basic elements of an OO program such as inheritance and class methods. Those things simply can't be represented in a relational database. That'll force us to make some ugly compromises when designing and implementing our OO code. I'm afraid that it'll lengthen our design and implementation phases and make future system upgrades much more difficult."

"I've done some research and there are quite a few commercial OO data packages available," said Sharon. "None of them have the track record of Oracle, but then the technology is fairly new. An OO database management system would be the best fit with Java, and it would open the door to other new technologies such as components."

Vince added, "It would also shorten the time we need to complete design since an OO database is directly based on the class diagram."

Sharon continued, "I think that this project presents a good opportunity for us to make the leap to the next generation of database software."

Bill was taken aback by the suggestion, but quickly replied, "Are you asking me to support two redundant databases based on two completely different database technologies? Management is already breathing down my neck about my budget. I've managed to save some money by switching to cheaper server hardware, but it's only a marginal improvement. Supporting another database management system will be a major effort. And we'd need new hardware to isolate the new DBMS from our existing database. I can't risk having a buggy new piece of software crash our operational databases. I'd also need to train my people to bring them up to speed on the new software. And how will we get data back and forth between the two databases?"

Sharon let the air clear for a bit for replying. "There's no question that it'll be a major undertaking. I'm not trying to downplay that fact or stretch your people and equipment to the breaking point. But there comes a time when we need to move on to newer technologies, and I think that time has arrived. I've already discussed the idea with Maria, and she thinks it deserves serious consideration."

Vince added, "I can do the database design either way. But we won't be building a base for the future if we use a relational database. I could design it to interface with indexed files on an old IBM mainframe if I had to. But why would we want to go backward instead of forward?"

OVERVIEW

Databases and database management systems are important components of a modern information system. Databases provide a common repository for data so that it can be shared among many organizational units and locations. Database management systems provide designers, programmers, and end users with sophisticated capabilities to store, retrieve, and manage data. Sharing and managing the vast amounts of data needed by a modern organization would simply not be possible without a database management system.

In Chapter 5, you learned to construct conceptual data models. You also learned to develop entity-relationship diagrams (ERDs) for traditional analysis and class diagrams for object-oriented (OO) analysis. To implement an information system, a project team must transform a conceptual data model into a more detailed database model, and implement that model within a database management system.

The process of developing a database model depends on the type of conceptual model and the type of data management software that will be used to implement the system. This chapter will describe the design of relational and OO data models and their implementation using database management systems. We will use examples from Rocky Mountain Outfitters to show how information collected during analysis is used in database design.

Databases and Database Management Systems

database (DB)
An integrated collection of stored data that is centrally managed and controlled

A **database (DB)** is an integrated collection of stored data that is centrally managed and controlled. A database typically stores information about dozens or hundreds of entity types or classes. The information stored includes entity or class attributes (for example, names, prices, and account balances) as well as relationships among the entities or classes (for example, which orders belong to which customers). A database also stores descriptive information about data such as field names, restrictions on allowed values, and access controls to sensitive data items.

database management system (DBMS)
System software that manages and controls access to a database

A database is managed and controlled by a **database management system (DBMS)**. A DBMS is a system software component that is generally purchased and installed separately from other system software components (for example, operating systems). Examples of modern database management systems include Microsoft Access, Oracle, DB/2, ObjectStor, and Gemstone.

Figure 10-1 illustrates the components of a typical database and its interaction with a DBMS, application programs, users, and administrators. The database consists of two related stores of information: the physical data store and the schema. The **physical data store** contains the raw bits and bytes of data that are created and used by the information system. The **schema** contains additional information about the data stored in the physical data store, including the following:

physical data store
The storage area used by a database management system to store the raw bits and bytes of a database

schema
A description of the structure, content, and access controls of a physical data store or database

- Access and content controls, including allowable values for specific data elements, value dependencies among multiple data elements, and lists of users allowed to read or update data element contents
- Relationships among data elements and groups of data elements (for example, a pointer from data describing a customer to orders made by that customer)
- Details of physical data store organization including type and length of data elements, the locations of data elements, indexing of key data elements, and sorting of related groups of data elements

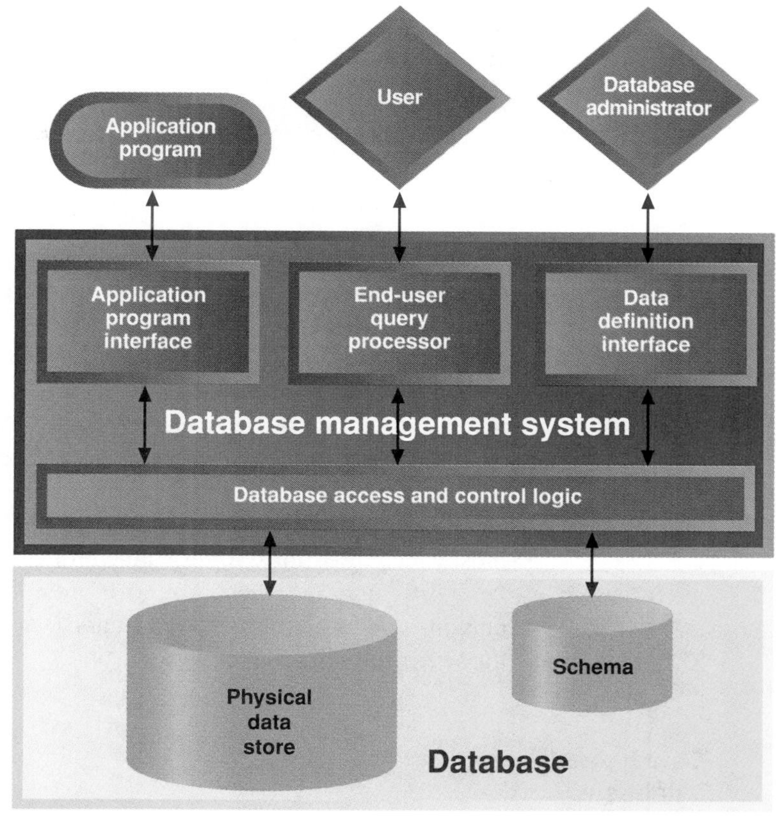

FIGURE 10-1
The components of a database and database management system and their interaction with application programs, users, and database administrators.

A DBMS has four key components: an application program interface (API), a query interface, an administrative interface, and an underlying set of data access programs and subroutines. Application programs, users, and administrators never access the physical data store directly. Instead, they tell an appropriate DBMS interface what data they need to read or write, using names defined in the schema. The DBMS accesses the schema to verify that the requested data exist and that the requesting user has appropriate access privileges. If the request is valid, then the DBMS extracts information about the physical organization of the requested data from the schema and uses that information to access the physical data store on behalf of the requesting program or user.

Databases and database management systems provide several important data access and management capabilities, including the following:

- Allowing simultaneous access by many users and/or application programs
- Allowing access to data without writing application programs (that is, via a query language)
- Managing all of the data used by an information system or organization as an integrated whole (that is, applying uniform and consistent access and content controls)

For these and other reasons, databases and DBMSs are widely used in modern information systems.

Database Models

DBMSs have evolved through a number of stages since their introduction in the 1960s. These stages have been characterized by a number of technology changes. The most significant change has been the type of model used to represent and access the content of the physical data store. Four such model types have been widely used:

- Hierarchical
- Network
- Relational
- Object-oriented

The hierarchical model was developed in the 1960s. It represented data using sets of records organized into a hierarchy. The network model also grouped data elements into sets of records but allowed those records to be organized into more flexible network structures. Few new databases have been developed with the hierarchical model since the early 1970s, and few new databases have been developed with the network model since the early 1980s. However, some older hierarchical and network databases are still in use today, especially in large-scale batch transaction processing applications.

The remainder of this chapter discusses in detail design issues for the relational and object-oriented database models. DBMSs based on these models are currently the most widely used for both existing and newly developed systems. Design issues for the hierarchical and network models are not described, since few students of information systems are likely to encounter DBMSs based on these models by the time they enter the work force.

Relational Databases

relational database management system (RDBMS)
A database management system that stores data in tables

table
A two-dimensional data structure containing rows and columns; also called a **relation**

row
The portion of a table containing data that describe one entity, relationship, or object; also called a **tuple** or **record**

field
A column of a relational database table; also called an **attribute**

The relational database model was first developed in the early 1970s, although it took nearly two decades to become the dominant database model. Relational databases were slow to be adopted because of the difficulties inherent in converting systems implemented with hierarchical and network DBMSs and because of the amount of computing resources required to successfully implement them. As with other theoretical advances in computer science, it took many years for the cost-performance characteristics of data storage and processing hardware to catch up to the new theory. Relational DBMSs now account for the vast majority of DBMSs currently in use.

A **relational database management system (RDBMS)** is a DBMS that organizes stored data into structures called **tables**, or **relations**. Relational database tables are similar to conventional tables—that is, they are two-dimensional data structures of columns and rows. However, relational database terminology is somewhat different from conventional table and file terminology. A single row of a table is called a **row**, **tuple**, or **record**, and a column of a table is called a **field**, or **attribute**. A single cell in a table is called a **field value, attribute value**, or **data element**.

Figure 10-2 shows the content of a table as displayed by the Microsoft Access relational DBMS. Note that the first row of the table contains a list of field names (column headings) and that the remaining rows contain a collection of field values that each describe a specific product. Note also that each row contains the same fields in the same order.

FIGURE 10-2
A partial display of a relational database table.

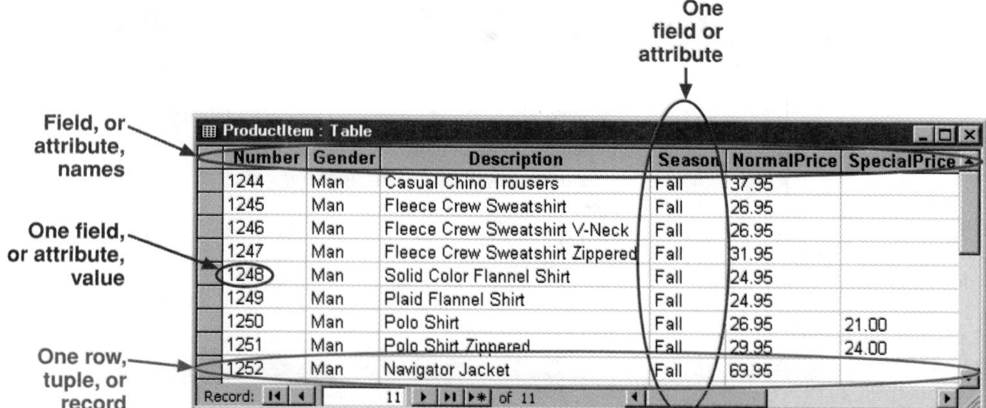

Number	Gender	Description	Season	NormalPrice	SpecialPrice
1244	Man	Casual Chino Trousers	Fall	37.95	
1245	Man	Fleece Crew Sweatshirt	Fall	26.95	
1246	Man	Fleece Crew Sweatshirt V-Neck	Fall	26.95	
1247	Man	Fleece Crew Sweatshirt Zippered	Fall	31.95	
1248	Man	Solid Color Flannel Shirt	Fall	24.95	
1249	Man	Plaid Flannel Shirt	Fall	24.95	
1250	Man	Polo Shirt	Fall	26.95	21.00
1251	Man	Polo Shirt Zippered	Fall	29.95	24.00
1252	Man	Navigator Jacket	Fall	69.95	

field value
The data value stored in a single cell of a relational database table;also called an **attribute value** or **data element**

key
A field that contains a value that is unique within each row of a relational database table

primary key
A key used to uniquely identify a row of a relational database table

Each table in a relational database must have a unique **key**. A key is a field or set of fields, the values of which occur only once in all the rows of the table. If only one field (or set of fields) is unique, then that key is also called the table's **primary key**. If there are multiple unique fields (or sets of fields), then one of the possible keys must be chosen as the primary key.

Key fields may be natural or invented. An example of a natural key field in chemistry is the atomic weight of an element in a table containing descriptive data about elements. Unfortunately, in business, few natural key fields are useful for information processing. Most key fields in a relational database are invented. Your wallet or purse probably contains many examples of invented keys, including your social security number, driver's license number, credit card numbers, and ATM card number. Some invented keys are externally assigned (for example, a Federal Express tracking number) and some are internally assigned (for example, Number in Figure 10-2). Invented keys are guaranteed to be unique because unique values are assigned by a user, application program, or the DBMS as new rows are added to the table.

Keys are a critical element of relational database design because they are the basis for representing relationships among tables. Keys are the "glue" that binds rows of one table to rows of another table—in other words, keys relate tables to each other. For example, consider the ERD fragment from the Rocky Mountain Outfitters example shown in Figure 10-3 and the tables shown in Figure 10-4. The ERD fragment shows an optional one-to-many relationship between the entities ProductItem and InventoryItem. The upper table in Figure 10-4 contains data representing the entity type ProductItem. The lower table contains data representing the entity type InventoryItem.

FIGURE 10-3
A portion of the RMO entity-relationship diagram.

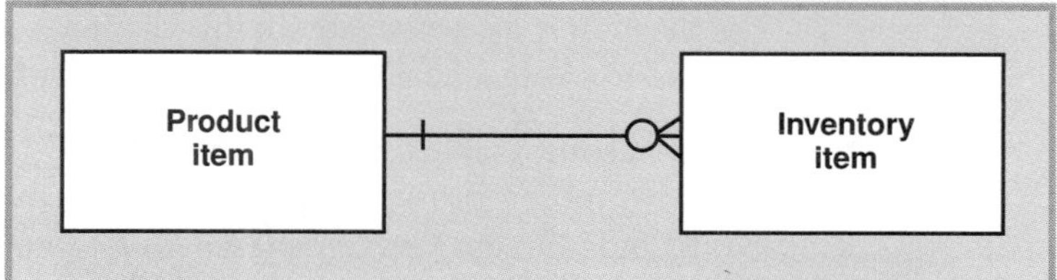

Number	Gender	Description	Season	NormalPrice	SpecialPrice
1244	Man	Casual Chino Trousers	Fall	37.95	
1245	Man	Fleece Crew Sweatshirt	Fall	26.95	
1246	Man	Fleece Crew Sweatshirt V-Neck	Fall	26.95	
1247	Man	Fleece Crew Sweatshirt Zippered	Fall	31.95	
1248	Man	Solid Color Flannel Shirt	Fall	24.95	
1249	Man	Plaid Flannel Shirt	Fall	24.95	
1250	Man	Polo Shirt	Fall	26.95	21.00
1251	Man	Polo Shirt Zippered	Fall	29.95	24.00
1252	Man	Navigator Jacket	Fall	69.95	

Number	ProductItemNu	Size	Color	Options	QuantityOnHand	Average Cost
86779	1244	30/30	Khaki		45	$12.75
86780	1244	30/30	Slate		10	$12.75
86781	1244	30/30	LightTan		17	$12.75
86782	1244	30/31	Khaki		22	$12.75
86783	1244	30/31	Slate		6	$12.75
86784	1244	30/31	LightTan		31	$12.75
86785	1244	30/32	Khaki		120	$12.75
86786	1244	30/32	Slate		28	$12.75
86787	1244	30/32	LightTan		21	$12.75

FIGURE 10-4
A relationship between data in two tables; the foreign key ProductItemNumber in the InventoryItem table refers to the primary key Number in the ProductItem table.

foreign key
A field value stored in one relational database table that also exists as a primary key value in another relational database table

The relationship between the entity types ProductItem and Inventory Item is represented by a common field value within their respective tables. The field Number (the primary key of the Product table) is also present within the InventoryItem table as the field ProductItemNumber. The field ProductItemNumber is called a **foreign key**. A foreign key is a field that duplicates the primary key of a different (or foreign) table. In Figure 10-4, the existence of the value 1244 as a foreign key within the InventoryItem table indicates that the values of Description, Season, NormalPrice, and SpecialPrice in the first row of the ProductItem table also describe inventory items 86779 through 86787.

Designing Relational Databases

Relational database design begins with either an ERD or a class diagram. This section explains how to create a schema based on an ERD. Schema creation based on a class diagram is discussed later in this chapter.

To create a relational database schema from an ERD, follow these steps:

1. Create a table for each entity type.
2. Choose a primary key for each table (invent one, if necessary).
3. Add foreign keys to represent one-to-many relationships.
4. Create new tables to represent many-to-many relationships.
5. Define referential integrity constraints.
6. Evaluate schema quality and make necessary improvements.
7. Choose appropriate data types and value restrictions (if necessary) for each field.

The following subsections discuss each of these steps in detail.

Representing Entities

The first step to creating a relational DB schema is to create a table for each entity on the ERD. Figure 10-5 shows the ERD for the RMO customer support system. Eleven entities are represented, and a table is created for each entity. The data fields of each table will be the same as those defined for the corresponding entity on the ERD. To avoid confusion, table and field names should match the names used on the ERD and/or in the project data dictionary. Initial table definitions for the RMO case are shown in Figure 10-6.

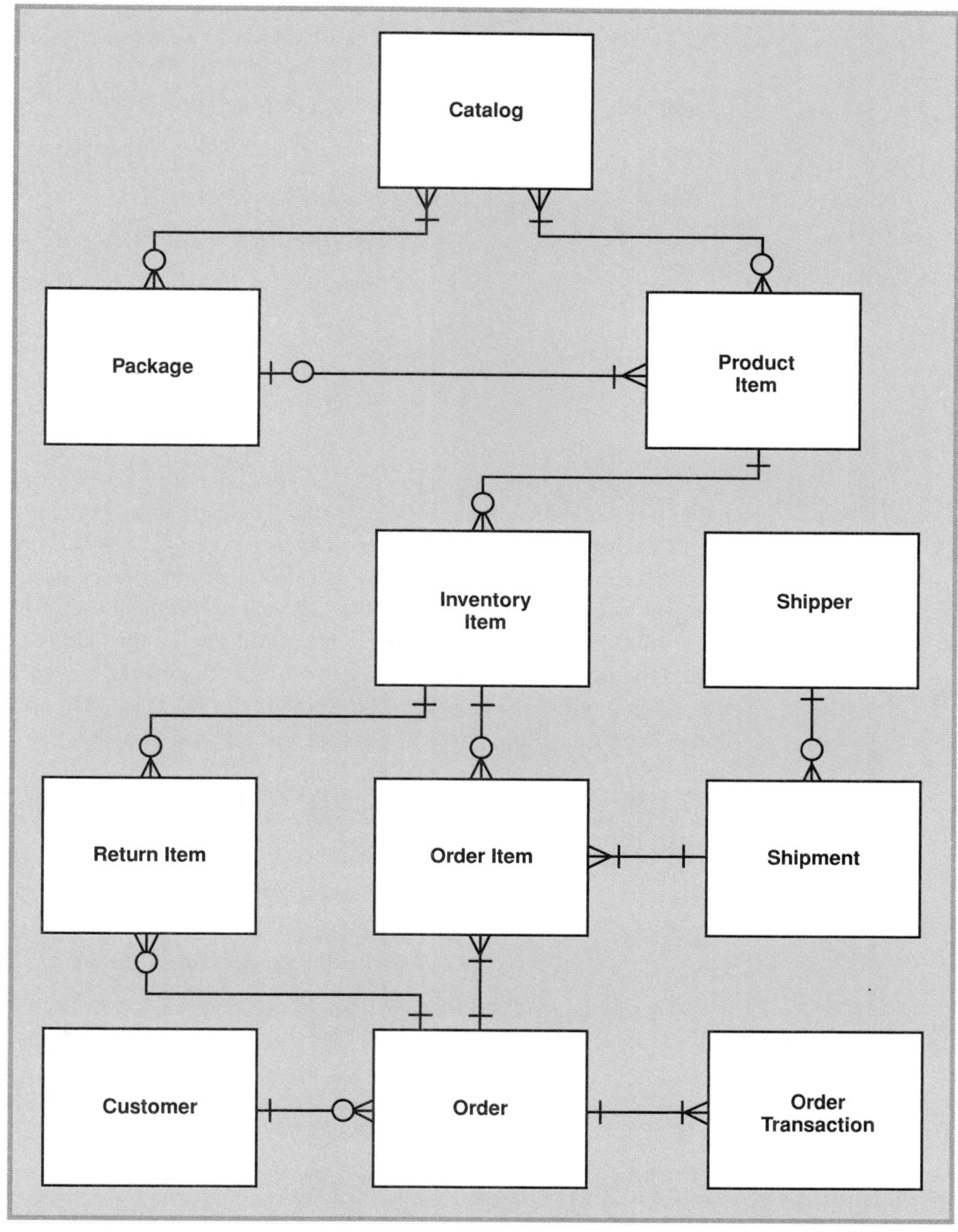

FIGURE 10-5
The RMO entity-relationship diagram.

FIGURE 10-6
An initial set of tables representing the entities on the ERD.

Table	Attributes
Catalog	Season, Year, Description, EffectiveDate, EndDate
Customer	Name, BillingAddress, ShippingAddress, DayTelephoneNumber, NightTelephoneNumber
InventoryItem	Size, Color, Options, QuantityOnHand, AverageCost, ReorderQuantity
Order	Date, PriorityCode, ShippingAndHandling, Tax, GrandTotal, DateReceived, ProcessorClerk, PhoneClerk, CallStartTime, LengthOfCall, EmailAddress, ReplyMethod
OrderItem	Quantity, Price, BackOrderStatus
OrderTransaction	Date, TransactionType, Amount, PaymentMethod
Package	Description, SalePrice
ProductItem	Vendor, Gender, Description, Season, NormalPrice, SpecialPrice
ReturnItem	Quantity, Price, Reason, Condition, Disposal
Shipment	TrackingNumber, DateSent, TimeSent, ShippingCost, DateArrived, TimeArrived
Shipper	Name, Address, ContactName, Telephone

After creating tables for each entity, the designer must select a primary key for each table. If a table already has a field or set of fields that are guaranteed to be unique, then the designer can choose that field or set of fields as the primary key. If the designer cannot choose a primary key from among the existing fields, then the designer must invent a new key field. Any name can be chosen for an invented key field, but the name should convey some indication that the field is a unique key field. Typical names include "Code," "Number," and "Identifier," possibly combined with the table name (for example, "ProductCode" and "OrderNumber"). Figure 10-7 shows the entity tables and identifies the primary key of each.

Table	Attributes
Catalog	**Number**, Season, Year, Description, EffectiveDate, EndDate
Customer	**AccountNumber**, Name, BillingAddress, ShippingAddress, DayTelephoneNumber, NightTelephoneNumber
InventoryItem	**Number**, Size, Color, Options, QuantityOnHand, AverageCost, ReorderQuantity
Order	**Number**, Date, PriorityCode, ShippingAndHandling, Tax, GrandTotal, DateReceived, ProcessorClerk, PhoneClerk, CallStartTime, LengthOfCall, EmailAddress, ReplyMethod
OrderItem	**Number**, Quantity, Price, BackOrderStatus
OrderTransaction	**Number**, Date, TransactionType, Amount, PaymentMethod
Package	**Number**, Description, SalePrice
ProductItem	**Number**, Vendor, Gender, Description, Season, NormalPrice, SpecialPrice

(*continued*)

FIGURE 10-7
Entity tables with the primary keys identified in bold.

Table	Attributes
ReturnItem	**Number**, Quantity, Price, Reason, Condition, Disposal
Shipment	**TrackingNumber**, DateSent, TimeSent, ShippingCost, DateArrived, TimeArrived
Shipper	**Number**, Name, Address, ContactName, Telephone

Representing Relationships

Relationships are represented within a relational database by foreign keys. The process of determining which foreign keys should be placed in which tables depends on the type of relationship being represented. The RMO ERD in Figure 10-5 contains three many-to-many relationships (at the top of the diagram) and nine one-to-many relationships. The rules for representing each relationship type are as follows:

- One-to-many relationship—add the primary key field(s) of the "one" entity type to the table that represents the "many" entity type.
- Many-to-many relationship—create a new table that contains the primary key field(s) of both related entity types.

Figure 10-8 shows the results of representing the nine one-to-many relationships within the tables from Figure 10-7. Each foreign key represents a single relationship between the table containing the foreign key and the table that uses that same key as its primary key. For example, the field AccountNumber was added to the Order table as a foreign key representing the one-to-many relationship between the entities Customer and Order. The foreign key ShipperNumber was added to the Shipment table to represent the one-to-many relationship between Shipper and Shipment. Note that when representing one-to-many relationships, foreign keys do not become part of the primary key of the table to which they are added.

Table	Attributes
Catalog	**Number**, Season, Year, Description, EffectiveDate, EndDate
Customer	**AccountNumber**, Name, BillingAddress, ShippingAddress, DayTelephoneNumber, NightTelephoneNumber
InventoryItem	**Number**, *ProductItemNumber*, Size, Color, Options, QuantityOnHand, AverageCost, ReorderQuantity
Order	**Number**, *AccountNumber*, Date, PriorityCode, ShippingAndHandling, Tax, GrandTotal, DateReceived, ProcessorClerk, PhoneClerk, CallStartTime, LengthOfCall,
OrderItem	**Number**, *OrderNumber, InventoryItemNumber, ShipmentNumber*, Quantity, Price, BackOrderStatus
OrderTransaction	**Number**, *OrderNumber, Date*, TransactionType, Amount, PaymentMethod
Package	**Number**, *CatalogNumber*, Description, SalePrice
ProductItem	**Number**, *PackageNumber*, Vendor, Gender, Description, Season, NormalPrice, SpecialPrice

(continued)

FIGURE 10-8
Represent one-to-many relationships by adding foreign key attributes (shown in italics).

Table	Attributes
ReturnItem	**Number**, *OrderNumber, InventoryItemNumber*, Quantity, Price, Reason, Condition, Disposal
Shipment	**TrackingNumber**, *ShipperNumber*, DateSent, TimeSent, ShippingCost, DateArrived, TimeArrived
Shipper	**Number**, Name, Address, ContactName, Telephone

Figure 10-9 expands the table definitions in Figure 10-8 by adding three new tables to represent the many-to-many relationships between Catalog and Package, Catalog and ProductItem, and Package and ProductItem. For example, the table CatalogProduct contains the foreign keys CatalogNumber and ProductNumber to represent the many-to-many relationship between Catalog and ProductItem. The primary key of the table is the combination of the foreign key fields. CatalogNumber and ProductNumber are shown in boldface italic to indicate that they are both foreign keys and parts of the primary key. Price is the only non-key field in the CatalogProduct table. Two more tables (CatalogPackage and PackageProduct) were also added to represent the other two many-to-many relationships in Figure 10-5.

FIGURE 10-9
The tables CatalogPackage, CatalogProduct, and PackageProduct are added to represent many-to-many relationships.

Table	Attributes
Catalog	**Number**, Season, Year, Description, EffectiveDate, EndDate
CatalogPackage	***CatalogNumber***, ***PackageNumber***
CatalogProduct	***CatalogNumber***, ***ProductNumber***, Price
Customer	**AccountNumber**, Name, BillingAddress, ShippingAddress, DayTelephoneNumber, NightTelephoneNumber
InventoryItem	**Number**, *ProductItemNumber*, Size, Color, Options, QuantityOnHand, AverageCost
Order	**Number**, *AccountNumber*, Date, PriorityCode, ShippingAndHandling, Tax, GrandTotal
OrderItem	**Number**, *OrderNumber, InventoryItemNumber, ShipmentNumber*, Quantity, Price, BackOrderStatus
OrderTransaction	**Number**, *OrderNumber*, Date, TransactionType, Amount, PaymentMethod
Package	**Number**, *CatalogNumber*, Description, SalePrice
PackageProduct	***PackageNumber***, ***ProductNumber***
ProductItem	**Number**, *PackageNumber*, Vendor, Gender, Description, Season, NormalPrice, SpecialPrice
ReturnItem	**Number**, *OrderNumber, InventoryItemNumber*, Quantity, Price, Reason, Condition, Disposal
Shipment	**TrackingNumber**, *ShipperNumber*, DateSent, TimeSent, ShippingCost, DateArrived, TimeArrived
Shipper	**Number**, Name, Address, ContactName, Telephone

Enforcing Referential Integrity

referential integrity
A consistent relational database state in which every foreign key value also exists as a primary key value

Now that we've described how foreign keys are used to represent relationships we need to describe how to enforce restrictions on the values of those foreign key fields. The term **referential integrity** describes a consistent state among foreign key and primary key values. Each foreign key is a reference to the primary key of another table. In most cases, a database designer wants to ensure that these references are consistent. That is, foreign key values that appear in one table must also appear as the primary key value of the related table. Referential integrity represents constraints in database content—for example, "an order must be from a customer" and "an order item must be something that we normally stock in inventory."

The DBMS usually enforces referential integrity automatically once the schema designer identifies primary and foreign keys. Automatic enforcement is implemented as follows:

- When a row containing a foreign key value is created, the DBMS ensures that the value also exists as a primary key value in the related table.
- When a row is deleted, the DBMS ensures that no foreign keys in related tables have the same value as the primary key of the deleted row.
- When a primary key value is changed, the DBMS ensures that no foreign key values in related tables contain the same value.

In the first case, the DBMS will simply reject the addition of any row containing an unknown foreign key value. In the latter two cases, a database designer usually has some control over how referential integrity is enforced. When a row containing a primary key is deleted, the DBMS can be instructed to delete all rows in other tables with corresponding keys. Or, the DBMS can be instructed to set all corresponding foreign keys to NULL. A similar choice is available when a primary key value is changed. The DBMS can be instructed to change all corresponding foreign key values to the same value or to set foreign key values to NULL.

Evaluating Schema Quality

After creating a complete set of tables, the designer should check the entire schema for quality. Ironing out any schema problems at this point ensures that none of the later design effort will be wasted. A high-quality data model has the following features:

- Uniqueness of table rows and primary keys
- Lack of redundant data
- Ease of implementing future data model changes

Unfortunately, there are few objective or quantitative measures of database schema quality. Database design is the final step in a modeling process, and as such, it depends on the analyst's experience and judgment. Various formal and informal techniques for schema evaluation are described in the following sections. No one technique is sufficient by itself, but the combination of these techniques can provide assurance that a database design is of high quality.

Row and Key Uniqueness A fundamental requirement of all relational data models is that primary keys and table rows be unique. Since each table must have a primary key, uniqueness of rows within a table is guaranteed if the primary key is unique. Data access logic within programs is usually dependent on the assumption that keys are unique. For example, a programmer writing a program to view customer records will generally assume that a database query for a specific customer number will return one and only one row (or none if the customer doesn't exist in the database). The program will be designed around this assumption and will probably fail if the DBMS returns two records.

A designer evaluates primary key uniqueness by examining assumptions about key content, the set of possible key values, and the methods by which key values are assigned. Internally invented keys are relatively simple to evaluate in this regard because the system itself creates them. That is, an information system that uses invented keys can guarantee uniqueness by implementing appropriate procedures to assign key values to newly created rows.

It is common for several different programs in an information system to be capable of creating new database rows. Thus, each such program needs to be able to assign keys to newly created database rows. However, the importance of key uniqueness requires that those procedures be consistently applied throughout the information system.

Because key issuance and management are such pervasive problems in information systems, many relational DBMSs automate key issuing services. Such systems typically automate a special data type for invented keys (for example, the AutoNumber type in Microsoft Access). The DBMS automatically assigns a key value to newly created rows, and communicates that value to the application program for use in subsequent database operations. Embedding this capability in the DBMS frees the IS developer from designing and implementing customized key assignment software modules.

Invented keys that aren't assigned by the information system must be given careful scrutiny to ascertain their uniqueness and the usefulness over time. For example, social security numbers are commonly used as keys in employee databases in the United States. Since the U.S. government has a strong interest in guaranteeing their uniqueness, the assumption that they will always be unique seems safe. But other assumptions concerning their use deserve closer examination. For example, will all employees who are stored in the database have a social security number? What if the company opens a manufacturing facility in Europe or South America?

Invented keys assigned by nongovernmental agencies deserve even more careful scrutiny. For example, Federal Express, UPS, and most shipping companies assign a tracking number to each shipment they process. Tracking numbers are guaranteed to be unique at any given point in time, but are they guaranteed to be unique forever (that is, are they ever reused)? Could reuse of a tracking number cause a primary key duplication in the RMO database? And what would happen if two different shippers assigned the same tracking number to two different shipments?

Uncertainties such as these make internally invented keys the safest long-term strategy in most cases. Although internally invented keys may initially entail additional design and development, they prevent one possible source of upheaval once the database is installed. Few changes have the pervasive and disruptive impact of a database key change in a large information system with terabytes of stored data and thousands of application programs and stored queries.

Data Model Flexibility Database flexibility and maintainability were primary goals in the original specification of the relational database model. A database model is considered flexible and maintainable if changes to the database schema can be made with minimal disruption to existing data content and structure. For example, adding a new entity to the schema should not require redefining existing tables. Adding a new one-to-many relationship should require only the addition of a new foreign key to an existing table. Adding a new many-to-many relationship should require that only a single new table be added to the schema.

Data redundancy plays a key role in determining the flexibility and maintainability of any database or data model. A truism of database processing is that "redundant storage requires redundant maintenance." That is, if data are stored in multiple places, then each of those places must be found and manipulated when data are added, changed, or deleted. Of course, performing any of those actions on multiple data storage locations is more complex (and less efficient) than performing them on a single location. Failure to update, modify, or delete multiple copies of the same information creates a condition called inconsistency. By definition, inconsistency cannot occur if information is stored only once.

The relational data model deliberately stores key fields multiple times (that is, redundantly) and non-key fields only once. Keys are stored once as a primary key and again each time they are used as a foreign key. Redundant key values are required because correspondence between the primary and foreign key is the only way to represent relationships among tables. But using redundant key values adds complexity to processes that manipulate key fields.

Relational DBMSs ensure consistency among primary and foreign keys by enforcing referential integrity constraints. But there are no automatic mechanisms for ensuring consistency among other redundant data items. Thus, the best way to avoid inconsistency in a relational database is to avoid redundancy in non-key fields. Database designers can avoid such redundancy by never introducing it into a schema—but it is all too easy to let redundancy slip in. If data redundancy is somehow introduced into the schema, then a process is required to identify the redundancy and remove it. The most commonly used process to detect and eliminate redundancy is database normalization.

normalization
A process that ensures relational database schema quality by minimizing data redundancy

Database Normalization **Normalization** is a formal technique used to evaluate the quality of a relational database schema. It determines whether a database schema contains any of the "wrong" kinds of redundancy and defines specific methods to eliminate that redundancy. Normalization is based on a concept called functional dependency and on a series of normal forms:

first normal form (1NF)
A relational database table structure in which there are no repeating fields or groups of fields

functional dependency
A one-to-one correspondence between two field values

second normal form (2NF)
A relational database table structure in which every non-key field is functionally dependent on the primary key

third normal form (3NF)
A relational database table structure in which no non-key field is functionally dependent on any other non-key field(s)

- **First normal form (1NF).** A table is in first normal form if it contains no repeating fields or groups of fields.
- **Functional dependency.** A functional dependency is a one-to-one relationship between the values of two fields. The relationship is formally stated as follows: *Field A is functionally dependent on field B if for each value of B there is only one corresponding value of A.*
- **Second normal form (2NF).** A table is in second normal form if it is in first normal form and if each non-key element is functionally dependent on the entire primary key.
- **Third normal form (3NF).** A table is in third normal form if it is in second normal form and if no non-key element is functionally dependent on any other non-key element.

Let's explain these concepts further.

First normal form defines a structural constraint on table rows. Repeating fields such as Dependent in Figure 10-10 are not allowed within any table in a relational database. Repeating groups of fields are also prohibited. In practice, this is not a difficult constraint to enforce since relational DBMSs do not allow a designer to define a table containing repeating fields.

SSN	Name	Department	Salary	Dependent1	Dependent2	Dependent3 ...	DependentN
111-22-3333	Mary Smith	Accounting	40,000	John	Alice	Dave	
222-33-4444	Jose Pena	Marketing	50,000				
333-44-5555	Frank Collins	Production	35,000	Jane	Julia		

FIGURE 10-10
An employee table with a repeating field.

Functional dependency is a difficult concept to describe and apply. The most precise way to determine whether functional dependency exists is to pick two fields in a table and insert their names in the italicized portion of the definition shown previously. For example, consider the fields Number and NormalPrice in the ProductItem table (see Figure 10-4). Recall that Number is an internally invented primary key that is guaranteed to be unique within the table. To determine whether NormalPrice is functionally dependent on Number, substitute NormalPrice for A and Number for B in the italicized portion of the functional dependency definition:

> "NormalPrice is functionally dependent on Number if for each value of Number there is only one corresponding value of NormalPrice."

Now ask yourself whether the statement is true for all rows that could possibly exist in the ProductItem table. If the statement is true, then NormalPrice is functionally dependent on Number. As long as the invented key Number is guaranteed to be unique within the ProductItem table, then the preceding statement is true. Therefore, NormalPrice is functionally dependent on Number.

A less formal way to analyze functional dependency of NormalPrice on Number is to remember that the ProductItem table represents a specific product sold by RMO. If that product can have only a single NormalPrice in the database, then NormalPrice is functionally dependent on the key of the table that represents products (Number). If it is possible for any product to have multiple normal prices, then the field NormalPrice is not functionally dependent on Number.

To evaluate whether the ProductItem table is in second normal form, we must first determine whether it is in first normal form. Since it contains no repeating fields, it is in first normal form. Then we must determine whether every non-key field is functionally dependent on Number (that is, consider each field in turn by substituting it for A in the functional dependency definition). If all the non-key fields are functionally dependent on Number, the ProductItem table is in 2NF. If one or more non-key fields are not functionally dependent on Number, then the table is not in 2NF.

Verifying that a table is in 2NF is more complicated when the primary key consists of two or more fields. For example, consider the RMO table CatalogProduct shown in Figure 10-11. Recall that this table represents a many-to-many relationship between Catalog and ProductItem. Thus, the table representing this relationship has a primary key consisting of the primary keys of Catalog (CatalogNumber) and ProductItem (ProductNumber). The table also contains one non-key field called Price.

FIGURE 10-11
The RMO CatalogProduct table.

CatalogProduct : Table

CatalogNumber	ProductNumber	Price
23	1244	$15.00
23	1245	$15.00
23	1246	$15.00
23	1247	$15.00
23	1248	$14.00
23	1249	$14.00
23	1252	$21.00
23	1253	$21.00
23	1254	$24.00
23	1257	$19.00

Record: 1 of 10

If this table is in 2NF, then the non-key field Price must be functionally dependent on the *combination* of CatalogNumber and ProductNumber. We can verify this functional dependency by substituting terms in the functional dependency definition and determining the truth or falsity of the resultant statement:

Price is functionally dependent on the combination of CatalogNumber and ProductNumber if for each combination of values for CatalogNumber and ProductNumber there is only one corresponding value of Price.

Analyzing the truth of the preceding statement is a tricky matter, since you must consider all the possible combinations of key values that might occur in the CatalogProduct file. A simpler way to approach the question is to think about the underlying entities represented in the table. A product can appear in many different catalogs. If price can be different in different catalogs, then the preceding statement is true. If a product's price is always the same, regardless of the catalog in which it appears, then the preceding statement is false and the table is not in 2NF. The correct answer doesn't depend on any universal sense of truth. Instead, it depends on RMO's normal conventions for setting product prices in different catalogs.

If a non-key field is functionally dependent on only part of the primary key, then the non-key field must be removed from its present table and placed in another table. For example, consider another version of the CatalogProduct table as shown in the upper half of Figure 10-12. The non-key field CatalogIssueDate is functionally dependent only on CatalogNumber, not on the combination of CatalogNumber and ProductNumber. Thus, the table is not in 2NF. To correct the problem, you must remove CatalogIssueDate from the CatalogProduct table and place it in a table that uses CatalogNumber alone as the primary key. Since the Catalog table in Figure 10-9 uses CatalogNumber alone as its primary key, CatalogIssueDate should be added to that table. If a Catalog table did not already exist, then you would need to create one to hold CatalogIssueDate, as shown in Figure 10-12.

FIGURE 10-12
Decomposition of a first normal form table into two second normal form tables.

CatalogProduct1NF : Table

CatalogNumber	ProductNumber	Price	CatalogIssueDate
23	1244	$15.00	8/1/2000
23	1245	$15.00	8/1/2000
23	1246	$15.00	8/1/2000
23	1247	$15.00	8/1/2000
23	1248	$14.00	8/1/2000
23	1249	$14.00	8/1/2000
23	1252	$21.00	8/1/2000
23	1253	$21.00	8/1/2000
23	1254	$24.00	8/1/2000
23	1257	$19.00	8/1/2000

Record: 11 of 11

Convert to second normal form

CatalogProduct : Table

CatalogNumber	ProductNumber	Price
23	1244	$15.00
23	1245	$15.00
23	1246	$15.00
23	1247	$15.00
23	1248	$14.00
23	1249	$14.00
23	1252	$21.00
23	1253	$21.00
23	1254	$24.00
23	1257	$19.00

Record: 1 of 10

Catalog : Table

Number	IssueDate
19	11/1/1999
20	2/1/2000
21	4/1/2000
22	6/1/2000
23	8/1/2000
24	9/15/2000

Record: 6

To verify that a table is in 3NF, we must check the functional dependency of each non-key element on every other non-key element. This can be cumbersome for a large table since the number of pairs that must be checked grows quickly as the number of non-key fields grows. The number of functional dependencies to be checked is $N \times (N-1)$ where N is the number of non-key fields. Note that functional dependency must be checked in both directions (that is, A dependent on B, and B dependent on A).

Consider the simple table shown in Figure 10-13. Assume that number is the primary key and that all customers live in the United States. Since there are three non-key fields, there are six functional dependencies to check:

- Is State functionally dependent on Street Address?
- Is Street Address functionally dependent on State?
- Is Zip Code functionally dependent on Street Address?
- Is Street Address functionally dependent on Zip Code?
- Is Zip Code functionally dependent on State?
- Is State functionally dependent on Zip Code?

The answer to the first five statements is no, but the answer to the last is yes. In the United States, all zip codes are wholly contained within a single state. Thus, for each value of Zip Code there is only one corresponding value of State (for example, 87123 is always in New Mexico). Including both fields in this table is a form of redundancy. For example, if the addresses of 100 customers who live in the 87123 zip code are stored in the table, the fact that 87123 is located in New Mexico is (redundantly) stored 100 times.

FIGURE 10-13
A table containing customer addresses.

CustomerAddressUSA : Table

Number	Street Address	State	Zip Code
134425	123 Main Street	NM	87123
187763	456 Oak Street	MO	65701
214435	678 Poplar Avenue	UT	84697

Record: 2 of 3

Because State is functionally dependent on Zip Code, the table is not in 3NF. To correct this problem, you must remove State from the table. But the correspondence between states and zip codes must be maintained somewhere in the database if the information system needs to generate complete mailing labels. The solution is to create a new table containing only Zip Code and State (see Figure 10-14). Zip Code is the primary key of this new table, and State is its only non-key field. Programs or methods that need to print or display a complete mailing address must now use the value of Zip Code in the CustomerAddress table to look up the corresponding value of State in the newly created table.

FIGURE 10-14
Conversion of a second normal form table to two third normal forms.

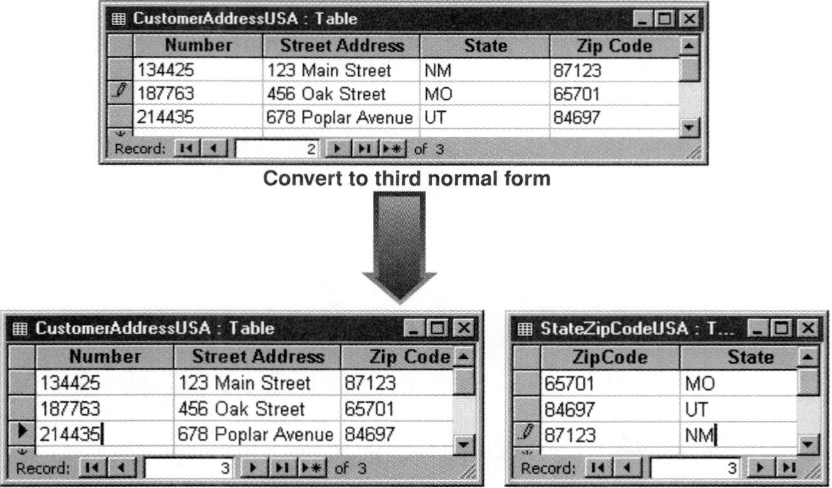
CustomerAddressUSA : Table

Number	Street Address	State	Zip Code
134425	123 Main Street	NM	87123
187763	456 Oak Street	MO	65701
214435	678 Poplar Avenue	UT	84697

Record: 2 of 3

CustomerAddressUSA : Table

Number	Street Address	Zip Code
134425	123 Main Street	87123
187763	456 Oak Street	65701
214435	678 Poplar Avenue	84697

Record: 3 of 3

StateZipCodeUSA : T...

ZipCode	State
65701	MO
84697	UT
87123	NM

Record: 3

Another type of 3NF violation occurs when a field can be computed from the value(s) of one or more other fields. For example, consider an order table with the fields Subtotal, ShippingCost, SalesTax, and GrandTotal. If GrandTotal is computed as

GrandTotal = Subtotal + ShippingCost + SalesTax

then GrandTotal is functionally dependent on the combination of the other three fields. Computational dependencies are a form of redundancy because a change to the value of any "variable" in the computation (for example, ShippingCost) also changes the "result" of the computation (for example, GrandTotal).

The method of correcting this type of 3NF violation is simple: Remove the computed field from the database. This has the desirable effect of eliminating redundancy in the database at the expense of programs that want to read the value of the computed variable. But just because the computed field is eliminated from the database doesn't mean that its value is lost. For example, any program or method that needs GrandTotal can simply query the database for the Subtotal, ShippingCost, and SalesTax and then add them together to compute GrandTotal.

Entity-relationship modeling and normalization are complementary techniques for relational database design. Note that the tables generated from the RMO ERD (see Figure 10-8) do not contain any violations of first, second, or third normal form. This is not a chance occurrence. Fields of an entity are functionally dependent on any unique identifier (primary key) of that entity. Fields of a many-to-many relationship are functionally dependent on unique identifiers of both participating entities. Thus, while creating an ERD, an analyst must directly or indirectly consider issues of functional dependency when deciding which fields belong with which entities or relationships.

We now turn our attention to the second type of database commonly in use today—the object-oriented database.

Object-Oriented Databases

object database management system (ODBMS)
A database management system that stores data as objects or class instances

Object database management systems (ODBMSs) are a direct extension of the OO design and programming paradigm. ODBMSs are designed specifically to store objects and to interface with object-oriented programming languages. Although it is possible to store objects in files or relational databases, there are many advantages to using a DBMS specifically designed for objects. These advantages include direct support for method storage, inheritance, nested objects, object linking, and programmer-defined data types.

ODBMSs first appeared as research prototypes in the 1980s and fledgling commercial products in the early 1990s. ODBMSs are maturing rapidly and are currently represented by many successful and widely used products including GemStone, ObjectStor, and NeoAccess. ODBMSs are the database technology of choice for newly designed systems implemented with OO tools, especially for scientific and engineering applications. ODBMSs are expected to supplant RDBMSs in more traditional business applications gradually over the next decade as OO technology becomes more widely used and as better tools for interfacing object and relational databases are developed.

Object Definition Language (ODL)
A standard object database description language promulgated by the Object Database Management Group

Because ODBMSs are relatively new, there are few widely accepted standards for specifying an object database schema. Some object database standards that are gaining wide acceptance are those proposed by the Object Database Management Group. One of these standards is the **Object Definition Language (ODL)**. ODL is a language for describing the structure and content of an object database. The schema examples in the sections that follow use ODL.

Designing Object Databases

To create an object database schema from a class diagram, follow these steps:

1. Determine which classes require persistent storage.
2. Define persistent classes.
3. Represent relationships among persistent classes.
4. Choose appropriate data types and value restrictions (if necessary) for each field.

Each of these steps is discussed in detail in the following sections.

Representing Classes

transient object
An object that doesn't need to store any attribute values between instantiations or method invocations

Objects can be classified into two broad types for purposes of data management. A **transient object** exists only during the lifetime of a program or process. Examples of transient objects include objects created to implement user-interface components (such as a view window or pop-up menu). Transient objects are created each time a program or process is executed and then destroyed when a program or process terminates.

persistent object
An object that must store one or more attribute values between instantiations or method invocations

A **persistent object** is not destroyed when the program or process that creates it ceases execution. Instead, the object continues to exist independently of any program or process. Storing the object state to persistent memory (such as a magnetic or optical disk) ensures that the object exists between process executions. Objects can be persistently stored with a file or database management system.

An object database schema includes a definition for each class that requires persistent storage. ODL class definitions can derive from the corresponding UML class diagram. Thus, classes already defined in UML are simply reused for the database schema definition.

For example, an ODL description of the RMO Customer class is

```
class Customer {
  attribute string Name
  attribute string BillAddress
  attribute string ShippingAddress
  attribute string HomePhone
  attribute string WorkPhone
}
```

This ODL class definition corresponds to the Customer table in Figures 10-6, 5-27, and 9-28. A similar ODL class definition would be created for each class shown in Figure 9-28. Once each class has been defined, then relationships among classes must be defined.

Representing Relationships

Each object stored within an ODBMS is automatically assigned a unique object identifier. This identifier may be a physical storage address or a reference that can be converted to a physical storage address at run time. In either case, each object has a unique identifier that can be stored within another object to represent a relationship.

Object identifiers are used to relate objects of one class to objects of another class. An ODBMS represents relationships by storing the identifier of one object within related objects. Object identifiers are the "glue" that binds related objects together. For example, consider a one-to-one relationship between the classes Employee and Workstation as shown in Figure 10-15. Each Employee object has an attribute called Computer that contains the object identifier of the Workstation object assigned to that employee. Each Workstation object has a matching attribute called User that contains the object identifier of the Employee that uses that workstation.

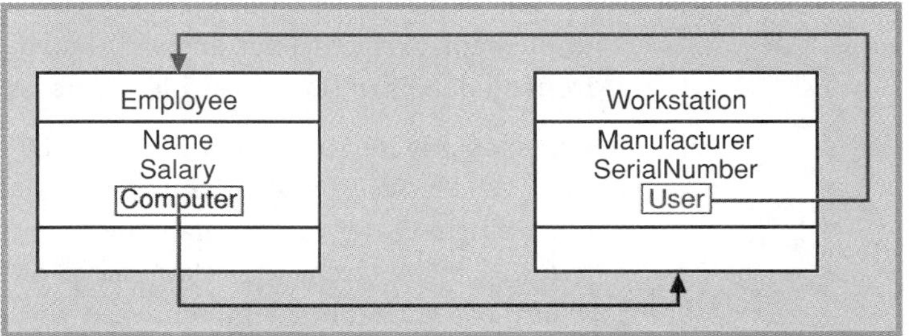

FIGURE 10-15
A one-to-one relationship represented with attributes (shown in color) containing object identifiers.

navigation
The process of accessing an object by extracting its object identifier from another (related) object

The ODBMS uses attributes containing object identifiers to find objects that are related to other objects. The process of extracting an object identifier from one object and using it to access another object is sometimes called **navigation**. For example, consider the following query posed by a user:

List the manufacturer of the workstation assigned to employee Joe Smith.

A query processor can find the requested employee object by searching all employee objects for the Name attribute Joe Smith. The processor can find Joe Smith's workstation object by using the object identifier stored in Computer. Note that a query processor can answer the opposite query (list the employee name assigned to a specific workstation) by using the object identifier stored in User. A matched pair of attributes is required to allow a relationship to be navigated in both directions.

Attributes that represent relationships are not usually specified directly by an object database schema designer. Instead designers specify them indirectly by declaring relationships between objects. For example, consider the following class declarations for the ODL schema language:

```
class Employee {
  attribute string Name
  attribute integer Salary
  relationship Workstation Uses
     inverse Workstation::AssignedTo
}
class Workstation {
  attribute string Manufacturer
  attribute string SerialNumber
  relationship Employee AssignedTo
     inverse Employee::Uses
}
```

The keyword *relationship* is used to declare a relationship between one class and another. The class Employee has a relationship called Uses with the class Workstation. The class Workstation has a matching relationship called AssignedTo with the class Employee. Note that each relationship includes a declaration of the matching relationship in the other class using the keyword *inverse*. The ODBMS is thus informed that the two relationships are actually mirror images of one another.

There are two advantages to declaring a relationship as shown above instead of creating an attribute containing an object identifier:

- The ODBMS assumes responsibility for determining how to implement the connection among objects. In essence, the schema designer has declared an attribute of type *relationship* and left it up to the ODBMS to determine an appropriate type for that attribute.
- The ODBMS assumes responsibility for maintaining referential integrity. For example, deleting a workstation will cause the Uses link of the related Employee object to be set to NULL or undefined.

The ODBMSs will automatically create attributes containing object identifiers to implement declared relationships. But the user and programmer will be shielded from all details of how those identifiers are actually implemented and manipulated.

One-to-Many Relationships Figure 10-16 shows a one-to-many relationship between the RMO classes Customer and Order. A Customer can make many different Orders, but a single Order can be made by only one Customer. A single object identifier is required to represent the relationship of an Order to a Customer. Multiple object identifiers are required to represent the relationship between one Customer and many different Orders, as shown in Figure 10-17.

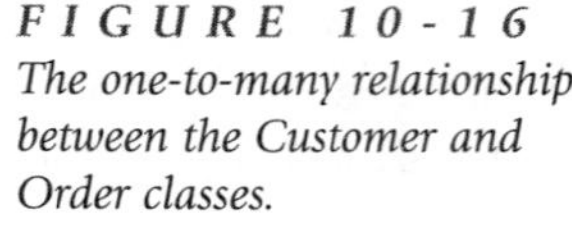
FIGURE 10-16
The one-to-many relationship between the Customer and Order classes.

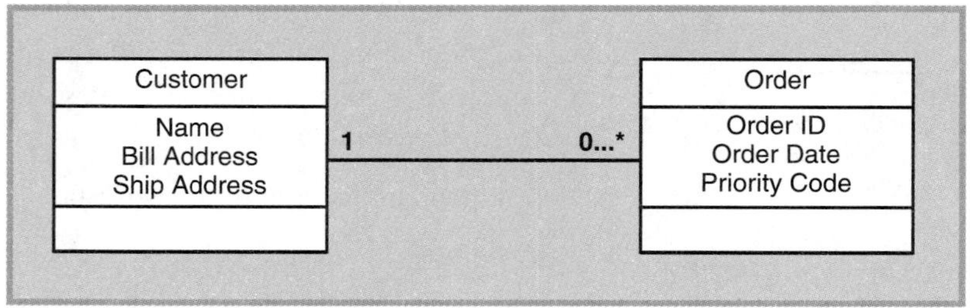

FIGURE 10-17
A one-to-many relationship represented with attributes containing object identifiers.

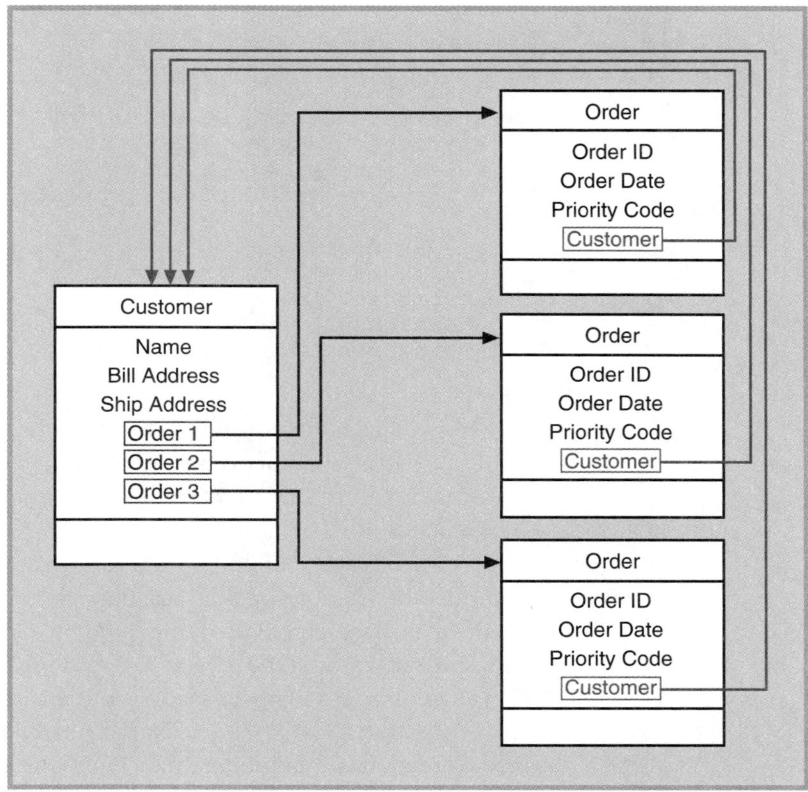

Partial ODL class declarations for the classes Customer and Order are as follows:

```
class Customer {
  attribute string Name
  attribute string BillAddress
  attribute string ShippingAddress
  attribute string HomePhone
  attribute string WorkPhone
  relationship set<Order> Makes
     inverse Order::MadeBy
}
class Order {
  attribute string OrderID
  attribute string Date
  attribute string PriorityCode
  attribute real TotalS&H
  attribute real TotalTax
  attribute real GrandTotal
}
```

The relationship Makes is declared between a single Customer object and a set of Order objects. By declaring the relationship as a set, you instruct the ODBMS to allocate multiple Order object identifier attributes dynamically to each Customer object.

Many-to-Many Relationships A many-to-many relationship is represented differently depending on whether the relationship has any attributes. Many-to-many relationships without attributes are represented similarly to one-to-many relationships. For example, assume that the many-to-many relationship between the RMO classes

Catalog and ProductItem has no attributes. In this case, the relationship can be represented as follows:

```
class Catalog {
  attribute string Description
  attribute integer Year
  relationship set<ProductItem> Contains
     inverse ProductItem::AppearsIn
}
class ProductItem {
  attribute string ProductID
  attribute string Description
  relationship set<Catalog> AppearsIn
     inverse Catalog::Contains
}
```

Note that the relationship is declared as a set within both class declarations. This allows an object of either class to be related to multiple objects of the other class.

Now let's turn our attention to many-to-many relationships with attributes. We normally think of an attribute as part of a class, not as part of a relationship. But in theory relationships among objects can (and often do) have attributes. For example, consider the relationship marriage between two person objects. It has attributes such as the date, time, and place of the wedding. Note that these are attributes of the relationship itself, not of the objects participating in the relationship.

Representing a many-to-many relationship with attributes requires a more complex approach. For example, assume that the relationship between Catalog and ProductItem has an attribute called Price. The approach needed to represent the relationship and its attribute is shown in Figure 10-18.

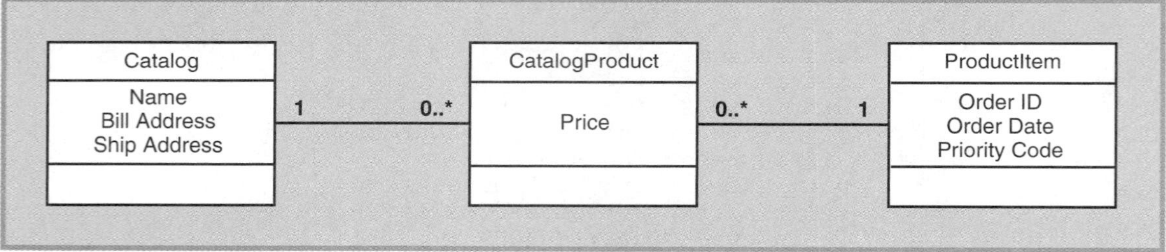

FIGURE 10-18
A many-to-many relationship represented with two one-to-many relationships.

association class
A class that represents a many-to-many relationship among other classes

An **association class** is a class that represents a relationship and stores the attributes of that relationship. In Figure 10-18, the association class CatalogProduct has been created to represent the many-to-many relationship between Catalog and ProductItem. The many-to-many relationship between Catalog and ProductItem has been decomposed into a pair of one-to-many relationships between the original classes and the new association class. The schema class declarations are as follows:

```
class Catalog {
  attribute string Description
  attribute integer Year
  relationship set<CatalogProduct> Contains1 inverse
     CatalogProduct::AppearsIn1
}
class ProductItem {
  attribute string ProductID
  attribute string Description
  relationship set<CatalogProduct> AppearsIn2 inverse
     CatalogProduct::Contains2
}
class CatalogProduct {
  attribute real Price
```

```
  relationship Catalog AppearsIn1 inverse Catalog::Contains1
  relationship ProductItem AppearsIn2 inverse
     ProductItem::Contains2
}
```

Generalization Relationships Figure 10-19 shows a generalization hierarchy from the RMO class diagram. Web Order, Telephone Order, and Mail Order are each more specific versions of the class Order. The ODL class definitions that represent these classes and their interrelationships are as follows:

```
class Order {
  attribute string OrderID
  attribute string Date
  attribute string PriorityCode
  attribute real TotalS&H
  attribute real TotalTax
  attribute real GrandTotal
}

class WebOrder extends Order
  attribute EmailAddress
  attribute ReplyMethod
}
class TelephoneOrder extends Order {
  attribute string PhoneClerk
  attribute string CallStartTime
  attribute integer LengthOfCall
}

class MailOrder extends Order {
  attribute string DateReceived
  attribute ProcessorClerk
}
```


The keyword extends indicates that WebOrder, TelephoneOrder, and MailOrder derive from Order. When stored in an object database, objects of the three derived classes will inherit all of the attributes, methods, and relationships defined for the Order class.

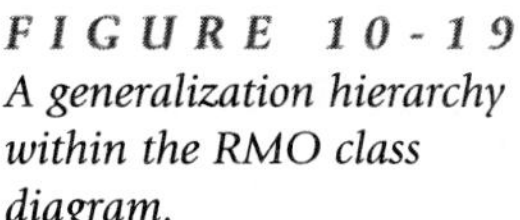

FIGURE 10-19
A generalization hierarchy within the RMO class diagram.

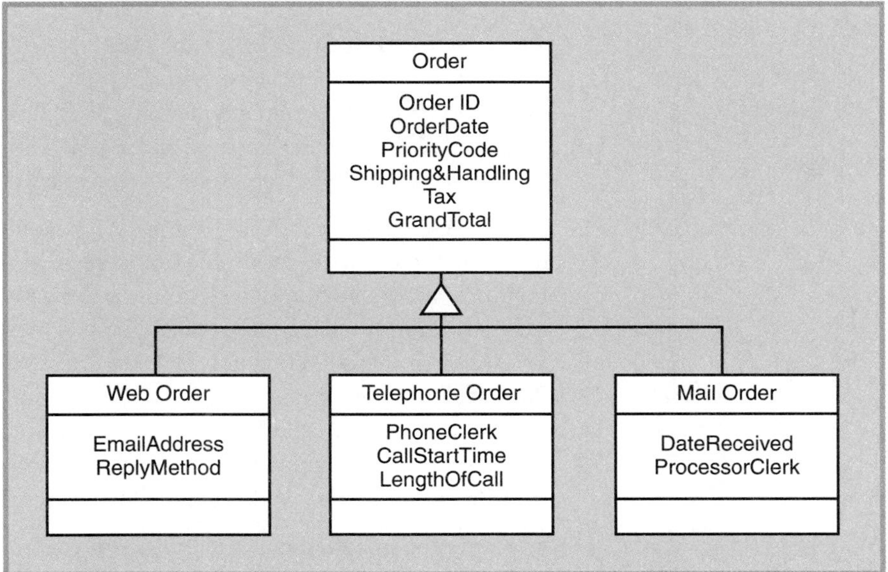

Key Attributes are not required in an object database since referential integrity is implemented with object identifiers. However, key attributes are useful in object databases for a number of purposes including guaranteeing unique object content and providing a means of querying database contents. The ODBMS automatically enforces uniqueness of key attributes in an object database. Thus, declaring an attribute to be a key guarantees that no more than one object in a class can have the same key value.

In addition to relational and object-oriented databases, a third type exists that mixes elements of both the relational and OO approaches. We discuss this hybrid next.

Hybrid Object-Relational Database Design

OO development tools were first widely employed in the mid- to late-1980s. During this same time period, RDBMSs were widely used and had reached a mature stage of development. Many OO tool developers exploited the existing base of RDBMS tools and knowledge by using an RDBMS to store persistent object states. This made sense both as an economy measure (there was one less OO tool to develop) and because many newer OO systems needed to manipulate data stored in existing relational databases. There is no widely accepted name to describe object storage using an RDBMS, so we will invent one to use for the remainder of the text: **hybrid object-relational DBMS** (or, simply **hybrid DBMS**).

hybrid object-relational DBMS
A relational database management system used to store object attributes and relationships; also called **hybrid DBMS**

The hybrid DBMS approach is currently the most widely employed approach to persistent object storage. Designing a hybrid database is essentially two design problems in one. That is, a complete relational database schema must be designed as described earlier in this chapter. The designer must also develop an equivalent set of classes to represent the relational database contents within the OO programs. This is a complex task because the database designer must bridge the differences between the object-oriented and relational views of stored data.

The following are the most important mismatches between the relational and OO views of stored data:

- Data stored in an RDBMS are static. Programmer-defined methods cannot be stored nor automatically executed.
- ODBMSs can represent a wider range of relationship types than RDBMSs, including classification hierarchies and whole-part aggregations. Relationships in an RDBMS can only be represented using referential integrity.

Because ERDs were developed prior to the OO paradigm, they have no features that can represent methods. Programs that access the database must implement methods internally. Inheritance cannot be directly represented in an RDBMS because a classification hierarchy cannot be directly represented.

Although the relational and OO views of stored data have significant differences, they also have significant overlaps. Recall from Chapter 5 that "things" within a system can be conceptually modeled using an ERD (the basis for a relational database schema), a class diagram (the basis for an OO database schema), or both. There is a considerable degree of overlap among the two representations, including the following:

- Grouping of data items into entities or classes
- Defining one-to-one, one-to-many, and many-to-many relationships among entities or classes

This overlap provides a basis for representing classes and objects within a relational database.

Classes and Attributes

Designers can store classes and object attributes in an RDBMS by defining appropriate tables in which to store them. For a completely new system, a relational schema can be designed based on a class diagram in essentially the same fashion as for an ERD. Table 10-1 describes the correspondence between OO, ER, and relational database concepts. A table is created to represent each class, and the fields of each table are the same as the attributes of the corresponding class. Each row holds the attribute values of a single object.

TABLE 10-1 Correspondence among concepts in the object-oriented, entity-relationship, and relational database views of stored data.

Object-Oriented	Entity-Relationship	Relational Database
Class	Entity Type	Table
Object	Entity Instance	Row
Attribute	Attribute	Column

A key field (or group of fields) must be chosen for each table. As described earlier, a designer can choose a natural key field from the existing attributes or add an invented key field. Primary key fields are needed to guarantee uniqueness within tables and to represent relationships using foreign keys.

Figure 10-20 shows a set of relational database tables that represent classes from the RMO class diagram in Figure 10-21. Note that the table definitions are identical to those in Figure 10-7, except for the addition of tables to represent the specialized classes Mail Order, Telephone Order, and Web Order.

FIGURE 10-20 Class tables with primary keys identified in bold.

Table	Attributes
Catalog	**Number**, Season, Year, Description, EffectiveDate, EndDate
Customer	**AccountNumber**, Name, BillingAddress, ShippingAddress, DayTelephoneNumber, NightTelephoneNumber
InventoryItem	**Number**, Size, Color, Options, QuantityOnHand, AverageCost, ReorderQuantity
MailOrder	**Number**, DateReceived, ProcessorClerk
Order	**Number**, Date, PriorityCode, ShippingAndHandling, Tax, GrandTotal
OrderItem	**Number**, Quantity, Price, BackOrderStatus
OrderTransaction	**Number**, Date, TransactionType, Amount, PaymentMethod
Package	**Number**, Description, SalePrice
ProductItem	**Number**, Vendor, Gender, Description, Season, NormalPrice, SpecialPrice
ReturnItem	**Number**, Quantity, Price, Reason, Condition, Disposal
Shipment	**TrackingNumber**, DateSent, TimeSent, ShippingCost, DateArrived, TimeArrived
Shipper	**Number**, Name, Address, ContactName, Telephone
TelephoneOrder	**Number**, PhoneClerk, CallStartTime, LengthOfCall
WebOrder	**Number**, EmailAddress, ReplyMethod

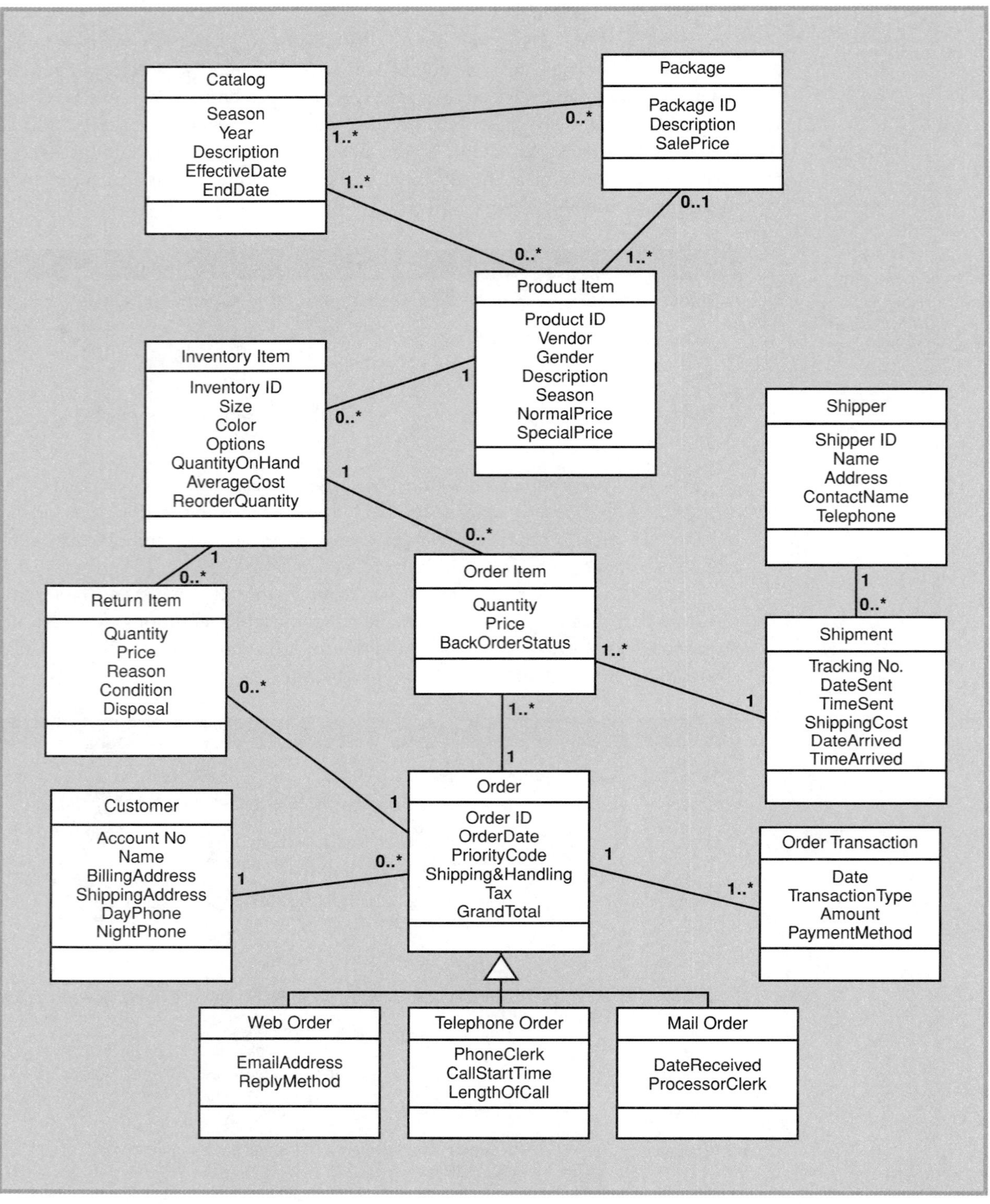

380

FIGURE 10-21
The RMO class diagram.

Relationships

ODBMSs use object identifiers to represent relationships among objects. But object identifiers are not created by RDBMSs, so relationships among objects stored in a relational database must be represented using foreign keys. Foreign key values serve the same purpose as object identifiers in an ODBMS. That is, they provide a means for one "object" to refer to another.

To represent one-to-many relationships, designers add the primary key field of the class on the one side of the relationship to the table representing the class on the many side of the relationship. To represent many-to-many relationships, designers create a new table that contains the primary key fields of the related class tables and any attributes of the relationship itself. Note these methods of representing relationships among objects are the same as previously described for representing relationships among entities.

Figure 10-22 extends the table definitions in Figure 10-20 by adding foreign keys representing the relationships shown in Figure 10-21. For example, the one-to-many relationship between the Customer and Order classes is represented by the foreign key AccountNumber stored in the Order table. The many-to-many relationship between the Catalog and ProductItem classes is represented by the table CatalogProduct that contains the foreign keys CatalogNumber and ProductNumber.

FIGURE 10-22 Relationship information added to the class tables by adding foreign key attributes (shown in italics) to represent one-to-many relationships and new tables (CatalogPackage, CatalogProduct, and PackageProduct) to represent many-to-many relationships.

Table	Attributes
Catalog	**Number**, Season, Year, Description, EffectiveDate, EndDate
CatalogPackage	***CatalogNumber***, ***PackageNumber***
CatalogProduct	***CatalogNumber***, ***ProductNumber***, Price
Customer	**AccountNumber**, Name, BillingAddress, ShippingAddress, DayTelephoneNumber, NightTelephoneNumber
InventoryItem	**Number**, *ProductItemNumber*, Size, Color, Options, QuantityOnHand, AverageCost, ReorderQuantity
MailOrder	***OrderNumber***, DateReceived, ProcessorClerk
Order	**Number**, *AccountNumber*, Date, PriorityCode, ShippingAndHandling, Tax, GrandTotal
OrderItem	**Number**, *OrderNumber, InventoryItemNumber, ShipmentNumber*, Quantity, Price, BackorderStatus
OrderTransaction	**Number**, *OrderNumber*, Date, TransactionType, Amount, PaymentMethod
Package	**Number**, *CatalogNumber*, Description, SalePrice
PackageProduct	***PackageNumber***, ***ProductNumber***
ProductItem	**Number**, *PackageNumber, CatalogNumber*, Vendor, Gender, Description, Season, NormalPrice, SpecialPrice
ReturnItem	**Number**, *OrderNumber, InventoryItemNumber*, Quantity, Price, Reason, Condition, Disposal
Shipment	**TrackingNumber**, *ShipperNumber*, DateSent, TimeSent, ShippingCost, DateArrived, TimeArrived
Shipper	**Number**, Name, Address, ContactName, Telephone
TelephoneOrder	***OrderNumber***, PhoneClerk, CallStartTime, LengthOfCall
WebOrder	***OrderNumber***, EmailAddress, ReplyMethod

Note that the tables in Figure 10-22 are identical to those in Figure 10-9 except for the content of the tables MailOrder, TelephoneOrder, and WebOrder (these tables will be discussed shortly). The similarity is no accident—it follows from the similarity between the RMO class and entity-relationship diagrams (Figures 10-5 and 10-21). The diagrams are similar because they represent the same underlying reality. Thus, it should be no surprise that the relational database schemas derived from a class diagram and an ERD representing that underlying reality are similar. In fact, it would be surprising (and probably indicative of an error) if they weren't similar.

Classification relationships such as the relationship among Order, MailOrder, TelephoneOrder, and WebOrder are a special case in relational database design. Just as a child class inherits the data and methods of a parent class, a table representing a child class inherits some or all of its data from the table representing its parent class. This inheritance can be represented in two ways:

- Combine all the tables into a single table containing the superset of all class attributes but excluding any invented key fields of the child classes
- Use separate tables to represent the child classes and substitute the primary key of the parent class table for the invented keys of the child class tables

Either method is an acceptable approach to representing a classification relationship.

Figure 10-9 shows the definition of the Order table under the first method. All of the non-key fields from MailOrder, TelephoneOrder, and WebOrder have been added to the Order table. For any particular order, some of the field values in each row will be null. For example, a row representing a telephone order would have no values for the fields EmailAddress, ReplyMethod, DateReceived, and ProcessorClerk.

Figure 10-22 shows the table definitions for the RMO case using the second method for representing inheritance. The relationship among the three child order types and the parent Order table is represented by the foreign key OrderNumber in all three child class tables. Note that the invented key of each table has been removed. Thus, in each case, the foreign key representing the inheritance relationship also serves as the primary key of the table representing the child class.

Now that we've covered the different approaches to database schema design, we consider the data types that are stored within that schema.

Data Types

data type
The storage format and allowable content of a field, class attribute, or program variable

primitive data type
A storage format directly implemented by computer hardware or a programming language

complex data type
A data type not directly supported by computer hardware or a programming language; also called **user-defined data type**

A **data type** defines the storage format and allowable content of a program variable, object state variable, or database field or attribute. **Primitive data types** are data types that are supported directly by computer hardware and programming languages. Examples include memory address (a pointer), boolean, integer, unsigned integer, short integer (one byte), long integer (multiple bytes), single characters, real numbers (floating point numbers), and double precision (double-length) integers and real numbers. Some procedural programming languages (such as C) and most OO languages enable programmers to define additional data types using the primitive data types as building blocks.

As information systems have become more complex, the number of data types used to implement them has increased. Examples of modern data types include dates, times, currency (money), audio streams, video images, motion video streams, and uniform resource locators (URLs, or web links). Such data types are sometimes called **complex data types** because they are usually defined as complex combinations of primitive data types. They may also be called **user-defined data types** because they may be defined by users during analysis and design or by programmers during design and implementation.

Relational DBMS Data Types

For each field in a relational database schema, the designer must choose an appropriate data type. For many fields, the choice of a data type is relatively straightforward. For example, designers can represent customer names and addresses using a set of fixed- or variable-length character arrays. Inventory quantities and item prices can be represented as integers and real numbers, respectively. A color can be represented by a character array containing the name of the color or by a set of three integers representing the intensity of the primary colors red, blue, and green.

Modern RDBMSs have added an increasing number of new data types to represent the data required by modern information systems. Table 10-2 contains a partial listing of some of the data types available in the Oracle RDBMS. Complex data types available in Oracle include DATE, LONG, and LONGRAW. LONG is typically used to store large quantities of formatted or unformatted text (such as a word processing document). LONGRAW can be used to store large binary data values, including encoded pictures, sound, and motion video.

TABLE 10-2
A subset of the data types available in the Oracle relational DBMS.

Type	Description
CHAR	Fixed-length character array
VARCHAR	Variable-length character array
NUMBER	Real number
DATE	Date and time with appropriate checks of validity
LONG	Variable-length character data up to 2 gigabytes
LONGRAW	Binary large object (BLOB) with no assumption about format or content
ROWID	Unique six-byte physical storage address

Modern RDBMSs can also perform many validity and format checks on data as they are stored in the database. For example, a schema designer can specify that a quantity on hand cannot be negative, that a U.S. zip code must be five or nine digits long, and that a string containing a URL must begin with *http://*. The validity and format constraints are then automatically shared by all application programs that use the database. Each program is simpler, and the possibility for errors due to mismatches among data validation logic is eliminated. Application programs still have to provide program logic to recover from attempts to add "bad" data, but they are freed from actually performing validity checks.

Object DBMS Data Types

ODBMSs typically provide a set of primitive and complex data types comparable to those of an RDBMS. ODBMSs also allow a schema designer to define format and value constraints. But ODBMSs provide an even more powerful way to define useful data types and constraints—a schema designer can define a new data type and its associated constraints as a new class.

A class is a complex user-defined data type that combines the traditional concept of data with processes (methods) that manipulate that data. In most OO programming languages, programmers are free to design new data types (classes) that extend those already defined by the programming language. Incompatibility between system requirements and available data types is not an issue, since the designer can design classes to specifically meet the requirements. To the ODBMS, instances of the new data type are simply objects to be stored in the database.

Class methods can perform many of the type and error checking functions previously performed by application program code and/or by the DBMS itself. In essence, the programmer constructs a "custom designed" data type and all of the programming logic required to use it correctly. The DBMS is freed from direct responsibility for managing complex data types and the values stored therein. It indirectly performs validity checking and format conversion by extracting and executing programmer-defined methods stored in the database.

The flexibility to define new data types is a chief reason that OO tools are so widely employed in non-business information systems. In fields such as engineering, biology, and physics, stored data is considerably more complex than simple strings, numbers, and dates. OO tools enable database designers and programmers to design custom data types that are specific to a problem domain.

Another issue that must be considered during database design is the locations where data are stored and accessed. In today's networked information systems, organizations often use distributed databases.

Distributed Databases

It is rare for all of an organization's data to be stored in a single database. Instead, data are typically stored in many different databases, often under the control of many different DBMSs. Reasons for employing a variety of databases and DBMSs include the following:

- Information systems may have been developed at different times using different DBMSs.
- Parts of an organization's data may be owned and managed by different organizational units.
- System performance is improved when data are physically close to the applications that use them.

The information system inventory of most organizations was not developed and installed at one time. Rather, information systems of varying size and function are developed over a period of decades. These systems are typically developed for different purposes, using different tools and support environments, and under the direction and control of different parts of an organization. As a result, the data of a single large organization are typically fragmented across a number of hardware, software, organizational, and geographic boundaries.

Database access is a significant performance factor in most information systems. Much of the activity of a typical information system is querying and updating database contents. Thus, delays in processing or answering an application program's database requests largely determine the application's throughput and response time.

Distributed Database Architectures

Chapter 7 described various approaches to organizing computing and information processing resources in a networked environment. There are several possible architectures for distributing database services, including the following:

- Single database server
- Replicated database servers
- Partitioned database servers
- Federated database servers

Combinations of these architectures are also possible.

Single Database Server Figure 10-23 shows a typical single database server architecture. Clients on one or more LANs share a single database located on a single computer system. The database server may be connected to one of the LANs or directly to the WAN backbone (as shown in the figure). Connection directly to the WAN ensures that no one LAN is overloaded by all of the network traffic to and from the database server.

FIGURE 10-23
A single database server architecture.

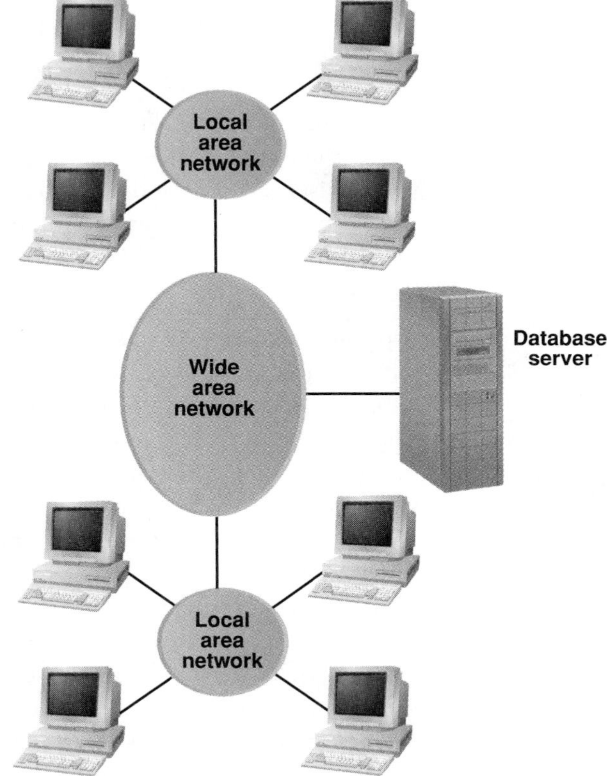

The primary advantage of a single database server architecture is its simplicity. There is only one server to manage, and all clients are programmed to direct requests to that server. Disadvantages of the single database server architecture include susceptibility to server failure and possible overload of the network or server. A single server provides no backup capabilities in the event of server failure. All application programs that depend on the server are disabled whenever the server is unavailable (such as during a crash or during hardware maintenance). Thus, single database server architecture is poorly suited to applications that must be available on a seven day, 24-hour basis.

Performance bottlenecks can occur within a single database server or in the network segment to which the server is attached. As transaction volumes grow, the capabilities of a single database server may become insufficient to respond quickly to all of the service requests it receives. In an attempt to improve performance, a designer may employ a more powerful computer system as the database server. But in an era of multiterabyte databases, it is not unusual for the size and transaction volume of larger databases to exceed the capabilities of any single computer system. Employing the largest mainframes may also be impractical due to cost, system management, or network performance considerations.

Requests to and responses from a database server may traverse large distances across local and wide area networks. Database transactions must also compete with other types of network traffic (such as voice, video, and web site accesses) for available transmission capacity. Thus, delays in accessing a remote database server may result from network congestion or propagation delay from client to server.

One way to reduce network congestion is to increase capacity within an entire network. But this approach is expensive and often impractical. Another approach specifically geared to improving database access speed is to locate database servers physically close to their clients (for example, on the same LAN segment). This approach minimizes the distance-related delay for requests and responses and removes a large amount of traffic from the WAN.

Moving a database server closer to its clients is a relatively simple matter when all of the clients are located close to one another. But what happens when clients are widely dispersed, as in today's global corporations? In this case, no single location for the database server can possibly improve database access performance for all clients at the same time. Thus, the "distant" clients must pay a greater performance penalty for database access.

Replicated Database Servers Designers can eliminate delay in accessing distant database servers by using a replicated database server architecture (see Figure 10-24). Each server is located near one group of clients and maintains a separate copy of the needed data. Clients are configured to interact with the database server on their own LAN. Database accesses are eliminated from the WAN and propagation delay is minimized. Local network and database server capacity can be independently optimized to local needs.

Replicated database servers also make an information system more fault tolerant. Applications can be programmed to direct access requests to any available server, with preference to the nearest server. When a server is unavailable, client access can be automatically redirected to another available server. Thus, accesses are redirected across a WAN when a local database server is unavailable. Designers can also achieve load-balancing by interposing a transaction server between clients and replicated database servers. The transaction server monitors loads on all database servers and automatically directs client requests to the server with the lowest load.

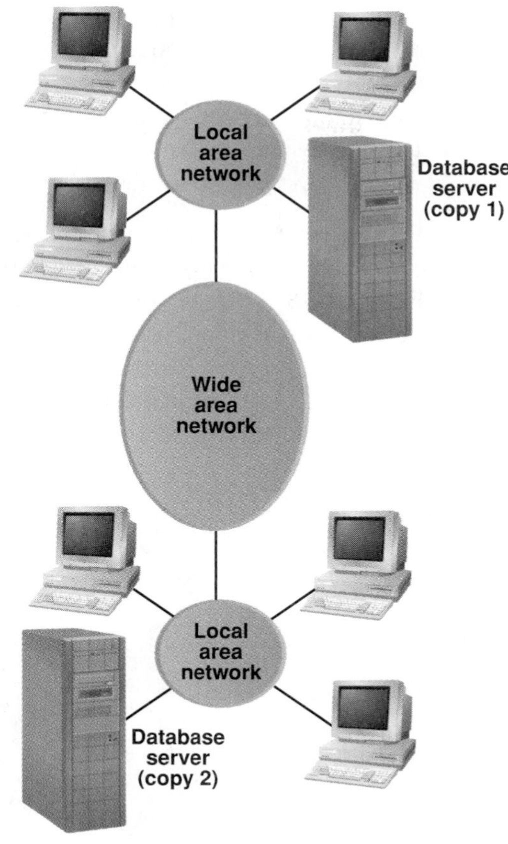

FIGURE 10-24
A replicated database server architecture.

In spite of their advantages, replicated database servers do have some drawbacks. Data inconsistency is a problem whenever multiple database copies are in use. When data are updated on one database copy, clients accessing that same data from another database copy receive an outdated response. To counteract this problem, updates to each database copy must periodically be made to all other copies of the database. This process is called **database synchronization**.

database synchronization
The process of ensuring consistency among two or more database copies

Designers can implement synchronization by developing customized synchronization programs or by using synchronization utilities built into the DBMS. Custom application programs are seldom employed because they are difficult to develop and because they would need to be modified each time the database schema or number and location of database copies change. Many DBMSs provide utilities for automatic or manual synchronization of database copies. Such utilities are generally powerful and flexible but also expensive. Incompatibilities among methods used by DBMSs to perform synchronization make using DBMSs from different vendors impractical.

The time delay between an update to a database copy and the propagation of that update to other database copies is an important database design decision. During the time between the original update and the update of database copies, application programs that access outdated copies aren't receiving responses that reflect current reality.

Designers can address this problem by reducing the synchronization delay. But shorter delays imply more frequent (or possibly continuous) database synchronization. Synchronization then consumes a substantial amount of database server capacity, and a large amount of network capacity among the related database servers must be provided. The proper synchronization strategy is a complex trade-off among cost, hardware and network capacity, and the need of application programs and users for current data.

Partitioned Database Servers Designers can minimize the need for database synchronization by partitioning database contents among multiple database servers. Figure 10-25 shows the division of a hypothetical database schema into two partitions. Each partition is accessed by a different group of clients. Figure 10-26 shows a partitioned database server architecture that maintains each partition on a separate database server. Traffic among clients and the database server in each group is restricted to a local area network.

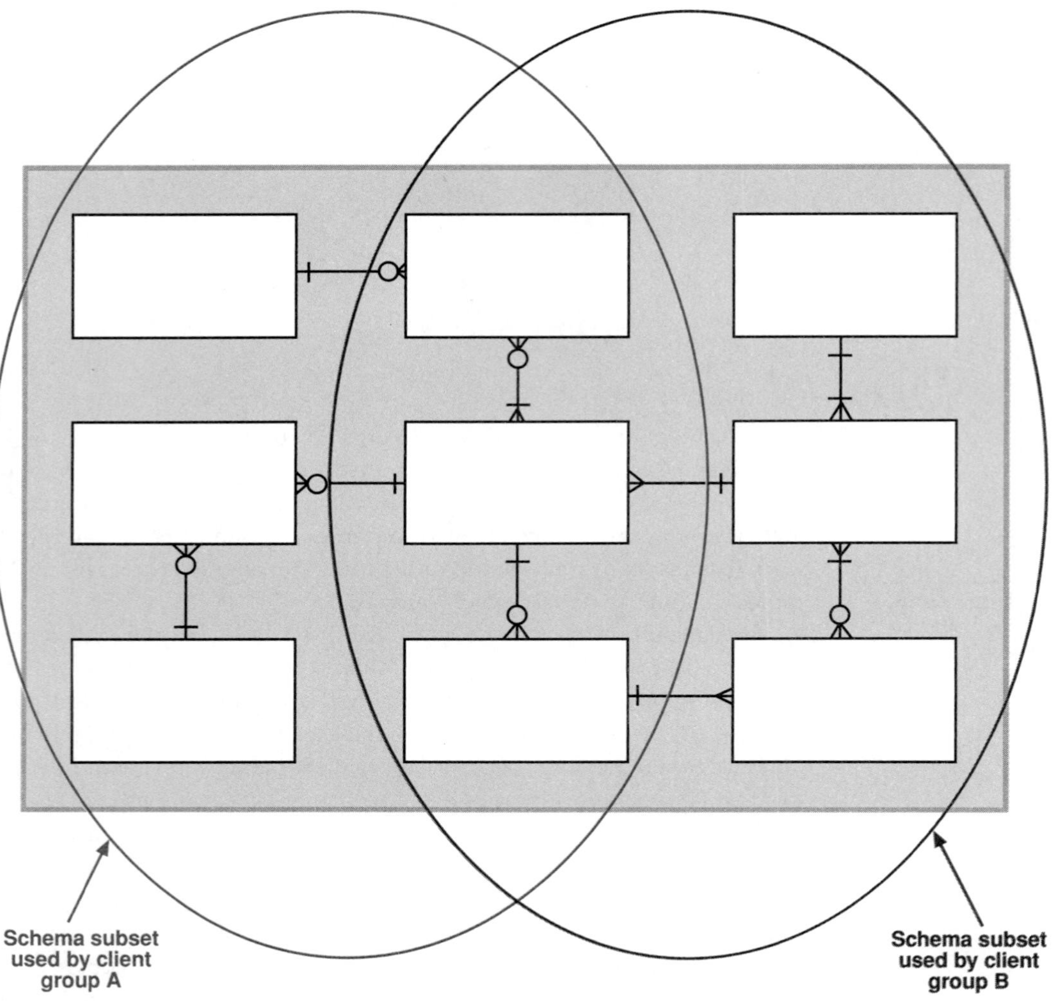

FIGURE 10-25
Partitioning a database schema into client access subsets.

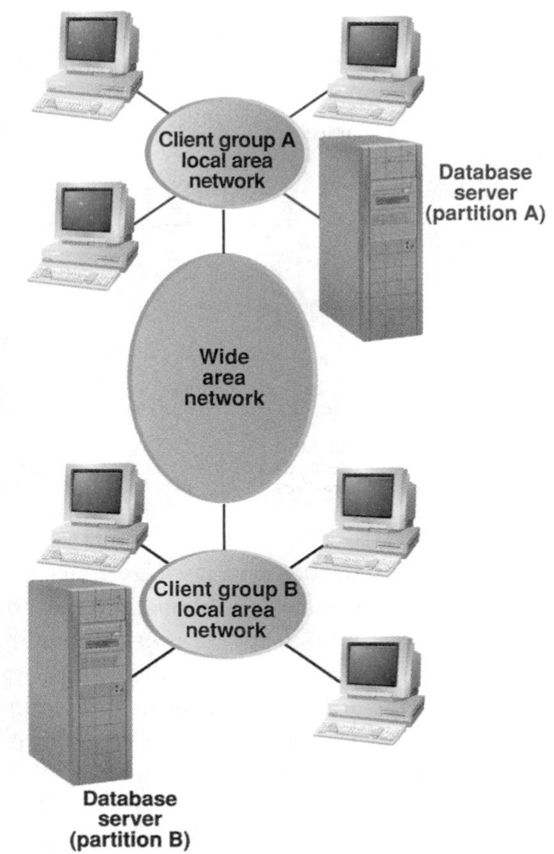

FIGURE 10-26
A partitioned database server architecture.

Partitioned database server architecture is feasible only when a schema can be cleanly partitioned among client access groups. Client groups must require access to well-defined subsets of a database (for example, marketing data rather than production data). In addition, members of a client access group must be located in small geographic regions. When a single access group is spread among multiple geographic sites (for example, order processing at three regional centers), then a combination of replicated and partitioned database server architecture is usually required.

It is seldom possible to partition a database schema into mutually exclusive subsets. Some portions of a database are typically needed by most or all users, and those portions must exist in each partition. Note the region of overlap in Figure 10-25. Database contents within that region would exist on each server, and those contents would need to be synchronized periodically. Thus, partitioning can reduce the problems associated with database synchronization, but seldom eliminates them entirely.

Federated Database Servers Some information systems are best served by a federated database architecture, as shown in Figure 10-27. This architecture is commonly used to access data stored in databases with incompatible storage models (for example, network and relational models) or DBMSs. A single unified database schema is created on a combined database server. That server acts as an intermediary between application programs and the databases residing on other servers. Database requests are first sent to the combined database server, which in turn makes appropriate requests of the underlying database servers. Results from multiple servers are combined and reformatted to fit the unified schema before the system returns a response to the client.

FIGURE 10-27
A federated database server architecture.

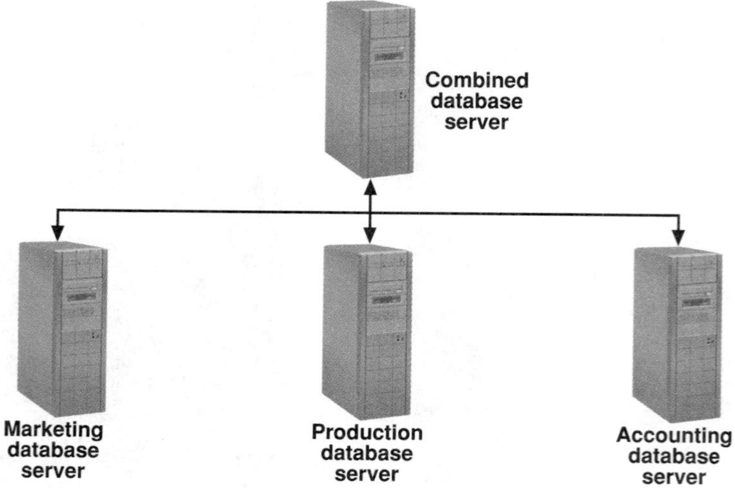

Federated database server architecture can be extremely complex. A number of DBMS products are available to implement such systems, but they are typically expensive and difficult to implement and maintain. Federated database architectures also tend to have high requirements for computer system and network capacity. But this expense and management complexity are generally less than would be required to implement and maintain application programs that interact directly with all of the underlying databases.

data warehouse
A collection of data used to support structured and unstructured managerial decisions

A common use of a federated database server architecture is to implement a data warehouse. A **data warehouse** is a collection of data used to support structured and unstructured managerial decisions. Data warehouses typically draw their content from operational databases within an organization and multiple external databases (for example, economic and trade data from databases maintained by governments, trade industry associations, and private research organizations). Because data originate from a large number of incompatible databases, a federated architecture is typically the only feasible approach for implementing a data warehouse.

Now that we've discussed the basic issues underlying distributed database design, we show how they come into play when making decisions for Rocky Mountain Outfitters' new customer support system.

The RMO Distributed Database Architecture

The starting point for designing a database architecture is information about the data needs of geographically dispersed users. Some of this information for RMO was gathered during the analysis phase (see Figures 6-35, 6-36, and 6-37) and is summarized here:

- Warehouses (Portland, Salt Lake City, and Albuquerque) need the ability to check inventory levels, query orders, record back orders and order fulfillment, and record order returns.
- The phone-order center (Salt Lake City) needs the ability to check inventory levels; create, query, update, and delete orders; query customer account information; and query catalogs.

- The mail-order center (Provo) needs the ability to check inventory levels, query orders, query catalog information, and update customer accounts.
- Direct customer access (location not yet determined) needs the same abilities as the phone-order center.
- Headquarters (Park City) needs the ability to query and adjust orders, query and adjust customer accounts, and create and query catalogs and promotions.

RMO has already decided to manage its database using the existing mainframe computer in the Park City data center. Thus, a WAN will be required to connect the server to LANs in the warehouses, phone-order center, mail-order center, headquarters, and data center. A connection will eventually be required for the web servers used for direct customer ordering, although these will probably be located at an existing site (such as the data center).

A single-server architecture for RMO is shown in Figure 10-28. This architecture would require sufficient WAN capacity to carry database (and other) traffic from all locations. The primary advantage of this architecture is its simplicity. There are no partitions or database copies to manage, and only a single server must be maintained. The primary disadvantages are relatively high WAN capacity requirements and the susceptibility of the entire system to failure of the single server.

FIGURE 10-28
A single-server database architecture for RMO.

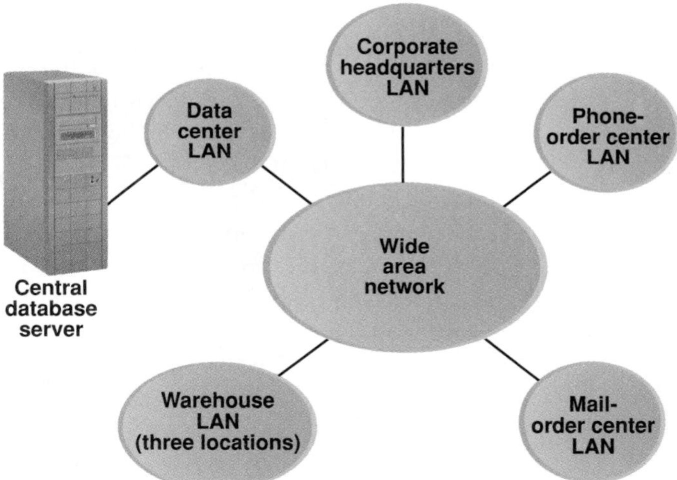

A more complex alternative is shown in Figure 10-29. A combination of database partitioning and replication is employed at each remote location. Each warehouse maintains a local copy of the order and inventory portions of the database. The phone- and mail-order centers maintain local copies of a larger subset of the database. Corporate headquarters relies on the central database server in the data center.

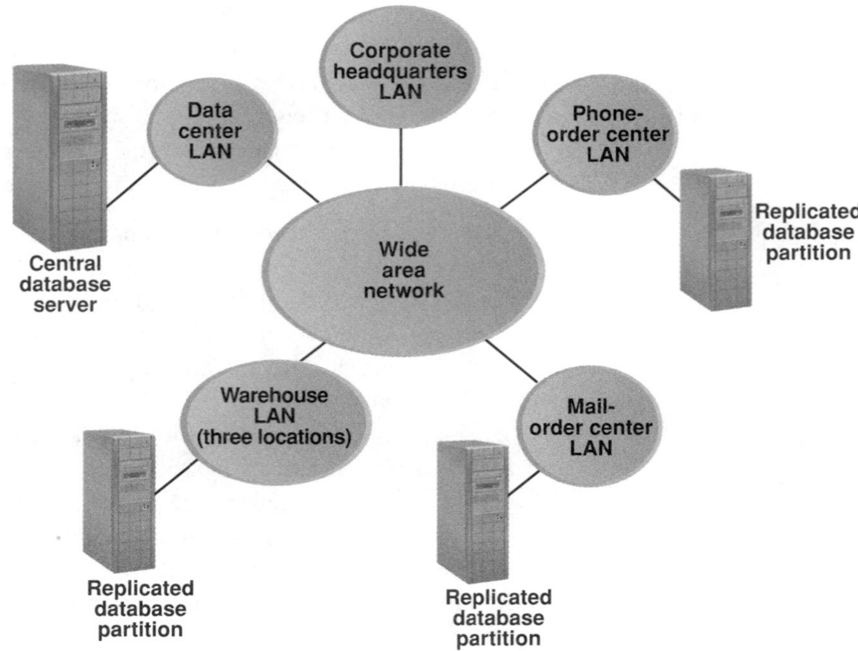

FIGURE 10-29
A replicated and partitioned database server architecture for RMO.

The primary advantages of this architecture are fault tolerance and reduced WAN capacity requirements. Each location could continue to operate if the central database server failed. However, as the remote locations continued to operate, their database contents would gradually drift out of synchronization. A synchronization strategy must be implemented to address both regular database updates and recovery from server failure. The strategy could vary by location.

The primary disadvantages to the distributed architecture are cost and complexity. The architecture saves WAN costs through reduced capacity requirements but adds costs for additional database servers. The cost of acquiring, operating, and maintaining the additional servers would probably be much higher than the cost of additional WAN capacity.

So which alternative makes the most sense for RMO? The answer depends on some data that hasn't yet been gathered and on answers to some questions about desired system performance. RMO management must also determine its goals for system performance and reliability. The distributed architecture would provide higher performance and reliability but does so at substantially increased cost. Management must determine whether the extra cost is worth the expected benefits.

Additional data about network traffic is needed to precisely determine LAN and WAN communication requirements between clients and database servers. Estimates of transaction and query volume are required for each location. These estimates must cover both normal and peak demand periods. Such data may be gathered during the analysis or the design phases. The data are required to determine an optimal configuration of LAN, WAN, and database server capacity. Note that the analysis of the data and the actual design of the networks and database architecture are complex endeavors that require highly specialized knowledge and experience.

Summary

Most modern information systems store data in a database and access and manage that data using a DBMS. Relational databases and DBMSs are most commonly used today, but object databases and DBMSs are increasing in popularity. One of the key activities of systems design is the development of a relational or object database schema.

A relational database is a collection of data stored in tables. A relational database schema is normally developed from an entity-relationship diagram. Each entity is represented as a separate table. One-to-many relationships are represented by embedding foreign keys in entity tables. Many-to-many relationships are represented by creating additional tables containing foreign keys of the related entities.

An object database stores data as a collection of related objects. The design class diagram is the starting point for developing an object database schema. The database schema defines each class, and the ODBMS stores each object as an instance of a particular class. Each object is assigned a unique object identifier. Relationships among objects are represented by storing the object identifier of an object within related objects.

Objects can also be stored within a relational database. Object attributes and relationships among objects—including one-to-many, many-to-many, and generalization hierarchies—can be represented. However, methods cannot be stored and inheritance cannot be directly represented.

Medium- and large-scale information systems typically use multiple databases or database servers in various geographic locations. A replicated database architecture employs multiple database copies on different servers, usually in different geographic locations. A partitioned database architecture employs partial database copies stored on different servers in close proximity to a distinct user subset. A federated database architecture employs multiple databases (possibly of different types) and a special-purpose DBMS that provides a unified view of the databases and a single point of access.

Key Terms

association class, p. 376
attribute, p. 358
attribute value, p. 359
complex data type, p. 382
data element, p. 359
data type, p. 382
data warehouse, p. 390
database (DB), p. 356
database management system (DBMS), p. 356
database synchronization, p. 387
field, p. 358
field value, p. 359
first normal form (1NF), p. 367
foreign key, p. 360
functional dependency, p. 367
hybrid DBMS, p. 374
hybrid object-relational DBMS, p. 378
key, p. 359
navigation, p. 373
normalization, p. 367
object database management system (ODBMS), p. 372
Object Definition Language (ODL), p. 372
persistent object, p. 373
physical data store, p. 356
primary key, p. 359
primitive data types, p. 382
record, p. 358
referential integrity, p. 365
relational database management system (RDBMS), p. 358
relation, p. 358
row, p. 358
schema, p. 356
second normal form (2NF), p. 367
table, p. 358
third normal form (3NF), p. 367
transient object, p. 372
tuple, p. 358
user-defined data types, p. 382

Review Questions

1. List the components of a DBMS and the purpose or function of each.
2. What is a database schema? What information does it contain?
3. Why have databases become the preferred method of storing data used by an information system?
4. List four different types of database models and DBMSs. Which are in common use today?
5. With respect to relational databases, briefly define the terms *row, tuple, column,* and *attribute.*
6. What is a primary key? Are duplicate primary keys allowed? Why or why not?
7. What is the difference between a natural key and an invented key? Which type is most commonly used in business information processing?
8. What is a foreign key? Why are foreign keys used or required in a relational database? Are duplicate foreign key values allowed? Why or why not?
9. List the steps used to transform an ERD into a relational database schema.
10. How is an entity on an ERD represented in a relational database?
11. How is a one-to-many relationship on an ERD represented in a relational database?
12. How is a many-to-many relationship on an ERD represented in a relational database?
13. What is referential integrity? Describe how it is enforced when a new foreign key value is created, when a row containing a primary key is deleted, and when a primary key value is changed.
14. What types of data (or fields) should never be stored more than once in a relational database? What types of data (or fields) usually must be stored more than once in a relational database?
15. What is relational database normalization? Why is a database schema in third normal form considered to be of higher quality than an unnormalized database schema?
16. Describe the process of relational database normalization. Which normal forms rely on the definition of functional dependency?
17. List the steps used to transform a class diagram into an object database schema.
18. What is the difference between a persistent object and a transient object? Provide at least one example of each object type.
19. What is an object identifier? Why are object identifiers required in an object database?
20. How is a class on a class diagram represented in an object database?
21. How is a one-to-many relationship on a class diagram represented in an object database?
22. How is a many-to-many relationship without attributes represented in an object database?
23. What is an association class? How are association classes used to represent many-to-many relationships in an object database?
24. Describe the two ways in which a generalization relationship can be represented in an object database.
25. Are key fields or attributes required in an object database? Why or why not?
26. Describe the similarities and differences between an ERD and a class diagram that models the same underlying reality.
27. How is a class on a class diagram represented in a relational database?
28. How is a one-to-many relationship on a class diagram represented in a relational database?
29. How is a many-to-many relationship on a class diagram represented in a relational database?
30. What is the difference between a primitive data type and a complex data type?
31. What are the advantages of having an RDBMS provide complex data types?
32. Does an ODBMS need to provide predefined complex data types? Why or why not?
33. Why might all or part of a database need to be replicated in multiple locations?
34. Briefly describe the following database replication methods: replicated database servers, partitioned database servers, and federated database servers.
35. What additional database management complexities are introduced any time database contents are replicated in multiple locations?
36. Why might a replicated database server be required? A partitioned database server? A federated database server?

Thinking Critically

1. The Universal Product Code (UPC) is a bar-coded number that uniquely identifies many products sold in the United States. For example, all copies of this textbook sold in the United States have the same UPC bar code on the back cover. Now consider how the design of the RMO database might change if all items sold by RMO were required by law to carry a permanently attached UPC (for example, on a label sewn into garments). How might the RMO relational database schema change under this requirement?
2. Assume that RMO plans to change its pricing policy. If two or more catalogs are in circulation at the same time, then all item and package prices in the catalogs must be the same. Prices can still rise or fall over time and those changes will be recorded in the database and printed in newly issued catalogs. Any customer who makes an order will always be given the lower of the current price or the price in the current catalog. What changes to the tables shown in Figure 10-9 will be required to ensure that the RMO database is in 3NF after the pricing policy change?
3. Assume that RMO will begin asking a random sample of customers who make orders by telephone about purchases made from competitors. RMO will give customers a 15 percent discount on their current order in exchange for answering a few questions. To store and use this information, the ERD and class diagram will be expanded with two new entities (classes) and three new relationships. The new entities (classes) are Competitor and ProductCategory. Competitor has a one-to-many relationship with ProductCategory, and the existing Customer entity (class) also has a one-to-many relationship with ProductCategory. Competitor has a single field (attribute) called Name. ProductCategory has four fields (attributes): Description, DollarAmountPurchased, MonthPurchased, and YearPurchased. Revise the relational database schema shown in Figure 10-9 to include the new entities and relationships. All tables must be in 3NF.
4. Assume that RMO is developing its database using object-oriented methods. Assume further that the database designers want to make some changes to the class diagram in Figure 10-21. Specifically, they want to make ProductItem an abstract parent class from which more specific product classes are specialized. Three specialized classes will be added: ClothingItem, EquipmentItem, and OtherItem. ClothingItem will add the attribute Color, and that same attribute will be removed from the InventoryItem class. Equipment item will also add an attribute called Color but will not have an attribute called Gender. OtherItem will have both the Color and Gender attributes. Revise the relational database schema in Figure 10-22 to store the new ProductItem generalization hierarchy. Use a separate table for each of the specialized classes.
5. Assume that RMO will use a relational database as shown in Figure 10-9. Assume further that a new catalog group located in Milan, Italy, will now create and maintain the catalog. To minimize networking costs, the catalog group will have a dedicated database server attached to its LAN. Develop a plan to partition the RMO database. Which tables should be replicated on the catalog group's local database server? Update Figure 10-29 to show the new distributed database architecture.
6. Revisit the issues raised in the Nationwide Books (NB) case at the beginning of the chapter. Should NB adopt an ODBMS for the new web-based ordering system? Why or why not?

Experiential Exercises

1. Network databases were not discussed in detail in this chapter, but they are discussed in many database textbooks. Investigate the network database model and its use of pointers to represent relationships among record types. In what ways is the use of pointers in a network database similar to the use of object identifiers in an object database? Does the similarity imply that object databases are little more than a renamed version of an older DBMS technology?
2. Access the Object Database Management Group web site (www.odmg.org) and gather information on changes and additions to the ODMG standards that are currently under consideration. Will the proposed changes and additions alter the approaches to object or distributed database design discussed in this chapter?
3. Investigate the student records management system at your school to determine what database management system is used. What database model is used by the DBMS? If the DBMS isn't object-oriented, find out what plans, if any, are in place to migrate to an ODBMS. Why is the migration being planned (or not being planned)?

4. Visit the web site of an on-line catalog vendor similar to RMO (such as www.llbean.com) or an on-line vendor of computers and related merchandise (such as www.cdw.com). Browse the on-line catalog and note the various types of information contained therein. Construct a list of complex data types that would be needed to store all of the on-line catalog information.

Case Studies

The Real Estate Multiple Listing Service System

Refer to the description of the Real Estate Multiple Listing Service system in the Chapter 5 case studies. Using the ERD and class diagram for that system as a starting point:

1. Develop a relational database schema in 3NF.
2. Develop an ODL database schema.

The State Patrol Ticket Processing System

Refer to the description of the State Patrol ticket processing system in the Chapter 5 case studies. Using the ERD and class diagram for that system as a starting point:

1. Develop a relational database schema in 3NF.
2. Develop an ODL database schema.

The Reliable Pharmaceutical Service

Refer to the description of the Reliable Pharmaceutical Service (RPS) system in the Chapter 6 case studies. Using the ERD (Figure 6-40) as a starting point:

1. Develop a relational database schema in 3NF.
2. Convert the ERD into a class diagram and develop an ODL database schema.

Further Resources

C. J. Date. *An Introduction to Database Systems*, 4th edition. Addison-Wesley, 1986. ISBN 0-201-14201-5.

Fred R. McFadden, Jeffrey A. Hoffer, and Mary B. Prescott. *Modern Database Management*, 5th edition. Addison-Wesley, 1998. ISBN 0-805-36054-9.

The Object Database Management Group web site, http://www.odmg.org.

Robert Orfali, Dan Harkey, and Jeri Edwards. *The Essential Client/Server Survival Guide*, 2nd edition. John Wiley & Sons. 1996. ISBN 0-471-15325-7.

Designing Inputs, Outputs, and Controls

Learning Objectives

After reading this chapter, you should be able to:

- Explain the importance of integrity controls
- Identify required integrity controls for inputs, outputs, and processing
- Understand the range of inputs and outputs necessary for a system
- Define inputs and outputs based on the requirements of the application program
- Design printed and on-screen reports appropriate for system users

Chapter Outline

Integrity Controls

Design of System Inputs

Design of System Outputs

CUSTOMIZED CARS CLEARINGHOUSE: MOVING INTO ELECTRONIC COMMERCE

Customized Cars Clearinghouse (CCC) is a California firm that specializes in parts for customizing automobiles. CCC carries a wide range of high-performance equipment, not only for new models but also for vintage automobiles. The company has built a loyal clientele based on one primary principle: "There is not a part that CCC cannot provide." Customers know that if they need a piece of equipment, CCC will find it for them.

Six months ago, the senior executives of CCC decided to expand their successful business into the world of electronic commerce. It seemed like a good move since automobile buffs currently had to travel long distances to come to one of the CCC stores for a part. CCC began and is now in the middle of a large development project to provide e-commerce as one of its sales channels.

Nathan Lopez, the project manager for the e-commerce project, was putting together a small team to review the security and control issues for the new system. Even though CCC had been very advanced in its use of IT, it had never opened up its system to direct input from customers and other people outside the company. Nathan wanted to make sure that there were adequate controls to protect both the customers and the CCC assets. The purpose of the development team is to design adequate system controls to make sure that only valid, correct transactions are entered into the system, that the transactions are protected, secure, and correctly processed. Obviously, the level of risk increases when a company's systems are opened up to the general public.

The team's first job is to identify all of the possible sources of fraudulent or incorrect information. The team then must ensure that the system includes adequate controls to make it as secure as possible. Nathan had been to several presentations on security and had heard that no system is 100 percent secure. He was carefully considering who would be on this team to make sure that all the correct controls were developed. He did not want the team to be too large, but he also wanted to make sure that it represents the correct mix of problem domain knowledge, system development experience, and security expertise. He finally identified three key people who would compose the team.

Mary Ellen Hamilton was a longtime CCC employee who had worked in almost every part of the company. She had worked both with sales and with accounting. She would bring a lot of knowledge about the company and about company systems. William Handspaw was the lead systems analyst on the project. He had also been with CCC a considerable length of time and had been involved in many successful development projects. Nathan had a lot of trust in William. Finally, Eric Snow, who was new to the company, would be the third member of the team. Eric had recently been hired into the systems department and brought extensive expertise in security and online systems. The team would depend on Eric for the detailed technical knowledge of how to integrate the controls into the system. However, the experience of Mary Ellen and William would provide the broad knowledge necessary to identify the necessary controls. Nathan was satisfied that this team would provide a good design for the necessary integrity controls.

OVERVIEW

In this chapter, you will learn about integrity controls for the system and about designing the interfaces to the system. The objective of this chapter is to give you a foundation in the issues and requirements to develop the controls and the interfaces. Chapter 12 focuses more specifically on human-computer interaction (HCI) and the design of the user interface to the system.

Designing system controls is crucial because more and more computer systems built today exist in an open environment. That is, they are part of a configuration or network that provides broad and wide access to many different people both within and outside an organization. One of the major considerations in the design of these systems is how to provide access to the required information and at the same time protect the information against damage—accidental or intentional. The user interface is especially vulnerable to damage since it is the method of access to the system. To set the groundwork for the design of the inputs and outputs in an open, yet controlled, environment, we begin this chapter with an introduction to controls.

Following the section on controls, the chapter covers inputs and outputs. In each of these sections, we discuss the principles and tasks that delineate the design of the inputs and outputs. These design tasks are closely related to the application architecture design and database design tasks explained in Chapters 9 and 10. In essence, the tasks described in this chapter define the architectural design of the system interfaces.

Integrity Controls

integrity control
Mechanisms and procedures that are built into the system to safeguard both the system and the information contained within it

Integrity controls are mechanisms and procedures that are built into the system to safeguard both the system and the information within it. Since most violations occur from inappropriate access, the design of the interface must carefully consider mechanisms to protect the system's integrity. However, integrity controls are not limited to input and output. We also discuss the broader issues associated with all types of integrity controls. Let's describe a few scenarios to illustrate why controls might be necessary.

- A furniture store sells merchandise on credit with internal financing. An error was made to a customer balance. How do we ensure that only a manager, someone with authority to make adjustments to credit balances, can make the correction?
- A person in accounts payable uses the system to write checks to suppliers. How does the system ensure that the check is correct and that it is made out to a valid supplier? How does the system ensure that no one can commit fraud by writing checks to a bogus supplier? How does the system know that a given payment has been authorized?
- Many companies now have internal LAN networks or intranets. How does a company protect its sensitive data from being accessed by outsiders or even from disgruntled employees?
- Electronic commerce is expanding at exponential rates, and many companies are now providing e-commerce sites. How does the company ensure that the financial transactions of its customers are protected and secure? How does a company make sure that its systems and databases are protected from hackers who use the Internet access paths to break in?

As you can see, all of these situations involve normal business and system activities. Developers of a new system must consider how to protect and maintain information integrity. In this section, you will learn about the different types of controls and how to include them in a newly developed information system.

Objectives of Integrity Controls

The objectives of integrity controls are to

- Ensure that only appropriate and correct business transactions occur
- Ensure that the transactions are recorded and processed correctly
- Protect and safeguard the assets (including information) of the organization

The first objective, to ensure that only appropriate and correct business transactions occur, focuses on the identification and capture of input transactions. Integrity controls must make sure that all important business transactions are included—that is, that none are lost or missing and that no fraudulent or erroneous transactions are entered.

The second objective, to ensure that the transactions are recorded and processed correctly, also relates to errors and fraudulent activities. Controls need to detect and alert users to data-entry errors and system bugs that cause problems in processing and recording data. An example of a fraudulent activity is a user who changes the dollar amount on an otherwise valid transaction.

The third objective, to protect and safeguard the assets of the organization, addresses loss of information due to computer crashes or catastrophes. It also includes protection of important information on computer files that could be destroyed by a disgruntled employee or a hacker.

Frequently, system developers are so focused on designing the system software itself that they forget to develop the necessary controls. Because computer systems are so pervasive and companies depend on information systems so heavily, a development project that does not specifically include integrity controls is inviting disaster. The system will be subject to errors, fraud, and deceptive practices, making it unusable.

System Access Controls

One general area of controls that is becoming increasingly important is access to the system. Modern operating systems, networking software, and Internet access all need to implement control mechanisms. These mechanisms can be used to control access to any resource managed by the operating system or network—including hardware, application programs, and data files.

system access control
An integrity control that determines who has access to a system and its data

System access controls are mechanisms that are established to restrict or control what portions of the computer system a person can use. This category includes controls to limit access to certain applications or functions within an application, restrict access to the computer system itself, and limit access to certain pieces of data.

With proper design and implementation, an information system can use access control functions embedded in system software. The advantage to this approach is that a consistent set of access controls is then applied to every resource on a hardware platform or network. Thus, the systems designer can implement a single access control scheme and apply it to every resource or information system.

The systems designer can also add controls over and above those already provided by system software. However, designing and implementing effective application-based access controls require technical expertise. Operating and network software developers expend considerable energy and resources to develop reliable and efficient access controls, and it is difficult and expensive for a typical organization to duplicate these efforts. For these reasons, most information systems build on the access control already within system software.

To begin development of access controls, we first must identify categories of users. The designers must consider all three of these categories: (1) unauthorized users, (2) registered users, and (3) privileged users. **Unauthorized users** are people who are not allowed access to any part or functions of the system. Such users include employees who are prohibited from accessing the system, former employees who no longer are permitted to access the system, and outsiders such as hackers and intruders. Controls must be able to identify and exclude access to these people.

unauthorized user
A person who does not have authorized access to a system

registered user
A user who is registered or known to the system and is authorized to access some part of it

Registered users are those who are authorized to access the system. Normally, various levels of registered users are set up depending on what they are authorized to view and update. The different levels of access are defined during the design of the new system. For example, some users may be allowed to view data but not update it, and other users can update only certain data fields. Some screens and functions of the new system may be hidden for other levels of registered users. The important point for systems designers is to recognize that there may be multiple levels of registered users.

privileged user
A user who has special security access privileges to a system

Privileged users include people who have access to the source code, executable program, and database structure of the system. These people include system programmers, application programmers, operators, and system administrators, and they may also have differing levels of security access. Usually system programmers have full access to all components of the systems and data. Application programmers have access to the applications themselves, but often not to the secure libraries and data files used for the systems in production. System administrators have access to all functions of the system, and can control and establish the various levels of registration and register users. A system administrator also usually has software programs to help control access and to monitor access attempts.

The idea of the different types of users is fundamental to the design of access controls. Figure 11-1 illustrates the various types of users and the kind of access that is appropriate. With this starting point, let's identify various access controls.

FIGURE 11-1
Users and access roles to computer systems.

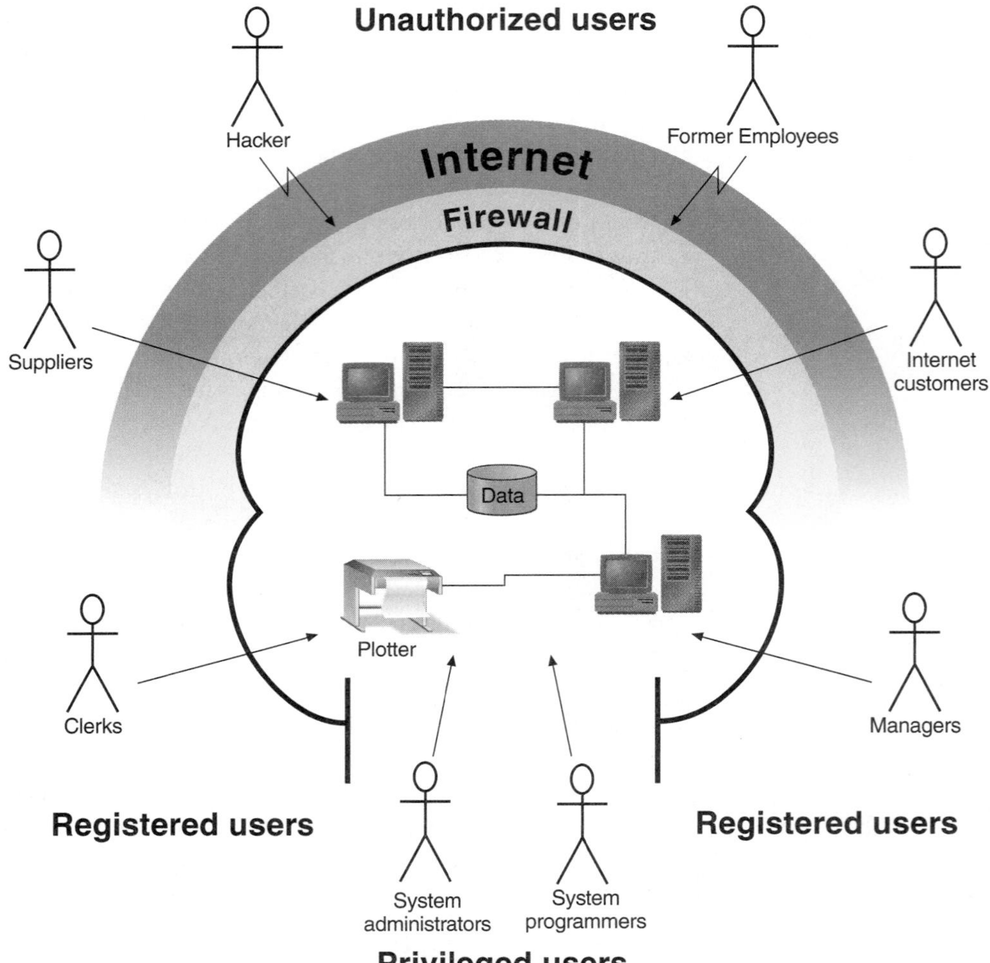

Physically Securing Locations Physically secure locations primarily protect physical equipment such as computers, hard disk storage devices, and backup data tapes. Securing a location cannot protect software or the on-line data files on networked systems. However, some mainframe computers require that supervisory functions be done at a specific master console. For that situation, protection of the physical console also increases security of the operating system and application programs.

Managing User Access The most common technique for managing user access to information systems is a user identifier (ID) and password. Ideally, each user has his or her unique logon identifier and a unique, secret password. User IDs are defined by a standard algorithm. Most techniques take a combination of the user name or initials to develop the ID.

Two techniques are used to define passwords. The computer can randomly generate and assign passwords, or each user can define his or her own password. There are advantages to both techniques. The first creates passwords that are usually longer and more random, but they tend to be hard for users to remember. Most users would have a hard time trying to remember a password such as a3x7869bts21. User-developed passwords are easier to remember, but they are usually not as complex and therefore not quite as secure. Some restrictions can be placed on the syntax of the password to ensure at least a minimum level of security. Typical restrictions are that the password must be at least 8 or 10 characters long, contain both numbers and letters, contain both lowercase and uppercase letters, and so forth. Also, easily guessed passwords such as names or birthdays should be avoided.

The security system should be organized so that all resources can be accessed using the same unique identifier and password combination. In other words, only one user ID/password combination should be required for access to the different systems throughout the organization. When users have to remember different IDs and passwords to access different systems, they often write them down and post them near the computer. Obviously, this defeats the purpose of user verification security.

One other policy to ensure ongoing security is that of changing passwords periodically. Some companies require users to change their passwords every 30 or 90 days. This can help avoid the potential security breach that occurs when someone discovers a user's password. There are two major disadvantages to this policy. First, users more easily forget their current password and consequently start writing it down and placing it where it is visible. The other problem is that all systems that require password access must be updated concurrently, which may not happen. Some systems may accept the new password, but others may still require the old password. Users then become confused trying to remember which password to use for which set of resources.

A final security step is to make sure the system keeps a record of attempted logons, especially unsuccessful ones. An unsuccessful logon may simply indicate that the user mistyped or forgot a password, but it may also indicate an attempted breach of security, which should be investigated.

Controlling Access with Visibility Physical security and user management controls impact the entire organization and may be administered by a centralized group. Another type of control determines what each user actually sees, and must be integrated into the application system itself. Designers define and develop the various levels of visibility and access.

Visibility controls are defined by the various identifying classes of registered users and which functions, screens, forms, fields, and reports will be available to those classes of users. For example, in a sales reporting system, salespeople may be able to see their own orders, sales, and commissions. A supervisor may be able to view the performance of all salespeople within his or her jurisdiction. The supervisor may also be able to enter and update information on specific orders. A higher-level manager may see the data and also be able to update individual commission rates, bonuses, and sales targets. The amount of visibility thus increases with the level of authority and responsibility in an organization.

Now that we have explored access and security issues, we turn to details about integrity controls for inputs and outputs that the system designer should consider. One of the most critical objectives to maintain system integrity is to prevent erroneous and misleading information from entering the system. Likewise, important information within the system should be safeguarded so that access is controlled.

Input Integrity Controls

Input integrity controls are used with all input mechanisms, from specific electronic devices to standard keyboard inputs. Input controls are an additional level of verification that helps reduce errors on input data. For example, a system may need a certain amount of information for a valid entry, but an input device cannot ensure that all the necessary fields have been entered. An additional level of verification, which we will call a *control,* is necessary to check for completeness.

Error Detection and Elimination The old computer systems adage of "garbage in, garbage out" relates to input controls, where the objective is to reduce bad data within the system due to erroneous input. Historically, the most common control method to ensure correct input was to enter data twice. This technique, called keypunch and verify, was first developed for batch entry of large amounts of data. One person would enter the data, and a second person would reenter it on equipment that would then verify that the two inputs were the same. Today, that technique is not used as much because many high-volume transactions are scanned for data. On-line systems also validate input as it is being entered. Here are the more common control techniques that are used today:

field combination control
An integrity control that verifies the data in one field based on data in another field or fields

value limit control
An integrity control that identifies when a value in a field is too large or too small

completeness control
An integrity control to ensure that all necessary fields on an input form have been entered

data validation control
An integrity control to validate the input data for correctness and appropriateness

- **Field combination controls** review various combinations of fields to ensure that the correct data is entered. For example, on an insurance policy, the application date must be prior to the date the policy is placed in force.
- **Value limit controls** check numeric fields to make sure that the amount entered is reasonable. For example, the amount of a sale or the amount of a commission usually falls within a certain range of values.
- **Completeness controls** ensure that all the necessary fields are completed. This kind of check can be executed as input occurs so that, depending on which fields are entered, additional required fields must also be entered. For example, if a dependent is entered on an insurance form, then that person's birthday must also be entered.
- **Data validation controls** ensure that numeric fields that contain codes are correct. For example, bank account numbers might be created with a seven-digit field and a trailing check digit to make an eight-digit account number. The check digit is calculated based on the previous seven digits, and the system recalculates it as the data-entry person enters the account number with check digit. If results do not match, then an input error has occurred. Other data validation can be done online against internal tables or files. For example, a customer number can be validated against the customer file at the time a new order is entered. The systems designer can reduce the need for this type of control by designing a system to obtain the data for a particular field from other information already in the system.

transaction logging
A technique whereby all updates to a database are recorded with the information of who, when, and how the update was performed

Transaction Logging **Transaction logging** is a technique by which any update to the database is logged with audit information such as user ID, date, time, input data, and type of update. The fundamental idea is to create an audit trail of all updates to the database that can trace any errors or problems that occur. The more advanced database systems—such as those that run on servers, workstations, and mainframes—include this as part of the DBMS software. However, several smaller DBMSs, particularly those that run on personal computers, do not include this capability, so design teams must add this capability directly to those applications.

Transaction logging achieves two objectives. First, it helps discourage fraudulent transactions. If a person knows that every transaction is logged, then that person is less apt to attempt a fraudulent transaction. For example, if a person knows his or her ID will be associated with every check request, then that person is not likely to request a bogus payment.

The second objective of a logging system is to provide a recovery mechanism for erroneous transactions. A mid-level logging system maintains the set of all updates. The system can then recover from errors by "unapplying" the erroneous transactions. More sophisticated logging systems can provide a "before" and "after" image of the fields that are changed by the transaction, as well as the audit trail of all transactions. These sophisticated systems are typically used only for highly sensitive or critical data files, but they do represent an important control mechanism that is available when necessary.

Output Integrity Controls

Output from a system comes in various forms, such as output that is used by other systems, printed reports, and data output on computer screens. The purpose of output controls is to ensure that output arrives at the proper destination, and is correct and accurate, current, and complete. It is especially important that reports with sensitive information arrive at the proper destination and that they not be accessed by unauthorized persons.

Destination Controls In the past, when most output was in printed form, a distribution control desk collected all the printed reports from the nightly processing and distributed them to the correct departments and people. This control desk was important because some of the reports had sensitive, confidential information, and it was important that those reports be kept secure. Systems with good controls printed destination and routing information on a report cover page along with the report. Today, businesses accomplish the same function of a control desk by placing printers in each of the locations that need printed reports. It is still a good idea, however, to print a cover sheet with destination and report heading information. Controlling access to these reports then becomes an issue of physical access. Destination codes and routing capabilities are included during the design process to handle the distribution of reports to separate printing facilities. These types of controls are called **destination controls**.

destination controls
Integrity controls to ensure that output information is channeled to the correct persons

Electronic output to other systems is usually provided in two forms: either an on-line transaction-by-transaction output, or a single data file with a batch of output transactions. Each form has its own type of controls. If the system produces on-line transactions, then it must ensure that each transaction includes the routing codes identifying the correct destination. Both systems will need to work together to ensure that each transaction is sent and received correctly. The output transaction will have verification codes and bits to permit the receiving system to verify the accuracy of the transaction. The receiving system also responds with an acknowledgment of a successful receipt of the transaction. Many of these controls are now built into the network transmission protocols. However, during the design activities, the systems designers need to be aware of the network and operating system capability and supplement it where necessary to ensure that the data are received successfully.

Controls for output data files carefully identify the contents, version, date, and time of the file. Normally, a system produces a data file, either on magnetic tape or disk, and another system must find that data file and use it. The major control issue is to ensure that the second system uses the correct data file. For example, to avoid serious problems, we want to make sure that Friday's transactions don't get run twice. Or, if, by some processing quirk, two data files are produced for the same day—one for the first half of the day and another for the second—the system must use both data files. Or, if the second system had processing errors and needs to be rerun, it must be able to find the correct file to use on its rerun. Controls for this situation generally have special beginning and ending records that contain date, time, version, record counts, dollar control totals, processing period, and so forth. During systems design, provision must be made to accumulate the appropriate totals and to produce the necessary control records.

Destination controls for computer screen output are not as widely used as for printed reports. Normally, the availability of information on computer screens is controlled by the user access controls discussed earlier. In some instances, however, destination controls limit what information can be displayed on which terminal. This is true primarily for military or other systems that house computer terminals in secure areas and provide access to the system's information to anyone who has access to the area. The design of these systems requires close coordination between the application program and the network security control system.

Completeness, Accuracy, and Correctness Controls The completeness, accuracy, and correctness of output information are a function primarily of the internal processing of the system rather than any set of controls. We ensure completeness and accuracy by printing control fields on the output report. For example, every report should have a date and time stamp, both for the time the report was printed and the date of the underlying data. Frequently, they are the same, but not always, especially when a report is reprinted due to a previous error. The following items are controls that should be printed on reports:

- Date and time of report printing
- Date and time of data on the report
- Time period covered by the report
- Beginning header with report identification and description
- Destination or routing information
- Pagination in the form "page — of —"
- Control totals and cross footings
- "End of Report" trailer
- Report version number and version date (such as for special printed forms)

Now that we have given an overview of input and output integrity controls, we can turn to the process of identifying and developing the architectural structure for the system inputs.

Design of System Inputs

When designing inputs for the system, the system developer must perform four tasks:

- Identify the devices and mechanisms that will be used to enter input
- Identify all system inputs and develop a list with data content of each

- Determine what kinds of controls are necessary for each system input
- Design and prototype the electronic forms (the windows the user works with) and other inputs

In this chapter, we discuss the first three tasks of the preceding list. Chapter 12 explains various techniques for designing and prototyping the electronic forms associated with direct user input. The first three items are architectural issues, which should be addressed first, whereas the fourth item is part of detailed, or low-level, design.

The first task, identifying devices and mechanisms, is a high-level review of the most up-to-date method of entering information into a system. In nearly all business systems, some input is done through electronic forms by end users. However, in today's high-technology world, there are numerous ways to enter information into a system. There are all types of scanning, reading, and transmitting devices that are faster, more efficient, and less error-prone.

The second task, developing the list of required inputs, provides the link between the design of the application software (explained in Chapter 9) and the design of the system and user interfaces. As described earlier, the design of the system inputs and interfaces must be integrated with the design of the application. This activity accomplishes that purpose.

The third task is to identify the control points and the level of security required for the system being developed. A statement of policy and control requirements should be done before beginning the detailed design of the electronic forms making up the user interface.

The fourth task, to design and prototype the electronic forms and other inputs, begins by creating user dialog scenarios and storyboard sketches of screens used by the users during each dialog. Sometimes working prototypes of electronic forms are created. Graphical programming languages—such as Visual Basic, Delphi, or Powerbuilder—can help in this activity. Building prototypes helps ensure that the forms are workable and acceptable to the users. Chapter 12 explains these various approaches.

Also included in the fourth task is the design of any required paper forms. Even though in the design of new systems we try to eliminate paper forms as much as possible, the requirements of data entry do not always make it feasible to eliminate paper forms completely. Thus, any forms that may be needed are designed as part of this task.

Identifying Devices and Mechanisms

Often when analysts begin developing a system, they assume that all input will be entered via electronic, graphical forms because they are now so common on personal computers and workstations. However, as the design of the user inputs commences, one of the first tasks is to evaluate and assess the various alternatives for entering information into the system.

The primary objective of any form of data input is to enter new, error-free data into the system or to update information with error-free data. The operative word here is error-free. Several rules of thumb help reduce input errors:

1. Capture the data as close to the originating source as possible.
2. Use electronic devices and automatic entry whenever possible.
3. Avoid human involvement as much as possible.
4. If the information is available in electronic form anywhere, use it instead of reentering the information.
5. Validate and correct information at the time and location it is entered.

Many firms design systems that enable the capture of data at the same point that the data are originally generated. For example, one way to sell a life insurance policy is to have the applicant, or the insurance agent, fill out a policy application. Then the agent sends the paper application to a central office to be entered into the system. In this method, numerous errors occur due to indecipherable writing, key entry errors, missing fields, and so forth. Another way to submit a new policy is to have the agent carry a laptop computer with easy-to-use electronic forms so that applicants can fill in the data themselves. Or better yet from a sales technique point of view, the agent can enter the data while the applicant looks over the agent's shoulder and verifies that the information is accurate and complete. In this last example, information about the applicant is entered and verified in the applicant's presence. In fact, a portable printer can even be attached to the laptop and the completed form can be printed out for the applicant to review immediately. This approach dramatically reduces the error rate, and speeds up the total business process, of new policy data entry.

The second and third points, automate data entry and avoid human involvement, are very closely related and in many instances are essentially different sides of the same coin, although using electronic devices does not automatically avoid human involvement. We list them separately to emphasize that you should think about both principles together. When developers think carefully about avoiding human input and using electronic input media, the resulting system has fewer electronic input forms and the problems associated with forms. One of the most pervasive sources of erroneous data is from users making mistakes in typing fields and numbers. Many methods and devices allow data to be captured without human keystroking. A few of the more prevalent devices are:

- Magnetic card strip readers
- Bar code readers
- Optical-character recognition readers and scanners
- Touch screens and devices
- Electronic pens and writing surfaces
- Digitizers, such as digital cameras and digital audio devices

We have all seen new systems that involve electronic devices. At the grocery store, electronic scanners identify and price each item from the printed UPC codes. Automatic weighing machines weigh and price the produce. Cash registers now read your check, including the amount and customer and bank information. New self-service checkout stations depend almost entirely on automated data entry devices.

Historically, legal requirements have required copies of paper contracts and signatures. Today, new precedents allow the digitizing of paper documents, including signatures. Credit card purchases now record digitized signatures to eliminate the need for paper credit card vouchers. This technique conforms to the principles stated above—the information is captured at its source in electronic form, which eliminates many of the sources of errors.

The next principle of error reduction is to reuse the information already in the computer. Some systems often require reentry of the same information multiple times. Not only does this generate errors, it creates multiple copies of the same information, which require an extra level of controls and programming to ensure the various copies are synchronized. When there is an error, it is very difficult to know which copy is correct. Also, when a change is required, it must be made to all copies of the data. One interesting example of using existing information, as well as other high-tech solutions, is found in car rental systems. When a car is rented, the customer, credit card, and car mileage and fuel information is captured. When the car is returned, an agent in the parking lot simply scans the contract ID and enters the return mileage and fuel-gauge level. Existing information is used to calculate the charges and print a credit card receipt for the customer, right in the parking lot. This solution was designed primarily to provide a higher level of customer service, but it also eliminates many problems and errors with data entry.

Another mechanism for system input is to have an interface directly from another system. Electronic data interchange (EDI) is one approach to reduce the need for user input, either with a scanning device or with the keyboard. Input information, such as purchase orders, invoices, inventory updates, and payments, all occur via one system sending transactions to another system. With EDI, this normally occurs between systems in separate organizations. However, the same principle applies to systems within the same organization. In fact, enterprise resource planning (ERP) systems are popular because they make companywide integrated systems possible. More information about ERP systems is provided in Chapter 14.

Another system-to-system interface that is gaining popularity is use of transactions based on the extensible markup language (XML). XML is much like hypertext markup language (HTML). HTML consists of embedding formatting information within the document itself. Thus, a program that can read HTML reads a text document and then, using the embedded markup codes, can process or format the document correctly. XML is an extension to HTML that permits textual messages to be sent with self-defining data structures embedded. Thus a transaction, which contains data fields, can be sent with XML codes to define the meaning of the data fields. Many newer systems are using this technique to provide a common system-to-system interface. Figure 11-2 illustrates a simple XML transaction that can be used to transfer customer information between systems.

FIGURE 11-2
A system-to-system interface based on XML.

```
<customer record>
        <accountNumber>RMO10989</accountNumber>
        <name>William Jones</name>
        <billingAddress>
                <street>120 Roundabout Road</street>
                <city>Los Angeles</city>
                <state>CA</state>
                <zip>98115</zip></billingAddress>
        <shippingAddress>
                <street>120 Roundabout Road</street>
                <city>Los Angeles</city>
                <state>CA</state>
                <zip>98115</zip></shippingAddress>
        <dayPhone>215.767.2334</dayPhone>
        <nightPhone>215.899.8763</nightPhone>
</customer record>
```

Developing the List of Inputs and Data Requirements of Each

The objective of this task is to ensure that the designer has identified all of the required inputs to the system and specified them correctly. As with other aspects of analysis and design, this task also provides a check on the quality of the analysis models. The fundamental approach used is to identify all information flows that cross the system boundary. The idea is the same for both traditional structured models and object-oriented models; however, the detailed techniques are unique to each model.

Using Structured Models During systems design using the structured techniques, one of the first tasks is to define the automation boundary. Figure 11-3, duplicated from Figure 9-2, is an example of an automation boundary on a data-flow diagram. Several of the inputs to the system based on this data-flow diagram are:

- Time-card information
- Updates to tax rate tables
- Updates to employee files

Even though it is possible to build an automation boundary on a high-level DFD and identify the inputs on this diagram, it usually is better to work from the DFD fragments or even more detailed DFDs. The high-level diagram frequently does not provide enough detail to discern many data flows and, hence, inputs to the system. For example, one of the processes on the diagram, *Correct errors,* is intersected by the system boundary. Thus, a view from a lower-level DFD is necessary to discern what processes to include within the automated system and what data flows cross the boundary.

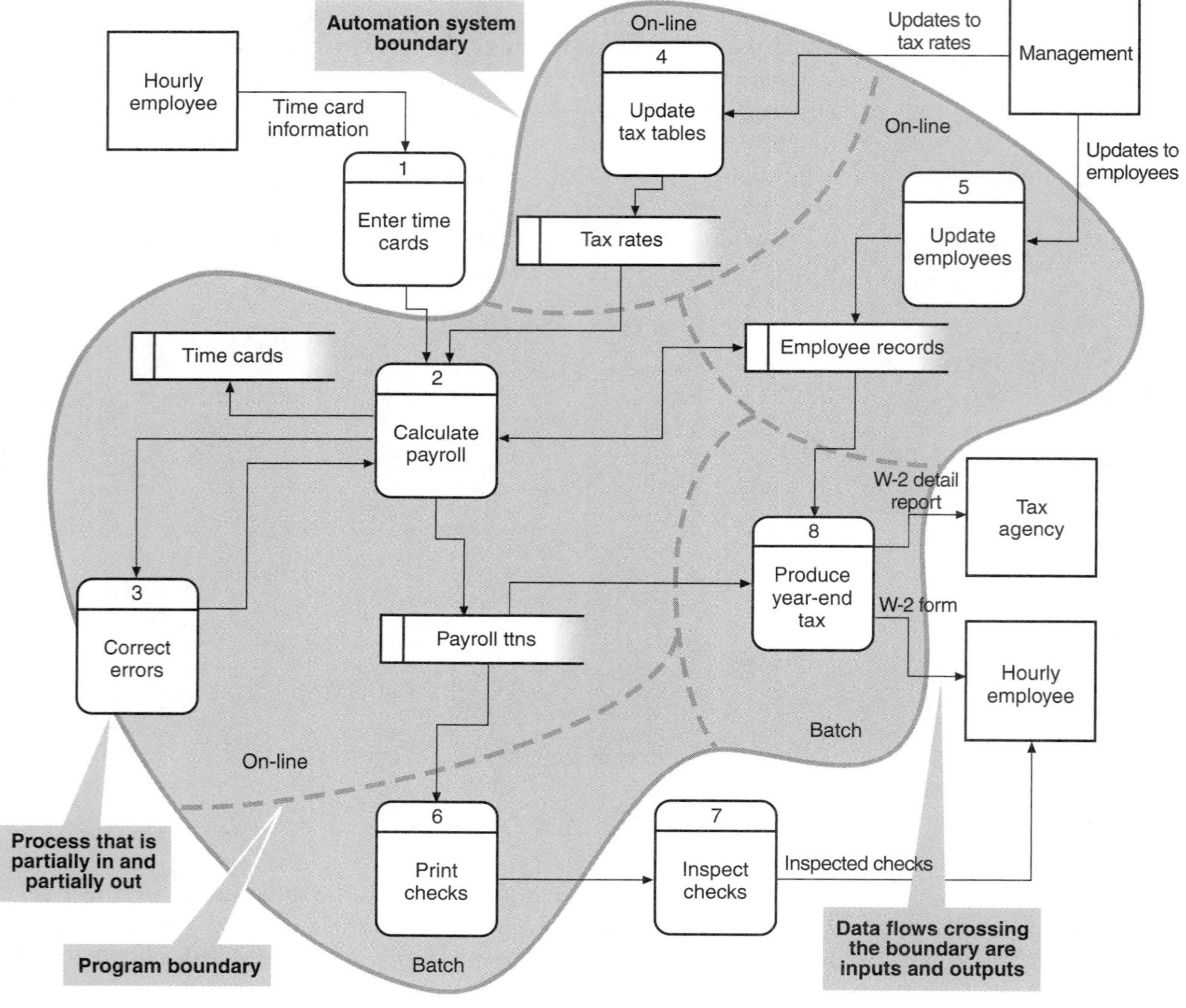

FIGURE 11-3
Automation boundary on a system-level DFD.

Chapter 6 explained how to build DFD fragments based on the events in the event table. For more complex models, you define system inputs by looking at each DFD fragment and creating the system boundary on each fragment. The high-level DFD with automation boundary gives a good overview, but the DFD fragments, or even the detailed DFD for each fragment, are easier to work with.

Figure 11-4 shows the *Create new order* detailed data-flow diagram with the automation boundary superimposed. On this diagram, all of the processes appear automated, along with the data files. A different version of the data-flow diagram, one for mail orders, might include a manual process to open the envelopes and enter the order into the computer. Figure 11-3 has an example of this kind of a manual process. The point is that not all processing is done with the computer. For this figure, however, the input data flows crossing the boundary are clearly defined, so required inputs will be the new order information data flow and the real-time link from the credit bureau. The input for the user interface will be the new order information, since the real-time link to the credit bureau will be an electronic system interface. The system interface to the credit bureau will also need to be designed, so it is important to remember that not all inputs originate with users.

FIGURE 11-4
The Create new order *DFD with an automation boundary.*

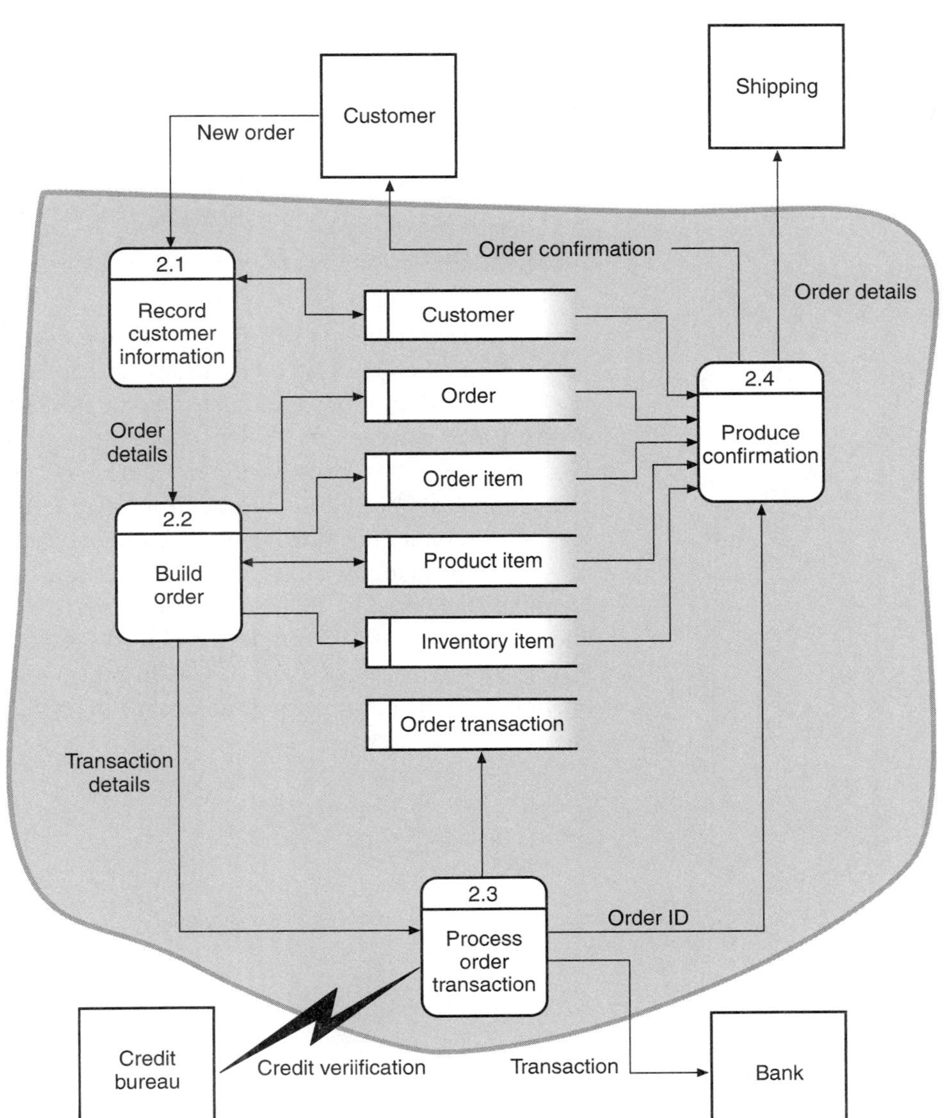

The designer analyzes each DFD fragment to determine the inputs required. The data flows that cross the boundary on the DFDs as inputs correspond to triggers for external events in the event table. The result of this task is a list of high-level inputs for the new system. Figure 11-5 is a list of inputs for RMO's customer support system as developed from the DFD fragments. To develop this list, the designer analyzed every DFD fragment in Figure 6-8 by drawing a system boundary. The purpose of this preliminary list is to provide a master control list of all system inputs that need to be designed. It does not, however, provide quite enough detailed information to design the input themselves. The additional information that is needed is obtained from the data flow definitions and structure charts.

FIGURE 11-5
List of inputs for the customer support system.

Item inquiry
New order information
Change order information
Order status inquiry
Order fulifillment notice
Back order notice
Order return notice
Catalog request
Customer account update
Promotion package information
Customer change adjustment
Catalog update information
Special promotion information
New catalog information

While developing the structure charts, the designer defines individual program modules and their associated data couples. Chapter 9 discussed the process of defining the detailed data content of each data couple. Each input data flow on a data flow diagram may translate into one or more physical inputs on the structure chart. In Figure 11-6, which is a duplication of Figure 9-18, input modules have been defined for getting customer information and for getting order information, including the details for several order line items. In this figure, the *New order* data flow on the DFD is expanded into four separate data couples on the structure chart. The structure chart identifies three modules that get data from outside the system. These three modules and their associated data couples are named *Get customer information, Get order information*, and *Get credit card information*. In other words, it requires three modules to provide all of the information from outside the system on the *Customer and new order information* data flow.

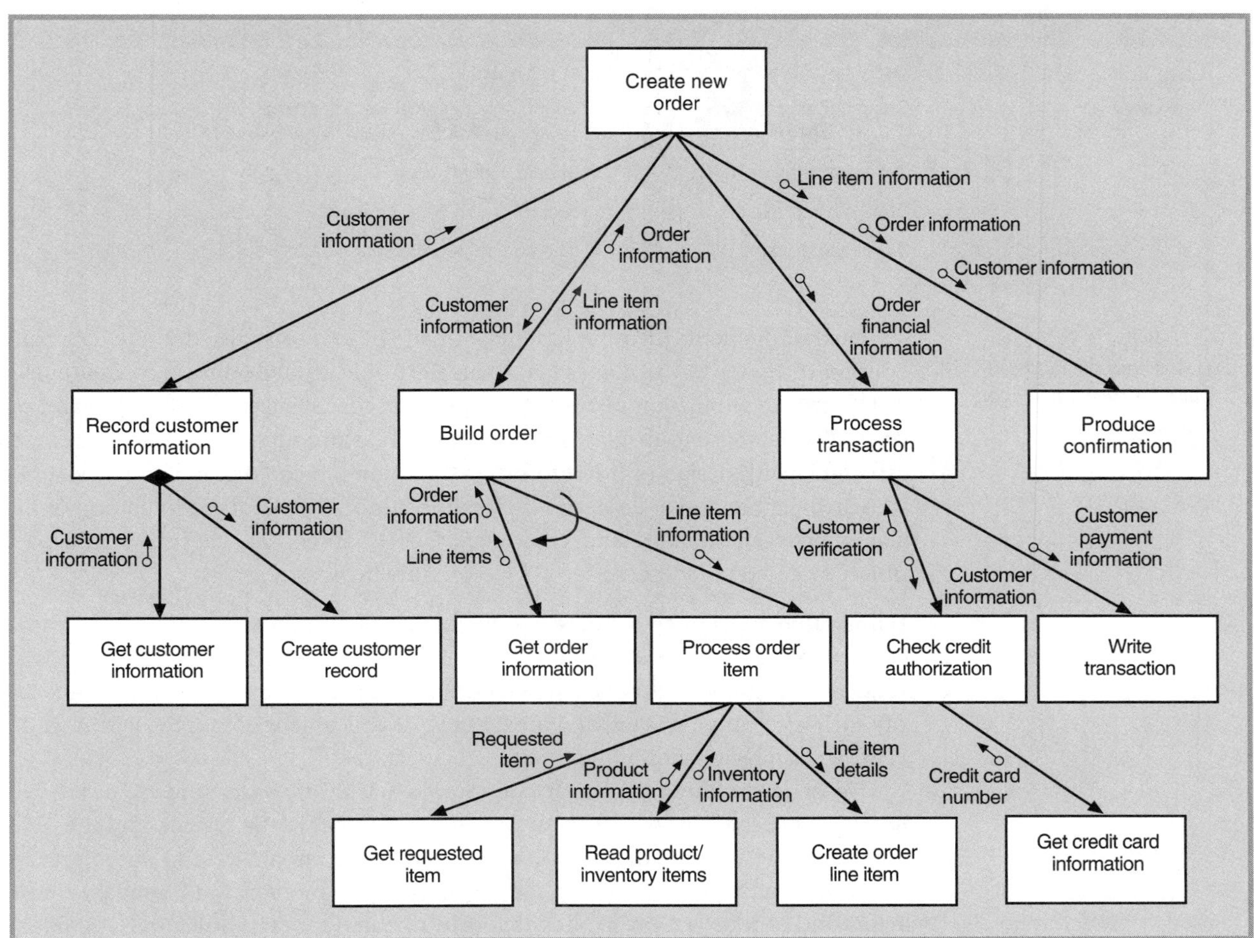

FIGURE 11-6
Structure chart for Create new order.

The next step is to analyze each module and data couple and list the individual data fields for each data couple. This analysis consists of reviewing the elements in the data stores to ensure that all elements on the data stores can be built based on the input data couples. Figure 11-7 expands Figure 11-5 to contain the data couples associated with each data flow as well as the data fields to be associated with each data couple.

Data flow	Data couples	Data elements
New order	Customer information	Account number, Name, Billing address, Shipping address, Day phone, Night phone
	Order information	Order date, Priority code
	Line item information	Product ID, Color, Size, Quantity
	Credit card number	Credit card number, Expiration date

FIGURE 11-7
A data flow and the data couples and data elements making up an input.

Each of the items identified as inputs made by end users listed in the data couple column of Figure 11-7 becomes part of an electronic input data form or an input/output form. An input/output form permits users to enter key blank fields and then query the database to display information such as product descriptions. These tasks associated with design of input forms are best done in conjunction with the application architecture design tasks and then refined through careful consideration of interaction between humans and computers (see Chapter 12). Let's now shift to the object-oriented paradigm to see how to identify the inputs.

Using Object-Oriented Models Identifying input with the object-oriented approach consists of the same tasks as with the structured approach. The difference is that sequence diagrams and design class diagrams are used. The sequence diagrams identify each incoming message, and the design class diagrams contain the pseudocode to verify the characteristics of the inputs.

Figure 11-8 is the sequence diagram for the telephone order scenario of the *Create new order* use case. In the figure, the actors are external to the system, and the objects are internal to the system. Every message that goes from an actor to an object represents an input to the system. In the object-oriented models, the boundary between actors and objects is more explicit than in structured models. In Figure 11-8, five separate messages go from the clerk (actor) to various objects. The messages, along with the actual message signatures as shown in the figure, are:

- Create new customer—CreateCustomer (Customer information)
- Check customer status—CheckStatus (CustomerName, PhoneNo)
- Create order—CreateOrder (Order information)
- Add item to order—AddToOrder (Product ID, Description, Qty)
- Finish order—FinishOrder ()

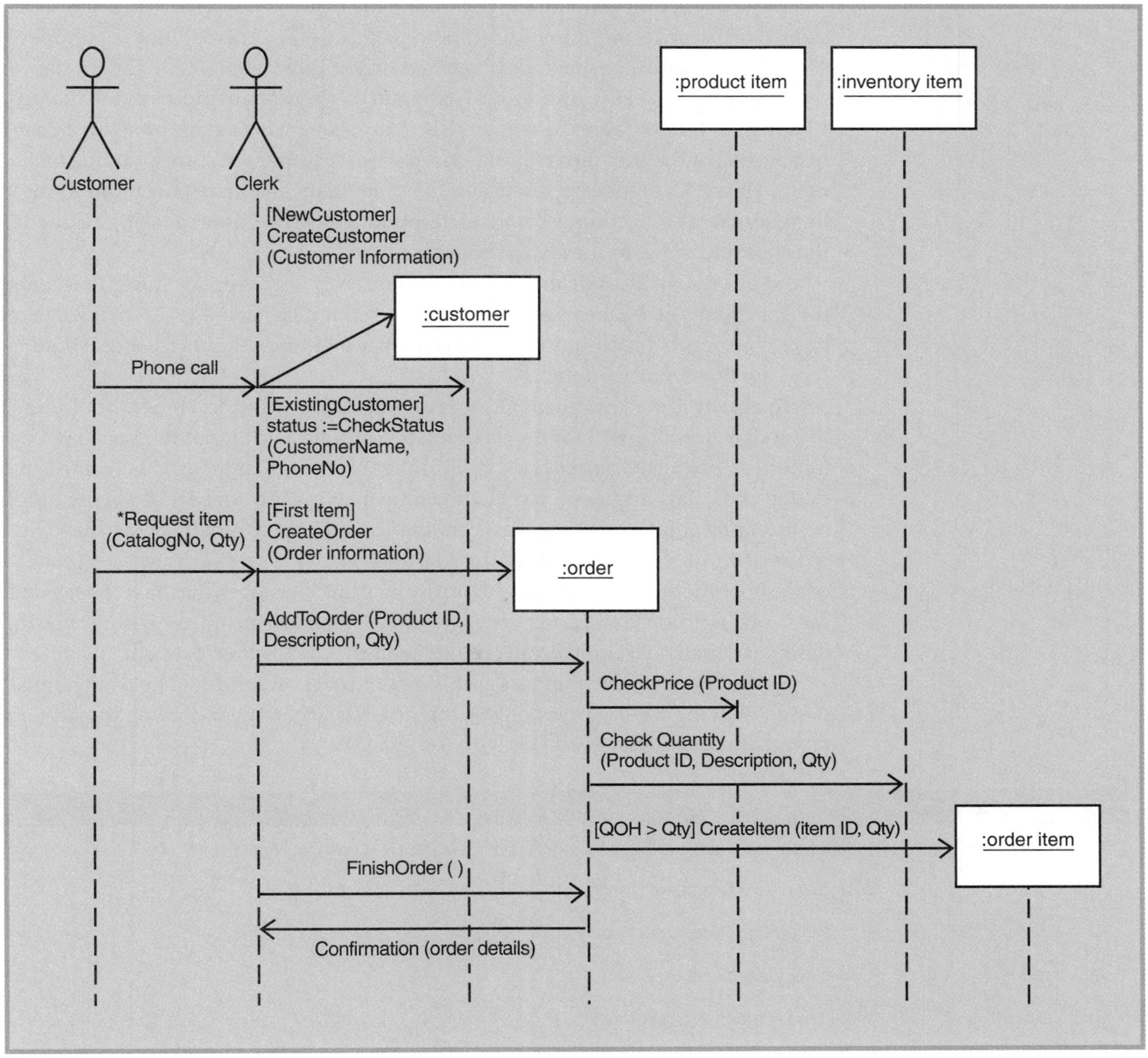

:product item
:inventory item
Customer
Clerk
[NewCustomer] CreateCustomer (Customer Information)
:customer
Phone call
[ExistingCustomer] status :=CheckStatus (CustomerName, PhoneNo)
*Request item (CatalogNo, Qty)
[First Item] CreateOrder (Order information)
:order
AddToOrder (Product ID, Description, Qty)
CheckPrice (Product ID)
Check Quantity (Product ID, Description, Qty)
[QOH > Qty] CreateItem (item ID, Qty)
:order item
FinishOrder ( )
Confirmation (order details)

FIGURE 11-8
Sequence diagram for Create new order.

The sequence diagram developed during the analysis phase identifies the series of steps that occur in the overall process to create a new order from a telephone call. Along with a sequence diagram, a detailed dialog is developed to highlight the communications between the user and computer. We explain more about dialogs in Chapter 12, but the point to note here is that a sequence diagram provides a detailed perspective of the user inputs to support the use case and the corresponding business event. The series of messages indicates that potentially five electronic input forms will be required. The five input forms correspond to the five messages originating from the clerk and going to the system objects.

Additional analysis of the messages themselves also supplies information about the data fields on the message. However, you will notice in Figure 11-8 that some messages are specified with more detail than other messages. In other words, some messages are not detailed enough.

To obtain a more thorough analysis of the messages, the developer should consult the design class diagram for the receiving object class. Each input message has a destination—a particular object class. Within the object class, a method is defined in the design class diagram to process the incoming message. An analysis of the pseudocode of the appropriate method assists in the identification of the fields that will be required to process the message. In addition, just as with the structured model, an analysis of the pseudocode will identify any other messages that may be needed to carry out the interchange between the system and the actor. As before, the input design is usually developed concurrently with the design class diagram.

Figure 11-9 is a table that lists each input message and the data fields that must be passed with the message. Not only is this analysis necessary to develop the user interface, it also provides a good check on the analysis.

Create customer	Account number, Name, Billing address, Shipping address, Day phone, Night phone
Check status	Customer name, Telephone number
Create order	Account number, Order date, Priority code
Add to order	Product ID, Color, Size, Quantity
Finish order	Credit card number, Expiration date

FIGURE 11-9
Input messages and data parameters from the sequence diagram.

Earlier chapters noted that the class diagram focused primarily on the application classes, such as *Customer* and *Order*. However, a complete design also requires user-interface classes to be added to the class diagram. It is possible to add the user-interface classes directly to the class diagram. In most of today's development environments, developers use programming languages with component libraries to develop the user interface. Such languages as Visual Basic and Java all have components to build windows and input screens. UML provides an additional notation specifically to identify user-interface classes. We use a small circle to denote an interface for a class. Figure 11-10 is a partial class diagram from RMO that shows the *Customer* class and five interface classes, one for each input window. The name for each interface begins with the letter *I* to indicate it is an interface class.

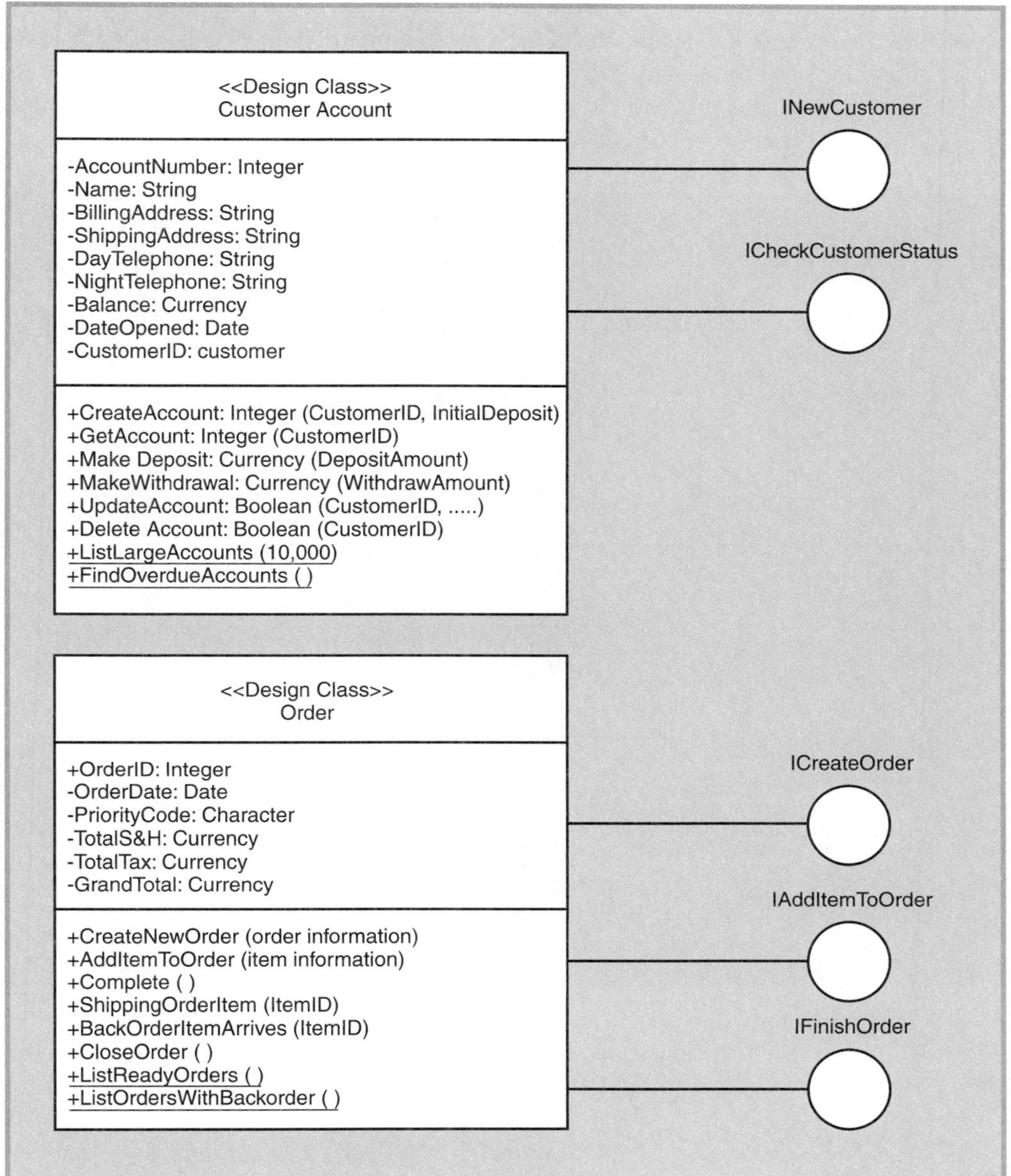

FIGURE 11-10
Customer and order classes with interfaces for the input forms from the Create new order *sequence diagram.*

Designing and Prototyping Input Forms

Paper forms are documents on which users record information that is either not entered into the computer or is entered later. In some cases, a paper form is needed to meet contractual or legal requirements. For example, some contracts such as life insurance policies have historically been documented with paper contracts. Although the courts have begun to accept some electronic media as legal contractual information, in many instances companies still prefer to use paper contracts with original signatures and guarantees. In other cases, information is captured at locations where computer data entry is not convenient. Good design principles dictate that the paper form and its electronic counterpart should have the same general layout and sequence of data fields. In other words, since the paper form will be used to enter the information into the system, the electronic and paper versions should mirror each other.

together. Figure 11-11 is an example of a customer order form for RMO. Many customers make purchases by mail through the catalog, so this form is essential. Even customers who order by telephone use the form as a working document to assist them in collecting information before they telephone.

Rocky Mountain Outfitters—Customer Order Form

Rocky Mountain Outfitters

Name and address of person placing order.
(Please verify your mailing address and make correction below.)
Order date __/__/__

Name

Address Apt. no.

City State Zip

Phone: Day ( )______ Evening ( )______

Gift Order or Ship To: (Use only if different from adress at left.)

Name

Address Apt. No

City State Zip

Gift ☐ Address for this Shipment Only ☐ Permanent Change of Address ☐

Gift Card Message

Delivery Phone ( )______

Item no.	Description	Style	Color	Size	Sleeve Length	Qty	Monogram Style	Price Each	Total

MERCHANDISE TOTAL ______

Total Monogramming charges ($5.00 per line per item) ______

Sales tax on merchandise delivered in Colorado and Utah ______

Regular FedEx shipping $4.50 per U.S. delivery address (Items are sent within 24 hours for delivery in 2 to 4 days) $4.50

Please add $4.50 per each additional U.S. delivery address ______

FedEx Standard Overnight Service
Add $6.00 for delivery in 2 business days to confirm U.S. address* ______

Any additional freight charges ______

International Shipping (see shipping information on back) ______

Method of Payment

Check/Money Order ☐ Gift Certificate(s) ☐ AMOUNT ENCLOSED $ ______

American Express ☐ MasterCard ☐ VISA ☐ Other ☐

Account Number

MO YR
__/__
Expiration Date

Signature ____________

FIGURE 11-11
RMO catalog order form.

The same general principles of good form design apply to both electronic forms and paper forms. We consider these principles in Chapter 12.

Design of System Outputs

The primary objective of system outputs is to present relevant information in the right place at the right time to the right people. Historically, the most common method of presenting output information has been in the form of printed textual reports. Although straight text or tables are still used extensively, new techniques, such as charts and diagrams, provide many more options for presentation, emphasis, and summarization of information.

As with input design, the tasks in this activity accomplish four objectives:

- Determine the type of each output.
- Make a list of specific outputs required based on application design.
- Specify any necessary controls to protect the information provided in the output.
- Design and prototype the report layout.

Historically, almost all output was produced in standard paper reports. Today there are many other alternatives to provide information. The purpose of the first task is to evaluate the various alternatives and design the most appropriate approach.

The list of required output reports is normally specified during the analysis phase as part of modeling system requirements. During design, the task is to coordinate the production of those outputs with the modules (structured techniques) and methods (object-oriented techniques) that are identified during the application architecture design.

The third task involves evaluating the value of the information to the organization and the need to protect it. Frequently, organizations will implement controls on the inputs and the systems but forget that output reports often have sensitive information that also must be protected and controlled.

ad hoc reports
Reports that are not predefined by a programmer but designed as needed by a user

It is possible today for users to develop their own reports using tools and preformatted templates. These reports, called **ad hoc reports**, are reports that have not been designed by programmer/analysts. An ad hoc report is the result of a new query to the database that a user of the system creates in response to a specific question or need. In Chapter 10, you learned about relational databases and structured query language (SQL). Many systems provide a simplified graphical tool to permit the users to formulate queries in SQL to produce ad hoc reports. Obviously, we do not design those reports during system development. However, the tools and capability to support user requests do need to be built into the system. The report design activity is a good time to ask whether the system requires an ad hoc reporting capability and to add such capability if necessary.

Determining the Type of Output

With the advent of office automation and other business systems, businesspeople thought that paper reports would no longer be needed. In fact, just the opposite has happened. Business systems have made information much more widely available, with a proliferation of all types of reports, both paper and electronic. In fact, one of the major difficulties with information systems today is the overwhelming amount of information. One of the most difficult aspects of output design is to decide what information to provide and how to present it to avoid a confusing mass of complex data.

Types of Reports Before looking at the different formats that analysts use in designing reports, let's discuss four types of output reports that users require: detailed reports, summary reports, exception reports, and executive reports.

detailed report
A report containing detailed transactions or records

Detailed reports are used to carry out the day-to-day processing of the business. They contain detailed information on business transactions. Sometimes a report may be for a single transaction, such as an order confirmation sheet that contains the details of a particular customer order. Other detailed reports may list a set of transactions. An example of this report may be a listing of all overdue accounts. Each line of the report will have information about a particular account. A clerk could use this report to research overdue accounts and determine actions to collect past-due amounts. The purpose of detailed reports is to provide working documents for people in the company.

summary report
A report that recaps or summarizes detailed information over a period of time or some category

exception report
A report that contains only information about nonstandard, or exception, conditions

Summary reports are often used to recap periodic activity. An example of this report is a daily or weekly summary of all sales transactions, with a total dollar amount of sales. This type of report is usually used by middle management to track departmental or division performance. **Exception reports** are also used to monitor performance. An exception report is produced only when a normal range of values is exceeded. When business is progressing normally, then no report is needed. But when something exceeds an expected range, a report is produced to alert staff. An example is a report from a production line that lists rejected parts. If the reject rate is above a set threshold, then a report is generated. Sometimes exceptions reports are produced regularly but they only highlight items that are out of range. Therefore, the rejected parts report might be produced every day if there are usually some rejected parts. A typical aged accounts receivable report might be produced each month showing the accounts that are past due. Unfortunately, there are always some accounts to list in such reports so they are produced regularly.

executive report
A summary report from various information sources that is normally used for strategic decisions

Executive reports are used by top management in the organization to assess overall health and performance. As such, these reports contain summary information from activities within the company. They may also contain comparative performance with industrywide averages. In this way, executives can assess competitive strengths or weaknesses of their company.

Chapter 1 discussed the various types of information systems that systems analysts develop. You will note that some types of information systems focus on producing a particular type of report. Although there is no strict requirement for a system to produce only one type of report, we often categorize a system based on the type of report it produces. The next section looks at some examples of printed reports.

internal output
A printed report or document produced for use inside of the organization

external output
Printed documents, like statements, notices, form letters, and legal documents, produced for use outside of the organization

turn-around document
An external output that includes a portion that is returned to the system as an input

control break report
A report that includes detailed and summary information

Internal versus External Outputs Printed outputs are classified as **internal outputs** and **external outputs**. Internal outputs are produced for use within the organization. The types of reports discussed previously fall under this category. External outputs include statements and notices and other documents that are produced for people outside of the organization. These need to be produced with the highest quality graphics and color. Some examples include monthly bank statements, late notices, order confirmation and packing slips (such as those provided to RMO customers), and legal documents such as insurance policies, discussed below. Some external outputs are referred to as **turn-around documents** because they are sent to a customer but include a tear-off portion that is returned for input later, such as a bill that contains a payment stub that is returned with a check. All of these printed outputs must be designed with care. Today's high-speed color laser printers enable all types of reports and other outputs to be produced.

An example of an internal output is shown in Figure 11-12. It is a report that includes both a detailed and summary section, sometimes called a **control break report**. The detailed section lists the transactions of records from the database, and the summary section provides totals and recaps of the information. This report is an inventory report for RMO. The report is sorted and presented by product. However, within each product is a list of each inventory item showing the quality currently on hand.

External outputs can consist of complex, multiple-page documents. A well-known example is the set of reports and statements that you receive with your car insurance statement. This statement is usually a multipage document consisting of detailed automobile insurance information and rates, summary pages, turnaround premium payment cards, and insurance cards for each automobile. Another example is a report of employment benefits with multiple pages of information customized to the individual employee. Sometimes the documents are printed in color with special highlighting or logos. All of this can be done as the documents are printed on laser printers. Figure 11-13 is one page of an example report for survivor protection from an employee benefit booklet. The text is standard wording, and the numbers are customized to the individual employee.

Rocky Mountain Outfitters — Products and Items

ID	Name	Season	Category	Supplier	Unit Price	Special	Special Price	Discontinued
RM0125	Outdoor Field	Spr/Fall	Mens C	8201	$39.00	▒	$0.00	No

Description Outdoor Nylon Jacket with Lining

Size	Color	Style	Units in Stock	Reorder Level	Units on Order
Large	Blue		1500	150	
Large	Green		1500	150	
Large	Red		1500	150	
Large	Yellow		1500	150	
Medium	Blue		1500	150	
Medium	Green		1500	150	
Medium	Red		1500	150	
Medium	Yellow		1500	150	
Small	Blue		1500	150	
Small	Green		1500	150	
Small	Red		1500	150	
Small	Yellow		1500	150	
Xlarge	Blue		1500	150	
Xlarge	Green		1500	150	
Xlarge	Red		1500	150	
Xlarge	Yellow		1500	150	

ID	Name	Season	Category	Supplier	Unit Price	Special	Special Price	Discontinued
RM0125	Hiking Walkers	All	Footwe	7993	$49.95	▒	$0.00	No

Description Hiking Walkers with Patterned Tread Durable Uppers

Size	Color	Style	Units in Stock	Record Level	Units on Order
10	Brown		1000	100	
10	Tan		1000	100	
11	Brown		1000	100	
11	Tan		1000	100	
12	Brown		1000	100	
12	Tan		1000	100	
13	Brown		1000	100	
13	Tan		1000	100	
7	Brown		1000	100	
7	Tan		1000	100	
8	Brown		1000	100	
8	Tan		1000	100	
9	Brown		1000	100	
9	Tan		1000	100	

FIGURE 11-12
RMO inventory report.

Survivor Protection

In the event of your death while working for a participating employer, your designated beneficiaries could receive:

Lump Sum Benefits

$50,000	Basic Life Insurance
$230,000	Supplemental Life Insurance
$148,677	Thrift Plan
$31,686	Tax Sheltered Annuity (TSA) Plan
$255	Social Security for your eligible dependents
$550,618	Total*

You have not elected Universal Life Insurance. If you would like more information on this plan, please call 1-800-555-7772.

*Refer to page 7 for additional information on the amount of coverage needed to provide ongoing replacement income.

Accidental Death Benefits
If your death is due to an accident, your designated beneficiaries will receive the above benefits plus:

$100,000	24-Hour Accidental Death and Dismemberment Insurance.
$100,000	Occupational Accidental Death and Dismemberment Insurance, if the accident is work related.

Monthly Death Benefits
If you die before receiving the Master Retirement Plan benefits and you are vested and have a surviving spouse, your spouse may be eligible for a Qualified Pre-Retirement Survivor Annuity.

In addition, your family may be eligible for the following estimated monthly benefits from Social Security, not to exceed a maximum of $2,591 based on:

$1,110	for each child under age 18;
$1,110	for a spouse with children under age 16; or
$1,058	for a spouse age 60 or older.

FIGURE 11-13
A sample employee benefit report.

drill down
The ability to link a summary field to its supporting detail and enable users to view the detail dynamically

Screen Output There are various types of screen output, each serving a different purpose, and each with its respective strengths and weaknesses. In most instances, screen output is formatted like a printed report, only displayed electronically. This kind of report can have detailed and summary sections, multiple pages, columns of data with headings, and so forth, just as a printed report does. However, an electronic report also has the advantage that it is dynamic—the user can have a real-time interchange. For example, an electronic report can have links to further information. One technique, called **drill down**, is the capability for the user to activate a "hot spot hyperlink" on the report, which tells the system to display a lower level, providing more detailed information. For example, Figure 11-14 contains a summary valuation report of inventory on hand. The report provides a summary valuation for each product item. However, if the user clicks on the hot link for any product, a detailed report pops up with the list of inventory items, the quantities on hand, and the valuation for each inventory item.

Monthly Sales Summary

Rocky Mountain Outfitters

Year **2001** *Month* **January**

Category	*Season Code*	*Web Sales*	*Telephone Sales*	*Mail Sales*	*Total Sales*
Footwear	All	$ 289,323	$ 1,347,878	$ 540,883	$ 2,178,084
Men's Clothing	Spring	$ 1,768,454	$ 2,879,243	$ 437,874	$ 4,691,484
	Summer	213,938	387,121	123,590	724,649
	Fall	142,823	129,873	112,234	384,930
	Winter	2,980,489	6,453,896	675,290	10,109,675
	All	1,839,729	4,897,235	349,234	7,086,198
Totals			14,747,368	$ 1,698,222	$ 23,391,023
Women's Clothing	Spring				965,610
	Summer				
	Fall				
	Winter				
	All				
Totals					

Monthly Sales Detail

Year **2001** *Month* **January** *Category* **Men's Clothing** *Season* **Winter**

Product ID	*Product Description*	*Web Sales*	*Telephone Sales*	*Mail Sales*	*Total Sales*
RMO12987	Winter Parka	$ 1,490,245	$ 3,226,948	$ 337,640	$ 5,054,833
RMO13788	Fur-Lined Gloves	149,022	322,695	33,765	505,482
RMO23788	Wool Sweater	596,097	1,290,775	135,058	2,021,930
RMO12980	Long Underwear	298,050	645,339	68,556	1,003,005
RMO32998	Fleece-Lined Jacket	447,075	1,258,079	100,271	1,805,425
	Total	$ 2,980,489	$ 6,453,896	$ 675,290	$ 10,109,675

FIGURE 11-14
A summary report with drill down to the detailed report.

linking
Connecting two or more reports electronically so that information from one links to information in another

Another technique, called **linking**, lets the user correlate information from one report to related information in another report. Most people are familiar with this concept on the pages that they browse on the Internet. This same capability can be very useful in a business report, which, for example, links the annual statements of key companies in a certain industry.

Another dynamic aspect of electronic reports is the capability to view the data from different perspectives. For example, it might be beneficial to view commission data by region, by sales manager, by product line, by time period, or compared to last season's data. Instead of printing all these reports, electronic format permits the different views to be generated only as needed. Sometimes long or complex reports include a table of contents with hot links to the various sections of the report. Some report-generating programs provide electronic reporting capability that includes all of the functionality that is found on pages on the Internet, including frames, hotlinks, graphics, and even animation.

Graphical and Multimedia Presentation The graphical presentation of data is one of the greatest benefits of the information age. Tools that permit data to be presented in charts and graphs have made information reporting much more user-friendly for printed and electronic formats. Information is being used more and more for strategic decision making by enabling businesspeople to look for trends and changes. In addition, today's systems frequently maintain massive amounts of data, much more than people can review. The only effective way to use much of this data is by summarizing it and presenting it in graphical form. Figure 11-15 illustrates a bar chart and a pie chart, which are two common ways to present summary data.

Men's Clothing Sales - January 2001

Web Sales
Telephone Sales
Mail Sales

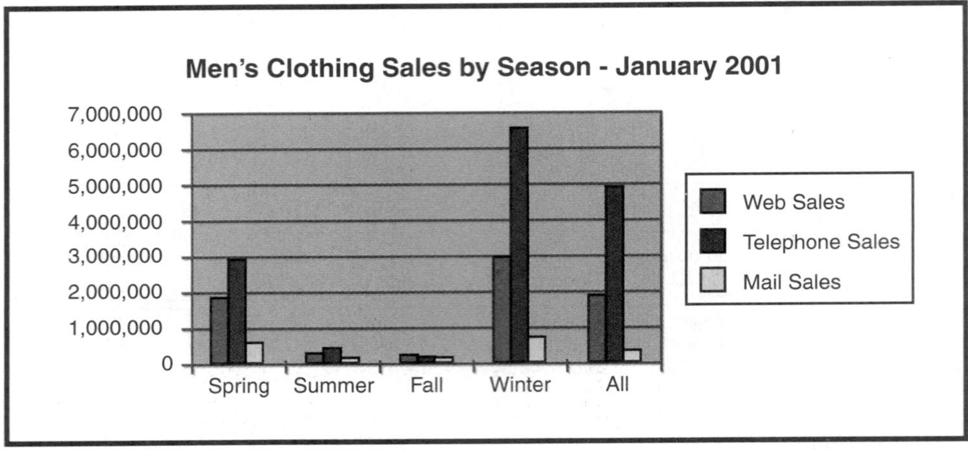

FIGURE 11-15
Sample bar chart and pie chart reports.

Multimedia outputs have become available only in the last several years as multimedia tool capabilities have increased. Today it is possible to see a graphical, and possibly animated, presentation of the information on a screen and have an audio description of the salient points. Combining visual and audio output is a powerful way to present information. (Of course, video games are pushing the frontier of virtual reality to include visual, audio, tactile, and olifactory outputs.)

As the design of the system outputs progresses, it is beneficial to evaluate the various presentation alternatives. Reporting packages can be designed into the system to provide a full range of reporting alternatives. Developers should carefully analyze each output report to determine the objective of the output and to select the form of the output that is most appropriate for the information and its use.

Making a List of Specific Reports Based on the Application Design

The objective of determining the list of reports is to ensure that each of the required outputs from the system is specified correctly. The basic approach is similar to the design of the system inputs. Outputs are responses in the event table—data that flow from the system to some destination. For structured techniques, the data flows that cross the system boundary are the outputs. For object-oriented techniques, messages that originate on internal classes and whose destination is an actor are the outputs. However, the uses of input and output are different. Whereas the data content of the input forms must support the needs of the database, the data content of the outputs must support the information requirements of the report users.

Using Structured Models The process of identifying the outputs is essentially the same as that of building the list of inputs from the data flows that cross the system boundary. In Figure 11-4, the *Create new order* DFD, there are three outputs: a confirmation to the customer, a notice to shipping, and a payment transaction report that goes to the bank. The same tasks as before—of building a table of the DFD outputs, defining exactly what reports are needed, and determining the data fields—should be done. Data-flow definitions for each of these outputs were created as part of the requirements model and should be available (see Chapter 6). The other input/output in Figure 11-4 is the electronic query going to the credit bureau. This also is an output, but it is usually not in the form of a report.

To verify that the structure chart modules are consistent with the structure of the output report, we look at the data couples and the report data requirements. For example, if the output is a single record output, such as the confirmation slip to a customer, then the module and data as shown in Figure 11-6 are correct to produce that information. An analysis of the data couple being sent to the module and the data fields on the output report will verify that the application has been designed correctly to generate the report.

However, with the other output from this module, the shipping order, the timing of the output can vary. Shipping orders can be produced one at a time as the order is prepared. In this case, the structure chart, as constructed, will function successfully. But the other way to prepare shipping orders is to produce a daily or semidaily report of items to be pulled from the shelves and boxed for shipping. This type of report is based on a set of orders, not a single order. To produce a combined shipping report, the system must have access to the database. It is not obvious from the structure chart in Figure 11-6 that the system will allow that type of shipping report. Comparing the data requirements of the report to the pseudocode in the module ensures that the output design is integrated successfully with the application design.

As we build the table of DFD outputs, report definitions, and data fields, we will add two more columns to the three already identified. These two additional columns list, first, the files or tables that will be required to produce the report and, second, an indication of whether the report is for a single instance of the file or includes a large set of records such as the entire file. Figure 11-16 is an example of the table of system outputs.

DFD data flow	Structure chart data couple	Data content of data couple	Database files required	Single or group order
Order confirmation	Produce confirmation	AccountNumber, Name BillingAddress, ShippingAddress, OrderNumber, Date PriorityCode, ShippingAndHandling, Tax, Total, *{ProductItemNumber, Description, Size, Color, Options, Quantity, Price}	Customer, Order, OrderItem	Single order
Shipping order details	Produce confirmation	AccountNumber, Name, ShippingAddress, OrderNumber, Date, PriorityCode	Customer, Order	Group of orders
Transaction details	Write transaction	AccountNumber, ShippingAndHandling, Tax, Total	Customer, Transaction	Single order

FIGURE 11-16
A table of system outputs with data requirements.

Using Object-Oriented Models Outputs are indicated by output messages in object-oriented sequence diagrams. That is, the message originates from an object internal to the system and has a destination of an external actor. In Figure 11-8, the output message *Confirmation ( )* is an example of an output message. In this case, it is generated as a result of the input *FinishOrder ( )* message to the internal order object. A review of all the output messages generated across all sequence diagrams provides the consistency check against the required outputs identified during the activities of the analysis phase.

Output messages that are based on an individual object (or record) are usually part of the methods of that object class. To report on all objects within a class, a class method is used. A class method is a method that works on the entire class of objects, not a single object. For example, a customer confirmation of an order is an output message that contains information for a single order object. However, to produce a summary report of all orders for the week, a class method looks at all the orders in the order class and sends output information for each one with an order date within the week's time period. In Figure 11-10, a class method such as *ListLargeAccounts ( )* sends output messages for a report that lists all accounts whose purchases have been greater than $10,000. You can add an interface class (the circle) to represent this output interface.

Designing and Prototyping Reports

Keep two principles in mind during the design of output reports:

- What is the objective of the report?
- Who is the intended audience?

Often designers must decide on the level of detail and format of the report. They can make these decisions correctly only if they know the objective of the report—that is, how the report is going to be used. It is often tempting to produce reports that mirror the structure and format of the data in the database. Newer systems, however, maintain a tremendous amount of detail in the database. Without careful consideration, designers can easily produce reports that suffer from **information overload**. Information overload occurs when so much data are provided that it becomes difficult for the user to find and focus on the information that is important. Many people have this problem when searching the Internet. Often when users search for a specific piece of information, the search engine returns an overwhelming number of results—such as 312,862 results for the search. Careful design and presentation are required to prevent this same problem with output reports.

information overload
The problem of providing too much information to users without providing techniques to organize and search the information

The format of the report is also important. Every report should have a meaningful title to indicate the data content. As we mentioned previously when discussing the accuracy of output controls, the report should include a heading that lists both the date the report was produced and a separate date indicating the effective date of the underlying information; sometimes the two dates may be different. Reports should also be paginated. In earlier systems, when reports were printed on continuous forms, page numbers were not as critical. With today's sheet-fed printers, however, it is easy for pages to be misplaced, and results can then be misinterpreted.

Labels and headings should be used to ensure the correct interpretation of the report data. Charts should be clearly labeled with the identification of the axes, units of measure, and a legend. In Figure 11-12, notice the headings and labels on the report to ensure that the data are not misinterpreted. Control breaks are used to divide the data into meaningful pieces that can be easily referenced. Use of lines, boldface, and different size fonts makes the report easy to read. Generally, report design is not difficult if you remember that the objective of any report is to provide meaningful information, not just data, and to provide it in a format that is easy to read.

Summary

This chapter began by introducing the concepts of integrity controls in systems. This introduction laid the groundwork for discussion of integrity controls associated with system inputs and outputs. The objectives of integrity controls are to:

- Ensure that only appropriate and correct business transactions occur
- Ensure that the transactions are recorded and processed correctly
- Protect and safeguard the assets (including information) of the organization

Integrity controls are concerned with defining who has access to the various components of the system and of the database. Security systems identify various classifications of users—such as unauthorized users, registered users, and privileged users—to set up access schemes. Additional integrity controls are concerned with reducing errors and maintaining the correctness of the data in the system.

Designing the inputs to the system is a four-step process:

1. Identify the devices and mechanisms that will be used to enter input.
2. Identify all system inputs and develop a list with data content of each.
3. Determine what kinds of controls are necessary for each system input.
4. Design and prototype the electronic forms and other inputs.

To develop a list of the inputs to the system, the designers use diagrams that were developed during the analysis and application design activity. For the structured approach, DFDs, data flow definitions, and structure charts are used. For the object-oriented approach, sequence diagrams are the primary source of information. The design class diagram is used to ensure that the correct data fields and the correct methods that produce the outputs are provided.

Changing technology now provides many options to enter data into a system. Input errors are reduced when electronic methods of inputting data are used instead of human keying of data. Such techniques as bar code scanners are effective ways to increase the speed of data capture as well as to reduce errors.

The process to design the outputs from the system consists of the same four steps as for input design. For output design, the DFDs and sequence charts are used to identify data flows and messages that exit the system. New technology provides numerous ways to present output with charts, graphs, and multimedia. Before deciding on an output media, the designer should carefully consider the intended audience and the purpose of the output.

Key Terms

ad hoc reports, p. 419
completeness control, p. 404
control break report, p. 420
data validation control, p. 404
destination controls, p. 405
detailed report, p. 419
drill down, p. 422
exception report, p. 419
executive report, p. 420
external output, p. 420
field combination control, p. 404
information overload, p. 426
integrity control, p. 400
internal output, p. 420
linking, p. 423
privileged user, p. 402
registered user, p. 401
summary report, p. 419
system access control, p. 401
transaction logging, p. 404
turn-around document, p. 420
unauthorized user, p. 401
value limit control, p. 404

Review Questions

1. What are the objectives of integrity controls in information systems? In your own words, explain what each of the three objectives means. Give an example of each.
2. Explain the three categories of user access privileges. Is three the right number, or should there be more or less than three? Why or why not?
3. What is your opinion about passwords? Should they be user selected or computer generated? Should they be changed periodically?
4. Explain what is meant by controlling access by visibility. Give some examples that you have seen.
5. Compare the strengths and weaknesses of using a DFD to define inputs with using a sequence diagram to define inputs. Which do you like the best? Why?
6. Explain the system boundary. Why was one used on a DFD but not used on a sequence diagram?
7. What additional information does the structure chart provide that is not obtained from a DFD in the development of input forms?
8. How are the data fields identified using the structured approach?
9. How are the data fields identified using UML and the object-oriented approach?
10. Explain four types of integrity controls for input forms. Which have you seen most frequently? Why are they important?
11. What protection does transaction logging provide? Should it be included in every system?
12. What are the different considerations between output screen design and output report design?
13. What is meant by drill down? Give an example of how you might use it in a report design.
14. What is the danger from information overload? What solutions can you think of?
15. Describe what kinds of integrity controls you would recommend be placed on all output reports. Why?

Thinking Critically

1. Scenarios presented at the beginning of the chapter described various situations that emphasized the need for controls. In the first, a furniture store sells merchandise on credit. Based on the descriptions of controls given in the chapter, identify the various controls that should be implemented in the system to ensure that corrections to customer balances are only made by someone with the correct authority.
2. Scenarios presented at the beginning of the chapter described various situations that emphasized the need for controls. In the second, an accounts payable clerk uses the system to write checks to suppliers. Based on the information in the chapter, what kinds of controls would you implement to ensure that checks are only written to valid suppliers, that check are written for the correct amount, and that all payouts have the required authorization? How would you design the controls if different payment amounts required different levels of authorization?
3. The executives of the company asked for a special decision support system type of report be made available to them on corporate financials. They want this report to be based on actual financial data for the past several years. The report is to have several input parameters so that they can do "what if" analysis of future sales, based on past performance. They would like the report to be viewable on-line as well as printable. What kinds of controls would you implement to (1) ensure that only authorized executives can request the report, (2) to ensure that the executives understand the basis (past and projected data) for a given report, and (3) that executives be aware of the sensitive nature of the information and treat it as confidential?
4. A payroll system has a data entry subsystem that is used to enter time card information for hourly employees. What kinds of controls would you implement to ensure that the data is correct and error free? What other controls would you include to ensure that a data entry clerk (who may be a friend of an employee) not inflate the hours on the time card (after it was approved by a supervisor)?
5. Based on the DFD given in Chapter 9 problem 3, Add a Class, and the structure chart you developed, identify the set of input and output screens that is necessary. Include the data fields that will be required.
6. Based on the DFD given in Chapter 9 problem 5, Special Order Purchasing, and the structure chart you developed, identify the set of inputs and outputs required. Develop the list of data fields for each screen and report.

7. Based on the collaboration and class diagrams shown in Chapter 9 problem 6, develop the list of inputs and the associated data fields for the portion of the system shown in the figure.
8. Using the statechart and class diagrams for Chapter 9 problem 8, develop the input forms that are needed for the Employee class. Be sure to list the data fields on each form.
9. You work for a grocery chain that always has large numbers of customers. To facilitate and speed checkout, the company would like to develop self-service checkout stands. Customers can check their own groceries and pay by credit card or cash. How would you design the checkout register and equipment? What kinds of equipment would you use to (1) make it easy and intuitive for the customers, (2) make sure that prices are entered correctly, and (3) that cash or credit card payments are done correctly? In other words, what kind of equipment would you have at the checkout station? In your solution, you may use existing state-of-the-art solutions, or invent new devices.

Experiential Exercises

1. Look on the Web for an electronic commerce site (for example amazon.com or Ebay.com). Evaluate the effectiveness of the screens on the user interface. What kind of security and controls are integrated into the system? Do you see potential problems with the integrity controls? Evaluate the design of the individual screens. How easy are they to read and use? What suggestions would you have to make them easier to use? How effective are they in minimizing data entry errors?
2. Many of you work for a local company or have a friend that works for a local company (fast food, doctor's office, video store, food store, and so forth). Look at one of these systems and evaluate the screens (and reports if possible) for ease of use and effectiveness. What kind of integrity controls are in place? How easy are the screens to use? What kinds of improvements would you make?
3. Research and find a system that is being constructed or has recently been constructed. You may work for a company that has a development in progress or have a friend who works for such a company. Another source of development projects is the university itself. Interview one of the developers. Ask about integrity controls, methodology for screen design, and guidelines to ensure consistency across the user interface. Ask about the size of the task (for example, how many screens, or hours required) and the method used to lay out the screens and reports (such as prototyping, case tools, and so forth).
4. Find out about the Abstract Windowing Toolkit (AWT) provided by Java or C++. Write a one-page description of the AWT, its purpose, and ways to use it. Your objective is to demonstrate that you understand the concept of the AWT and how it is used to build windows and input screens in a Windows environment.

Case Study

All-Shop Superstores

All-Shop Superstores is a regional chain of superstores in the Boston, New York, Washington, D.C. corridor. These stores compete with other giants, such as Wal-Mart, KMart, Target, and others. The stores contain both large grocery stores as well as domestics, clothing, automobile, and home improvement. Overall, the margins in this portion of the retail industry are very small. Grocery has always been small, in the range of 5 to 10 percent. Domestics, clothing, and so forth is a little higher, but to compete with Wal-Mart, all margins must be kept low.

In order to reduce operating costs as much as possible, All-Shop has decided to move very heavily into electronic data interchange (EDI) with its suppliers. All-Shop is aware that several of its more advanced competitors allow their suppliers to manage inventory levels in the stores themselves. For example, paper hygiene products such as disposable diapers and toilet paper are high-volume products that require very close monitoring of inventory levels. All-Shop has already installed sophisticated sales and inventory systems that track activity of each individual item (tracked by UPC code) on a daily basis. These systems not only capture daily activity but also maintain histories in a data warehouse of activities to support on-line data analysis.

The first step for All-Shop was to enable its major suppliers to have access to its daily sales and inventory database. In that way, the suppliers can monitor sales activities and check inventory to ensure that deliveries are made on a timely basis to be sure that inventory levels are maintained at an optimum level. The system should also permit each supplier to access and check the status of individual payables and a history of past payables. Obviously, all of this information must be controlled by supplier so that suppliers cannot observe other supplier's information.

Based on what you have learned in this and previous chapters, develop a use case diagram identifying the use cases that apply to the supplier as an actor. Even though this is really a system to system interface, the supplier system can be considered an actor. Identify two lists of controls that you consider to be necessary for this interface. On the first list, identify overall controls for the entire EDI interface that may be necessary. Then, for the second list, for each identified use case develop a specific set of controls that will be necessary. Base your analysis on the types of controls discussed in the chapter as well as the three primary objectives of integrity controls. In other words, your assignment is to develop a statement of required controls that can be used by the system developers to ensure that the assets and information of All-Shop are adequately protected.

As a second step, All-Shop is considering a plan to provide access to its data warehouse for its suppliers in order to permit them to analyze past trends and help design promotions to increase sales of not only individualo products, but overall sales. In other words, All-Shop is building partnerships with its suppliers to maximize its presence in the retail market place. One major concern of All-Shop executives is how to ensure that the suppliers treat this information with maximum security and not use it to damage All-Shop objectives. How can they ensure that this information is not used to benefit its competitors inadvertantly since suppliers also work with All-Shop competitors?

Do you think this second step is a wise move for All-Shop? If not, why not? If so, what kinds of controls and contractual arrangements should be made to protect All-Shop? In this second example, you may see how a narrow focus on integrity controls may be inadequate to protect proprietary information. A broader view and understanding of controls and their objectives is required in this instance.

Real Estate Multiple Listing Service System

Based on the DFD fragments you developed in Chapter 6 and the structure charts from Chapter 9, develop a table of inputs along with the associated data couples and data fields for each input. Also develop a table of outputs with the required data fields.

Reliable Pharmaceutical Service

Based on the collaboration and sequence diagrams you developed in Chapter 7, develop a list of inputs and outputs required for the system. Also using the class diagram you developed in Chapter 9, develop a list of parameters that need to be passed via each input or output. Also identify any specific controls that may be necessary to ensure that prescription information is entered accurately.

Downtown Video Rental System

Using the collaboration and sequence diagrams you developed in Chapter 7, develop a list of inputs and outputs, along with the necessary data fields, for the system.

Further Resources

Ben Shneiderman. *Designing the User Interface: Strategies for Effective Human-Computer Interaction*. Addison-Wesley Longman, 1998.

Brenda Laurel. *The Art of Human-Computer Interface Design*. Addison-Wesley, 1990.

Donald Warren, Jr., and J. Donald Warren. *The Handbook of IT Auditing*. Warren Gorham & Lamont, 1998.

Donald A. Wayne and Peter B. B. Turney. *Auditing EDP Systems*. Prentice Hall, 1990.

David Benyon, Diana Bental, and Thomas Green. *Conceptual Modeling for User Interface Development*. Springer-Verlag, 1999.

IS Audit and Control Association. *IS Audit and Control Journal*, Volume I. 1995.

Human-Computer Interaction

Learning Objectives

After reading this chapter, you should be able to:

- Explain why the user interface *is* the system to the users
- Discuss the importance of the three principles of user-centered design
- Describe the historical development of the field of human-computer interaction (HCI)
- Describe the three metaphors of human-computer interaction
- Discuss how visibility and affordance affect usability
- Apply the eight golden rules of dialog design when designing the user interface
- Define the overall system structure as a menu hierarchy
- Write user-computer interaction scenarios as dialogs
- Create storyboards to show the sequence of forms used in a dialog
- Use UML class diagrams, sequence diagrams, and collaboration diagrams to document dialog designs
- Design windows forms that are used to implement a dialog
- List the additional design considerations for direct customer interaction at a web site

Chapter Outline

TWELVE

INTERFACE DESIGN AT AVIATION ELECTRONICS

Bob Crain was admiring the user interface for the manufacturing support system recently installed at Aviation Electronics (AE). Bob was the plant manager for AE's Midwest manufacturing facility responsible for producing general aviation devices used in airplanes all over the world. The manufacturing support system was used for all facets of the manufacturing process, including product planning, purchasing, parts inventory, quality control, finished goods inventory, and distribution. Bob was involved extensively in the development of the system over a period of several years, including initial planning and development. The system reflected most everything he knew about manufacturing. The information services team that developed the system relied extensively on Bob's expertise. That was the easy part for Bob.

What particularly pleased Bob was the final user interface. Bob insisted that the development team think "outside the box." He did not want just another cookie-cutter transaction processing system. He wanted a system that acted like a partner in the manufacturing process—with a look and feel that really fit the work the users were doing. After all, the facility produced devices that had usability as a major design goal. Shouldn't the manufacturing support system be designed that way, too?

The first project manager assigned to the project didn't want to discuss usability at all. "We'll add the user interface later after we work out the accounting controls," was a typical comment. When Bob insisted that the project manager be replaced, the information services department sent Sara Robinson to lead the project.

Sara was a soul mate; she started out by asking about events affecting the manufacturing process and about cases where users need support from the system. Although she had a team of analysts working on the accounting transaction details right from the beginning, she always focused on how the system would be used from the user's point of view. Bob and Sara conducted meetings to involve users in discussions about how they might use the system, even asking users to act out the roles of the user and the system carrying on a conversation. That approach was outside the box.

At other meetings Sara presented sketches of screens and asked users to draw on them, placing information right on the sketches they wanted to see and options they wanted to be able to select. These sessions produced many ideas. For example, many users did not sit at their desks all day—they needed larger and more graphic displays they could see from across the room. Many users needed to refer to several displays, and they needed to be able to read them simultaneously. Several functions were best performed using graphical simulations of the manufacturing process. Users made sketches showing how the manufacturing process actually worked, and these sketches were later used to define much of the interface. Sara and her team kept coming back every month or so with more examples to show, asking for more suggestions.

When the system was finally completed and installed, most users already knew how to use it since they had been so involved in its design. Bob knew everything the system could do, but he had his own uses for it. He sat at his desk and clicked the *monitor ongoing processes* button on the screen, and the manufacturing support system gave him his morning briefing.

OVERVIEW

One of the key systems design activities is to design the user interface for a system. Designing the user interface means designing the inputs and outputs involved when the user interacts with the computer to carry out a task. This chapter emphasizes the interaction between the user and the computer—called *human-computer interaction*, or HCI. The inputs and outputs identified in Chapter 11 are the starting place. For every input, a developer must consider the interaction between user and computer and design an interface to process the input. Similarly, for every output produced at the request of a user (on-line reports, for example), the developer must design the interaction. Because the interaction is much like a dialog between the user and the computer, user-interface design is often referred to as *dialog design*.

This chapter begins with discussion of the user interface by providing background on user-centered design, the development of the field of human-computer interaction, and several metaphors used to describe the user interface. Many guidelines are available to help ensure usability of the system, and some of the most important guidelines are discussed. Next, approaches to documenting dialog designs are presented, including the use of UML diagrams from the object-oriented approach to development. Guidelines for designing windows forms and web pages are also provided. Examples are given throughout the chapter, including some dialog design examples for Rocky Mountain Outfitters.

The User Interface

Many people involved in developing a system think of the user interface as a component added to the system near the end of the development process. But the user interface to an interactive system is much more than that. The user interface is everything the end user comes into contact with while using the system—physically, perceptually, and conceptually (see Figure 12-1). To the end user of a system, the user interface *is* the system itself.

FIGURE 12-1
Physical, perceptual, and conceptual aspects of the user interface.

Desk, chair, light, keyboard, mouse, touch screen, keypad, manuals, printed documents, paper forms.
Windows, menus, dialog boxes, buttons, lines, shapes, textures, colors, fonts, sounds, speech.
Customers, products, orders, catalogs, adding, deleting, updating, printing, select-click-drag-drop, double-click-escape-click-click.

Many system developers, particularly those who work on highly interactive systems, echo this point of view in claiming that to design the user interface is to design the system. Therefore, consideration of the user interface should come very early in the development process. The term **human-computer interaction (HCI)** is generally used to refer to the study of end users and their interaction with computers.

human-computer interaction (HCI)
The study of end users and their interactions with computers

Physical Aspects of the User Interface

Physical aspects of the user interface include the devices the user actually touches, including the keyboard, mouse, touch screen, or keypad. But other physical parts of the interface include reference manuals, printed documents, data-entry forms, and so forth that the end user works with while completing tasks at the computer. For example, a mail-order data-entry clerk at Rocky Mountain Outfitters works at a computer terminal but uses printed catalogs and hand-written order forms when entering orders into the system. The desk space, the documents, the available light, and the computer terminal hardware all make up the physical interface for this end user.

Perceptual Aspects of the User Interface

Perceptual aspects of the user interface include everything the end user sees, hears, or touches (beyond the physical devices). What the user sees includes all data and instructions displayed on the screen, including shapes, lines, numbers, and words. The user might rely on the sounds made by the system, even a simple beep or click that lets the user know a keystroke or selection has been recognized by the system. More recently, computer-generated speech makes it seem that the system is actually talking to the user, and with speech recognition software, the user can talk to the computer. The user "touches" objects such as menus, dialog boxes, and buttons on the screen using a mouse, but the user also touches objects such as documents, drawings, or records of transactions with a mouse when completing tasks.

Conceptual Aspects of the User Interface

Conceptual aspects of the user interface include everything the user knows about using the system, including all of the problem domain "things" in the system the user is manipulating, the operations that can be performed, and the procedures followed to carry out the operations. To use the system, the end user must know all about these details—not how the system is implemented internally, but what the system does and how it can be used to complete tasks. This knowledge is referred to as the **user's model** of the system. Much of the user's model is a logical model of the system, as you learned in Chapters 5, 6, and 7. A logical model of the system requirements can be quite detailed, so the user must know quite a few details to operate the system. Recall also that a systems analyst relies on the end users to help define the requirements that the analyst captures in various models. The user's knowledge of the requirements for the system becomes the fundamental determinant of what the system is, and if the user's knowledge of the system is part of the interface, then the user interface must be much more than a component added near the end of the project.

user's model
What the user knows about using the system, including the problem domain "things" the user is manipulating, the operations that can be performed, and the procedures followed when carrying out tasks

User-Centered Design

Many researchers focus their attention on creating analysis and design techniques that place the user interface at the center of the development process because they recognize the importance of the user interface to system developers and system users. These techniques are often referred to collectively as **user-centered design**. User-centered design techniques emphasize three important principles:

user-centered design
A collection of techniques that place the user at the center of the development process

- Focus early on the users and their work
- Evaluate designs to ensure usability
- Use iterative development

The early focus on users and their work is consistent with the approach to systems analysis in this text—it is important to understand and identify the system users and the requirements they have for the system. The traditional approach to development focuses more on the requirements from the business point of view—what needs to be accomplished and what are the sources and destinations for data. The object-oriented approach, probably because object-oriented systems are more interactive, focuses more on users and their work by identifying actors, use cases, and scenarios followed when using the system. As discussed in Chapter 7, the automation boundary between user and the computer is defined very early during requirements modeling.

User-centered design goes much further in attempting to understand the users, however. What do they know? How do they learn? How do they prefer to work? What motivates them? The amount of focus on users and their work varies with the type of system being developed. If the system is a shrink-wrapped desktop application marketed directly to end users, the focus on users and their preferences is intense.

usability
The degree to which a system is easy to learn and use

The second principle of user-centered design is to evaluate designs to ensure usability. **Usability** refers to the degree to which a system is easy to learn and use. Ensuring usability is not easy—there are many different types of users with differing preferences and skills to accommodate. Design features that are easy to use for one person might be difficult for another. If the system has a variety of end users, how can the designer be sure the interface will work well for all of them? If it is too flexible, for example, some end users may feel lost. On the other hand, if it is too rigid, some users will be frustrated.

But there is more to consider for ease of use and learning. These concepts are often in conflict, because an interface that is easy to learn is not always easy to use. For example, menu-based applications with multiple forms and dialog boxes and extensive prompts and instructions are easy to learn—indeed, they are self-explanatory. Easy-to-learn interfaces are appropriate for systems that are used infrequently by an end user. But if office workers use the system all day, it is important to make the interface fast and flexible, with shortcuts, hot keys, and information-intensive screens. This second interface might be harder to learn, but it will be easier to use once it is learned. Office workers (with the support of their management) are willing to invest more time learning the system in order to become efficient users.

Developers use many techniques to evaluate interface designs to ensure usability. User-centered design requires testing all aspects of the user interface. Some usability testing techniques collect objective data that can be analyzed with statistics to compare designs. Some techniques collect subjective data about user perceptions and attitudes. To assess user attitudes, developers conduct formal surveys, focus group meetings, design walkthroughs, paper and pencil evaluations, expert evaluations, formal laboratory experiments, and informal observation.

The third principle of user-centered design is to use iterative development—doing some analysis, then some design, then some implementation, then repeating the processes. After each iteration, the project team evaluates the work on the system to date. Iterative development keeps the focus on the user by continually returning to the user requirements during each iteration and by evaluating the system after each iteration. Iterative development is discussed throughout this text as applying to both traditional and object-oriented approaches to development. Chapter 13 discusses iterative development in more detail.

Human-Computer Interaction as a Field of Study

human factors engineering (or ergonomics)
The study of human interaction with machines in general

User-interface design techniques and HCI as a field of study evolved from studies of human interaction with machines in general, referred to as **human factors engineering** or **ergonomics**. The formal study of human factors began during World War II, when aerospace engineers studied the effects of arranging controls in the cockpit on airplane pilots. Pilots are responsible for controlling many devices as they fly, and the effectiveness of the interaction between the pilot and the devices is critical. If the pilot makes a mistake (that is, if he or she can't correctly use a device), the plane can crash. What the pilot does is the "human factor" that engineers realized was often beyond their control.

One story about the importance of the human factor involved a minor change to the design of the cockpit of a plane. The designers switched the locations of the throttle and the release handle for the ejection seat. The result was a dramatic increase in the number of unexplained pilot ejections. When under pressure, the pilots grabbed what they thought was the throttle and ejected themselves from the plane. Initially, designers dismissed the problem as the need for better training. But even with training, pilots under pressure continued to grab the wrong handle. It became apparent that the key to the "human factor" was to change the machine to accommodate the human rather than trying to change the human to accommodate the machine.

The field of human factors was first associated with engineering, since engineers designed machines. But engineers, who are generally used to precise specifications and predictable behavior, often find the human factor frustrating. Gradually, specialists emerged who began to draw on many disciplines to understand people and their behavior. These disciplines include cognitive psychology, social psychology, linguistics, sociology, anthropology, and others, as shown in Figure 12-2. Information systems specialists with an interest in human-computer interaction study computers plus all of these disciplines.

FIGURE 12-2
The fields contributing to the study of HCI.

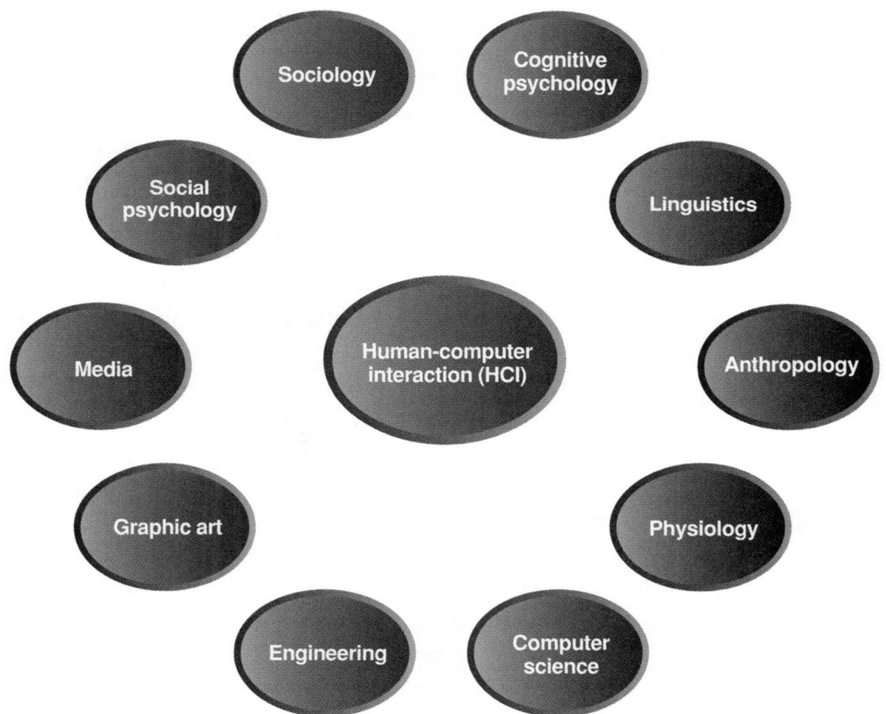

Although many human factors specialists began focusing on computers, an important contribution to the development of the field of human-computer interaction began with the Xerox Corporation in the 1970s. Xerox produced high-speed photocopy machines that provided an increasing number of special options and capabilities that the human operator could specify. The designers of the photocopy machines recognized the importance of making the complex machines easy for the operators to learn and use. Xerox customers wanted minimal training time for their operators, and operator errors could be costly. For example, if a clerk began a large photocopy job but made a mistake in specifying details, it would be wasteful and delay the distribution of important documents. Therefore, Xerox emphasized the usability of its machines.

Xerox established a research and development laboratory, called the Xerox Palo Alto Research Center (Xerox PARC), to study issues that affect how humans operate machines. As a result of this investment, Xerox eventually offered photocopy machines with touch screen, menu-driven interfaces that displayed icons representing objects like documents, stacks of paper, staples, and sorting bins.

Research and development at Xerox PARC also involved work on computers and object-oriented programming. The first pure object-oriented programming language, called Smalltalk, was created at Xerox PARC by Alan Kay and associates to facilitate the development of interactive user interfaces. In the early 1970s, Kay envisioned an advanced, portable personal computing platform (similar to today's ultralight notebook computers) called the Dynabook. Many researchers thought such a machine could not be built for three or four decades because the hardware required for the Dynabook was not available. Kay decided to work on the software that would run the machine in anticipation of the hardware, which led to Smalltalk.

Smalltalk includes classes that make up the key parts of windowing interfaces today—windows, menus, buttons, labels, text fields, and so forth. The design and programming philosophy used to describe and build these interfaces was developed along with the language—all 100 percent object-oriented.

Because of the work at PARC, Xerox eventually developed and marketed one of the first general-purpose personal computers with a graphical user interface—the Xerox Star—in the late 1970s. Although it was ahead of its time and far too expensive, it is considered a landmark development in computing. The key features were exploited in the early 1980s by a small company near Xerox PARC named Apple Computers. Apple first exploited the Xerox Star's features as the Apple Lisa and then as the Apple Macintosh. The work at Xerox PARC had a substantial impact on object-oriented programming, personal computers, and user-interface design.

Now that the object-oriented approach to system development is becoming more influential, user-interface design concepts and development techniques pioneered at labs like Xerox PARC are becoming better integrated into system development methodologies used for business systems. The field of HCI has grown and now sponsors many academic journals, conferences, and book series devoted to research and practice. Undergraduate and graduate degree programs are also available to train HCI specialists.

Metaphors for Human-Computer Interaction

There are many ways to think about human-computer interaction, referred to as *metaphors* or *analogies*. These are the direct manipulation metaphor, the document metaphor, and the dialog metaphor. Since each metaphor provides an analogy to a different concept, each provides implications for the design of the user interface.

direct manipulation
A metaphor of HCI in which the user interacts directly with objects on the display screen

The Direct Manipulation Metaphor **Direct manipulation** assumes the user interacts with objects on the screen instead of typing commands on a command line. Objects the user can interact with are made visible on the screen so the user can point at them and manipulate them with the mouse or arrow keys. The earliest direct manipulation interfaces were word processors that allowed the user to type in words directly where desired in a document. By the early 1980s, electronic spreadsheet applications (first VisiCalc, then Lotus 1-2-3) became available for IBM DOS PCs that used a direct manipulation approach—the user typed numbers, formulas, or text directly into cells on a spreadsheet. The spreadsheet on the screen was conceptually similar to a paper spreadsheet that was familiar to people working in accounting and finance. In that respect, they were easy to understand and natural to use, and end users found these applications useful because they could include formulas that would do the calculations on the spreadsheets automatically. These early direct manipulation DOS applications were an important reason for the success of the personal computer. Even though they did not have graphical user interfaces, they were very popular because they made interacting with a computer straightforward, natural, and useful.

The Smalltalk language developed at Xerox PARC extended direct manipulation to all objects on the screen. Some of these objects are interface objects like buttons, check boxes, scroll bars, and slider controls, but other problem domain objects such as documents, schedules, file folders, and business records were also displayed as objects the user could directly manipulate. For example, an interface might include a trash can object; to delete a document file, the user clicks on the document with the mouse and drags the document to the trash can. By directly manipulating the objects in this way, the user tells the computer to delete the document file.

desktop metaphor
A direct manipulation approach where the display screen includes an arrangement of common desktop objects

Direct manipulation coupled with object-oriented programming eventually evolved into the **desktop metaphor**, in which the display screen includes an arrangement of common desktop objects—a notepad, a calendar, a calculator, and folders containing documents. Many desktops now include a telephone, an answering machine, a CD player, and even a video monitor. Interacting with any of these objects is similar to interacting with the real-world objects they represent (see Figure 12-3). End users now expect all applications, including business information systems, to be as natural to work with as objects on the desktop.

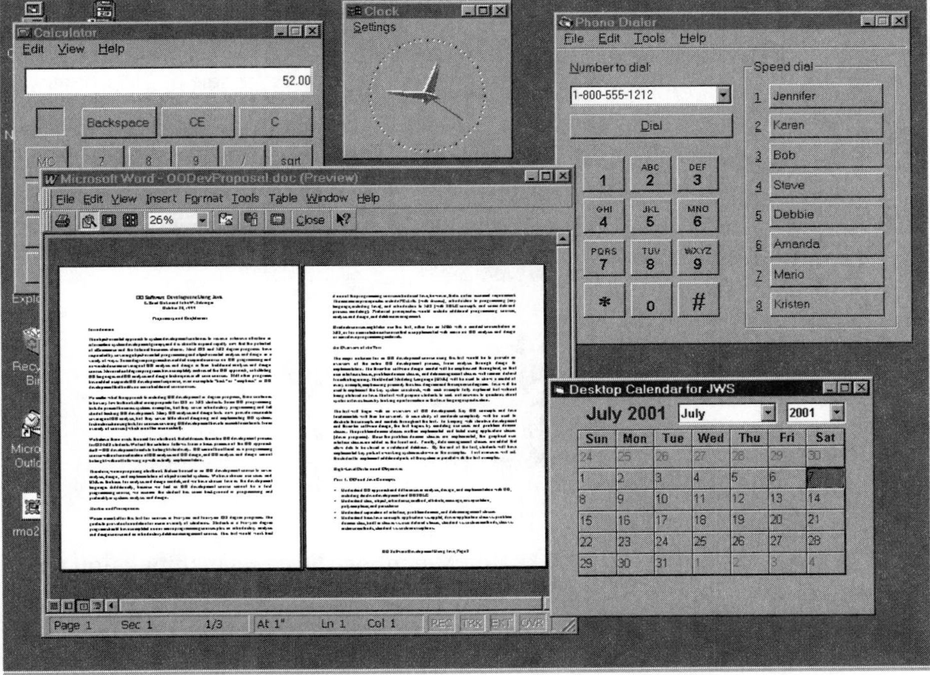

FIGURE 12-3
The desktop metaphor based on direct manipulation shown on a display screen.

document metaphor
A metaphor of HCI in which interaction with the computer involves browsing and entering data on electronic documents

The Document Metaphor Another view of the interface is the **document metaphor**, in which interaction with the computer involves browsing and entering data on electronic documents. These documents are much like printed documents, but because the documents are in electronic form, additional functionality is available that makes them more interactive. Electronic versions of documents can be organized differently

hypertext
Documents that allow the user to click on a link and jump to a different part of the document or to another document

hypermedia
Technology that extends the hypertext concepts to include multimedia content such as graphics, video, and audio

than paper versions because the reader can jump around from place to place. **Hypertext** documents allow the user to click on a link and jump to a different part of the document or to another document entirely.

Most common desktop applications create and edit electronic documents, which are not limited to text and usually include word processing, spreadsheets, presentations, and graphics. All of these applications produce documents, but any one document can contain words, numbers, and graphics produced by any of these applications, making documents collections of all sorts of interrelated media. **Hypermedia** extends the hypertext concept to include multimedia content such as graphics, video, and audio that can be linked together for navigation by the user in a document.

The World Wide Web is organized around the document metaphor because everything at a web site is organized as pages that are linked together as hypermedia (note that HTML means *hypertext* markup language). Even transactions are processed by selecting information on a web page document. The document metaphor and the browser interface as ways of describing and designing interactive systems will continue to affect user interface designs (see Figure 12-4).

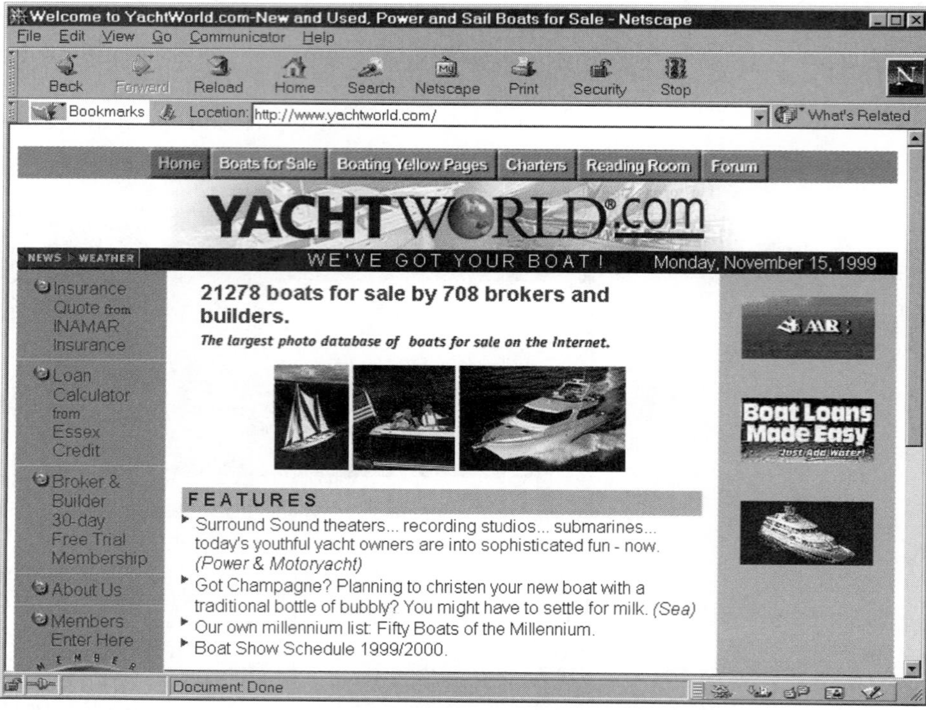

FIGURE 12-4
The document metaphor shown as hypermedia in a web browser.

dialog metaphor
A metaphor of HCI in which interacting with the computer is much like carrying on a conversation or dialog

The Dialog Metaphor The direct manipulation and document metaphors emphasize objects in the computer with which the user interacts. Another view of the interface is the **dialog metaphor**, in which interacting with the computer is much like carrying on a conversation or dialog. User-interface design is often referred to as dialog design. Carrying on a dialog or conversation with someone means that each person is listening to and responding to questions and comments from the other person, exchanging information in a sequence. The dialog metaphor is another way of thinking about human-computer interaction because the computer "listens to" and responds to questions or comments from the user, and the user "listens to" and responds to questions and comments of the computer. Like the direct manipulation metaphor, the dialog metaphor is based on an object-oriented view of the system because communication involves messages from one "object" to another. Figure 12-5 shows how the user and the computer communicate by sending messages to each other.

I am ready to work now
Okay, which task do you want to complete?
I want to check for messages
It looks like you have these three messages...

FIGURE 12-5
The dialog metaphor expresses that the user and computer interact by sending messages.

Consider the following dialog between a manager and an assistant:

Manager:	Did I get any messages while I was out?
Assistant:	Yes, you have three messages—from Bob, Mary, and Lim.
Manager:	What did Lim have to say?
Assistant:	Lim left a message at 8:15 P.M. last night regarding the meeting next Monday about the inventory management system. The message is, "Can we change the time for the meeting to 10:30? I'll be delayed by the testing session."
Manager:	I better respond. Say that the change is not a problem.
Assistant:	Okay, I'll leave him that message. Do you want the next message?
Manager	What did Mary have to say?
Assistant:	She left a message at 8:15 this morning regarding lunch. She said, "Lunch is still on, but Joe will not be able to join us."
Manager:	Okay, no response. And that's all for now. Thanks.
Assistant:	Okay, you still have one message from Bob. I'll remind you later.

This dialog involves the manager and the assistant carrying on a conversion about messages. The questions asked by the manager and the responses and follow-up questions asked by the assistant seem clear and natural. Would the basic dialog be any different if this were an automated phone answering service that responded to voice commands and replied in a computer-generated voice? Probably not. Would it be any different if this were a computer application simulating an intelligent "assistant"? Probably not. The basic dialog followed would be the same: a question, a response, another question, a response that might include a request for clarification, a response to the request for clarification, and a final response.

The basic dialog is also the same for a typical e-mail application, even though the user and computer send messages in different ways. The user selects a menu item for *read new mail*. The computer lists the new mail messages for the user to choose, the user chooses one message, then the computer displays the message. It might seem odd to think of interaction with an e-mail application as being similar to the dialog just presented, but the basic information exchanged and the sequence are the same.

The user and the computer both send messages, but each is forced to use a different language because of limitations in both the user and the computer. The user cannot understand cryptic binary codes, nor plug in directly to the computer to interpret the electrical impulses the computer uses to represent the binary codes. The natural language of the computer just won't work for people. The computer has to adapt to the user and provide its messages in a form that is natural for the user—text and graphics that the user can see and read.

Similarly, the computer cannot understand complex voice messages, facial expressions, and body language that are the natural language of the user, so the user has to adapt to the computer and provide messages by clicking the mouse, dragging objects, and typing words on the keyboard. Advances in computer technology are making it possible for the user to communicate in more natural ways, but the typical user interfaces today still rely on the mouse and keyboard. One reason is the need for silence and also privacy in the office, so it is not clear whether voice commands will become common in computer applications.

The challenge of user-interface design is to design a natural dialog sequence that allows the user and computer to exchange the messages needed to carry out a task. Then the designer needs to design the details of the language for the user to send the messages to the computer, plus the language for the computer to send messages to the user.

Figure 12-6 shows the dialog between manager and assistant translated into the languages used by the user and the computer. Interface designers use a variety of informal diagrams and written narratives to model human-computer interaction. This is one way the dialog design details can be modeled. Additional techniques are described later in this chapter.

	Message	User's language	Computer's language
Manager	Did I get any messages while I was out?	Click the *read messages* menu item on the main menu.	
Assistant	Yes, you have three messages—from Bob, Mary, and Lim.		Look up new messages for the user and display a new message form with message headers listed in a list box.
Manager	What did Lim have to say?	Double-click the message from Lim in the list box.	
Assistant	Lim left a message at 8:15 p.m. last night regarding the meeting next Monday about the inventory management system. The message is, "Can we change the time for the meeting to 10:30? I'll be delayed by the testing session."		Look up the message body for the selected message and display it in message detail form.
Manager	I better respond. Say that the change is not a problem.	Click the Reply button on the message detail form. Type in the message, "Okay, that is not a problem." Click the Send button.	Display the new message form addressed to the sender.
Assistant	Okay, I'll leave him that message. Do you want the next message?		Display the Message Sent dialog box and redisplay the new messages form with message headers listed in the list box.
Manager	What did Mary have to say?	Double-click the message from Mary in the list box.	
Assistant	She left a message at 8:15 this morning regarding lunch. She said, "Lunch is still on, but Joe will not be able to join us."		Look up the message body for the selected message and display it in message detail form.
Manager	Okay, no response. And that's all for now. Thanks.	Click the *close message* button. Click the *close new message form* button.	Redisplay a new messages form with message headers listed in the list box.
Assistant	Okay, you still have one message from Bob. I'll remind you later.		Display the Closing Read New Mail dialog box, showing one unread message remaining.

FIGURE 12-6
The user's language and the computer's language used to implement an e-mail application based on the natural dialog between manager and assistant.

Interface Design Guidelines

There are many published interface design guidelines to guide system developers. User-interface design guidelines range from general principles to very specific rules. Some of the guidelines and rules for designing outputs were described in Chapter 11. This section describes some well-known guidelines for designing the dialog between user and computer. Later this chapter presents some of the guidelines and rules for designing input forms. Some system development organizations adopt **interface design standards**—general principles and rules that must be followed for any system developed by the organization. Design standards help ensure that all user interfaces are usable and that all systems developed by the organization have a similar look and feel.

interface design standards
General principles and rules that must be followed for the interface of any system developed by the organization

Visibility and Affordance

Donald Norman proposes two key principles that ensure good interaction between a person and a machine: visibility and affordance. These two principles apply to human-computer interaction just as they do for any other device.

visibility
A key principle of HCI that states all controls should be visible and provide feedback to indicate the control is responding to the user's action

Visibility means that a control should be visible so users know it is available, and that the control should provide immediate feedback to indicate it is responding. For example, a steering wheel is visible to a driver, and when the driver turns it to the left, it is obvious that the wheel is responding to the driver's action. Similarly, a button that can be clicked by a user is visible, and when it is clicked, it changes to look as though it has been pressed to indicate it is responding. Sometimes buttons make a clicking sound to provide feedback.

affordance
A key principle of HCI that states the appearance of any control should suggest its functionality

Affordance means that the appearance of any control should suggest its functionality—that is, the purpose for which the control is used. For example, a control that looks like a steering wheel suggests that the control is used for turning. In the computer, a button affords clicking, a scroll bar affords scrolling, and an item in a list affords selecting. Norman's principles apply to any objects on the desktop, such as those shown in the examples in Figures 12-3 and 12-4 previously.

If user-interface designers make sure all controls are visible and clear in what they do, the interface will be usable. Most users are familiar with the Windows interface and the common Windows controls. However, designers should be careful to apply these principles of visibility and affordance when designing web pages. Many new types of controls are now possible at web sites, but these controls are not always as visible and their effects are not always as obvious as they are in a standard Windows interface. More objects are clickable, but it is not always clear what is clickable, when a control has recognized the click, and what the click will accomplish.

Eight Golden Rules

Ben Shneiderman proposes eight underlying principles that are applicable in most interactive systems (see "Further Resources" at the end of the chapter for Shneiderman's text). Although these are general guidelines rather than specific rules, he names them "golden rules" to indicate they are the key to usability (see Figure 12-7).

FIGURE 12-7
The eight golden rules for designing interactive interfaces.

1 Strive for Consistency
2 Enable Frequent Users to Use Short Cuts
3 Offer Informative Feedback
4 Design Dialogs to Yield Closure
5 Offer Simple Error Handling
6 Permit Easy Reversal of Actions
7 Support Internal Locus of Control
8 Reduce Short-Term Memory Load

Strive for Consistency Designing a consistent appearing and functioning interface is one of the most important design goals. The way information is arranged on forms, the names and arrangement of menu items, the size and shape of icons, and the sequence followed to carry out tasks should be consistent throughout the system. People are creatures of habit. Once we learn one way of doing things, it is difficult to change. When operating a computer application, many of the actions taken become automatic—we do not have to think about what we are doing. People who can touch type do not have to think about each key press—the fingers just respond automatically. Consider what would happen to touch typists if rows two and three on the keyboard were reversed. They would not be able to use the keyboard (and certainly wouldn't like it). If a new application comes along that has a different way of functioning, productivity will suffer and users will not be happy.

The Apple Macintosh first emphasized the benefits of consistency in the 1980s. Apple provided applications with the Macintosh that set the standard that developers should follow when creating new applications. If new applications were consistent with these applications, Apple claimed that learning them would be easy. Apple also published a standards document to explain how to be consistent with the Macintosh interface. Similar examples and standards documents followed for the Microsoft Windows interface.

Business information systems are different than desktop applications originally produced for the Macintosh. Sometimes an application needs to be inconsistent with the original guidelines. For example, the original standards specified to include menus on the menu bar for File, Edit, and Format, in every application. All document-oriented applications, such as word processors, spreadsheets, and graphics, need those menus. But many business systems do not have file, edit, and format functions. Most other guidelines and standards do apply, however.

Research has also shown that inconsistent interfaces sometimes are beneficial. If the user is interacting with multiple applications in separate windows, a different visual appearance may help the user differentiate them. Additionally, when the user is learning several applications in one session, some differences in the interfaces may help the user remember which application is which. Inconsistencies introduced for these reasons should be subtle and superficial. The basic operation of the applications should be the same.

Enable Frequent Users to Use Short Cuts Users who work with one application all day long are willing to invest the time to learn short cuts. They rapidly lose patience with long menu sequences and multiple dialog boxes when they know exactly what they want to do. Therefore, short-cut keys reduce the number of interactions for a given task. Also, designers should provide macro facilities for users to create their own short cuts.

Sometimes the entire interface should be designed for frequent users who do not need much flexibility. Consider the mail-order data-entry clerks for Rocky Mountain Outfitters. They enter orders into the system all day long from paper forms mailed by customers. These users need an interface that is simple, fast, and accurate. Long dialogs, multiple menus, and multiple forms would slow these users down.

Offer Informative Feedback Every action a user takes should result in some type of feedback from the computer so the user knows the action was recognized. Even keyboard clicks help the user, so the click is included deliberately. If the user clicks a button, the button should visually change and perhaps make a sound.

Feedback of information to the user is also important. If the Rocky Mountain Outfitters mail-order clerk enters a customer ID number in an order screen, the system should look up the customer to validate the ID number, but it should also display the name and address to the clerk so the clerk is confident the number is correct. Similarly, when the clerk enters a product ID for the order, the system should display a description of the product. As the clerk's attention shifts back and forth from the mail-order form to the computer screen, he compares the name and product description from the system to the information on the form to confirm everything is okay. This sense of confirmation and the resulting confidence in the system is very important to users, particularly when they work with a system all day. But the system should not slow the user down by displaying too many dialog boxes to which the user must respond.

Sometimes feedback is provided to help the user in other ways. The phone-order representative at Rocky Mountain Outfitters needs information from the system just as the mail-order clerk does, but he also needs more information. Phone customers may ask questions, so the information provided as feedback for the phone-order representative is more detailed and flexible. We discuss some designs for the phone-order representative at RMO later in this chapter.

Design Dialogs to Yield Closure Each dialog with the system should be organized with a clear sequence—a beginning, middle, and end. Any well-defined task has a beginning, middle, and end, so user tasks on the computer should also feel this way. If the user is thinking, "I want to check my messages," as in the earlier manager and assistant dialog example, the dialog begins with a request, involves an exchange of information, and then ends. The user can get lost if it is not clear when a task starts and ends. Additionally, the user often focuses intently on a task, so when it is confirmed that the task is complete, the user can clear her mind and get ready to focus on the next task.

If the system requirements are defined initially as events to which the system responds, each event leads to processing of one specific, well-defined activity. In the traditional, structured approach, each activity is defined by data flow diagrams and structured English. With the object-oriented approach, each activity (a use case) might be further defined as multiple scenarios, each with a flow of events. Each scenario is a well-defined interaction; therefore, event decomposition sets the stage for dialogs with closure in both the traditional approach and the object-oriented approach.

Offer Simple Error Handling User errors are costly, both in the time needed to correct them and in the resulting mistakes. If the wrong items are sent to a customer at Rocky Mountain Outfitters, it is a costly error. Therefore, the systems designer must prevent the user from making errors whenever possible. A chief way to do this is to limit available options and allow the user to choose from valid options at any point in the dialog. Adequate feedback, as discussed previously, also helps reduce errors.

If an error does occur, the system needs mechanisms for handling it. The validation techniques discussed in Chapter 11 are useful for catching errors, but the system must also help the user correct the error. When the system does find an error, the error message should state specifically what is wrong and explain how to correct it. Error messages should not be judgmental. It is not appropriate to blame the user or make the user feel inadequate.

The system also should make it easy to correct the error. For example, if the user typed in an invalid customer ID, the system should tell the user that and then place the cursor in the customer ID text box with the previously typed number displayed and ready to edit. This way, the user can see the mistake and edit it rather than retype the entire ID. Consider the following error message that occurs after a user has typed in a full screen of information about a new customer:

```
The customer information entered is not valid. Try again.
```

This message doe not explain what is wrong or what to do next. Further, after displaying this message, what if the system cleared the data-entry form and redisplayed it? The user would have to reenter everything previously typed, yet still have no idea what is wrong. The error message did not explain it, and now that the typed data has been cleared, the user cannot tell what might have been wrong. A better error message would say:

```
The date of birth entered is not valid. Check to be sure
only numeric characters in appropriate ranges are entered in
the date of birth field...
```

The input form should redisplay with all fields still filled in and the cursor should be at the field with invalid data, ready for the user to edit it.

Permit Easy Reversal of Actions Users need to feel that they can explore options and take actions that can be canceled or reversed without difficulty. This is one way users learn about the system—by experimenting. It is also a way to prevent errors; as users recognize they have made a mistake, they cancel the action. In the game of checkers, a move is not final until the player takes his fingers off the game piece; it should be the same when a user drags an object on the screen. Additionally, be sure to include cancel buttons on all dialog boxes and allow users to go back one step at any time. Finally, when the user deletes something substantial—a file, a record, or a transaction—the system should ask the user to confirm the action.

Support Internal Locus of Control Experienced users want to feel that they are in charge of the system and that the system responds to their commands. They should not be forced to do anything or made to feel as if the system is controlling them. Systems should make users feel they are deciding what to do. Much of this comfort and control can be provided by the wording of prompts and messages. Writing out a dialog like the manager and assistant message dialog given previously will lead to a design that conveys the feeling of control.

Reduce Short-Term Memory Load People have many limitations, and short-term memory is one of the biggest. As discussed earlier in this book, people can remember only about seven chunks of information at a time. The interface designer cannot assume the user will remember anything from form to form, or dialog box to dialog box, during an interaction with the system. If the user has to stop and ask, "Now what was the filename? the customer ID? the product description?" then the design places too much of a burden on the user's memory.

With these eight golden rules in mind, an interface designer can help ensure that user interactions are efficient and effective. We now turn to some basic techniques for documenting the design of the dialog.

Documenting Dialog Designs

Many techniques are available to help the designer think through and document dialog designs. The dialogs that must be designed are based on the inputs and outputs requiring user interaction, as defined in Chapter 11. These are used to define a menu hierarchy that allows the user to navigate to each dialog. Storyboards, prototypes, and UML diagrams can be used to complete the designs.

Events, Subsystems, and Menu Hierarchy

Chapter 11 described how inputs and outputs are obtained from data-flow diagrams (in the traditional approach) and sequence diagrams (in the object-oriented approach). Generally, each input *obtained interactively from a user* requires a dialog design. Additionally, each output produced *at the request of a user* requires a dialog design. The events documented early during the analysis process are the key to listing the dialogs that need to be designed. External events are triggered by inputs, and temporal events produce outputs.

Dialog design must be done simultaneously with other design activities. As shown in Chapter 9, the structure charts for subsystems (transaction analysis) include details about menu structure of the interactive parts of the system. Additionally, the structure chart for each event (transform analysis of each DFD fragment) also includes details about the dialog with the user. The object-oriented approach also integrates dialog design very early, even during analysis tasks. Sequence diagrams and collaboration diagrams include details about the dialog. When we discuss menu design and dialog design, remember that these tasks are not done in isolation.

The overall system structure from the standpoint of the user is reflected in the menus available. Each menu contains a hierarchy of options, and they are often arranged by subsystem or by actions on objects. Rocky Mountain Outfitters' customer support system includes the order entry subsystem, order fulfillment subsystem, customer maintenance subsystem, catalog maintenance subsystem, and a reporting subsystem added during design. Menus might also be arranged based on objects—customers, orders, inventory, and shipments. Within each menu, there might be duplicate functions, such as *Look up past orders*, under customers and under inventory.

Sometimes several versions of the menus are needed based on the type of user. For example, mail-order clerks at RMO do not need many of the options available—they process new orders only. The phone-order sales representatives need many more options, but they still do not need all system functions. And some options should be available only to managers, such as management reports and price adjustments.

Menus should also include options that are not in the event list—most important are options related to the controls discussed in Chapter 11. These include backup and recovery of databases in some cases, plus user account maintenance. Additionally, user preferences are usually provided to allow the user to tailor the interface. Finally, menus should always include help facilities.

All events that lead to dialogs in the RMO customer support system are listed and grouped by subsystem in Figure 12-8. These groupings form one set of menu hierarchies. In addition, there are menu hierarchies for utilities, preferences, and help. The list in the figure is only one of many possible menu hierarchy designs—a starting place.

FIGURE 12-8
One overall menu hierarchy design for the RMO customer support system (not all users will have all of these available).

Five menu hierarchies grouped by subsystem and based on events/use cases

Order Entry
- Check Availability
- Create New Order
- Update Order

Order Fulfillment
- Check Order Status
- Record Fulfillment
- Record Back Order
- Create Return

Customer Account Maintenance
- Provide Catalog
- Update Customer Account
- Adjust Customer Charges
- Distribute Promotional Package

Catalog Maintenance
- Update Catalog
- Create Special Promotion
- Create New Catalog

Reporting and Queries
- Order Summary Report
- Transaction Summary Reports
- Fulfillment Summary Reports
- Prospective Customer Lists
- Customer Adjustment Reports
- Catalog Activity Reports
- Ad Hoc Query Facility

Three menu hierachies added during design for controls, preferences, and help

System Utilities
- Printers and Devices
- Back Up and Recovery
- User Accounts
  - Maintain Accounts
  - Change Current User Password

User Preferences
- Dialog Style
- Color and Font
- Shortcuts
- Macro Facility

Help
- Contents and Index
- Search for Help
- Task List
- About the RMO System

A dialog design is created for each of these menu options. After completing the dialog designs for all options, the designer can define the structure of the menus for different types of users. Menu hierarchies can be rearranged easily as the design evolves.

Dialogs and Storyboards

Once all required dialogs have been identified, dialog designs can be documented in many ways, and there are no *de facto* standards. One approach is to list the key steps followed for the dialog with a written description of what the user and computer do at each step. The format for writing these steps can follow the flow of events model for scenarios, described in Chapter 7 for the object-oriented approach (see Figure 7-3), or they can be more free-form.

Designers can also document dialog designs by writing out a dialog like the one presented earlier in this chapter (for the manager and assistant), then documenting how the user and computer express themselves, as shown previously in Figure 12-6. Sometimes the designer can go right from the dialog to sketching a series of screens that make the design more visible. Consider the DownTown Videos rental system described in a case study at the end of Chapter 7. One event is *Customer rents videos*, and the dialog for renting videos to a customer that goes on between the clerk and the system would follow this pattern:

System:	What DownTown Videos option do you want?
User:	I need to rent some videos.
System:	Okay, what is the customer's ID number for this rental?
User:	It is 201564.
System:	Fine, your customer is Nancy Wells at 1122 Silicon Avenue. Note that she has two kids, Bob and Sally, who can only rent PG-13 videos. Correct?
User:	Yes.
System:	What's the ID number of the first video she wants to rent?
User:	It is 5312.
System:	Fine, that is *Titanic* (the 1953 version with Barbara Stanwyck) with rental charge of $1.00 for three days, rated PG. What's the ID number of the next video, or is that all?
User:	It is 8645.
System:	Fine, that is *Titanic* (the 1998 version) with a rental charge of $3.00 for three days, rated PG-13. What's the ID number of the next video, or is that all?
User:	That's all.
System:	Okay, collect $4.00 plus $.24 tax, for a total of $4.24 and let me know when you are done.
User:	Yes, I'm done.
System:	Remind Nancy Wells that these videos are due back Wednesday the 9th before 11 P.M. closing.
User:	Okay. Thanks.

The designer thinks through the sequence of screens the user would need to carry on this dialog with the computer. The user needs to enter a customer ID and several video IDs. The computer would have to look up and display information on the customer and each video rented. Then the computer would calculate the total due and remind the user of the due date so the user can pass that information on to the customer.

storyboarding
A technique used to document dialog designs by showing a sequence of sketches of the display screen

One technique used to show the screens is called **storyboarding**—showing a sequence of sketches of the display screen during a dialog. The sketches do not have to be very detailed to show the basic design concept. A storyboard can be implemented with a visual programming tool such as Visual Basic, but using simple sketches drawn with a graphics package can help keep the focus on the fundamental design ideas.

Figure 12-9 shows the storyboard for the rent videos dialog. The system has a menu hierarchy based on the event list plus needed controls, preferences, and help. The dialog uses one form and a few dialog boxes and adds more information to the form as the dialog progresses. Note that the questions the computer asks are shown in the prompt area at the bottom of the form, matching almost identically the phrases used in the written dialog. The user has a choice of either scanning or typing the few IDs that must be entered. Information provided to the user is shown in labels on the form. The information provided allows the user to confirm the identity of the customer, see any restrictions that might apply, and pass on information about cost and return dates to the customer.

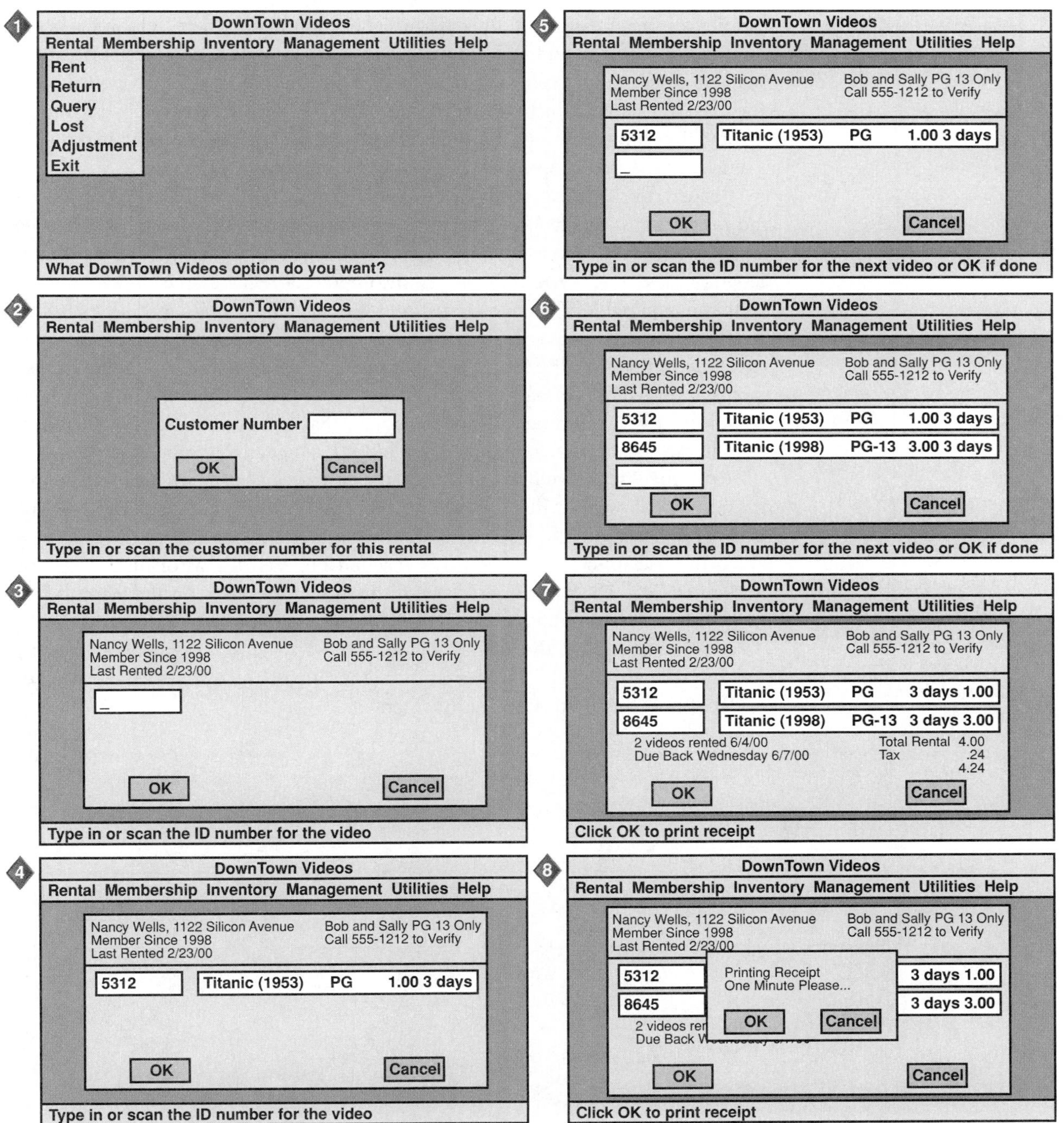

FIGURE 12-9
Storyboard for the DownTown Videos rent videos dialog.

These approaches to dialog design provide only a framework to work from, and the resulting design remains fairly general. As working prototypes are produced, many details still have to be worked out. As the design progresses, reviewing the golden rules and other guidelines will help you keep the focus on usability.

Dialog Documentation with UML Diagrams

The object-oriented approach provides specific UML diagrams that are useful for modeling user-computer dialogs. In the object-oriented approach, objects send messages back and forth, listening to and responding to each other in sequence. People also send messages to objects and receive messages back from the objects. The sequence diagram described in Chapter 7 includes an actor (a user) sending messages to objects in the system and the objects returning information in the form of messages, shown in sequence. It is much like a dialog between the user and various objects in the system.

The object-oriented approach involves adding more types of objects to class diagrams and interaction diagrams as the project moves from analysis to design, as discussed in Chapter 9. Interface objects can be added to these diagrams to show more detail about the design of the dialog between the user and computer. The first step is to determine what forms are required for the dialog based on informal dialog design techniques described previously. Next, the sequence diagram for the scenario is expanded to show the user (an actor) interacting with the forms. The user-interface classes that make up the forms can then be modeled using a class diagram. Finally, the sequence diagram is further expanded to show the user interacting with specific objects that make up the form. A collaboration diagram can also be used for another view of the dialog design for the scenario.

Recall the Rocky Mountain Outfitters use case *Look up item availability*. A sequence diagram was shown for this use case in Chapter 7 (Figure 7-12). Figure 12-10 shows an expanded version of the sequence diagram that includes a form (or window) with which the user interacts. The form is inserted between the actor and the problem domain objects. The actor interacts with the interface objects on the form to talk to the problem domain objects, which reply through the interface objects on the form. The analysis model of the interaction just shows the messages between actor and problem domain objects. The design model inserts the interface to create a physical model of how the interaction is implemented.

FIGURE 12-10
A sequence diagram for the RMO Look up item availability *dialog with an item search form added.*

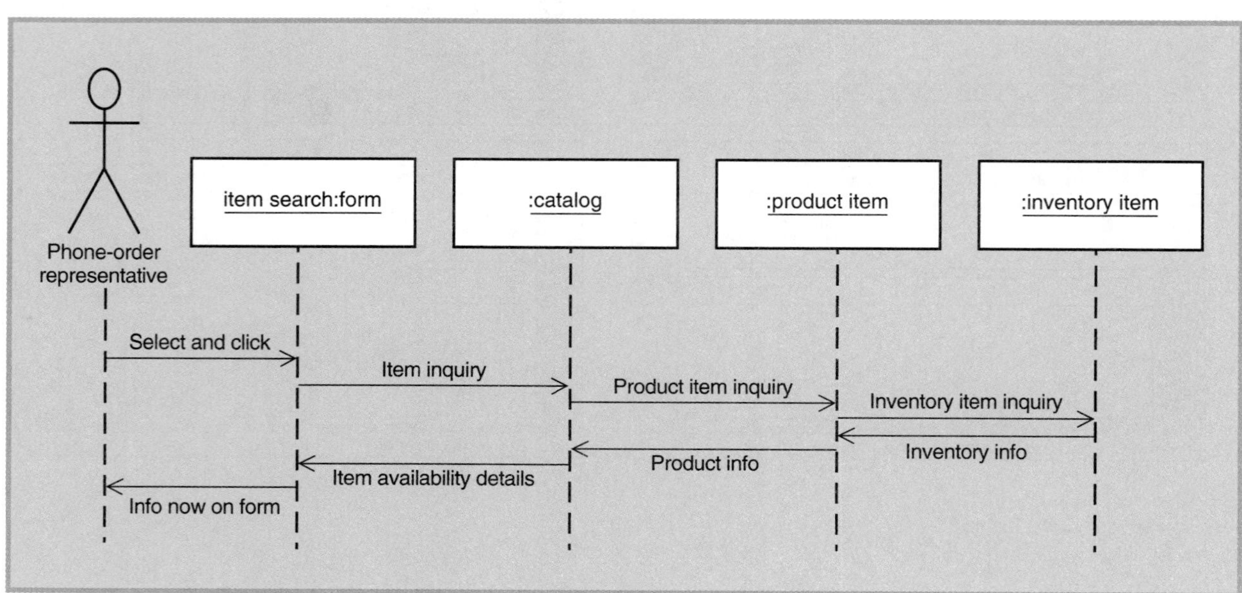

The form shown on the sequence diagram includes specific interface objects. Figure 12-11 shows a class diagram with the interface classes used to make up the form. (Recall that the class diagram was also shown in Chapter 11 to define the methods that interact with specific forms.) The Frame class in Figure 12-11 represents the basic structure that contains other interface objects. A Menu Bar is attached to the frame, a menu bar contains Menus, and a menu contains Menu Items. The associations between these classes are aggregations shown with the diamond symbol on the class diagram (whole-part or aggregation relationships are discussed in Chapter 5). Other classes shown are List, Button, and Label, which also are parts of the frame. This example, based loosely on Java, enables the frame to "listen" for events that occur to interface objects, such as clicking a menu item or a button. The frame's actionListener() method is invoked when the frame "hears" that an event has occurred.

FIGURE 12-11
A class diagram showing interface classes making up the item search *form.*

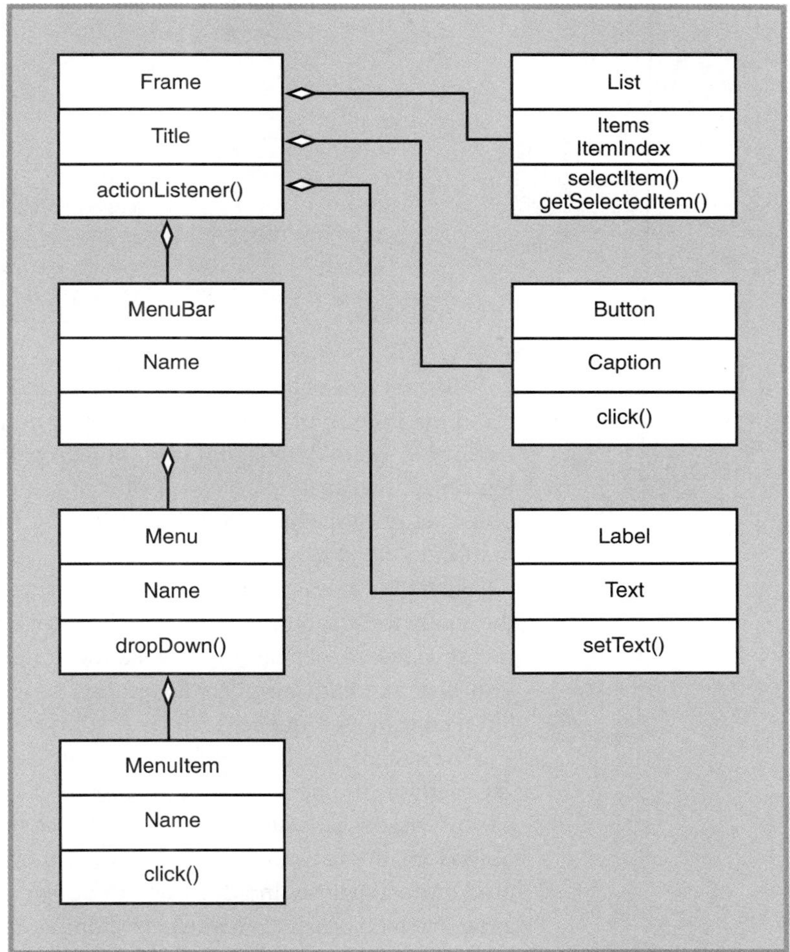

A sequence diagram can be used to model the messages between the user and the specific objects making up the form, including the messages the interface objects send to each other. Figure 12-12 shows the further expanded sequence diagram. This model emphasizes the details of the form's design, so the problem domain details can be omitted. Nothing has changed for the part of the sequence diagram where problem domain objects interact among themselves. The interface objects have simply been plugged in between them and the actor.

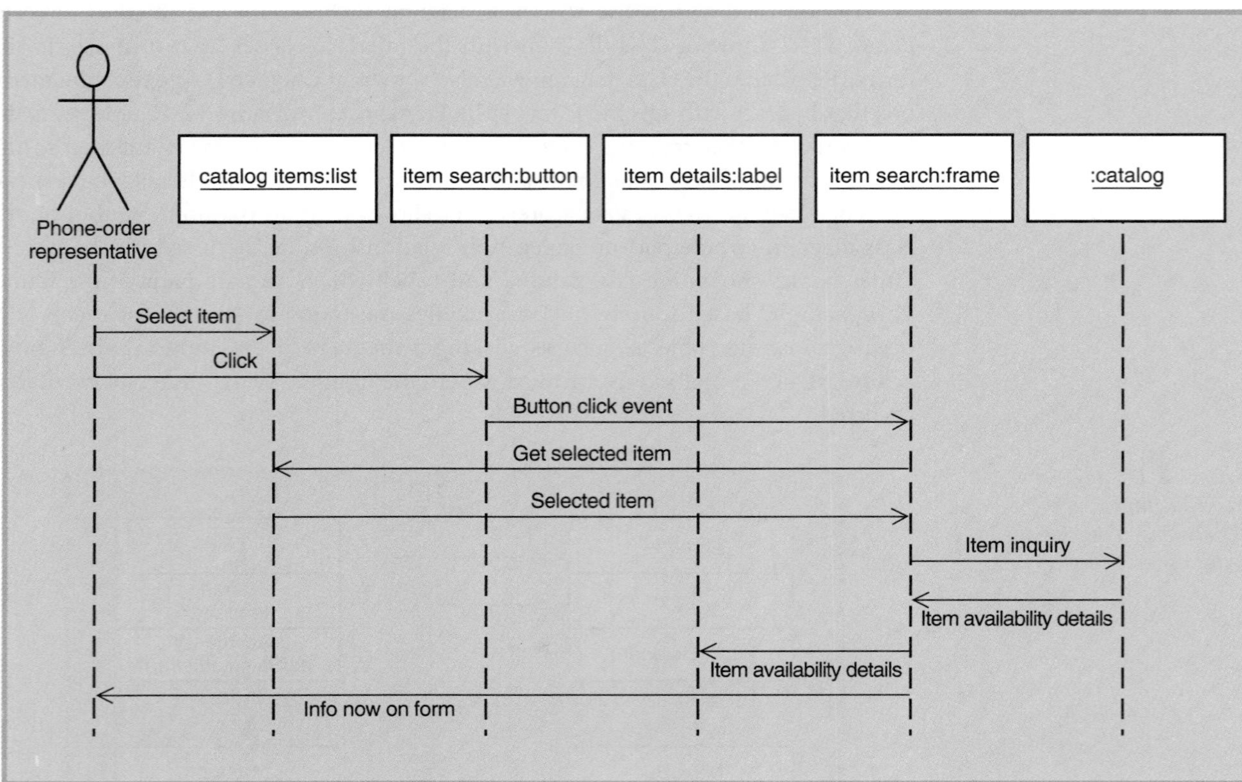

FIGURE 12-12
A sequence diagram showing specific interface objects for the Look up item availability *dialog (not all problem domain objects are shown).*

The user interacting with the *Look up item availability* form selects an item to look up in the list and then clicks the search button. The frame "hears" the click, asks the list for the selected item, and uses the selected item in the message to query the catalog object, which interacts with other problem domain objects. When the catalog object returns the requested information, the frame tells the label to display the information to the user.

Figure 12-13 shows a collaboration diagram containing the same information as the sequence diagram. Recall from Chapter 7 that either or both of these diagrams might be drawn, depending on what the designer wants to emphasize. The collaboration diagram, with numbered messages, emphasizes the messages between objects, but it still shows sequence.

The systems analyst needs to think through the natural sequence of interaction very early in the development process to create the sequence diagram for each scenario of each use case. This basic dialog is then extended to include user-interface objects for the user and computer when they interact. Since the same diagrams are used during analysis and design, with design adding more detail, the object-oriented approach lends itself to iterative development.

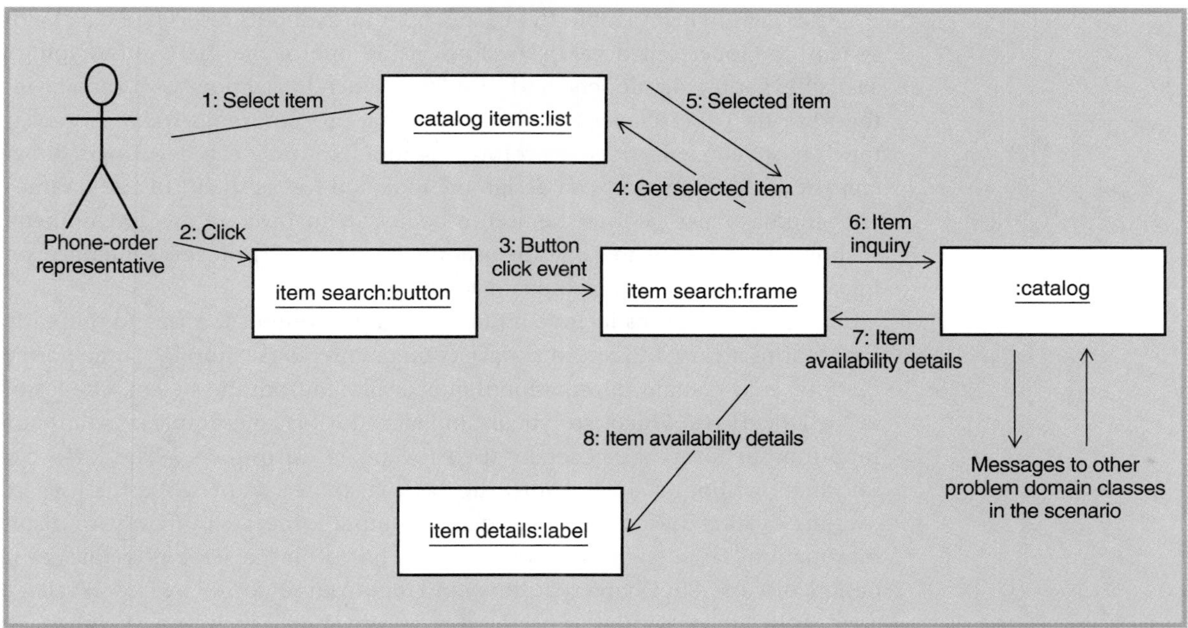

FIGURE 12-13
A collaboration diagram for the Look up item availability *dialog.*

Guidelines for Designing Windows Forms

Each dialog might require several windows forms. Each form must be carefully designed for usability. Almost all of the new business systems today are developed for an interactive Microsoft Windows, X-Windows (UNIX), or Macintosh environment. Form design principles are the same for any of these three environments. In this section, when we refer to forms or windows, we mean any one of these three environments. Within the windows environment, however, we should distinguish between standard input forms and browser input forms.

Standard input forms are those that are programmed in a programming language, such as Visual Basic or Java. The advantage of standard forms is that they are extremely flexible and can access data directly on the workstation. **Browser forms**, on the other hand, are programmed using HTML or XML and follow Internet conventions. Browser forms can be displayed using a markup language browser and must follow those conventions. Currently HTML and XML forms do not permit the same level of design flexibility as standard input forms. However, technology is changing rapidly, and markup languages may soon support all the desired features. The advantage of forms based on HTML and XML protocols is that the same forms can be used for internal staff as well as on the Internet. E-commerce is growing so rapidly that many firms are designing the user interface on all new systems based on HTML and XML formats.

browser forms
Forms programmed using HTML or XML that follow Internet conventions

After identifying the objective of the input form and its associated data fields, the system developer can develop the form using one of the many prototyping tools available. Earlier, developers used to spend tremendous amounts of time laying out the form on paper diagrams before beginning programming. Today, however, it is much more efficient to use a prototyping tool. Not only can developers design the content of the form, they can design the look and feel of the form at the same time. This approach also permits the user to be heavily involved in the development, and such involvement ensures that the user interface is in fact the *user's* interface, providing a strong sense of ownership and acceptance.

Categories of forms include input forms, input/output forms, and output forms. Input forms are used to record a transaction or enter data, although some portions of the form may contain information that is displayed from the system. The form used in the DownTown Videos storyboard in Figure 12-9 is an example of an input form. Input/output forms are generally used to update existing data. The form displays information about a single entity, such as a customer, and enables the user to type over the existing information to update it. Output forms are primarily for displaying information. The design of output forms is based on the same principles as report design, discussed in Chapter 11. Input and input/output forms are closely related and are designed using similar principles. Every form should include a careful analysis of the integrity controls required for data input, discussed in Chapter 11. There are four major issues to be considered in the design of these forms:

- Form layout and formatting
- Data keying and entry
- Navigation and support controls
- Help support

Form Layout and Formatting

Form layout and formatting are concerned with the general look and feel of the form. Everyone has encountered systems with hard-to-use input forms—the font was too small, the labels were hard to understand, the colors were abrasive, the navigation buttons were not obvious, and so forth. On the other hand, forms that are easy to use are well laid out, with the fields easily identified and understood. One of the best methods to ensure that forms are well laid out is to prototype various alternatives and let the user test them. The user will let you know which characteristics are helpful and which are distracting. As you design your input forms, you should think about the following:

- Consistency
- Headings, labels, and logos
- Distribution and order of data-entry fields and buttons
- Font sizes, highlighting, and colors

We place consistency at the top of the list of things to consider because of its importance, as discussed previously. Some large systems require several teams of programmer/analysts to develop the input forms, and periodically systems are developed without adequate communication between the teams. All of the forms within a system need to have the same look and feel. A consistent use of function keys, short cuts, control buttons, and even color and layout makes a system much more useful and professional looking. On the other hand, a system that is not consistent across all forms can be error-prone and difficult to use.

The headings, labels, and logos on the form help to convey the purpose and use of the form. A clear, descriptive title at the top of the form helps to minimize confusion about the use of the form. Labels should also be easily identified and easy to read.

The designer should also carefully place the data-entry fields around the form. Related fields can be placed next to each other and even isolated with a fine-lined box. The designer should also carefully consider the tab order. If input is coming from a paper form, then the tab order should follow the reading of the paper form—top to bottom, left to right. Blank space should be used throughout the form so that the fields do not appear crammed together and are easy to distinguish and read. Normally navigation buttons are at the bottom of the window. *De facto* traditions for the placement of buttons are developing based on standards of the large development firms such as Apple, Microsoft, Sun, Oracle, and others. It is a good idea to be sensitive to these traditions as they change with new technology upgrades.

The purpose of font size, highlighting, and color is to make the form easy to read. A careful mix of large and small fonts, bold and normal type, and different color fonts or background can help the user find important and critical information on the form. Too much variation causes the form to become cluttered and difficult to use. However, judicious use of these techniques will make the form more easily understood. Column headings and totals can be made slightly larger or boldfaced. The form can highlight negative or credit balances by changing font color. However, font color and background color should be used in concert to ensure that the field is readable. For example, red type on green or black backgrounds or a dark color on a dark background are not good choices—some people with colorblindness cannot distinguish red from green or black.

Figure 12-14 is an example of a form designed for the Rocky Mountain Outfitters customer support system. This form is used to look up information about a product and to add it to an order. Notice how the title and labels make the form easy to read. The natural flow of the form is top to bottom with related fields placed together. Navigation and close buttons are easily found but are not in the way of data-entry activity.

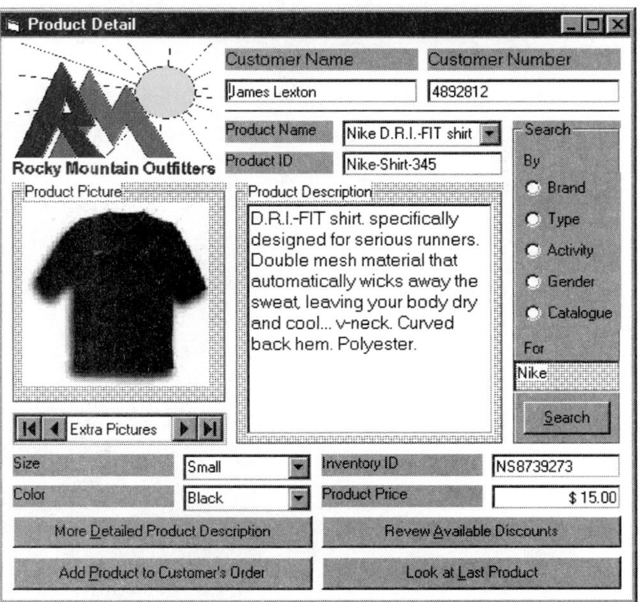

FIGURE 12-14
The RMO Product Detail form used to look up information about a product, select size and color, and then add it to an order.

Data Keying and Entry for Standard Windows

The heart of any input form is the entry of the new data. Even here, however, a primary objective is to require as little data entry as possible. Any information already in the computer, or that can be generated by the computer, should not be reentered. A generous use of selection lists, check boxes, automatic retrieval of descriptive fields, and so forth will speed up data entry and reduce errors. The Product Detail form for RMO (Figure 12-14) shows many examples that reduce the need to enter data.

text box
An input control that accepts keyboard data entry

Several types of data-entry controls are widely used in windows systems today. A **text box** is the most common element used for data entry. A text box consists of a rectangular box that accepts keyboard data. In most cases, it is a good idea to add a descriptive label to identify what should be typed in the text box. Text boxes can be designed to limit the entry to a specified length on a single line or to permit scrolling with multiple lines of data.

list box
An input control that contains a list of acceptable entries the user can select

spin box
A variation of the list box that presents multiple entries in a text box from which the user can select

combo box
Another variation of the list box that permits the user to enter a new value or select from the entries

Variations of a text box consist of a list box, a spin box, and a combo box. A **list box** contains a list of the acceptable entries for the box. The list usually consists of a predefined list of data values, and the user selects one from the list. The list can be presented either within the rectangular box or as a drop-down list. A variation of a list box is a spin box. A **spin box** presents the possible values within the text box itself. Two spinner arrows let the user scroll through all the values. A **combo box** also contains a predefined list of acceptable entries but also permits the user to enter a new value when the list does not contain the required value. The list of available data values can be provided either as hard-coded entries in the form itself, or as the values from some type of file or table. Both a list box and combo box facilitate data entry by minimizing keystrokes and the corresponding possibility of errors.

radio buttons (or option buttons)
Input controls that enable the user to select one option from a group

check boxes
Input controls that enable the user to select more than one option from a group

Two types of input controls are used in groups: radio buttons (sometimes called *option buttons*) and check boxes. **Radio buttons** are associated together as a group, and the user selects one and only one of the group. The system automatically turns off all the other buttons in the group when one is selected. Since all of the possible values appear on the form, this control is used only when the list of alternatives is small and the values never change. **Check boxes** also work together as a group. However, check boxes permit the user to select as many values as desired within the group. Sometimes radio buttons and check boxes are called Boolean fields because the selection for each control is either on or off—that is, checked or unchecked. Figure 12-15 shows a form with examples of these data-entry controls.

FIGURE 12-15
Examples of data-entry controls on an input form.

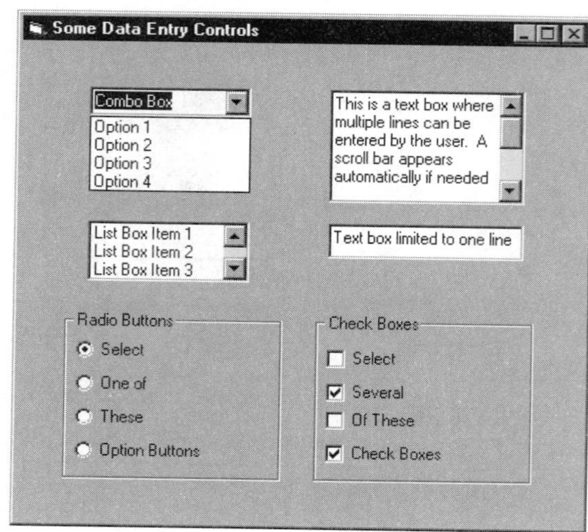

Navigation and Support Controls

Standard Windows interfaces provide several controls for navigation and window manipulation. For Microsoft applications, these controls consist of Minimize, Maximize, and Close buttons in the top-right corner, horizontal and vertical scroll bars, the record selection bar on the left panel, record navigation arrows at the bottom of the window, and so forth. To maintain consistency across systems, it is generally a good idea to utilize these navigation controls when available. A well-designed user interface, however, should also include other controls or buttons. You can place buttons on the form to enable users to move to other relevant screens, to search and find data, and to close the open window.

Data Keying and Entry for the Browser Interface

Browser interfaces are dependent on the capabilities available in markup languages such as HTML and the newer XML. Although originally these languages had limited capabilities, today's versions can support the same basic type of data entry controls. The major difference between a standard windows input form and a browser input form is that the windows form can easily perform edits field by field as the data are entered. In a browser input form, the edits are not performed until the entire form is transmitted to the server computer. However, as browser programs become increasingly more sophisticated, more and more capabilities will be provided while the data are being entered. With the growth and emphasis being placed on web technology, we expect in the future that windows input forms and browser input forms will have very similar capabilities.

Help Support

A primary objective in the design of each input form is for it to be intuitive that users will not need help. However, even well-designed forms will be misunderstood, and on-line help is always recommended. Three types of help are common in today's systems: (1) a tutorial that walks you through the use of the form, (2) an indexed list of help topics, and (3) context-sensitive help.

Most systems provide tutorial help to assist in training new users. Tutorials can be organized by task, which generally includes one dialog with a set of related forms. Every new system should have an indexed list of help topics. This list can be invoked either through a keyword search or, as with many Microsoft systems, with a help wizard. The help wizard is simply a program that does an automatic keyword search based on words found in a question or sentence. The wizard will return several alternative help topics based on the results of the keyword searches.

Context-sensitive help can be based on the indexed list of help, but it is invoked differently. Context-sensitive help automatically displays the appropriate help topic based on the location of the cursor. In other words, if the cursor is within a certain text field, and the user invokes context-sensitive help, then the help for that text field is displayed.

Dialog Design for Rocky Mountain Outfitters

Now that dialog design concepts and techniques have been discussed, we can demonstrate the process of designing one specific dialog for Rocky Mountain Outfitters. The Rocky Mountain Outfitters customer support system includes support for the phone-order sales representative processing an order for a customer. The dialog corresponds to the event *Customer places an order* and more specifically to the scenario *Phone-order representative creates new order*. The target environment for this part of the system is the phone-order representative's desktop PC on a Windows platform.

The designer starts by referring to the models produced during analysis, either the data-flow diagram fragment and corresponding detailed DFD for the activity or the sequence diagram for this scenario, depending on the approach used for analysis. The four basic steps followed in these models are as follows:

1. Record customer information.
2. Create a new order.
3. Record transaction details.
4. Produce order conformation.

With the traditional approach to development, a structure chart would be produced as shown in Chapter 9 to correspond to these steps. With the object-oriented approach, the sequence diagram for this scenario (Figure 7-17) could be updated to include forms the user will need for interaction. The specifics of identifying and detailing the inputs and outputs for this scenario were shown in Chapter 11.

A partial sequence diagram that includes a main menu form, a customer form, and an order form are added to the sequence diagram shown in Figure 12-16 (page 454). The actor (user) clicks a menu item on the main menu form that sends a display message to the customer form. The actor supplies customer information and a new customer object is created or the status is checked if it is an existing customer. The customer form sends a display message to the order form, and so on. Based on the sequence of processing required, a basic dialog can be written to convey in more detail how the dialog will flow from the user's perspective. The details of the interface objects needed on each form should wait until this dialog is refined. A new customer placing an order with a phone sales representative (the user) might resemble the following dialog:

Computer: What customer support system option do you want?

User: I need to create a new order.

Computer: Okay, is it a new or existing customer?

User: It's a new customer.

Computer: Fine, give me the customer's name, address, phone, and so on.

User: The customer is Ginny Decker, 11980 Visual Blvd....

Computer: Okay, what is the first item she wants to order?

User: Boots, Women's, Hiking Supreme Line in the Spring Fling catalog.

Computer: They come in these sizes and colors, and we have all in stock except size 9.

User: Okay, one pair of size 8 in tan.

Computer: That comes to $65.50 plus tax, shipping, and handling. Anything else?

User: Yes, Raincoat, Woman's, On the Run Line in the Spring Fling catalog.

Computer: They come in these sizes and colors, and we have all in stock except medium and large blue.

User: Okay, one coat size small in brown.

Computer: That comes to $87.95 plus tax, shipping, and handling. Anything else?

User: That's it.

Computer: How does the customer want the order shipped?

User: UPS two-day air.

Computer: Does the customer want to use a credit card, or is there another payment approach for this order?

User: Use the Mastercard number xx674-22-3388 expiring January 03.

Computer: Okay, the order is recorded. To summarize, for customer Ginny Decker, ship by UPS two-day mail Boots, Women's, Hiking Supreme Line, size 8, tan, at $65.50, and Raincoat, Woman's, On the Run Line, size small, brown, at $87.95. Total cost, $153.45 plus $9.20 tax and $13.40 shipping and handling—$176.05 charged to Mastercard xx674-22-3388.

User: Thanks.

While working on this dialog, the designer can begin to define the forms that will be required for the user and the computer. We identified some of these forms while designing the inputs in Chapter 11. A refined list of forms might include:

- Main menu form
- Customer form
- Item search form
- Product detail form
- Order summary form
- Shipping and payment options form
- Order confirmation form

The designer can then use the list of forms to define a design concept for the flow of interaction from form to form. One approach is to show potential forms in sequence, as shown in Figure 12-17. After the main menu form, the customer form appears first, which the user fills in or updates, and then the item search form displays to let the user search for an item. Product details are shown for the item, and then the item search form is shown again. When all items are selected, the order summary form is shown, and so on. The designer should concentrate on highlighting parts of the dialog that occur through each form rather than worrying about the physical design of each form. After considering what information is needed on each form, the designer can create a more detailed storyboard or implement prototype forms using a tool such as Visual Basic.

This initial design is very sequential but reasonable for a first iteration. Phone sales representatives at RMO evaluated the storyboard and the prototype, and suggested that the sequence was too rigid; it assumes the dialog always follows the same sequence. However, because phone-order representatives are on the phone with customers, they have to follow the customers' lead. Sometimes customers do not want to give out information about themselves until the order is processed and confirmed, for example. Sometimes customers want to know the totals for the order or to review details about something already included in the order. But the sequential design assumes the customer information always comes first. Although the sequential approach might work well for mail-order clerks, more flexibility is required for the phone-order representatives.

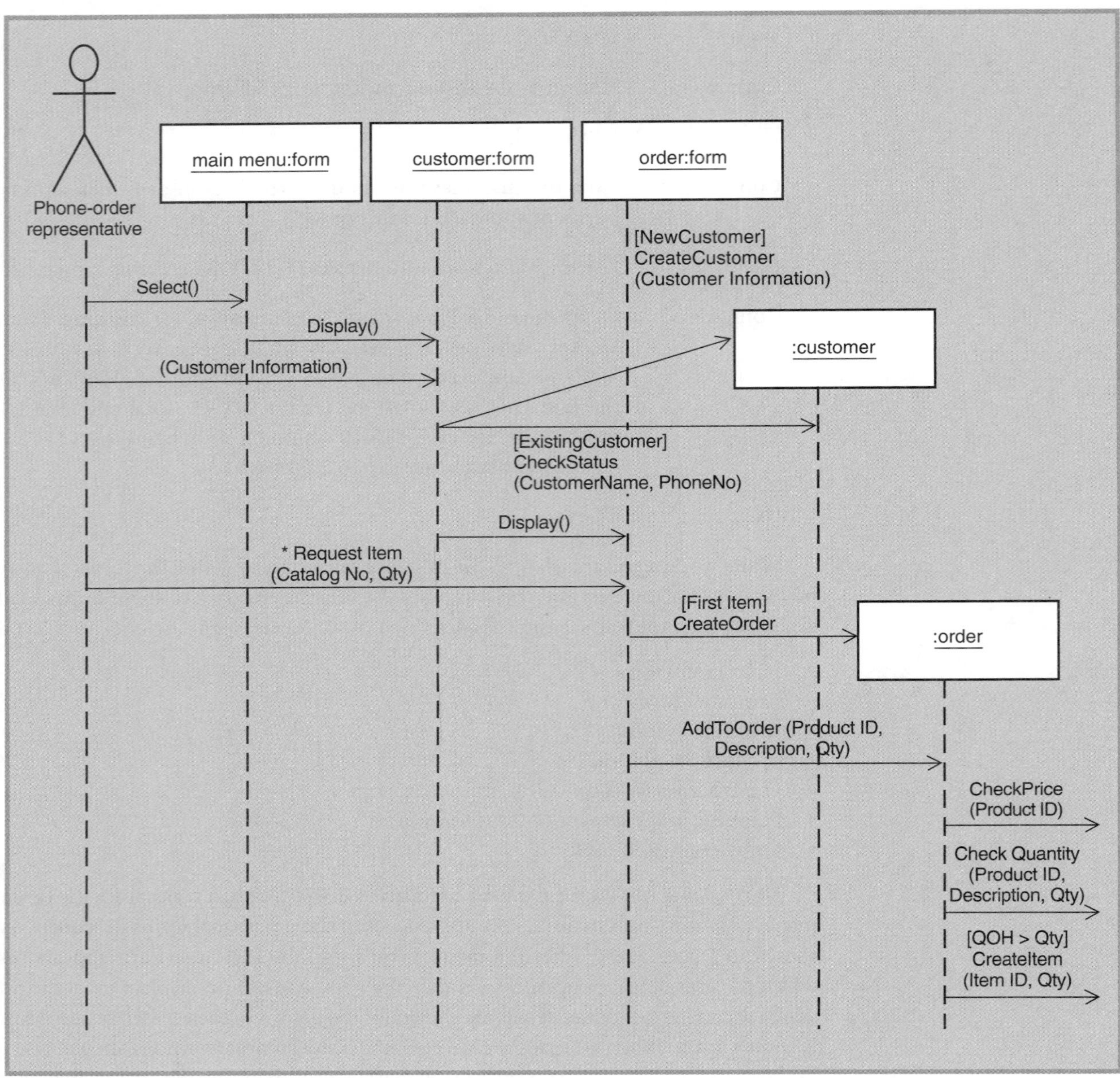

FIGURE 12-16
Partial sequence diagram for RMO's Create new order *dialog, adding the main menu, the customer form, and the order form.*

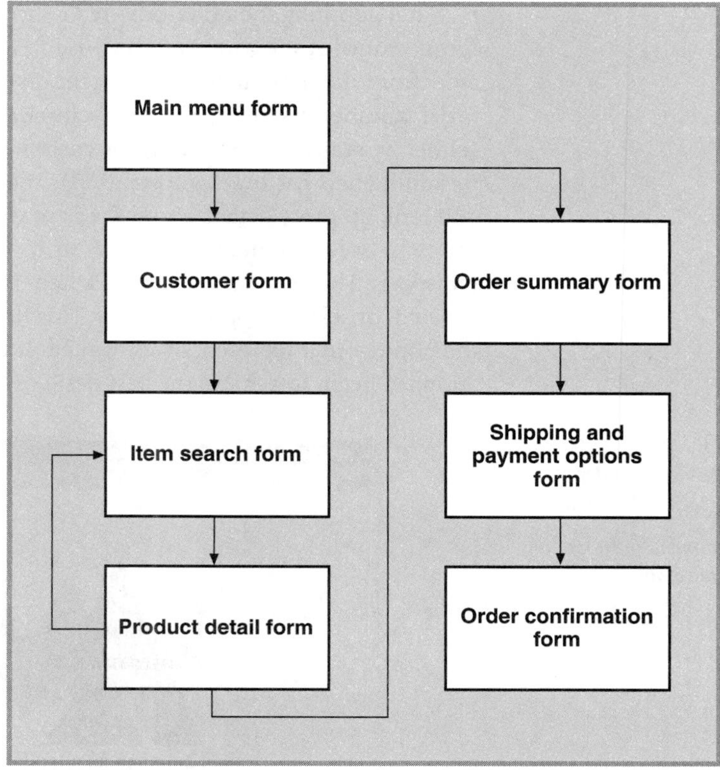

FIGURE 12-17
A design concept for the sequential approach to the Create new order *dialog.*

To address the flexibility and information needs of the users, the project team developed a second design concept. This design concept makes the order form the center of the dialog, with options to switch to other forms at will. After each action, the order form is redisplayed to the phone-order representative with the current order details. The order-centered design concept is shown in Figure 12-18. It allows the same sequence to be followed as the basic dialog, but it also allows flexibility when needed. It also shows the user more information about the order throughout the dialog in case the information is needed.

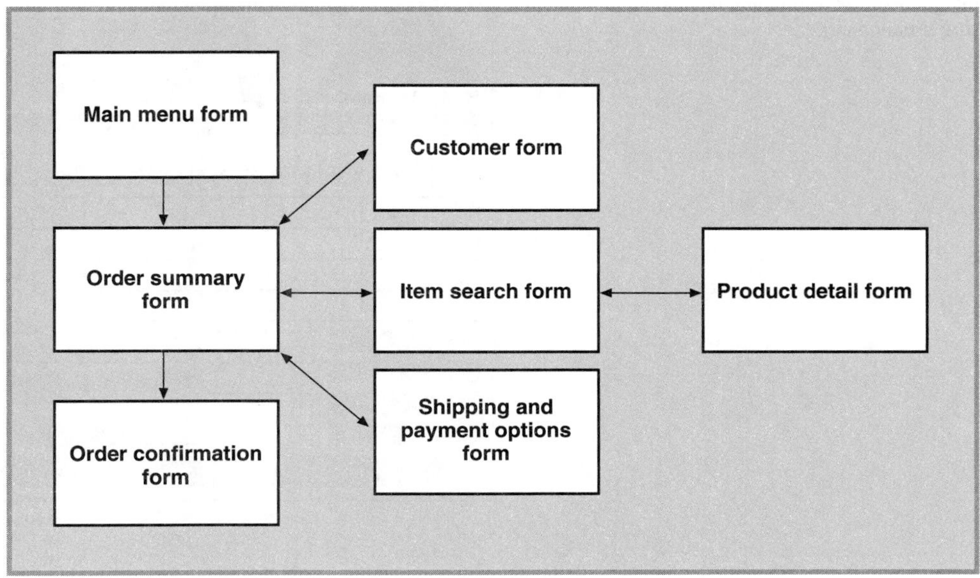

FIGURE 12-18
A design concept for an order-centered approach to the Create new order *dialog.*

After adopting the order-centered concept, the project team designed the detailed forms. Some of the forms are shown in Figure 12-19. After the user selects *Place an order* from the main menu, the system displays the order summary form with a new order number assigned. The user can add the customer information immediately (either by searching for a previous customer based on customer number or name or by adding new customer information). The user can then search for a requested item and look up more details about the item on the product detail form. If the customer wants to order the item, the user adds it to the order and the order summary form redisplays. The user can add another item to the order, make changes to the first item ordered, or select shipping options. This flexibility and information display are what the phone-order representatives wanted. It took quite a few iterations and user evaluations to begin to achieve the best design.

FIGURE 12-19
Prototype forms for an order-centered approach to the dialog.

(a) The main menu form for an order-centered approach to the dialog.

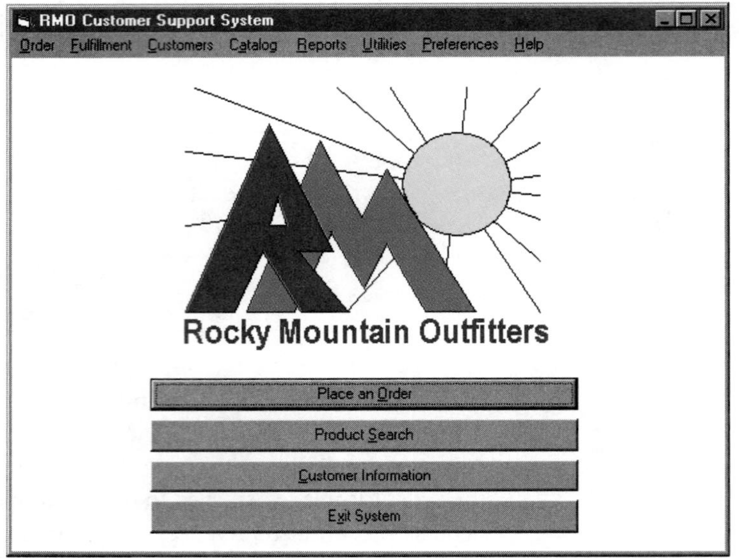

(b) The order summary form for an order-centered approach to the dialog beginning a new order.

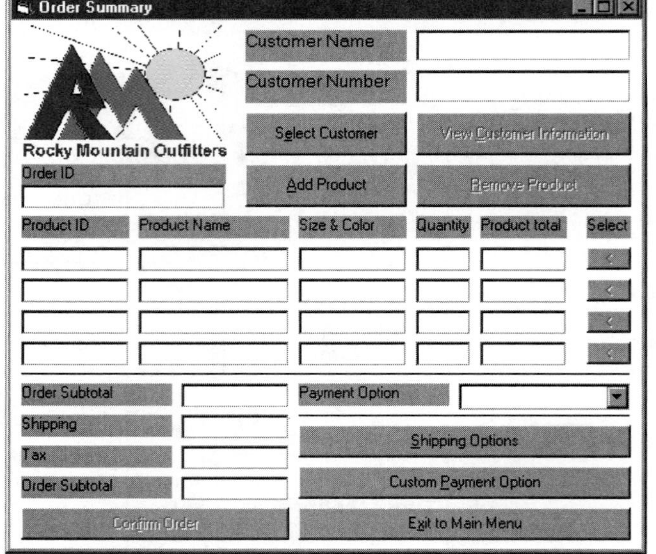

(c) The product detail form after the user selects the Add Product item.

Product Detail
Customer Name
Customer Number
James Lexton
4892812
Rocky Mountain Outfitters
Product Name
Nike D.R.I.-FIT shirt
Product ID
Nike-Shirt-345
Search
By
Brand
Type
Activity
Gender
Catalogue
For
Nike
Search
Product Picture
Product Description
D.R.I.-FIT shirt. specifically designed for serious runners. Double mesh material that automatically wicks away the sweat, leaving your body dry and cool... v-neck. Curved back hem. Polyester.
Extra Pictures
Size
Small
Inventory ID
NS8739273
Color
Black
Product Price
$ 15.00
More Detailed Product Description
Revew Available Discounts
Add Product to Customer's Order
Look at Last Product

(d) The order summary form after the user adds several products.

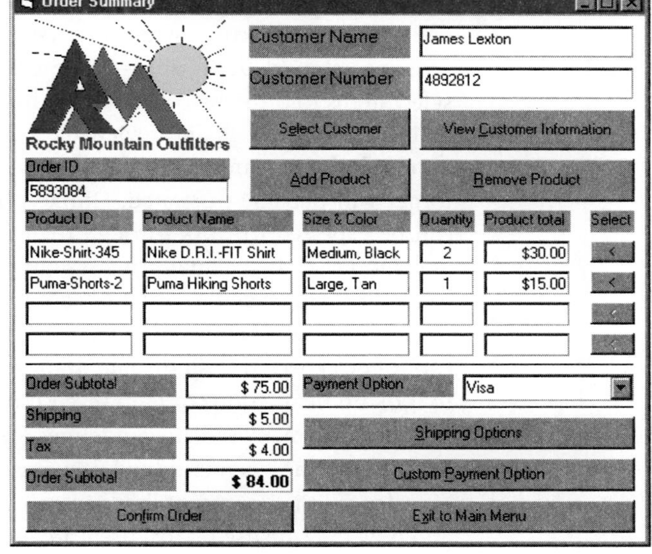

(e) The shipping option form for a completed order.

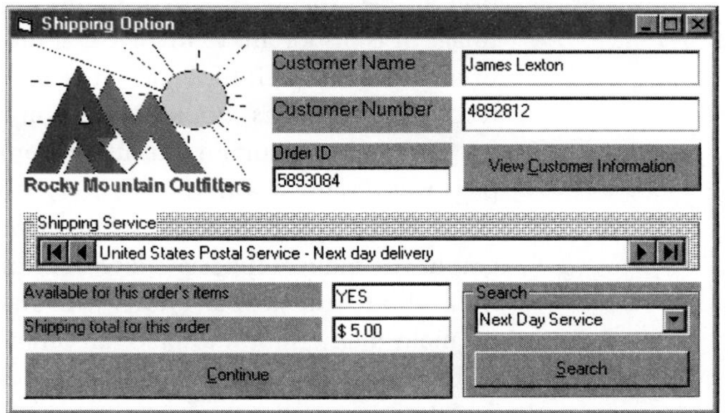

Dialog Design for Web Sites

The RMO customer support system event *Customer places order* requires several different dialog designs (one for each scenario), including a scenario for the phone-order representatives just discussed and a scenario for the mail-order clerks. But another key system objective is to allow customers to interact with the system directly to place orders via the World Wide Web. The design of a complete company web site is beyond the scope of this text, but some general rules and guidelines apply to direct customer orders.

The basic dialog between user and computer will be about the same as for the phone-order representative, but the web site will have to provide even more information for the user, be even more flexible, and be even easier to use. First, customers may want to browse through all possible information about a product. More pictures will be needed, including pictures showing different colors and patterns of items. Although the phone-order representative needs detailed information on the screen to be able to answer questions, the customer will probably want even more. The information will need to be displayed differently, too. For example, the phone-order clerk will be accustomed to a dense display of information and know where to look for the details, but customers will need the information organized to make it easy to locate details the first time they use the system.

The system will need to be very flexible because customers will have different preferences for interacting with the system. As discussed for the phone order, the sequence should be flexible—some customers will want to browse first and even select items to order before entering any information about themselves. Options like reviewing past orders and reviewing shipping and payment approaches will also be required. If a customer wants to do something that the system does not allow, the customer will become frustrated. Unlike phone-order and mail-order employees, who can be expected to work through their frustration, the customer can simply log off and shop elsewhere. Finally, the system must be so easy to learn and use that the customer does not even have to think about it. The initial dialog options need to be very clear, and once a sequence starts, all options should be self-explanatory. Customers cannot be expected to sit through training just to use a web site, nor should they have to look up instructions or help (even though they should be available).

There are many specific guidelines for the design of web sites, and they are evolving all the time. As mentioned previously regarding the guidelines for visibility and affordance, it is important that controls used on a web page be clear in what they do and how they are used. Most users are more concerned about speed than fancy graphics and animations, but novice designers often go overboard on graphics and animation at the expense of speed. Additionally, since the web site reflects the company image, it is important to involve graphic designers and public relations professionals in the design. A well-thought-out visual theme is important. Focus groups and other feedback techniques should be used on the design.

The Rocky Mountain Outfitters' home page is shown in Figure 12-20. The home page was designed with the RMO logo above an inspiring photograph of the mountains. Focus groups working with marketing consultants helped select the photograph. The main emphasis is direct customer interaction. The user can choose to learn more about RMO, contact RMO with an e-mail message, or request a catalog. There is no main menu option to place an order. Instead, the web site offers the opportunity to browse through pages of RMO products. The customer can search for products based on keywords or product ID numbers, or the customer can select a category of product from a list. Weekly specials are also offered. When customers find something they want, an order is created. They can change their mind at any time.

FIGURE 12-20
Rocky Mountain Outfitters' home page.

The RMO web site uses the shopping cart analogy. Once customers find a product they want, they select the quantity, size, and other options, then they add the item to their shopping cart. They can view the shopping cart at any time, then continue browsing and adding items. When they are done, they check out and the system confirms the order. Figure 12-21 shows a product detail page reached after a user navigates through the men's clothing option. Figure 12-22 shows the shopping cart with a summary of an order.

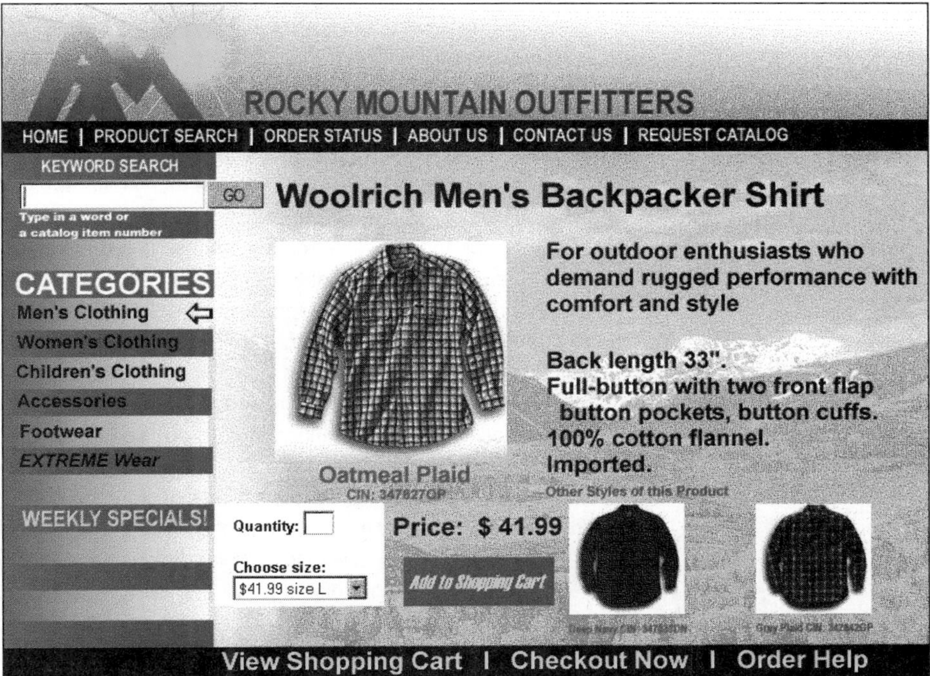

FIGURE 12-21
The product detail page from the Rocky Mountain Outfitters' web site.

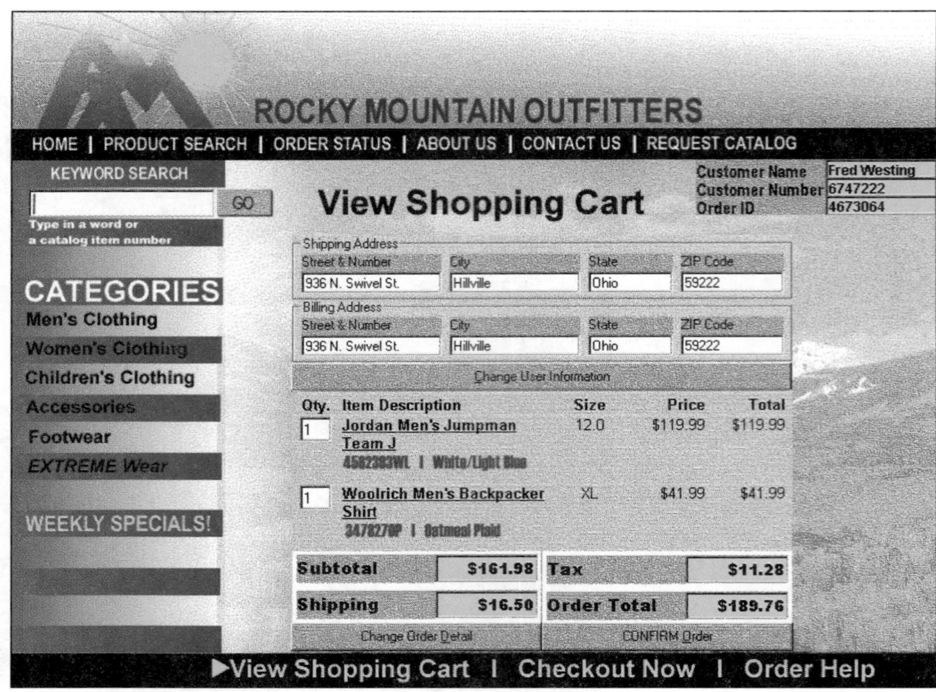

FIGURE 12-22
The shopping cart page from the Rocky Mountain Outfitters' web site.

Summary

This chapter described the key concepts and techniques for designing the interaction between the user and the computer—human-computer interaction (HCI)—following up on the material describing input design and output design in Chapter 11. The user interface is everything the user comes into contact with while using the system—physically, perceptually, and conceptually. To the user, the user interface *is* the system. The knowledge the user must have to use the system (the user's model) includes information about objects and functions available in the system—the kind of information that defines the requirements model the analyst works hard to uncover during systems analysis.

User-centered design refers collectively to techniques that focus early on users and their work, evaluate designs to ensure usability, and use iterative development. Usability refers to the degree to which a system is easy to learn and use. Ensuring usability is a complex task because design choices that promote ease of learning versus those that promote ease of use often conflict. Additionally, there are many different types of users to consider for any system. Human-computer interaction as a field of research grew out of human factors engineering (ergonomics) research that studies human interaction with machines in general.

There are many different ways to describe the user interface, including the desktop metaphor, the document metaphor, and the dialog metaphor. The dialog metaphor emphasizes the interaction between user and computer, and interface design is often called *dialog design* for that reason. Interface design guidelines and interface design standards are available from many sources. Norman's visibility and affordance guidelines state that controls should be visible, provide feedback indicating that they are working, and be obvious in their function. Shneiderman's eight golden rules are to strive for consistency, provide short cuts, offer informative feedback, design dialogs to yield closure, offer simple error handling, permit easy reversal of actions, support internal locus of control, and reduce short-term memory load.

Dialog design starts with identifying dialogs based on events. Additional dialogs are needed for integrity controls added during design, for user preferences, and for help. Menu hierarchies can be designed for different types of users, and once the dialogs are designed, menu hierarchies can be rearranged easily. Writing a dialog sequence much like a script can help a developer work out the key information that needs to be exchanged during the dialog.

A storyboard showing sketches of screens in sequence can be drawn to convey the design for review with users, or prototypes can be created using a tool such as Visual Basic. The object-oriented approach provides UML models that can document dialog designs, including sequence diagrams, collaboration diagrams, and class diagrams. Each form used in a dialog needs to be designed, and there are guidelines for the layout, selection of input controls, navigation, and help. Designing a dialog for a web site is similar to any other dialog, except users need more information and more flexibility. Additionally, because a web site reflects the company's image to customers, graphics designers and public relations professionals should be involved.

Key Terms

affordance, p. 444
browser forms, p. 455
check boxes, p. 458
combo box, p. 458
desktop metaphor, p. 439
dialog metaphor, p. 440
direct manipulation, p. 438
document metaphor, p. 439
human-computer interaction, p. 434
human factors engineering (ergonomics), p. 436
hypermedia, p. 440
hypertext, p. 440
interface design standards, p. 444
list box, p. 458
radio buttons (option buttons), p. 458
spin box, p. 458
storyboarding, p. 450
text box, p. 458
usability, p. 436
user-centered design, p. 435
user's model, p. 435
visibility, p. 444

Review Questions

1. Why is interface design often referred to as dialog design?
2. What are the three aspects of the system that make up the user interface for a user?
3. What is the term generally used to describe the study of end users and their interaction with computers?
4. What are some examples of physical aspects of the user interface?
5. What are some examples of perceptual aspects of the user interface?
6. What are some examples of conceptual aspects of the user interface?
7. What collection of techniques places the user interface at the center of the development process?
8. What are the three important principles emphasized by user-centered design?
9. What term refers to the degree to which a system is easy to learn and use?
10. What is it about the "human factor" that engineers find difficult? What is the solution to human factor problems?
11. What are some of the fields that contribute to the field of human-computer interaction?
12. What research center significantly influenced the nature of the computers we use today?
13. What are the three metaphors used to describe human-computer interaction?
14. A desktop on the screen is an example of which of the three metaphors used to describe human-computer interaction?
15. What type of document allows the user to click on a link and jump to another part of the document?
16. What type of document allows the user to click on links to text, graphics, video, and audio in a document?
17. What is the name for general principles and specific rules that developers must always follow when designing the interface of a system?
18. What two key principles are proposed by Norman to ensure good interaction between a person and a computer?
19. List the eight golden rules proposed by Shneiderman.
20. What is the technique that shows a sequence of sketches of the display screen during a dialog?
21. What UML diagram can be used to show how the interface objects are plugged in between the actor and the problem domain classes during a dialog?
22. What UML diagram can be used to show the interface objects that are contained on a window or form?

23. What are the three basic types of windows forms used in business systems?
24. What is the input control (interface object) used for typing in text?
25. What are some of the input controls that can be used to select an item from a list?
26. What two types of input controls are included in groups?
27. What are three requirements for usability of a direct customer access web site beyond those of a windows interface used by employees?
28. What popular analogy is used for direct customer access with a web site when customers shop on-line?

Thinking Critically

1. Think of all of the software you have used. What are some examples of ease of learning conflicting with ease of use?
2. Visit some web sites and identify all of the controls used for navigation and input. Are they all obvious? Discuss some differences in visibility and affordance of the controls.
3. Consider the human factor solution that states it is better to change the machine rather than try to change the human to accommodate the machine. Are there machines (or systems) that you use in your daily life that still have room for improvement? Are the current generations of Windows PC and Apple Mac as usable as they might be? If not, what improvements can you suggest? Is the World Wide Web as usable as it might be? If not, what improvements would help? Are we just beginning to see some breakthroughs in usability, or have most of the big improvements already been made?
4. Review the dialog between user and computer shown for DownTown Videos. Create a table like Figure 12-6 that shows how the dialog can be converted to the user's language and the computer's language. Discuss how moving from the dialog to the table takes a logical model and creates a physical model.
5. Refer again to the table shown in Figure 12-6. Create a storyboard for the e-mail system based on the information in the table.
6. Read through the following dialog that shows a user trying to place an order with the system. Critique the dialog in terms of ease of learning and ease of use.

User:	I want to order a product.
System:	Okay. Enter your name and address.
User:	My name is Timothy Mudd, 5139 North Center Street, Los Angeles, CA 98210.
System:	Oh, we have all of that information on file, but thank you for entering it again.
User:	I want to order the Acme Drill Press with adjustable belt drive.
System:	Sure, continue with your request.
User:	I want the blue color and rubber feet but not the foot brake.
System:	Sure, anything else?
User:	I want it shipped priority with the special discount you offer.
System:	We hear you.
User:	Okay, that's all.
System:	We are sorry, but your transaction violated transaction code 312 and we must terminate the session.

7. Refer to the form shown in Figure 12-15, showing data-entry controls. Draw a UML class diagram that represents the objects that make up the form.
8. Review all of the controls that come with Visual Basic that are visible when added to a form. Discuss how well each satisfies the visibility and affordance requirements.

Experiential Exercises

1. Review the sequential design concept and the order-centered design concept for the Rocky Mountain Outfitters *Create new order* dialog. Consider what information and navigation options are not needed by the mail-order clerks. Then design the forms that would be appropriate for clerks to use when entering mail orders for RMO. Either sketch the forms or implement the forms using a tool like Visual Basic. Use input controls that minimize data entry.

2. Evaluate the course registration system at your university. List the basic steps followed through a dialog with the system. What are some of the problems with the system from the perspective of ease of learning and ease of use? In what ways is the system inflexible? In what ways is needed information not available? Is too much information provided that distracts from the task at hand?
3. Evaluate the on-line catalog system at your university library (see some of the questions and issues in exercise 2). Write a dialog that shows the interaction between the user and the system. Rewrite the dialog to improve it. Create a storyboard to show how your design would look and feel.
4. Locate a system that users work with all day. Ask some of the users of the system what they like and dislike about it. Ask how they learned the system. Ask them about the typical errors that are made with the system and how they are prevented or corrected. Observe them working with the system. Does it appear that any of the interaction is "automatic"?
5. Find a web site with direct customer ordering. Browse through some product descriptions and note the design of the dialog and the web pages. What do you like and dislike about the design? Evaluate the web site based on visibility and affordance and on the eight golden rules.

Case Studies

The 21st Century University System

Consider the following dialog for a user interacting with the course registration system at 21st Century University. The user is a student registering for classes at the university.

System: What course registration function do you want to perform?
User: *Preregistration.*

System: Okay, what is your account ID and password?
User: *My account ID is SATZ791, and my password is xxxxxxxx.*

System: Okay, welcome back, Kevin. You last adjusted your schedule on 7/7. What preregistration function do you want to perform?
User: *To add courses.*

System: These are the courses you are preregistered for now:
(The system displays a list of course information for three courses.)

Since you are a CIS major, these are all of the CIS courses you might add that have no time conflicts with your present schedule (note, though, that prerequisites have not been checked for these).

(The system displays a list of CIS courses available.)

Your options are:

- Switch to a list of courses in another department
- List these courses again but ignore time conflicts
- List these courses again after checking prerequisites
- Provide details of the indicated course
- Add the indicated course to your schedule
- Cancel

User: *I want to switch to courses in another department.*
System: Fine. To what department do you want to switch?

User: *Art.*
System: These are the courses for which you are registered now:

(The system displays a list of course information for three courses.)
As requested, these are the art courses you might add that have no time

conflicts with your present schedule (note, though, that prerequisites have not been checked for these courses).

(The system displays a list of art courses available.)

Your options are:

- Switch to a list of courses in another department
- List these courses again, but ignore time conflicts
- List these courses again after checking prerequisites
- Provide details of the indicated course
- Add the indicated course to your schedule
- Cancel

User: *Add the second course in the list.*
System: (The system continues the dialog.)

1. Is this dialog a logical or a physical model? Could it be a phone registration system? Could it be handled manually by a clerk at a desk? Discuss.
2. Discuss how well this dialog follows the eight golden rules.
3. Complete the dialog by adding likely exchanges between the student and the system as the student adds one more course. The student should request more details about a course, then decide to add the course. Include a request from the student to see final schedule details.
4. Design and implement either a storyboard or prototype for this dialog, using a tool such as Visual Basic, and being as faithful to the dialog as possible. Make up some sample data to show in your design as needed.

The DownTown Videos Rental System

This chapter includes an example of a storyboard for DownTown Videos, a case study first introduced in Chapter 7. The storyboard showed the *Rent out videos* dialog. Revisit the DownTown Videos case and complete the following.

1. Implement the storyboard in this chapter as a prototype using a tool such as Visual Basic.
2. Write a dialog and then create a storyboard for the event *Customer returns videos*. Consider that one or more videos might be returned and that one or more of them might be late, requiring a late charge.
3. Using a tool such as Visual Basic, implement the storyboard as a prototype, then ask several people to evaluate it. Discuss the suggestions made.

The Waiters on Wheels System

Review the Chapter 5 opening case study that describes Waiters on Wheels, the restaurant meal delivery service. The analyst found at least 14 events for the system, which are listed in the case.

1. Create a set of menu hierarchies for the system based on the events listed. Then add additional menu hierarchies for system utilities based on controls, user preferences, and help.
2. The most important event listed in the case is *Customer calls in to place an order*. Write out a dialog between the user and the computer with a natural sequence and appropriate exchange of information.
3. Sketch a storyboard of the forms needed to implement the dialog.
4. Ask several people to evaluate the design and discuss any suggested changes.
5. Implement a prototype of the final dialog design using a tool such as Visual Basic.

The State Patrol Ticket Processing System

Review the State Patrol ticket processing system introduced as a case study at the end of Chapter 5.

1. Create a set of menu hierarchies for the system based on the events in the case. Then add additional menu hierarchies for system utilities based on controls, user preferences, and help.
2. For the event *Officer sends in new ticket*, write out a dialog between the user and the computer for the dialog *Record new ticket* with a natural sequence and appropriate exchange of information.
3. Sketch a storyboard of the forms needed to implement the dialog. Be sure to use input controls that minimize data entry, such as list boxes, radio buttons, and check boxes.
4. Ask several people to evaluate the design and discuss any suggested changes.
5. Implement a prototype of the final dialog design using a tool such as Visual Basic.

Another Alternative for Rocky Mountain Outfitters

A few of the phone-order representatives at Rocky Mountain Outfitters were not completely satisfied with the order-centered dialog design presented in this chapter. They suggested that the design could be streamlined if only one form was displayed throughout the dialog—the order summary form. They thought the order summary form could expand to show additional information instead of switching to separate forms for customer information, product information, or shipping information. When the additional information is not needed, the form could contract. They thought this approach would be easier on the eyes, with less effort being expended to focus and refocus on multiple forms that pop up. They requested that a prototype of the one-form design concept be created for their review. The interface might include both options, allowing users to choose the one they prefer.

1. Draw a storyboard to show how the one-form design concept might look as customer information, product information, and shipping information are added to the expanded form.
2. Implement a prototype of the storyboard using a tool such as Visual Basic.
3. Ask several people to evaluate the design, specifically comparing it to the order-centered design in the text, and discuss the results.
4. Can you describe or implement yet another alternative?

Further Resources

Patrick J. Lynch and Sarah Horton. *Web Style Guide: Basic Design Principles for Creating Web Sites*. Yale University Press, 1999.

Deborah J. Mayhew. *Principles and Guidelines in Software User Interface Design*. Prentice Hall, 1992.

Donald Norman. *The Design of Everyday Things*. Doubleday, 1990.

Jenny Preece, Yvonne Rogers, David Benyon, Simon Holland, and Tom Carey. *Human Computer Interaction*. Addison Wesley Publishing Company, 1994.

Jeffery Rubin. *Handbook of Usability Testing*. John Wiley & Sons, 1994.

Ben Shneiderman. *Designing the User Interface* (2nd ed.). Addison Wesley, 1992.

PART 4

Implementation and Support

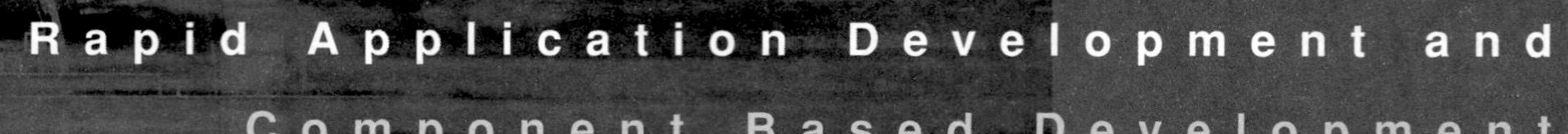

Rapid Application Development and Component Based Development

Learning Objectives

After reading this chapter, you should be able to:

- *Describe rapid application development and the prototyping and spiral development approaches.*
- *Compare the prototyping and spiral development approaches to more traditional development approaches.*
- *Choose an appropriate development approach to match project characteristics.*
- *Implement a risk management process.*
- *Describe additional rapid development techniques, including joint application design, tool-based design and development, and code reuse.*
- *Describe object frameworks and how they are used to speed software development.*
- *Describe components and the process by which they are developed and deployed.*

Chapter Outline

Rapid Application Development

Rapid Development Approaches

Rapid Development Techniques

Object Frameworks

Components

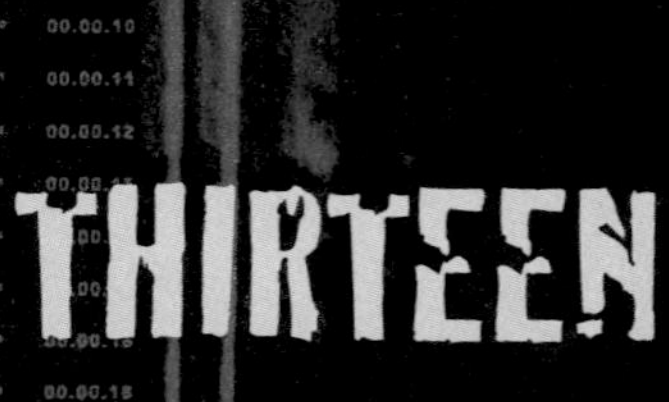

CDs DIRECT 2U: SPEEDING DEVELOPMENT OF A CUSTOMER SUPPORT SYSTEM

Tony Griego, Trude Schultz, Yvonne Young, and Jim Blake were meeting to discuss design, development, and deployment options for a new customer service system for CDs Direct 2U. The system will be used by customer support telephone staff and also provide a web-based interface for direct customer interaction. The analysis phase was nearly finished, and the team was working on design and implementation proposals to present to the steering committee.

Tony started the meeting. "Well, the requirements certainly paint a bigger picture than we initially thought. And some of them seem likely to change as the project progresses. I don't see how we can possibly meet the original delivery target without significantly scaling back system scope or drastically improving our typical design and development speed. But the steering committee is adamant that this project be finished by the end of the year. Our competitors are already using systems like this and we need to roll ours out soon just to survive."

Trude spoke up, "I ran some time and cost estimates yesterday, and it appears that we're looking at a 14-month project at 140 percent of our original design and development budget. Boosting the budget another 20 percent might shave a couple of months off the projected schedule, but I think that's our shortest possible schedule."

Tony replied, "That's pretty close to my seat-of-the-pants guess. But I don't think that the committee will agree to more money *and* a later delivery date."

Yvonne said, "I think that some options might buy us the schedule reductions we need. For one, we've never really tried a prototyping approach on a large system, and this project seems like a good candidate. Also, there are many new tools available for developing Internet-based applications."

Jim added, "I've been looking into component-based development, and this project looks like a good candidate. We may be able to purchase some components to cover some of the requirements and plug them into the parts we develop. Inventory management and credit verification seem like good candidates. And component-based design and implementation would also give us greater flexibility to update pieces of the system as Internet and database technology change."

Tony replied, "Those suggestions are all good, and we should investigate them. I don't know if any one will be enough to solve our problems. But maybe two or three of them in combination can be the silver bullet we need to bring this project in on schedule."

OVERVIEW

The pace of change in business and other organizations has accelerated throughout the 20th century, and no one expects that pace to slow during the 21st century. Rapid change in both business practices and information technology has created a significant management problem: how to quickly develop and deploy information systems that implement the latest business practices and employ cutting-edge technologies. Yet, changing technology is a two-edged sword. One edge provides system developers with new tools and techniques to improve their own productivity. The other increases user expectations for system functionality. Organizations that can quickly develop systems to effectively exploit new technologies thrive and prosper. Those that don't stagnate and flounder.

The tools and techniques described in earlier chapters provide a foundation for rapid system development, but there are additional tools and techniques that can be applied to speed up the process. This chapter presents a small but important subset of those tools and techniques. The first half of the chapter concentrates on rapid development techniques. The chapter begins with a discussion of what rapid application development is and isn't. We then describe an alternative approach to system development (the spiral life cycle). Then we discuss a set of techniques that can be used with the spiral and other life cycles to increase development speed or reduce the risk of schedule overruns.

The latter half of the chapter focuses on two technologies—object frameworks and components. Both technologies can significantly increase development speed through code reuse and flexible software construction. We will describe the technologies and how they can be successfully integrated into the system development life cycle (SDLC) to speed software development.

Rapid Application Development

Rapid application development (RAD) is an overused and poorly understood term. RAD is something that most software developers claim they do but cannot precisely define. At various times (including the present), RAD has been equated with tools and techniques such as prototyping, fourth-generation programming languages, CASE tools, and object-oriented (OO) analysis, design, and development. Methodologies have also been developed that are either called RAD or claim to result in RAD. Given the blizzard of competing and confusing claims, it's little wonder that few people can precisely define what RAD is.

For the moment, we'll avoid defining the term. Instead we'll first describe some of the things that influence development speed. After we have provided that background, we'll return to a discussion of RAD.

Reasons for Slow Development

The reasons for slow development are far too numerous to list or describe exhaustively in this text. Steve McConnell, in his recent book *Rapid Development: Taming Wild Software Schedules*, provides many pages of reasons. We focus on three broad categories that have delayed many a project: rework, shifting requirements, and inadequate or inappropriate tools and techniques. The following sections briefly describe the problem areas. Methods of addressing the problems are discussed in later sections of the chapter.

Rework Think of the process of building a house. A foundation is laid, the house is framed, the roof and exterior walls are covered, windows and doors are installed, interior mechanical systems are installed, interior fixtures are installed, and finally finishing touches such as paint and flooring are added. Now imagine that after each stage is initially completed, the builder decides that the result produced is poor quality. So the work is ripped out and done over again. The result is that each construction stage is performed twice and the project schedule doubles.

This scenario may sound contrived, but similar scenarios are relatively common in system development. It is not uncommon for software construction tasks to be performed multiple times to correct problems with earlier attempts. Clearly, one way to avoid lengthening the development schedule is to ensure that each construction activity is performed only once.

Avoiding rework requires two things:

- The right software (and only the right software) must be constructed or procured.
- The development process must always produce software that meets minimum quality standards.

The only way to ensure that the right software is built is to ensure that requirements and overall design constraints are fully known before design and construction begins. There are many ways to ensure that construction always produces quality results. Some of the more effective ways are fully specifying important design parameters before beginning construction, assigning well-trained and motivated people to construction tasks, and providing system builders with the right tools for the job.

Shifting Requirements One reason that many projects fall behind schedule is that requirements change during the project. Changes to requirements require corresponding changes to design and construction. The later in the project a change occurs, the more costly it is to incorporate it into the system. For example, changes to a house design (such as enlarging a room and putting in additional windows) before detailed design require relatively little effort. Changes after detailed design require drawing up new blueprints and modifying the materials list and construction schedule. Changes during construction usually require undoing some already completed construction (that is, rework).

Changes to software requirements are also subject to increasing costs as development progresses. A commonly used estimate is that a change that could be implemented for $1 during analysis costs $10 during architectural design, $100 during detailed design, $1,000 during coding and testing, and $10,000 after the system is operational (see Figure 13-1).

FIGURE 13-1
The cost of implementing a requirements change increases in each life cycle phase.

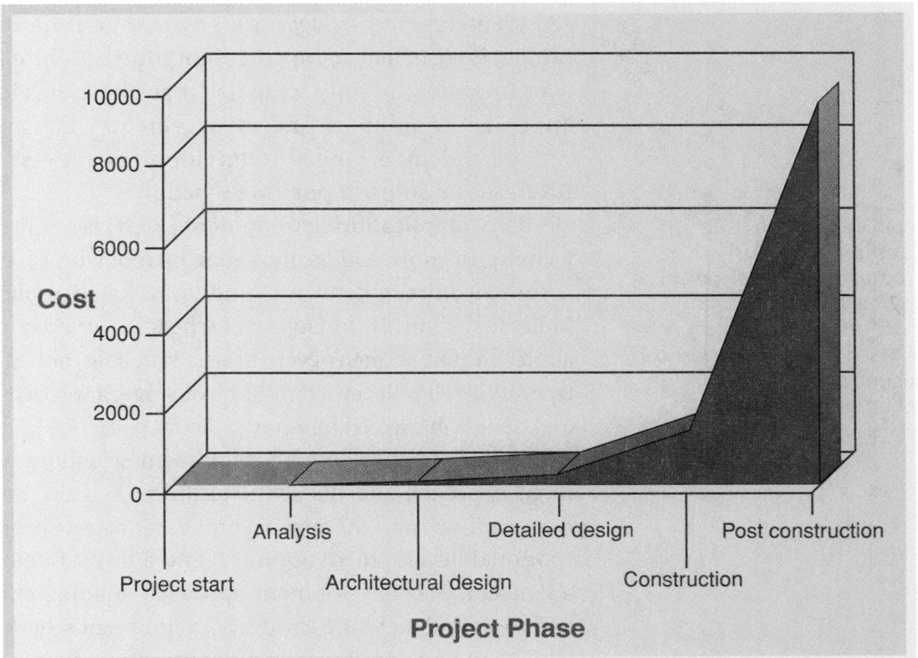

Some shift in requirements is to be expected in most software development projects. Users aren't always sure what they want at the beginning of the project. Also, a rapidly changing environment may necessitate changes before the system is fully developed. Ignoring important requirements changes only guarantees that the system will not meet the users' needs when it's delivered. To avoid schedule extensions while satisfying user requirements, a developer should anticipate and incorporate changes into the development process.

Tools and Techniques No one set of tools and techniques is best suited to all system development projects. Differences in system requirements, development methodologies, and target operating environments may require different tools for analysis, design, and construction. For example, programming languages such as assembly and C are typically used for systems that require maximum execution speed. Object-oriented development and programming techniques are generally best for systems that will evolve over relatively long periods of time (such as many business applications). Sequential development is usually most efficient when requirements are well known and relatively static. Evolutionary development is generally best when requirements are poorly understood or rapidly shifting.

Using the wrong tool or technique for a given project can reduce quality, increase development time, or both. Building a house with power tools is fast, and the result is generally high quality. Building a house with nothing more than hammers and hand saws will take much longer and probably result a lower-quality product. Finishing a 100-story building on schedule requires not only the right tools and skills but also efficient construction and project management techniques. The parallels to software development are obvious. A development team can avoid slow development and poor quality by matching tools and techniques to the project at hand.

What Is RAD?

The time required to develop a system depends on a number of factors, including user requirements, budget, development approach, development tools and techniques, and management of the development process. Assume that both user requirements and budget are fixed (they usually are). Assume further that the budget is sufficient to buy the right tools and hire a well-trained staff. Under these assumptions, the only variables that affect development speed are development approach, techniques, and management. Choosing the right approach and techniques and managing the development process efficiently and effectively will result in the shortest possible schedule.

rapid application development (RAD)
A collection of development approaches, techniques, tools, and technologies, each of which has been proven to shorten development schedules under some conditions

Rapid application development (RAD) is a collection of development approaches, techniques, tools, and technologies, each of which has been proven to shorten development schedules under some conditions. Rapid application development is not a silver bullet that shortens development schedules for every project. There is no universal RAD approach that shortens every project schedule, nor is there any one technique, tool, or technology that fits every project. No single approach or technique by itself is sufficient, and none will succeed for every project.

The key to shortening development schedules is to identify the overall development approach and the set of techniques, tools, and technologies most suitable to that approach and the specific project. For some projects, RAD may require an unconventional development approach and a host of newer techniques. For other projects, a conventional development approach supplemented by a few specific techniques and tools will yield the shortest development schedule.

The remainder of this chapter explores a diverse set of RAD concepts. As you read the material, keep in mind that the approaches, tools, techniques, and technologies need to be mixed and matched to specific projects. They are all related by their ability to speed development under some circumstances. But they are a set of alternatives that must be individually evaluated for their effectiveness with a given development project.

Rapid Development Approaches

Previous chapters presented a relatively conventional approach to system development. We divided the SDLC into phases, and each phase into a set of activities. We described a linear flow of activities and a series of models, moving from requirements, to architectural design, and then to detailed design. Presenting development activities in sequential order is a logical way to *teach* systems analysis and design. But it isn't necessarily the best way to *do* analysis and design for a particular project.

The conventional approach to system development tends to be sequential. In particular, it stresses completing requirements definition before design, and design before implementation. The conventional approach does provide some opportunities for parallel activities, particularly within major life cycle phases. But significant use of parallelism and iteration is not the norm in conventional development.

The oldest approaches to software development were purely sequential. They had their roots in an era when:

- Systems were relatively simple and independent of one another.
- Computer hardware resources were very expensive.
- Software development tools were relatively primitive.

A sequential approach to system development is well suited to such an environment.

Underlying older development approaches are these assumptions: Simple systems are simple to analyze and model, and it is reasonable to assume that requirements can be captured fully and accurately before any design or construction activity begins. Complete analysis prior to design and construction is not only reasonable, it's an economic necessity. The high cost of computer resources and the labor-intensive nature of programming make building the "wrong" system a very expensive and time-consuming mistake. Complete design prior to construction is also necessary. Issues of technical feasibility and operational efficiency are paramount when computing resources are expensive, and those issues are best addressed by designing the entire system all at once.

Now consider the current validity of the assumptions just listed. Information systems today often consist of millions of lines of code and are interconnected with many other systems. Computer power has doubled every two to three years at no increase in cost. Labor costs have risen and there is a shortage of skilled personnel in most areas of software development. Software development tools leverage greater hardware power to allow programs to be developed quickly and cheaply. Clearly, none of the original assumptions that motivated sequential system development apply today. But does that mean that sequential development should never be used? The answer is no, but the reasons are complex and interdependent.

As described earlier, shifting requirements are common due to changes in the external environment and from uncertainties about requirements at the beginning of a project. For example, consider the following scenario. A competitor has successfully implemented a new customer support system based on new technology. Another firm responds by initiating a project to develop a similar system. But it has no experience with that type of system or with the underlying technology.

In this scenario, it isn't reasonable to assume that all system requirements can be specified completely and accurately before design and construction begin. Also, technical, operational, and economic feasibility cannot be completely determined at the beginning of the project. Thus, a sequential development approach is poorly suited to the project. An evolutionary approach with experimentation and learning is more appropriate. Such an approach allows requirements to be discovered and gradually refined. It also allows both users and developers to work their way up the learning curve of the new technology gradually.

Now let's revisit an earlier premise—the shortest possible schedule cannot be achieved if software is constructed multiple times. To ensure that we don't build things more than once, we must ensure that we build the right things and that our efforts produce quality products. The first of those requirements is essentially a restatement of the need for complete and accurate analysis and design prior to construction. But how can we do this if requirements are complex and shifting or feasibility is uncertain?

Obviously, we can't. Thus, we must accept some rework as an unavoidable consequence of uncertainty about requirements or feasibility. And we must restructure our approach to system development to accommodate the uncertainties. The exact way in which we restructure will depend on:

- Project size
- The degree of (un)certainty of requirements or feasibility at the start of the project

- The expected rate of change in user requirements during the life of the project
- The experience and confidence that developers have in the proposed implementation technology

Large project size, uncertain or shifting requirements, and new technologies are all indicators of a need to depart from sequential development. Projects combining several of these characteristics require the most significant SDLC modifications, as summarized in Figure 13-2.

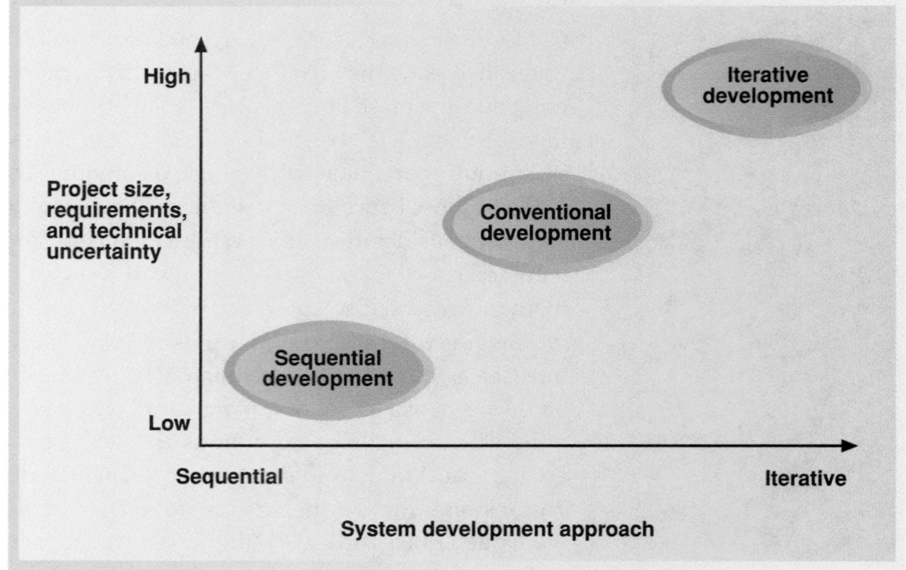

FIGURE 13-2
The development approach as a function of project characteristics.

Note that we're not suggesting that sequential or conventional development be completely abandoned. Sequential or conventional development is still the fastest way to develop software *when uncertainty about requirements and feasibility is low.* We're also not suggesting that the techniques described in earlier chapters be abandoned. All of the techniques described in earlier chapters are useful and efficient for many types of projects (otherwise, we wouldn't have spent 12 chapters describing them!). But when technology is new or requirements can't be completely specified, then techniques must be reorganized and possibly supplemented within a nonsequential development approach.

The Prototyping Approach to Development

Prototyping was described earlier in Chapter 4 in the section "Build Protoypes." Here we briefly review that material and then describe a complete life cycle based on prototyping. Prototyping is the process of building a partially or fully functioning system model that looks and acts as much as possible like a real system. A prototype is a system in its own right, but it may be incomplete or may not exactly match the final set of user requirements.

discovery prototype
A prototype system used to discover or refine system requirements or design parameters

developmental prototype
A prototype system that is iteratively developed until it becomes the final system

Two types of prototypes are commonly used in software development. A **discovery prototype** is often used during analysis and occasionally during design. A **developmental prototype** is a prototype system that is not intended to be thrown away. Developmental prototypes become all or part of the final system. Developmental prototypes are primarily used in iterative software development. A simple prototype is developed, and additional features are added in subsequent development phases. Or, a series of independent prototypes may be developed and later combined to form a complete system.

prototyping development approach
A development approach based on iterative refinement of a developmental prototype

Steps in the Prototyping Development Approach Figure 13-3 shows a **prototyping development approach**. The planning phase is as described in Chapter 2. The analysis phase can be implemented with traditional or OO techniques, provided that the techniques are compatible with those used for design and implementation. Analysis models may be less detailed than with conventional development approaches, under the assumption that some requirements will be discovered or fully specified by prototyping.

FIGURE 13-3
A system development approach based on developmental prototypes.

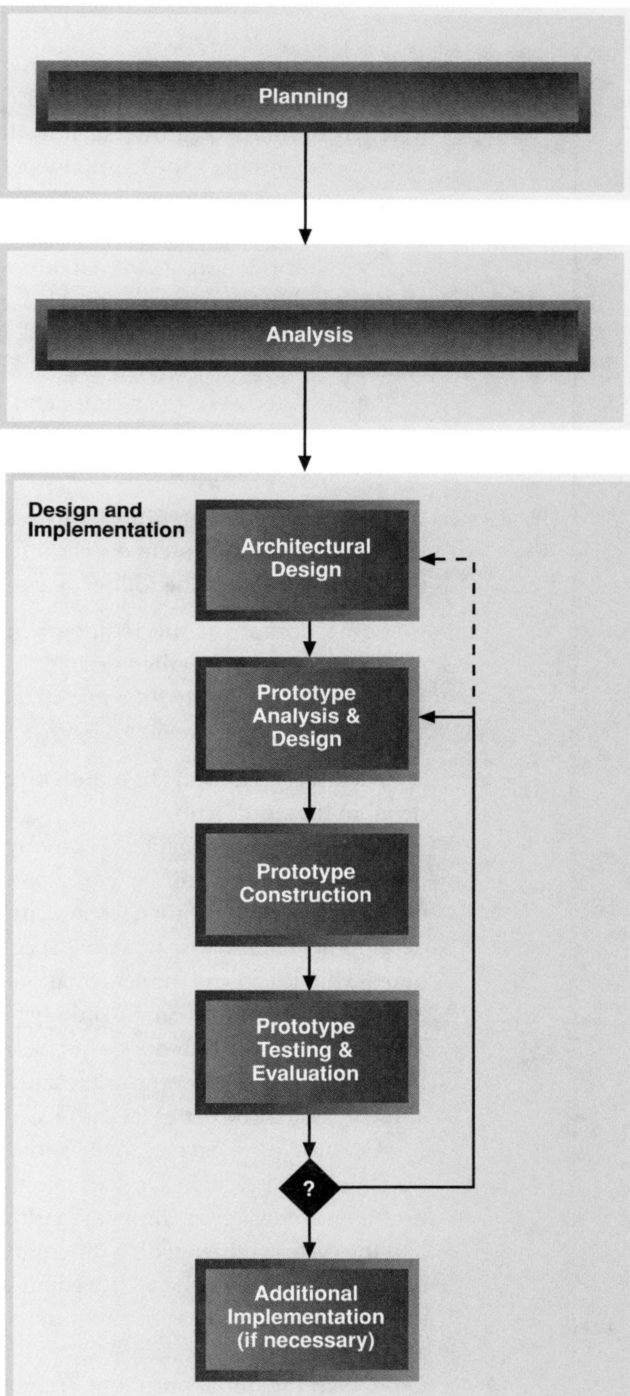

Design and implementation start out similarly to conventional design phases but differ substantially after architectural design is completed. Architectural design establishes the overall top-level design parameters and implementation environment. The analyst plans the remaining design and implementation activities by defining a series of prototypes or modifications to them. Typically, the requirements are divided into subsets based on specific functions (for example, order entry, order fulfillment, and purchasing) and architectural system boundaries (such as client and server). These subsets may be further divided into a series of implementation steps, starting with a core set and adding functions in each step.

Once the prototype series has been defined, detailed design and implementation proceed in an iterative fashion. Each cycle begins by defining the requirements and design of a single prototype or prototype version, including aspects of analysis that were left unfinished earlier (such as detailed user-interface specifications). Once the developer determines the requirements and detailed design of a prototype, it is developed, tested, and evaluated. Testing and evaluation determine whether the prototype meets the objectives for the current cycle and whether any additional cycles are required. Information uncovered during testing and evaluation may require the developer to modify the development plan or, in unusual cases, the architectural design.

Some requirements of the system may be left out of the prototype development cycle and completed after the final cycle. Examples of such requirements include backup and recovery operations and systemwide security measures. Requirements with no interactive interfaces and those with a systemwide impact are generally good candidates for postprototype implementation.

When to Use a Prototyping Approach The prototyping development approach usually results in faster system development than a conventional development approach under some or all of the following conditions:

- Some portion of the requirements cannot be fully specified independently of architectural or detailed design.
- Technical feasibility for some system functions is unknown or uncertain.
- Prototype development tools are powerful enough to create a fully functional system.

The third condition is a requirement for successful prototyping and is discussed in detail in the next section.

As discussed in Chapter 4, prototyping is a good means of solidifying uncertain requirements. Users can generally provide better feedback about requirements when examining a prototype than when examining graphical or textual models. A prototyping development approach allows full specification of some requirements that can be combined with design and implementation. Some of the requirements take longer to specify than others, and those can be prototyped to avoid delays in the entire project. However, a prototyping approach does work best when most of the requirements are understood in advance. If significant uncertainty exists, other techniques and development approaches may be better for the project.

A prototyping development approach also works well to ascertain the technical feasibility of a design. Development and testing of early prototypes help sort out problems in design or implementation. Unsuccessful attempts are simply discarded, and the cycle is retried with a different set of parameters. Once success is achieved, the developer can revisit the architectural design and revise the development plan of subsequent prototypes. As with requirements, successful prototyping requires that most aspects of architectural design and technical feasibility be specified in advance. Otherwise, the initial prototype is simply too big and cumbersome for rapid development and testing.

Some functions or requirements may not be suited to prototyping. Examples include systems or portions of a system that are:

- Noninteractive (such as a program that automatically generates orders to suppliers)
- Internally complex (such as a module that schedules deliveries for fastest delivery time or minimal cost using a complex algorithm)
- Subject to stringent performance or security requirements (such as a program that generates thousands of electronic payments per hour)

Noninteractive programs and systems exhibit little observable behavior to test directly or validate. Thus, a developer can derive little value from prototyping their requirements. Internally complex programs or systems often have relatively well-defined processing requirements that can be formally stated (for example, mathematically). Such requirements are more easily implemented by a conventional development approach based on structured or OO requirements and design models. Iterative development actually wastes development time for such systems because the repeated design, testing, and evaluation are unnecessary. Finally, software with stringent performance requirements must be constructed with tools that are optimized to produce efficient executable code. Unfortunately, such tools are often not optimal for development speed.

Prototyping Tool Requirements Successful prototyping requires system developers to employ development tools with power, flexibility, and developmental efficiency. Many of the modern "visual" programming tools (such as Microsoft Visual Basic and PowerBuilder) satisfy these requirements. Many more specialized tools (such as Oracle Forms) satisfy the requirements for specialized applications and technical environments.

The key requirement for a successful prototyping tool is development speed. The ability to build and examine many prototypes is crucial to elicit complete and accurate user requirements and to determine technical feasibility through experimentation. The development tools must make it possible to construct, modify, or augment a prototype in a matter of minutes, hours, or at most a few days.

Flexibility and power are necessary for rapid development. Successful prototyping tools employ a highly interactive approach to application development. Other typical techniques and capabilities include:

- WYSIWYG (what you see is what you get) development of user-interface components
- Generation of complete programs, program skeletons, and database schema from graphical models or application templates
- Rapid customization of software libraries or components
- Sophisticated error-checking and debugging capabilities

Figure 13-4 shows an example of WYSIWYG development of a user-interface prototype for the RMO customer support system. The dialog box appears exactly as it would when displayed by an application program. To add user-interface components to the dialog box, you drag templates from the small window labeled Controls. You can customize features by clicking on the component and altering values in its properties list. To edit the underlying application program code, you can double-click the component and edit the pop-up code editor window that appears.

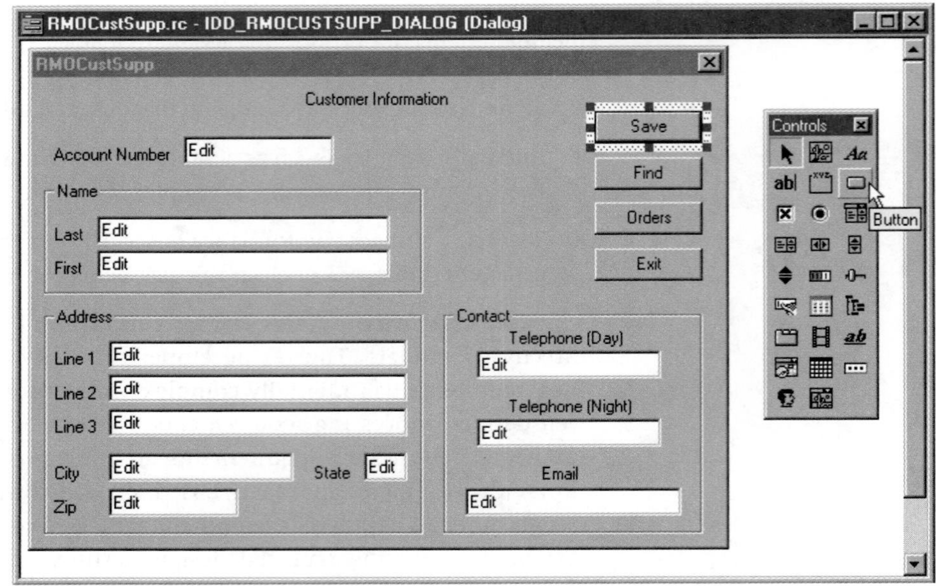

FIGURE 13-4
Development of a user interface prototype using a WYSIWYG dialog box editor.

Different prototyping tools provide varying capabilities. No one tool is best in all categories, and many tools are specialized for specific technical environments (such as operating or database management systems) or application types (such as data entry and query to or from a DBMS and web site development). Thus, it is important to choose the tool that best matches the characteristics of the project at hand.

Prototyping must also satisfy conditions specific to each project, including:

- Suitability to the technical environment in which the system will be deployed
- Ability to implement all system requirements
- Ability to interface with software developed with other tools

The technical environment in which the prototyping tool is used may be quite different from the technical environment ultimately used to deploy the system. Differences may include hardware, operating system, DBMS, network environment, and many other factors. It is not always possible to choose a prototyping tool that develops software that works in both the prototyping environment and the target environment.

For example, a developer may use a simple PC software package such as Microsoft Office to develop the prototype. But if the deployment environment is an IBM mainframe supporting thousands of interactive users, then it's unlikely that the prototype will function reliably in that environment. When prototyping and deployment environments differ substantially, it is typically more feasible to redevelop the application from scratch, using the prototype only as a tangible representation of system requirements.

Some prototyping tools are designed to develop software components that can be easily incorporated into (or extended by) other software modules. Others are designed to produce stand-alone software, which is fine as long as the tool is capable of building software for all of the required system features. If there are any gaps in the tool capabilities, then using that tool to build the final system is infeasible because the prototype will not be able to function with software developed by other tools.

Fortunately, in recent years, incompatibility of tools has increasingly become less of a problem. In particular, object-oriented and component-based technologies and standards have made software interoperability more possible and may someday become the norm. For the present, incompatibility is a potential problem that must be recognized and planned against.

The Spiral Approach to Development

spiral development approach
An iterative development approach in which each iteration may include a combination of planning, analysis, design, or development steps

The spiral development approach is an iterative development approach in which each iteration may include a combination of planning, analysis, design, or development steps. First described by Barry Boehm (see "Further Resources"), the approach has since become widely used for software development. There are many different ways of implementing the spiral development approach. Figure 13-5 illustrates one version that uses developmental prototypes (for other versions, see the Boehm and McConnell references in the "Further Resources" section). It is a more radical departure from traditional development than the prototyping development approach described earlier.

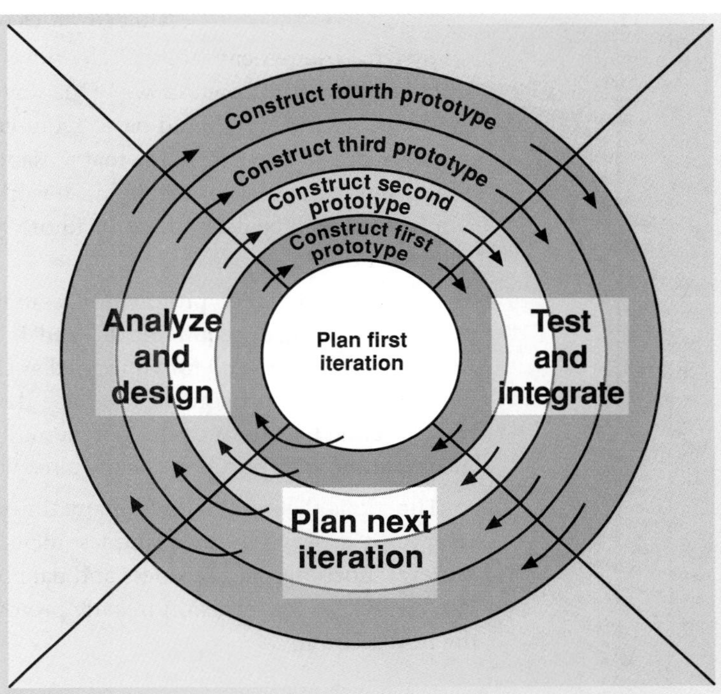

FIGURE 13-5
The spiral life cycle.

Steps in the Spiral Development Approach The spiral development approach begins with an initial planning phase, as shown in the center of Figure 13-5. The purpose of the initial planning phase is to gather just enough information to begin developing an initial prototype. Planning phase activities include a feasibility study, a high-level user requirements survey, generation of implementation alternatives, and choice of an overall design and implementation strategy. A key difference between the spiral approach and the prototyping approach is the detail of user requirements and software design. The spiral approach uses less detail than the prototyping approach.

As in the prototyping approach, preliminary plans for a series of prototypes may also be developed. But these plans must be very flexible because analysis and design activities are very limited. Because many analysis and design details are bypassed, plans often need to be altered later in the project. Thus, detailed planning for future prototypes is usually avoided.

After the initial planning is completed, work begins in earnest on the first prototype (the blue ring in Figure 13-5). For each prototype, the development process follows a sequential path through analysis, design, construction, testing, integration with previous prototype components, and planning for the next prototype. When planning for the next prototype is completed, the cycle of activities begins again.

Although the figure shows four prototypes, the spiral development approach can be adapted for any number of prototypes.

Which features to implement in each cycle's prototype is an important decision. Developers should choose features to include based on a number of criteria including:

- User priorities
- Uncertain requirements
- Function reuse
- Implementation risk

How these criteria are evaluated varies widely from one project to another.

System requirements are typically prioritized into categories such as "must have," "should have," and "nice to have." One way to minimize schedule length is to include the "must have" and "should have" requirements in the earliest prototypes. Doing so greatly increases the probability that a usable system can be delivered to the customer while minimizing development time. If work takes longer than expected, lower-priority requirements can be delayed until a future system upgrade, added after installation, or ignored entirely.

As described earlier, prototyping is an excellent tool for firming up uncertain or poorly defined user requirements, and including them in early prototypes allows them to be explored and fully specified as soon as possible. That knowledge can then be used to guide subsequent prototype development. Delaying the development of poorly defined portions of the system may result in unanticipated rework if they are incompatible with portions that are already implemented.

Functions that will be used many times are excellent candidates for inclusion in early prototypes. Typical examples include data-entry screens, data lookup and retrieval functions, and business and data objects. High-risk functions are also excellent candidates for inclusion in early prototypes. Risk analysis is discussed further in the next section.

Benefits and Risks of Spiral Development The spiral development approach has many advantages over traditional and prototyping approaches, including:

- High parallelism. There are many opportunities to overlap activities both within and among prototyping cycles. For example, planning of the next prototype can usually overlap testing and integration of the previous prototype.
- High user involvement. Users can be involved at each planning, analysis, and testing stage. Frequent and continual user participation produces a better system and higher user satisfaction.
- Gradual resource commitment. Resource consumption is much more evenly spread out over a spiral life cycle. This may lead to more efficient utilization of some resources (such as personnel). However, the total development cost is generally higher. See Figure 13-6 for a comparison of typical expenditures over traditional and spiral development approaches.
- Frequent product delivery. Every prototype is a working system in its own right. With sufficient planning, these prototypes can be immediately put to work. Frequent product delivery also leads to more testing, thus improving the product by catching more bugs.

FIGURE 13-6
Cumulative cost plotted against time for spiral and sequential development.

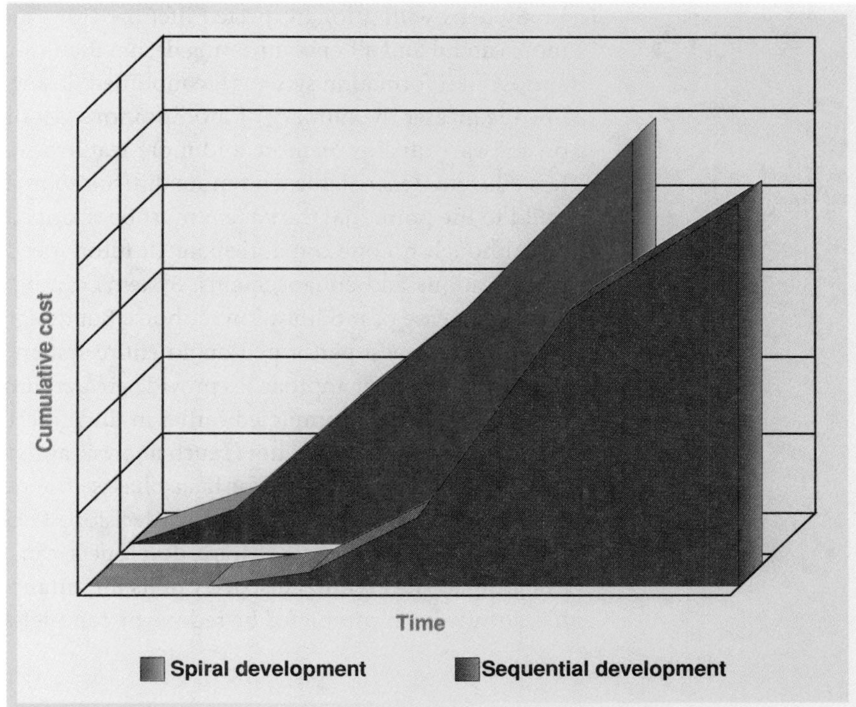

So why isn't every system developed with the spiral development approach? The primary drawbacks of the spiral approach are management and design complexity. Projects that use spiral development are more complex to manage than traditional projects because more activities occur in parallel and more people are working on the project at earlier stages. Also, because not all analysis and design occurs before construction, there is a higher probability that some rework will be necessary.

In a sequential development approach, most or all design activities are completed before construction begins. When a design is specified completely for an entire system, the result is generally of higher quality than if the same system were designed one piece at a time. In the spiral approach, fewer high-level design activities are performed before construction begins. Design decisions that may seem optimal when looking at a portion of the system may be less desirable when looking at the entire system. In mathematics and management science, this situation is sometimes described by the phrase "locally optimal but globally suboptimal."

A designer using the spiral development approach runs a much greater risk of making decisions that are globally suboptimal (possibly resulting in lower performance, a greater number of bugs, or more difficult maintenance after installation). Whether or not these problems are significant depends on several factors, including:

- Interdependence among system functions
- Expected life of the system
- Design team experience and skill
- Luck!

Systems with highly interdependent functions run a greater risk of design problems when the spiral approach is employed. Design decisions for one function also impact many other parts of the system. Parts of the system that are constructed in later iterations may inherit a large number of design constraints that prove difficult or impossible to resolve.

Systems with a long expected lifetime (for example, more than five years) need more careful and all-encompassing design than those with short lives. This is because almost all information systems are modified during their lifetimes, and the longer the life, the greater the number of modifications. Modifications have a cumulative effect on software quality. As more and more features are added to the system, it becomes less efficient, less reliable, and more difficult to maintain. Eventually, these problems build to the point that the system must be scrapped and reimplemented.

Although no one can anticipate all future needs, a designer can allow for future modifications and enhancements. Systems can be designed for rapid development, maximum ease of modification, or both. But designing for future upgrades and flexibility is a task best performed on an entire system, not iteratively on its subsystems. Other development approaches provide greater flexibility in the design because more design activities are completed earlier in the project life cycle.

It is difficult to decide how much analysis and design to do in the initial planning phase and what can be left for later phases. The experience and skill of the analysis and design team are critical to this decision. Luck is also a factor. Even the most skilled and experienced software developers can misidentify a critical analysis or design area. The eventual discovery of its importance may require that some work that has already been completed be redone or completely thrown away.

Rapid Development Techniques

Chapter 3 defined a technique as a collection of guidelines that help an analyst complete a system development activity or task. Many techniques have been developed over the years to speed up development. Some of them live up to their claims, although none can do so for all projects and development scenarios.

This section presents a small group of techniques proven to shorten project schedules. None is unique to a particular system development approach, although some work better with (or are required by) a particular approach.

Risk Management

All software development projects face risks that can result in schedule delays. Examples include changing user requirements, failure of hardware, support software, or tools, and loss of needed development resources such as funding or personnel. Risks are present regardless of what approach is taken to system development. When problems or unforeseen changes occur, they can add weeks, months, or years to a development schedule. Thus, identifying and dealing with risks are important and necessary parts of all approaches to system development.

risk management
A systematic process of identifying and mitigating software development risks

Risk management is a systematic process of identifying and mitigating software development risks. There are many specific ways to implement a risk management process. However, the underlying principles of risk management are the same regardless of the specific process used. These principles are based on the following ideas:

- Most risks can be identified if specific attention is directed to them.
- Risks appear, disappear, increase, and decrease as the development process proceeds.
- Small risks should be monitored, whereas large risks should be actively mitigated.

risk
A condition or event that can cause a project to exceed its shortest possible schedule

Risk Assessment For this chapter, we define a **risk** as a condition or event that can cause a project to exceed its shortest possible schedule. This definition is narrow, but sufficient for our purposes. Conditions are things that currently exist but are unknown or poorly understood at the present time. For example, when upgrading an existing system, a developer may become aware of bugs in critical software components only after a new feature is implemented. Events are things that may occur but haven't yet. Examples include user requirement changes, reassignment of key personnel, and failure of a hardware vendor to meet a delivery deadline. Figure 13-7 describes several important categories of software development risks. A longer list with more specifics can be found in McConnell (see the "Further Resources" section at the end of the chapter).

FIGURE 13-7
Major categories of development schedule risk.

System requirements
- Requirements are poorly understood when the schedule is created
- Customer adds or changes features late in development
- IS staff insists on unnecessary features (e.g., security or robustness)

Tools and technology
- Target deplolyment environment is poorly suited to the application
- Development tools are poorly suited to the target deployment environment
- New tools have unknown productivity and/or learning curve
- New technology is unstable
- Technology standards are not yet fully developed

People
- Users are not adequately involved in all stages of development
- Project management is poor
- Developers are poorly trained or underperforming
- Developer or manager skills are poorly matched to specific tasks

Environment
- Upper management's commitment to the project is weak or variable
- Competitive environment is rapidly changing
- Technological environment is rapidly changing
- Governmental action (or inaction)

Each risk has a probability and one or more outcomes. The risk probability is the chance that event will occur or the condition will be discovered. An outcome is the action that will be required if the event occurs or the condition is discovered. For example, for a new system with specific performance requirements (for example, a certain number of transactions processed per minute), there is a risk that the compiled software will not be able to meet those requirements. Possible outcomes include the need to spend time optimizing the software for greater speed and the need to install and test the software on newly acquired higher-performance hardware. In either outcome, additional resources—including time—will be required to remedy the problem. When examining schedule overrun risks, a developer usually expresses all outcomes in similar units of time (for example, four weeks to redesign and implement, and two weeks to install and test on more powerful hardware).

Developers need to determine the probability of each risk based on their experience. For example, a developer may consider the risk of not meeting performance requirements to be low because similar systems on similar hardware platforms have been successfully designed and implemented in the recent past. Thus, the developer may assign a relatively low numerical probability (such as 10 percent) to the risk. Other risks may have much higher probabilities. For example, a developer may know that certain support software required by a new system has a long history of bugs and revisions to fix them. Thus, the developer may assign a relatively high probability (such as 60 percent) to the risk that a bug will be discovered and that some portion of the system will have to be redesigned or reimplemented.

If a risk has multiple possible outcomes (as most do), then each outcome also has a probability. Outcome probabilities are actually conditional probabilities. That is, they are probabilities that the outcome will be required *if* the risk event occurs or the risk condition is discovered. The sum of all outcome probability risks for a specific risk is 100 percent if the outcomes are mutually exclusive and collectively exhaustive. Table 13-1 summarizes risks, outcomes, and their probabilities for a hypothetical project.

Risk	Risk probability	Possible outcomes	Outcome probability	Outcome delay	Expected delay	
					Outcome	Risk
Critical bug in support software	60%	Redesign and reimplement	70%	4 weeks	1.68	
		Reimplement only	30%	2 weeks	0.36	2.04
Significant user requirements change	30%	Redesign and reimplement existing code	50%	6 weeks	0.90	
		Redesign and implement additional code	50%	4 weeks	0.60	1.50
Key staff member reassigned during implementation	20%	Hire and orient a replacement	30%	4 weeks	0.24	
		Reassign tasks to remaining personnel	70%	12 weeks	1.68	1.92
System fails performance test	10%	Optimize design or code	80%	4 weeks	0.32	
		Upgrade hardware	20%	2 weeks	0.04	0.36
Hardware not delivered on time	10%	One week delay	60%	1 week	0.06	
		Two weeks delay	30%	2 weeks	0.06	0.15
		Three weeks delay	10%	3 weeks	0.03	
			Total expected schedule delay			**5.97**

TABLE 13-1
An analysis of risks, outcomes, and expected schedule delays.

Developers multiply outcomes, probabilities, and schedule delays to create an expected delay for each outcome. For example, Table 13-1 shows that the expected delay associated with outcome "optimize design or code" is 0.32 weeks. This was computed by multiplying the probability of failing the performance test, the conditional probability that optimization would be sufficient to fix the problem, and the expected amount of time required to do the optimization ($0.1 \times 0.8 \times 4 = 0.32$ weeks). The expected delays for each outcome of a single risk are added to determine the expected delay for the risk (for example, $0.32 + 0.04 = 0.36$ weeks, for failing the performance test). The expected delay for the entire project is the sum of the expected delays for each risk. Note that these calculations assume that risks and their outcomes are independent of one another.

Steps in Risk Management Figure 13-8 contains a flow chart for risk management. The process begins at the start of the development project and continues until it is completed. The first step is to identify all risks that are likely to affect the project. This task is relatively difficult but critically important since risks that aren't identified can't be managed. A group of project participants (including users) usually identifies risks, since more heads are better than few to generate and evaluate a large number of ideas.

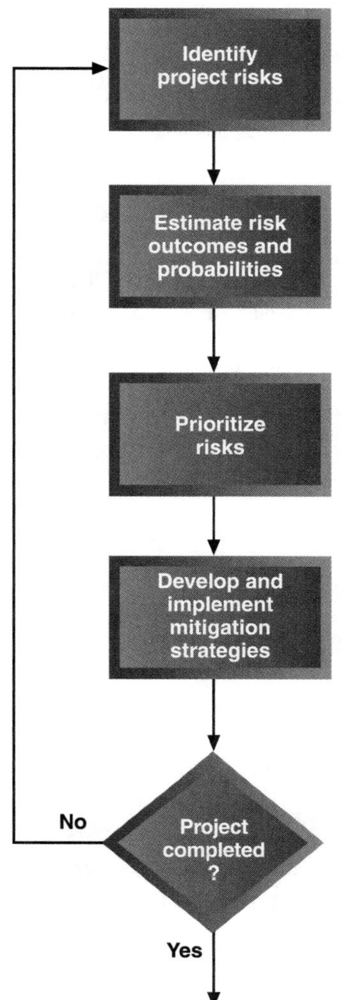

FIGURE 13-8
Steps in risk management.

Outcomes and probabilities for each risk are then determined. They may be difficult to estimate. For example, a tool-related failure may have a relatively well-defined outcome (such as time needed to reimplement with another tool) but a poor probability estimate, particularly if the development team doesn't have much experience with the tool. Sometimes both outcomes and the probability may be difficult to estimate, such as the risk of a user adding significant new requirements late in the project. As when identifying risks, a group generally is used so that a large base of knowledge and experience can be applied to the task.

Once outcomes and probabilities have been estimated, the risks must be prioritized. Highest priority should go to risks with long expected delays. Risks with long outcomes and low probabilities are also given high priority since probability estimates are inherently inaccurate. The goal of prioritizing risks is to generate a list of "high" risks and a list of "low" risks. Low-priority risks (for example, the last two rows in Table 13-1) are simply monitored to ensure that they don't become higher-priority risks. High-priority risks (for example, the first three risks in Table 13-1) call for more active management approaches.

risk mitigation
Any step(s) taken to minimize the expected schedule delay of a risk

Risk mitigation is a generic term that describes any steps taken to minimize the expected schedule delay of a risk. Specific mitigation steps can vary widely. For example, the risk of a customer adding significant requirements late in the project might be mitigated by developing an early throwaway prototype to validate requirements, actively involving the customer in all phases of project planning, or requiring the customer to sign a contract rigidly specifying the system features. Risk mitigation for a failed performance test might include detailed benchmarking of the tool and hardware, development of a test prototype, or parallel development with multiple tools.

As a project proceeds, more information about risks becomes available. The additional information may arise as a natural by-product of development activities, or it may arise from specific risk mitigation efforts. In either case, new and revised information should be periodically evaluated. Thus, the entire risk management process is actually a loop whose frequency varies from project to project. In projects with few risks and moderately long schedules, developers may reevaluate risks infrequently (for example, after completion of each phase of the development life cycle or after one or more iterations of the spiral development approach). Projects with many risks and longer development schedules may require formal reexamination of risks more frequently (for example, at the end of every week).

When to Implement Risk Management Risk management should always be used regardless of which approach is employed for a project. But inherently riskier projects call for more thorough and active risk management. Unknown requirements and feasibility are a principal element of overall project risk. Since such unknowns normally motivate developers to use iterative development, active and thorough risk management should be key elements of those approaches.

When a spiral development approach is employed, risk management should be a key component both of the initial planning phase and of planning for every prototype iteration. When possible, developers should incorporate mitigation measures for the highest priority risks into the activities of the earliest development cycles. For example, developers should address uncertainties about requirements, tools, or technology by incorporating them into the first development cycle. This provides further information about risks early in the project and allows the maximum possible time for risk mitigation.

Joint Application Design

Joint application design (JAD) was described in Chapter 4 in the section "Conduct Joint Application Development Design." JAD is a popular and very effective technique for defining system requirements. It is primarily used as a systems analysis technique and occasionally to specify some higher-level design parameters. JAD shortens the time needed to specify system requirements by including all key decision makers in one or more intensive sessions. This approach shortens the project schedule by minimizing communication overhead and concentrating the efforts of many people in a short space of time. Participation by all key decision makers ensures that the project gets off to a speedy start with solid planning and clearly defined objectives.

JAD can be incorporated into any development approach, but it is especially well suited to iterative development because both emphasize prototypes and development speed. JAD can be used to implement most or all of the initial planning phase of the spiral development approach. The user-interface prototypes developed during the JAD session can form the starting point for the first full-fledged developmental prototype in the spiral development approach.

When used with a conventional development approach, a JAD session does not normally address all of the design issues that the initial planning phase of the spiral development approach does, such as choice of implementation tools, overall architectural design, and initial definition of the evolutionary prototypes. However, these decisions are next in line after those normally considered in a JAD session. The JAD session can be extended to include these decisions, or the results of the JAD session can be the input to a later phase.

Tool-Based Development

The term *tool* was defined in Chapter 3 as software that helps create models or other deliverables required in the project. The intervening chapters have discussed a large number of tools, including CASE tools, project management tools, database management systems, and integrated development environments. For this chapter, we limit our attention to tools that are used directly to build software. Such tools include compilers, code generators, and object frameworks but exclude tools such as project management and model development software, unless they are included within a CASE tool or integrated development environment.

No one tool is best for producing all types of software. Tools have different strengths and weaknesses. Some parts of a system can be built very quickly because they match development tool capabilities very well. Other parts of a system are much more difficult to build because of limitations or gaps in tool capabilities. Building these parts of the system requires much more time because human ingenuity and experimentation are required to get the tool to do something it does poorly or was never designed to do.

tool-based development
A development approach that chooses a tool or tools that best match system requirements and doesn't develop any requirements that aren't easily implemented with the tool(s)

The premise of **tool-based development** is simple—choose a tool or tools that best match the requirements and don't develop any requirements that aren't easily implemented with the tool(s). In essence, tool-based development is an application of the generic 80/20 or 90/10 rule. Resources are best used to construct a system that satisfies the 80–90 percent of the requirements most easily implemented, and the 10–20 percent of the requirements that are difficult to implement with the tool(s) are discarded.

Although tool-based development is easy to describe, it is very difficult to implement in an organization because it requires a developer to say no early and often. The developer and the customer must agree on what requirements will or won't be included in the system. The developer must be willing and able to say no to future requests to add difficult requirements. Saying no so often may not be possible due to user needs or project politics.

User satisfaction may be much lower than with other development methods because some desired functions are left out of the system. The developer can mitigate dissatisfaction by clearly describing the schedule and cost impacts of all requirements that don't directly match tool capabilities. On the plus side, design and development can proceed much more quickly, and reductions in coding and testing time are dramatic. Thus, tool-based implementation forces a clear and direct trade-off between development speed and system functionality.

Figure 13-9 shows a simple process for tool-based development in the context of a sequential development approach. New activities are added to the end of the analysis phase, including:

- Investigating development tools that support the system requirements
- Selecting tools and modifying system requirements to match the tool set

Note that requirement priorities are assumed to drive the tool selection, not the other way around. Subsequent development phases are driven by the underlying capability and methodology of the chosen tool set. The developer can also incorporate tool-based development into the prototyping or spiral development approaches by embedding a tool selection activity within the analysis or initial planning phase of those approaches.

FIGURE 13-9
Tool-based development applied within a traditional sequential life cycle.

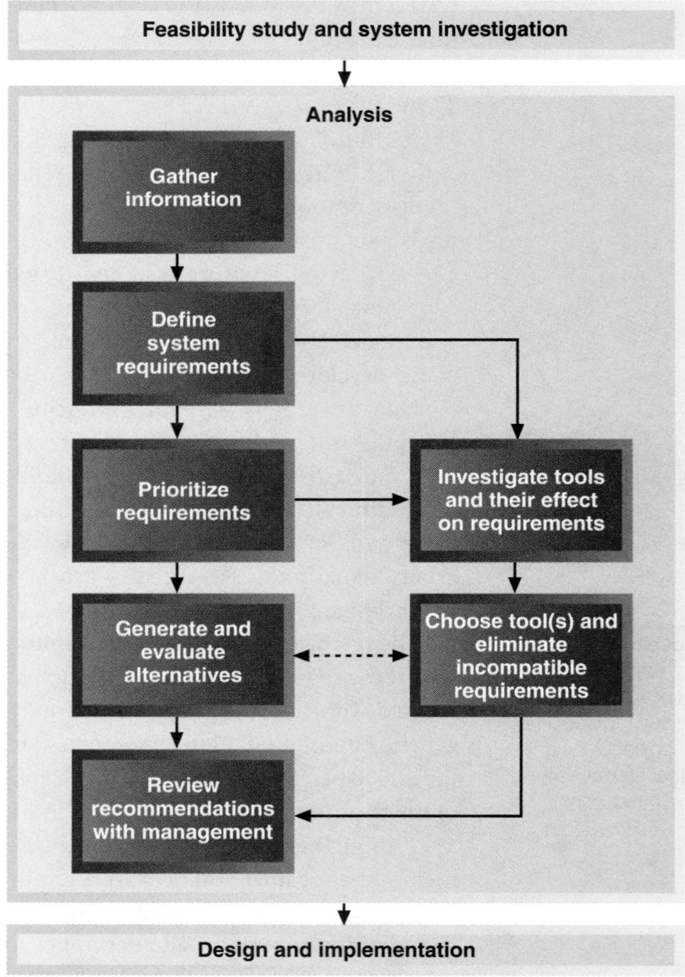

If a new tool is chosen, then project costs will increase because the new tool must be purchased and developers must be trained. Project managers must plan carefully to avoid having a new tool (or tools) lengthen the project schedule. Tool acquisition and installation must be completed before implementation activities begin. Training should be concurrent with systems analysis, and it should be completed prior to architectural or detailed design.

A developer must take great care when choosing multiple tools to build a new system. Tools that aren't specifically designed to work together may have hidden incompatibilities that don't become apparent until late in the implementation phase. For example, development tools may use incompatible methods for representing data. Such incompatibilities may not become apparent until integration testing, when data stored in a program developed with one tool are passed via a function call or method invocation to software developed with another tool. Smoothing over these incompatibilities can add as much work and time to a project as adding system features that aren't well matched to a specific tool.

Another danger of tool-based development is the tendency to choose a different tool set for every project. This is particularly problematic when users hear success stories about a new tool set and they want a system similar to the one described in the success story. The problem is that each new tool adopted by a development team dramatically decreases productivity. Every tool has a learning curve, and productivity gains can't be realized until the development team has worked its way up the learning curve. High productivity is typically not achieved during the first project that uses the new tool. Tools less suited to the project requirements may actually deliver a better system on a shorter schedule if the development team has extensive experience with the tool.

Software Reuse

software reuse
Any mechanism that allows software used for one purpose to be reused for another purpose; also called **code reuse**

Software reuse (or **code reuse**) is a general term that describes any mechanism that allows software used for one purpose to be reused for another purpose. Software reuse can significantly shorten a development schedule by reducing the effort, time, and money required to develop or modify information systems. Whether software reuse actually reduces development effort depends on many specifics, including:

- The effort required to identify potentially reusable software
- The extent to which existing software functions require modification to suit a new purpose
- The extent to which existing software must be repackaged into a form that can be plugged into a new system

There are many different ways of reusing software. At one extreme, an existing system (for example, an accounting software package) can be purchased, thus avoiding the vast majority of implementation phase activities. At the other extreme, small portions of code from previously written programs can be reused in another program. Of course, there are many kinds of software reuse between these extremes, and many different approaches, techniques, tools, and technologies can be used.

OO design and programming techniques have provided new tools to address software reuse. One of the reasons that OO development methods have become so popular is the premise that classes and objects are easier to reuse than programs built with traditional programming tools. Note the use of the word *premise* in the previous sentence. There is still too little hard evidence to say definitively that OO development is faster than traditional development. But most software developers do believe that OO development is faster, and ongoing research should soon close the gap between belief and knowledge.

Figure 13-10 shows two particular methods of OO software reuse. A project team can reuse OO source code by deriving new classes from existing classes or entire object frameworks. Executable objects are sometimes called components. A component can be a single object, an entire system, or anything in between. The last two sections of this chapter describe object frameworks and components in detail.

Software reuse is a technique applicable to any system development approach. Analysts and developers must actively search for reusable software. For large-scale reuse, the search must begin before architectural design. However, the search can be time-consuming and expensive. The effort needed to search, evaluate, adapt, and integrate existing software to a new purpose may be greater than the effort to build a new system from scratch.

The search for smaller-scale reusable software can occur during systems design or development and is usually less time intensive. In many cases, developers need look no further than their programming toolkit. Most modern programming toolkits reuse software in one or more forms. Examples include object frameworks, program templates, code generators, and component libraries. Modern programming toolkits also provide tools for adapting and integrating reusable code.

FIGURE 13-10
A comparison of various object-oriented code reuse methods.

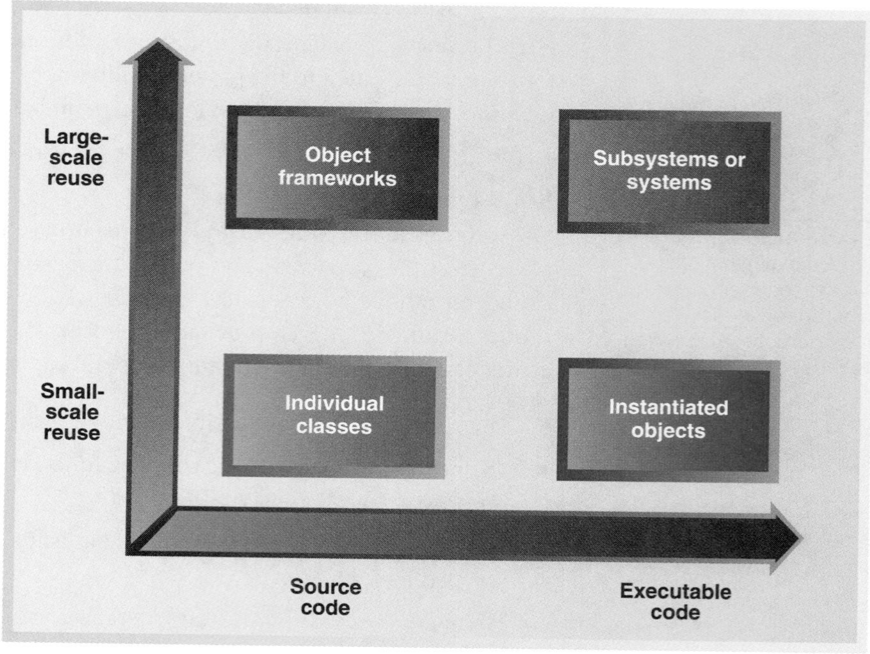

Object Frameworks

Similar functions are embedded in many different types of systems. For example, the graphical user interface (GUI) is nearly ubiquitous in modern software. Many features of a GUI—such as drop-down menus, help screens, and drag-and-drop manipulation of on-screen objects—are used in many or most GUI applications. Other functions—such as search, sorting, and simple text editing—are also common to many applications. Reusing software to implement such common functions is a decades-old development practice. But such reuse was awkward and cumbersome with older programming languages. Object-oriented programming languages provide a simpler method of software reuse.

object framework
A set of classes that are specifically designed to be reused in a wide variety of programs or systems

foundation classes
The classes within an object framework

An **object framework** is a set of classes that are specifically designed to be reused in a wide variety of programs. The object framework is supplied to a developer as a precompiled library or as program source code that can be included or modified in new programs. The classes within an object framework are sometimes called **foundation classes.** Foundation classes are organized into one or more inheritance hierarchies. Programmers develop application-specific classes by deriving them from existing foundation classes. Programmers then add or modify class attributes and methods to adapt a "generic" foundation class to the requirements of a specific application.

A Sample Object Framework

The Microsoft Foundation Classes were developed as an aid to programming applications that run within the Windows operating system. Classes are provided for a number of general-purpose processing needs, including file and database manipulation, network I/O, graphic images, general-purpose data structures, and user-interface objects. The classes are typically used within C++ programs developed with the Microsoft Visual C++ toolkit.

Examples include the Save As dialog box (see Figure 13-11), print dialogs, menus, toolbars, window frames, and window control menus. These user-interface features are similar from application to application because they execute common user-interface functions within the operating system service layer. The Microsoft Foundation Classes make it easy to reuse these common functions by encapsulating them within classes.

FIGURE 13-11
A standard Windows Save As dialog box, an object derived from the Microsoft Foundation Class CFileDialog.

Figure 13-12 shows a portion of the Microsoft Foundation Classes hierarchy. The root of the hierarchy is a generic class called CObject. CObject encapsulates a minimal set of functionality for all Windows objects, including access to class information during program execution and diagnostic dumps for tracing program errors. The class CCmdTarget derives from CObject and adds functionality to allow the object to respond to external messages (such as keyboard input and mouse clicks).

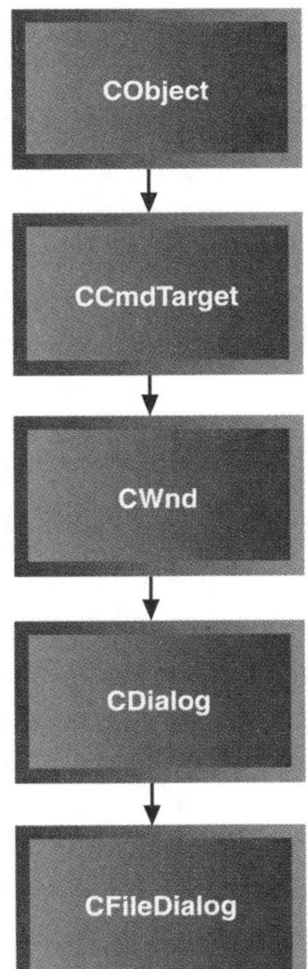

FIGURE 13-12
The derivation tree of the class CFileDialog within the Microsoft Foundation Classes.

The CWnd class encapsulates the behavior of windows on the user's display device. Its methods implement functionality common to all windows, including frames, window resizing, window movement, and minimization and maximization. The classes CObject, CCmdTarget, and CWnd are generic classes that are not intended to be used directly by application programs. Instead, application programs typically use classes from lower levels of the class hierarchy.

The CDialog class encapsulates behavior common to all dialog boxes. This behavior includes the OK button, the Cancel button, and the feature of most dialog boxes that prevents further actions by the user until the OK or Cancel button has been clicked. The class CFileDialog provides a template for the behavior of dialog boxes used to save files (such as the dialog box shown in Figure 13-11).

An application program can create a Save As dialog box by creating an object of the class CFileDialog. The class provides some ability to customize class behavior through parameters used initially to create an object or by sending messages to internal class methods that define default behaviors. For example, a programmer can specify the default directory shown in the "Save in:" drop-down list at the top of Figure 13-11. He or she can also define the types of files displayed in the "Save as type:" drop-down list at the bottom of Figure 13-11. The programmer can create a more customized version of a Save As dialog box by deriving a new class from CFileDialog. The programmer is then free to add or modify data and methods of the new class to suit the needs of the particular application.

An object framework such as the Microsoft Foundation Classes provides a large set of reusable software to programmers. The class library provides tools for programmers to deal with the complexity of modern operating systems and application programs. Application development without such an object framework would be much more tedious, time-consuming, error-prone, and expensive.

Object Framework Types

Object frameworks have been developed for a variety of programming needs. Examples include:

- User-interface classes—classes for commonly used objects within a graphical user interface, such as windows, menus, toolbars, and file open and save dialogs.
- Generic data structure classes—classes for commonly used data structures such as linked lists, indices, and binary trees and related processing operations such as searching, sorting, element insertion, and element deletion.
- Relational database interface classes—classes that allow OO programs to create database tables, add data to a table, delete data from a table, or query the data content of one or more tables.
- Classes specific to an application area—classes specifically designed for use in application areas such as banking, payroll, inventory control, and shipping.

General-purpose object frameworks such as the Microsoft Foundation Classes typically contain classes from the first three categories. Classes in these categories can be reused in a wide variety of application areas. Application-specific object frameworks provide a set of classes for use in a specific industry or type of application. Application-specific application frameworks are usually designed by third parties as extensions to a general-purpose object framework (such as a set of banking classes designed to be used with the Microsoft Foundation Classes). Application-specific frameworks may also be developed by IS departments in larger companies.

At one time, software industry watchers assumed a thriving market for application-specific frameworks would develop. They thought third-party developers and software consulting firms would design and sell general-purpose and industry-specific object frameworks, and these frameworks would become the basis of many information systems. The market does exist, but it is much smaller than originally thought. Many factors may account for this situation:

- The market for ready-to-use information systems, especially those for enterprisewide information systems (for example, SAP and PeopleSoft), has grown quickly. Such systems eliminate the need for an application-specific object framework market.
- Programming staff are in short supply. Competing demands for skilled personnel have left few programmers available for long-term development and maintenance of application-specific frameworks.
- There are several competing OO programming languages, including C++, Java, and Smalltalk, and none has captured the market. Programmers would need to develop equivalent frameworks in all three languages to reach a large portion of the market.
- Rapid maturation of component technology is simplifying and standardizing reuse of executable software.

Many large organizations have moved aggressively to develop their own application frameworks. Smaller firms usually do not do so because the resource requirements are substantial. An application- or company-specific framework requires a significant development effort typically lasting several years. The effort is repaid over time through continuing reuse of the framework in newly developed systems. The effort is also repaid through simplified maintenance of existing systems. But the payoffs occur far in the future, often making them less valuable than the current funds required to build the frameworks.

The Impact of Object Frameworks on Design and Implementation Tasks

Developers need to consider several issues when determining whether to use object frameworks. Object frameworks impact the process of systems design and development in several different ways:

- Frameworks must be chosen before detailed design begins.
- Systems design must conform to specific assumptions about application program structure and operation that the framework imposes.
- Design and development personnel must be trained to use a framework effectively.
- Multiple frameworks may be required, necessitating early compatibility and integration testing.

The process of developing a system using one or more object frameworks is essentially one of adaptation. The frameworks supply a template for program construction and a set of classes that provide generic capabilities. System designers adapt the generic classes to the specific requirements of the new system. Frameworks must be chosen early so that designers know the application structure imposed by the frameworks, the extent to which needed classes can be adapted from generic foundation classes, and the classes that cannot be adapted from foundation classes and must be built from scratch.

Of the three object layers typically used in OO system development (view, business logic, and data), the view and data layers most commonly derive from foundation classes. User interfaces and database access tend to be the areas of greatest strength in object frameworks, and they are typically the most tedious classes to develop from scratch. It is not unusual for 80 percent of a system's code to be devoted to view and data classes. Thus, constructing view and data classes from foundation classes provides significant and easily obtainable code reuse benefits. Adapting view classes from foundation classes has the additional benefit of ensuring a similar look and feel of the user interface across systems and across application programs within systems.

Object frameworks can be divided into two types—those that impose a specific structure on the programs that use them and those that do not. Many of the larger and more widely used application frameworks are in the former category. For example, the Microsoft Foundation Classes require application programs to be objects derived from one of the application classes defined by the hierarchy (such as CWinThread or CWinApp). These classes implement a program organization based on a core code module that waits for messages passed in from the operating system and then dispatches responsibility for responding to those messages to specific objects within the program.

The task of an application program using this framework is to create objects and methods that respond to specific messages delivered by the operating system. A programmer can decide which messages to process and how to process them, but the basic structure and operation of the program cannot be altered. Because of this imposed structure, programs are not designed from the ground up as self-contained entities. Rather, a template imposed by the object framework is the starting point for all program design activities. This fundamentally alters the nature of program design and the specific tasks that must be completed.

Successful use of an object framework requires a great deal of up-front knowledge about its class hierarchies and program structure. That is, designers and programmers must be familiar with a framework before they can successfully use it. Thus, a framework should be selected as early as possible in the SDLC, and developers must be trained in use of the framework before they begin to implement the new system.

Pitfalls to Consider for Object Frameworks

The complexity of employing object frameworks is multiplied when more than one framework is used in the same system. This situation is increasingly common as more frameworks become available for specific business needs and application areas. For example, consider a system development project for a complete banking system in which separate frameworks from different vendors are chosen for the view, business logic, and data layers of the application. To compound the complexity further, consider the use of two different view layer frameworks—one for Windows-based workstations and another for mainframe-based applications using dumb terminals.

Several potential problems must be considered when integrating multiple frameworks within a single system, including naming overlaps and inconsistent assumptions or templates for program structure. Overlaps in naming are typically resolved by modifying names in the framework. But this is possible only if the framework source code is available. Frameworks delivered as compiled libraries cannot be easily edited to implement name changes.

Inconsistent assumptions about program structure are a far more serious integration problem—one that frequently defies solution. If two frameworks each insist on dictating the overall structure of a program by implementing a master application program object, then it is difficult or impossible to combine the two frameworks within a single program. Typically, the only way to integrate them is to modify the source code of one framework to bypass the assumed program structure template. But this requires an intimate knowledge of the internal structure and function of the framework, and the resulting product is generally far less reliable than the original framework.

Components

component
A standardized and interchangeable software module that is fully assembled and ready to use

In addition to using object frameworks, developers often use components to speed system development. A **component** is a software module that is fully assembled and ready to use. Components may be single executable objects or groups of interacting objects. A component may also be a non-OO program or system "wrapped" in an OO interface. Components implemented with non-OO technologies must still implement objectlike behavior. In other words, they must implement a public interface, respond to messages, and hide their implementation details.

Components are standardized and interchangeable software parts. They differ from objects or classes because they are binary (compiled and linked) programs, not symbolic (source code) programs. This distinction is important because it makes components much easier to reuse and reimplement than source code programs.

For example, consider the grammar-checking function in most word processing programs. A grammar-checking function can be developed as an object or as a subroutine. Other parts of the word processing program can call the subroutine or object methods via appropriate source code constructs (for example, a C++ method invocation or a BASIC subroutine call). The grammar-checking function source code is integrated with the rest of word processor source code during program compilation and linking. The executable program is then delivered to users.

Now consider two possible changes to the original grammar-checking function:

1. The developers of another word processing program want to incorporate the existing grammar-checking function into their product.
2. The developers of the grammar-checking function discover new ways to implement the function that result in greater accuracy and faster execution.

To integrate the existing function into a new word processor, the source code of the grammar-checking function must be provided to the word processor developers. They then code appropriate calls to the grammar checker into their word processor source code. The combined program is then compiled, linked, and distributed to users.

When the developers of the grammar checker revise their source code to implement the faster and more accurate function, they deliver the source code to the developers of both word processors. Both development teams integrate the new grammar-checking source code into their word processors, recompile and relink the programs, and deliver a revised word processor to their users.

So what's wrong with this scenario? Nothing in theory, but a great deal in practice. The grammar-checker developers can provide their function to other developers only as source code, which opens up a host of potential problems concerning intellectual property rights and software piracy. Of greater importance, the word processor developers must recompile and relink their entire word processing programs to update the embedded grammar checker. The revised binary program must then be delivered to users and installed on their computers. This is an expensive and time-consuming process. Delivering the grammar-checking program in binary form would eliminate or minimize most of these problems.

A component-based approach to software design and construction solves both of these problems. Component developers such as the developers of the grammar checker can deliver their product as a ready-to-use binary component. Users such as the developers of the word processing programs can then simply plug in the component. Updating a single component doesn't require recompiling, relinking, and redistributing the entire application. Perhaps applications already installed on user machines could query an update site via the Internet each time they started and automatically download and install updated components.

At this point, you may be thinking that component-based development is just another form of code reuse. But structured design, object-oriented inheritance, and client-server architecture all address code reuse in different ways. What makes component-based design and construction different are the following:

- Components are binary (executable) units of code. Structured design and object-oriented inheritance are methods of reusing source code.
- Components are executable objects that advertise a public interface (that is, a set of methods and messages) and hide (encapsulate) the implementation of their methods from other components. Client-server architecture is not necessarily based on OO principles. Component-based design and construction are an evolution of client-server architecture into a purely OO form.

Components provide an inherently flexible approach to systems design and construction. Developers can design and construct many parts of a new system simply by acquiring and plugging in an appropriate set of components. They can also make newly developed functions, programs, and systems more flexible by designing and implementing them as collections of components. Component-based design and construction has been the norm in the manufacturing of physical goods (such as cars, televisions, and computer hardware) for decades. However, it is only now emerging as a viable approach to designing and implementing information systems.

Component Standards and Infrastructure

Interoperability of components requires standards to be developed and readily available. For example, consider the video display of a typical IBM-compatible personal computer. One video display unit cable is connected to a power source. The plug at the end of this cable has a specific physical configuration (for example, three prongs of a certain size and placement). The video display unit expects certain functions from this plug—namely, that electrical power of a known type (for example, 120-volt, 60-cycles-per-second alternating current) will flow through it. All other video display units expect the same functions from the plug. Thus, the electrical power plug is an interface standard.

The plug on the end of the other cable used by the video display unit is also an interface standard. The plug has a specific form, and each connector in the plug carries a well-defined electrical signal. The physical form of the plug and the type of signals carried through each connector were defined years ago by a group of computer and video display manufacturers. Adherence to this standard guarantees that any video display unit will work with any IBM-compatible personal computer and vice versa. The existence of an interface standard encourages manufacturers to develop compatible components. This, in turn, creates interoperability.

Components may also require standard support infrastructure. For example, video display units are not internally powered. Thus, they require not only a standard power plug but also an infrastructure to supply power to the plug. A component may also require specific services from an infrastructure. For example, a cellular telephone requires the cellular service provider to assign a transmission frequency with the nearest cellular radio tower, transfer the connection from one tower to another as the user moves among telephone cells, establish a connection to another person's telephone, and relay all voice data to and from the other person's telephone via the public telephone grid. All cellular telephones require these services.

Software components have a similar need for standards. Components could be hard-wired together, but this reduces their flexibility. Flexibility is enhanced when components can rely on standard infrastructure services to dynamically find other components and establish connections with them.

Connecting Components In the simplest systems, all components execute on a single computer under the control of a single operating system. Interprocess communication facilities provided by the operating system can be used to connect components. Connection is more complex when components are located on different machines running different operating systems and when components can be moved from one location to another. In this case, a network protocol independent of the hardware platform and operating system is required. In fact, using a network protocol is desirable even when components all execute on the same machine because such a protocol guarantees that systems can be used in different environments—from a single machine to a network of computers.

Modern networking standards have largely addressed the issue of a common hardware and communication software to connect distributed software components. Internet protocols are a nearly universal standard and, thus, provide a ready means of transmitting messages among components. Internet standards can also be used to exchange information among two processes executing on the same machine. However, Internet standards alone do not fully supply a component connection standard. The missing pieces are:

- Definition of the format and content of valid messages and responses
- A means of uniquely identifying each component on the Internet and routing messages to and from that component

Many organizations have attempted to define the missing standards for connecting distributed components. Currently, only two of these standards are well developed and widely implemented: CORBA and DCOM. The **Common Object Request Broker Architecture (CORBA)** was developed by the **Object Management Group (OMG)**, a consortium of computer software and hardware vendors. The standard is currently in its third revision, and it is just now being widely implemented. The **Distributed Component Object Model (DCOM)** is a Microsoft-developed standard for component interoperability. It is widely implemented in Windows-based application software, and it is increasingly used in three-tier distributed applications based on Microsoft Internet Information Server and Transaction Server.

Common Object Request Broker (CORBA)
A standard for software component connection and interaction developed by the Object Management Group (OMG)

Object Management Group (OMG)
A consortium of computer companies that developed the CORBA standard

Distributed Component Object Model (DCOM)
A standard for software component connection and interaction developed by Microsoft

For at least the near future, the decision to use or implement components requires adoption of CORBA, DCOM, or both. If one standard is chosen then suitable development tools and server software must be acquired and installed. Development personnel typically require weeks or months of training to become proficient with the tools and server software. Choosing both standards is an expensive and complicated proposition because the standards are incompatible. Each standard requires unique server software, tools, and developer skills. An information system adapted to both standards is really two different systems.

Components and the Development Life Cycle

Purchasing and reusing components is a viable approach to speed completion of a system. There are two development scenarios that involve components:

- Purchased components can form all or part of a newly developed or reimplemented system.
- Components provide one model for designing and deploying a newly developed or reimplemented system.

Each scenario has different implications for the system development life cycle. The following sections explore these implications and their effects on various phases of the SDLC.

System Performance Component-based software is usually deployed in a distributed environment. Components typically are scattered among client and server machines and among local area network (LAN) and wide area network (WAN) locations. Distributing components across machines and networks raises the issue of performance. System performance depends on where components are located (that is, component topology), the hardware capacity of the computers on which they reside, and the communication capacity of the networks that connect computers. Performance also depends on demands for network and server capacity made by other applications and communication traffic such as telephone, video, and interactions among traditional clients and servers.

The details of analyzing and fine-tuning computer and network performance are well beyond the scope of this text. But anyone planning to build and deploy a distributed component-based system should be aware of the performance issues. These issues must be carefully considered during systems design, implementation, and support.

Steps developers should take to ensure adequate performance include the following:

- Examine component-based designs to estimate network traffic patterns and demands on computer hardware.
- Examine existing server capacity and network infrastructure to determine their ability to accommodate communication among components.
- Upgrade network and server capacity prior to development and testing.
- Test system performance during development and make any necessary adjustments.
- Continuously monitor system performance after installation to detect emerging problems.
- Redeploy components, upgrade server capacity, and upgrade network capacity to reflect changing conditions.

Implementing these steps requires a thorough understanding of computer and network technology as well as detailed knowledge of existing applications, communications needs, and infrastructure capability and configuration. Applying this knowledge to real-world problems is a complex task typically performed by highly trained specialists.

Object-Oriented Techniques It is possible to develop component-based applications without using OO analysis, design, or development techniques, but it isn't recommended. A component is a distributed object that passes messages to (or calls methods of) other components. Since objects are the basis of component construction and interaction, the proper design tools and techniques are object-oriented. OO design, in turn, requires OO analysis. Traditional structured analysis and design techniques are poorly matched to component-based systems.

From the users' viewpoint, the requirements for a component-based system are no different from the requirements of a system implemented with more traditional technology. That is, the behavioral aspects of the system (that is, data inputs, the user interface, and basic processing functions) are the same. Thus, analyzing and documenting user requirements for a component-based system is the same as for any other OO system.

Design phase issues for a newly developed component-based system are similar to those of any system developed with object-oriented techniques. A suitable class hierarchy must be defined, and the messages and methods must be fully specified. The techniques and models used to do this (such as class diagrams, interaction diagrams, and statecharts) are the same as for any other OO analysis and design effort.

Purchased Components If purchased components are considered as an alternative, then a developer must search for suitable software during the latter part of systems analysis. The developer must also determine the correspondence between the capabilities of candidate software and user requirements. The process is similar to that undertaken when evaluating purchased application software.

Factors that make choosing purchased components a more complex problem than choosing complete application software include the following:

- Components may address narrow problems.
- Components require a suitable infrastructure in which to operate.
- Installing components and managing the component infrastructure require new skills.

The market for components that address specific business needs is relatively young. Some types of components (such as electronic payment processing) are available from many vendors, but others (such as basic accounting functions) are not. The components that are available tend be highly specialized. This situation will no doubt change as the market matures. But for the near future, building a large-scale system from components requires integrating solutions from multiple vendors with custom development to fill in the gaps. As with any multiple-vendor solution, compatibility can be difficult to ascertain at the outset and problematic to maintain over the long term.

Design for Reuse Most of the advantages of component-based systems arise from their reusability. Thus, object (component) reuse deserves extra attention in the analysis and design phases of the SDLC. Component reuse also has significant implications for maintaining information systems.

OO analysis and design concentrates on developing generally useful objects by modeling real-world objects and their behavior. To the extent the real world is accurately modeled, the developed objects can be reused in other systems and across programs within a system. A component-based system exploits object reuse by sharing executable objects among many applications.

When a system reuses objects by sharing source code, a perfect fit between an old object and a new system is not required. The designers of the new system are free to modify existing objects as needed to suit new requirements. Appropriate use of inheritance and overriding existing methods allow such modifications to be made relatively quickly, with maximum reuse of existing code.

In contrast, modifying an existing component to accommodate the requirements of a new system is much more complex. An existing component has already been installed and is already in use as a part of one or more information systems. Any change to the component for use in a new system may require:

- Modifying the existing component
- Developing a new component with substantial redundant functions

- Developing a new component that uses the existing component but adds some new functions

Modifying an existing component can be risky because it may break existing applications. The modified component may have subtle differences that cause errors in older applications. Thorough testing can determine whether such problems exist. But the effort required to test older systems can be substantial.

Both CORBA and DCOM support the deployment of multiple component versions. Thus, two components can use the same name but a different version number. Older applications can use the older version, and new applications can use the newer version. This ensures that each application sees the component it was originally designed to use. It also eliminates the possibilities of introducing new bugs into old applications.

But component versions have their own set of problems. Developers have more components to keep track of, and redundancy among their functions can create maintenance problems. If a redundant function is found to have a bug, then multiple components must be modified to fix the problem. One way around this problem is to design a new component version to call methods directly in the older version. The new component thus uses (wraps) the old one and adds any new or modified functions. However, this approach creates a complex chain of interdependence among component versions. The interdependence chains can be difficult to manage if there are many versions. They can also be inefficient due to excessive interaction among component versions.

The best long-term strategy is usually to modify or wrap existing components, thus avoiding the problem of redundant maintenance. Old systems can be tested with the new components and converted to use them if testing succeeds. Applications that do not function can be forced to use older components until they (or the new components) are repaired. Once all applications using the old component have been repaired, the old component can be deleted or merged with the newer version.

Designers are sometimes tempted to bypass future maintenance problems by designing components with forward-looking features. They try to guess what features not currently needed might be useful in future applications. They then add these features in the hope that future applications will be able to use existing components without modification.

Designing for the future may sound like a good strategy, but it introduces several problems, including:

- Inaccuracy and waste. It's difficult to guess what component behavior will be needed in future applications. If the designer guesses wrong, then the extra design and implementation effort is wasted.
- Cost. The extra effort adds cost to the current project and lengthens its schedule.

Component designers must walk a fine line between developing components that are general enough to be reused in the future but specific enough to be useful and efficient in a current application.

Summary

Rapid application development (RAD) is a term that covers a variety of tools, techniques, and development approaches that speed application development. RAD tools include compilers for advanced programming languages, integrated development environments, object frameworks, and components and their supporting infrastructure. RAD techniques include risk management, joint application design, tool-based design, and software reuse. RAD development approaches include prototyping, spiral development, and (under the right circumstances) traditional development life cycles.

All RAD tools, techniques, and development approaches do not speed development for all projects. Developers must carefully examine project characteristics to determine which RAD concepts are mostly likely to speed development. Some parts of RAD (such as risk management and software reuse) will speed development for most projects. Other parts of RAD (such as tool-based design and spiral development) are applicable to far fewer projects.

Software reuse is a fundamental approach to rapid development. It has a long history, although it has been applied with greater success since the advent of object-oriented programming, object frameworks, and component-based design and development. Object frameworks provide a means of reusing existing software through inheritance. Object frameworks provide a library of reusable source code, and inheritance provides a means of quickly adapting that code to new application requirements and operating environments.

Components are units of reusable executable code that behave as distributed objects. They are plugged into existing applications or combined to make new applications. Like the concept of software reuse, component-based design and implementation are not new. But the standards and infrastructure required to support component-based applications have only recently emerged. Thus, components are only now entering the mainstream of software development techniques.

Key Terms

Common Object Request Broker Architecture (CORBA), p. 505
code reuse, p. 497
component, p. 503
developmental prototype, p. 482
discovery prototype, p. 482
Distributed Component Object Model (DCOM), p. 505
foundation classes, p. 498
object framework, p. 498
Object Management Group (OMG), p. 505
prototyping development approach, p. 483
rapid application development (RAD), p. 480
risk, p. 490
risk management, p. 490
risk mitigation, p. 493
software reuse, p. 497
spiral development approach, p. 487
tool-based development, p. 495

Review Questions

1. During what life cycle phase is it least expensive to implement a requirements change? During which phase is it most expensive?
2. Is RAD a single approach or technique? Why or why not?
3. What factors determine the fastest development approach for a specific project?
4. Under what condition(s) is the sequential development approach likely to be faster than alternative development approaches?
5. Under what condition(s) is the prototyping development approach likely to be faster than alternative development approaches?
6. Under what condition(s) is the spiral development approach likely to be faster than alternative development approaches?
7. How does the process (composition and order of phases and activities) of prototype-based system development differ from a sequential development approach? How does it differ from a spiral development approach?
8. How does the process (composition and order of phases and activities) of spiral system development differ from a sequential development approach?
9. Which approach to system development (conventional, prototyping, or spiral) is likely to result in the shortest possible development time when user requirements *or* technical feasibility are poorly understood at the start of the project? Why?
10. Which approach to system development (conventional, prototyping, or spiral) is likely to result in the shortest possible development time when user requirements *and* technical feasibility are poorly understood at the start of the project? Why?
11. What development tool characteristics are required for successful use of the prototyping or spiral development approaches?
12. Why isn't the spiral development approach used for all projects?
13. Define the following terms and phrases: risk, risk management, expected delay for a single outcome, expected delay for a single risk.
14. With which development approaches (conventional, prototyping, or spiral) should risk management be used?
15. How should JAD be incorporated into the prototyping or spiral approaches to software development?
16. Describe tool-based development. With which development approaches (conventional, prototyping, or spiral) can it be used?
17. What is an object framework? How is it different from a library of components?
18. What are the differences between general-purpose and application-specific foundation classes?
19. For which layers of an OO program are off-the-shelf components most likely to be available?
20. What is a software component?
21. Why have software components only recently come into widespread use?
22. In what ways do components make software construction and maintenance faster?
23. Name two interoperability standards for software components. Compare and contrast the standards.

Thinking Critically

1. Reread the opening case at the beginning of the chapter. Describe the pros and cons of developing the system described therein with a conventional development approach, the prototyping development approach, and the spiral development approach. Which approach do you think would work best for this project?
2. Consider the capabilities of the programming language and development tools used in your most recent programming or software development class. Are they sufficiently powerful to implement developmental prototypes for single-user software on a personal computer? Are they sufficiently powerful to implement developmental prototypes in a multiuser, distributed, database-oriented, and high-security operating environment? If they were being used with a tool-based development approach, what types of user requirements might be sacrificed because they didn't match language or tool capabilities?

3. The Object Data Management Group (ODMG) was briefly described in Chapter 10. Visit the web sites of the OMG (www.omg.org) and ODMG (www.odmg.org) and gather information to answer the following questions. What is the goal or purpose of each standard? What overlap, if any, exists among the two standards? Are the standards complementary?
4. Read the article by Scott Lewandowski listed in the "Further Resources" section. Compare and contrast the CORBA and Microsoft DCOM approaches to component-based development. Which approach appears better positioned to deliver on the promises of component-based design and development? Which approach is a true implementation of distributed objects? Which approach is likely to dominate the market in the near future?
5. Compare and contrast object frameworks and components in terms of ease of modification before system installation, ease of modification after system installation, and overall cost savings due to code reuse. Which approach is likely to yield greater benefits for a unique application system (such as a distribution management system that is highly specialized to a particular company)? Which approach is likely to yield greater benefits for general-purpose application software (such as a spreadsheet or virus protection program)?
6. Assume that a project development team has identified risks, outcomes, and probabilities for a new web-based insurance pricing system as summarized in the following table. Compute the expected schedule delay for each risk and for the entire project. Which risks should be actively managed? What mitigation measures might be appropriate? Why do the outcome probabilities for some risks sum to more than 100 percent?

Risk	Risk probability	Possible outcomes	Outcome probability	Outcome delay—weeks
Change in state insurance laws or regulations requires changes to pricing algorithms	100	Update pricing for 1–10 states	60	1
		Update pricing for 11–20 states	25	2
		Update pricing for 21–30 states	10	3
		Update pricing for 31–40 states	4	4
		Update pricing for 41–50 states	1	5
Significant change in Internet security standards	30	Update server software	100	2
		Minor content update	70	4
		Major content update	30	10
Management decides to negotiate preferred service provider agreement with a major ISP (for example AOL)	25	No format/content changes required	70	0
		Cosmetic format/content changes required	20	2
		Significant format/content changes required	10	6
		Minor capacity upgrade	50	5
		Major capacity upgrade	50	12
Change in federal insurance law or regulations requires changes in pricing algorithms	10	Minor recoding	70	2
		Major recoding	30	12

TABLE 13-2

7. Consider the similarities and differences between component-based design and construction of computer hardware (such as personal computers) and design and construction of computer software. Can the "plug-compatible" nature of computer hardware ever be achieved with computer software? Does your answer depend on the type of software (for example, system or application software)? Do differences in the expected lifetime of computer hardware and software impact the applicability or desirability of component-based techniques?

Experiential Exercises

1. Talk to someone at your school or place of employment about a recent development project that was canceled due to slow development. What development approach was employed for the project? Would a different development approach have resulted in faster development?
2. Talk to someone at your school or place of employment about a recent development project that failed due to inability to satisfy user or technical requirements. Review the reasons for failure and the risk management processes that were used. What changes (if any) in the risk management process might have prevented the failure or mitigated its effects?
3. Consider a project to replace the student advisement system in use at your school with one that employs modern features (for example, web-based interfaces, instant reports of degree progress, and automatic course registration based on a long-term degree plan). Now consider how such a project would be implemented using tool-based development. Investigate alternative tools such as Visual Studio, PowerBuilder, and Oracle Forms, and determine (for each tool) what requirements would need to be compromised for the sake of development speed if the tool were chosen.
4. Examine the capabilities of a modern programming environment such as Visual C++, Visual Basic, Visual Works, or Delphi. Is an object framework or component library provided? Does successful use of the programming environment require a specific development approach? Does successful use require a specific development methodology?
5. Examine the technical description of a complex end-user software package such as Microsoft Office. In what ways, if any, was component-based software development used to build the software?

Case Study

Midwestern Power Services (MPS) provides natural gas and electricity to customers in four Midwestern states. Like most power utilities, MPS is facing the prospect of significant federal and state deregulation over the next several years. Federal deregulation has opened the floodgates of change but provided little guidance or restriction on the future shape of the industry. State legislatures in two of the states MPS serves have already begun deliberations about deregulation, and the other two states are expected to follow shortly.

The primary features of the proposed state-level deregulation are:

- Separating the supply and distribution portions of the current regulated power utility business
- Allowing customers to choose alternate suppliers regardless of what company actually distributes electricity or natural gas

The deregulation proposals seek to increase competition in electricity and natural gas by regulating only distribution. Natural gas extraction and electricity generation would be unregulated, and consumers would have a choice of wholesale suppliers. The final form of deregulation is unknown, and its exact details will probably vary from state to state.

MPS wants to get a head start on preparing its systems for deregulation. Three systems are most directly affected—one for purchasing wholesale natural gas, one for purchasing wholesale electricity, and one for billing customers for combined gas and electric services. The billing system is not currently structured to separate supply and distribution charges, and it has no direct ties to the natural gas and electricity purchasing systems. MPS's general ledger accounting system is also affected because it is used to account for MPS's own electricity-generating operations.

MPS plans to restructure its accounting, purchasing, and billing systems to match the proposed deregulation framework:

- Customer billing statements will clearly distinguish between charges for supply and distribution of both gas and electricity. Prices for supply will be determined by the wholesale suppliers of each power commodity. Revenues will be allocated to appropriate companies (such as distribution charges to MPS and supply charges to wholesale providers).
- A new payment system for wholesale suppliers will be created to capture per-customer revenues and generate payments from MPS to wholesale suppliers. Daily payments will be made electronically based on actual payments by customers.
- MPS's own electricity-generating operations will be restructured into a separate profit center, similar to other wholesale power providers. Revenues from customers who choose MPS as their electricity supplier will be matched to generation costs.

MPS's current systems were all developed internally. The general ledger accounting and natural gas purchasing systems are mainframe-based. They were developed in the mid-1980s, and incremental changes have been made ever since. All programs are written in COBOL, and DB/2 (a relational DBMS) is used for data storage and management. There are approximately 50,000 lines of COBOL code.

The billing system was rewritten from the ground up in the mid-1990s and has been slightly modified since that time. The system runs on a cluster of servers using the UNIX operating system. The latest version of Oracle (a relational DBMS) is used for data storage and management. Most of the programs are written in C++, although some are written in C and some use Oracle Forms. There are approximately 80,000 lines of C and C++ code.

MPS has a network that is used primarily to support terminal-to-host communications, Internet access, and printer and file sharing for microcomputers. The billing system relies on the network for communication among servers in the cluster. The mainframe that supports the accounting and purchasing systems is connected to the network, although that connection is primarily used to back up data files and software to a remote location. The company has experimented with web-browser interfaces for telephone customer support and on-line statements. However, no functioning web-based systems have been completed or installed.

MPS is currently in the early stages of planning the system upgrades. It has not yet committed to specific technologies or development approaches. It has also not yet decided whether individual systems will be upgraded or entirely replaced. The target date for completing all system modifications is three years from now, but the company is actively seeking ways to shorten that schedule.

1. Describe the pros and cons of the traditional, prototyping, and spiral development approaches to upgrading the existing systems or developing new ones. Do the pros and cons change if the systems are replaced instead of upgraded? Do the pros and cons vary by system? If so, should different development approaches be used for each system?

2. Is tool-based development a viable development approach for any of the systems? If so, suggest tools that might be appropriate. For each tool suggested, identify the types of requirements likely to be sacrificed due to a poor match to tool capabilities.

3. Assume that all systems will be replaced with custom-developed software. Will an object framework be valuable for implementing the replacements? Is an application-specific framework likely to be available from a third party? Why or why not?

4. Assume that all systems will replaced with custom-developed software. Should component-based design and development be actively pursued? Why or why not? Does MPS have sufficient skills and infrastructure to implement a component-based system? If not, what skills and infrastructure are lacking?

5. List the risks and risk outcomes that may affect the planned upgrade or replacement. Which risks require active management? What mitigation measures should be pursued?

Further Resources

Barry W. Boehm. "A Spiral Model of Software Development and Enhancement," *Computer*, May 1988, pp. 61–72.

Frederick P. Brooks, Jr. "No Silver Bullet: Essence and Accidents of Software Engineering," *Computer*, April 1987, pp.10–19.

Scott M. Lewandowski. "Frameworks for Component-Based Client/Server Computing," *ACM Computing Surveys*, volume 30:1, March 1998, pp. 3–27.

Steve McConnell. *Rapid Development: Taming Wild Software Schedules*. Microsoft Press, 1996, ISBN 1-55615-900-5.

Robert Orfali, Dan Harkey, and Jeri Edwards. *The Essential Client/Server Survival Guide* (2nd ed.). John Wiley & Sons, 1996, ISBN 0-471-15325-7.

Steve Sparks, Kevin Benner, and Chris Faris. "Managing Object-Oriented Framework Reuse," *Computer*, volume 29:9, September 1996, pp. 53–61.

Object Data Management Group (developer of an object database management standard) home page, http://www.odmg.org.

Jane Wood, and Denise Silver. *Joint Application Development* (2nd ed.). John Wiley & Sons, 1995, ISBN 0-471-04299-4.

Packaged Software and Enterprise Resource Planning

Learning Objectives

After reading this chapter, you should be able to:

- Discuss three major analysis and design issues associated more often with packages than with custom-developed software
- Name the three types of package customization and explain when each would be necessary
- Identify and explain the four major characteristics of ERP systems and discuss several reasons for adopting ERP
- Identify and explain several critical success factors for ERP implementation
- Distinguish the three major approaches to ERP system development
- Discuss how methods used for ERP systems deployment differ from those for conventional system development
- Identify and explain the four major steps to ERP package selection
- Discuss the major areas of expansion for ERP
- Identify the four high-level work areas of SAP R/3
- Identify and discuss the five types of master data in SAP R/3
- Describe the function of R/3 Basis software
- Describe how transactions are entered in SAP R/3

Chapter Outline

Packaged Software

Enterprise Resource Planning

A Closer Look at One ERP Package: SAP R/3

FOURTEEN

PIZZA AMORE: REVAMPING SYSTEMS TO MANAGE MORE INFORMATION

Pizza Amore began as a simple pizza parlor in 1965. In 15 years, it has expanded to deliver 338 million pizzas from more than 6,200 stores in 65 international markets, totaling $3.2 billion in sales (more than 10 percent of the pizza market). To handle this enormous size and growth, the chain determined it had to undertake several business process reengineering (BPR) and enterprise resource planning (ERP) initiatives.

Much of the emphasis of BPR and ERP focuses on Pizza Amore's supply chain management. Most people would not think of a pizza parlor as a manufacturing company, but Pizza Amore produces some 4.2 million pounds of dough in 18 distribution centers each week, and delivers this dough (with another 150 kinds of food and paper products) to 4,500 U.S. franchises using a fleet of 160 trucks. Many kinds of software packages are employed to manage this complexity. Pizza Amore uses forecasting software from Prescient Systems, Inc., to anticipate demand. An application from Manugistics, Inc., schedules and routes delivery trucks. An on-board computer system from Kronos, Inc., feeds data into a time-and-attendance application. Pizza Amore also uses extensive data warehousing and data mining techniques with tools from Cognos, Inc., and Hyperion Solutions Corp. to anticipate changes in its market. As Pizza Amore's vice president of supply chain management, explains, "We're trying to shift the way we manage our business, from an entrepreneurial, seat-of-the-pants way to a more information-based way."

One challenge Pizza Amore's staff faced was to get all these applications to seamlessly interact within one comprehensive system. To achieve this goal, the company launched an initiative to (1) replace its legacy systems with enterprise technology, (2) ensure Year 2000 compliance of financial systems, and (3) streamline its product distribution throughout the organization. The company's existing systems had limited functionality and were difficult to manage and synchronize. Alex Fernandez was appointed head of the team charged with the design and installation of Pizza Amore's ERP system. Fernandez and his team investigated many alternatives, eventually settling on PeopleSoft because of the tools available to manage Pizza Amore's supply chain and overall business environment.

By integrating with Pizza Amore's current in-store systems and providing on-line analytical processing (OLAP), PeopleSoft provides increased information access and analysis for better decision making across the enterprise. The first phase of PeopleSoft 7.5 implementation concentrates on core transactional functionality and was scheduled for startup in October 1998. Additional phases will be geared toward improving the efficiency of its many subprocesses. Pizza Amore will continually improve upon its solution with additional enhancements and new releases from PeopleSoft, including advanced client-server and network computing technologies, and utilization of its corporate intranet.

But the path to complete ERP is not easy. Some of the problems encountered by Pizza Amore stem from the fact that PeopleSoft still has some catching up to do with the rest of the ERP field. An analyst with Forrester Research says that rivals SAP, Oracle, and Baan offer more mature, functionally rich manufacturing products. Pizza Amore was already using many diverse application packages before starting its PeopleSoft ERP rollout. Systems analysts discovered that PeopleSoft often didn't accommodate the data fields used by the other packages. For example, the truck routing system used a data field containing the delivery stop sequence that needed to be conveyed to the warehousing system for loading products on trucks. Pizza Amore decided to take the risky step of modifying the core PeopleSoft modules to include such fields, depending heavily on the integration experience of its IT staff to build one-to-one interfaces between its ERP system and its bolt-ons. Fernandez chose not to rely on PeopleSoft to perform the integration, explaining, "If you keep control over the developers and you know what's been done, when you need to upgrade, it's not that tough."

OVERVIEW

The information systems of large companies are often very complex. Even though Rocky Mountain Outfitters (RMO) is a relatively small company, it has many of the same needs as Pizza Amore, needs that can possibly be addressed with stand-alone software packages, or even enterprise resource planning (ERP), instead of custom-developed software. This chapter begins by looking at the key issues of working with application packages in general, followed by a survey of ERP: what it is, why it should be considered important, and how it is implemented. Finally, a particular ERP package, SAP R/3, will be examined to give you a better understanding of how ERP really works.

Figure 14-1 illustrates a computer-based information system consisting of many tightly linked components (hardware, software, data, networks, procedures, and people) working together to provide the information needs of one or more business functions. The process of systems analysis and design must adequately address all these components to satisfy the needs of users—the ultimate judges of system success. The focus of this text has been on the analysis and design of business application software—computer programs created to solve specific business problems. At issue in this chapter are two major questions related to business application software. First, what is the best way to procure this software? Second, after the software is obtained, how is it best utilized to achieve the objectives of the firm? The way in which individual companies, large or small, answer these questions has a tremendous impact on the future of IS organizations, the future of the enterprises they serve, and the future of the global economy.

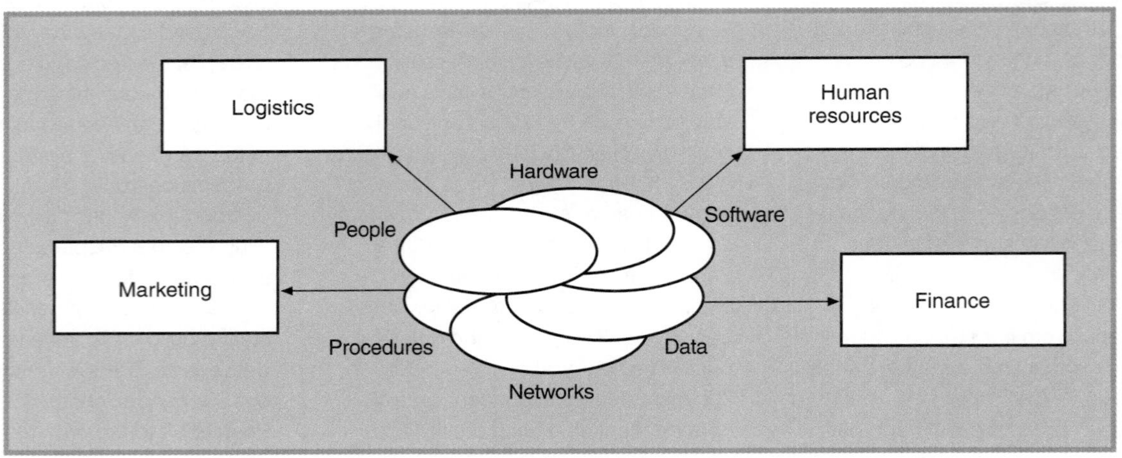

FIGURE 14-1
IS components serving multiple business functions.

Let's consider the first question: How is software best procured? Throughout the early history of business computing, applications consisted almost entirely of custom-developed software—software created on demand according to the unique specifications of the intended users. The two parties primarily involved in system creation were (1) in-house developers and (2) in-house users. As the field of information systems matured, enterprising individuals realized that certain business applications, such as financial accounting functions, were fairly standardized. They began to create, as third-party developers outside the firm, packaged software designed to perform common or standardized functions and to be sold to a large number of potential customers with little or no modification. The continuing demand for rapidly deployable, functional, and affordable software still drives third-party developers to explore new applications for software packages, applications founded on the "best practices" (a set of fairly standard procedures commonly accepted as the optimum way of performing a given function) in various industries. Such packages enable firms to avoid "reinventing the wheel" every time a new application is needed. Therefore, some systems analysts maintain that the best way to procure software is by purchasing a package, provided the package can meet most, if not all, user requirements.

Now for the second question: How is application software best utilized in a business environment? Should software be used only to satisfy the needs of individual, isolated departments, or is the entire organization best served when the information produced in one department is readily available for use in another? The **integration** of application software, a process in which separate components or subsystems are combined and problems in their interactions are addressed, has long been a goal of many organizations. A rudimentary form of integration is often achieved through batch processing, but that approach delays the availability of critical information. Recent advances in client-server technology and telecommunications have fueled a trend toward integrating business applications in real time across departments, functions, and divisions of corporations, and even between the firm and its external suppliers and customers. The combined migration toward packaged software and real-time integration has spawned a new paradigm in management information systems—enterprise resource planning (ERP), the process of using integrated software applications to achieve departmental and enterprisewide objectives. Several ERP system development firms—such as SAP, PeopleSoft, Oracle, Baan, and J.D. Edwards—are currently using this paradigm.

integration
The process of combining separately produced components or subsystems and addressing problems in their interactions

The key players in enterprise computing that will be germane to our exploration of packages and ERP are shown in Figure 14-2. As head of IS at Rocky Mountain Outfitters, Mac Preston is keenly aware that an ERP approach to system development is a definite possibility for RMO.

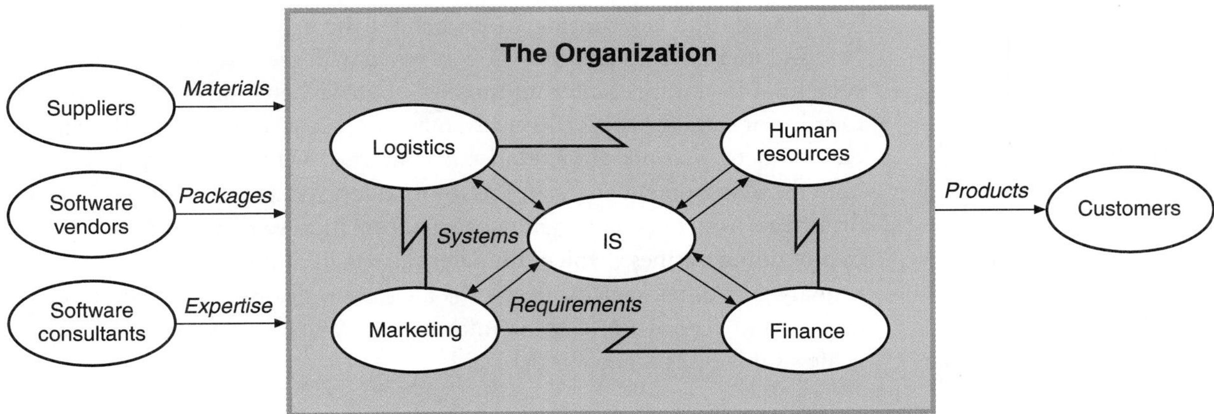

FIGURE 14-2
The key players in enterprise computing.

Packaged Software

As we've already discussed, packaged software traditionally has been focused on performing isolated business functions. More recently, the emergence of packaged suites of integrated application software, better known as enterprise resource planning (ERP) systems, has completely reengineered the information systems function in many large corporations. Stand-alone packages cannot share information unless

system integrator
An individual who is specially trained in integrating application software across various computing platforms

forced to do so through the efforts of **system integrators**, individuals who are specially trained in integrating application software across various computing platforms. With an ERP system, however, the advanced application software for all functional departments in an organization can easily share information, enabling tremendous improvement in effectiveness and efficiency. Organizations are now faced with critical decisions concerning the futures of their information systems. Should they continue to maintain their existing legacy systems or scrap them entirely? If they plan to acquire new application software, should it be through continued custom development or by purchasing packages? If packages are involved, should firms integrate separate vendors' products, or should they purchase a comprehensive ERP system? Or, should organizations pursue a combination of custom development, packages, and a purchased ERP system? In the case of RMO, it plans to develop custom software and purchase packaged software. The company also plans to integrate several of its key applications. However, RMO has not yet addressed a possible need for ERP. Obviously, the choices are not easy or simple.

The Trend toward Packaged Software

As discussed in Chapter 8, custom development may be preferred when a package cannot provide the required functionality. However, current trends appear to favor the packaged software solution. A recent survey reported that during a one-year period, spending on new in-house application development had dropped 34 percent, while spending on software packages had doubled. If this trend continues, how will in-house system development be affected? Obviously, the decision to purchase a software package, in lieu of custom development, can have far-reaching consequences for the information technology (IT) function within an organization and for the organization as a whole.

What are the key considerations behind the acquisition of packaged software? First and foremost, an enterprise is concerned with the *effectiveness* of its information systems. A firm wants *quality* information to make better business decisions, perhaps to create or sustain a competitive advantage, or just to survive in a turbulent economic environment. Second, an enterprise is concerned with the *efficiency* associated with both the development and use of its information systems. The efficiency with which a firm creates, operates, and maintains its information systems contributes greatly to its cost of doing business, which ultimately affects its ability to compete or survive. The primary goal of packaged software, vis-à-vis custom software, is both to improve the quality and reduce the cost of information. Whether packaged software actually achieves this two-part goal is debatable.

Implementation and Support of Packaged Software

Chapter 8 introduced the issues of analysis and design when packaged software is part of the solution. In the analysis phase, user requirements for the system are defined. If a package can be found that meets all the critical requirements and the vast majority of the users' preferences, it will be seriously considered. In the design phase, many more detailed issues are addressed, such as what computing platform is required for the package, how users will be trained to use the new package, and how changeover from the existing system will take place. These types of issues actually differ little from those that arise from custom development.

However, three critical considerations of analysis and design are unique to packages: customization, integration, and future upgrades. First, a software package quite often fails to meet all the requirements or preferences identified in the analysis phase.

Therefore, the package may require some type of customization before it is acceptable for installation and use. Second, a software package, as part of a larger system, must often receive data from several systems, either custom or packaged, and provide data to still others. The package may therefore require integration with existing legacy applications or databases, or with other new packages. Third, a package will undoubtedly undergo at least one upgrade by the vendor during the life span of the overall system. Customization, integration, and upgrades of packages must be addressed in planning and analysis, thoroughly planned in design, flawlessly executed in implementation, and continually monitored in support. These relationships are illustrated in Figure 14-3. For example, RMO is planning to acquire accounting/finance and human resources packages. The staff should address the cost of future package upgrades in the planning phase and then must successfully install the upgrades in the support phase. The following sections discuss the three major considerations of packaged software in detail.

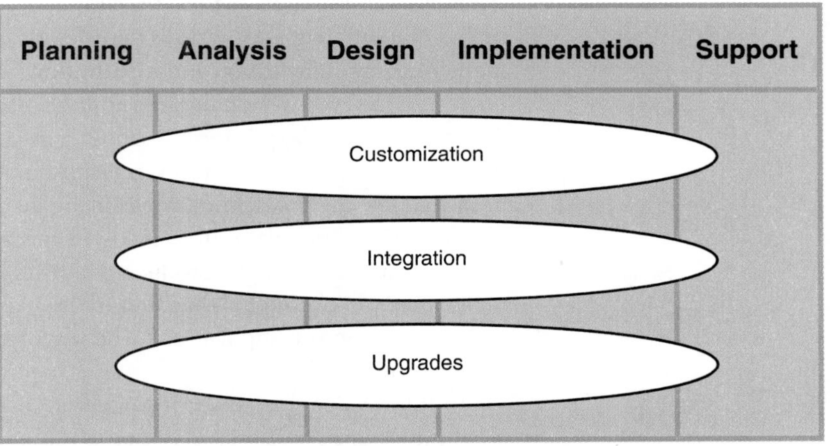

FIGURE 14-3
Three critical issues of packaged software spanning all phases of the SDLC.

Customizing Packages Packages are usually designed with the hope that little or no customization will be required, since the best practices of the industry or business function and an abundance of options and features are often built in. However, users may desire to customize a package to meet personal preferences or satisfy unique functional requirements. Since excessive customization can lead to problems later in the system development life cycle, it should be considered only with foresight and careful planning.

There are three types of customization: configuration, modification, and enhancement. **Configuration** involves selecting from an assortment of options, such as assembling software modules into a complete system or defining values of parameters in a software package. Configuration can be as simple as changing the format of interfaces to suit a user's preference or adding company-specific details to printed invoices. On the other hand, configuration can be as complex as integrating several different package modules, possibly from different vendors. The in-house developer can usually perform simple configuration tasks through utilities provided with the package. The assistance of a package's vendor or an experienced third-party consultant may be needed for more complex configuration, especially in the case of ERP systems.

configuration
The process of selecting from an assortment of options, such as assembling software modules into a complete system or defining values of parameters in a software package

Modification of a package involves changing program code to alter the existing form of processing. In many cases, the in-house developer may not have access to the program code of the package. In other cases, the program code may be accessible, but the package vendor may discourage and even fail to support modification. Modifications made to an existing package can cause problems when the user later decides to install an upgraded version.

modification
The process of changing program code to alter the previously existing form of processing

enhancement
The process of adding program code or program modules to provide additional functionality to software

Enhancement involves adding program code or program modules to provide additional functionality to software. Enhancement does not alter the existing processes in the package—it just adds to them. For example, a comprehensive ERP system may be enhanced with a custom-developed forecasting module.

Some form of customization is nearly always required for large, complex packaged solutions. Most of the customization involves configuration or enhancement, not modification. Consultants who install ERP systems often do extensive configuration of various modules as requested by the customer. To facilitate the process, a package should support interoperability with legacy and other systems through standards such as Microsoft's Distributed Component Object Model (DCOM) or OMG's Common Object Request Broker Architecture (CORBA). Firms that develop and market packaged solutions usually emphasize creating reusable components that are easily customizable by deployment teams.

Integrating Packages Information systems of the past were implemented primarily as "islands of automation" within functional departments. A company would ordinarily run several different applications in different areas, such as human resources, finance, marketing, manufacturing, and distribution, without giving much thought to shared interests. However, it soon became apparent that an application in one area may need data created by an application in another area. For example, a manufacturing department in Mexico and a distribution facility in Australia may both need sales forecasts produced by the marketing department at corporate headquarters in the United States. Sharing such data often involved having the IS department create and maintain separate programs to produce data files with specific contents and formats, to mail data tapes overnight when and where needed, to load data into local computers, and then to run the applications. The disadvantages of such an approach to sharing information are obvious.

To improve performance, organizations are attempting to integrate many of their most important applications so that all functions can access needed data in a standardized format in real time. Doing so should significantly improve the efficiency of operations and the effectiveness of decision making. The need for real-time application integration is even greater when firms are seeking to expand into electronic commerce, extend supply chains, compete globally, and enhance customer service. A recent survey reports four major obstacles to integration, as shown in Figure 14-4. Note that the most formidable barriers are technical, due to a shortage of knowledgeable system integrators and the inherent difficulty of integrating disparate applications. However, other obstacles (getting executive buy-in and building consensus) are organizational in nature.

enterprise application integration (EAI)
The process of linking applications to support the flow of information across multiple business units and IT systems

The current trend toward sharing data among what were once islands of automation is called **enterprise application integration** (EAI), the process of linking applications to support the flow of information across multiple business units and IT systems. There are essentially two ways to integrate applications: (1) Use middleware to link individual packages, or (2) purchase a comprehensive ERP package that already has the middleware built in. (ERP packages are discussed in depth later in this chapter.)

As we discussed previously, middleware is special software that sits between different applications and translates and routes data between them. Middleware can be obtained by purchasing a packaged integration tool or through custom development. A recent survey indicated that as many as 73 percent of large companies are currently involved in real-time application integration using a combination of commercial integration tools and custom coding. Custom development of middleware has its drawbacks, since it is very costly and the resulting programs are very difficult to maintain.

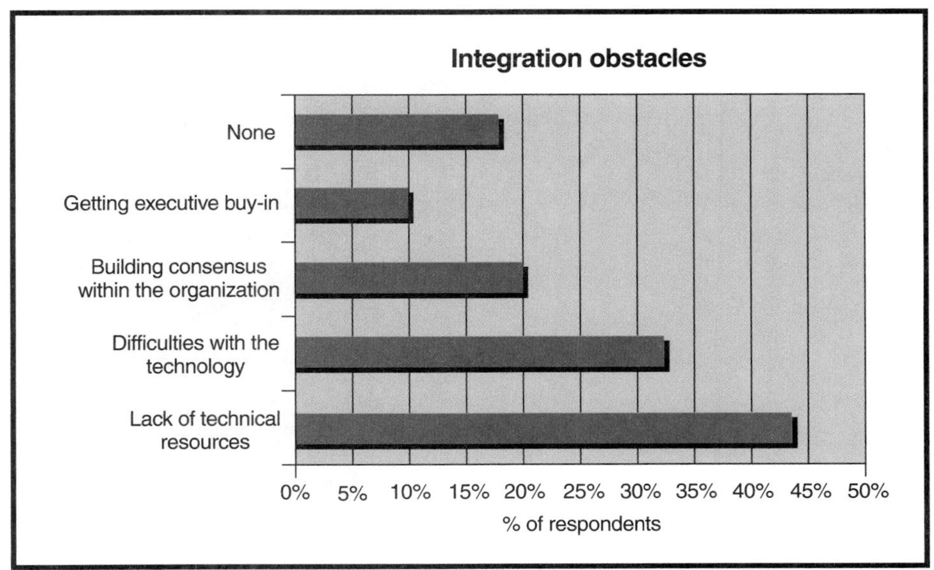

FIGURE 14-4
Obstacles to integration.

Upgrading Packages Software vendors are notorious for releasing frequent upgrades of their packages. An upgrade usually results from the addition of functions or from improved ease of use. Sometimes a new release, or patch, is needed to fix a bug in the previous version of the package. Upgrades are often fairly straightforward for in-house staff or vendors, because the upgrade is designed to be compatible with the previous version. However, when extensive customization and integration have occurred with previous versions at the user's site, upgrades can be particularly challenging. This is why customization and integration must be carefully planned and documented during system development.

Enterprise Resource Planning

The previous section discussed general issues of software packages. Once a package is acquired as part of an information system, it may be customized, integrated, or upgraded within a single function or department of an organization. This section explores information systems that are designed to reach beyond the needs of a single department. The goal of such systems is to improve the operation and competitiveness of an entire organization, not just one or a few departments. Such software is at the hub of what is called enterprise resource planning (ERP). This section explores what ERP is, why firms choose ERP, how ERP is implemented, and how ERP functions.

RMO plans to integrate its customer support system, inventory management system, and retail store system, but it has not addressed whether to integrate these systems with its new accounting/finance and human resources packages. Successfully doing so would essentially constitute an ERP system.

What Is ERP?

Enterprise resource planning (ERP) is the process of using integrated application software to improve the effectiveness and efficiency of the entire enterprise, thereby contributing to a firm's competitive advantage. ERP is thought by many to be the way that virtually all firms will operate business information systems—such as transaction processing systems, management information systems, decision support systems, and artificial intelligence/expert systems—in the new millennium. Figure 14-5 illustrates the basic differences between nonintegrated ISs and ISs in an ERP system: In the former, only one function segregates and accesses data; in the latter, all functions can quickly and easily share a common database.

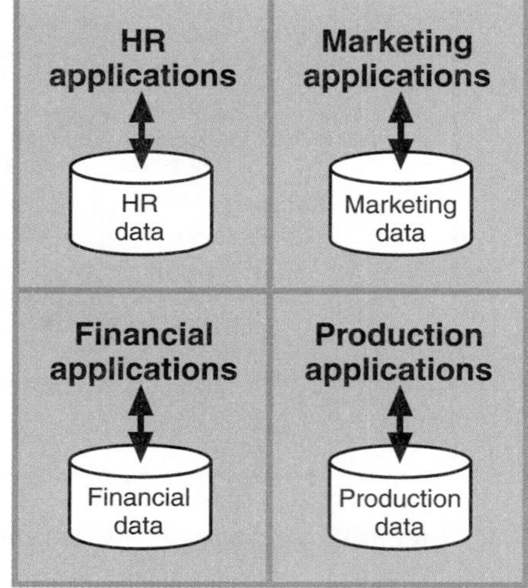

FIGURE 14-5a
Nonintegrated IS.

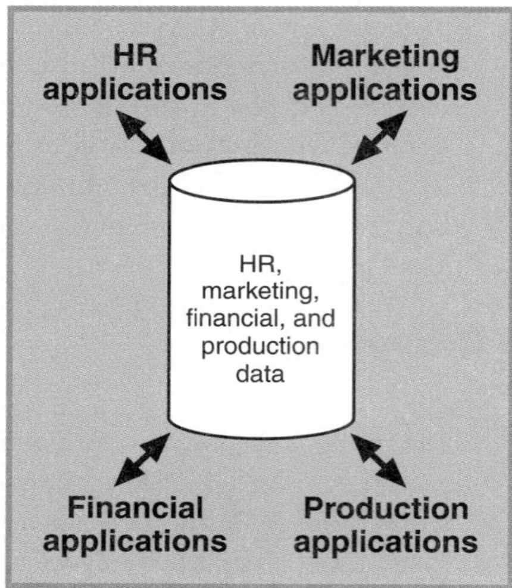

FIGURE 14-5b
Integrated IS in an ERP.

The Purpose of ERP ERP is designed to provide best practice functionality within departments and a high level of integration across all enterprise functions, such as human resources, finance, and manufacturing. Not only is ERP designed to integrate functions, it is also designed to integrate organizational units, such as plants and divisions, across the country or around the world. Going one step further, ERP has the capability of integrating previously separate and independent organizations—that is, integrating a firm with its suppliers and customers. Thus, ERP is intended to help integrate the entire **value chain**, the sequence of all processes within an organization that add value to the product or service.

value chain
The sequence of all processes within an organization that add value to the product or service offered

Additionally, ERP can bridge the gaps in the value chain in real time, making critical data available to a host of different users (inside or outside the firm) exactly when needed. Thus, ERP is designed to break down significant barriers of function, space, and time. RMO could use ERP to link its financial, human resource, manufacturing, inventory, retail, and distribution systems. Managers could instantly see sales and inventory at the Park City and Denver retail outlets or examine the performance of mail order, phone order, or Internet sales. ERP could tie manufacturing plants in Salt Lake City and Portland directly to all these sales systems and to the inventory systems for the Salt Lake City, Portland, and Albuquerque distribution centers. To plan the training and movement of personnel within the various manufacturing, distribution, and sales functions, RMO could integrate these functions with the human resource system, and instantly update the bottom line by integrating all these systems with the financial system. An ERP approach to doing business can be a powerful competitive weapon.

ERP and Business Information Systems There are many types of business information systems within organizations, including transaction processing systems (TPS) (for example, order processing, payroll, and inventory control), management information systems (MIS) (sales forecasting, production planning, and assembly line scheduling), and decision support systems (DSS) (facilities planning, product promotion, and distribution). ERP is often used first to integrate the core TPS in human resources, finance, and manufacturing and later to integrate the MIS and DSS of various departments. Not only does ERP integrate functional departments, but it does so on-line in real time, not a day or week later. ERP can span functional departments within a single facility as well as geographical boundaries of large, multinational corporations. ERP can reach beyond the boundaries of the firm to include external suppliers and customers in the information system, further extending the firm's supply chain. These aspects of ERP are illustrated in Figure 14-6. The integrated, real-time, global, and extended characteristics of the information obtained through the TPS of an ERP system make the MIS and DSS of the organization more valuable. Through ERP, the firm can make faster and better decisions to become more effective, efficient, and, ultimately, competitive.

FIGURE 14-6
Four major characteristics of an ERP system.

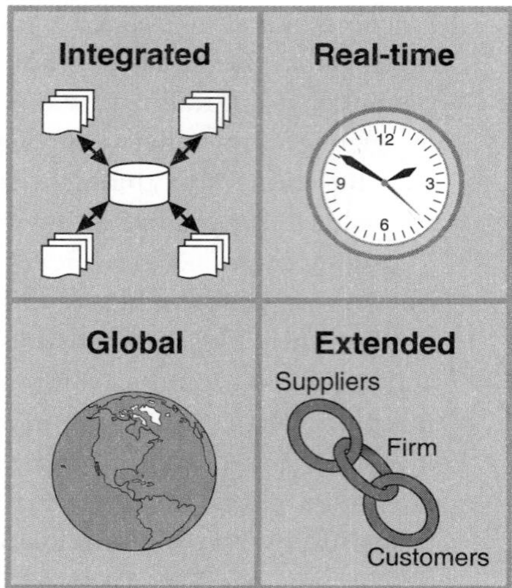

customer-relationship management
The process of meeting the needs of all customers from initial inquiry to support after the sale

supply-chain management
Managing all resources and processes involved from raw-material acquisition to finished goods delivery

sales-force automation
Processes that make it easier for sales personnel to service customer accounts

e-commerce
The buying and selling of goods and services on the Internet, especially the World Wide Web

business intelligence
The acquisition and use of information designed to improve competitive advantage

total quality management (TQM)
The process of achieving the highest levels of quality throughout the entire organization

Major ERP Vendors The major players in the packaged ERP market are (in descending total revenue order) SAP AG, PeopleSoft, Oracle, J.D. Edwards, and Baan. These companies have exhibited phenomenal growth in recent years, although that growth now appears to be slowing somewhat. In general, these ERP vendors began offering integrated packages for managing the three core business functions: finance, manufacturing, and human resources. Recently, they have expanded by offering added functions in **customer-relationship management** (meeting the needs of all customers from initial inquiry to support after the sale), **supply-chain management** (managing all resources and processes involved from raw-material acquisition to finished goods delivery), **sales-force automation** (helping sales personnel to service customer accounts), **e-commerce** (buying and selling goods and services on the Internet, especially the World Wide Web), and **business intelligence** (acquiring and using information designed to improve competitiveness).

Why Consider ERP?

The changing climate of business and advances in information technology are leading many firms away from the legacy systems of the past and toward the packaged systems of today. Recently, businesses have experienced trends toward mergers, government deregulation, globalization, and business process reengineering (including downsizing). These business pressures cause firms to search for new ways to compete, such as through **total quality management** (TQM), the process of achieving the highest levels of quality throughout the entire organization, and business process reengineering (BPR), the process of making radical, large-scale improvements in business processes to improve efficiency and effectiveness significantly. Also, recent advancements in client-server technology, object-orientation, telecommunications, and the Internet have opened new opportunities for applying software solutions to business problems. All these changes have combined to open the door for advanced software packages to take over wherever homespun legacy systems may be falling short.

Reasons for Implementing ERP Firms are constantly looking for ways to survive and thrive in a complex competitive environment. Many firms are losing their competitive edge by relying on systems that are approaching obsolescence. From a broad perspective, there are several reasons that firms are turning to ERP. First, many believe that information technology (IT) is an excellent way to enhance their competitiveness. They may, however, question their ability or desire to continue to maintain a staff that is expert in all areas of increasingly complex technologies. ERP is one means of updating a firm's information systems without investing in in-house software development. Second, some firms go beyond the goal of simply improving IT and subscribe to the philosophy of business process reengineering. These firms may choose to implement ERP and reengineer their processes to match the model of the ERP system. Third, reorganization through mergers or expansion into global markets has motivated some firms to integrate geographically dispersed systems within one ERP system. All these reasons essentially point to the need to replace outdated technologies.

From a business's perspective, there are several reasons to embrace ERP. Based on several case studies, the most common reasons given for ERP projects in major companies are:

- Reducing the workforce in core transaction processing systems (such as order processing and inventory management) by eliminating manual tasks and duplication
- Supporting global business operations (such as one order-processing system for all customers worldwide)
- Achieving economies of scale by replacing division-level systems (such as corporatewide versus plantwide purchasing)
- Reducing information system development staff by using packaged solutions
- Improving customer service through better logistics based on more complete and accessible information
- Improving data integrity through a common database
- Improving decision support via more timely reports and drill-down capability

Costs and Benefits Investing in an ERP package may cost several hundred thousand dollars for a midsize firm (one with annual sales of about $200 million to $1 billion) and several million dollars for a large corporation (one with annual sales greater than about $1 billion). The cost of the software itself is usually very competitive among the major vendors, but the cost of consulting services can be quite high, primarily due to the shortage of qualified people. The payback is usually estimated at between 6 and 30 months, depending on the size and complexity of the system. Savings are usually based on an estimated 30 percent reduction in administrative and information systems costs. Payback can also be calculated in terms of the ability to continue to compete in the long term against firms that are currently upgrading their information systems with ERP. An enterprise should be able to generate revenue more effectively by providing better data more quickly for decision support, by creating products at lower cost, and by offering better service to suppliers and customers.

The firm must ultimately look at the justification for ERP by comparing the current information system to the proposed ERP system. The annual cost of operation and maintenance of the existing system should be compared to the initial cost of installing the ERP system plus its annual cost of operation and maintenance. Although its benefits are difficult to quantify, the ERP system should add to revenue by enabling better and faster decisions and providing better customer service. RMO may be unable to survive in the highly competitive outerwear industry without extremely fast, effective, and integrated information systems, such as those provided by ERP.

Implementing ERP

Since ERP systems are so complex and they affect the entire organization, it is critical to implement them systematically and carefully. So, what are the factors that can help ensure success?

Critical Success Factors of ERP ERP projects can be extremely complex, time consuming, and expensive, and there are always high levels of risk involved in overhauling an information system. Based on several case studies to date, some of the most important factors critical for successful implementation are:

- Strong top management support
- Centralized project management with a *business* leader (as opposed to an IT leader) in charge
- Strong IT management and staff support
- Heavy user involvement (from *all* affected areas) in project management and implementation
- Business process reengineering to standardize on the capabilities of the software, rather than customization of the software to existing business processes
- Retraining of existing software developers in ERP implementation and maintenance
- Extensive training of end users in the new system
- Use of consultants (knowledgeable in specific business functions and experienced in specific ERP modules) to lead in implementation and training
- A respected and effective champion for ERP within the organization
- Effective and continuous communication among all parties involved in the ERP project
- Top-notch systems analysts with excellent business knowledge and excellent technical skills
- Retaining of these top-notch systems analysts throughout the project and beyond
- Sensitivity to user resistance to new systems

Approaches to ERP In addition to recognizing the critical success factors for ERP implementation, an organization should understand that there are essentially three approaches (see Figure 14-7). One option is to purchase a comprehensive packaged system, such as those offered by the leading ERP vendors (SAP, PeopleSoft, Oracle, J.D. Edwards, Baan, and so on). Such systems attempt to cover the information needs of all critical business functions (finance, manufacturing, human resources, marketing, and distribution). This approach may be best for the enterprise as a whole, but some modules that attempt to provide needed functionality to certain departments may be lacking. The comprehensive package may also lack the desired flexibility to accommodate changes rapidly in the way a firm prefers doing business.

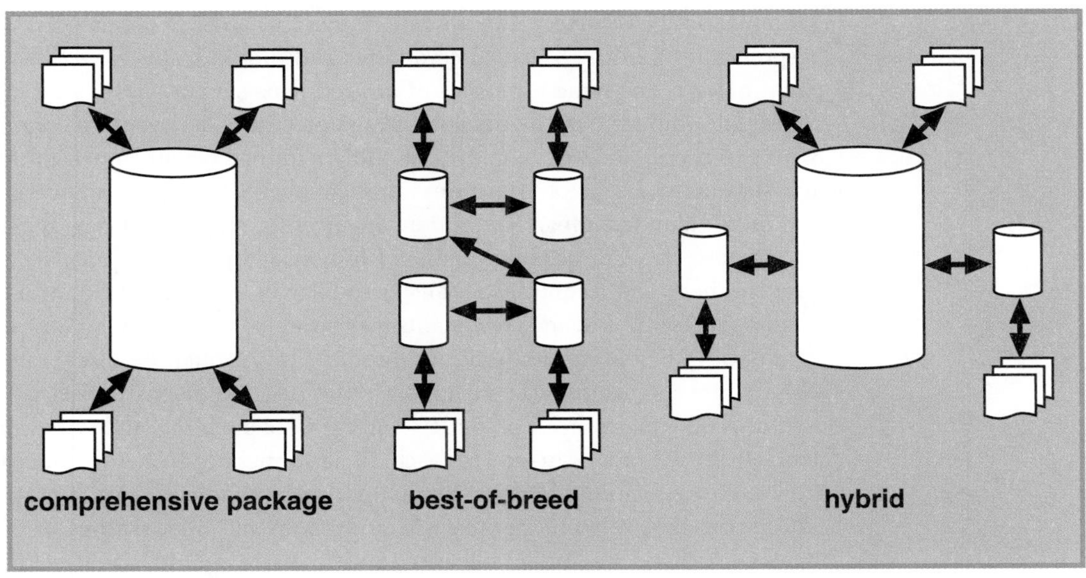

FIGURE 14-7
Three approaches to ERP.

best-of-breed
A software package developed for a particular application that is usually considered to incorporate the most current and best practices within the industry

A second approach is known as **best-of-breed**, in which the presumed best available software (custom or packaged) is chosen for a particular application (such as production scheduling or e-commerce) and is integrated with other best-of-breed solutions in other functional areas. This approach provides each individual function or department with the latest and greatest software solution. Although each business function may be extremely well served under the best-of-breed approach, the company as a whole may suffer due to a lack of integration, since disparate packages are often not designed to integrate easily with each other. However, the availability of middleware (software tools to link applications) can make best-of-breed a viable option. With the use of middleware and a skilled systems integrator (either an in-house developer or a consultant), functional areas can use the best software for their application and still share important data with other functions in the firm. Currently, RMO is banking on this approach by purchasing packages for accounting/finance and human resources; developing its own systems for customer support, inventory management, and retail stores; and integrating the latter three systems.

A third approach is a hybrid of the comprehensive package and best-of-breed approaches. A particular comprehensive package may have many desirable features but may not provide all the functionality needed by the organization. For example, a firm may choose a well-known ERP vendor to provide a backbone system for, human resources, manufacturing, and finance, but plug in various best-of-breed solutions for supply-chain or sales forecasting applications. Instead of having myriad best-of-breed applications attempting to interconnect, the firm can connect a smaller number of applications to the ERP backbone system, thus simplifying integration. The hybrid approach may be necessary if a company thinks that it cannot afford to modify certain key business processes significantly to fit the process supported by a comprehensive package. The hybrid approach is also recommended if a firm believes it needs to develop its own software in certain areas to provide unique effectiveness and achieve a competitive edge. This custom software can be added to the package of core applications that provide basic efficiencies.

Organizational Structure for an ERP Project Chapter 2 discussed the roles and responsibilities of the oversight committee and project teams for a system development project. The same principles of project management apply to an ERP project. Since ERP would usually be considered mission critical, an oversight committee, consisting of clients and key executives, should be formed. The members of the committee should represent all functions affected by ERP: finance, marketing, human resources, manufacturing, distribution, engineering, and information systems.

The project team usually consists of in-house IS personnel (managers, analysts, programmers, and technical specialists) and key in-house users. For an ERP project, vendor representatives and consultants may be added to the team because of the complexity of ERP software. In-house IS personnel must often be hired and trained to help ensure successful implementation. With the comprehensive package approach, the focus of IS personnel is on understanding, implementing, and supporting the system. Under the best-of-breed approach, IS personnel need a better understanding of how to create custom middleware or use packaged middleware to integrate the application packages. A hybrid approach requires both an understanding of the comprehensive package and of middleware. System users should also be educated and trained in all aspects of ERP that affect their jobs.

Since ERP projects are often large and extensive, an organization should provide comfortable meeting rooms to accommodate committee and team meetings; work spaces for project managers, team members, and consultants; computers, telephones, desks, chairs, bookshelves, and file cabinets; and administrative support to make arrangements for travel, lodging, meals, and meetings. A hypothetical organizational structure for an ERP project is provided in Figure 14-8.

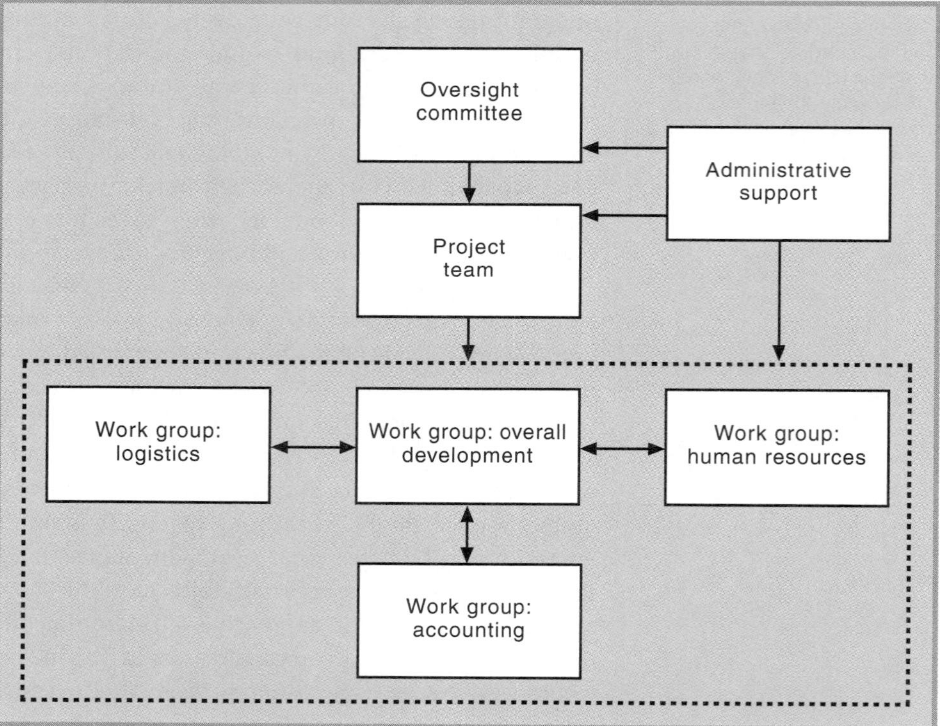

FIGURE 14-8
Organizational structure for an ERP project.

RMO, with its $100 million annual sales, would be considered a small company. If RMO considers an ERP project, the oversight committee would consist of its president and three vice presidents. The project team could include the assistant vice president of production, director of operations, director of catalog sales, assistant vice president of accounting and finance, and assistant vice president of systems. Various work groups could include systems personnel and managers and operators from manufacturing, distribution, purchasing, warehousing, retail sales, telephone sales, human resources, and accounting. The project team and work groups could also include vendor representatives and consultants.

An ERP System Development Methodology

There are many horror stories of failed attempts to implement ERP packages, regardless of vendor. Such failures are usually due to poor project management practices. As with any other system development project, a well-defined methodology is essential for successful implementation and continued, problem-free operation. The system development life cycle, presented in Chapter 2, is just as applicable to a comprehensive or best-of-breed ERP system as to any custom-developed system.

An ERP project differs from a more typical system development project in two important ways. First, an ERP project has a much greater *scope* than a typical project, involving virtually every function or department in the organization. Second, the ERP project often uses packaged software extensively, either a comprehensive package or several best-of-breed packages, instead of custom software. These important differences affect the system development life cycle in many ways.

Phase 1: Planning An oversight committee is formed to initiate the project. The primary task of the oversight committee is to identify the general purpose and overall scope of the ERP project and to appoint a project manager and project team members to develop the system.

The project team clearly defines the problem to be addressed by the ERP system and determines the detailed scope of the project. The team must then evaluate alternative approaches to ERP—custom solution, comprehensive package, integrated best-of-breed packages, or some combination—and make at least an initial selection. Once an approach is selected, the project team develops a proposed project schedule and budget, addresses feasibility, and presents its findings to the oversight committee, both orally and in writing.

Phase 2: Analysis Once the analysis phase has been reached, the oversight committee has decided to proceed with the project and has probably committed to a specific approach. However, specific vendors are not yet selected. The project team begins creating work groups within the different functions of the organization to gather information and define requirements. In-house IS personnel or outside consultants may be employed to assist the work groups in this effort.

After all requirements have been defined, a major responsibility of the project team is to evaluate the vendors that could meet the requirements and to make its recommendation to the oversight committee. A comprehensive package approach requires evaluation of the major vendors. A best-of-breed package approach requires the evaluation of several competing packages for a given function or department. A custom approach may require the evaluation of only a few software development tools. (Keep in mind that a typical ERP project could involve all three approaches simultaneously.) The oversight committee makes the final vendor selections. Details on the vendor evaluation process are presented later.

Another major responsibility of the project team during analysis is the identification of any business process reengineering initiatives that will be required, based on the selection of software packages. Again, work groups within various functions may be involved in the step. The amount and type of BPR may depend on the ERP approach and packages selected.

Although doing so is not highly recommended, a firm may choose to change a package instead of changing its business procedures. Areas where such customization is most likely to occur and its impact on project schedule and cost are determined during analysis.

Once a vendor or vendors have been selected, representatives and possibly new consultants may be added to the analysis process. If a comprehensive package is selected, the oversight committee and project team members should receive intensive training in its purpose and operation. If several best-of-breed packages are chosen, then committee and team members representing certain functions should receive training on the respective packages. If custom development is involved, technical specialists should begin training on the specific tools to be used.

Finally, prototype ERP systems may be developed in all areas to demonstrate the integration of modules to users and further refine requirements. At this point, the alternatives considered earlier (that is, which ERP approach and which ERP vendor) are revisited and reevaluated. During this reevaluation, the project team may swing from comprehensive package to best-of-breed, or from one vendor to another. If so, the project team presents its recommendations, written and orally, to the oversight committee for approval and continuation.

The analysis phase should take less time with a comprehensive package approach and more time with a custom development approach. Regardless of the approach, however, analysis for an ERP project usually takes much longer than for a stand-alone system serving only one department or function.

Phase 3: Design The design phase begins after the final selection of an ERP vendor. The extent of design depends on the ERP approach. If packages are chosen, user interfaces are largely predetermined, although some minor customization may be desired. The design of user interfaces would be most extensive under a custom approach. Comprehensive packages usually require minimal use of middleware since they need to interface with only a few legacy systems or third-party packages. The best-of-breed approach may require a complex system interface design since many different packages from different vendors must communicate. A custom approach would require a normal degree of system interface design.

Designing applications, prototypes, and databases is very involved under a custom approach, minimal for a comprehensive approach, and somewhere in-between for a best-of-breed approach. Prototype development is highly beneficial to continue the refinement of requirements, even when using a comprehensive package.

Hardware design is very important in ERP systems since several business functions will be sharing many resources. A distributed client-server architecture may be recommended over an existing mainframe system. Networks connecting departments, buildings, or cities must be considered. Certain vendors' software may be designed to run on platforms not currently in use.

End users should receive intensive training in ERP packages at this point in the project. In addition to preparing them for the new system, training helps refine requirements during prototyping and eases the transition to design.

More of the details of BPR are fleshed out during design. From a process perspective, new procedures for business activities are documented. From a personnel perspective, new job descriptions are created, and plans are made to reorganize and retool the workforce, especially for those whose jobs are changing dramatically.

Phase 4: Implementation When custom software is chosen, this subphase is known as construction. However, for a comprehensive package, programs are already constructed in modules for such functions as accounts payable, purchasing, and inventory control. However, for a particular function, such as manufacturing, modules vary based on the type of business process used (discrete versus continuous). Such modules must be configured into a cohesive system. With best-of-breed packages, programs from different vendors must be integrated into a complete system using middleware. With comprehensive or best-of-breed, integration of new components with legacy components takes priority. For example, existing data may need to be reformatted for compatibility with new programs.

Once team members have configured and integrated modules with other programs and components, the remaining subphases of implementation are very similar to those of a conventional custom software project. First, a prototype of the system is usually created, validated over several iterations, and revised until a production-ready system is obtained. Second, the completed system is thoroughly verified and tested, then tuned for performance. Third, the system is thoroughly documented and end users receive final training. Finally, the new system is installed throughout the organization according to the rollout plan created in the design phase (pilot implementation, parallel implementation, or total cutover).

During this phase, all plans for BPR are implemented. Since hardware, software, data, and networks have already been implemented, the only components of the new system that remain are people and procedures. Organizational structures are changed, placing personnel in their new positions. The new procedures, in which all personnel have been previously trained, are also implemented.

Phase 5: Support The purpose of this phase is to ensure the short-term and long-term success of the new system. End-user support is critical for such an enterprisewide, integrated system. Extensive training can prevent catastrophes, but a certain degree of ongoing staff support is necessary with such a drastic, comprehensive change. A well-staffed and well-managed Help Desk should help to ensure a smooth transition.

Maintenance of the ERP system is also a key element of success. Corrective maintenance involves fixing bugs discovered by users. The extreme care taken in analysis, design, and implementation should minimize these bugs. When they do occur, a rapid response by knowledgeable consultants will ensure good public relations. Adaptive maintenance is needed when current versions of packages are upgraded or when comprehensive package modules, best-of-breed packages, or custom software are added or removed to meet changing requirements. Perfective maintenance may be needed to fine-tune the system for optimum performance. A system audit could be undertaken periodically to determine whether objectives of the ERP system are being achieved.

Special Topics on ERP Development

Now that we have presented general ERP system development methodology, we turn to some details critical to ERP development: selecting a comprehensive package, customizing a package, linking applications within a system, and using packages for application development.

Selecting a Comprehensive ERP Package Guidelines for selecting individual software packages were presented earlier in Chapter 8. There are many additional considerations when a firm is contemplating the purchase of a comprehensive ERP package. Both the costs and the risks are much higher in the case of the latter. The selection process for such a major investment as an ERP system should be thorough. We now provide more detailed information about the steps of evaluation.

Step 1: Preliminary Evaluation

To begin, a firm evaluates ERP packages by looking at broad, general requirements. Possibly four or five packages are considered at this preliminary stage. The decision makers could be members of the project team or the oversight committee, depending on a firm's management style. Package vendors and consultants are involved to provide needed information. Independent sources of information such as trade journals and personal contacts are consulted. Some of the key items to consider in a preliminary evaluation of an ERP package follow:

- Critical functionality, such as ability to deal with multiple currencies, to do simulations, and to support e-commerce.
- Up-front cost of the package. Initial quotes may vary widely among vendors, but they are often willing to negotiate.
- Cost of ongoing support for the package (upgrades, add-ons, and so on).
- The long-term viability of the vendor. How long will a company be around?
- The fit between the business model of the package and the actual business processes of the firm. Often it is easier for a firm to consider reengineering its processes before significantly altering the software.
- Satisfactory, *guaranteed* implementation time.
- A sizable installed base (at least 20 successful installations) so the firm is not one of the "guinea pigs."
- Local support (in multiple countries, if applicable) with a guaranteed, acceptable level of service.
- Proven success in the international environment (if applicable).
- Availability of customization tools to bolt on third-party packages or link to legacy systems.
- The best accommodation of the firm's management style (for example, preference for seeing the big picture or for drilling down to detailed data).

Sometimes, a given package exhibits what may be called a showstopper—a missing feature or unsupported business process that transforms an otherwise good fit into a complete mismatch.

The result of this preliminary evaluation is a short list of perhaps two competing ERP packages.

Step 2: Detailed Evaluation

Once the field of packages is narrowed, lower-level managers, staff, or even operating personnel may assist in the selection process through meetings or workshops. Such workshops could be organized by major function (for example, finance or manufacturing) or by department (such as accounts payable within finance or production scheduling within manufacturing) to consider dozens of detailed evaluation criteria, again based on feedback from vendors and consultants. Just a few examples of the detailed criteria are the package's ability to accomplish the following:

- Set prices according to different criteria
- Provide financial information in multiple currencies
- Control commissions of sales representatives
- Support just-in-time (JIT) production

Such workshops provide two important benefits. First, the candidate's ability to address successfully both mandatory and optional functionality is accurately determined.

Second, users of the new system are involved in the evaluation process, thereby increasing their understanding and acceptance of the package and ERP in general. The results of the detailed evaluation process should also be documented.

Step 3: Vendor Presentations

After the detailed selection criteria have been applied to all candidate ERP packages, the project team can schedule vendor presentations for both the oversight committee and for the lower-level managers, staff, and operators. Both the form and substance of these presentations may be evaluated to assist with the final recommendation.

Step 4: On-Site Visits

An additional method for evaluating competing ERP vendors is on-site visits to different firms that use the packages under consideration. The names of such firms may be supplied by the vendor or, better yet, uncovered by the project team. During an on-site visit, information about the quality and quantity of vendor support, the amount of customization required for the package, the performance and reliability of the software, requirements for hardware and additional software, and implementation time can be obtained and added to the set of evaluation criteria.

Step 5: The Final Decision

After all previous steps have been completed, the project team usually performs a formal and structured evaluation based on (1) weights assigned to the criteria and (2) vendor scores on these criteria. The result is a total weighted score for each ERP candidate. Other factors may also influence the decision on a particular package, especially where total scores are extremely close. One vendor may have a very high total score but may also have one glaring weakness that could make the package unsuitable. Certain intangible factors may also come into play. At this point, however, the project team must select a vendor to continue the project. It is very important to achieve consensus to help ensure the future success of the project. The project team's recommendation of a vendor is usually submitted to the oversight committee for final approval.

Linking Applications to an ERP Package A popular approach to ERP is the hybrid approach: implementing a comprehensive ERP package from a major vendor and "bolting on" various best-of-breed software solutions to the main ERP backbone. One strategy for increasing chances of success with a hybrid approach is to select a package that is already partnered with the ERP vendor. However, true partnering implies actual shared development, not just joint marketing, to ensure that the applications will work together smoothly.

Once a best-of-breed package is selected, some firms have the vendor integrate the package with the ERP system. Other firms have their own developers create the interfaces between the ERP system and the third party application, often using tools provided by the third party. This approach gives the organization more control and the ability to test the links thoroughly.

An alternative to creating interfaces is purchasing enterprise application integration (EAI) tools (that is, middleware). Since this technology is relatively new, a team must be careful in selecting a middleware vendor. An additional strategy is to hire a third-party consultant with experience in the industry and its applications to oversee the integration.

Using ERP Packages for Application Development One potential advantage of a comprehensive ERP package approach is that the package can serve as a platform for continued application development and integration. Project teams can extend functions by incorporating custom code to meet special requirements not provided by the package. A number of software vendors, including the major ERP package vendors, provide tools ranging from high-end development environments to specialized integration technologies. For example, SAP offers connectors to such tools as Microsoft's Visual Studio and IBM's Visual Age to make it easier to link, for example, a Visual Basic application with SAP R/3. Other vendors have built their ERP packages using certain 4GL tools, making it easier for the customer to use the same tool to create custom applications.

This approach to application development is certainly not traditional. Developers in an ERP environment often become system integrators, adding new packages to the ERP framework or extending the functionality of an ERP application through custom development. However, linking custom applications to ERP packages is complex since many such packages were not developed using standard architectures and much of the ERP package's inner workings may be proprietary. Also, many of the application development tools offered by the major vendors are seriously lacking. For example, a given package may include a low-level programming language and may reveal its underlying data schema, but it may not provide the testing and debugging functionality that developers have come to expect. Using an ERP package for true application development is, therefore, a distant dream.

The Future of ERP

ERP systems were originally intended to integrate the information systems of departments in an organization to drive down costs. As many firms have largely accomplished these objectives, they are looking for new integration capabilities beyond the walls of the factory or corporate headquarters. The major areas of expansion for ERP systems that are currently undergoing rapid development and growth are supply chain management and business intelligence.

Supply Chain Management Companies essentially compete on the basis of supplying quality products and services faster and cheaper through superior management of materials, information, cash, and work processes. By integrating each functional area into a comprehensive strategy, an organization can address the key issues of cost, speed, complexity, flexibility, and customer needs. An ERP platform is ideal for driving down costs and improving response along the entire supply chain, from raw materials to finished goods.

The supply chain involves three major links: suppliers, in-house processes, and customers. Suppliers and customers can be linked to the firm's ERP system via e-commerce. In-house processes are improved through the integrated, real-time, and global characteristics of an ERP system. One often-overlooked way that ERP can accomplish better response to demand is through improved human resource (HR) management. Internal human resources are an extremely vital link in any supply chain. HR management is more than just handing out paychecks or helping an employee navigate through a cafeteria-style benefit plan. An ERP system can, for example, provide intranet-based learning to employees and maintain information on employee skills to determine the best way to deploy the workforce to meet rapidly changing requirements. At present, RMO has no plans to integrate its HR package with other systems, but perhaps it should reconsider.

Business Intelligence Businesses can collect all types of information on customers, operations, competitors, products, and employees, into what is often called a data warehouse. However, the key to using this information for competitive advantage is integrating it into a robust decision-support tool. ERP systems often provide such tools, helping a firm to identify strengths, pinpoint weaknesses, uncover new opportunities, and capitalize on trends by combining historical and predictive views of its performance with information about the business. RMO could capitalize on such business intelligence applications since it is in the highly competitive retail sales industry.

A Hypothetical ERP Scenario An ideal scenario, enabled by an integrated, real-time, global, extended ERP system, could go something like this:

- A consumer orders a car (or major appliance, computer, books, clothing, or so on) at home from a web-based retailer.
- The retailer's web-based order-entry system is also linked to one or more banks to provide the consumer with on-line loan approval.
- The banks' loan approval systems are linked to one or more credit bureaus' expert systems to determine the creditworthiness of the individual and the appropriate interest rate and term of the loan.
- Upon approval, the consumer's order of the product automatically triggers a shipping order at the distributor to deliver the goods.
- The shipping order at the distributor automatically triggers an assembly order at the assembly plant to replace the shipped item in inventory.
- The assembly order at the assembly plant automatically triggers subassembly orders at sister plants.
- The subassembly orders at the sister plants automatically trigger inventory withdrawals.
- The inventory withdrawals at the subassembly plants automatically trigger purchase orders directed to their suppliers of raw materials.
- The purchase orders at the raw materials plants automatically trigger shipping orders, inventory withdrawals, production orders, purchase orders, and so on.

This entire hypothetical process could actually take only a few minutes (or seconds) in real time, linking the customer's order backward through the value chain to the most basic raw materials, using ERP and computer networking technologies. The rapid, real-time response of the ERP system helps to ensure a smooth, constant flow of goods and materials, minimizing the need for excessive inventories and providing excellent customer service as well. Firms that use ERP systems to their fullest capabilities could gain significant competitive advantage in the marketplace of the future.

A Closer Look at One ERP Package: SAP R/3

The first part of this chapter discussed packaged software in general, and then discussed packaged software in the context of ERP. This section examines a particular ERP package, SAP R/3. The purpose of this section is not to empower the reader to become an SAP consultant, but simply to provide a better understanding and appreciation of ERP systems through the exploration of SAP R/3.

What Is SAP?

SAP AG is a German software development firm (see Figure 14-9) that created and markets the SAP R/3 enterprise resource planning system (pronounced S-A-P, not "sap"). SAP began creating software in 1972. In 1983, SAP introduced the R/2 (meaning real-time system, version 2) system of integrated business application software for mainframe architectures. In 1989, SAP R/3 was released utilizing the increasingly popular client-server architecture and relational database technologies. SAP is currently the fourth-largest software vendor in the world and the market leader in ERP systems. There are now more than 8,000 installations of R/3 in over 50 countries.

FIGURE 14-9
SAP's home page.

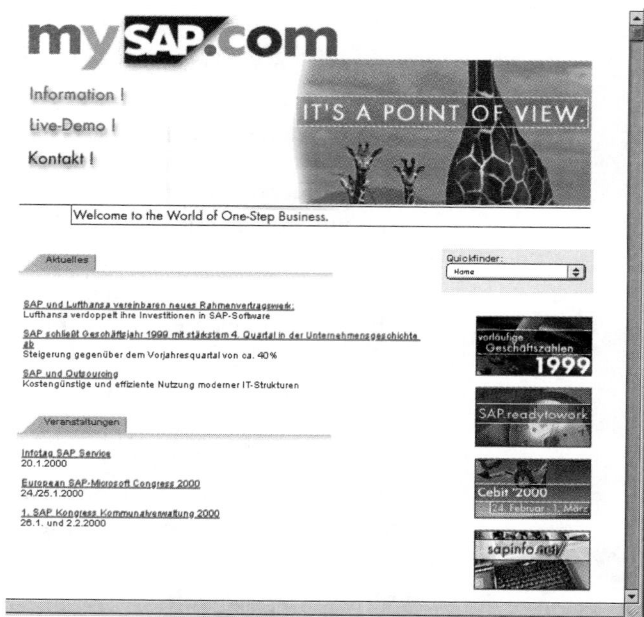

ERP systems are perhaps the most complicated information systems that a firm can implement, and SAP R/3 is considered to be the most complex ERP system available. SAP R/3 requires more than 15 gigabytes of disk space just to install and contains nearly 10 million lines of code. Although the complexity of SAP R/3 provides it with a high degree of flexibility and utility, it also makes for challenging implementation.

Organization of SAP R/3 Implementation of SAP R/3 within a company consists of a set of modules tailored to the needs of the enterprise. Tailoring means configuring several related application modules, not modifying or enhancing existing code. The modules within SAP R/3 are organized into four high-level work areas:

- Logistics
- Accounting
- Human Resources
- Business Tools

The first three areas represent the major functional groupings within most enterprises. Within Logistics are functions such as materials management, production scheduling, and plant maintenance. The Accounting area contains modules such as Financial Accounting, Controlling, and Project Management. Under the Human Resources area are modules such as Personnel Administration, Payroll, and Benefits. Not every firm, however, would choose to implement all available modules in these three areas. The Business Tools area provides the systems integrator the means to perform R/3 system installation and administration, including configuration and programming activities.

SAP R/3's integrated software modules are designed to meet information needs across the entire enterprise. The modules selected by a given firm become its SAP R/3 configuration. A firm just starting out with SAP R/3 may choose to implement only the core modules: Financial Accounting, Sales and Distribution, Materials Management, and Production Planning. After a firm successfully implements and uses these modules, it may choose to add others over time, such as Controlling, Warehouse Management, Plant

Maintenance, Quality Management, Project System, Treasury Management, and Asset Management.

SAP Industry Solutions SAP R/3 is a highly configurable system designed to meet the ERP needs of a wide variety of industrial segments. The SAP system includes the basic core functionality common to most organizations: finance, human resources, and logistics. To better serve the needs of different industries, SAP has created Industrial Business Units (IBUs) to provide system solutions for many industries, based on what are generally considered "best practices." A brief overview of some of these industry-specific solutions can provide a better understanding of how SAP R/3 may be applied.

- **The automotive industry**. SAP's solution for the automotive industry includes a customer-pulled supply chain and a global infrastructure for the value chain including component suppliers, after-market parts suppliers, and wholesalers. Additional functionality includes flow manufacturing, Kanban, shipping scheduling, and global available-to-promise capability.
- **The electronics industry**. Time-to-market is a key to success in this industry. New products with high quality and low prices are in demand. A real-time, integrated system can provide many benefits. SAP R/3 offers product data configuration management, configure-to-order quoting and estimating, product data management, bill-of-material, and routing development. The SAP High-Tech module provides a knowledge- and rule-based product configuration engine to forecast and monitor dynamic pricing and supply scenarios.
- **The public sector industry**. Public organizations should operate as efficiently and effectively as private enterprises. The SAP solution for this industry includes a Human Resources module that provides employee self-service (ESS), benefits administration, and event management capabilities, among others. The Purchasing and Materials Management module enables automatic day-to-day purchasing, supplier selection, and electronic data interchange (EDI).
- **The retail industry**. This industry is characterized by a wide variety of distribution possibilities (walk-in traffic, catalog sales, e-commerce) and small margins. Retailers must carefully manage inventory and monitor customer preferences. SAP provides Merchandise Logistics Tools to track customer buying behavior. The Retail module includes software for goods receipt, labeling, inventory management, and promotions management. RMO falls within this category and would be most interested in SAP's offerings within this IBU.

Other SAP IBUs include insurance, telecommunications, aerospace and defense, banking, chemicals, consumer products, healthcare, oil and gas, pharmaceuticals, and utilities. Obviously, SAP is endeavoring to tailor its ERP system to an extremely wide variety of business applications.

Master Data The SAP R/3 system revolves around several standardized databases utilized by various application programs (see Figure 14-10). At the core of SAP R/3 are five master databases: the Material Master, the Vendor Master, the Customer Master, the Human Resources Master, and the General Ledger. The Material Master is used to store data relating to all materials tracked in the system. Logistics modules such as Materials Management, Sales and Distribution, Production Planning, Plant Maintenance, and Quality Management use these data. Materials within a firm are defined in SAP R/3 as one of several types such as Finished Products, Semi-finished

Products, Raw Materials, Operating Supplies, and Packaging. The types of fields associated with an item of material may include material number, unit of measure, procurement type, temperature conditions, shelf life, and price.

FIGURE 14-10
The five master databases for SAP R/3.

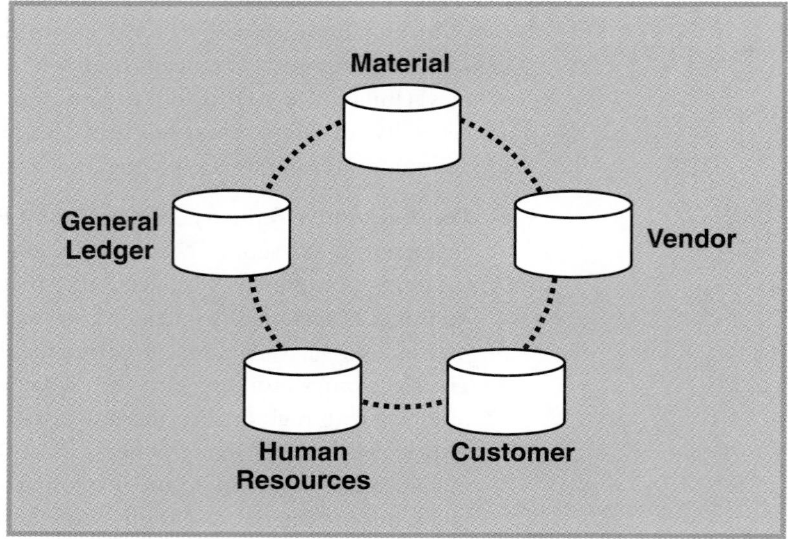

The Vendor Master contains information about external suppliers, such as address, telephone, tax codes, bank accounts, payment terms, and order currencies. The Customer Master stores data about a firm's business partners that buy products or services. Customers may be one of several types, including sold-to partners, ship-to partners, payers, or bill-to partners. For example, when you buy a gift for a friend, you may be the sold-to partner, your friend may be the ship-to partner, and your credit card company may be the payer. All the information associated with these various parties, such as names, goods receiving hours, preferred shipping method, payment terms, and tax classifications, must be stored in the Vendor Master.

The Human Resources Master contains information about a firm's employees. This information is stored in various infotypes—logical groupings of similar data fields. For example, the Personal Data infotype contains name, birth date, marital status, and number of children. The Address infotype consists of various subtypes such as Permanent, Temporary, and Mailing. The Events infotype tracks important events for an employee, such as hire date, pay change, or change in organizational assignment.

The General Ledger accounts are the backbone of the entire SAP system and are used to produce reports to meet legal requirements. All transactions that affect external reporting requirements are reflected in General Ledger accounts. For example, billing a customer will increase the receivables account, and shipping an item will reduce the inventory account.

Implementing SAP R/3

As mentioned earlier, implementing SAP R/3 can be complex and time consuming. Here we highlight general issues of architecture, middleware, configuration, and accelerated installation.

Architecture of SAP R/3 The R/3 system is designed for a three-tier client-server architecture (see Figure 14-11). The three tiers, or levels, are (1) presentation, (2) application, and (3) database. The user of R/3 works on a client machine at the presentation level with SAPGUI (SAP graphical user interface) software, enabling the user to work in a familiar Windows environment. Through SAPGUI, the user sends and receives data from what is called R/3 System software on various SAP application servers. It is at this application level that specific modules, such as Payroll or Materials Management, function. The R/3 System at the application level communicates with myriad SAP data tables within a relational database management system (RDBMS) on a database server. The language SAP applications use to interface with the RDBMS is standard SQL.

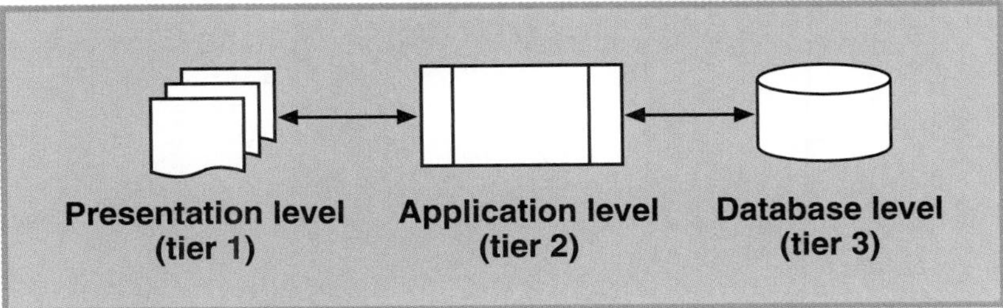

FIGURE 14-11
The three-tier architecture of SAP R/3.

R/3 Basis R/3 Basis is the middleware that enables the integration of various SAP R/3 application programs on various computing platforms. Basis also provides R/3 administration tools (the data dictionary and the function library). Basis allows the R/3 system to run on a variety of platforms (such as AS/400, UNIX, or Windows NT) and to utilize different database systems (such as Oracle or DB2). There are three different types of interfaces that Basis controls: user, programming, and communications.

The user interface is a graphical user interface (GUI) that makes work easier for users, developers, and administrators. For example, a developer may use the GUI of the data dictionary to create a table by pointing and clicking, rather than using SQL directly. The programming interface utilizes the ABAP/4 Development Workbench, which is required to create or change database tables and programs in the R/3 system. The communications interface provides the link among internal programs, as well as between R/3 and third-party programs. SAP uses standard relational database technology to store all three types of R/3 data: master, transactional, and configuration.

Accelerated SAP (ASAP) Over the past decade, SAP and many other vendors have acquired a reputation for very expensive and slow ERP implementations. To address this issue, SAP has developed a methodology for a quicker, more efficient implementation called Accelerated SAP, or just ASAP. ASAP is designed for small to mid-sized companies and should require only about six months to implement, as opposed to one to two years for a normal SAP implementation in a large firm. ASAP follows the same basic phases just presented but with added emphasis on implementing the system with virtually no modification. Under this plan, the company may be forced to change its processes to fit the software. However, such BPR should be much more feasible for a small to mid-sized company than for a large one. RMO would be a very good candidate for ASAP.

Using SAP R/3

This section gives you a feel for how a user would navigate through SAP R/3. SAPGUI is loaded on a client computer that is networked to the server on which the SAP R/3 system has been installed. In a Microsoft Windows environment, the user starts the SAP R/3 program like any other program (by selecting *Start, Programs, SAP Frontend, SAP R3*). The logon screen would then appear requiring the user to provide a user ID and password. Menus and buttons are available on this window for such functions as logging off or creating a new password (see Figure 14-12).

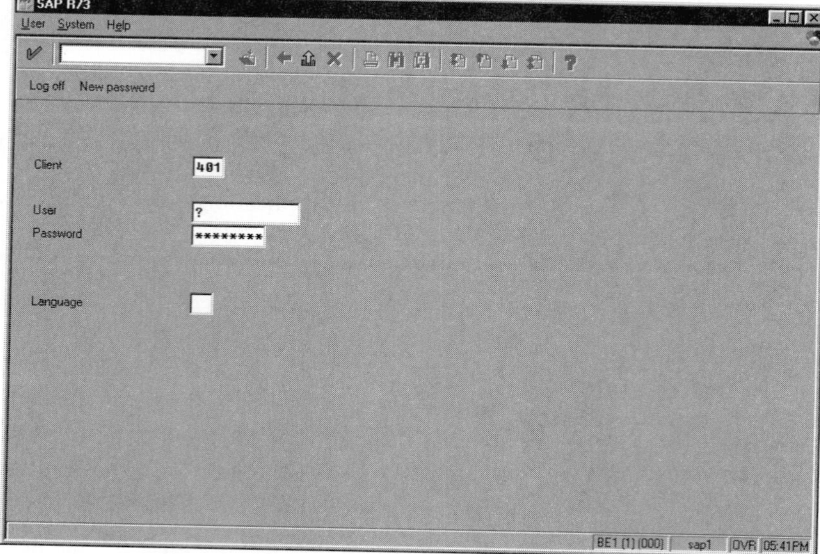

FIGURE 14-12
SAP R/3's logon screen.

Once logged on, the user will see the main SAP R/3 screen (see Figure 14-13). The menu bar on this screen contains several options: *Office, Logistics, Accounting, Human resources, Information systems, Tools, System,* and *Help*. The *Logistics, Accounting,* and *Human resources* options represent the basic core of the SAP R/3 system. Selecting the *Logistics* option displays a pull-down menu with the following options: *Materials management, Sales/distribution, Production, Plant maintenance, Service management, Quality management,* and *Project management*. Selecting *Accounting* displays the following options: *Financial accounting, Treasury, Controlling, Enterprise control, Capital investmt mgt, and Project management.*

Selecting *Human resources* displays the following options: *Personnel administration, Time management, Incentive wages, Payroll, Benefits, Planning, Recruitment, Travel expenses,* and *Information system.*

Of course, the availability of all such menu options is contingent on having the associated modules configured and installed in the system.

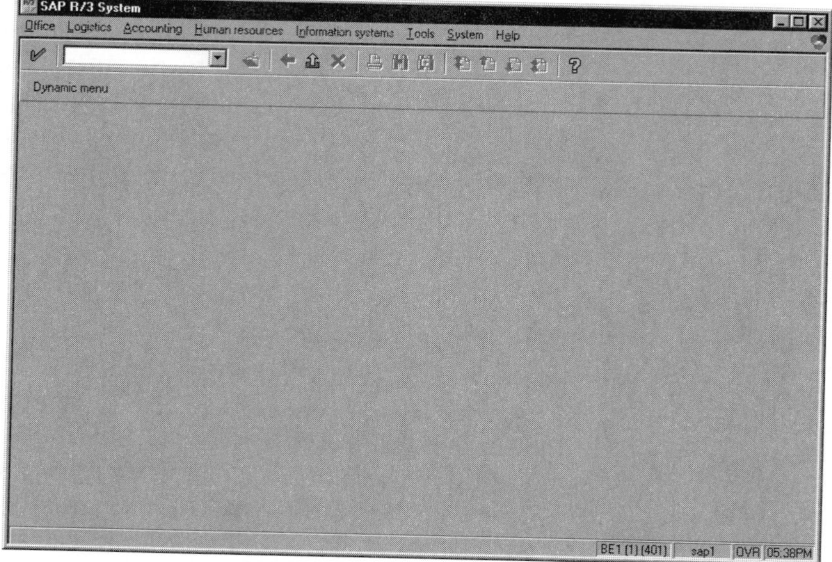

FIGURE 14-13
SAP R/3's main screen.

At the heart of the SAP R/3 system is the entry of business transaction data, followed by the use of that data to operate the firm efficiently and effectively. At the main SAP R/3 screen, the user can enter a four-digit transaction code in lieu of traversing a lengthy menu path to enter a valid SAP transaction. For example, to obtain a screen for confirming production work performed, the user can enter the code C011 instead of selecting *Logistics, Production, Production control, Confirmation, Enter, For operation,* and *Time ticket*. Obviously, a user well acquainted with transaction codes in a particular functional area would much rather enter the codes than use the menu. Figure 14-14 shows a screen called the Stock/Requirements List, which shows a real-time version of the MRP process. This output shows forecast demand and planned orders plus any dynamic changes such as completed production orders or material withdrawals.

FIGURE 14-14
SAP R/3's transaction screen for confirming production.

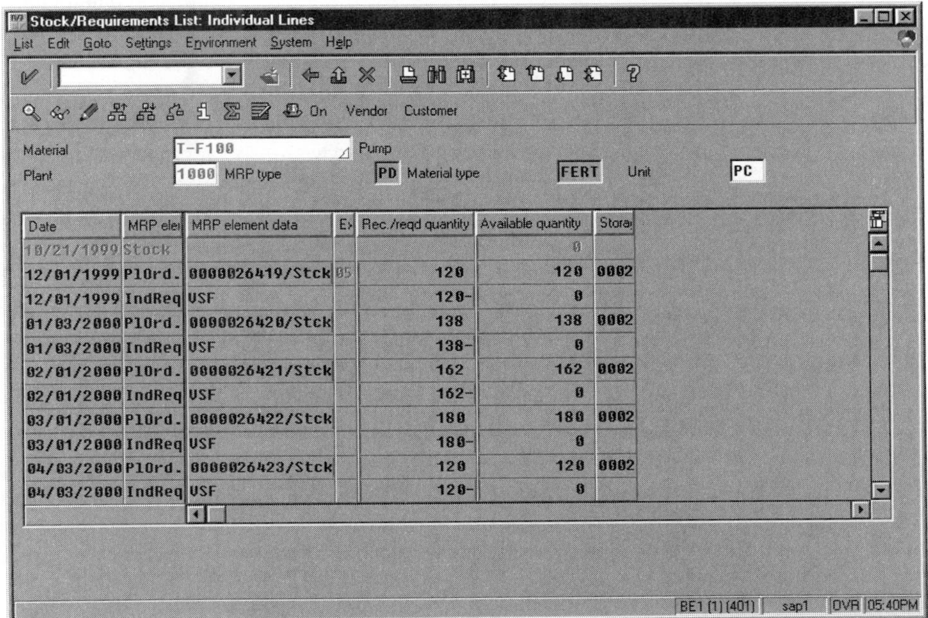

Date	MRP elei	MRP element data	E>	Rec./reqd quantity	Available quantity	Stora
10/21/1999	Stock				0	
12/01/1999	PlOrd.	0000026419/Stck	05	120	120	0002
12/01/1999	IndReq	USF		120-	0	
01/03/2000	PlOrd.	0000026420/Stck		138	138	0002
01/03/2000	IndReq	USF		138-	0	
02/01/2000	PlOrd.	0000026421/Stck		162	162	0002
02/01/2000	IndReq	USF		162-	0	
03/01/2000	PlOrd.	0000026422/Stck		180	180	0002
03/01/2000	IndReq	USF		180-	0	
04/03/2000	PlOrd.	0000026423/Stck		120	120	0002
04/03/2000	IndReq	USF		120-	0	

A button bar is also available in SAP R/3. Clicking the Back button will move you backward through the path just traversed. The Exit Session button closes down the current session. The Help button provides on-line help. The Dynamic Menu button displays a listing of menu bar options in a tree structure until it truncates a particular task with its transaction code. The user can click that option or enter the transaction code to access the desired task.

After a user has logged on to SAP R/3, he or she may create up to six different sessions in SAP R/3 at the same time. To create a session, the user selects *System, Create Session* from the main screen menu. Multiple sessions, available on the taskbar in Windows, allow the user to work on different tasks without spending extra time moving around through menus. A user trying to access the same record in a database in different sessions will be locked out of that record until the original update is completed.

To log off of SAP R/3 when finished for the day, the user either selects *System, Log off* from the menu system or clicks the Close button in the SAP R/3 window.

Summary

The issue of how best to procure business application software—custom-developed versus packaged—is extremely critical for the future of IS organizations and business enterprises. Equally important is the question of how best to deploy business application software—integrated versus stand-alone. The most recent trend in IS development is the deployment of integrated packages to increase both the efficiency and effectiveness of information technology as it serves to boost the competitiveness of the entire enterprise.

Packaged software is designed by software vendors to be very easy to use and highly functional for its intended users. Application packages are now becoming more popular as both business processes and information systems requirements are becoming more complex. The debate rages whether a firm should custom-develop its own applications to match many of its unique system requirements, or whether the firm should alter its processes, through business process reengineering, to match many of the "best practices" supported by recent packages. The three major issues that differentiate application packages from custom software are (1) customization, (2) integration, and (3) upgrades. The systems analyst must address these three issues during the systems analysis and design phases.

A major class of packaged software is the comprehensive enterprise resource planning (ERP) system. ERP is the process of running virtually the entire organization with highly integrated application software. The approaches to deploying an ERP system include (1) creating the system in-house, (2) assembling the system from a diversified set of "best-of-breed" application packages, and (3) purchasing the system as a comprehensive package. One of the most popular approaches to ERP is the comprehensive package. ERP is purported to result in increased efficiency throughout the organization, reduced IS staff, and improved customer service. However, implementing such a comprehensive, enterprisewide system can be extremely challenging, with many potential pitfalls. Properly organizing people (employees, vendors, and consultants) for an ERP project is a major key for success.

The current market leader in comprehensive ERP systems is the German company SAP, which offers its R/3 system worldwide. SAP R/3 software is organized around three major application areas (Logistics, Accounting, and Human Resources) plus Business Tools, which allows for the continuing administration of the system. SAP R/3 can be configured for a wide range of industries, such as automotive, electronics, public sector, and retail. SAP R/3 is usually installed on a client-server platform by system integrators with the assistance and guidance of in-house systems analysts. While very large corporations typically install more of the available modules with some customization (requiring more time and expense), SAP offers the scaled-down ASAP system to mid-sized and small firms for more rapid deployment.

Key Terms

best-of-breed, p. 529
business intelligence, p. 526
configuration, p. 521
customer-relationship management, p. 526
e-commerce, p. 526
enhancement, p. 522
enterprise application integration, p. 522
modification, p. 521
sales-force automation, p. 526
supply-chain management, p. 526
system integrators, p. 520
total quality management, p. 526
value chain, p. 525

Review Questions

1. What is meant by a "best-of-breed" software package?
2. What are the two major critical success factors for the development of software packages?
3. What are the three most common types of software packages?
4. What are the three types of package customization?
5. What are the four major characteristics of ERP systems?
6. What are the three types of employee teams or committees that would usually be involved in an ERP project?
7. What are the major areas into which ERP vendors are expanding?
8. What does the *R/3* in SAP R/3 stand for?
9. What are the four major high-level work areas of R/3?
10. What are the five master data sets found in R/3?
11. What are the three tiers found in the R/3 architecture?

Thinking Critically

1. Under what conditions would a firm prefer to develop custom software rather than purchase one or more packages?
2. How have the purposes of business information systems changed over the past 50 years?
3. Discuss when BPR would and would not be necessary when implementing an ERP system.
4. Discuss how the role of a traditional systems analyst will change when his or her firm adopts an ERP system.
5. Give several examples of how both efficiency and effectiveness are improved with the implementation of ERP.
6. Discuss how ERP could make substantial improvements in all four types of business information systems.
7. Which should come first: ERP or BPR? Defend your position.
8. Discuss the major differences between a system development methodology for a custom project and that for an ERP project. Cover all five phases of the SDLC.
9. What does "ASAP" stand for in the context of SAP software? Under what conditions would you recommend ASAP to a company?

Experiential Exercises

1. Do a search on the web for *ERP* or *enterprise resource planning*. Find an article that discusses a recent development in ERP or provides more detail about a topic introduced in this chapter. Prepare a written or oral presentation of this article.
2. Form two teams of students to debate the issue of custom versus packaged software development. Provide arguments and evidence, pro and con, for the statement, "Custom development is essential for creating and sustaining competitive advantage."
3. Form two teams of students to debate the issue of best-of-breed versus comprehensive ERP systems. Provide arguments and evidence, pro and con, for the statement, "A comprehensive ERP package is superior to a best-of-breed approach."
4. Do some web research on the term *middleware*. Prepare a written or oral report on some recent developments or other interesting aspects of this subject.
5. Perform the following role play for the class: You are the IS manager. It is your job to explain to four other students in the class (who are playing the role of IS analysts and programmers) why the company is adopting ERP and how it will affect their jobs. The IS staff have heard rumors that at least 50 percent of them will probably be laid off.
6. Prepare a hypothetical cost/benefit analysis for an ERP system, showing the expected payback.
7. Select a particular type of enterprise (such as electronics manufacturer, governmental agency, or pharmaceutical company). Assume that your company is either small (less than $200 million sales), mid-sized ($200 million to $1 billion sales), or large (greater than $1 billion sales), and operates either domestically or globally. By simply reviewing the web sites of several (at least five) ERP vendors, draw up a preliminary evaluation reducing the number of candidates to two. Make sure you provide sound reasons and evidence for your decision.

Case Study

Rayco Products Corporation

Dave Chen works as a production planner for Rayco Products Corporation (RPC), a very successful multinational, Fortune 500 company. It is Dave's responsibility to create monthly production plans for five plants in the United States and Canada. To be closer to the action, Dave is located at one of the five plants in Texas instead of at corporate headquarters in Ohio. One of the primary inputs to Dave's production planning process is an 18-month, item-level sales forecast that groups literally thousands of distinct part numbers into dozens of product families. The marketing group in Ohio prepares this forecast each month. One of Dave's major problems is creating the production plans in a timely manner so that the plants can adjust their material, manpower, and equipment requirements to meet expected customer demand.

The process works something like this: Corporate IS in Ohio runs the sales forecast program based on historical sales figures and input received from marketing personnel (regarding promotions, new products, and so on). Several copies of the forecast data tape are produced and mailed overnight to the plant IS departments where the three production planners are located, including Dave in Texas. When the IS department at the Texas plant receives the tape, it runs a program on the mainframe to produce a very large printout, and a data file on disk, of the forecast for all product families and their individual part numbers. When the file is created, Dave runs a program to download it to his PC. He then imports the text file into an Excel spreadsheet. From there, Dave runs several macros to manipulate the data into an initial production plan.

The process of creating a final production plan continues as Dave receives additional input on current finished goods inventory levels at several distribution centers and manpower and equipment availability at the five plants. After a lot of manual input and several iterations, Dave has five final production plans covering the next 12 months. Dave and the other production planners then fly to corporate headquarters to attend a full day of meetings with representatives from marketing, distribution, and production to discuss any necessary changes to the plans. After the meeting, Dave incorporates the changes in the spreadsheets and mails floppy disks to the five plants overnight. From there, production schedulers at the plants take the spreadsheets and manually input the production plans for the various product families into their MRP systems to generate detailed material requirements. Manpower and equipment requirements are estimated manually.

The entire process of creating, mailing, and running the forecast tapes, then downloading the data to the PC and running the spreadsheet macros, takes at least three days (barring any glitches in computer operations or in shipping tapes). This puts Dave in an extreme bind to interpret the data and create the production plans in time for the scheduled monthly meeting. The plants are also in a hurry to receive the final production plans so they can begin ordering material, scheduling overtime, and shifting production to other plants, if necessary. The entire process—from the monthly close of business to the creation of material, manpower, and machine requirements at the plants—takes at least seven business days, or about one-third of a business month.

No one, especially plant personnel, seems to like the current process of creating production plans. Nearly everyone involved in the process has serious doubts about the integrity of the data since it has been handled so many times. Dave thinks there must be a more timely and accurate way of getting the information from marketing and distribution, creating and approving production plans, and getting the final adjusted plans to the plants.

1. What major courses of action would you recommend for RPC in terms of ERP, and why?
2. What specific benefits would RPC derive for production planning via an ERP implementation?
3. What potential problems would RPC face during and after an ERP implementation?

Further Resources

Erin Callaway. *Enterprise Resource Planning: Integrating Applications and Business Processes across the Enterprise*. Computer Technology Research Corporation, 1999.

Bradley D. Hiquet. *SAP R/3 Implementation Guide: A Manager's Guide to Understanding SAP*. Macmillan Technical Publishing, 1998.

Sergio Lozinsky. *Enterprisewide Software Solutions: Integration Strategies and Practices*. Addison-Wesley, 1998.

Avraham Shtub. *Enterprise Resource Planning (ERP): The Dynamics of Operations Management*. Kluwer Academic Publications, 1999.

Norbert Welti. *Successful SAP R/3 Implementation: Practical Management of ERP Projects*. Addison Wesley Longman, Inc., 1999.

Liane Will. *SAP R/3 System Administration: The Official SAP Guide*. Sybex, 1999.

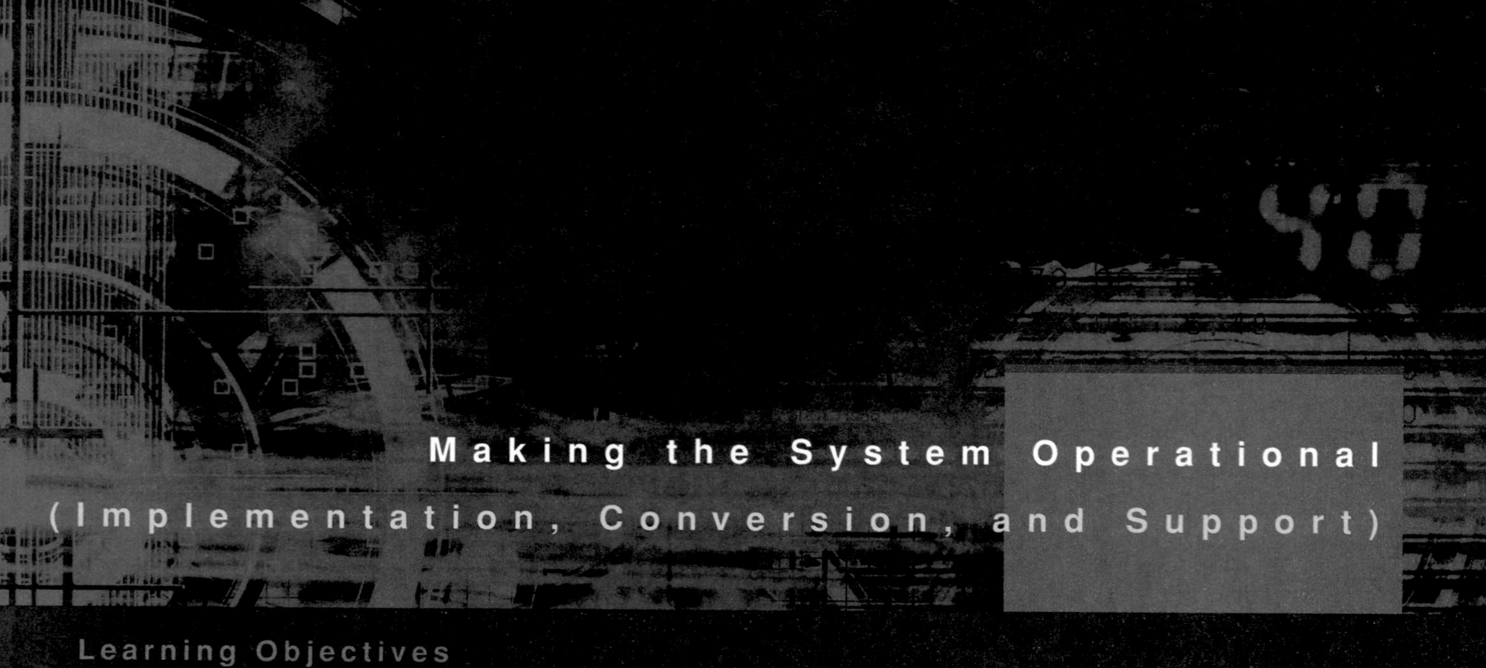

Making the System Operational (Implementation, Conversion, and Support)

Learning Objectives

After reading this chapter, you should be able to:

- Describe implementation and postimplementation activities.
- Choose an appropriate approach to program development.
- Describe various types of software tests and how and why each is used.
- List various approaches to system implementation and conversion and describe the advantages and disadvantages of each.
- Describe different types of documentation and the processes by which they are developed and maintained.
- Describe training and support requirements for a new system.

Chapter Outline

TRI-STATE HEATING OIL: JUGGLING PRIORITIES TO BEGIN OPERATION

It was 8:30 A.M. on Monday morning, and Maria Grasso, Kim Song, Dave Williams, and Rajiv Gupta were just about to begin the weekly project status review meeting. Tri-State Heating Oil had started developing a new customer order and service call scheduling system five months ago. The target completion date was 10 weeks away, but the project was behind schedule. Analysis and design had taken eight weeks longer than anticipated because key users had disagreed on what new system requirements to include and the system scope was larger than expected.

Maria began the meeting by saying, "We've gained a day or two since our last meeting due to better-than-expected unit testing results. All of the methods developed last week sailed through unit testing, so we won't need any time this week to fix errors in that code."

Kim spoke, "I wouldn't get too cocky just yet. All of the nasty surprises in my last object-oriented project came during integration testing. We're completing the user-interface classes this week, so we should be able to start integration testing with the business classes sometime next week."

Dave nodded enthusiastically and said, "That's good! We have to finish testing those user-interface classes as quickly as possible because we're scheduled to start user training in three weeks. I need that time to develop the training materials and work out the final training schedule with the users."

Rajiv replied, "I'm not sure that we should be trying to meet our original training schedule with so much of the system still under development. What if integration testing shows major bugs that require more time to fix? And what about the unfinished business and database classes? Can we realistically start training with a system that's little more than a user interface with half a system behind it?"

Dave replied, "But we have to start training in three weeks. We contracted for a dozen temporary workers so that we could train our staff on the new system. Half of them are scheduled to start in two weeks and the rest two weeks after that. It's too late to renegotiate their start dates. We can extend the time they'll be here, but delaying their starting date means we'll be paying for people who we aren't using."

Maria spoke up and said, "I think that Rajiv's concerns are valid. It's not realistic to start training in three weeks with so little of the system completed and tested. We're at least five weeks behind schedule, and there's no way we'll recapture more than four or five days of that during the next few weeks. I've already looked into rearranging some of the remaining coding to give priority to the work most critical to user training. There are a few batch processes that can be back-burnered for awhile. Kim, can you rearrange your test plans to handle all of the interactive applications first?"

Kim replied, "I'll have to go back to my office and take another look at the dependencies among those programs. Offhand, I'd say yes, but I need a few hours to make sure."

Maria replied, "Okay, let's proceed under the assumption that we can rearrange coding and testing to complete a usable system for training in five weeks. I'll confirm that by e-mail later today as soon as Kim gets back to me. I'll also schedule a meeting with the CIO to deliver the bad news about temporary staffing costs."

After a few moments of silence Rajiv asked, "So what else do we need to be thinking about?"

Maria replied, "Well, let's see. ... There's user documentation, hardware delivery and setup, operating system and DBMS installation, the network upgrade, and stress testing for the distributed database accesses."

Rajiv smiled and said to Maria, "You must have been a juggler in your youth, and it was good practice for keeping all of these project pieces up in the air. Does management pay you by the ball?"

Maria chuckled and replied, "I do think of myself as a juggler sometimes. And if management paid me by the ball, I could retire as soon as this project is finished!"

OVERVIEW

This chapter focuses on the activities that occur after a system has been designed. Activities that occur before the system is turned over to its users are collectively called *implementation*. Activities that occur after the system becomes operational are described by the terms *maintenance*, *support*, or both.

Implementation, maintenance, and support activities are often considered to be straightforward and dull—they don't attract the same attention or enthusiasm as analysis and design activities. The situation is analogous to the difference between architecture and construction. An architect gets most of the credit for creating a new building, even though his or her job essentially ends with the blueprints. Yet, the vast majority of the effort that goes into making the building a reality occurs after the blueprints are finished. The same is true of information system development—care and quality of implementation are critical to the success of a system.

Implementation consumes substantially more time and resources than earlier phases of the system development life cycle (SDLC). A large number of people are required to build and test software, acquire hardware and other operational resources, and assemble them into a functioning system. Myriad interdependent activities must be completed, including program development, quality assurance, physical installation, documentation, and training. Managing the implementation phase is much more complex than managing earlier phases because so many people and tasks must be coordinated.

Information systems are the lifeblood of a modern organization. Thus, maintaining and supporting those systems are two of the most important jobs in an organization. Maintenance and support activities ensure that the system and its users function efficiently and effectively for years after installation. Most organizations spend much more money maintaining and supporting existing systems than they do building new ones.

Program Development

Developing a complex system is an inherently difficult process. Consider the complexity of manufacturing automobiles. Tens of thousands of parts must be fabricated or purchased. Parts are assembled into small subcomponents (such as dashboard instruments, wiring harnesses, and brake assemblies), which are in turn assembled into larger subcomponents (such as instrument clusters, engines, and transmissions) and eventually into a complete automobile. Parts and subcomponents must be constructed, tested, and passed on to subsequent assembly steps. There are tens or hundreds of thousands of individual production steps. The effort, timeliness, cost, and output quality of each step depend on all of the preceding steps.

Program development (or construction) is similar in many ways to automobile manufacturing. Requirements and design specifications have already been determined. What remains is a complex production and assembly process that must ensure efficient resource use, minimal construction time, and maximum product quality. But unlike automobile manufacturing, the process is not designed once and then used to build thousands of similar units. Instead, a software manufacturing process must be redeveloped for each new project to match that project's unique characteristics.

When most people think of system development, they primarily think of programming. Programming isn't the only development phase activity, but it is clearly one of the most important. Its importance arises from several factors, including:

- Required resources
- Managerial complexity
- System quality

Program development consumes more resources than any other system development activity. Program development (including unit testing) typically accounts for at least one-third of all development labor. Program development also accounts for between one-third and one-half of the project development schedule. The magnitude of resources and time consumed during program development clearly warrants careful planning and management attention.

Order of Implementation

One of the most basic decisions to be made about program development is the order in which program components will be developed. Several orderings are possible, including:

- Input, process, output
- Top-down
- Bottom-up

Each project must adapt one or a combination of these approaches to specific project requirements and constraints.

input, process, output (IPO) development
A development order that implements input modules first, process modules next, and output modules last

Input, Process, Output (IPO) Development This order is based on data flow through a system or program. Programs or modules that obtain external input are developed first. Programs or modules that process the input (that is, transform it into output) are developed next. Programs or modules that produce output are developed last.

For structured designs and programs, an analyst can determine IPO ordering by examining the system flow chart and structure charts. For example, consider the payroll system flow chart in Figure 15-1. The programs *Maintain tax tables* and *Maintain employee database* obtain and modify data inputs for other programs, so they would be the first programs implemented. The Payroll program combines input and processing, so it would be implemented next. The *Check printing* and *Year-end tax* programs produce system outputs, so they would be implemented last.

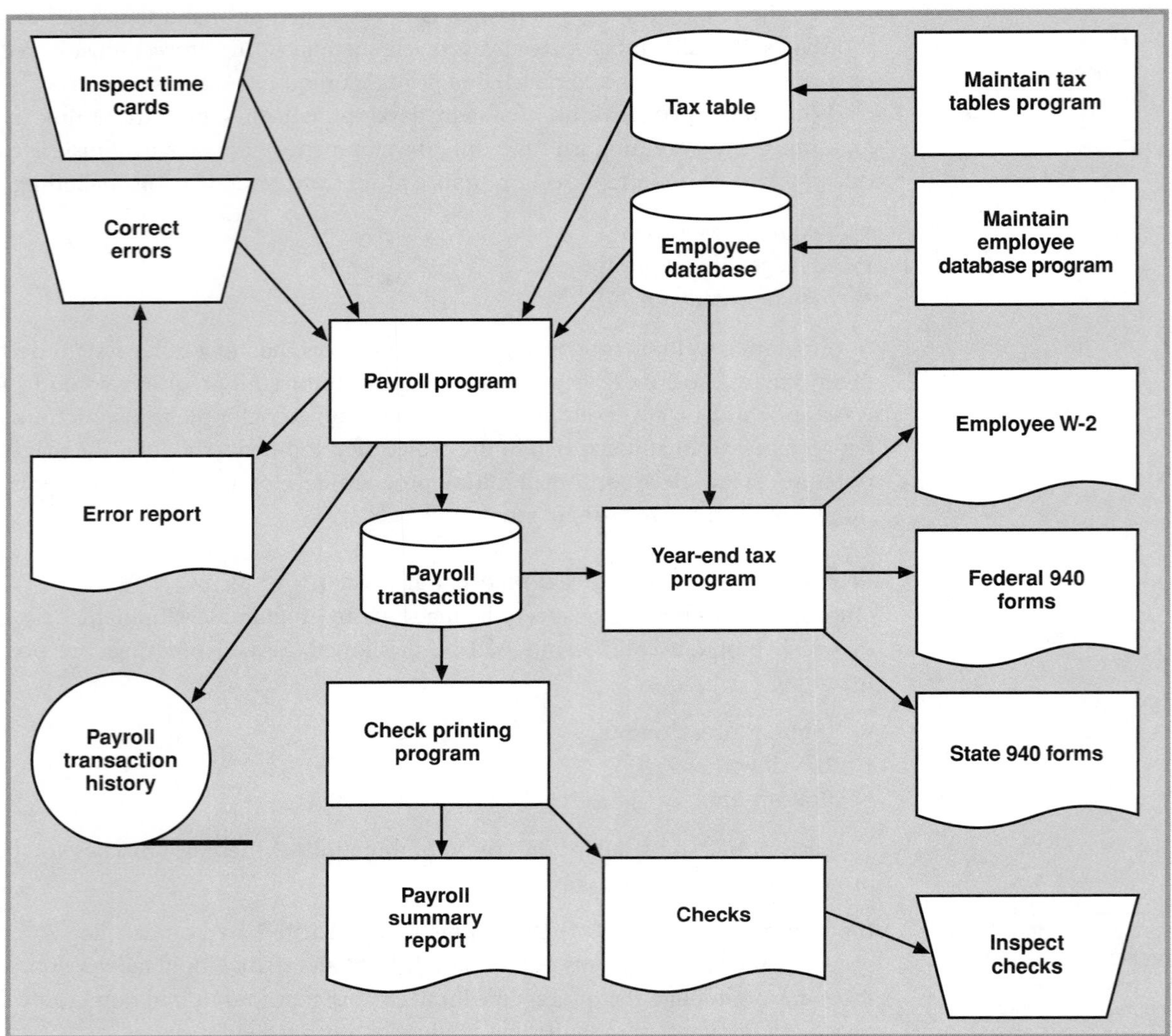

FIGURE 15-1
A system flow chart for a payroll system.

Figure 15-2 shows a structure chart for the Payroll program. An analyst can apply IPO implementation order to modules within a program by classifying them as input, process, and output modules. If the structure chart was constructed using transform analysis, then modules will be clearly organized into afferent (input), central transform (process), and efferent (output) "legs." (See the section "Developing a Structure Chart" in Chapter 9 for a review of structure chart organization.) For this program, the modules in the afferent leg of the structure chart (those below and including *Enter time cards*) would be implemented first, followed by the modules in the transform leg (those below and including *Calculate amounts*), and finally the modules in the efferent leg (those below and including *Output payroll*).

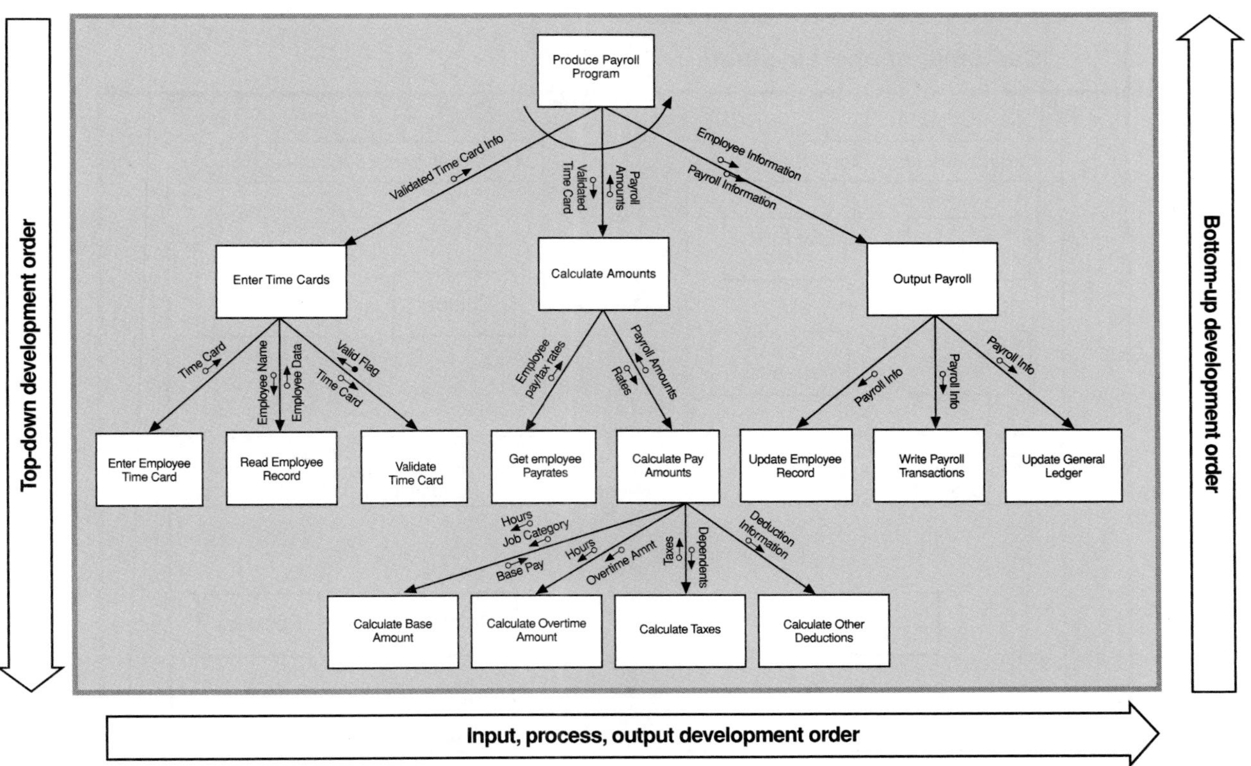

FIGURE 15-2
A structure chart for the Payroll program in Figure 15-1.

IPO development order can also be applied to object-oriented (OO) designs and programs. The key issue to analyze is data dependency—that is, which classes and methods capture or generate data that are needed by other classes or methods. One source of information about data dependency among classes is the package diagram. For example, the package diagram in Figure 15-3 shows that the customer and catalog maintenance subsystems are not dependent on each other or on either of the other two subsystems. The order-entry subsystem is dependent on both the customer and catalog maintenance subsystems, and the order fulfillment subsystem is dependent on the order-entry subsystem.

Data dependency among the packages (subsystems) implies data dependency among their embedded classes. Thus, the classes Customer, Catalog, and Package have no data dependency on the remaining RMO classes. Under IPO development order, those three classes are implemented first.

The chief advantage of IPO development order is that it simplifies testing. Because input programs and modules are developed first, they can be used to enter test data for process and output programs and modules. The need to write special-purpose programs to generate or create test data is reduced, thus speeding the development process.

IPO development order is also advantageous because important user interfaces (for example, data-entry routines) are developed early. User interfaces are more likely to require change during development than other portions of the system. Early development allows for early testing and user evaluation. If changes are needed, there is still plenty of time to make them. Early development of user interfaces also provides a head start for related activities such as training users and writing documentation.

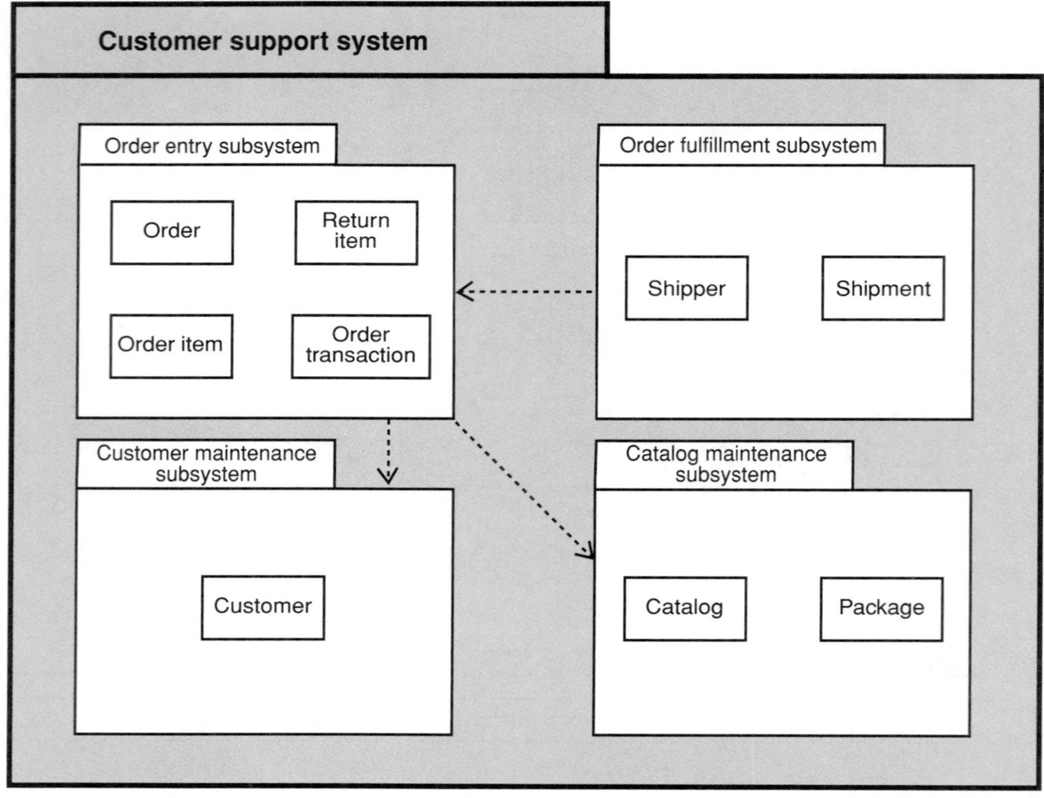

FIGURE 15-3
A package diagram for the four RMO subsystems.

A disadvantage of IPO development order is the late implementation of outputs. Output programs are useful for testing process-oriented modules and programs; analysts can find errors in processing by manually examining printed reports or displayed outputs. IPO development defers such testing until late in the development phase. However, analysts can usually generate alternate test outputs by using the query processing or report writing capabilities of a DBMS. If such outputs can be quickly and easily defined, then the disadvantage of late implementation of output routines is substantially mitigated.

Top-Down and Bottom-Up Development Order The terms *top-down* and *bottom-up* have their roots in structured design and structured programming. Both terms describe the order of implementation with respect to a module's location within a structure chart. For example, consider the structure chart in Figure 15-2. **Top-down development** begins with the module at the top of the structure chart (Produce Payroll Program). **Bottom-up development** begins with the set of modules at the lowest level of the structure chart.

top-down development
A development order that implements modules at the top of a structure chart first

bottom-up development
A development order that implements modules at the bottom of a structure chart first

Top-down and bottom-up program development can also be applied to OO designs and programs, although there is no obvious visual analogy as with structure charts. The key issue is method dependence—that is, which methods call which other methods. Information about method dependence can be obtained from method logic descriptions, sequence diagrams, and class diagrams. Method logic and sequence diagrams explicitly show messages passed between objects. The class diagram shows inheritance relationships among classes. Subclasses have method dependence on their parent classes.

The primary advantage of top-down development is that there is always a working version of a program. For example, top-down development of the program in Figure 15-2 would begin with a complete version of the topmost module and dummy (or stub) versions of its three subordinate modules (stub modules are discussed later in the "Testing" section). This set of modules is a complete program that can be compiled, linked, and executed, although at this point it wouldn't do very much when executed.

Once the topmost module is completed, construction moves down to the next level of the structure chart. As each of these modules is implemented, stubs for the modules on the next lower level are added. At every stage of development, the program should be complete (that is, it should be able to be compiled, linked, and executed). Its behavior becomes more complex and realistic as development proceeds, however.

The primary disadvantage of top-down development order is that it doesn't use programming personnel very efficiently at the beginning of construction. Development has to proceed through two or three iterations before a significant number of modules or methods are being developed simultaneously. However, if the first few iterations of the program can be completed quickly, then the disadvantage is minimal.

The primary advantage of bottom-up development is that many programmers can be put to work immediately. In addition, lower-level modules are often the most complex and difficult to write, so early development of those modules allows more time for development and testing. Unfortunately, bottom-up development also requires writing a large number of driver programs to test bottom-level modules, which adds additional complexity to the development and testing process (this issue is discussed further in the "Testing" section). Also, the entire system isn't assembled until the topmost modules are written. Thus, testing of the system as a whole is delayed.

Other Development Order Considerations IPO, top-down, and bottom-up development are only a starting point for creating a program development plan. Other factors that must be considered include user feedback, training, documentation, and testing. User feedback, training, and documentation all depend heavily on the user interfaces of the system. Early implementation of user interfaces allows user training and the development of user documentation to begin early in the development process. It also allows the users to provide feedback on the quality and usability of the interface. Note that interface construction is substantially advanced if user-interface prototypes were constructed and validated during analysis or design.

Testing is also an important consideration when determining development order. As individual software components (such as modules or methods) are constructed, they must be tested. Programmers must find and correct errors as soon as possible because they become much harder to find and more expensive to fix as the construction process proceeds. It's important both to identify portions of the software that are susceptible to errors and to identify portions of the software where errors can pose serious problems that affect the system as a whole. These portions of the software need early construction and testing regardless of where they fit within the basic approaches of IPO, top-down, or bottom-up development.

Testing and construction are highly interdependent. For this reason, a formal plan covering both testing and construction is normally created before either activity begins. The construction and test plan covers many specifics, including:

- Development order
- Testing order
- Data used to test modules, module groups, programs, and subsystems

- Acceptance criteria
- Personnel assignments (construction and testing)

Testing is discussed in detail later in this chapter. But for now, keep in mind that construction and testing go hand in hand. Their interdependence and complexity necessitate formal planning and regular comparisons between the plan and actual performance.

Framework Development

When constructing a large OO system, it is not unusual to build an object framework (or set of foundation classes) that covers most or all of the application-specific classes. For example, when constructing an OO account maintenance system for a bank, developers might design and construct a set of classes to represent customers and various types of bank accounts (for example, savings, checking, and certificates of deposit). (Refer to the "Object Frameworks" section of Chapter 13 for a discussion of object frameworks.)

Foundation classes are typically reused in many parts of the system and across many different applications. Because of this, they are a critical system component. Errors in a foundation class can affect every program in the system. In addition, later changes to foundation classes may require significant changes throughout the system.

Foundation classes are typically implemented first to minimize the impact of errors and changes. Foundation classes are typically assigned to the best programmers and are tested more thoroughly than other classes. Early and thorough testing guarantees that bugs or other problems will be discovered before other code that depends on the foundation classes has been written.

Team-Based Program Development

A team of programmers normally works on program development. Using multiple programmers compresses the development schedule by allowing many portions of the system to be developed simultaneously. However, team-based program development introduces its own set of management issues. These include:

- Organization of programming teams
- Task assignment to specific teams or members
- Member and team communication and coordination

There are many different ways of organizing an implementation team. Some commonly used organizational models include:

- Cooperating peer
- Chief developer
- Collaborative specialist

Table 15-1 summarizes the characteristics of each team type, and the types of projects and tasks best suited to each.

Team type	Team characteristics	Task and project types
Cooperating peers	Equal skill levels Overlapping specialties Consensus-based decision making	Experimentation Creative problem solving
Chief developer	Organized as a military platoon or squad One leader makes all important decisions	Well-defined objectives Well-defined path to completion
Collaborative specialists	Wide variation in skill and experience Minimal overlap in technical specialties Leader is primarily an administrator Consensus-based decision making	Diagnosis or experimentation Creative and integrative problem solving Wide range of technology

TABLE 15-1
A comparison and summary of development team types.

cooperating peer team
A team with members of roughly equal skill and experience with overlapping areas of specialization

chief developer team
A team with a single leader who makes all important decisions

collaborative specialist team
A team with members who have wide variation in and minimal overlap of skills and experience

A **cooperating peer team** includes members of roughly equal skill and experience with overlapping areas of specialization. Members are considered equals, although they may be assigned tasks of varying importance or complexity. Decisions are primarily made by consensus, and the team frequently meets to exchange information and build consensus.

A **chief developer team** is similar to a small military unit. An assigned leader performs a number of functions, including technical consulting, team coordination, and task assignment. In this type of team, there is much less communication than with a cooperating peer team. The chief developer makes most of the important decisions, although he or she may seek input from members individually or collectively.

A **collaborative specialist team** is similar to a cooperating peer team, but its members have wide variation in and minimal overlap of skills and experience. Such teams are often composed of members from different organizational subunits. There may be an appointed leader, but that member typically performs only administrative functions (such as scheduling, coordinating, and interacting with external constituencies). Technical decisions are generally made by consensus, although member opinions usually carry extra weight within the member's own area of expertise.

Some common principles of team organization underlie all development projects and organizational structures. One is that team size should be kept relatively small (no more than 10 members). Larger team sizes tend to be inefficient due to the inherent complexity of communication and coordination in large groups. When more than 10 developers are assigned to a project, it is best to break them up into small teams (approximately five members each). Each team should be assigned a relatively independent portion of the project. One member of each team should be designated to handle coordination and communication with other teams. Having a single point of contact simplifies communication and provides for some specialization of functions within each team.

Another common principle of team organization is that team structure should be matched to the task and project characteristics. Teams with a well-defined implementation task that does not push the limits of member knowledge or technical feasibility are usually best organized as chief programmer teams. A chief programmer team operates very efficiently in such an environment.

Teams assigned to tasks that require experimentation or high creativity are better served by a cooperating peer or collaborative specialist model. Because of their more open communication, cooperating peer teams are especially well suited to tackling tasks that require generating and evaluating a large number of ideas. Overlap in skill, speciality, and experience allows thorough evaluation of each idea.

Collaborative specialist teams are well suited to projects that span a large range of cutting-edge technology. They are also well suited to tackling projects that require integrated problem solving (for example, diagnosing and fixing bugs in an existing complex system). However, success depends on a true collaborative process, which is sometimes difficult to achieve among members with wide variation in skill and experience.

Member skills must be appropriately matched to the tasks at hand. Skill matching is a fairly obvious requirement with respect to technical skills such as database management, user interfaces, and numeric algorithms. Skill matching is less obvious but no less important for nontechnical skills. Teams need a mix of nontechnical skills and traits, including the ability to generate new ideas, build consensus, manage details, and communicate with external constituencies.

Source Code Control

source code control system (SCCS)
An automated tool for tracking source code files and controlling changes to those files

Development teams need tools to help coordinate their programming tasks. A **source code control system (SCCS)** is an automated tool for tracking source code files and controlling changes to those files. An SCCS stores project source code files in a repository. The SCCS acts like a librarian—it implements check-in and checkout procedures, tracks which programmer has which files, and ensures that only authorized users have access to the repository.

Programmers can manipulate files in the repository as follows:

- Check out a file in read-only mode.
- Check out a file in read/write mode.
- Check in a modified file.

A programmer checks out a file in read-only mode when he or she wants to examine the code without making changes (for example, to examine a module's interfaces to other modules). When a programmer needs to make changes to a file, it is checked out in read/write mode. The SCCS allows only one programmer to check out a file in read/write mode. The file must be checked back in before another programmer can check it out in read/write mode.

Figure 15-4 shows the main display of Microsoft Visual SourceSafe. Various source code files from the RMO customer support system are shown in the display. Some files are currently checked out by programmers. For each file checked out in read/write mode, the program lists the programmer who checked it out, the date and time of checkout, and the current location of the file. The icon for each checked-out file is displayed with a red border and check mark.

An SCCS prevents multiple programmers from updating the same file at the same time, thus preventing inconsistent changes to the source code. Source code control is an absolute necessity when programs are developed by multiple programmers. It prevents inconsistent changes and automates coordination among programmers and teams. The repository also serves as a common facility for backup and recovery operations.

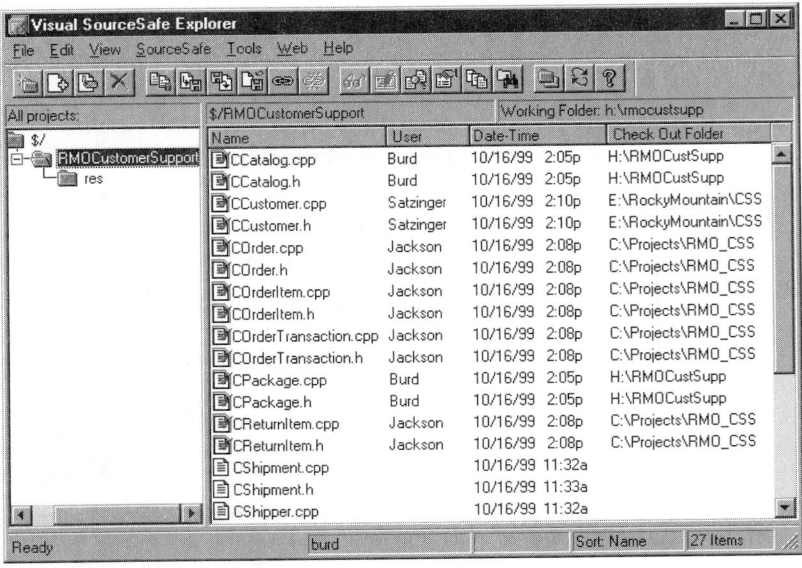

FIGURE 15-4
Project files managed by a source code control system.

Versioning

Medium- and large-scale systems are complex and constantly changing. Changes occur rapidly during implementation and slower afterward. System complexity and rapid change create a host of management problems—particularly for testing and support. Testing is problematic in such an environment because the system is a moving target. By the time an error is discovered, the code that caused the error may already have been moved, altered, or deleted. Support is complex for similar reasons. Support personnel need to know the state of the system as it is installed on a user's computer system to respond properly to a bug report or request for help.

Complex systems are developed, installed, and maintained in a series of versions to simplify testing and support. A system version created during development is called a test version. A test version contains a well-defined set of features and represents a concrete step toward final completion of the system. Test versions provide a static system snapshot and a checkpoint to evaluate the project's progress.

alpha version
A system that is incomplete but ready for some level of rigorous testing

beta version
A system that is stable enough to be tested by end users

production version, production release, or release version
A system that is formally distributed to users or made operational

An **alpha version** is a test version that is incomplete but ready for some level of rigorous testing. Multiple alpha versions may be defined depending on the size and complexity of the system. The lifetime of an alpha version is typically short (usually measured in days or weeks).

A **beta version** is a test version that is stable enough to be tested by end users. A beta version is produced after one or more alpha versions have been tested and known problems have been corrected. End users test beta versions by using them to do real work. Thus, beta versions must be more complete and less prone to disastrous failures than alpha versions. Beta versions are typically distributed to end users and evaluated over a period of weeks or months.

A system version created for long-term release to users is called a **production version**, **release version**, or **production release**. A production version is considered a final product, although software systems are rarely "finished" in the traditional sense

maintenance release
A system update that provides bug fixes and small changes to existing features

of that term. Multiple production versions are used to add features and fix bugs discovered after installation. Minor production releases (sometimes called **maintenance releases**) provide bug fixes and small changes to existing features. Major production releases add significant new functionality and may be the result of rewriting an older release from the ground up.

Figure 15-5 shows a series of possible test and production versions for the RMO customer support system. Each version is described in Figure 15-6. The system is delivered in two major production releases—versions 1.0 and 2.0. Each initial production release is preceded by one or more alpha and beta test versions. Each version adds or updates functionality and includes bug fixes for the previous version. Version 1.1 is a maintenance or minor production release of version 1.0.

FIGURE 15-5
A timeline of test and production versions for the RMO customer support system.

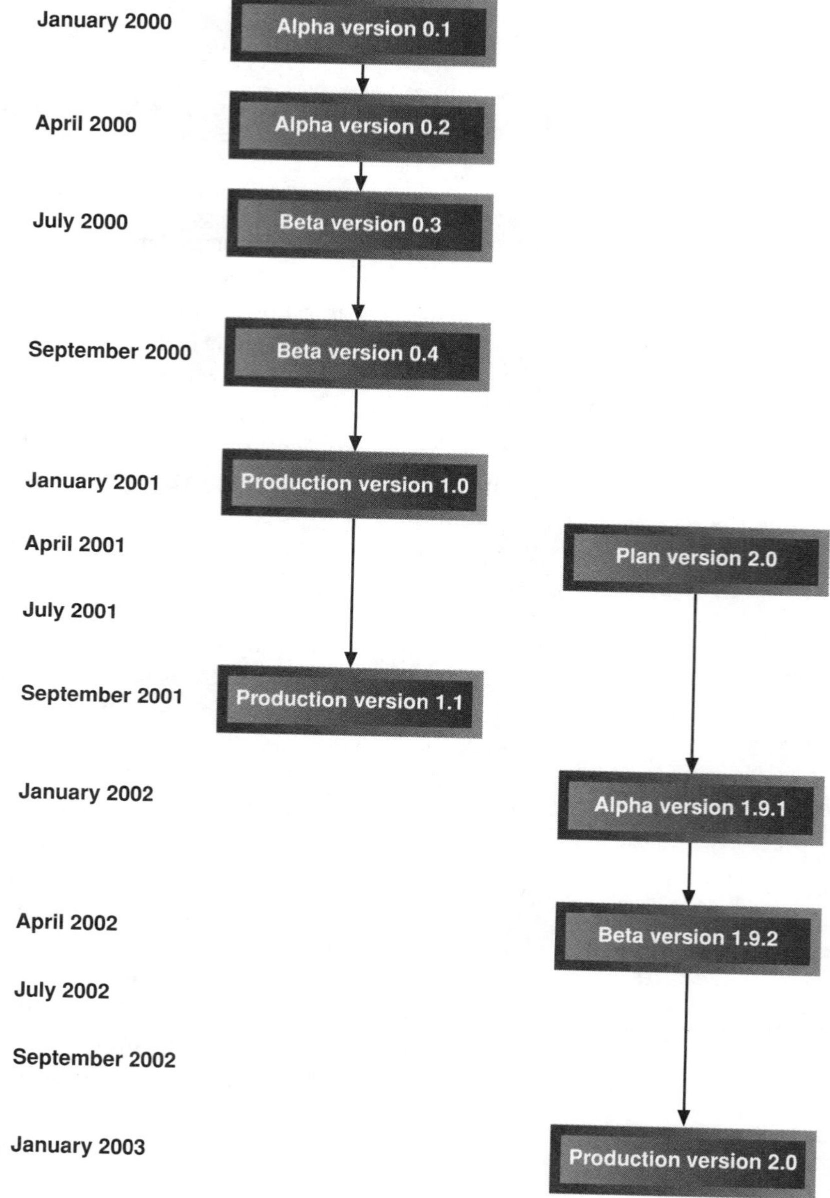

FIGURE 15-6
Descriptions of versions in Figure 15-5 for the RMO customer support system.

Alpha 0.1—Basic database functionality with simple CRUD capabilities. Handles regular transactions only, no reports or printing capability.

Alpha 0.2—Full CRUD for all transaction types with screens in near final form, no reports or printing capability. Includes bug fixes for version 0.1.

Beta 0.3—Screens in final form with simple on-line help, simple printing of screen contents. Includes bug fixes for version 0.2.

Beta 0.4—Adds reports and formatted printing. Includes bug fixes for version 0.3.

Production 1.0—Includes all bug fixes for version 0.4.

Prduction 1.1—Adds keystroke shortcuts for experienced users, includes all bug fixes for version 1.0.

Alpha 1.9.1—Adds simple database extraction into a DSS tool to version 1.0.

Beta 1.9.2—Adds user-friendly database navigation and downloads into a DSS tool to version 1.1. Includes all bug fixes for version 1.9.1.

Production 2.0—Includes all bug fixes for version 1.9.2.

Note that the time line for developing version 2.0 overlaps maintenance changes to version 1.0. Alpha version 1.9.1 is based on production version 1.0, but beta version 1.9.2 is based on production version 1.1. Overlapping older production versions with test versions of future production releases is typical.

Keeping track of versions is complex. Each version needs to be uniquely identified for users and testers. In applications designed to run under Windows, the version information is typically viewed by choosing the About item of the standard help menu, as shown in Figure 15-7. Users seeking support or reporting errors in a beta version use this feature to report the system version to testers or support personnel.

FIGURE 15-7
The About box of a typical Windows application.

Controlling multiple versions of the same system requires sophisticated version control software. Version control capabilities are normally built into an SCCS. Programmers and support personnel can extract the current version or any previous version for execution, testing, or modification. Modifications are saved under a new version number to protect the accuracy of the historical snapshot.

Beta and production versions must be stored as long as they are installed on any user machines. Stored versions are used to evaluate future bug reports. For example, when a user reports a bug in version 1.0, support personnel will extract that release from the archive, install it, and attempt to replicate the user's error. Feedback provided to the user will be specific to version 1.0 even if the most recent production release is a higher-numbered version.

Quality Assurance

quality assurance (QA) The process of ensuring that an information system meets minimal quality standards

As with any business procedure or system, quality is a major concern with information systems. **Quality assurance (QA)** is the process of ensuring that an information system meets minimal quality standards as determined by users, implementation staff, and management. QA is sometimes equated with finding bugs in program code, but this is a narrow and incomplete view. QA is a series of activities that are performed throughout the SDLC to build systems correctly from the start or detect and fix errors as soon as possible. Integrating quality assurance into early project phases allows many programming errors to be completely avoided. It also ensures that the system that is actually developed meets the needs of the users and the organization.

QA activities during analysis concentrate on identifying gaps or inconsistencies in system requirements. QA activities during design concentrate on satisfying stated requirements and on making design decisions that will lead to easily implemented bug-free programs. QA activities during implementation consist primarily of testing. However, design and implementation overlap in many projects. Thus, quality assurance activities for design are typically integrated with testing activities.

QA activities are often shortchanged during design and especially during implementation. There are several reasons this occurs, including the following:

- Schedule pressures can build with each successive phase. QA and testing activities may be bypassed in an ill-fated attempt to speed up the project.
- QA activities require development personnel to open their work to thorough examination and criticism by others. Many people are reluctant to do this.
- Many people view testing and test personnel as the bearers of bad news. They mistakenly believe that no news is good news and that bypassing testing is a way of avoiding bad news.

There is a simple way to prevent human nature and schedule pressure from derailing quality assurance activities: Integrate QA into the project and schedule from the beginning and never abandon it. Quality standards should be clearly stated, be measurable, and require a product that doesn't meet those standards to be fixed no matter what the effect on the schedule or budget.

Another key factor in firmly establishing QA in the development process is to build an environment of openness, collegiality, and mutual respect among project participants. Personnel must be receptive to suggestions and constructive criticism and be willing to provide suggestions and criticisms to others. QA cannot be effective if it is allowed to devolve into an exercise in destructive criticism, finger-pointing, and blame assignment.

The cost of fixing an error rises during each development phase. Errors are best detected during analysis or design and, thus, never committed to program code. Errors in programs are much easier to fix in the early stages of implementation than during later acceptance testing or, in the worst case, after the system is operational. This economic reality makes QA efforts throughout the development life cycle well worth their cost.

Technical Reviews

Most programmers have had many experiences in which they were unable to correct an error because they "couldn't see it." But when the source code containing the error is shown to another programmer, the other programmer spots it immediately. Common examples include misspelled keywords, malformed *if* statement conditions, and illegal or spurious characters in source code. Such errors happen to programmers at all skill and experience levels.

technical review
A formal or informal review of design or construction details by a group of developers

A **technical review** opens the design and construction process to input from other people. Technical reviews provide an opportunity for other people to find problems and offer constructive criticism. Technical reviews vary widely from one organization to another and sometimes among projects within an organization. Some organizations use informal processes, while others adopt formal procedures.

A *walkthrough* is a review by two or more people of the accuracy and completeness of a model or program. Walkthroughs are most often used during analysis and design, although they can be used during implementation. During design and implementation, a walkthrough is a technical review in which two or more developers review a design or implementation to assess and improve its quality. Typically, one of the participants has already created a model or module before the walkthrough. The developer describes its underlying assumptions and operation, and the other participants provide comments and suggestions.

inspection
A formal review of design or construction details by a group of developers, where each person plays a specific role

An **inspection** is a more formal version of a walkthrough. Participants review and analyze materials before they meet as a group. Review materials include the model or code to be inspected, related models (for example, a structure chart or class diagram), and notes on specific types of errors that could occur. Group meetings usually follow a standard format.

When the group meets, participants play specific roles, including presenter, critic, and secretary. The presenter (usually the developer of the model or code) summarizes the material being inspected. The critics describe errors or concerns they found before the meeting, and the errors are discussed by all members of the group. Additional errors or problems may be uncovered during the discussion. The participants discuss possible solution strategies and agree on a specific approach. The secretary records all of the errors and the agreed-upon solution strategies.

Walkthroughs and inspections are important QA processes because they can detect errors *before* code has been written. Studies have shown that technical reviews accomplish the following:

- Reduce the number of errors that reach testing by a factor of 5 to 10
- Reduce testing costs by approximately 50 percent

Technical reviews reduce development costs and shorten the development schedule because a large number of errors never are passed along to be coded, tested, diagnosed, or fixed.

Testing and technical reviews each find between 50 and 75 percent of errors. But some errors are more easily detected by one method or the other. Some errors are rarely found by one technique but easily found by the other. Thus, the two techniques are more effective jointly than individually.

Testing

Testing is the process of examining a product to determine what defects it contains. To conduct a test, programmers must have already constructed the software and there must be well-defined standards for what constitutes a defect. The developers can test products by reviewing their construction and composition or by exercising their function and examining the results. This section concentrates on the latter type of testing. This process is shown in Figure 15-8.

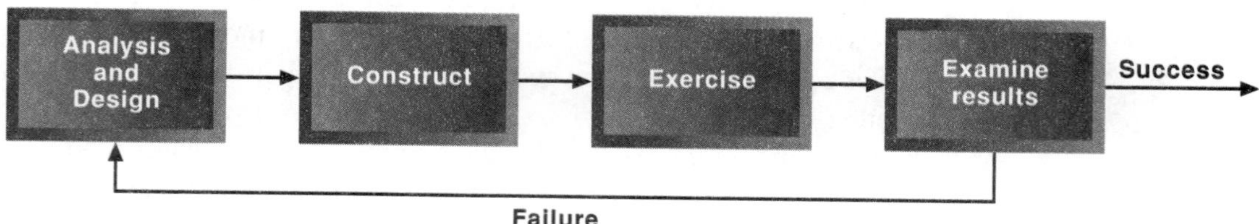

FIGURE 15-8
A generic model of software testing.

Software Testing An information system is an integrated collection of software components. Components can be tested individually or in groups, or the entire system can be tested as a whole. Testing components individually is called unit testing. Testing components in groups is called integration testing. Testing entire systems is called system testing. Each type of testing is described in detail later in this section.

The three testing types are each correlated to a specific phase of the SDLC, as shown in Figure 15-9. A system test examines the behavior of an entire system with respect to technical and user requirements. These requirements are determined during the analysis phase of the SDLC. During high-level design, the division of the system into high-level components and the structural design of those components are determined. Integration testing tests the behavior of related groups of software components. Low-level design is concerned with the internal construction of individual components. Unit testing tests each individual software component in isolation.

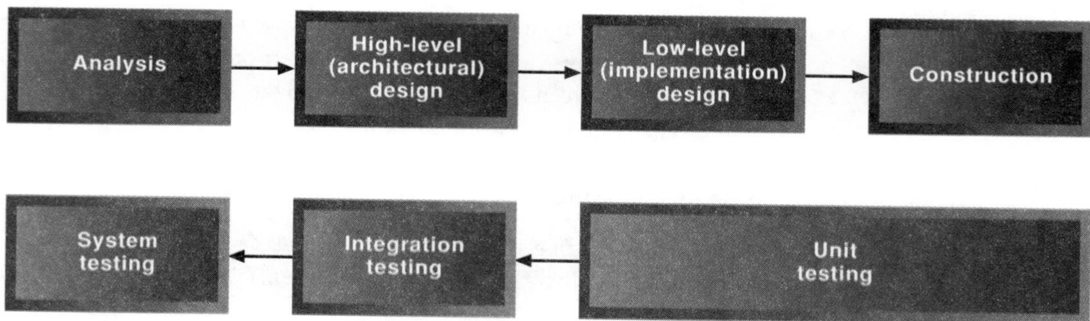

FIGURE 15-9
The correspondence between SDLC phases and various types of testing.

Because each testing level is related to a specific phase of the SDLC, testing activities can be spread out through the life cycle, as illustrated in Figure 15-10. Planning for each type of testing can occur with its related SDLC phase, and development of specific tests can occur once the planning is complete. Tests cannot be conducted until relevant portions of the system have been constructed, however.

Analysis	High-level (architectural) design		Low-level (implementation) design		Construction		
Plan system tests	Plan integration tests	Develop system tests	Plan unit tests	Develop integration tests	Implement unit tests	Implement integration tests	Implement system tests

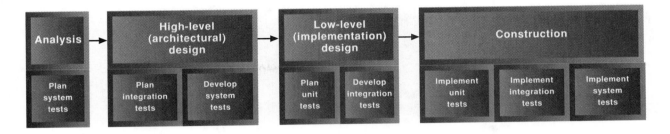

FIGURE 15-10 SDLC phases and the testing activities that can be performed within each phase.

test case
A formal description of a starting state, one or more events to which the software must respond, and the expected response or ending state

test data
A set of starting states and events used to test a module, group of modules, or entire system

unit testing, or module testing
Testing of individual code modules or methods before they are integrated with other modules

An important part of developing tests is specifying test cases and data. A **test case** is a formal description of the following:

- A starting state
- One or more events to which the software must respond
- The expected response or ending state

Both starting state and events are represented by a set of **test data**. For example, the starting state of a system may represent a particular set of database contents (such as the existence of a particular customer and order for that customer). The event may be represented by a set of input data items (such as a customer account number and order number used to query order status). The expected response may be a described behavior (such as the display of certain information) or a specific state of stored data (such as a canceled order).

Preparing test cases and data is a tedious and time-consuming process. At the program or module level, every instruction must be executed at least once. Ensuring that all instructions are executed during testing is a complex problem. Fortunately, there are automated tools based on proven mathematical techniques to generate a complete set of test cases. See the Watson and McCabe article in the "Further Resources" section for a thorough discussion of this topic.

Unit Testing This is the process of testing individual code modules before they are integrated with other modules. Unit testing is sometimes called **module testing**, although that term implies that software units are structured programming modules. In fact, unit testing can be applied to structured or OO software, and the unit being tested may be a function, subroutine, procedure, or method (for the remainder of this section, we will use the term *module* to refer to any of these programming constructs). Units may also be relatively small groups of interrelated modules, that are always executed as a group. The goal of unit testing is to identify and fix as many errors as possible before modules are combined into larger software units (such as programs, classes, and subsystems). Errors become much more difficult and expensive to locate and fix when many modules are combined.

Few modules are designed to operate in isolation. Instead, groups of modules are designed to execute as an integrated whole. Modules may call other software units to perform tasks or may be called by other modules. This relationship is easily seen in a structure chart (see Figure 15-11), although it also exists among methods in OO software. For example, the module *Calculate pay amount* is called by the module *Calculate amounts*, which in turn calls the three modules below it in the structure chart.

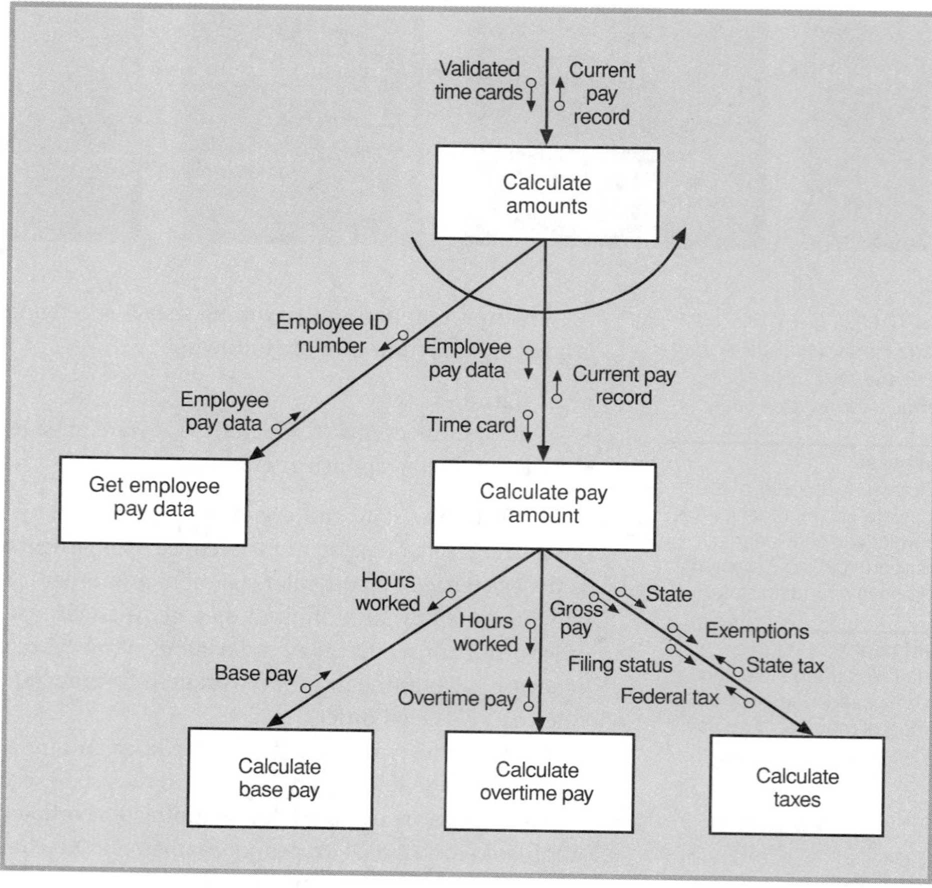

FIGURE 15-11
A portion of a structure chart for a program to calculate payroll.

driver
A module, developed for unit testing, that simulates the calling behavior of a module that hasn't yet been developed

If *Calculate pay amount* is being tested in isolation, then two types of testing modules are required. The first module type is called a driver. A **driver** simulates the calling behavior of a module. A driver module implements the following functions:

- Sets the value of input parameters to the tested function
- Calls the tested module, passing it any required input parameters
- Accepts return parameters from the tested module and prints or displays them

Figure 15-12 shows a simple driver module for testing *Calculate pay amount*. A more complex driver module might use test data consisting of hundreds or thousands of module inputs and correct outputs stored in a file or database. The driver would loop through the test data and repeatedly call *Calculate pay amount*, check the return parameter against the expected value, and print or display warnings of any discrepancy.

```
module main()

// Driver Module to Test CalculatePayAmount()
{
   // Declare Module Parameters

   record EmployeePayData   {
      integer EmployeeIDNumber;
      boolean SalariedEmployee;
      real PayRate;
      char[2] State;
      integer FilingStatus;
      integer Exemptions;
   }
   record TimeCard   {
      integer EmployeeIDNumber;
      date StartDate;
      array[7] of real HoursWorked;
   }
   record CurrentPayRecord   {
      real BasePay;
      real OvertimePay
      real FederalTax;
      real StateTax;
   }

   // Set Input Parameter Values

   EmployeeData.EmployeeNumber=123456789;
   EmployeeData.SalariedEmployee=false;
   EmployeeData.PayRate=32.50;
   EmployeeData.State="AZ";
   EmployeeData.FillingStatus=1;
   EmployeeData.Exemptions=5;
   TimeCard.EmployeeIDNumber=123456789;
   TimeCard.StartDate=05/21/2000;
   TimeCard.HoursWorked[0]=0.0;
   TimeCard.HoursWorked[1]=0.0;
   TimeCard.HoursWorked[2]=8.5;
   TimeCard.HoursWorked[3]=7.5;
   TimeCard.HoursWorked[4]=8.0;
   TimeCard.HoursWorked[5]=8.0;
   TimeCard.HoursWorked[6]=9.0;

   // Call Tested Module

   call CalculatePayAmount (EmployeeData, TimeCard, CurrentPay Record);

   // Print Results

   print(EmployeeData, TimeCard, CurrentPayRecord);
}
```

FIGURE 15-12
A driver module for testing Calculate pay amount.

Using a driver allows a subordinate module to be tested before modules that call it have been written. Drivers are used extensively in bottom-up development since child modules (or methods) are developed and unit-tested before their parents are developed.

stub
A module, developed for testing, that simulates the execution or behavior of a module that hasn't yet been developed

The second type of testing module used to perform unit tests is called a stub. A **stub** simulates the behavior of a called module that hasn't yet been written. A unit test of *Calculate pay amount* would require three stub modules, one for each of the modules that appear below it in Figure 15-11. Stubs are relatively simple modules that usually have only one or two lines of executable code. Each of the stubs used to test *Calculate pay amount* can be implemented as a statement that simply returns a constant regardless of the parameters passed as input. Figure 15-13 shows sample code for each of the three stub modules.

```
Module CalculateBasePay(HoursWorked,BasePay)

// Stub Module

array[7] of real HoursWorked;
real BasePay;
{
   BasePay=1000.00;
   return;
}

Module CalculateOvertimePay(HoursWorked,OvertimePay)

// Stub Module

array[7] of real HoursWorked;
real OvertimePay;
{
   OvertimePay=125.00;
   return;
}

Module CalculateTaxes(GrossPay,FilingStatus,State,Exemptions,FederalTax,StateTax)

// Stub Module

real GrossPay;
integer FilingStatus;
char[2] State;
integer Exemptions;
real FederalTax;
real StateTax;
{
   FederalTax=275.00;
   StateTax=75.00;
   return;
}
```


FIGURE 15-13
Stub modules used for testing Calculate pay amount.

Stubs are needed for top-down development. In fact, top-down development often begins by writing a stub for every module or method in a program or class. Individual stub modules and methods are then replaced with fully implemented code as it is developed.

integration test
A test of the behavior of a group of modules or methods

Integration Testing An **integration test** tests the behavior of a group of modules or methods. The purpose of an integration test is to identify errors that were not or could not be detected by unit-testing individual modules or methods. Such errors may result from a number of problems, including:

- Interface incompatibility. For example, a caller module passes a variable of the wrong data type to a subordinate module.
- Parameter values. A module is passed or returns a value that was unexpected (such as a negative number for a price).
- Run-time exception. A module generates an error such as "out of memory" or "file already in use" due to conflicting resource needs.
- Unexpected state interactions. The states of two or more modules interact to cause complex failures (such as an order class method that operates correctly for all possible customer object states except one).

These are some of the most common integration testing errors, but there are many other possible errors and causes.

Once an integration error has been detected, the responsibility for incorrect behavior must be traced to a specific module or modules. The person responsible for performing the integration test is generally also responsible for identifying the cause of the error. Once the error has been traced to a particular module, the programmer who wrote the module is asked to rewrite it to correct the error.

Integration testing of structured software is straightforward but not necessarily easy to implement. Most structured modules are called by only a single parent module. In addition, most structured modules do not store a permanent state within themselves. Internal variables are always reinitialized to the same values each time the module is called. The combination of these characteristics allows test personnel (often with the assistance of automated testing tools) to generate test cases and data that exercise all possible control paths through the software being tested. Confidence that testing has revealed important errors increases with the number of control paths that are tested.

In contrast, integration testing of OO software is much more complex and not as well understood. There is no clear hierarchical structure to an OO program. An OO program consists of a set of interacting objects that can be created or destroyed during execution. Object interactions and control flow are dynamic and complex.

Additional factors that complicate OO integration testing include the following:

- Methods can (and usually are) called by many other methods, and the calling methods may be distributed across many classes.
- Classes may inherit methods and state variables from other classes.
- The specific method to be called is dynamically determined at run time based on the number and type of message parameters.
- Objects can retain internal variable values (that is, the object state) between calls. The response to two identical calls may be different due to state changes that result from the first call or occur between calls.

The combination of these factors makes it difficult to determine an optimal testing order. The factors also make it difficult to predict the behavior of a group of interacting methods and objects. Thus, developing and executing an integration test plan for OO software are much more complex than for structured software. Specific methods and techniques for dealing with that complexity are well beyond the scope of this textbook. See the "Further Resources" for OO software testing references.

system test
A test of the behavior of an entire system or independent subsystem

build and smoke test
A system test that is performed daily

acceptance test
A system test performed to determine whether the system fulfills user requirements

System Testing A system test tests the behavior of an entire system or independent subsystem. System testing is normally first performed by developers or test personnel to ensure that the system does not malfunction in obvious ways and that the system fulfills the developers' understanding of user requirements. Later testing by users confirms whether the system does indeed fulfill their requirements.

A **build and smoke test** is a system test that is typically performed daily. The system is completely compiled and linked (built) and a battery of tests is performed to see whether anything malfunctions in an obvious way ("smokes"). Build and smoke tests are commonly associated with iterative or rapid development. However, build and smoke tests can also be used in more traditional projects if top-down development is employed.

Build and smoke tests are valuable because they provide rapid feedback regarding significant problems. Any problem that occurs during a build and smoke test must be the result of code modified or added since the previous test. Daily testing ensures that errors are found quickly and that they can be easily tracked to their source. Less frequent testing provides rapidly diminishing benefits because more code has changed and errors are more difficult to track to their source.

An **acceptance test** is a system test performed to determine whether the system fulfills user requirements. Acceptance testing is typically the last round of testing before a system is handed over to its users. Acceptance testing is a very formal activity in most development projects. Details of acceptance tests are sometimes included in the RFP and procurement contract when a new system is built by or purchased from an external party.

Who Tests Software? There are many participants in the testing process. Their exact number and role depend on the size of the project and other project characteristics. Specific participants include:

- Programmers
- Users
- Quality assurance personnel

testing buddy
A programmer assigned to test code written by another programmer

Programmers are generally responsible for unit-testing their own code prior to integrating it with modules written by other programmers. In some organizations, programmers are assigned a **testing buddy** to help them test their own code. Programmers are assigned specific responsibility for testing their buddy's code prior to integration testing. Having a different programmer test the code usually finds more errors.

Users are primarily responsible for beta testing and acceptance testing. When beta versions are developed, they are distributed to a group of users for testing over a period of days, weeks, or months. Volunteers are frequently used, although this is not always desirable since volunteers are not a random sample of users. Volunteer beta testers tend to be more computer literate and have a higher tolerance for bugs and malfunctions than ordinary users. These characteristics may result in higher-quality feedback for some problems but a lack of feedback for other problems.

Acceptance testing is normally conducted by users in partnership with IS development or operations personnel. The rigor and importance of acceptance tests require participation by a large number of users across a wide range of user levels (for example, data-entry clerks and the managers who will "own" the system). Although IS personnel may perform system setup and troubleshooting functions, it is ultimately up to the users to accept or reject the system.

In a large system development project, a separate quality assurance group or organization is usually formed. The QA group is responsible for all aspects of testing except unit testing and acceptance testing. The QA group's responsibilities and activities typically include:

- Developing a testing plan
- Performing integration and system testing
- Gathering and organizing user feedback on alpha and beta software versions and identifying needed changes to the system design or implementation

To maintain objectivity and independence, the QA group normally reports directly to the project manager or to a permanent IS manager.

Installation

Once a new system has been developed and tested, it must be installed and placed into operation. Installing a system and making it operational are complex because there are many conflicting constraints, including cost, customer relations, employee relations, logistical complexity, and overall exposure to risk. Some of the more important issues to consider when planning installation include:

- Incurring costs of operating both systems in parallel
- Detecting and correcting errors in the new system
- Potentially disrupting the company and its IS operations
- Training personnel and familiarizing customers with new procedures

Different approaches to installation represent different trade-offs among cost, complexity, and risk. The most commonly used installation approaches are:

- Direct installation
- Parallel installation
- Phased installation

Each approach has different strengths and weaknesses, and no one approach is best for all systems. Each approach is discussed in detail here.

Direct Installation

direct installation, or immediate cutover
An installation method that installs a new system, quickly makes it operational, and immediately turns off any overlapping systems

In a **direct installation**, the new system is installed and quickly brought to operational status, and any overlapping systems are then turned off. Direct installation is also sometimes called **immediate cutover**. Both systems are concurrently operated for only a brief time (typically a few days or weeks) while the new system is being installed and tested. Figure 15-14 shows a time line for direct installation.

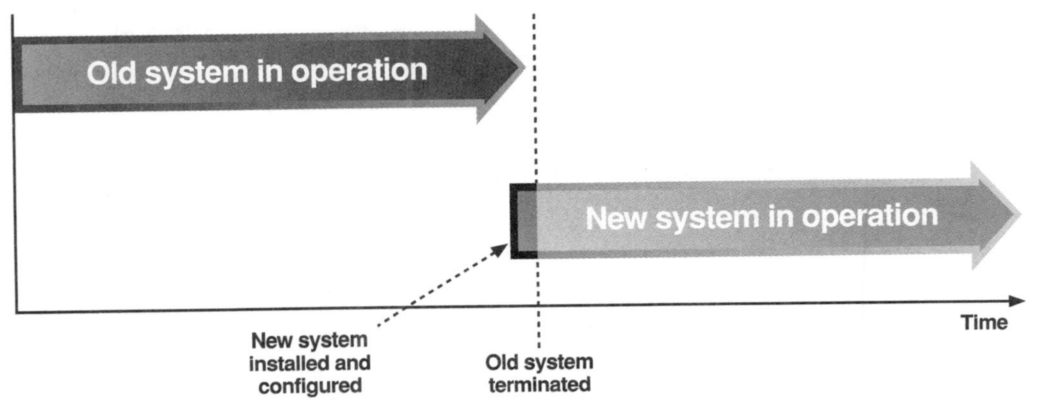

FIGURE 15-14
Direct installation and cutover.

The primary advantage of direct installation is its simplicity. Since the old and new systems aren't operated in parallel, there are fewer logistical issues to manage and fewer resources required. The primary disadvantage of direct installation is its risk. Because older systems are not operated in parallel, there is no backup in the event of a new system failure. The magnitude of the risk depends on the nature of the system, the cost of workarounds in the event of a system failure, and the cost of system unavailability or less-than-optimal system function.

Direct installation is typically used under one or both of the following conditions:

- The new system is not replacing an older system (automated or manual).
- Downtime of days or weeks can be tolerated.

If neither condition applies, then parallel or phased installation is usually preferable to minimize the risk of system unavailability.

Parallel Installation

parallel installation
An installation method that operates both the old and new systems for an extended time period

In a **parallel installation**, the old and new systems are both operated for an extended period of time (typically weeks or months). Figure 15-15 illustrates the time line for parallel installation. Ideally, the old system continues to operate until the new system has been thoroughly tested and determined to be error-free and ready to operate independently. As a practical matter, the time allocated for parallel operation is often determined in advance and limited to minimize the cost of dual operation.

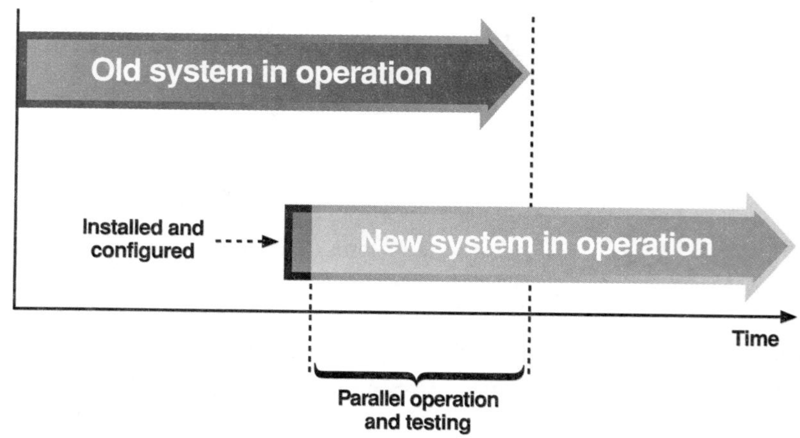

FIGURE 15-15
Parallel installation and operation.

The primary advantage of parallel installation is a relatively low risk of system failure and the negative consequences that might result from that failure. If both systems are operated completely (that is, using all data and exercising all functions), then the old system functions as a backup for the new system. Any failure in the new system can be mitigated by relying on the old system.

The primary disadvantage of parallel installation is cost. During the period of parallel operation, the organization pays to operate both systems. There are many extra costs associated with operating two systems in parallel, including:

- Hiring temporary personnel (or temporarily reassigning existing personnel)
- Acquiring extra space for computer equipment and personnel
- Increasing managerial and logistical complexity

Unless the operational costs of the new system are substantially less than that of the old system, the combined operating cost is typically 2.5 to 3 times the cost of operating the old system alone.

Parallel operation is generally best when the consequences of a system failure are severe. Parallel operation substantially reduces the risk of a system failure through redundant operation. The risk reduction is especially important for "mission-critical" applications such as customer service, production control, basic accounting functions, and most forms of on-line transaction processing. Few organizations can afford any significant downtime in such important systems.

Full parallel operation may be impractical for any number of reasons, including the following:

- Inputs to one system may be unusable by the other, and it may not be possible to use both types of inputs.
- The new system may use the same equipment as the old system (for example, computers, I/O devices, and networks), and there may not be sufficient capacity to operate both systems.
- Staffing levels may be insufficient to operate or manage both systems at the same time.

When full parallel operation is not possible or feasible, a partial parallel operation may be employed instead. Possible modes of partial parallel operation include the following:

- Processing only a subset of input data in one of the two systems. The subset could be determined by transaction type, geography, or sampling (for example, every 10th input transaction).

- Performing only a subset of processing functions (such as updating account history but not printing monthly bills).
- Performing a combination of data and processing function subsets.

Partial parallel operation always entails the risk that significant errors or problems will go undetected. For example, parallel operation with partial input increases the risk that errors associated with untested inputs will not be discovered.

Phased Installation

phased installation
An installation method that installs a new system and makes it operational in a series of steps or phases

In a **phased installation**, the system is installed and brought into operation in a series of steps or phases. Each phase adds components or functions to the operational system. During each phase, the system is tested to ensure that it is ready for the next phase. Phased installation can be combined with parallel installation, particularly when the new system will take over the operation of multiple existing systems.

Figure 15-16 shows a phased installation with both direct and parallel installation of individual phases. The new system replaces two existing systems. The installation is divided into three phases. The first phase is a direct replacement of one of the existing systems. The second and third phases are different parts of a parallel installation that replace the other existing system.

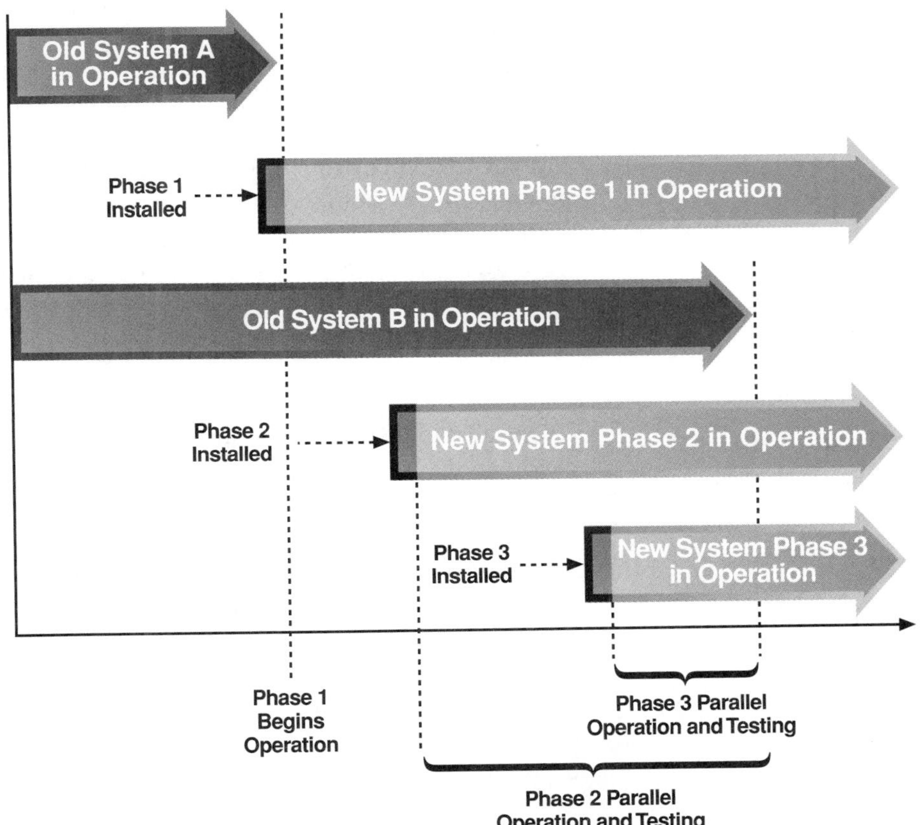

FIGURE 15-16 Phased installation with direct cutover and parallel operation.

There is no single method for performing phased installation. Installation specifics such as the composition of specific phases and their order of installation vary widely from one system to another. These specifics determine the number of installation phases, the order of installation, and the parts of the new system that are operated in parallel with existing systems.

The primary advantage of phased installation is reduced risk. Risk is reduced because failure of a single phase is less problematic than failure of an entire system. The primary disadvantage of phased installation is increased complexity. Dividing the installation into phases creates more activities and milestones, thus making the entire process more complex. However, each individual phase contains a smaller and more manageable set of activities. If the entire system is simply too big or complex to install at one time, then the reduced risks of phased installation outweigh the increased complexity inherent in managing and coordinating multiple phases.

Phased installation is most useful when a system is large, complex, and composed of relatively independent subsystems. If the subsystems are not relatively independent, then it is difficult or impossible to define independent installation phases. System size and complexity may also be too great for an "all at once" installation to be feasible. In that case, there is really no choice but to use phased installation.

Personnel Issues

Installing a new system places significant demands on personnel throughout an organization. Installation is typically a time of demanding schedules, rapid learning and adaptation, and high stress. Planning should anticipate these problems and take appropriate measures to mitigate their effects.

New system installation typically stretches IS personnel to their limits. Many tasks must be performed in little time. The problem is most acute in parallel installation, where personnel must operate both the old and new systems. Often, development and customer support personnel must be temporarily reassigned to provide sufficient manpower to operate both systems. Reassignment may reduce progress on other ongoing projects and also reduce support and maintenance activities for other systems.

Temporary and contract personnel may be hired to increase available manpower during an installation. Two types of personnel are particularly useful:

- Personnel with experience in hardware and software installation and configuration
- Personnel with experience (or who can be trained) to operate the old system

Installation may require technical skills that are in short supply within the organization. In that case, hiring contractors to assist in hardware and software installation is a necessity. Hardware and software specialists can be contracted directly from vendors or from IS consulting firms.

Hiring temporary employees to operate the old system during a parallel installation has several benefits. First, it provides the extra manpower needed to operate both systems. Second, it frees permanent employees for training and new system operations. Temporary personnel are typically hired several months in advance, trained in the operation of the old system, and employed for the expected duration of parallel operation. If problems are encountered with the new system, then employment contracts can be extended until the old system can be safely phased out.

Another personnel issue that must be considered is employee productivity. All new systems have a learning curve for users and system operators. Both require training before the installation begins. But no matter how good the training, users and operators will require a period of time (typically a few months) after system installation to reach their peak efficiency. Manpower requirements will be higher during that time, as will the general level of employee stress.

Documentation

Preparing documentation is an important but frequently overlooked activity during implementation. Documentation provides information to users on how to operate and maintain the system. Documentation also provides information needed for future modifications or reimplementation.

The nature of both system and user documentation has changed substantially in the last few decades. Change has been driven by rapidly advancing technology. Prior to the 1980s, most documentation was printed on paper and organized into bound or loose-leaf books. Automated documentation is now the norm. Automated documentation formats include:

- Electronic manuals, such as documents in Microsoft Word or Adobe Acrobat format
- Hyperlinked documents, such as documents formatted for web browser viewing with embedded links among documentation components
- On-line help—descriptive text, graphics, tutorials, and definitions embedded within an application
- Electronic system models—text and graphics formatted and stored in graphics file formats such as GIF, JPEG, and Viseo
- Tool-specific system models—models developed with system development tools such as integrated programming environments, DBMSs, and CASE tools

Electronic manuals can be distributed in a number of standard formats such as Adobe Acrobat and Windows help files. Choosing a standard format ensures that users will have or can easily obtain appropriate software to view the documentation. Hyperlinked documents allow users to navigate rapidly among related topics. Most electronic document formats allow hyperlinks. Using web page formats allows users to share a single copy. Web pages make documentation changes simpler to distribute since only the server copy needs to be updated.

Electronic and tool-specific system models are primarily intended for software developers' use. Generic model formats (such as ordinary text and GIF images) can be formatted as any type of electronic format. Tool-specific models must generally be accessed via specific software tools (for example, viewing a model generated by a CASE tool usually requires a viewer supplied by the CASE tool). However, some development tools allow models to be exported to other formats (such as Acrobat or Microsoft Word).

Documentation can be loosely classified into two types:

system documentation
Descriptions of system functions, architecture, and construction details, as used by maintenance personnel and future developers

user documentation
Descriptions of how to interact with and maintain the system, as used by end users and system operators

- **System documentation**—descriptions of system functions, architecture, and construction details
- **User documentation**—descriptions of how to interact with and maintain the system

System documentation is generated throughout the SDLC as outputs of each life cycle phase and activity. User documentation is created during the implementation phase of the SDLC. The development team cannot create user documentation earlier because many details of the user interface and system operation either haven't yet been determined or are subject to change during development.

System Documentation

System documentation serves one primary purpose: providing information to designers and developers who will maintain or reimplement the system. Most or all of the documentation needed for this purpose is generated as a byproduct of analysis, design, and implementation activities. Table 15-2 shows three phases of the SDLC and the system documentation produced or modified in each phase. The documentation produced in each phase is useful for future maintenance or upgrades, although documentation produced in later phases is used more frequently than documentation produced in earlier phases.

TABLE 15-2
Life cycle phases and system documentation generated in each phase.

Life cycle phase	System documentation	
	Traditional approach	Object-oriented approach
Analysis	Entity-relationship diagram Data-flow diagram Process description Data flow and element definition	Class diagram Use case Sequence diagram Collaboration diagram Statechart diagram
	Event list	
Design	System flow chart Structure chart	Design class diagram Package diagram
	Module or method pseudocode Database schema diagram	
Implementation	Program source code Database schema source code Test data	

Source code is the most frequently used documentation since it is the most direct link to the system's executable software. Direct changes to binary code are complex and expensive, so changing and recompiling source code is the only realistic method of altering a system's behavior. Test data is needed to test the system after change. Rerunning old tests with old test data helps determine whether a change in one part of the system has accidentally "broken" some other part of the system.

Source code can be difficult (and thus inefficient) for humans to use as documentation because it is entirely textual and often poorly commented. Important types of system information—such as how programs interact and what user needs a program satisfies—are usually not documented within source code. Such information is needed when evaluating significant changes in system design or function and when tracing errors that flow from one program to another via shared data. Information needed to perform these tasks is readily available in analysis and design models.

Design models tend to be used more often than analysis models because design parameters change more often than system requirements. Examples of maintenance changes that require design models but not analysis models include redeploying existing programs or databases to new hardware, fixing bugs in individual programs, and optimizing the performance of an existing distributed system. Such changes alter the corresponding design models (for example, the system flow chart or package diagram) but do not change analysis phase models.

Analysis models do change when user requirements are altered. For example, adding a new transaction type or an entirely new processing subsystem changes the data-flow diagram or class diagram. It also changes other analysis phase models such as the entity-relationship diagram, event list, use cases, and statecharts.

System documentation must be actively managed. It must be stored in an accessible location and form, retrieved when necessary for maintenance changes, and updated once changes have been implemented. In large organizations with many information systems, managing the documentation is a very formal process. Large organizations typically have one person responsible for archiving and retrieving documentation and for enforcing documentation standards.

Failure to maintain system documentation adequately compromises the value of a system. Systems with inadequate documentation are difficult or impossible to maintain, thus increasing the likelihood that a system will be prematurely scrapped or reimplemented. Maintaining documentation is an activity that extends the useful life of a productive asset.

System documentation is redundant with the system itself. That is, any information contained within a separate store of system documentation can also be obtained by directly examining the system. For example, programmers can determine the entities and relationships of a relational database by examining the SQL statements that describe the database schema. Programmers can also determine the modular structure of a traditional program or the classes within an OO program by directly examining the program source code. If the source code is unavailable, then program structure can be also be determined from executable code, although the process is much more difficult.

As with databases, redundancy creates problems of maintainability and possible inconsistency. As changes are made to the system, its documentation must also be updated. If documentation is not updated, then it is inconsistent with the system and useless to future designers and maintenance programmers. Making documentation an integral part of the installed system minimizes or eliminates inconsistency because updates to the system automatically update the documentation.

Two approaches can be used to create a self-documenting system. The first is to build the system using a full life cycle CASE tool. With a CASE tool, the system is built automatically from design models and stored by the CASE tool. Design models, in turn, are built automatically (or nearly so) based on analysis models. To implement a system change, a programmer modifies an analysis or design model and then regenerates the installed system. The CASE tool automatically maintains consistency among the installed system and the models. As long as only the models are changed (instead of the source or executable code), the models and system will always be consistent.

The second approach to implementing a self-documenting system is to use a reverse engineering tool. A reverse engineering tool can generate system models by examining source code. For example, such a tool can generate a class diagram by examining OO programs, and generate a structure chart by examining a program written in a procedural programming language. If a reverse engineering tool is powerful and reliable enough to generate all types of system documentation, then there is no need to maintain a separate store of documentation. The source code itself is the documentation, and the reverse engineering tool generates other forms of documentation on demand.

Both CASE and reverse engineering tools tend to be highly specialized to specific operating environments (such as programming languages, database management systems, and operating systems). They also tend to be expensive and to have steep learning curves. As a result, they aren't used as often as you might think. Thus, for many systems, system documentation must still be maintained separately and redundantly.

User Documentation

User documentation provides ongoing support for end users of the system. It primarily describes routine operation of the system, including functions such as data entry, output generation, and periodic maintenance. Topics typically covered include:

- Software startup and shutdown
- Keystroke, mouse, or command sequences required to perform specific functions

- Specific program functions required to implement specific business procedures (for example, the steps followed to enter a new customer order)
- Common errors and how to correct them

Topical coverage is usually supplemented by a table of contents, a general description of the purpose and function of the program or system, a glossary, and an index.

User documentation for modern systems is almost always electronic and is usually an integral part of the application. Most modern operating systems provide standard facilities to support embedded documentation. Figures 15-17 and 15-18 show portions of the electronic user documentation of a typical Windows application. Figure 15-17 shows a single page of the documentation with embedded hyperlinks (in red) and pop-up definitions (in blue). Figure 15-18 shows the table of contents that would appear after the user chooses the appropriate item from the application help menu. The user can access an index or search engine by clicking the appropriate tab at the top of the screen.

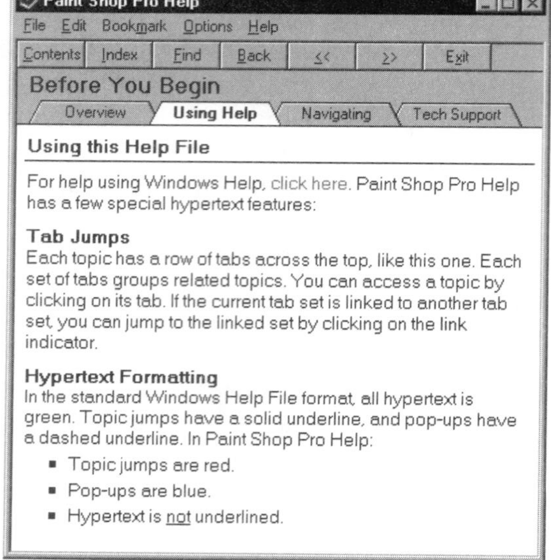

FIGURE 15-17
A sample page of electronic user documentation.

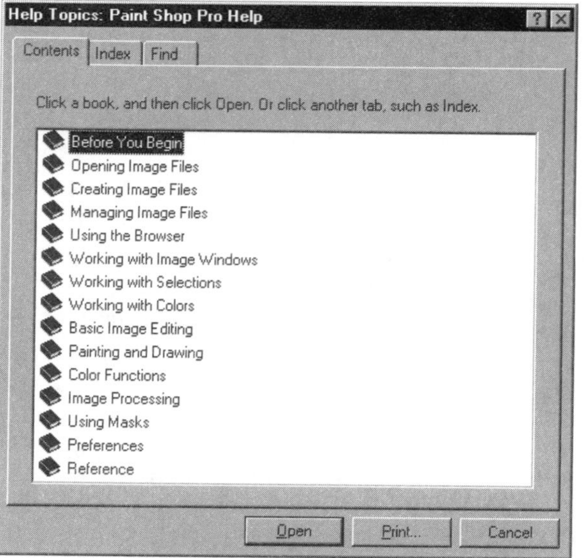

FIGURE 15-18
A sample table of contents, index, and search screen for electronic user documentation.

User documentation is an important organizational asset. Unfortunately, many organizations fail to prepare comprehensive high-quality user documentation for internally developed systems. Some of the reasons include the following:

- The assumption that trained programmers can examine source code, figure out how the system works, and train users on an as-needed basis
- The assumption that the users trained during system implementation will informally pass on their knowledge to future users
- The resources and special skills required to develop documentation and keep it up to date

As discussed in the previous section, source code is a poor form of system documentation. For similar reasons, it is an even worse form of indirect user documentation. Source code provides a detailed instruction-by-instruction view of how pieces of a system work. It provides little or no information about how those pieces interact and how the entire system functions within a specific context. Supplementing source code with other forms of system documentation does improve the situation. But even then, the process of figuring out how a system works based only on system documentation is slow and error-prone.

Knowledge of how to use a system is as important an asset as the system itself. Once initial training is completed, that knowledge is stored in the minds of end users. But there is no guarantee that the knowledge will be maintained or effectively transferred to subsequent users. Employee turnover, reassignment, and other factors make direct person-to-person transfer of operational knowledge difficult and uncertain. In contrast, written or electronic documentation is easier to access and far more permanent.

Developing good user documentation requires special skills and considerable time and resources. Writing clearly and concisely, developing effective presentation graphics, organizing information for easy learning and access, and communicating effectively with a nontechnical audience are skills with high demand and limited supply. Development takes time, and high-quality results are achieved only with a thorough review and testing process. Unfortunately, preparing user documentation is often left to technicians lacking in one or more necessary skills. Also, preparation time, review, and testing are often shortchanged due to schedule overruns and the last minute rush to tie up all the loose ends of implementation.

Training and User Support

Good documentation can reduce training needs as well as the frequency of support requests. But some amount of training before and support after installation is almost always required. Remember, users are part of the system, too! Without training, users would slowly work their own way up the learning curve, error rates would be high, and the system would operate well below peak efficiency. Training allows users to be productive as soon as the system becomes operational. Support activities ensure continuing user productivity long after installation.

There are two classes of users—end users and system operators—who must be considered for purposes of documentation, training, and support. End users are people who use the system day to day to achieve the system's business purpose. System operators are people who perform administrative functions and routine maintenance to keep the system operating. Table 15-3 shows representative activities for each role. In smaller systems, a single person may fill both roles.

TABLE 15-3
Typical activities of end users and system operators.

End user activities	System operator activities
Creating records or transactions	Starting or stopping the system
Modifying database contents	Querying system status
Generating reports	Backing up data to archive
Querying database	Recovering data from archive
Importing or exporting data	Installing or upgrading software

Training and support activities vary with the target audience. Audience characteristics that affect training include the following:

- Frequency and duration of system use
- Need to understand the system's business context
- Existing computer skills and general proficiency
- Number of users

In general, end users use the system frequently and for extended periods of time and system operators interact with the system infrequently and usually for short time periods. End users use the system to solve a particular business problem or to implement specific business procedures, and system operators are usually computer professionals with limited knowledge of the business processes that the system supports. End-user computer skill levels vary widely, while system operators typically have higher and more uniform skill levels. The number of end users is generally much larger than the number of system operators.

Training for end users must emphasize hands-on use and the application of the system within a specific business context, such as order entry, inventory control, or accounting. If the users are not already familiar with those procedures, then training must also include them. Widely varying skill and experience levels call for at least some hands-on training including practice exercises, questions and answers, and one-on-one tutorials. Self-paced training materials can fill some of this need, but complex systems typically require at least some face-to-face training. The relatively large number of end users makes group training sessions feasible, and it is also possible to train a subset of well-qualified end users who then can train other users (that is, to train the trainers).

System operator training can be much less formal when the operators are not end users. Experienced computer operators and administrators can learn most or all they need to know by self-study. Thus, formal training sessions may not be required. Also, the relatively small number of system operators makes one-on-one training feasible, if it is necessary.

Determining the best time to begin formal training can be difficult. On one hand, starting relatively early in the implementation phase provides plenty of time for learning and can ensure that users "hit the ground running." On the other hand, starting early can be frustrating to both users and trainers because the system may not be stable or complete. Little is more frustrating to an end user than trying to learn on a buggy, crash-prone system with features and interfaces that are constantly changing.

In an ideal world, training doesn't begin until the interfaces are finalized and a test version has been installed and fully debugged. But the typical end-of-project crunch makes that approach a luxury that is often sacrificed. Instead, training materials are normally developed as soon as the interfaces are reasonably stable, and end-user training begins as soon as possible thereafter. It is much easier to provide training if system interfaces are developed early and if the top-down modular development approach is employed.

Ongoing Training and User Support

The term *user support* covers training and user assistance that occur after the system is up and running. Some of the activities are the same as preinstallation training activities. For example, training of new users must occur periodically due to employee turnover (for example, new hires and transfers). Other activities such as refresher training and help desk operation are unique to support.

User support can be provided by a number of methods, including the following:

- On-line documentation and troubleshooting
- Resident experts
- A help desk
- Technical support

On-line documentation and troubleshooting have surged as a support method in recent years. Much of this support is built into the application, although increasingly web sites are being used to provide it. The goal of on-line support is to minimize the need for human support by putting useful information into the hands of end users when they need it. Achieving that goal requires well-designed support materials that are comprehensive and easy to use.

Resident experts are the most commonly used form of user support, and they are usually provided informally. A resident expert can be an on-site IS staff member or (more frequently) a business area staff member or user who provides support to other users. The position of resident expert is often an informal one. The person in that position frequently grows into it simply by displaying exceptional computer literacy or knowledge of software. Over time, all other users tend to approach that person first with questions or problems.

A help desk is a permanent IS department that provides end-user support for a wide range of systems and software. Help desks are staffed by personnel trained to install, operate, and troubleshoot application software including commonly used off-the-shelf products (such as word processors). A help desk serves as a central contact point for users. Help desk staff are trained to handle the majority of user problems and questions. Those who require further assistance are forwarded to technical support.

Technical support is typically a specific function or department within IS maintenance. It is located within maintenance because of the close relationship between user support, change requests, and system error reporting. If help desk personnel can't solve a user's problem, then there's a good chance that an error has been discovered or that there is a gap between system capability and user needs. If the problem is a system error, then maintenance needs to be notified quickly to investigate the cause and correct it if it is critical. Noncritical errors and unmet user needs must also be brought to the attention of maintenance, but timeliness is less critical. In either case, technical support is the bridge between users and maintenance activities.

Maintenance

software maintenance
Modification of a software product after delivery to correct faults, improve performance or other attributes, or adapt the product to a changed environment

The Institute of Electrical and Electronics Engineers (IEEE) and the American National Standards Institute (ANSI) have defined **software maintenance** as modification of a software product after delivery to accomplish at least one of the following objectives:

- Correct faults
- Improve performance or other attributes
- Adapt the product to a changed environment

The term *maintenance* covers virtually everything that happens to a system after delivery except total replacement or abandonment.

In most organizations, the cost of maintaining existing systems is at least as great as the cost of developing new ones. Existing systems are an organizational asset and must be actively managed to preserve their value and utility. In that sense, maintaining software is similar to maintaining other types of capital assets such as buildings and equipment.

Maintenance involves change—to adapt to a new environment, to adapt to changing user requirements, and to fix problems as they occur or are discovered. But change is risky. Making changes to an operational system is much more difficult than making changes to a system under development. When a change causes a developmental system to crash, there are no users calling the support desk and no immediate financial impact. But changes to operational systems have an immediate impact on users, customers, and the organization as a whole.

Failure of an operational system can have disastrous consequences. Thus, software maintenance differs greatly from new system development. New system development generally occurs in a relatively open environment where change is expected, new ideas are tried out, and risk taking is tolerated, if not encouraged. In contrast, maintenance occurs in a very conservative environment where change is tolerated as a necessary evil and risk taking is strongly discouraged.

Maintenance activities include:

- Tracking modification requests and error reports
- Implementing changes
- Monitoring system performance and implementing changes to improve performance or increase capacity
- Upgrading hardware and system software
- Updating documentation to reflect maintenance changes

Maintenance and new system development do have many activities in common, including analysis, design, construction, testing, and documentation. However, implementation of those activities differs in many ways, including scope and detail, triggering events, and implementation constraints. Each maintenance activity is described in detail in the following sections.

Submitting Change Requests and Error Reports

To manage the risks associated with change, most organizations adopt formal control procedures for all operational systems. Formal controls are designed to ensure that potential changes are adequately described, considered, and planned before being implemented. Typical change control procedures include:

- Standard change request forms
- Review of requests by a change control committee
- Extensive planning for design and implementation

Figure 15-19 shows a sample change request form that has been completed by a user or system owner and submitted to the change control committee for consideration. The change control committee reviews the change request to assess the impact on existing computer hardware and software, system performance and availability, security, and operating budget. The recommendation of the change control committee is formally recorded in a format such as the sample shown in Figure 15-20. Approved changes are added to the list of pending changes for budgeting, scheduling, planning, and implementation.

Change Request				
Request Date	2/1/2000	**Change Type**	☐	**Error Correction**
Requested By	Wen-Hsu Chang, Comptroller		☒	**Modification**
Target System	Customer Accounts - Refunds		☐	**New Function**

Change (or Error) Description

U.S. check formats will soon change due to a recently enacted federal law. The new format reserves an area to the right of the current routing number to be used for a security bar code checksum.

The law requires the new checksum to be printed on checks dated on or after January 1, 2001.

We currently use a portion of the area in question to print a multicolored security symbol. The security symbol will need to be moved or eliminated, and the security bar code checksum will have to be added.

Change Request ID	2000-11		
Date Received	2/2/2000	**Review Date**	2/7/2000, 0930-1100
Review Participants	W. Chang (Comptroller), R. Brooks (IS Operations), J. Hernandez (IS Security), G. Weeks (IS Change Coordinator)		

FIGURE 15-19
A sample change request form.

Change Review				
Change Request ID	2000-11	**Date Reviewed**	2/7/2000	
Priority	☐ **Critical**	☒ **Necessary**		☐ **Optional**
Hardware Implications need to verify ability of current printers to write a security bar code in mandated area				
Software Implications database will need to be modified to store the security bar code with other check information check writing program must be modified to generate and print the security bar code				
Performance Implications none				
Operating Budget Implications none				
Other Implications none				
Disposition	☒ **Approved**	☐ **Rejected**		☐ **Suspended**
Reason				
Latest Implementation Date	12/31/2000			
Reevaluation Date	n/a	**Signature**		

FIGURE 15-20
A sample change review form.

Bugs can be reported using a standard change request form, but many computing organizations use a different form and procedure due to the immediate need to fix them. Bug reports can come from many sources, including end users, computer operators, or IS support staff. Bug reports are typically routed to a single person or organization for logging and follow-up.

Implementing a Change

Change implementation follows a miniature version of the system development cycle. Most of the same activities are performed, although they may be reduced in scope or completely eliminated. In essence, a maintenance change is an incremental development project in which the user and technical requirements are fully known in advance. Thus, analysis phase activities are typically skimmed or skipped.

Planning for a change includes the following activities:

- Identify what parts of the system must be changed
- Secure resources (such as personnel) to implement the change
- Schedule design and implementation activities
- Develop test criteria and a testing plan for the changed system

Identifying affected parts of the existing system requires access to the existing system documentation. The documentation is reviewed by design, development, and operations staff to determine the scope of the change. Test criteria and plans for the existing system are the starting point for testing the new system. The testing plan is simply updated to account for changed or added functions. The modified plan and test data are archived for use in future change projects.

Design may be combined with planning if the change is relatively simple. For more complex changes, a separate design phase is used. The existing system design is evaluated and modified as necessary to implement the proposed changes. As with test plans and data, the revised design is archived for use in future change projects.

production system
The version of the system used day to day

test system
A copy of the production version that is modified to implement changes

Implementation activities are normally performed on a copy of the operational system. The **production system** is the version of the system used day to day. The **test system** is a copy of the production version that is modified to implement changes. The test system may be developed and tested on separate hardware or on a redundant system. Redundant systems are commonly used for round-the-clock transaction processing applications. The test system becomes the operational system only after complete and successful testing.

Upgrading Computing Infrastructure

Computer hardware, system software, and networks must be periodically upgraded for many reasons, including:

- Software maintenance releases
- Software version upgrades
- Declining system performance

Like application software, system software (for example, operating and database management systems) must periodically be changed to correct errors and add new functions. System software developers typically distribute maintenance releases several times per year. The frequency of maintenance update distribution has increased in recent years in part due to the advent of Internet-based software distribution. In some cases (such as virus checkers and operating system security subsystems), updates may be released weekly or even more frequently.

As with internally generated changes, system software updates are risky. Application software that worked well with an older software version may fail when that software is updated. For this reason, system software updates are extensively tested before they are applied to operational systems. In many cases, maintenance and version updates are simply ignored to reduce risk. Unless errors related to system software have already been encountered, there is little immediate benefit to an upgrade. Operational system maintenance usually follows the old engineering maxim, "If it isn't broken, don't fix it!"

An infrastructure upgrade may be required to add capacity or address a performance-related problem. Increases in transaction volume or support of new systems on existing hardware and networks may reduce performance to unacceptable levels. Infrastructure upgrades are implemented like any other type of change. The primary difference is how a performance upgrade is initiated.

Input from user or IS staff may indicate the need for a performance upgrade. But a final determination of whether an upgrade is needed and what exactly should be upgraded requires thorough investigation and research. Computer and network performance is complex and highly technical. What is a perceived as a performance problem may have little to do with hardware or network capacity. If the problem is due to hardware or networks, then the specific problem must be identified and a suitable upgrade chosen.

Performance problems require complex diagnosis to determine the best approach to address the problem. Staff with sufficient technical background to understand all of the relevant trade-offs should diagnose the problem. Larger IS organizations may have permanent staff with such skills, but many organizations must rely on contract personnel or consultants to diagnose performance problems, recommend corrective measures, and install or configure those measures. The skills come at a high price, but they can prevent an organization from wasting larger amounts of money buying hardware or network capacity that isn't really needed.

Summary

System development activities that occur after design and before system delivery are collectively called implementation. Implementation is complex because it consists of many interdependent activities including programming, quality assurance, hardware and software installation, documentation, and training. Implementation is difficult to manage because activities must be properly sequenced and progress must be continually monitored. Implementation is risky because it requires significant time and resources and because it often affects systems vital to the daily operation of an organization.

Programming and testing are two of the most interdependent implementation activities. Software components must be constructed in an order that minimizes development resources and maximizes the ability to test the system and correct errors. Unfortunately, those two goals are often in conflict. Thus, a program development plan is a trade-off among available resources, available time, and the desire to detect and correct errors prior to system installation.

Installation, documentation, and training are activities that normally follow program development. They are highly interdependent since an installed and documented system is a prerequisite for complete training. Manpower utilization and the number of directly affected personnel generally peak during these activities.

The terms *support* and *maintenance* describe activities that occur after a system is made operational. Support activities assist users in realizing the full benefit of the system. Maintenance activities ensure that the system functions at peak efficiency and that needed changes are implemented with minimal disruption to the organization. For most systems, the resources required for support and maintenance are greater than the resources required to develop the system. Because of high resource requirements and greater operational risk, support and maintenance are normally implemented as formal and carefully managed activities.

Key Terms

acceptance test, p. 569
alpha version, p. 559
beta version, p. 559
bottom-up development, p. 554
build and smoke test, p. 569
chief developer team, p. 557
collaborative specialist team, p. 557
cooperating peer team, p. 557
direct installation, or immediate cutover, p. 571
driver, p. 566
input, process, output (IPO) development, p. 551
inspection, p. 563
integration test, p. 568
maintenance releases, p. 560
parallel installation, p. 571
phased installation, p. 573
production system, p. 585
production version, release version, or production release, p. 559
quality assurance (QA), p. 562
software maintenance, p. 582
source code control system (SCCS), p. 558
stub, p. 567
system documentation, p. 575
system test, p. 569
technical review, p. 563
test case, p. 565
test data, p. 565
test system, p. 585
testing buddy, p. 570
top-down development, p. 554
unit testing, or module testing, p. 565
user documentation, p. 575

Review Questions

1. List and briefly describe the three basic approaches to program development order. What are the advantages and disadvantages of each?
2. How can the concepts of top-down and bottom-up development order be applied to object-oriented software?
3. Describe three approaches to organizing programming teams. For what types of projects or development activities is each approach best suited?
4. What is a source code control system? Why is such a system necessary when multiple programmers build a program or system?
5. Define the terms *alpha version, beta version,* and *production version.* Are there well-defined criteria for deciding when an alpha version becomes a beta version or a beta version becomes a production version?
6. List and briefly describe implementation phase QA activities other than software testing. What is the effect on software testing of not performing nontesting QA activities?
7. What are the characteristics of good test cases?
8. Define the terms *acceptance test, integration test, system test,* and *unit test.* In what order are these tests normally performed? Who performs (or evaluates the results of) each type of test?
9. What is a driver? What is a stub? With what type of test is each most closely associated? With what development order is each most likely to be used?
10. What factors make testing object-oriented programs more complex than testing structured programs?
11. Briefly describe direct, parallel, and phased installation. What are the advantages and disadvantages of each installation approach?
12. Why are additional personnel generally required during the later stages of system implementation?
13. What are the differences between documentation for end users and system operators?
14. How or why is system documentation redundant with the system itself? What are the practical implications of this redundancy?
15. List the types of documentation needed to support maintenance activities. Which documentation types are needed most frequently? Which are needed least frequently?
16. How do training activities differ between end users and system operators?
17. How does implementing a maintenance change differ from a new system development project? How are they similar?
18. Why might system software upgrades not be installed? What are the costs of not installing them?

Thinking Critically

1. Examine the system flow chart for Rocky Mountain Outfitters in Figure 9-8. Develop a preliminary development plan based on IPO development order. Which programs are difficult to classify as input, process, or output? Is a straightforward application of IPO development order appropriate for this system? If not, what changes should be made to the preliminary development plan?
2. This chapter discussed top-down and bottom-up development order for transform-oriented structure charts. Can these development orders also be applied to transaction-oriented structure charts such as the one shown in Figure 9-13? If so, how?
3. Describe the process of testing software developed using both top-down and bottom-up development order. Which method results in the fewest resources required for testing? What types of errors are likely to be discovered earliest under each development order? Which development order is best as measured by the combination of required testing resources and ability to capture important errors early in the testing process?
4. Assume that the Rocky Mountain Outfitters customer support system will be developed as described in your answer to question 1. Assume that 14 people are available for programming and testing. What sizes and types of teams are best suited to the project?
5. Consider the issue of documenting a system using only electronic models developed with a full life cycle CASE tool. The advantages are obvious (for example, the analyst modifies the models to reflect new requirements and automatically generates an updated system). Are there any disadvantages? Hint: The system may be maintained for a decade or more.
6. Some types of system documentation (such as models developed during the analysis phase) are seldom looked at once the system is made operational. What are the advantages and disadvantages of not keeping such documentation types?

Experiential Exercises

1. Assume that you and five of your classmates are charged with developing the *Create new order* subsystem shown in Figure 9-19. Create a development and testing plan to write and test the required modules. Assume that you have three weeks to complete all tasks.
2. Implement a formal QA process in one of your programming or system development classes (be sure to obtain permission from your instructor first). Form a group of students to implement an inspection process. Have one or all members prepare presentation materials for code that they've written and distribute them prior to a group meeting. If possible, create a buddy system for testing each others' code. Evaluate the results in terms of time required for code development and quality of the final product.
3. Examine the end-user documentation supplied with a typical personal or office productivity package such as Microsoft Office. Compare the documentation to the categories described in this chapter. Which categories of documentation are supplied? How might documentation for a business application differ in content or format?
4. Talk with a computer center or IS manager about the testing process recently used to test hardware and software for Y2K bugs. What types of tests were performed? How were test cases and test data generated? What types of teams were used to develop and implement the tests?
5. Talk with an end user at your school or work about the documentation and training provided with a recently installed or distributed business application. What type of training and documentation were provided? Did the user consider the training to be sufficient? Does the user consider the documentation to be useful and complete?

Case Studies

The HudsonBanc Billing System Upgrade

Two regional banks with similar geographic territories merged to form HudsonBanc. Both banks had credit card operations and operated billing systems that had been internally developed and upgraded over three decades. The systems performed similar functions, and both operated primarily in batch mode on IBM mainframes. Merging the two billing systems was identified as a high-priority cost-saving measure.

HudsonBanc initiated a project to investigate how to merge the two billing systems. Upgrading either system was quickly ruled out because the existing technology was considered old, and the costs of upgrading the system were estimated to be too high. HudsonBanc decided that a new system should be built or purchased. Management preferred the purchase option since it was assumed that a purchased system could be brought on-line more quickly and cheaply. An RFP was prepared, many responses were received, and after months of analysis and investigation, a vendor was chosen.

Hardware for the new system was installed in early January. Software was installed the following week, and a random sample of 10 percent of the customer accounts was copied to the new system. The new system was operated in parallel with the old system for two months. To save cost involved with complete duplication, the new system computed but did not actually print billing statements. Payments were entered into both systems and used to update parallel customer account databases. Duplicate account records were checked manually to ensure they were the same.

After the second test billing cycle, the new system was declared ready for operation. All customer accounts were migrated to the new system in mid-April. The old system was turned off on May 1, and the new system took over operation. Problems occurred almost immediately. The system was unable to handle the greatly increased volume of transactions. Data entry slowed to a crawl, and payments were soon backed up by several weeks. The system was not handling certain types of transactions correctly (for example, charge corrections and credits for overpayment). Manual inspection of the recently migrated account records showed errors in approximately 50,000 accounts.

It took almost six weeks to adjust the incorrect accounts manually and to update functions to handle all transaction types correctly. On June 20, the company attempted to print billing statements for the 50,000 corrected customer accounts. The system refused to print any information for transactions more than 30 days old. A panicked consultation with the vendor concluded that fixing the 30-day restriction would require more than a month of work and testing. It was concluded that manual entry of account adjustments followed by billing within 30 days was the fastest and least risky way to solve the immediate problem.

Clearing the backlog took two months. During that time, many incorrect bills were mailed. Customer support telephone lines were continually overloaded. Twenty-five people were reassigned from other operational areas, and additional phone lines were added to provide sufficient customer support capacity. System development personnel were reassigned to IS operations for up to three months to assist in clearing the billing backlog. Federal and state regulatory authorities stepped in to investigate the problems. HudsonBanc agreed to allow customers to spread payments for late bills over three months without interest charges. Setting up the payment arrangements further aggravated the backlog and staffing problems.

1. What type of installation did Hudson Banc use for its new system? Was it an appropriate choice?
2. How could the operational problems have been avoided?

The DownTown Videos Rental System

Using the design class diagram you developed in Chapter 9 for the DownTown Videos rental system, develop an implementation and testing plan. Specify the order in which classes and their methods will be implemented and the groups of methods and classes that will be tested during integration testing.

The Reliable Pharmaceutical Service System

Using the structure chart you developed in Chapter 9 for the Reliable Pharmaceutical Service, develop an implementation and testing plan. Specify the order in which modules will be implemented and the groups of modules that will be tested during integration testing.

Further Resources

Robert V. Binder. "Object-Oriented Software Testing," *Communications of the ACM*, volume 37:9, September, 1994, pp. 29–101.

Barry Boehm. *Software Engineering Economics*. Prentice-Hall, 1981, ISBN 0-13-822122-7.

Robert G. Ebenau, and Susan H. Strauss. *Software Inspection Process*. McGraw-Hill, 1994, ISBN 0-07-062166-7.

William Horton. *Designing and Writing Online Documentation: Hypermedia for Self-Supporting Products* (2nd ed.). John Wiley & Sons, 1994, ISBN 0-471-31635-5.

William Horton. *Illustrating Computer Documentation: The Art of Presenting Information Graphically on Paper and Online*. John Wiley & Sons, 1991, ISBN 0-471-53845-0.

International Association of Information Technology Trainers (ITrain) web site—http://itrain.org/.

Edward Kitt. *Software Testing in the Real World: Improving the Process*. ACM Press, 1995, ISBN 0-201-87756-2.

David Kung, Jerry Gao, Pei Hsia, Yasufumi Toyoshima, Chris Chen, Young-Si Kim, and Young-Kee Song. "Developing an Object-Oriented Software Testing and Maintenance Environment," *Communications of the ACM*, volume 38:10, October, 1995, pp. 75–87.

Steve McConnell. *Code Complete*. Microsoft Press, 1995, ISBN 1-55615-484-4.

Arthur H. Watson, and Thomas J. McCabe. "Structured Testing: A Testing Methodology Using the Cyclomatic Complexity Metric," NIST Special Publication 500-235, National Institute of Standards and Technology, September 1996, http://hissa.ncsl.nist.gov/HHRFdata/Artifacts/ITLdoc/235/mccabe.html.

The Responsibilities of a Project Manager

Information system development is a complex process, and success is not guaranteed. The industry abounds with case after case in which a new system was never completed or did not satisfy the requirements when it was installed. One of the most important contributors to successful completion of a new information system is the quality of the project management.

Every project has a senior person who is responsible for the project and functions as the project manager. Titles may range from project manager, to senior analyst, to senior systems developer. However, every person on the project team helps with project management responsibilities. As a new systems analyst, you will very likely participate in project management tasks such as estimating your work, assisting in cost/benefit analyses, making presentations, and providing status and milestone completion information. There is a shortage of system analysts who have good project management skills, and one of the most highly remunerated career paths is to become a competent, successful project manager.

The project manager has responsibilities both within and outside of the project. This appendix focuses primarily on those tasks directly related to managing the project. We discuss the fundamental principles and practices for good project management in six primary areas:

- Providing leadership and vision
- Planning and organizing the project
- Organizing and managing the project team
- Estimating costs and developing the project budget
- Monitoring and controlling the project schedule
- Ensuring the quality of the final result

Providing Leadership and Vision

It has been said that "management without leadership is bureaucracy and leadership without management is chaos." Project management entails both leadership skills and management skills. What is leadership? Sometimes we think that leadership is an innate characteristic or personal charisma. Although there may be some truth to that belief, leadership is much more than personality. There are specific tasks that good leaders do, and leadership skills can be learned. We do not attempt to teach leadership here—there are many books and leadership training courses available. We do address two aspects of leadership that are important to project management—namely, providing vision and motivating team members.

During a development project, there are a multitude of tasks and literally hundreds of details that must be understood and documented. It is easy to become so involved in these details that the overall goals of the project are forgotten. Without a project vision, it is difficult to maintain balance among the various tasks and decisions. The project manager should first develop a vision for the project and then share that vision with the project team. There are two aspects of a project vision. First, the vision should define the primary objectives of the project by answering questions such as, "Why are we undertaking this project? What is it expected to achieve? What benefit is it to the sponsoring organization?" Second, there should be a vision of how the processes of the project are to be carried out, such as, "How do the members of the team relate to each other? What standards of conduct are expected? How are coordination and cooperation achieved? What are the standards of performance? What kind of development methodology will be used?"

Each project manager has a unique style of leadership, and there are many effective styles. Sometimes a vision statement is developed at the beginning of the project to help solidify the vision and to keep it visible. Other managers are less formal and communicate the vision through informal meetings and visits with the team members. One common aspect of leadership, however, is frequent communication between the project manager and team members. Instilling a vision requires frequent and continual communication with the project team.

The other primary aspect of leadership is to help motivate members of the project team. Often novice project managers assume that all team members are motivated by money: wage increases and bonuses. Yet, different people are motivated in different ways. There are many different methods to provide job satisfaction and high motivation, including assignment of responsibilities, rank advancements, performance recognition, and satisfaction with the job itself. A strong project manager often works closely with the human resource director to develop a comprehensive approach to motivation and recognition for the members of the team.

In all contexts, however, there is one pitfall to avoid. People become demotivated very rapidly if they do not know what is going on. The old military approach to inform people based on a "need to know" does not instill a high level of motivation. It is much more effective to have frequent and broad-based communication within the team. Obviously, balance must be achieved between too much detail and not enough. An effective guideline is to keep the team informed on the progress of the individual working groups, decisions on the direction of the project, and areas of difficulty within the project. A central repository of the materials produced by the various working groups also can assist in sharing information about the system itself. On very large projects, which last a long time and involve many people, a project newsletter or status sheet is sometimes used to disseminate important information.

Planning and Organizing the Project

The first set of defined tasks of project management consists of those that plan and organize the project. Even though the project manager has primary responsibility for planning and organizing, systems analysts are also heavily involved. Project planning continues throughout the life of the project. During the project planning phase, an

baseline project plan
The initial project plan that serves as the baseline for plan modifications

initial plan, frequently called the **baseline project plan**, is developed. The baseline project plan identifies and details the various tasks and activities of the analysis phase. Later phases of design and implementation are planned only at a high level. As the project progresses, and as the analysis activities are completed, the baseline project plan is updated to reflect changes in the scope of the project and decisions that are made concerning design and implementation.

There are five components of planning and organizing the project:

- Estimating the size of the project
- Defining project milestones
- Developing a detailed list of tasks
- Assigning resources to the tasks
- Developing and maintaining a project schedule

The first four components are so interrelated that they are usually done together. They provide information to develop the final project schedule. To estimate the size of the project, the team studies the business need and makes a preliminary assessment concerning the size of the final system. Based on the difficulty of the problem and the complexity of the final system, a detailed task list is developed for activities of the analysis phase. Task lists for design and implementation activities are less detailed. The resources required for each task are identified and an estimate of the number of person-days required to complete the task is developed. During this initial planning, the resources are often defined only by type, such as systems analyst, programmer, user, and technician. As the project is staffed, specific people are assigned to the tasks.

As the list of tasks is developed, but before the final schedule is completed, a set of project milestones is defined. Milestones should be items that can be precisely measured as completed.

Finally, the list of tasks, their time estimates, resource requirements, and identified milestones are all combined into a project schedule. Most project managers use a charting technique such as a PERT chart or a Gantt chart to display the project schedule graphically. During the life of the project, the project schedule is continually updated. As portions are completed, or when milestones are reached, the project schedule often needs to be revised. Additional tasks may be identified or the estimated size of existing tasks may need to be modified.

Organizing and Managing the Project Team

One of the key elements of project management is getting work done through other people. It follows then that one of the most critical components of project management is organizing and managing the project team. We are concerned here only with employee responsibilities for project tasks, not with general employment issues such as raises, performance reviews, vacations, sick leave, and so forth. In fact, many team members report to other administrative units over which the project manager has no administrative responsibility.

There are four important areas of focus in organizing and managing the project team:

- Recruiting and obtaining staff for the team
- Assigning team members to work groups and tasks
- Ensuring team members are well trained
- Ensuring that the teams develop into effective workteams

The first two items, obtaining staff and assigning them to work groups and tasks, are direct results of planning and organizing the project. Under the third item, ensuring that team members are well trained, project managers may need to provide training for new development tools, unfamiliar operating environments, new programming languages, or new technology. If only a few members of the team lack training, then they can be enrolled in commercial courses or in-house courses. If the technology is new to the entire team, then special training sessions are usually scheduled for all.

The final item, developing effective workteams, is frequently ignored. However, team-building exercises enhance the level of trust among team members and help in the day-to-day working associations. Generally, the project manager works with the human resources director to establish a plan for training and team building for members of the project.

Estimating Costs and Developing the Project Budget

Cost estimation and the development of the corresponding budget are two of the most difficult components of project management. Development projects often go over budget. The primary problem with estimating the cost of the project is that the estimate must be done before the project is accepted—that is, the sponsoring organization wants to know how much it will cost before it commits to the project. The project manager must estimate the cost before completely understanding the requirements and complexities involved.

Estimating the cost of the project is an ongoing task. Frequently, costs of remaining tasks must be reestimated as major phases are completed or near completion. Thus, a cost estimate is provided as part of project planning, a reestimate is done as alternatives are evaluated during analysis, and a final estimate is performed as the design is completed.

Various quantitative estimation techniques have been tried. Many work well on projects for well-defined applications such as inventory management or sales entry. However, any time new technology or different problem domains are investigated, such as large integration projects or electronic data interchange systems across multiple companies and channels of distribution, standard estimating techniques quickly become inadequate.

The sources of project costs vary depending on the approach taken to develop the new system. Some companies do all development work in-house. Other hire consultants or software development firms to assist in the development. For projects done completely in-house, the largest cost is usually the cost of salaries and wages. One fairly effective method (no method is completely effective) to estimate salaries and wages is to utilize the detailed list of tasks generated for the project schedule. A common problem is underestimating the amount of work to be done. However, if a detailed task list was developed and resource requirements were estimated, then they can provide a fairly accurate estimate of the human resources required. Since the costs for salaries and wages are directly proportional to the number of people involved and the time to completion, scope changes and project delays directly affect this cost. It is almost impossible to foresee everything that impacts the length of the project.

For projects done with outside assistance, the provider should help with the estimate by providing estimates for the parts of the project it will complete. Other categories—such as software licenses, training, and facilities—are smaller components and can normally be estimated fairly accurately.

The project budget is simply a timeline of the cash flow based on the project estimates. Salaries and wages are spread throughout the life of the project based on staffing levels. Software licenses are paid at the beginning of the project. Computer equipment is a small component as the project begins, and a large component immediately prior to rollout of the new system. Installation of network cabling and facilities will also occur in the months immediately prior to rollout. Normally the more difficult part is estimating the costs. Building the budget from those costs is rather straightforward.

Monitoring and Controlling the Project Schedule

Once the project begins, the project manager must see that the plan is executed and that the project progresses on schedule. Managers use many different techniques to control the project, depending on the manager's style of management. Some follow a philosophy of "management by walking around," which is an informal technique to measure the pulse of the project. Other managers have regular, formal status reports and meetings. There are also numerous training programs to teach skills necessary for project control. Regardless of the manager's chosen style, five ideas are important to assist in the control of an ongoing project:

- Tracking progress using the schedule and defined milestones
- Maintaining an open items control log
- Understanding and utilizing corrective procedures carefully
- Implementing consistent status review and communication techniques
- Managing client expectations

The ease with which the project manager can monitor the project depends on the quality of the project schedule. A detailed and well-thought-out schedule not only assists in the assignment of tasks, but also provides the milestones and steps that assist in monitoring progress. The saying, "planning your work, and working your plan," is a good guideline for monitoring schedules.

Measuring completion of a certain task by estimating the percentage completed is highly subjective and risky. How can you evaluate exactly the amount of work left? Percentage completion may seem useful, but a better method is to measure progress with defined milestones such as "completed" or "not completed."

During every project, some problems cannot be resolved the first time they are encountered. Frequently, senior management must address new business rules. Technical issues also may require additional research before final decisions can be made. One technique to ensure that these open items are not forgotten is to maintain an open items control log. As each item is identified, it is logged, and a person or group is assigned to resolve it. Each item can then be tracked for its status, final resolutions, and the date it is closed. Figure A-1 illustrates an open items control log.

Item number	Short name	Date added	Description	Person responsible	Date resolved	Resolution
1	Comm rate	7/8/2000	Change rate based on specials	George Young	9/1/2000	See writeup on commissions
2	Partial ship	9/5/2000	Ship full item or allow partial item shipments	Sean Kroff	Open	
3	History report	9/5/2000	Define level of detail for transaction history	Michelle Hensen	10/25/2000	See history report definition
4						
5						

FIGURE A-1
A sample open items control log.

Unanticipated problems can also arise, ranging from missed user requirements to technology that does not work. A good project has only minor unexpected problems. Some projects have problems that are so complex and pervasive that the project is canceled. Experienced systems analysts and project staff can help avoid major problems, but all projects require corrective measures when small and even medium-size problems occur. The most common corrective procedures include rearranging or reassigning tasks, reducing the scope of the project, or adding more staff to address deficiencies. Any corrective procedure must be taken very carefully to assess the various options and their results before they are chosen. It is impossible to identify standard solutions to problems since each problem is unique to the specific project, system, resource, management, and business environment.

Finally, a good project manager knows how to maintain a good working relationship with the user and the client. In project manager jargon, we say a good manager knows how to "manage the client expectations." There is a fine line between being overly optimistic and being a "dooms-dayer." One of the most important skills of a good project manager is to be able to communicate with the client in a way that maintains optimism but does not allow the client to expect more than is realistically feasible.

Ensuring the Quality of the Final Result

An old adage for developing systems says, "Build the system right, but build the right system." The first idea, build the system right, is concerned with building a system that runs correctly, that gives the right answers, and that is robust. The second idea, build the right system, is concerned with building a system that meets the user requirements—that is, one that accomplishes the business objectives. The project manager has a responsibility to ensure the quality of the final product in both areas.

In processes such as assembly lines or manufacturing plants that produce the same product over and over again, the current approach to quality control is, "Build it right instead of inspecting and correcting after the fact." For software development, however, each project is one of a kind. So, even though we always try to do it right the first time, quality control requires that we test and inspect to verify that completed work was done correctly. The inspection process to ensure that the system meets the user requirements is called **validation**. The system is validated against the user requirements. The inspection process that ensures the system is correct and consistent is called **verification**.

validation
Proving that the system satisfies the user requirements

verification
Proving that the system is internally consistent and correct

walkthrough
A method to review completed documents or diagrams to test for accuracy

beta testing
Testing of the system by the users to validate that the requirements are met

system testing
Testing of the system by technical staff to eliminate internal processing errors

Two primary techniques are used to ensure quality, both for validation and verification: walkthroughs and testing. **Walkthroughs** are used primarily in the early phases of the project when there are documents and diagrams that can be reviewed. Walkthroughs are used to validate the diagrams against the user requirements and to verify that the diagrams are consistent and correct. Testing is used primarily in the later parts of the project when there are computer programs that can be executed and tested. **Beta testing** is the term used for validating the system against the user requirements. **System testing** is the term for testing the entire system for errors. The project manager's responsibility is to ensure that the project includes adequate walkthrough and testing sessions.

Key Terms

baseline project plan, p.593
beta testing, p.597
system testing, p.597
validation, p.596
verification, p.596
walkthrough, p.597

Review Questions

1. What are the six major areas of internal responsibility for a project manager?
2. Describe the difference between leadership and management.
3. Define a *baseline project plan.*
4. Explain the difference between validation and verification.
5. Explain the difference between beta testing and system testing.
6. Discuss the issues related to controlling a project after it has begun.

Application Questions

1. List several ways that a baseline project plan is helpful for managing a development project.
2. The discussion in this appendix focuses on the internal responsibilities of a project manager. Discuss what other kinds of responsibilities a project manager might have, particularly external responsibilities or those that are focused on the sponsoring organization.
3. Discuss the importance of walkthroughs in ensuring the quality of the new system.
4. Discuss the importance of team-building exercises. Find and report on an article on team building.
5. Find an article or report on a development project that has gone off course. Discuss the reasons and corrective actions that might have been taken to prevent the problem.

Calculating Net Present Value, Payback Period, and Return on Investment

A capital investment is an expenditure by an organization in equipment, land, or other assets that are used to carry out the objectives of the organization. Since money—including money available for capital investment—is a scarce resource, one important responsibility of senior-level management of any organization is to decide which of the available alternative investments to choose. Organizations use several different methods to determine which investment is best. The three methods this appendix discusses are net present value (NPV), payback period, and return on investment (ROI). Definitions and brief explanations of these terms are given in Chapter 2. This appendix explains these measures in more detail and shows how to calculate them. More in-depth discussion of the principles behind these measurements, including important concepts about selecting alternatives, can be found in books on finance. You can calculate these measurements by hand using tables for rates and factors. In addition, financial calculators now have built-in function keys to calculate these measures. This appendix demonstrates the manual method to provide an understanding of the principles.

Net Present Value
Calculations

The two basic concepts of net present value are (1) that all benefits and costs are calculated in terms of today's dollars (that is, present value) and (2) that benefits and costs are combined to give a net value. Hence the name *net present value*.

discount rate
The annual percentage rate that an amount of money is discounted to bring it to a present value

The first step in calculating the NPV is to decide on the **discount rate** to use. Most organizations have a policy, determined by the chief financial officer, that states the standard discount rate used for investments. This rate may vary by type of investment—we will use the discount rate for software development. The discount rate equates future values to current values. For example, $100 received one year in the future is only worth $94.34 today with a discount rate of .06 (6 percent). The equation that represents this is

$$\textit{present value} = \textit{amount received in future} \,/\, (1 + \textit{discount interest rate})^{\textit{number of years}}$$

discount factor
The accumulation of yearly discounts based on the discount rate

For dollar amounts received more than one year in the future—for example, in three years—the discount rate is applied three times to give a **discount factor**. Let's use PV for present value, FV for future value, i for discount interest rate, F for discount factor, and *n* for the number of years in the future. The equations that apply to find a present value for a future value across multiple years are

$$F_n = 1/(1 + i)^n \text{ and } PV = FV * F_n$$

For example, $100 received three years hence at a discount rate of 6 percent is

$$F_n = 1/[(1+ .06) * (1+ .06) * (1+ .06)] = .8396$$
$$PV = \$100.00 * .8396 = \$ 83.96$$

Finance books provide tables of discount factors by discount rate for multiple years, so you do not have to calculate them. Financial calculators have the preceding equations built-in, so you simply enter the future value, the discount rate, and the number of years to get the present value.

Figure B-1 shows the table containing the NPV calculations that were done for Rocky Mountain Outfitters (RMO) in Chapter 2. Notice that the NPV calculation is done over a period of five years. Most organizations have an investment policy that determines the time period allowed for NPV calculations. In this example, RMO policy states that an investment in a new information system must have a positive value within at most five years. Thus, the NPV calculations cover a five-year period. In the figure, year 0 is considered the present year.

	RMO cost benefit analysis	Year 0	Year 1	Year 2	Year 3	Year 4	Year 5	Total
1	Value of benefits	$ -	$ 889,000	$ 1,139,000	$ 1,514,000	$ 2,077,000	$ 2,927,000	
2	Discount factor (10%)	1	0.9091	0.8264	0.7513	0.6830	0.6209	
3	Present value of benefits	$ -	$ 808,190	$ 941,270	$ 1,137,468	$ 1,418,591	$ 1,817,374	$6,122,893
4	Development costs	$(1,336,000)						$(1,336,000)
5	Ongoing costs		$ (241,000)	$ (241,000)	$ (241,000)	$ (241,000)	$ (241,000)	
6	Discount factor (10%)	1	0.9091	0.8264	0.7513	0.6830	0.6209	
7	Present value of costs	$ -	$ (219,093)	$ (199,162)	$ (181,063)	$ (164,603)	$ (149,637)	$(913,559)
8	PV of net of benefits and costs	$(1,336,000)	$ 589,097	$ 742,107	$ 956,405	$ 1,253,988	$ 1,667,737	
9	**Cumulative NPV**	**$(1,336,000)**	**$(746,903)**	**$ (4,796)**	**$ 951,609**	**$2,205,597**	**$ 3,873,334**	
10	**Payback period**		**2 years + 4796 / (4796 + 951,609) = 2 + .005 or 2 years and 2 days**					
11	**5-year return on investment**		**6,122,893 - (1,336,000 + 913,559) / (1,336,000 + 913,559) = 172.18%**					

FIGURE B-1
Net present value, payback, and return on investment calculations for RMO.

The anticipated benefits are also associated with the year in which they accrue. Line 1 in Figure B-1 shows the annual benefits of the new system. Notice that these benefits begin in year 1. RMO anticipates that these benefits will increase over time with increased usage and utility from the system. Year 0 is considered to be the period at the beginning of the useful life of the system—that is, the point in time of the initial investment. All of the development costs are assigned to year 0 even if the development project spans multiple years. Line 4 shows the development costs.

Annual operating costs, as shown on line 5, are applied to the year in which they occur. Operating costs normally increase over time, since as the system gets older, it frequently requires more maintenance.

You can apply the next two steps in either order: You can either combine the benefits and costs (net them together) and then apply the discount factor, or you can apply the discount factor to the benefits and costs individually and then combine the two discounted amounts. This example uses the second approach and discounts the benefits and costs separately before combining them. (You will see the reason later when we discuss ROI.)

In the example, we have used a discount rate of 10 percent. Notice that in year 0, which is the present, the discount factor is 1.00. Each year thereafter, the factor reduces since it is a continual multiplication of the previous year's factor by 1/(1 + discount rate). The discount factors are given in line 2 and repeated on line 6. On line 3 is the present value of each of the future annual benefits, and line 7 has the present value of the future operating costs. Finally, line 8 combines lines 3 and 7 for each year of the benefits and costs.

To obtain the net present value of across all the years, you combine the amounts in each column of row 8. Figure B-1 shows a running accumulation, year by year, in row 9. As the figure shows, at the end of five years, the total NPV for this investment is $3,873,334.

Payback Period
Calculations

Payback is a method to determine the point in time at which the initial investment is paid off. Various methods are used to calculate payback. You can calculate the payback period using either present value amounts or cash flow amounts. Cash flow amounts do not take the time value of money into consideration and only calculate based on nondiscounted dollar amounts. Another approach is to consider only the initial investment and not include annual operating costs in the calculations. In our example, we use the time value of money (that is, the discount rate), and also use net benefits (that is, benefits minus operating costs). The year when this value goes positive is the year in which payback occurs. In the RMO example, this happens within the third year.

You can obtain a more precise answer by determining the fraction of a year when payback actually occurs. In this example, the third year began with a deficit of $4,796 and ended the year with a positive value of $951,609. The equation for the fraction of a year is

(|beginning-year amount| / (year-end amount + |beginning-year amount|))

where the vertical bars indicate absolute value. For RMO, the fraction of a year before payback is

(4796 / (951,609 + 4796)) = .005

Payback thus occurs in 2.005 years. Therefore, the payback period is two years and two days (.005 * 365).

Return on Investment

The objective of the NPV is to determine a specific value based on a predetermined discount rate. The objective of the return on investment is to calculate a percentage return (like an interest rate) so that the costs and the benefits are exactly equal over the specified time period. Figure B-1, illustrates a return on investment calculation for RMO. The time period can be the expected life of the investment (that is, the productive life of the system) or it can be an arbitrary time period. This example simply calculates the ROI for the five-year period. In another situation, RMO might want to consider a 10-year ROI, assuming the system will be functional for at least 10 years.

Line 11 of Figure B-1 contains the calculations for a five-year return on investment. The process is to use the amounts for the discounted benefits and the discounted costs. These numbers are given on the last column in the table. The equation to calculate ROI is

ROI = (Estimated time period Benefits – Estimated time period Costs) / Estimated time period Costs

For RMO, assuming a five-year benefit period, the ROI is

ROI = 6,122,893 - (1,336,000 + 913,559) / (1,336,000 + 913,559) = 172.18 %

In other words, the investment in the development costs returned over 172 percent on the investment for a period of five years. Obviously, since the system is generating benefits at that point in time, if we assume the lifetime was longer, we would get a much higher ROI.

Key Terms

discount rate, p.598
discount factor, p.598

Review Questions

1. What is the difference between the following two procedures?
   a. Discount benefits and discount costs first, then net these values together to get a net present value.
   b. Net benefits and costs together first, then discount the combined figure to get a net present value.
2. Explain the difference between the discount rate and the discount factor.
3. How do you determine the time period for NPV? For ROI?
4. What is meant by payback period?

Thinking Critically

1. Calculate the net present value, the payback period, and the return on investment using a discount rate of 8 percent for the following systems development project. The development costs for the system were $225,000.

Year	Anticipated Annual Benefits	Expected Annual Operating Costs	Discount Factors at 8 Percent
1	$55,000	$5,000	.9259
2	$60,000	$5,000	.8573
3	$70,000	$5,500	.7938
4	$75,000	$5,500	.7349
5	$80,000	$7,000	.6805
6	$80,000	$7,000	.6301
7	$80,000	$7,000	.5833
8	$80,000	$8,000	.5401

FIGURE B-2
Benefits, costs, and discount factors.

Developing a Project Schedule with PERT/CPM Charts

Over the last 50 years, the rapid increase in technology has permitted the design and construction of very complex systems. Some of the early complex systems were developed during the 1960s as the United States strove to place a man on the moon. To coordinate and control the thousands of detailed tasks required to build and test the rockets as well as to design, build and test all of the environmental support—such as command modules, landing vehicles, and space suits—a sophisticated scheduling and control system was needed. Many project teams that developed these sophisticated systems used PERT (Program Evaluation Review Technique) and CPM (Critical Path Method) charts, which are two slightly different methods of graphically scheduling and monitoring the activities of complex projects. Today the two techniques have essentially merged into a single method. Chapter 2 provides a brief introduction to PERT/CPM charts, and this appendix explains the techniques to build a PERT/CPM chart and demonstrates how to use Microsoft Project to develop PERT/CPM charts.

Building PERT/CPM Charts

As indicated in Chapter 2, developing a PERT/CPM chart is a four-step process:

1. Identify all the tasks for the project.
2. Determine the amount of work to complete each task.
3. For each task, identify the immediate predecessor tasks.
4. Chart the tasks on the PERT/CPM chart, calculating start and finish times for each task.

work breakdown structure The hierarchy of tasks, activities, and phases of a project; this hierarchy is used as method to estimate and schedule the tasks of a project

Recall from Chapter 2 that phases, activities, and tasks were identified as one way to organize the work to be done. Sometimes this is referred to as a **work breakdown structure**. A work breakdown structure simply refers to the concept of building a hierarchy of the work to be done. Chapter 2 identified only three levels: phases, activities, and tasks. However, in very large projects, three levels may not be sufficient. To add more than three levels, an analyst can subdivide any activity into smaller activities before subdividing them into the lowest-level tasks.

You can develop the work breakdown structure using either a top-down approach or a bottom-up approach. A top-down approach starts with a phase, then identifies the activities, then identifies the individual tasks. In a bottom-up approach, the project team, in a brainstorming session, simply tries to think of all the individual tasks that need to be done. Then the team organizes those tasks into activities, higher-level activities (if necessary), and finally the major phases of the project. Figure C-1 is an example of a work breakdown structure for the Rocky Mountain Outfitters (RMO) customer service system. As can be seen, the project planning phase is broken down into very detailed tasks. The other phases of the project are given only at the activity level.

FIGURE C-1
Work breakdown structure for the project planning phase for RMO.

1.0 Project planning phase
- 1.1 Define the problem
  - 1.1.1 Meet with users
  - 1.1.2 Determine scope
  - 1.1.3 Write statement of business benefits
  - 1.1.4 Write statement of need
  - 1.1.5 Define statement of system capabilities
  - 1.1.6 Develop context diagram
- 1.2 Confirm project feasibility
  - 1.2.1 Identify intangible costs and benefits
  - 1.2.2 Estimate tangible development and operational costs
  - 1.2.3 Estimate tangible benefits and do cost/benefit analysis
  - 1.2.4 Evaluate organizational and cultural feasibility
  - 1.2.5 Evaluate technical feasibility
  - 1.2.6 Evaluate required schedule
  - 1.2.7 Evaluate resource availability
- 1.3 Produce the project schedule
  - 1.3.1 Develop work breakdown schedule
  - 1.3.2 Estimate resources, durations, and precedences
  - 1.3.3 Develop a PERT chart and a Gantt chart
- 1.4 Staff the project
  - 1.4.1 Develop a resource plan for the project
  - 1.4.2 Identify and request specific technical staff
  - 1.4.3 Meet with users, and identify and request user staff
  - 1.4.4 Organize the project team
  - 1.4.5 Conduct team-building exercises
  - 1.4.6 Conduct initial skills training
- 1.5 Launch the project
  - 1.5.1 Prepare presentation materials
  - 1.5.2 Make executive presentation
  - 1.5.3 Set up project facilities and support resources
  - 1.5.4 Conduct official kickoff meeting

2.0 Analysis phase
- 2.1 Gather information
- 2.2 Define system requirements
- 2.3 Build prototypes for discovery of requirements
- 2.4 Prioritize requirements
- 2.5 Generate and evaluating alternatives
- 2.6 Review recommendations with management

3.0 Design phase
- 3.1 Design and integrate the operating environment
- 3.2 Complete the application design
- 3.3 Design the user interface(s)
- 3.4 Design the system interface(s)
- 3.5 Design and integrate the database
- 3.6 Prototype for design alternatives
- 3.7 Design and integrate the system controls

4.0 Implementation phase
- 4.1 Construct software components
- 4.2 Verify and test
- 4.3 Prototype for tuning
- 4.4 Convert the data
- 4.5 Conduct training
- 4.6 Complete the documentation
- 4.7 Install the system

The second step is to identify the amount of work required for each task. The work required for each of the activities is composed of the work for the individual tasks. Frequently, for the first iteration of the PERT/CPM chart, we may estimate only the length of time for the task. More advanced scheduling, however, will require us to identify and assign resources so that people are not double-scheduled for concurrent tasks.

There are various techniques for estimating the amount of time required for each task. Often three estimates are made: optimistic, pessimistic, and expected. A schedule can be built using each set of estimates to give an overall optimistic completion date, a pessimistic, one and an expected completion date. Another alternative is to combine each of the three estimates for each task and use one estimate, which we call the most probable duration. The three individual estimates can be combined by a simple averaging (pessimistic + optimistic + expected / 3) or by some type of weighted average. One popular weighting method is to weigh the estimate for the expected duration two to four times more than the pessimistic or optimistic estimate. A weighted average simply ensures that any large deviation—either optimistic or pessimistic—will not drastically affect the most probable duration.

Figure C-2 continues with the RMO example given in Figure C-1. For each task of the project planning phase, we have estimated an optimistic, pessimistic, and most probable duration. Estimates for the number of analysts assigned to each task are also given. For the other phases of the project, we have estimated durations at the activity level. In other words, since the detailed tasks for these activities have not been identified, the analysis phase has been estimated at the activity level. Later phases are not shown.

TASK LIST WITH RESOURCE ESTIMATES

	Phases, activities, and tasks	Optimistic duration (days)	Pessimistic duration (days)	Most probable (days)	Number resources
1.0	Project planning phase				
1.1	Define the problem				
1.1.1	Meet with users	1	4	2	2
1.1.2	Determine scope	1	3	2	2
1.1.3	Write statement of business benefits	1	2	1	1
1.1.4	Write statement of need	1	2	1	1
1.1.5	Define statement of system capabilities	1	2	1	1
1.1.6	Develop context diagram	1	2	1	1
1.2	Confirm project feasibility				
1.2.1	Identify intangible costs and benefits	1	1	1	2
1.2.2	Estimate tangible development and operational costs	1	3	2	2
1.2.3	Estimate tangible benefits and do cost/benefit	1	3	2	2
1.2.4	Calculate NPV, ROI, and payback	1	1	1	1
1.2.5	Evaluation organizational and cultural feasibility	1	1	1	1
1.2.6	Evaluate technical feasibility	1	2	2	2
1.2.7	Evaluate required schedule	1	2	1	1
1.2.8	Evaluate resource availability	1	1	1	1

(continued)

	TASK LIST WITH RESOURCE ESTIMATES				
	Phases, activities, and tasks	Optimistic duration (days)	Pessimistic duration (days)	Most probable (days)	Number resources
1.3	Produce the project schedule				
1.3.1	Develop work breakdown schedule	2	4	3	2
1.3.2	Estimate resources, durations, and precedences	1	3	2	2
1.3.3	Develop PERT chart and Gantt chart	1	3	2	2
1.4	Staff the project				
1.4.1	Develop a resource plan for the project	1	1	1	2
1.4.2	Identify and request technical staff	1	2	1	1
1.4.3	Meet with users, identify and request user staff	1	3	2	1
1.4.4	Organize project team	1	1	1	1
1.4.5	Conduct team-building exercises	2	5	3	2
1.4.6	Conduct initial skills training	2	6	4	2
1.5	Launch the project				
1.5.1	Prepare presentation materials	1	1	1	1
1.5.2	Make executive presentation	1	1	1	1
1.5.3	Set up project facilities and support resources	1	5	3	2
1.5.4	Conduct official kick-off meeting	1	1	1	1
2.0	Analysis phase				
2.1	Gather information	30	120	60	5
2.2	Define system requirements	30	100	60	5
2.3	Build prototypes for discovery of requirements	25	80	50	2
2.4	Prioritize requirements	5	15	10	2
2.5	Generate and evaluate alternatives	5	15	8	2
2.6	Review recommendations with management	2	5	3	2

FIGURE C-2
Phases, activities, and tasks with duration and resource estimates.

The third step is the one that lays the foundation for the development of the PERT/CPM chart. In this step, the immediate predecessor tasks for each task are identified. Immediate predecessor tasks either produce a result that is needed by the identified task or need the same human resources and so must be scheduled sequentially. Figure C-3 organizes the predecessor tasks and prepares for the transfer of information to a PERT/CPM chart. In Figure C-3, the summary activities are removed, only the most probable duration is included, and the predecessor activities for each task are identified.

TASK LIST WITH RESOURCE ESTIMATES				
	Phases, activities, and tasks	**Most probable (days)**	**Number resources**	**Predecessor**
0.0	Start			
1.1.1	Meet with users	2	2	0.0
1.1.2	Determine scope	2	2	1.1.1
1.1.3	Write statement of business benefits	1	1	1.1.2
1.1.4	Write statement of need	1	1	1.1.3
1.1.5	Define statement of system capabilities	1	1	1.1.4
1.1.6	Develop context diagram	1	1	1.1.5
1.2.1	Identify intangible costs and benefits	1	2	1.1.6
1.2.2	Estimate tangible development and operational costs	3	2	1.2.1
1.2.3	Estimate tangible benefits	2	2	1.2.1
1.2.4	Calculate NPV, ROI, and payback	1	1	1.2.2, 1.2.3
1.2.5	Evaluation organizational and cultural feasibility	1	1	1.2.4
1.2.6	Evaluate technical feasibility	2	2	1.2.5
1.2.7	Evaluate required schedule	1	1	1.2.6, 1.3.3
1.2.8	Evaluate resource availability	1	1	1.2.7
1.3.1	Develop work breakdown schedule	3	2	1.2.8
1.3.2	Estimate resources, durations, and precedences	2	2	1.3.1
1.3.3	Develop PERT chart and Gantt chart	2	2	1.3.2
1.4.1	Develop a resource plan for the project	1	2	1.3.3, 1.2.6
1.4.2	Identify and request specific technical staff	1	1	1.4.1
1.4.3	Meet with users, identify and request user staff	2	1	1.4.1
1.4.4	Organize project team	1	1	1.4.2, 1.4.3
1.4.5	Conduct team building exercises	3	2	1.4.4
1.4.6	Conduct initial skills training	4	2	1.4.5
1.5.1	Prepare presentation materials	1	1	1.4.1
1.5.2	Make executive presentation	1	1	1.5.1
1.5.3	Set up project facilities and support resources	3	2	1.5.2
1.5.4	Conduct official kick-off meeting	1	1	1.5.1

(continued)

TASK LIST WITH RESOURCE ESTIMATES				
Phases, activities, and tasks		Most probable (days)	Number resources	Predecessor
2.1	Gather information	60	5	1.5.3, 1.5.4, 1.4.6
2.2	Define system requirements	60	5	2.1
2.3	Build prototypes for discovery of requirements	50	2	2.1
2.4	Prioritize requirements	10	2	2.2, 2.3
2.5	Generate and evaluating alternatives	8	2	2.4
2.6	Review recommendations with management	3	2	2.5

FIGURE C-3
Tasks with predecessor tasks identified.

In the final step, we graph the tasks based on the most probable duration and the predecessor relationships. Figure C-4 is a partial PERT/CPM chart that illustrates the tasks of the activity defining the tasks numbered 1.1.1 through 1.1.6. We have chosen this set of tasks first to show the basic format of a PERT/CPM chart. The tasks are represented by the rectangles, and the predecessor relationships are indicated by the connecting arrows. The top compartment within the rectangle is the task name. The middle-leftmost compartment is the task ID, the middle-rightmost compartment is the duration, and the bottom-left corner is the start time for each task. The calculation of the start times is done left to right, starting with the leftmost task. You detemine the start time of each task by adding the predecessor's duration to the start time of the predecessor task.

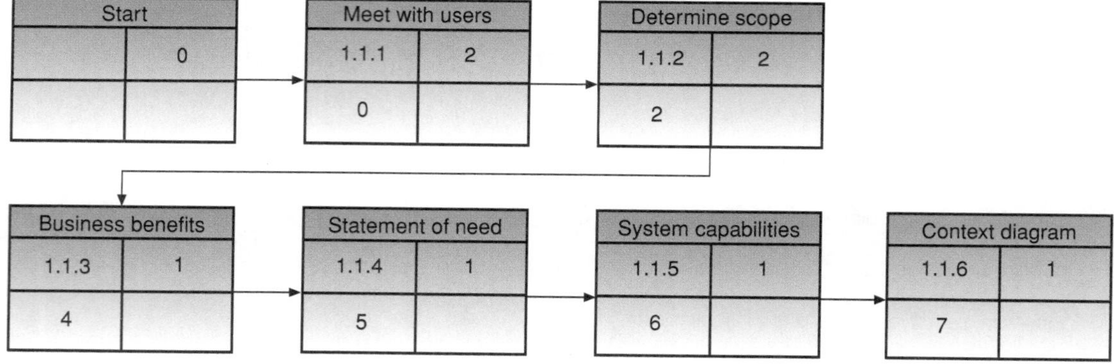

FIGURE C-4
A PERT/CPM chart for the first six tasks for the RMO project.

Figure C-5 presents a more complex PERT/CPM chart to illustrate some additional characteristics. As in the previous figure, the middle-rightmost compartment represents the duration, and the bottom-leftmost compartment is the start time. However, in this figure, two tasks, 1.2.2 and 1.2.3, are on parallel paths because they can be done concurrently. Whenever there are two or more concurrent paths, one can possibly finish before the other. In fact, in this case, task 1.2.2 has a duration of three days, and task 1.2.3 only requires two days. To account for this difference, the definition of start time needs to be expanded. The bottom-leftmost compartment is defined as the **early start time**, and the bottom-rightmost compartment is called the **late start time**.

early start time
The earliest time that a task can begin as determined by completion times of predecessor tasks, calculated as the latest completion time of all immediate predecessor tasks; early start times are calculated left to right

late start time
The latest time that a task can begin and still keep the project on schedule, calculated as the required start time of a successor task minus the task duration; late start times are calculated right to left, starting at the end of the project.

slack time
The difference between the early start time and the late start time for a task; the amount of time that a task can delay starting and still not have a negative impact on the project

The early start time is the earliest that a task can begin since it depends on the completion of the prior task. The late start time is the latest time that an activity can start and still keep everything on schedule. For example, as shown in the chart, task 1.2.4 is scheduled to begin on day 12, so both tasks 1.2.2 and 1.2.3 must finish before day 12. Task 1.2.2 starts on day 9 and requires three days. However, task 1.2.3 is scheduled to begin on day 9 but only requires two days. Hence, if task 1.2.3 did not start until day 10, then it would still finish on time for task 1.2.4 to begin on schedule on day 12. Thus, task 1.2.3 has a late start of day 10. The difference between the early start and the late start is called **slack time** for the task.

Early start time is the earliest possible time that a given task can begin. The early start times for the tasks are computed from left to right. The early start time for a task is the finish time for the latest predecessor task. In other words, the earliest that a task can start is after the completion of the longest predecessor task. Of course, if a task has only one predecessor task, then the early start time is easily calculated as the finish time of that single predecessor.

Late start times are calculated from right to left beginning with the ending task. Once all of the early start times have been calculated, a total project completion time (using the longest dates) can be determined. The project completion date is the time that the final task is completed. To calculate the latest start time, you need to subtract the duration of that task from the total project time. The computation continues to progress right to left by subtracting the task's duration from the late start time of the successor task. Since this computation goes from right to left, only tasks that have multiple successor tasks can possibly have multiple answers. If a task has multiple successors, the latest it can start is in time for the earliest late start time of all of the successors. If there is a single successor task, then the late start time is easily calculated as the late start of the successor minus the duration of the current task.

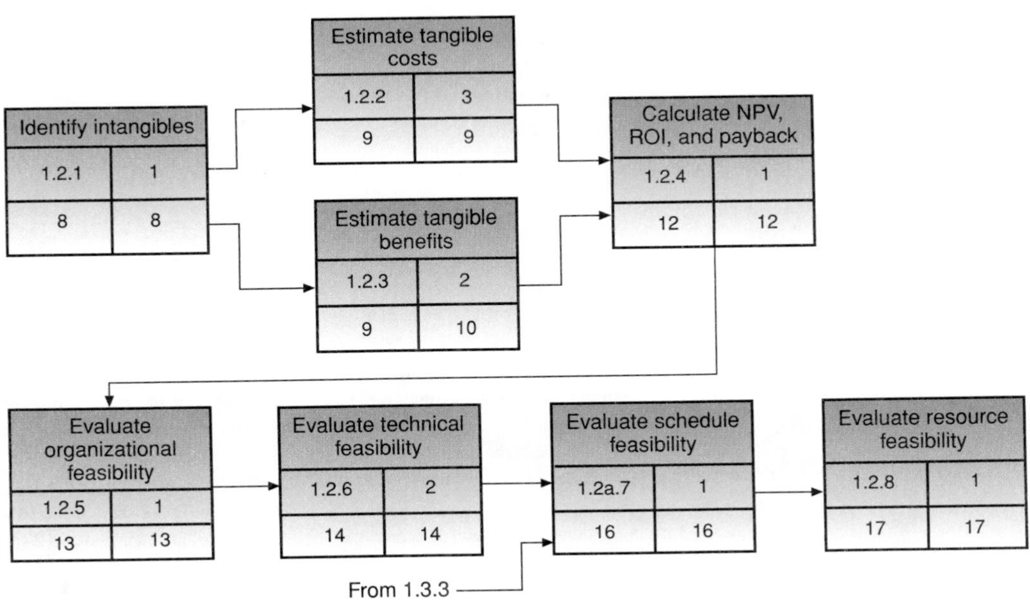

FIGURE C-5
A PERT/CPM chart showing early start and late start times.

As you look at the chart, notice that some tasks have an early start time and a late start time that are the same. In other words, the earliest and the latest day that the task can begin is the same. There is no slack time for that task. Those tasks that have this characteristic are considered to be on the critical path. Chapter 2 defined the critical path as the shortest possible time for the project to be completed. If any task on the critical path falls behind schedule, then the entire project will fall behind schedule. This process gives another, more precise definition of the critical path: Any task with the same early start time and late start time is on the critical path. Every project will have at least one path, and maybe more, of connected tasks with the early start time and late start time that are equal.

Creating the RMO Project Schedule
with Microsoft Project

There are several tools to help with the construction of PERT/CPM charts. Obviously, one of the problems with building a schedule is that it may be constantly changing. It requires a tremendous amount of work to try to keep such a chart current. Microsoft Project is an automated project scheduling tool that is fairly widely used. The following figures illustrate some of the capability of MS Project.

Figure C-6 is a snapshot of the MS Project page that contains the list of tasks. In our earlier examples, the phases and activities were eliminated and only the tasks were shown on the chart. MS Project permits the definition of multiple levels of summary tasks, which permits the activity-level entries to be included. MS Project also automatically assigns a sequence number to each task. Thus, the work breakdown structure numbering scheme developed in the earlier examples is used only as a comment field. The format of the rectangles that represent the tasks is the same as the earlier examples, with the inclusion of the MS Project ID numbers instead of the work breakdown numbers.

FIGURE C-6
A task list for RMO from MS Project.

	Text1	Task Name	Duration	Start	Finish	Predecessor
1	**1.0**	⊟ **Planning Phase**	**30 days**	**Mon 1/10/00**	**Fri 2/18/00**	
2	**1.1**	⊟ **Define the problem**	**8 days**	**Mon 1/10/00**	**Wed 1/19/00**	
3	1.1.1	Meet with users	2 days	Mon 1/10/00	Tue 1/11/00	
4	1.1.2	Determine Scope	2 days	Wed 1/12/00	Thu 1/13/00	3
5	1.1.3	Write Statement of Business Benefi	1 day	Fri 1/14/00	Fri 1/14/00	4
6	1.1.4	Write Statement of Need	1 day	Mon 1/17/00	Mon 1/17/00	5
7	1.1.5	Define Statement of System capabil	1 day	Tue 1/18/00	Tue 1/18/00	6
8	1.1.6	Develop Context Diagram	1 day	Wed 1/19/00	Wed 1/19/00	7
9	**1.2**	⊟ **Confirm project feasibility**	**13 days**	**Thu 1/20/00**	**Mon 2/7/00**	
10	1.2.1	Identify Intangible Costs and Benefit	1 day	Thu 1/20/00	Thu 1/20/00	8
11	1.2.2	Estimate tangible development and	3 days	Fri 1/21/00	Tue 1/25/00	10
12	1.2.3	Estimate tangible benefits and do c	2 days	Fri 1/21/00	Mon 1/24/00	10
13	1.2.4	Calculate NPV, ROI and Payback	1 day	Wed 1/26/00	Wed 1/26/00	11,12
14	1.2.5	Evaluation Organizational & Cultural	1 day	Thu 1/27/00	Thu 1/27/00	13
15	1.2.6	Evaluate Technical Feasibility	2 days	Thu 1/27/00	Fri 1/28/00	13
16	1.2.7	Evaluate Required Schedule	1 day	Wed 2/2/00	Wed 2/2/00	13,20
17	1.2.8	Evaluate Resource Availability	1 day	Mon 2/7/00	Mon 2/7/00	13,23
18	**1.3**	⊟ **Produce the project schedule**	**7 days**	**Wed 1/26/00**	**Thu 2/3/00**	
19	1.3.1	Develop Work Breakdown Structure	3 days	Wed 1/26/00	Fri 1/28/00	11
20	1.3.2	Estimate resources, durations, prec	2 days	Mon 1/31/00	Tue 2/1/00	19
21	1.3.3	Develop PERT chart and Gantt char	2 days	Wed 2/2/00	Thu 2/3/00	20
22	**1.4**	⊟ **Staff the project**	**11 days**	**Fri 2/4/00**	**Fri 2/18/00**	
23	1.4.1	Develop a resource plan for the proje	1 day	Fri 2/4/00	Fri 2/4/00	21
24	1.4.2	Identify and request specific technic	1 day	Mon 2/7/00	Mon 2/7/00	23
25	1.4.3	Meet with users, identify and reques	2 days	Mon 2/7/00	Tue 2/8/00	23
26	1.4.4	Organize project team	1 day	Wed 2/9/00	Wed 2/9/00	24,25
27	1.4.5	Conduct team building exercises	3 days	Thu 2/10/00	Mon 2/14/00	26
28	1.4.6	Conduct initial skills training	4 days	Tue 2/15/00	Fri 2/18/00	27

Figure C-7 is a snapshot of part of the MS Project Pert/CPM chart page. It is read in the same way as the earlier examples. The bold red tasks connected with the bold connecting lines identify the critical path.

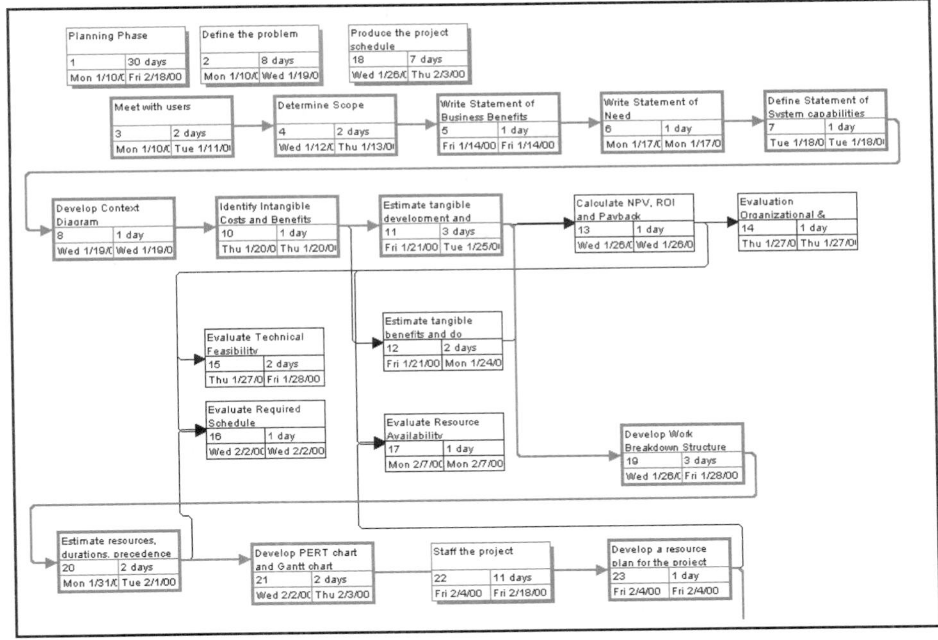

FIGURE C-7
A snapshot of an MS Project PERT/CPM chart.

Building Gantt Charts

As indicated in Chapter 2, a Gantt chart is a bar chart of the project tasks. The horizontal axis represents calendar dates, which makes a Gantt chart useful to note the progress of the various tasks over time. Figure C-8 presents the Gantt chart for the same RMO tasks. This example demonstrates several useful features.

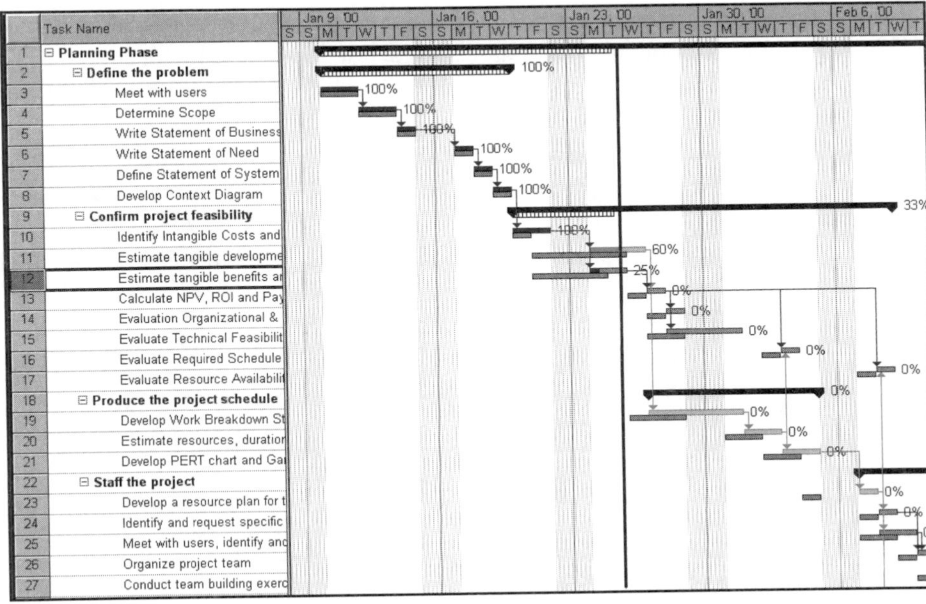

FIGURE C-8
A snapshot of a partial Gantt chart from MS Project for RMO.

When the original schedule is developed, it can be stored as the baseline project schedule. The baseline schedule remains constant throughout the project. During the project, a current schedule can be maintained to reflect changes and additions to the original schedule. A comparison between the two schedules illustrates where the updated schedule has deviated from the original.

The Gantt chart can also indicate completion of tasks. As shown in Figure C-8, each task has a double bar; the bottom bar is the baseline schedule, and the upper bar is the current schedule. Bars that are darkened are completed tasks, and bars that are partially darkened are partially completed. The vertical line on the chart indicates today's date. Since the Gantt chart is plotted against actual calendar dates, the reader can see which tasks are on schedule and which are not on schedule. Any tasks to the left of today's date that are not completely darkened are behind schedule. Any tasks that have parts of their bars darkened to the right of today's date are ahead of schedule. Of course, those tasks that are darkened to the left and not darkened to the right are exactly on schedule.

Key Terms

early start time, p. 609
late slack time, p. 609
late start time, p. 609
work breakdown structure, p. 603

Review Questions

1. Why is it important to develop a detailed list of tasks when preparing to build a project schedule.
2. Explain in your own words the process of calculating the late start times on a PERT/CPM chart.
3. Discuss the difference between a PERT/CPM chart and a Gantt chart. List comparative advantages and disadvantages of each.
4. Can a project have two critical paths? Explain why or why not. Give an example if possible.

Thinking Critically

1. You have a younger friend who is getting ready to go to college. Make an activity list of all the activities you can think of for her to do to start college. How did you decide how detailed to make your list?
2. Now expand the list to individual tasks. Build a work breakdown structure showing activities and tasks.
3. Draw a PERT/CPM chart of the tasks listed in question 2.
4. Draw a Gantt chart for the tasks listed in question 2.
5. Draw a PERT/CPM chart for getting ready for class in the morning. Identify the critical path and the total time it takes.
6. Remember one of your classes where you had a team research project. Develop a work breakdown structure and a PERT/CPM chart for the project. Since this was a team project, you should be able to have multiple concurrent paths. Show the critical path.
7. Illustrate on a Gantt chart the tasks for question 6.

APPENDIX D

Presenting the Results to Management

After an analyst has generated various system development alternatives and done analysis to select the most likely candidates, he or she needs to present the results to upper management. The purpose of this presentation is to review the analysis of the candidate alternatives and secure final approval from management on the approach to be used for the rest of the project. This approval is necessary to ensure continued support and funding for the project. Usually the attendees at this presentation include key client personnel, those involved in directing and funding the project; management of the information systems group in the organization; and the managers and team leaders of the project.

When you are required to present your results to management, be sure you precisely understand the following:

- What is the objective of the presentation?
- Who will attend the presentation and what is their background, interest, expectations, and knowledge of the subject?
- When is the presentation scheduled and how much time is allotted?
- Where will the presentation be given? What is the room like? What facilities and equipment are available?
- What other related issues, distractions, meetings, attendees, and so forth might affect the delivery and outcome of the presentation?

Of the preceding items, the first two are the most critical. You should not even begin to prepare the presentation until you have firmly established the primary objective and you know who your audience is. These first two items will determine both the content and the format of the presentation.

In some instances, the objective may be to make a firm recommendation for action and have that recommendation accepted. In other cases, the objective may be to establish a framework in which the most viable alternatives can be discussed and a selection made. Sometimes the project is already approved and the objective is to inform the client about the status and next steps. In each case, the content and format of the presentation are different.

The characteristics of the audience are also an important consideration. Most senior-level executives are interested in summary information: how much it will cost, how long it will take, what the costs/benefit trade-off is, how the system fits with the organization's strategic direction, what the impact on organization and operations will be, and so forth. Middle-level managers are more likely to be interested in how the new system will work and how it will affect their staff, budget, and operations. If the audience is too disparate, it may be better to make multiple presentations and hold an executive-level decision meeting and a mid-level information meeting. In any event, to have a successful presentation, you need to know your audience and its characteristics.

Preparing the Presentation

There are myriad books and techniques on making presentations. We recommend a method called the OABC format: Opening, agenda, body, and closing. The method is not only easy to remember but effective.

The Opening

In the opening part of your presentation, you must get the attention of your audience and prepare it to receive your message. There are two primary items you may want to include in this opening:

- Establish the rationale for the project—that is, recap the problem or the need that initiated the project. You should keep this recap short in the opening, however.
- Present your final recommendation, unless you want to build up to it as a conclusion or climax of your presentation.

Basically these two items say, "This is the problem," and, "Here is the solution. Now the rest of the presentation will fill in the details." By presenting the problem and solution together, you reinforce that the solution does, in fact, solve the stated problem. The rest of the presentation then elaborates the details. In this opening, you want to connect with your audience, by establishing rapport, and grabbing the audience's interest. Otherwise, you are fighting an uphill battle in the remainder of the presentation.

The Agenda

In the agenda portion, you should briefly describe the major elements and sequence of your presentation. Doing so will help your audience stay connected to you and in tune with your presentation. During the body of your presentation, you will describe the details of your analysis. A brief agenda beforehand will help the audience understand how the details fit together to build to your conclusion. Developing an agenda also helps you organize your thoughts. As you write the agenda, you should ensure that all the items you include in the presentation contribute to your objectives. If some do not, omit them.

The Body

The body of the presentation develops each of the major elements in the agenda. As you develop the details, keep in mind your objectives and the audience. Make sure each element contributes to the objective and climax of the presentation. Also, consider whether each element is discussed at the right level of detail for your audience.

After you are sure these first two points are satisfied, then you must determine how best to present your material so that it is clear, accurate, and easily understood. You want your audience to understand the details of the topic you are presenting. It should also be obvious to the audience how this part of the presentation contributes to the final objective. Your task is not to keep the audience guessing, but to make everything in the presentation crystal clear.

Numerous approaches can be taken to develop major points: cause and effect, comparisons, details, tables, explanations, stories, visuals, multimedia, examples, and so forth. You should be sure to give the audience some way to remember what you say. Because people forget 90 percent of what they hear, you must select a strategy that will help them to remember. However, be sure to make your strategy both unforgettable and effective, not just unforgettable.

People tend to remember information better under the following conditions:

- The information is interesting.
- The information is important to them.
- The information has an immediate application or fills an immediate need.
- They have prior knowledge about the subject.
- They take notes or write down points from the presentation.

Finally, it is important that you build credibility with your audience during your presentation. That means you must do your homework before you stand up before the audience. Every premise and conclusion that you draw must be supported by facts, analysis, and logic. You may not want to include all of the detailed analysis and calculations as part of the presentation, but you should be able to provide additional detail when needed. In some cases, this may mean having some backup visuals or materials handy in case questions are asked. You must be prepared to field the types of questions that your audience is likely to ask. The better you know your audience, the better you can anticipate the types of questions and information that will be requested. Your credibility with the audience heavily influences its acceptance of your conclusions.

The Closing

The closing provides an opportunity to emphasize your major points and to highlight conclusions and recommendations. Your closing remarks should end with a positive tone. Obviously, you do not want to sound overbearing or pushy. Seek the right mix of confidence without sounding like a know-it-all.

If you field questions at the end of the presentation, you should include a concluding remark after the questions. You want the audience to remember your conclusion and recommendation. Questions can take the topic to many different areas that have little to do with the conclusions. You want the audience to remember your conclusion even after questions have been answered.

Giving the Presentation

All of us get nervous when we have to give a presentation. You can minimize your nervousness if you are well prepared and have practiced your presentation. As you prepare to give your presentation, remember the following:

- **Time:** Stay within the allotted time. If you practice, you will know how to adjust your presentation to fit the time and still make your essential points.
- **Room layout:** Be sure the room is arranged so that everyone can participate in the presentation. Be sure the facilities and equipment all work correctly.
- **Dress:** Dress appropriately for your audience. Although there are exceptions, normally you should dress to the same level as the ranking individual in the audience will dress. Remember, your objective is to gain the respect of your audience, so you should dress to fit in with them.

- **Body language:** Eye contact is a necessary component to building rapport with your audience. As much as possible, maintain eye contact with the audience, including people at the ends of the rows. Use arm gestures as appropriate to emphasize your points and to add vitality to your presentations. It is appropriate to move around from time to time if the setting permits it. This will help maintain the attention of the audience.

Use of Visual Aids

Visual aids are an important tool to assist you. Not only will they help keep the audience's attention, but they can help clarify important points. However, remember that you are the most important visual. Use other visuals to enhance your presentation, not to be the center of attention and draw attention away from you.

Common forms of audiovisual aids in use today include:

- Overhead projection
- Posters and flip charts
- Computer-projected shows and images
- Projected slides
- Handouts
- Artifacts

When you are preparing audiovisual aids, there are several things you should keep in mind:

- **Professional image:** Above all, make your aids professional looking. This means they should be professionally drawn using some type of automated tool. Do not use hand-drawn diagrams. They will undermine your credibility with the audience. Take the time to make your visual aids look crisp and clean. Add color and logos when appropriate. Double- and triple-check to make sure there are no misspelled words.
- **Audibility and visibility:** When you are selecting the type of aid to use, make sure that it can be heard or seen easily. The size and layout of the room sometimes dictate what is feasible to use. Check both the lighting and the speaker system to ensure that they work as you expect.
- **Understandability and readability:** As you prepare your materials, remember that they should be used to emphasize your presentation, not replace you. Visual or audio materials that are too extensive tend to detract from rather than enhance the presentation. Make them succinct and to the point.
- **Ease of use:** Whatever media you choose should be easy for you to access and use. You should be able to flow from one point to the next in both your presentation and your audiovisual aids without having to stop and readjust or reconfigure your materials. Otherwise, you may lose the attention and focus of the audience. The impression you want to make is that the audiovisual components flow naturally with the presentation. Nobody should notice how they happen to appear.
- **Technology:** We are all moving increasingly toward high-technology aids. However, do not choose to use high-technology aids just because they are high tech. In many instances, the older, more standard tools will serve you better. However, if you choose to use technology, we strongly recommend two principles. First, check out everything beforehand. Do a walkthrough of your presentation using the high-tech materials in the same room and with the same equipment to

be used for the real presentation. Second, have a plan B. It seems that high tech is especially subject to Murphy's law, "If something can go wrong, it will," and to its corollary, "If nothing can go wrong, it still will."

As you use visual materials, remember the following principles:

- **Referencing:** Point to the specific part of the visual aid you are referencing. However, do not turn away from the audience to do so. Glance or point briefly to the element you want to reference, then return your attention to the audience.
- **Interpretation:** Interpret the visuals. Do not expect the audience to read between the lines of the point you are trying to make. Make explicit explanations of the materials on the visual.
- **Flow:** Think about and practice the flow of the visuals. Remove them and turn off the source light if you are finished with a visual and plan to present topics not related to the visual. Make sure you know how to flow from one to the next. Be sure your visuals are in the correct order.
- **Handouts:** You can use handouts to assist the audience to follow your presentation. They can also be useful to encourage the audience to take notes on your presentation. If you plan to have detailed handouts, consider distributing them at the end of the presentation. Handouts that can be read quickly or that you will use step by step in your presentation can be distributed as you begin. In that case, explain what you plan to do so that the audience can follow along with you.

Questions and Answers

Frequently, you should allot time for questions and answers. You may either invite questions during the presentation or at the end. Questions during the presentation tend to make the presentation more informal and interactive. The danger, of course, is that they may take you off track. They can also take excessive time so you are not able to make your conclusions. However, if an individual does not understand part of your presentation, it may be better to clear up the misunderstanding before moving on. Often a workable solution is to accept questions during the presentation to clarify a point but to hold other related questions until the end.

Ideally, the question-and-answer period assists you in achieving your presentation objective. The way you handle questions can either enhance or weaken your credibility with the audience. Even questions that seem to detract can enhance the presentation if you handle them in a professional manner. Here are some guidelines:

- Make sure everyone has heard the question. Repeat it, if necessary.
- Listen carefully to the question and do not start answering it before the person is through asking.
- Make sure you understand the question before answering. If you are unsure, ask for clarification.
- When you answer, begin talking to the person asking the question. Then include the rest of the audience.
- After you answer, make sure you answered the right question. You can ask whether the answer was satisfactory.
- If you do not know the answer, admit it and, if appropriate, indicate that you will find out the answer. Do not be defensive. There are always more questions than there are answers. If necessary, find out how you can later get the answer to the person who asked the question.

- Use your visuals to support your answers, especially if someone requests more detail.
- Keep your answers short and concise. Clearly answer the question and then move on to the next.
- If you have a hostile questioner, listen to the question carefully. Let the person express his or her views. Show respect for that person's views, and then honestly express your opinion without being hostile. If disagreement still exists, then indicate in fairness to the rest of the audience that it may be appropriate to continue the discussion of those points later.
- Remember to end the question period with a repetition of your major conclusion or recommendation before ending the presentation.

A

D

E

F

G

H

J

K

L

O

P

S

U

V

W

CANTERBURY CHRIST CHURCH UNIVERSITY COLLEGE
LIBRARY